# Encyclopedia of
# Plant Physiology

New Series Volume 14 A

Editors

A. Pirson, Göttingen
M. H. Zimmermann, Harvard

# Nucleic Acids and Proteins in Plants I

## Structure, Biochemistry and Physiology of Proteins

Edited by

D. Boulter and B. Parthier

Contributors

L. Beevers J. D. Bewley D. D. Davies J. W. Hart C. F. Higgins
R. C. Huffaker P. J. Lea R. Manteuffel A. Marcus Ph. Matile
M.-N. Miège B. J. Miflin K. Müntz D. H. Northcote
B. Parthier J. W. Payne P. I. Payne J. A. M. Ramshaw
A. P. Rhodes D. D. Sabnis J. L. Stoddart H. Thomas J. H. Weil

With 135 Figures

Springer-Verlag Berlin Heidelberg New York 1982

Professor Dr. DONALD BOULTER
University of Durham
Department of Botany
Science Laboratories, South Road
Durham, DH1 3LE/UK

Professor Dr. BENNO PARTHIER
Akademie der Wissenschaften der DDR
Institut für Biochemie der Pflanzen Halle
Weinberg 3, Postfach 250
401 Halle (Saale)/GDR

ISBN-13: 978-3-642-68239-1 e-ISBN-13:978-3-642-68237-7
DOI: 10.1007/978-3-642-68237-7

Library of Congress Cataloging in Publication Data. Main entry under title: Nucleic acids and proteins in plants. (Encyclopedia of plant physiology; new ser.; v. 14, pt. A–). Bibliography: p. Includes index. Contents: pt. A. Structure, biochemistry, and physiology of proteins. 1. Nucleic acids. 2. Plant proteins. 3. Botanical chemistry. I. Boulter, D. II. Parthier, Benno. III. Series. QK711.2.E5 new ser., vol. 14, pt. A, etc. 81-18256 [QK898.N8] 581.1s [581.19′245] AACR2.


Softcover reprint of the hardcover 1st edition 1982



2131/3130-543210

# Contents

## 4 Post-Translational Modifications

L. BEEVERS (With 3 Figures)



## 5 Protein Degradation

PH. MATILE

## 6 Physiological Aspects of Protein Turnover

D.D. Davies (With 9 Figures)

## 7 Structures of Plant Proteins

J.A.M. Ramshaw (With 31 Figures)



## 8 Protein Types and Distribution

M.-N. Miège (With 14 Figures)

## 10 Biochemistry and Physiology of Leaf Proteins

R.C. HUFFAKER (With 10 Figures)



## 11 Microtubule Proteins and P-Proteins

D.D. SABNIS and J.W. HART (With 3 Figures)

## II. Nucleic Acids and Proteins in Relation to Specific Plant Physiological Processes

### 14 Seed Development

K. MÜNTZ (With 12 Figures)

## 15 Protein and Nucleic Acid Synthesis During Seed Germination and Early Seedling Growth

J.D. BEWLEY (With 13 Figures)



## 16 Leaf Senescence

J.L. STODDART and H. THOMAS (With 6 Figures)

# List of Contributors


L. Beevers
Dept. of Botany and Microbiology
University of Oklahoma
770 Van Vleet Oval
Norman, Oklahoma 73019/USA

J.D. Bewley
Dept. of Biology
University of Calgary
Calgary, Alberta
Canada T2N 1N4/Canada

D.D. Davies
School of Biological Sciences
University of East Anglia
Norwich, NR4 7TJ/UK

J.W. Hart
Dept. of Botany
Aberdeen University
St. Machar Drive
Aberdeen AB9 2UD/UK

C.F. Higgins
Dept. of Biochemistry
University of Dundee
Dundee DD1 4HN/UK

R.C. Huffaker
Plant Growth Laboratory
University of California
Davis, California 95616/USA

P.J. Lea
Biochemistry Department
Rothamsted Experimental Station
Harpenden, Herts. AL5 2JQ/UK

R. Manteuffel
Zentralinstitut für Genetik
und Kulturpflanzenforschung
der Akademie der Wissenschaften
der DDR
Correnssstr. 3
4325 Gatersleben/GDR

A. Marcus
Institute for Cancer Research
7701 Burholme Avenue
Philadelphia, Pennsylvania 19111/USA

Ph. Matile
Dept. of General Botany
Swiss Federal Institute of Technology
Sonneggstr. 5
8092 Zurich/Switzerland

M.-N. Miège
Laboratoire de Chimie Taxonomique
Université de Genève
1 chemin de l'Impératrice
1292 Chambesy – Genève/Switzerland

B. Miflin
Biochemistry Department
Rothamsted Experimental Station
Harpenden, Herts. AL5 2JQ/UK

K. Müntz
Zentralinstitut für Genetik
und Kulturpflanzenforschung
der Akademie der Wissenschaften
der DDR
Correnssstr. 3
4325 Gatersleben/GDR

D.H. Northcote
Dept. of Biochemistry
University of Cambridge
Tennis Court Road
Cambridge CB2 1QW/UK

B. Parthier
Akademie der Wissenschaften
der DDR
Institut für Biochemie der Pflanzen
Halle
Weinberg 3, Postfach 250
401 Halle (Saale)/GDR

J.W. Payne
Dept. of Botany
University of Durham
Durham DH1 3LE/UK

P.I. Payne
The Plant Breeding Institute
Maris Lane
Trumpington, Cambridge CB2 2LQ/UK

J.A.M. Ramshaw
The Agassiz Museum
Harvard University
Cambridge, Massachusetts 02138/USA

*Present address:*
C.S.I.R.O.
(Division of Protein Chemistry)
343 Royal Parade
Parkville 3052/Australia

A.P. Rhodes
The Miln Marsters Group Limited
Plant Breeding and Research Centre
Docking, Kings Lynn
Norfolk PE 31 8LR/UK

D.D. Sabnis
Dept. of Botany
Aberdeen University
St. Machar Drive
Aberdeen AB9 2UD/UK

J.L. Stoddart
Plant Biochemistry Department
Welsh Plant Breeding Station
Plas Gogerddan
Aberystwyth, Dyfed SY23 3EB/UK

H. Thomas
Plant Biochemistry Department
Welsh Plant Breeding Station
Plas Gogerddan
Aberystwyth, Dyfed SY23 3EB/UK

J.H. Weil
Institut de Biologie Moléculaire et Cellulaire
15, rue Descartes
67084 Strasbourg/France

# List of Abbreviations

| | |
|---|---|
| A | adenosine (likewise: C, cytidine; G, guanosine; I, inosine; U, uridine; T, thymidine; ψ, pseudouridine) |
| $i^6A$ | isopentenyl-adenosine |
| $ms^2i^6A$ | 2-methyl-thio-iso pentenyl-adenosine |
| Aa-RS | amino acyl-tRNA synthetases; (Thr-RS, threonine-tRNA synthetase, other amino acids correspondingly) |
| ABA | abscisic acid |
| ADP | adenosine 5′-diphosphate (likewise CDP, GDP, UDP) |
| AMP | adenosine 5′-monophosphate, adenylic acid (likewise CMP, GMP, UMP, TMP) |
| cAMP | cyclic 3′,5′AMP |
| Ap | adenosine 3′-monophosphate (likewise Cp, Gp, Up, Tp) |
| ATP | adenosine 5′-triphosphate (likewise CTP, GTP, UTP, TTP) |
| BA | benzyladenine |
| BANA | 2-N-benzoyl-D-arginine-β-naphthylamide |
| bp | basepairs |
| BSA | bovine serum albumin |
| 2C | diploid (likewise 4C, tetraploid, etc.) |
| CER | cisternal endoplasmic reticulum |
| Con A | Concanavalin A |
| CPMV | cowpea mosaic virus |
| CTN | N-carbobenzoxy-L-tyrosine-P-nitrophenyl ester |
| D | green area duration |
| 2,4-D | dichlorophenoxyacetic acid |
| dA | deoxyadenosine (likewise: dC, dG, etc.) |
| DAHP | 3-deoxy-D-arabino heptulosonate-7-phosphate |
| DCMU | 3-(3′,4′-dichlorophenyl)-1,1-dimethyl urea |
| DEAE | diethylaminoethyl |
| cDNA | complementary DNA |
| ctDNA | chloroplast DNA |
| EC | energy charge |
| EDTA | ethylene-diamintetra-acetic acid |
| EF | elongation factor in ribosomal translation (e.g. EF-1; $EFT_u$, $EFT_s$, EFG) |
| EGTA | ethyleneglycolbis-tetra-acetic acid |
| eIF | eucaryotic initiation factor |
| ER | endoplasmic reticulum |
| FMN | flavin mononucleotide |
| GO, G1, G2 | phases of the cell cycle |
| GA | gibberellin |
| $GA_3$ | gibberellic acid |
| GlcNAc | N-acetylglucosamine |
| GOGAT | glutamine oxoglutarate amino-transferase |
| GP II | glycoprotein II |
| GPD | glyceraldehyde-3-phosphate dehydrogenase |
| GS | glutamine synthetase |
| HMW | high molecular weight |
| Hyp | hydroxyproline |
| IAA | indole-3-acetic-acid |
| ICL | isocitrate lyase |
| IF | initiation factor |
| IGP | indole-3-glycerol-phosphate |
| Kbp | kilobasepairs |
| Kd | kilodalton |

| | |
|---|---|
| Ku | unfolding equilibrium constant |
| LAP | leucine aminopeptidase |
| Lg | leghaemoglobin |
| LMW | low molecular weight |
| LNA | L-leucyl-β-naphthylamide |
| LSD | least significance difference |
| MAK | methylated albumin kieselguhr |
| MDH | malate dehydrogenase |
| MDMP | D-2-(4-methyl-2,6-dinitroanilo)-N-methylproprionamide |
| $M_r$ | molecular mass |
| mRNA | messenger RNA |
| mRNP's | messenger ribonucleic acid particles |
| MSO | methionine sulphonamine |
| MT | microtubule |
| MTOC | microtubule-organising centres |
| MW | molecular weight (mol. wt.) |
| N | nucleoside (usually in connection with p, pp or ppp) and chemical symbol for nitrogen |
| NAA | naphthaleneacetic acid |
| NAD | nicotinamide adenine dinucleotide (oxidized form) |
| NADH | nicotinamide adenine dinucleotide (reduced form) |
| NiR | nitrite reductase |
| NR | nitrate reductase |
| NTP | nucleoside triphosphate |
| pA | adenosine 5′-monophosphate (in polynucleotide chains); likewise pC, pG. pU, etc. |
| PAL | phenylalanine ammonia lyase |
| pBR322 | plasmid of E. coli frequently used for gene transformation |
| P5C / P2C | = pyrroline-5-carboxylate / pyrroline-2-carboxylate |
| $P_i$ | inorganic phosphate |
| poly(A) | polyadenylic acid, polyadenylated 3′ terminus of (virus or messenger) RNA; likewise: poly(U), poly(dA), poly(dC), poly(dA-T), etc. |
| poly(A)$^+$RNA | RNA containing terminal poly(A) |
| poly(A)$^-$RNA | RNA lacking terminal poly(A) |
| Q-MAP's | quantitative ratio MT-associated proteins |
| NQ-MAP's | non-quantitative ratio MT-associated proteins |
| RCA | Ricinus communis agglutinin |
| RER | rough endoplasmic reticulum |
| rRNA | ribosomal RNA |
| $tRNA^{Val}_{Euglena}$ | valine-specific transfer RNA of Euglena (correspondingly for other amino acids and other species) |
| RQ | respiratory quotient |
| RuBPCase (RuBPC) | ribulose-1,5-bisphosphate carboxylase |
| S | sedimentation coefficient in Svedberg units |
| SDS-PAGE | sodium dodecylsulphate polyacrylamide gel electrophoresis |
| SF | senescence factor |
| TAP | tubulin assembly protein |
| TER | tubular endoplasmic reticulum |
| Thr-RS | threonine-tRNA synthetase (other amino acids correspondingly) |
| $T_m$ | melting point |
| TMV | tobacco mosaic virus |
| TRV | tobacco rattle virus |
| WGA | wheat germ agglutinin |

# Introduction

D. BOULTER and B. PARTHIER

At the time of the former edition of the Encyclopedia of Plant Physiology, approximately 25 years ago, no complete plant protein amino acid sequences or nucleic acid sequences had been determined. Although the structure of DNA and its function as the genetic material had just been reported, little detail was known of the mechanism of its action, and D.G. CATCHSIDE was to write in the first chapter of the first volume of the Encyclopedia: "There is a considerable body of evidence that the gene acts as a unit of physiological action through the control of individual enzymes". No cell-free transcription and protein-synthesizing systems were available and the whole range of powerful methods of recombinant DNA technology was still to be developed. Today for the first time with plant systems, it is possible not only to describe their molecular biology but also to manipulate it, i.e., to move from a description to a technological phase.

The properties of living systems are inscribed by those of the proteins and nucleic acids which they synthesize. Proteins, due to their very large size, occur as macromolecules in colloidal solution or associated in supra-molecular colloidal form. The colloidal state confers low thermal conductivity, low diffusion coefficients and high viscosity, properties which buffer a biological system from the effects of a changing environment. Biological systems not only have great stability, but also the capacity to reproduce. The information to specify the structures and processes from birth to death is stored and transmitted to offspring by nucleic acids which are also transcribed and translated into the proteins of the system. Thus the biochemical processes of biological systems only proceed at the necessary fast rate via the agency of enzyme (protein) action. Many other biological processes also show specificity, e.g., cell interaction, intracellular transport and compartmentation, defence mechanisms, etc. and these too are mediated by proteins. Clearly, not all of these protein types could be dealt with in one volume, and accounts of some will be found in several forthcoming volumes of this series, e.g., membrane proteins, light harvesting proteins, phytochrome etc.; the decision as to which protein and nucleic acid data should be collected together in a separate volume has, of necessity, had to be somewhat arbitrary and depended primarily on whether sufficient was known about a topic to warrant a separate chapter elsewhere in the series.

For production reasons, it has been necessary to make a somewhat artificial division of the field into two volumes. Furthermore, as is inherent in a multi-author book, some overlapping of information has been unavoidable but this has been kept to a minimum and has been concentrated, wherever possible, to give the different viewpoints of the authors. Authors in their turn have had to be selective due to the limited space available to them and this will have inevitably introduced an aspect of heterogeneity.

With all of these methodological and conceptual advances, it is not surprising that the arrangement of the topics in this volume is very different to that in the corresponding Volume 8 of the old series. This volume covers our present knowledge of the plant genomes, ribonucleic acids, proteins and viruses, as well as selected aspects of developmental plant physiology, i.e., seed development, germination, chloroplast differentiation, leaf senescence and cell wall differentiation where the investigations have been conducted primarily from an analysis of the changes in nucleic acids and proteins. The explosive advances in molecular biology and their application in plant investigation has led to a predominance of molecular data over plant physiological data in the present volume.

At present, plants supply more than 70% of the world's food and this percentage will tend to increase with increasing world population and as energy becomes more expensive and in short supply. Even in developed countries where the amount of animal protein consumed predominates, plant protein is very slowly being substituted as analogues, extenders, and even as novel foods, and this trend is forecast to continue, especially if a better understanding of plant proteins and their functional properties is forthcoming. In agriculture, conventional breeding methods and agronomic practices have made very important improvements in crop productivity and will continue to do so in the near future. However, the plant breeder is seeking to use the powerful methods of somatic genetics and recombinant DNA technology in order to overcome present barriers to genetic recombination. Continued progress in all these areas depends upon a better understanding of the structure and function of plant proteins and nucleic acids and their interaction in gene organisation and expression, cell differentiation, and plant development. These topics form the core of the data presented in Volume 14.

It can confidentally be predicted that, as well as contributing to crop improvements, studies on the proteins and nucleic acids of plants will continue to be a source of important problems in cell biology, intra-cellular transport, structure/function relationships, molecular evolution and developmental biology for many years to come.

We are grateful to our author colleagues for their co-operation in the difficult task of selection and for the patience of those many authors in connection with the somewhat extended publication date. We are also indebted to Professor A. Pirson, Series Editor, for his invaluable advice and support and would like to thank the staff of Springer-Verlag for their indispensible help with this publication.

Durham and Halle
October 1981

D. BOULTER
B. PARTHIER

# I. Biosynthesis and Metabolism of Protein Amino Acids and Proteins

# 1 Ammonia Assimilation and Amino Acid Metabolism

B.J. MIFLIN and P.J. LEA

## 1 Introduction

The field of amino acid biosynthesis from inorganic nitrogen has been one of the most active in plant biochemistry in the last five years and a recent book has been devoted entirely to it (MIFLIN 1980a). To provide a reasonable coverage of the field in a short chapter is difficult; here we present an outline of the biochemical pathways involved in the flow of nitrogen from ammonia into (and in some cases out of) amino acids, and give some indication of how the pathways may be regulated and how the reactions are integrated into the plant's metabolism, particularly in relation to subcellular localization. We have attempted to support this outline with a reasonable number of citations from which the interested reader can derive much further information; nevertheless the citations are by no means exhaustive and to the authors of relevant papers that have been omitted we offer our regrets and ask for their understanding.

## 2 Ammonia Assimilation and Transamination

### 2.1 Introduction

Ammonia assimilation in plants has been covered in detail recently (MIFLIN and LEA 1976, 1977, 1980). This review will consider the evidence for the operation of the glutamine synthetase/glutamate synthase pathway of ammonia assimilation in higher plants with detailed coverage of very recent investigations.

Ammonia is produced from an external source by the reduction of nitrate or the fixation of nitrogen gas (VENNESLAND and GUERRERO 1979, STEWART W.D.P. 1979); but it also arises from the catabolism of an internal source, e.g., glycine, asparagine or arginine (see Sect. 5). Regardless of the origin it is essential that the ammonia is rapidly converted into a form where it is not toxic.

### 2.2 Enzymes Involved in the Glutamate Synthase Cycle

Glutamine synthetase (GS)[L-glutamate:ammonia ligase (ADP forming) EC 6.3.1.2] catalyzes reaction (1): –

$$\text{L-Glutamate} + \text{ATP} + NH_3 \rightarrow \text{L-glutamine} + \text{ADP} + \text{Pi} + H_2O \qquad (1)$$

There are a number of problems in assaying the enzyme (MIFLIN and LEA 1977). The most reliable assay is that involving hydroxylamine in the so-called "biosynthetic" reaction (2): –

$$\text{L-Glutamate} + \text{ATP} + \text{hydroxylamine} \rightarrow$$
$$\xrightarrow{Mg^{2+}} \text{glutamyl-}\gamma\text{-hydroxamate} + \text{ADP} + P_i + H_2O \qquad (2)$$

The hydroxamate product reacts with acidified ferric chloride to give a red-brown complex usually measured at 540 nm. The enzyme is universally distributed throughout the plant kingdom and is usually present in large amounts in all tissues. All the enzymes so far studied have a great affinity for ammonia ($K_m$ $10^{-5}$–$10^{-4}$ M), which should prevent any build-up of ammonia.

The ammonia assimilated into the amide group of glutamine is transferred to the α-amino position of glutamate via the action of glutamate synthase (3). (L-Glutamate:$NADP^+$ oxidoreductase (transaminating) EC 1.4.1.13 or L-Glutamate:ferredoxin oxidoreductase (transaminating) (EC 1.4.7.1)

$$\text{2-Oxoglutarate} + \text{L-glutamine} + \text{NADPH (or NADH or ferredoxin red.)} \rightarrow$$
$$2\ \text{L-glutamate} + NADP^+ \text{ (or } NAD^+ \text{ or ferredoxin ox.)} \qquad (3)$$

Glutamine synthetase and glutamate synthase are able to act in conjunction to form the glutamate synthase cycle (Fig. 1). The key point is that one molecule of glutamate continuously recycles whilst the second may be transaminated to other amino acids or converted to proline and arginine (Sect. 4.3).

Glutamate may also be regenerated when the amide amino group of glutamine is transferred to form other compounds such as asparagine or carbamoyl phosphate. This glutamate may then be used for further $NH_3$ assimilation.

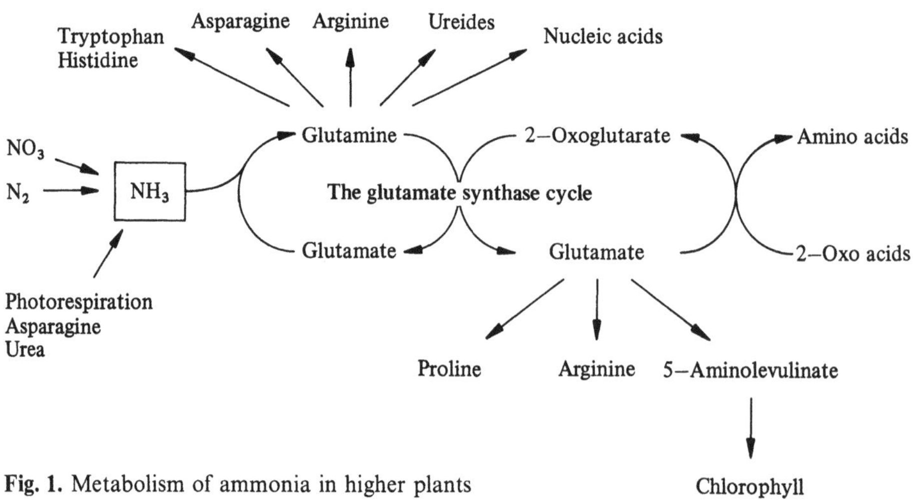


**Fig. 1.** Metabolism of ammonia in higher plants

## 2.3 Evidence for the Glutamate Synthase Cycle

Three kinds of evidence for the operation of the cycle in various plant tissues will be presented: presence of the two requisite enzymes, labelling studies, and inhibitor studies.

### 2.3.1 Assimilation in Leaves

Glutamine synthetase (GS) has been isolated from leaves of *Pisum* (O'NEAL and JOY 1973b, 1974, 1975) and from *Lemna* (STEWART and RHODES 1977a, b). In most cases the specific activity of the enzyme is higher in the leaf than in the root, although if a large proportion of the nitrogen is assimilated in the root then the situation may be reversed (LEE and STEWART 1978). In maize (*Zea mays*) initial evidence suggested that a major proportion of the leaf GS was present in the mesophyll cells (RATHNAM and EDWARDS 1976), but later investigations suggested the enzyme was equally distributed between the mesophyll and bundle-sheath cells of $C_4$ plants (HAREL et al. 1977, MOORE and BLACK 1979). Two separate forms of GS have recently been isolated from soyabean (*Glycine max*) hypocotyls (STASIEWICZ and DUNHAM 1979), barley (*Hordeum vulgare*) (MANN et al. 1979, 1980), rice (*Oryza sativa*) (GUIZ et al. 1979) and pea (*Pisum sativum*) leaves (ESTIGNEEVA et al. 1977) by ion exchange chromatography. BARRATT (1980) has shown by staining for GS activity on starch gels, that in most leaves tested there are two distinct isoenzymes which can be separated by electrophoresis. The two forms present in barley leaves have similar molecular weights but differ in stability, pH optima and response to thiol reagents. The apparent $K_m$ values for ammonia and ATP are similar but the isoenzyme present only in leaves shows negative cooperativity in the binding of glutamate (MANN et al. 1979; GUIZ et al. 1979).

Ferredoxin-dependent glutamate synthase was initially isolated from pea leaves (LEA and MIFLIN 1974), but since that time has been detected in a wide range of higher plants (RHODES et al. 1976, NICKLISH et al. 1976, WALLSGROVE et al. 1977, LEE and STEWART 1978, STEWART and RHODES 1977b, 1978). The enzyme from *Vicia faba* leaves has been partially purified and is specific for ferredoxin, glutamine and 2-oxoglutarate (WALLSGROVE et al. 1977). Recently MATOH et al. (1980) have shown the presence of a separate NADH-specific GOGAT in etiolated pea shoots. In green expanded pea leaves this activity is only about 5% of the total GOGAT activity (R.M. WALLSGROVE unpublished).

In $C_4$ plants initial studies suggested that glutamate synthase was not present in bundle-sheath cells (RATHNAM and EDWARDS 1976), but later studies (HAREL et al. 1977, MOORE and BLACK 1979) indicated that a higher proportion of leaf GOGAT was present in the bundle-sheath cells.

$[^{15}N]$-labelling data obtained by LEWIS and PATE (1973) indicated that $[^{15}N]O_3$, $[^{15}N]$-glutamate and $[^{15}N$-amide$]$-glutamine were all incorporated into the amino-N of amino acids, suggesting the direct conversion of the amide-N of glutamine to the 2-amino position. BAUER et al. (1977) added $[^{14}N]O_3^-$ to pea leaves from plants previously grown on $[^{15}N]O_3^-$, and showed that although there was evidence of multiple pools the $^{15}N$ in ammonia decreased rapidly followed by glutamine and glutamate.

CANVIN and ATKINS (1974) found that although [$^{15}$N]-ammonia assimilation in leaves was stimulated by light, a considerable amount also occurred in the dark; this was considered at the time to be due to the action of glutamate dehydrogenase in the mitochondria. Later studies by ITO et al. (1978) showed that in the dark [$^{15}$N]-ammonia was incorporated into glutamine but not into glutamate, suggesting that although ferredoxin-dependent glutamate synthase could not function in the dark, ATP was available for glutamine synthetase.

Recently RHODES et al. (1980) analyzed [$^{15}$N]$H_3$ assimilation in *Lemna* under steady-state conditions and subjected the results to careful computer analysis. Two models, involving two separate compartments, were found to fit the data. Both had in compartment (1) the glutamate synthase cycle and in compartment (2) a second site of glutamine synthesis. The models could not exclude the possibility that 10% of the total N influx may be assimilated directly into glutamate via a second reaction. Confirmation as to the exact route of [$^{15}$N]$H_3$ assimilation was then obtained using methionine sulphoximine (MSO) and azaserine, respectively inhibitors of glutamine synthetase and glutamate synthase, which totally inhibited incorporation. Initial work by LEWIS and PROBYN (1978a) suggested that when *Datura* leaves were fed high levels of [$^{15}$N]$O_3$ there was direct incorporation into glutamate. However, later LEWIS and PROBYN (1978b) in *Datura* and KAISER and LEWIS (1980) in *Helianthus* showed that MSO completely blocked $^{15}$N incorporation into amino acids, irrespective of the $NO_3$ level applied, indicating that glutamine synthesis was necessary prior to glutamate formation. A number of other studies using MSO and azaserine have been carried out in *Lemna* (STEWART and RHODES 1976, STEWART CR 1979) and spinach cells (WOO and CANVIN 1980), all confirming the operation of the glutamate synthase cycle in leaves.

### 2.3.2 Green Algae

Glutamine synthetase from *Chlorella* has been studied by a number of Russian workers (KRETOVICH and EVSTIGNEEVA 1974) and shown to be composed of six subunits of molecular weight 53,000 (RASULOV et al. 1977).

Glutamate synthase was initially isolated from the green alga *Chlorella,* where it could be readily assayed using a freezing and thawing technique (LEA and MIFLIN 1975). The enzyme was also demonstrated in *Platymonas striata* (EDGE and RICKETTS 1978) *Chlorella* (SHATILOV et al. 1978) and *Caulerpa simpliciuscula* (MCKENZIE et al. 1979). In the alga *Chlamydomonas* both ferredoxin and NAD(P)H-dependent glutamate synthase activities have been detected (CULLIMORE and SIMS 1981a). It is probable that the pyridine nucleotide enzyme is utilized for heterotrophic growth.

BAKER and THOMPSON (1961) showed that in *Chlorella,* after very short time periods, [$^{15}$N]$H_3$ was incorporated predominantly into the amide position. Later work by BASSHAM and KIRK (1964) suggested that [$^{15}$N]$H_3$ was directly incorporated into glutamate although their data is open to a different interpretation (MIFLIN and LEA 1976). More recent experiments with the brown alga *Macrocystis* (O.A.M. LEWIS personal communication) and the marine alga *Caulerpa* (CH'NG 1980) have both suggested that [$^{15}$N]$H_3$ can be assimilated via the glutamine synthetase/glutamate synthase pathway at low levels of exogenous nitrogen.

Some 15 years before the glutamine synthetase/glutamate synthase pathway was shown to operate in plants, VAN DER MEULEN and BASSHAM (1959) fed diazo-oxo-norleucine and azaserine (both inhibitors of glutamine-amide transfer reactions) to cultures of the green algae *Scenedesmus* and *Chlorella*. Incorporation of $[^{14}C]O_2$ into glutamate was inhibited and there was an accumulation of label in glutamine and 2-oxoglutarate as would be expected if glutamate synthase had been inactivated; similar inhibition of $[^{15}N]$-ammonia assimilation in *Chlamydomonas* has also been shown by CULLIMORE and SIMS (1981b).

### 2.3.3 Roots and Tissue Culture

Glutamine synthetase has been purified from rice roots (KANAMORI and MATSUMOTO 1972), and the enzyme isolated from roots by a number of other workers (MIFLIN 1974, WEISSMAN 1976, EMES and FOWLER 1978, STEWART and RHODES 1978, SAHULKA and LISA 1979). WASHITANI and SATO (1977b, 1978) have isolated the enzyme from tissue culture cells.

Glutamate synthase was first isolated from pea root tissue by FOWLER et al. (1974) and in a somewhat different form by MIFLIN and LEA (1975). Although the activity tends to be lower in roots than in other tissues its presence there has been confirmed by STEWART and RHODES (1978), LEE and STEWART (1978), ARIMA (1978), EMES and FOWLER (1978), OAKS et al. (1979, 1980). There is some confusion over the electron donor in root systems, as the majority of enzymes can utilize NADH, NADPH or ferredoxin, although in rice roots only the ferredoxin enzyme has been detected (ARIMA 1978). No attempt, however, has been made to separate or purify the enzymes.

The first demonstration of a NAD(P)H-dependent glutamate synthase in tissue cultures was by DOUGALL (1974) and was subsequently extended by DOUGALL and BLOCH (1976), FOWLER et al. (1974), WASHITANI and SATO (1978) and EMES and FOWLER (1978). Recently CHIU and SHARGOOL (1979) have purified glutamate synthase 430-fold from soyabean suspension cell cultures, using the enzyme's affinity for Blue Sepharose; this enzyme preferentially uses NADH over NADPH but not ferredoxin.

Early work by COCKING and YEMM (1961) and later by YONEYAMA and KUMAZAWA (1975) on feeding $^{15}N$ to roots is difficult to interpret, although it is possible to deduce some evidence for the GS/glutamate synthase pathway after the roots have been returned to $^{14}N$ (MIFLIN and LEA 1976). ARIMA and KUMAZAWA (1977) showed that if rice roots were pretreated with MSO to such an extent that 90% of the GS was inactivated, the incorporation of $[^{15}N]H_3$ into glutamine was inhibited by 90% and into glutamate by 77%. PROBYN and LEWIS (1979) showed that the addition of MSO to the roots of *Datura* fed $[^{15}N]O_3$ almost completely blocked incorporation into amino acids, but had no effect on the rate of nitrate reduction. $[^{13}N]$-feeding studies have been carried by SKOKUT et al. (1978) on tobacco (*Nicotiana tabacum*) tissue culture cells. The labelling kinetics of cells fed $[^{13}N]H_3$ or $[^{13}N]O_3^-$ indicated that glutamine was the primary assimilation product, with glutamate formed second. Inhibition studies with MSO, using cells grown on ammonium salts and fed [$^{13}$N]$H_3$, could not exclude that some nitrogen was being incorporated directly into glutamate. The data could, however, be explained (see MIFLIN and LEA 1980) by the presence of two sites of glutamine synthesis, only one of which was involved in glutamate synthesis. This suggestion is clearly the same as that proposed by RHODES et al. (1980) and discussed in Section 2.3.1 to explain their [$^{15}$N]-feeding data with *Lemna*.

### 2.3.4 Maturing Seeds

The primary assimilation of ammonia is not normally considered to be of major importance in maturing seeds actively engaged in protein synthesis. However the three major transport compounds (arginine, asparagine and ureides, see Sect. 5.2) have to be metabolized via ammonia and the GS/glutamate synthase pathway.

Although GS was first purified from dry pea seeds (ELLIOTT 1953) and its properties studied since (WEBSTER 1964, KINGDON 1974), there is no data available on the purified enzyme from seeds at earlier stages of maturation. The enzyme has been demonstrated in maturing legume cotyledons (LEA and FOWDEN 1975, STOREY and BEEVERS 1978, SODEK et al. 1980) and in maturing cereal grains (DUFFUS and ROSIE 1978, MIFLIN and SHEWRY 1979, OAKS et al. 1979).

A ferredoxin-dependent glutamate synthase was first isolated from maturing lupin (*Lupinus albus*) seeds (LEA and FOWDEN 1975) and later an NAD(P)H-dependent enzyme was found in pea (BEEVERS and STOREY 1976) and soyabean (STOREY and REPORTER 1978). MIFLIN and LEA (1976) reported unpublished work demonstrating the presence of a pyridine nucleotide-dependent enzyme in barley (later published in MIFLIN and SHEWRY 1979). SODEK and DA SILVA (1977) showed that in the developing endosperm of maize, NAD(P)H-dependent glutamate synthase gave a well-defined peak of activity coinciding with the period of most active N accumulation. OAKS et al. (1979) carried out a comparison of maize root and immature maize endosperm glutamate synthase. They demonstrated that $K^+$ ions were essential to detect maximum activity of the enzyme from both sources. DUFFUS and ROSIE (1978) were unable to find glutamate synthase in barley endosperms, probable because the enzyme was damaged by the techniques they employed.

Labelling studies employing $^{15}$N (LEWIS and PATE 1973, LEWIS 1975, ATKINS et al. 1975) have all suggested that amide nitrogen is readily transferred to protein amino-nitrogen via glutamine in developing seeds. Inhibitor studies (LEA et al. 1979a) with isolated pea cotyledons grown in vitro showed that azaserine, but not MSO, inhibited growth on glutamine, suggesting a requirement for the glutamate synthase reaction, however, the possibility of long-term effects of azaserine on nucleic acid synthesis cannot be excluded.

### 2.3.5 Legume Root Nodules

Discussion of ammonia assimilation in root nodules is complicated by the difficulties involved in the separation of the bacteroid from the plant cytoplasm. A number of the apparent discrepancies in the literature can probably be explained by incomplete separation or total ignorance of the many organelles present in root nodules. The following discussion has been centred on the plant portion of the root nodule, usually termed "the cytoplasm". In the bacteroids there is a consensus that the amounts of assimilatory enzymes are too small to account for all the ammonia that is produced by nitrogen fixation (BROWN and DILWORTH 1975, KURZ et al. 1975, ROBERTSON et al. 1975b, PLANQUE et al. 1978, STRIPF and WERNER 1978, UPCHURCH and ELKAN 1978).

GS is found in the "plant" portion of the legume nodule (DUNN and KLUCAS 1973, ROBERTSON et al. 1975a, PLANQUE et al. 1978). The GS activity in this portion increases markedly during lupin nodule development in conjunction with leghaemoglobin and nitrogenase (ROBERTSON et al. 1975a). GS has been extensively purified from the cytosol of the root nodules of soyabean (MCPARLAND et al. 1976) where it comprises 2% of the soluble protein. The authors concluded that the enzyme consisted of eight identical subunits of 45,000–47,000 molecular weight, arranged in two sets of planar tetramers 1 nm apart forming a cube of dimensions of 10 nm across each side.

Despite problems in assaying glutamate synthase in root nodules, large amounts of a pyridine nucleotide enzyme have been demonstrated in lupins (ROBERTSON et al. 1975b, RADYWKINA et al. 1977, RATAJCZAK et al. 1979), *Lathyrus* (LEE and STEWART 1978), *Phaseolus* (AWONAIKE 1980) and soyabean (SEN and SCHULMAN 1980). BOLAND et al. (1978) measured the activity of glutamate synthase and GS in the plant fraction of 12 different legumes. Specific activities of glutamate synthase reported varied between 7% and 100% of those of GS. In *Phaseolus* glutamate synthase activity was very high in young nodules, but decreased with age (AWONAIKE 1980). The enzyme has been purified extensively from lupin (BOLAND and BENNY 1977) and *Phaseolus* (AWONAIKE 1980). In both plants the enzyme is specific for NADH, glutamine and 2-oxoglutarate, has a high affinity for 2-oxoglutarate and a molecular weight of 220,000–235,000. The mechanism of the glutamate synthase action has been shown (BOLAND 1979) to involve compulsory binding of NADH as first substrate, followed by random-order binding of glutamine and 2-oxoglutarate, but AWONAIKE (1980) proposed that after NADH, glutamine bound second and 2-oxoglutarate third.

The initial $^{15}N$ studies of KENNEDY (1966) suggested that ammonia was incorporated directly into glutamate; however, the data can be interpreted in terms of the GS/glutamate synthase pathway (MIFLIN and LEA 1976). $^{13}N_2$ has now also been used in soyabean nodules; the radioactivity is initially rapidly incorporated into glutamine and then glutamate, the kinetics of labelling showed a typical precursor-product relationship (MEEKS et al. 1978). OHYAMA and KUMAZAWA (1978, 1980) have followed $^{15}N_2$ incorporation into amino acids in intact soyabean nodules. Addition of MSO inhibited amino acid synthesis and promoted the accumulation of $[^{15}N]H_3$, azaserine also decreased amino acid synthesis, but there was accumulation of $^{15}N$ in the amide group of glutamine as well as in ammonia.

## 2.4 Alternative Pathways of Ammonia Assimilation

The previous section has centred solely on the GS/glutamate synthase pathway as we consider that this is the major route of ammonia assimilation in plants.

However, there are a number of enzymes in the literature which have been proposed as possible assimilatory mechanisms.

glutamate dehydrogenase (EC 1.4.1.3)

$$\text{2-Oxoglutarate} + NH_3 + NAD(P)H_2 \leftrightarrows \text{L-Glutamate} + H_2O + NAD(P) \quad (4)$$

alanine dehydrogenase (EC 1.4.1.1)

$$\text{Pyruvate} + NH_3 + NAD(P)H_2 \leftrightarrows \text{Alanine} + H_2O + NAD(P) \quad (5)$$

aspartate dehydrogenase

$$\text{Oxaloacetate} + NH_3 + NAD(P)H_2 \leftrightarrows \text{Aspartate} + H_2O + NAD(P) \quad (6)$$

Two other possible mechanisms have been proposed involving the direct incorporation of ammonia into asparagine and carbamoyl phosphate, however as they do not involve the synthesis of a 2-amino group they will not be considered in this section. Alanine dehydrogenase has been purified from cyanobacteria (ROWELL and STEWART 1976) and aspartate dehydrogenase can be shown to operate at low pH values in vitro (KRETOVICH et al. 1978), but there is no evidence of direct assimilation of ammonia into either alanine or aspartate in vivo.

Until 1970 glutamate dehydrogenase (GDH) [reaction (4)] was considered to be the major enzyme involved in ammonia assimilation in all organisms. In bacteria and fungi at high ammonia concentrations (1 mM and above) the enzyme appears to be active, however, in algae and higher plants there is little evidence of an assimilatory function in vivo. The presence of GDH, which can be assayed readily in the assimilatory direction, is the most frequently used piece of evidence for its operation in ammonia assimilation. A comprehensive review of GDH in plants has recently been published by STEWART et al. (1980); only a brief summary will be given here.

GDH activity has been detected in a wide variety of higher plant tissues including seeds (THURMAN et al. 1965, NAGEL and HARTMANN 1980), roots (PAHLICH and JOY 1971, JOY 1973, OAKS et al. 1980), root nodules (BROWN and DILWORTH 1975, STONE et al. 1979, AWONAIKE 1980, DUKE et al. 1980), hypocotyls (YUE 1968), epicotyls (DAVIES and TEIXEIRA 1975), germinating cotyledons (CHOU and SPLITTSTOESSER 1972), maturing cotyledons (SODEK et al. 1980) and in green shoots (LEECH and KIRK 1968, LEA and THURMAN 1972, BARASH et al. 1975, NAUEN and HARTMANN 1980). The enzyme has been extensively studied in the aquatic plant *Lemna* (JOY 1971, SHEPARD and THURMAN 1973, EHMKE and HARTMANN 1976, 1978, STEWART and RHODES 1977b, SCHEID et al. 1980). The ability of GDH to utilize either NADH or NADPH as a coenzyme has frequently been demonstrated in plant extracts, although the NAD-dependent activity is usually much greater in higher plants. The molecular weight of NAD-dependent GDH varies from 208,000 (PAHLICH and JOY 1971) to 270,000 (STONE et al. 1979) with subunits of 46,000 (STEWART et al. 1980) to 58,000 (SCHEID et al. 1980). There is therefore some discrepancy in the literature as to whether the enzyme exists as a tetramer or hexamer.

Isoenzymes of GDH are often present, seven being the most common number (THURMAN et al. 1965, YUE 1968, LEE 1973, NAUEN and HARTMANN 1980, SCHEID et al. 1980). Their number can be increased by the addition of ammonia to the plant (BARASH et al. 1975, NAUEN and HARTMANN 1980) and 17 separate isoenzymes have been demonstrated in ammonium-infiltrated pea shoots (NAUEN and HARTMANN 1980). Evidence with purified enzyme preparations suggest that the isoenzymes all have the same molecular weight but different charge characteristics (PAHLICH 1972, NAGEL and HARTMANN 1980). Further evidence obtained by PAHLICH et al. (1980), using monospecific antibodies against pea seed GDH, confirms that the seven bands are immunologically identical.

Divalent cations are necessary for optimum activity; $Ca^{2+}$, $Zn^{2+}$, $Mn^{2+}$, $Co^{2+}$ and $Fe^{2+}$ all activate, whilst EDTA strongly inhibits, GDH (KING and WU 1971, JOY 1973, EHMKE and HARTMANN 1978, KINDT et al. 1980, SCHEID et al. 1980). The isolation and assay of GDH in EDTA buffers, although often attempted, is not recommended.

The apparently high $K_m$ value of GDH for ammonia is one of a number of pieces of evidence used by MIFLIN and LEA (1976) to suggest that it is unlikely to have a role in ammonia assimilation in vivo. A table produced by STEWART et al. (1980) indicates that values in higher plants lie between 5.2 and 70 mM, although values determined in the green algae may be as low as 0.5–0.67 mM (BROWN et al. 1974, MIFLIN and LEA 1976, GAYLER and MORGAN 1976). However, care must be taken in quoting one value for the $K_m$ for ammonia, as it may be influenced by pH and concentration of the substrates. PAHLICH and GERLITZ (1980) have recently suggested that pea seed GDH exerts strong negative cooperativity with ammonia, and that the catalytic efficiency of the enzyme increases at low ammonia concentrations. The $K_m$ value for ammonia was shown to decrease 25-fold at low ammonia levels. Utilizing GDH isolated from *Medicago sativa,* NAGEL and HARTMANN (1980) were able to show that by reducing the concentration of NADH from 100 to 5 μM the apparent $K_m$ was reduced almost tenfold.

The reaction mechanism of plant GDH was initially proposed by KING and WU (1971) to be similar to that of microbial enzymes, i.e., the compulsory ordered binding of NADH, 2-oxoglutarate and ammonia, followed by the ordered release of glutamate and NAD. However, two papers in 1977 suggested that the reaction may be fully random (GROAT and SOULEN 1977) or at least a partially random sequential mechanism (GARLAND and DENNIS 1977). Two more recent publications (STONE et al. 1980, NAGEL and HARTMANN 1980) confirm the initial proposal of an ordered mechanism.

## 2.5 Localization of Ammonia Assimilation

### 2.5.1 Enzyme Distribution

Much of the evidence of subcellular localization is of poor quality and must be viewed with scepticism (for further discussion see LEECH 1977 and QUAIL

1979). Studies with differential centrifugation suggest that glutamine synthetase (O'NEAL and JOY 1973a, HAYSTEAD 1973) glutamate synthase (LEA and MIFLIN 1974) and GDH (LEECH and KIRK 1968, LEA and THURMAN 1972) are present in chloroplasts and that GDH is also present, in much larger quantities, in mitochondria (see references quoted in STEWART et al. 1980). These localizations have been confirmed by density gradient centrifugation techniques using mechanically prepared homogenates (MIFLIN 1974, LEA and THURMAN 1972, WALLSGROVE et al. 1979a). However, the use of protoplasts as starting material allows a much improved recovery of intact chloroplasts and WALLSGROVE et al. (1979a) showed that probably all of the ferredoxin-dependent glutamate synthase of a leaf is in the chloroplast but that the glutamine synthetase occurs in the cytoplasm and the chloroplast. NAD-dependent GDH was almost entirely in the mitochondria. Further studies using this technique showed that the chloroplast isoenzyme of glutamine synthetase differs from the cytoplasmic one (MANN et al. 1979, 1980). Similarly, where GDH is found in chloroplasts it differs in properties, i.e., ratio of activity of NADPH/NADH (LEA and THURMAN 1972) or $K_m$ for ammonia (GAYLER and MORGAN 1976, MCKENZIE et al. 1979) from the mitochondrial enzyme.

A claim has been made that glutamine synthetase is present in the mitochondria (JACKSON et al. 1979) although the enzyme was only tested for by an indirect technique and no evidence of a product (or the product-related compound glutamyl hydroxamate) reported. Subsequent studies by our group (WALLSGROVE et al. 1980) and by NISHIMURA et al. (1980) have failed to substantiate the claim. In *non-green* tissue glutamine synthetase (WASHITANI and SATO 1977a, b, MIFLIN 1974, EMES and FOWLER 1978) and glutamate synthase (EMES and FOWLER 1978) have been shown to be present in plastids. Whether or not glutamine synthetase is present in the cytoplasm is subject to some doubt but the probability is high. Recent studies with the root nodule plant fraction have shown that the bulk of the glutamine synthase is cytoplasmic with a trace in plastids, whereas the glutamate synthase is definitely present in plastids (AWONAIKE et al. 1981). Whilst NAD-GDH is located in root mitochondria separated on density gradients (MIFLIN 1970), an NADP-dependent form has been found in tissue culture plastids separated by column techniques (WASHITANI and SATO 1977b).

### 2.5.2 Studies with Isolated Organelles

Various techniques have been used to show that isolated intact chloroplasts assimilate ammonia into amino acids in the light via the glutamate synthase cycle (see LEA and MIFLIN 1979 for detailed discussion). Thus SANTARIUS and STOCKING (1969) and MITCHELL and STOCKING (1975) have shown the ammonia-dependent formation of [$^{14}$C]-glutamine from [$^{14}$C]-glutamate; LEA and MIFLIN (1974) the net synthesis of glutamate from glutamine and 2-oxoglutarate and WALLSGROVE et al. (1979b) the formation of glutamate and glutamine from $NH_4^+$ and $NO_2^-$. Furthermore ANDERSON and DONE (1977a, b) have coupled the reaction to oxygen evolution, presumed to be dependent on the reoxidation of ferredoxin by glutamate synthase. The above groups have confirmed by the use of inhibitors, that the reactions are due to the action of the glutamate synthase cycle and not the presence of GDH. WASHITANI and SATO (1977a, b) have isolated plastids from non-green tissue cultures and showed them capable of ammonia assimilation in a manner consistent with the glutamate synthase cycle.

In contrast, there has been little evidence that mitochondria can assimilate ammonia although the reverse reaction, i.e., the oxidation of glutamate, is readily catalyzed. However, DAVIES and TEIXEIRA (1975) reported a small rate of assimilation under anaerobic conditions which, as they pointed out, is scarcely of widespread physiological relevance. Recently the question has been reconsidered because of the interest in the photorespiratory release of ammonia that occurs in mitochondria. Since some authorities consider that the electron transport chain may be blocked during photosynthesis, the state of the mitochondria may be analogous to that during anaerobiosis. WALLSGROVE et al. (1980) therefore attempted to see if mitochondria, whose electron transport was blocked with antimycin A, could assimilate ammonia released from glycine. No evidence was found in favour of this although DAVIES and TEIXEIRA's (1975) observations were confirmed when relatively large amounts of ammonia (16 mM) were supplied to the mitochondria.

## 2.6 Regulation

We have suggested elsewhere (MIFLIN and LEA 1980) that it is unlikely that green plants, grown in the light under conditions in which photosynthesis and $CO_2$ fixation are occurring normally, in any way limit their ammonia assimilation pathway. This is particularly likely to be true for plants growing in the field, where nitrate is the normal source of nitrogen and for which controls and limitations, if they exist, are likely to operate on the uptake and reduction of nitrate to ammonia. In favour of this assertion is the fact that most agricultural crops increase the total protein synthesized per unit area linearly with increased nitrogenous fertilizer supplied (e.g., see MIFLIN 1980b).

In the case of non-green plants, or plants grown under aquatic conditions with plentiful amounts of ammonia (and often limiting light conditions, at least in the laboratory), the situation may be different. With aquatic plants such as *Lemna* there is evidence that the amounts of assimilatory enzymes in the tissue can change according to nitrogen status (e.g., RHODES et al. 1976, STEWART and RHODES 1977a), and that the activity of glutamine synthetase may be altered by light–dark transitions (RHODES et al. 1979, STEWART et al. 1980). In general, we consider that the energy and carbohydrate status of the tissue is the most important factor in regulating the flux of N through the assimilatory pathways (including nitrate assimilation) and that the evolutionary pressure on green plants would have resulted in mechanisms which were extremely efficient at taking up N, when available, and preventing the build-up of toxic ammonia. Thus the light activation of glutamine synthetase, which parallels similar activation of other chloroplast enzymes (BUCHANAN 1980), is probably part of a general series of mechanisms (e.g., see MIFLIN 1977 for further discussion), ensuring that the assimilation of ammonia is as efficient as possible under conditions where nitrate is being reduced and there is a ready supply of carbon acceptors, or when ammonia is being generated during photorespiration (KEYS et al. 1978).

## 3 Transamination

Whichever of the two pathways of ammonia assimilation described in Section 2 is operating, glutamate is the first compound formed with a 2-amino group. Nitrogen is transferred to other compounds by a series of reactions catalyzed by aminotransferases (see WIGHTMAN and FOREST 1978, GIVAN 1980). The most widely investigated enzyme is aspartate aminotransferase (glutamate-oxoaloacetate transaminase, EC 2.6.1.1) which catalyzes the following reversible reaction:

$$\text{Glutamate} + \text{oxaloacetate} \rightleftharpoons \text{2-oxoglutarate} + \text{aspartate}$$

Using this route, aspartate is synthesized and 2-oxoglutarate can recycle to take part in the glutamate synthase cycle. Similar reactions are involved in the synthesis of glycine, alanine, isoleucine, leucine, valine, serine, and the aromatic amino acids.

The amino group of aspartate may be used directly in the synthesis of asparagine, lysine, threonine and methionine or transferred into arginine or nucleic acids. Aspartate:oxaloacetate:malate conversions are also thought to be important in the transport of reducing power out of the mitochondria (WOO et al. 1980) and chloroplasts (HEBER and WALKER 1979) and in the transport of $CO_2$ between the mesophyll and bundle-sheath cells of $C_4$ plants (RAY and BLACK 1979).

Transamination is also important in the metabolism of non-protein amino acids (see FOWDEN et al. 1979) in particular ornithine and $\gamma$-amino butyrate. The first step in amino acid breakdown is also often a transamination reaction (MAZELIS 1980). Asparagine in particular is now thought to be metabolized through this type of reaction in leaves (LLOYD and JOY 1978).

## 4 Biosynthesis of the Other Amino Acids

### 4.1 Introduction

The synthesis of the remainder of the amino acids will be dealt with below in groups according to the way in which the carbon skeleton of the amino acid is derived, following the well-established concepts of amino acid families. An attempt will be made to present evidence for the suggested pathways from both in vivo studies and the presence of the relevant enzymes. The regulation of synthesis of the amino acids will be discussed and, where suitable studies have been done, some indication of the subcellular compartmentation of the reactions will be given. For further reading and more detailed information the reader is referred to articles by GIVAN (1980), BRYAN (1980), GIOVANELLI et al. (1980) THOMPSON (1980), KEYS (1980), GILCHRIST and KOSUGE (1980) and ANDERSON (1980). Amino acid biosynthesis in plants has also been reviewed in recent years by BRYAN (1976) and MIFLIN and LEA (1977).

## 4.2 Synthesis of Amino Acids Derived from Pyruvate

Alanine (by direct transamination see above), the branched chain amino acids leucine, valine and isoleucine, and lysine derive all or part of their carbon from pyruvate; lysine, however, obtains the majority of its carbon from aspartate and its synthesis is thus covered in that section. The pathway to isoleucine, leucine and valine is given in Fig. 2. (In this and subsequent figures the reactions and the enzymes catalyzing them are keyed into the text by the use of numbers thus (*1*) is catalyzed by threonine dehydratase.) Evidence for this occurring

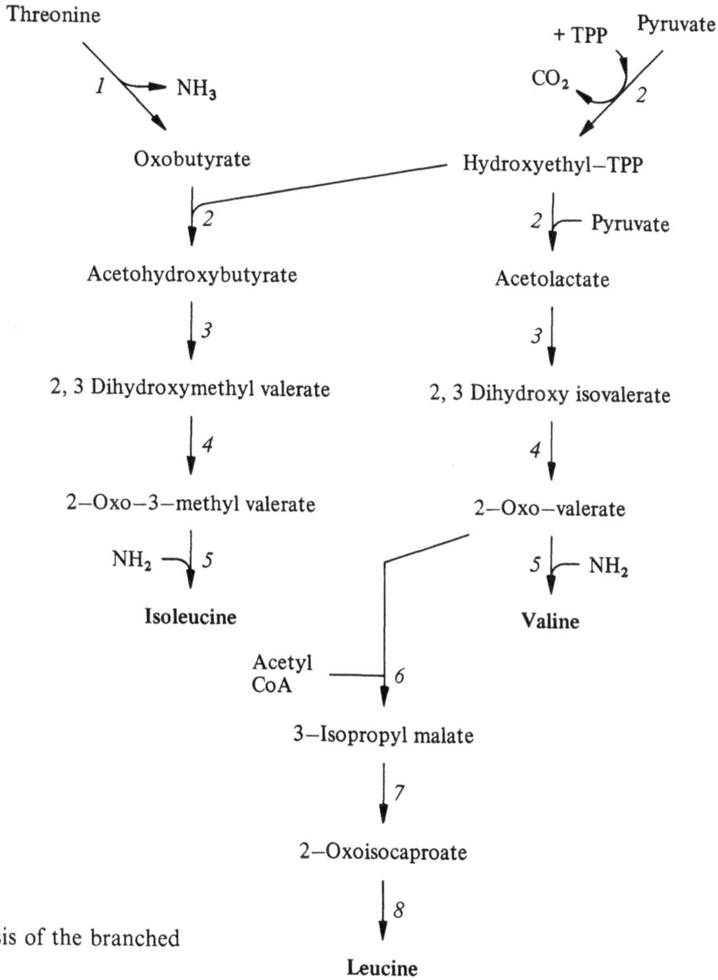


**Fig. 2.** The synthesis of the branched chain amino acids

in vivo has been obtained from isotope dilution experiments (Dougall and Fulton 1967a) and by feeding labelled precursors (Kagan et al. 1963, Butler and Shen 1963, Borstlap 1975).

### 4.2.1 Enzymic Evidence

The pathway shown in Fig. 2 does not have a common portion in terms of carbon flux but does appear to be common in the sense that the enzymes catalyzing the conversion of acetohydroxybutyrate to oxomethylvalerate are the same as those converting acetolactate to oxovalerate. Starting from threonine (whose synthesis is dealt with in Sect. 4.4) the first enzyme is threonine dehydratase EC 4.2.1.16 (*1*) which also deaminates threonine to give oxobutyrate. This enzyme has been found in several plant tissues and shows extremely complex kinetics, indicating its probable allosteric nature (Dougall 1970, Sharma and Mazumder 1970, Bleckman et al. 1971, Bryan 1980). The first "common" enzyme in the pathway is acetolactate synthase EC 4.1.3.18 (*2*). Although this has chiefly been studied by measuring the the production of acetolactate, it also produces acetohydroxybutyrate (Miflin 1971). The reaction involves the decarboxylation of pyruvate to form hydroxyethyl-thiamine pyrophosphate which is then accepted by either another molecule of pyruvate or by 2-oxobutyrate. The enzyme is found in a wide range of plant species (Miflin and Cave 1972). The enzymes catalyzing reactions *3* and *4* are acetohydroxyacid reductoisomerase EC 1.1.1.86 and dihydroxyacid dehydratase EC 4.2.1.9 and their presence is indicated in plant extracts; Kanamori and Wixom (1963) and Kiritani and Wagner (1970) have purified the latter enzyme from spinach. The final steps (*5*) of valine and isoleucine synthesis are catalyzed by a transaminase which may or may not be highly specific (see Givan 1980). The synthesis of isopropylmalate (*6*) has been demonstrated in plant extracts (Oaks 1965b), but no detailed enzymology of this branch of the pathway has been carried out; again transaminases active with leucine are well-documented in plants (Givan 1980).

### 4.2.2 Subcellular Localization

Evidence that at least part of the pathway occurs in chloroplast comes from the finding that acetolactate synthase is associated with plastids in green and non-green tissues (Miflin 1974). Furthermore there is evidence that threonine dehydratase is present in easily sedimenting particles (Bleckman et al. 1971). The light-dependent synthesis of isoleucine from [$^{14}$C]-labelled aspartate and threonine has been demonstrated in isolated chloroplasts (Mills and Wilson 1978, Mills et al. 1980) and recently Bassham et al. (1981) have found that isolated chloroplasts synthesize valine from [$^{14}$C]$O_2$ at a rate commensurate with the synthesis in whole cells.

### 4.2.3 Regulation

Studies with labelled precursors show that addition of very small amounts of exogenous leucine regulates the in vivo synthesis of this amino acid in maize root tips (OAKS 1965a); similar evidence also suggests that the synthesis of valine and isoleucine is regulated. MIFLIN (1969) has used an alternative approach in which the inhibitory effects of amino acids on the growth of seedlings has been measured; again these results suggest that regulation of the pathway occurs and probably involves cooperative effects between leucine and valine. Similar results have also been obtained with duckweed (*Spirodela*) (BORSTLAP 1970). When isolated enzymes are studied, feedback inhibition can be demonstrated for threonine dehydratase (*1*) and acetolactate synthase (*2*). Threonine dehydratase is inhibited by isoleucine but valine can partially antagonize this effect under certain conditions (DOUGALL 1970, SHARMA and MAZUMDER 1970, BLECKMAN 1971, BRYAN 1980). Acetolactate synthase is cooperatively inhibited by low concentrations of leucine and valine, although both amino acids are inhibitory on their own at higher concentrations (MIFLIN 1971, MIFLIN and CAVE 1972). OAKS (1965b), using relatively crude extracts of maize, has also provided evidence consistent with isopropylmalate being subject to feedback inhibition by leucine. Consequently there exists, at the enzyme level, a series of controls that should enable the plant to maintain the desired balance of leucine, valine and isoleucine synthesis.

### 4.3 Synthesis of Amino Acids Derived from Glutamate

Besides glutamine, proline and arginine derive all or part of their carbon from glutamate. The pathways involved are given in Fig. 3. Evidence for the pathway is given below.

#### 4.3.1 Enzymic Evidence

Amino-acid acetyltransferase, EC 2.3.1.1 (*1*) has been described for cell-free extracts of *Chlorella* and *Beta vulgaris* (MORRIS and THOMPSON 1975, 1977). In both cases the enzyme appears to copurify with glutamate acetyltransferase, EC 2.3.1.35. (*2*) but is much more unstable so that preparations with only the latter activity may be obtained. In higher plants enzyme (*2*) has about a tenfold lower $K_m$ for glutamate than (*1*) but the values are about the same for both in *Chlorella*. ATP: acetylglutamate kinase EC 2.7.2.8 (*3*) has been found in cell-free extracts of algae (FARAGO and DENES 1967, MORRIS and THOMPSON 1975) and higher plants (MORRIS and THOMPSON 1977, MCKAY and SHARGOOL 1977). The *Chlamydomonas* enzyme has been purified 120-fold and its characteristics studied in some detail (FARAGO and DENES 1967). The next steps involve the reduction of acetyl glutamylphosphate to acetylglutamic semialdehyde catalyzed by N-acetyl-$\gamma$-glutamyl phosphate reductase EC 1.2.1.38 (*4*). This reaction has only been demonstrated relatively rarely including in cell-free extracts of *Beta vulgaris* (MORRIS et al. 1969). Similarly little work has been done on acetyl-ornithine aminotransferase EC 2.6.1.11 (*5*) except for the demonstration of activity in cell-free extracts of rose tissue culture cells (DOUGALL and FULTON 1967b). The product of (*5*) is N-acetyl ornithine which is converted to ornithine by (*2*) above. Ornithine is converted to arginine via citrulline and

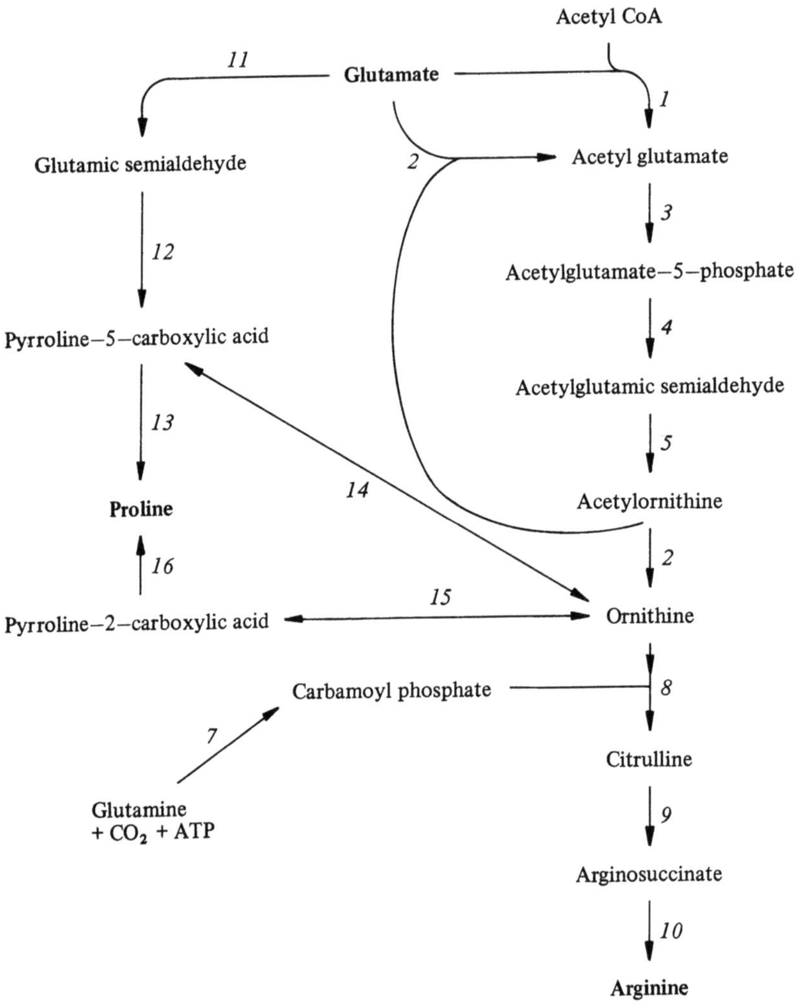


**Fig. 3.** Pathway for the synthesis of proline and arginine

requires carbamoyl phosphate as co-substrate. Carbamoyl phosphate is formed from glutamine ATP and $CO_2$ via the action of carbamoyl-phosphate synthetase EC 6.3.5.5 (*7*). This enzyme has been found in several plants (O'NEAL and NAYLOR 1969, 1976, ONG and JACKSON 1972, SHARGOOL et al. 1978, ROUBELAKIS and KLIEWER 1978a). Ornithine then combines with carbamoylphosphate to form citrulline catalyzed by ornithine carbamoyltransferase EC 2.1.3.3 (*8*). This enzyme has been purified to near homogeneity (2,000-fold) (KLECZKOWSKI and COHEN 1964) from peas. Two forms of this enzyme have been described in several plants (EID et al. 1974; GLENN and MARETZKI 1977, SPENCER and TITUS 1974) which differ in a number of properties (e.g., size, pH optima, chromatography on DEAE) and possibly in their physiological function. Arginosuccinate synthetase EC 6.3.4.5 (*9*) converts citrulline to arginosuccinate. Reports of the existence of this activity in cell-free extracts of germinating pea seeds and vine tissues have been made (SHARGOOL and COSSINS 1968, SHARGOOL 1971, 1975, SHARGOOL et al. 1978, ROUBELAKIS and KLIEWER 1978b). The enzyme has been purified 400-fold from germinating pea cotyledons and separated from (*10*) (SHARGOOL 1971). The arginosuccinate so formed is subsequently cleaved via the action of arginosuccinate lyase EC 4.3.2.1 (*10*) to give arginine and fumarate. Many reports of this enzyme activity in cell-free extracts have been made (WALKER and MYERS 1953, SHARGOOL 1971, 1975, ROUBELAKIS and KLIEWER 1978b). The activity is difficult to stabilize, the reason for this may be due to a protease which degrades the enzyme (SHARGOOL 1975).

In microorganisms the biosynthesis of proline is considered to occur via a non-acetylated pathway (UMBARGER 1978). The enzymic evidence for the first step however is not particularly detailed even in *Escherichia coli*. The reactions involved are postulated as the activation of the $\gamma$-carboxyl group of glutamate by the formation of glutamyl-5-phosphate and its subsequent hydrolysis and reduction to glutamyl-5-semialdehyde (*11*) (i.e., an analogous series of reactions to the first steps of the aspartate pathway). The intermediate glutamyl-5-phosphate, which is unstable, is considered to exist only as an enzyme-bound intermediate. The product glutamic semialdehyde spontaneously cyclizes to form pyrroline-5-carboxylate. The only evidence that we are aware of that reactions (*11, 12*) occur in plants is the report of MORRIS et al. (1969) that cell-free extracts of *Beta vulgaris* can convert [$^{14}$C]-glutamate to proline, dependent on the presence of ATP and NADH or NADPH. Much more evidence is available for the presence of pyrroline-5-carboxylate reductase EC 1.5.1.2 (*13*) in that extracts of several plants reduce pyrroline-5-carboxylate to proline (NOGUCHI et al. 1966, RENA and SPLITSTOESSER 1975, VANSUYT et al. 1979) with either NADH or NADPH, although the latter was the better donor for the purified (150-fold) enzyme from tobacco (NOGUCHI et al. 1966). Plant extracts have also been assayed for their ability to carry out the reverse reaction and the activity termed L-proline dehydrogenase. It has been suggested that the two activities are the reversible properties of the same protein (THOMPSON 1980, RENA and SPLITSTOESSER 1975) although the two reactions may differ in their pyridine nucleotide specificities.

As shown in Fig. 3 there are two pathways which are potentially capable of interconverting ornithine and proline. Ornithine has an amino group in the 2 and 5 positions; according to which is transferred it may thus be converted to either glutamic-5-semialdehyde or 2-oxo-5-amino valeric acid, both of which cyclize non-enzymically to give respectively pyrroline-5-carboxylate (P5C) and pyrroline-2-carboxylate (P2C). Evidence for the presence of ornithine: 2-oxo acid 5-aminotransferase EC 2.6.1.13 (*14*) in extracts of many different plants has been claimed (SUNI and DENES 1967, LU and MAZELIS 1975, SCHER and VOGEL 1957, SMITH 1962, MAZELIS and FOWDEN 1969), but the product has usually only been measured by the formation of a yellow complex with ortho-aminobenzaldehyde. This reaction is not specific for P5C since P2C gives a similar colour – the addition compounds can, however, be separated easily by electrophoresis (HASSE et al. 1967) or by paper chromatography (MCNAMER and STEWART 1974) which, in hindsight, should have been utilized in the above studies. SENEVIRATNE and FOWDEN (1968) did investigate the products of ornithine transamination by extracts of mung bean mitochondria and found evidence for the transamination of both 2 and 5 amino groups; however, they were unable to exclude the possibility that some or all of the 5-transaminase activity was due to a non-enzymic reaction of ornithine with pyridoxal phosphate. Stronger evidence for the presence of ornithine: 2-oxoglutarate 2-aminotransferase, comes from the work of HASSE et al. (1967), who showed that extracts of mung bean (*Phaseolus aureus*) and *Lupinus angustifolius* yielded an ortho-aminobenzaldehyde complex that co-electrophoresed with that obtained with authentic P2C but was separate from that obtained with P5C. Conversion of P5C to proline, from ornithine transamination, can be catalyzed by (*13*) as discussed above and plants also contain pyrroline-2-carboxylate reductase, EC 1.5.1.1 (*16*) as shown by MEISTER et al. (1957) and MACHOLAN et al. (1965).

### 4.3.2 In Vivo Studies

Many labelling studies have been carried out that show [$^{14}$C]-glutamate may be metabolized to proline and to arginine (see THOMPSON 1980) but these do not allow discrimination between the multiple pathways in Fig. 3. Specific evidence for the involvement of the acetylated pathway in arginine biosynthesis comes from the isotope competition experiments of DOUGALL and FULTON (1967b) in which they showed that, in the presence of N-acetylornithine, the incorporation of $^{14}$C from glutamate into arginine was decreased by 70%. Similar or greater reductions were found in the presence of ornithine and citrulline. In these experiments they found no reduction in the labelling of any other of the protein amino acids, thereby suggesting that N-acetylornithine and ornithine are not normally on the route between glutamate and proline. Despite this conclusion there is evidence that plants can convert ornithine to proline. By using specifically double-labelled ornithine MESTICHELLI et al. (1979) have shown convincingly that plants can convert ornithine to proline via a 2-transamination, presumably involving the formation and subsequent reduction of P2C. Although the authors claim, on the basis of their results and from criticism of some of the enzyme studies "that the time is ripe for a re-examination of proline biosynthesis" their results do nothing to disprove the accepted view that proline is normally formed via the non-acetylated pathway (Fig. 3 reactions *10–12*). MESTICHELLI et al. (1979) assume that exogenously supplied, labelled ornithine actually enters the pool of intermediates involved in endogenous proline synthesis. That this assumption is unjustified is shown by work in *Neurospora crassa* (DAVIS and MORA 1968) in which exogenous ornithine only serves as a source of arginine in mutants that have lost ornithine transaminase activity. In the presence of the transaminase, exogenously supplied ornithine is catabolized, rather than used as a source for the biosynthesis of arginine. Furthermore, MESTICHELLI et al. (1979) stopped their feeding experiments, after incubation in labelled ornithine, by placing their plants in an oven at 45–50 °C, however, since MORRIS et al. (1969), and BOGGESS et al. (1976a, b) have shown that wilting accelerates the conversion of ornithine to proline, the results observed by the authors may have been unduly influenced by the stress imposed prior to desiccation and may not necessarily relate to the conversions occurring in unstressed tissues.

Controversies regarding metabolic pathways in microorganisms have often been resolved by the use of auxotrophic mutants. In higher plants few such mutants exist, however GAVAZZI et al. (1975) have isolated proline auxotrophs of maize. Feeding experiments in which P5C was unable to support growth suggested that the block was at P5C reduction (RACCHI et al. 1978). However, preliminary enzyme experiments showed that the mutants contained as much P5C reductase (*13*) as the wild type (BERTANI et al. (1980). The possibility that ornithine could overcome the proline requirement was also tested but with negative results (RACCHI et al. 1978). Consequently no conclusion can yet be made regarding the nature of the mutation.

### 4.3.3 Sub-Cellular Localization

Almost all the studies on the localization of the enzymes in Fig. 3 fail to measure up to required standards of experimental procedure and thus fail to provide any conclusive data. The vast majority of authors using non-green tissue have ignored the possibility of the presence of any sedimentable organelles other than mitochondria, i.e., have taken no account of plastids or microbodies. Since both those latter organelles contain enzymes involved in amino acid metabolism (see MIFLIN 1974, MIFLIN and LEA 1977, LEA et al. 1981) it is not possible to conclude that because an enzyme sediments at 10,000 *g* for ca. 15 min it is present in the mitochondria; claims to this effect should be ignored, unless supplementary evidence is presented. Such evidence should involve assays of marker enzymes for all known organelles and preferably separations involving density gradient centrifugation (see LEECH 1977 for a further discussion of the problem). Of the evidence available there is the suggestion, supported by the light-stimulation of the reactions in vivo (NOGUCHI et al. 1968, HANSON and TULLY 1979), that proline synthesis occurs in the chloroplasts (NOGUCHI et al. 1966) and that the enzymes for the synthesis of carbamoyl phosphate (*7*) and citrulline (*8*) are present in plastids (SHARGOOL et al. 1978), and that the transamination of ornithine (SPLITTSTOESSER and FOWDEN 1973) can occur in mitochondria. The lack of knowledge of the compartmentation of the reactions of pathways as complex as those shown in Fig. 3 and of the possibility of multiple forms of the enzymes in separate compartments, severely limits our understanding. Such information is urgently required if progress is to be made in this important area of metabolism. Recently TAYLOR and STEWART (1981) have shown that (*2*), (*7*) and (*8*) are in the choroplast.

### 4.3.4 Regulation

Evidence for the in vivo regulation of the pathway has been provided by the $^{14}C$ feeding studies of NOGUCHI et al. (1968), OAKS (1965a), OAKS et al. (1970), and BOGGESS et al. (1976a) in which proline and arginine have been found to inhibit their own synthesis from carbon precursors. The degree of control of proline biosynthesis appears to change as a function of age; thus mature maize root segments show less control than root tips (OAKS et al. 1970). Wilting also appears to release the control mechanisms operating on proline synthesis, probably prior to the synthesis of P5C (BOGGESS et al. 1976a).

In vitro studies on proline biosynthesis are so fragmentary that it is not possible to have any knowledge of the biochemical mechanisms governing the rate of proline synthesis. However, the enzymic controls on arginine synthesis have been studied. Arginine has been found to inhibit the transacetylation of glutamate from acetyl CoA (*1*) but not from acetylornithine (*2*) (MORRIS and THOMPSON 1975, 1977). Arginine also appears to control the activity of acetylglutamate kinase (*3*) in all species studied (MORRIS and THOMPSON 1975, 1977, MCKAY and SHARGOOL 1977, FARAGO and DENES 1967). Both of these inhibitions, which occur at relatively low (mM or less) concentrations of arginine, may be physiologically relevant, and could potentially turn off the pathway independent of which type of acetylation reaction is occurring. The other source of carbon and nitrogen for arginine is carbamoyl phosphate which is also a precursor of pyrimidines. Control of carbamoyl phosphate synthetase might therefore be expected to be more complex so as to prevent one pathway shutting off the other. It has been found that the enzyme is inhibited by UMP but that this inhibition is partially relieved by ornithine (ONG and JACKSON 1972, O'NEAL and NAYLOR 1976) which should allow the plant to synthesize arginine in the presence of UMP. SHARGOOL (1973) has also reported the control of arginosuccinate synthetase by energy charge, a control that is modified by the level of arginine.

### 4.4 Synthesis of the Aspartate Family of Amino Acids

Apart from asparagine, aspartate also donates carbon via a series of reactions stemming from the same two initial steps to lysine, methionine, threonine and isoleucine. Because isoleucine also receives some of its carbon from pyruvate and has enzyme steps in common with the other branched chain amino acids (leucine and valine) its synthesis is covered above. The reactions of the aspartate pathway are summarized in Fig. 4. Evidence for this pathway operating in higher plants has come from many studies which are reviewed below and also elsewhere (MIFLIN et al. 1979, BRYAN 1980).

Evidence for the operation of the pathway in vivo as shown in Fig. 4 has come from a number of studies. Feeding radioactive aspartate leads to the formation of lysine (VOGEL 1959, DUNHAM and BRYAN 1971, MOLLER 1974). Studies with specifically labelled substrates and determination of the distribution of label in specific carbons of lysine have ruled out the participation of the fungal aminoadipic acid pathway and are entirely consistent with the reactions shown in Fig. 4 (MOLLER 1974). Isotope dilution experiments (DOUGALL and FULTON 1967a) have indicated that homoserine is an intermediate in threonine and methionine synthesis, that cystathionine and homocysteine (but not apparently O-succinyl or O-acetylhomoserine) are intermediates in methionine formation and that diaminopimelate is on the route to lysine synthesis.

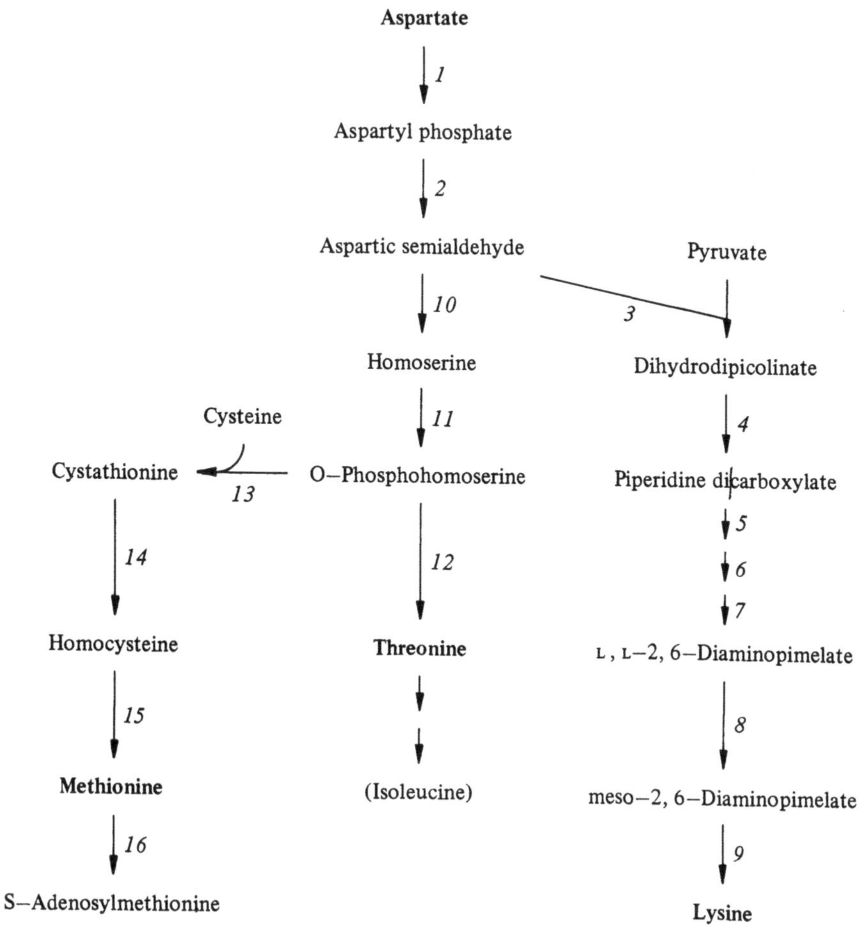


**Fig. 4.** The synthesis of amino acids derived from aspartate

### 4.4.1 Enzymic Evidence

The initial enzyme aspartate kinase EC 2.7.2.4 (*1*) has been demonstrated in many plants (for example see BRYAN et al. 1970, CHESHIRE and MIFLIN 1975, AARNES and ROGNES 1974, AARNES 1974, SHEWRY and MIFLIN 1977, GENGENBACH et al. 1978, LEA et al. 1979b, ROGNES et al. 1980). The enzyme has been generally found to have complex substrate kinetics with respect to aspartate and to require $Mg^{2+}$ ions. Levels of activity are often low and considerable care needs to be taken in assaying the enzyme, particularly in crude extracts, because of competing reactions (see DAVIES and MIFLIN 1977, MIFLIN et al. 1979 for further discussion of this problem). The enzyme exists in multiple forms (SAKANO and KOMAMINE 1978, DAVIES and MIFLIN 1978) which have different regulatory properties (see below). The product of the reaction, aspartyl-phosphate, is unstable and probably is not released free into the cell environment but rather immediately reduced by aspartate-semialdehyde dehydrogenase

EC 1.2.1.11 (*2*). However, it should be pointed out that there is no evidence for any association between these two enzymes. Aspartate semialdehyde dehydrogenase has been little studied but has been shown to be present in pea (SASAOKA 1961), maize (see BRYAN 1980) and barley extracts (DEGRYSE and MIFLIN unpubl.). Aspartate semialdehyde is at the first branch point of the pathway; on the route to lysine formation it is condensed with pyruvate to form dihydrodipicolinate catalyzed by the enzyme dihydrodipicolinate synthase, EC 4.2.1.52 (*3*). This is present in cell-free extracts of several plants (CHESHIRE and MIFLIN 1975, MAZELIS et al. 1977, MATTHEWS and WIDHOLM 1978) and has recently been purified and characterized (WALLSGROVE and MAZELIS 1981). Beyond this step there is no further report of enzymes from plants catalyzing the postulated reactions (*4–8*) until the last step, the decarboxylation of diaminopimelate. Early studies by SHIMURA and VOGEL (1966) indicated that green plants contained diaminopimelate decarboxylase EC 4.1.1.20 (*9*) which has been confirmed by VOGEL and HIROVONEN (1971), MAZELIS et al. (1976), MAZELIS and CREVELING (1978), and SODEK (1978). The enzyme appears to be specific for the meso form of the substrate.

Aspartic semialdehyde is also a precursor of threonine and methionine. It is first reduced via the action of homoserine dehydrogenase EC 1.1.1 (*10*) to homoserine, an enzyme which has been found in a wide range of plant species (e.g., see BRYAN 1969, AARNES and ROGNES 1974, DI MARCO and GREGO 1975, AARNES 1977a, MATTHEWS and WIDHOLM 1979a, b, WALTER et al. 1979) and considerably purified (DICAMELLI and BRYAN 1975, WALTER et al. 1979). The enzyme is generally stimulated by monovalent cations; it exists in a number of isoenzymic forms which fall into two classes distinguished by their size, location in the cell, response to $K^+$ ions and their regulatory properties (see below) and use both NADPH and NADH (but the former is generally a better substrate). This means that care must be taken to separate the isoenzymes before their properties, especially kinetics, can be determined.

Homoserine kinase EC 2.7.1.39 (*11*) catalyzes the addition of a phosphate group to homoserine and has been demonstrated in cell-free extracts of barley (AARNES 1976, 1978) and peas (THOEN et al. 1978a). Homoserine phosphate is then metabolized to threonine via the action of threonine synthase, EC 4.2.99.2 (*12*). Reports of the presence of this enzyme in plants were first made by SCHNYDER et al. (1975) and have been confirmed by the work of MADISON and THOMPSON (1976), AARNES (1978) and THOEN et al. (1978b).

Detailed studies by GIOVANELLI, DATKO and MUDD (GIOVANELLI et al. 1974, DATKO et al. 1974, 1977, GIOVANELLI et al. 1980) have established that phospho-homoserine is the 2-aminobutyryl donor for cystathionine biosynthesis in plants and thus is at the second branch point in the pathway. Cystathionine synthase (*13*) [EC unclassified since EC 4.2.99.9 O-succinylhomoserine (thiol)-lyase is obviously not an appropriate name for the plant enzyme] has been studied in crude extracts of several plants (DATKO et al. 1974, 1977) and is relatively non-specific in that it can use a range of 2-aminobutyryl donors. Cystathionine is cleaved by cystathionine-$\beta$-lyase EC 4.4.1.8 (*14*) to give homocysteine, pyruvate and ammonia. The enzyme has been purified (430-fold) from spinach (*Spinacia*) leaves (GIOVANELLI and MUDD 1971) and also shown to cleave cysteine. ANDERSON and THOMPSON (1979) have purified a cysteine lyase from turnips which is also capable of cleaving cystathionine; the true physiological role of this enzyme is not clear. The final step in methionine synthesis is the methylation of homocysteine catalyzed by tetrahydropteroylglutamate methyltransferase EC 2.1.1.14 (*15*), shown to be present in extracts of plants by BURTON and SAKAMI 1969, DODD and COSSINS 1970, CLANDININ and COSSINS 1974, SHAH and COSSINS 1970b.

Besides the above reaction certain other transformations can occur in plants which are relevant to methionine biosynthesis. Enzymic evidence (GIOVANELLI and MUDD 1967, DATKO et al. 1974, 1977) has shown that plants have the potential to directly sulphydrate O-phosphohomoserine (or other 2-aminobutyryl donors) to form homocysteine.

$$\text{O-phosphohomoserine} + H_2S \rightarrow \text{homocysteine} + \text{Pi}$$

This reaction would bypass reactions *13* and *14* in Fig. 4. Careful assessment of the relative contributions of the two pathways in *Chlorella* (GIOVANELLI et al. 1978) and *Lemna* (MACNICOL et al. quoted in GIOVANELLI et al. 1980) have indicated that the direct sulphydration pathway is unlikely to contribute more than about 10% of the homocysteine formed in plants. In considering methionine synthesis it is important to note that it is not only used for protein synthesis but is the precursor of S-adenosylmethionine which is used by plants in a variety of reactions (e.g., ethylene biosynthesis, methylation reactions etc. for further discussion see BRIGHT et al. 1980, GIOVANELLI et al. 1980). The enzyme responsible for its synthesis is methionine adenosyltransferase EC 2.5.1.6 (*16*) and has been shown to be present in barley and *Pisum* (MUDD 1960, CLANDININ and COSSINS 1974, AARNES 1977b).

### 4.4.2 Subcellular Localization

Two approaches have been used to attack this problem. Firstly, organelles have been isolated and incubated with radioactive aspartate and the products determined and, secondly, the distribution of component enzymes in various subcellular fractions has been studied. The results with intact organelles show that isolated, intact chloroplasts carry out a light-dependent synthesis of lysine, threonine, isoleucine and methionine from $[^{14}C]$-aspartate or $[^{14}C]$-malate (MILLS and WILSON 1978, MILLS et al. 1980). These reactions could not be demonstrated in mitochondria (MILLS et al. 1980). The results thus show that at least part of the total synthesis of the aspartate family of amino acids occurs in the chloroplast.

Subcellular distribution studies have also shown that some at least of the aspartate kinase and homoserine dehydrogenase activity of the cell is associated with the chloroplasts (LEA et al. 1979b, BURDGE et al. 1979, WAHNBAECK-SPENCER et al. 1979). More detailed studies with homoserine dehydrogenase have shown that about half of the enzyme is associated with the chloroplast and that the remainder appears in the cytosol (SAINIS et al. 1981). The chloroplastic form in peas and barley differs from the cytosol form in size, affinity characteristics, regulatory properties and inhibitor kinetics and thus the two forms are obviously isoenzymes of entirely different nature and correspond to the two classes of enzyme established by enzyme purification studies (BRYAN 1980, DICAMELLI and BRYAN 1975, 1980, AARNES and ROGNES 1974, AARNES 1977a, WALTER et al. 1979, SAINIS et al. 1981). Whether or not the cytosolic form plays any biosynthetic role remains to be determined.

The localization of two enzymes on the lysine branch has been studied. Initial results with mechanically isolated chloroplasts showed that diaminopimelate decarboxylase (*19*) was present in intact plastids (MAZELIS et al. 1976). Further studies with chloroplasts isolated from protoplasts indicate that all of this enzyme and dihydrodipicolinate synthase, in the leaf cell, is present in the chloroplasts (WALLSGROVE et al. 1979c, WALLSGROVE and MAZELIS 1980). These results suggest that the chloroplast (and probably, in non-green tissue, the plastid) is the sole site of lysine synthesis.

Of the other enzymes, there is suggestive evidence that threonine synthase (AARNES 1978) and a proportion of methionine adenosyltransferase and tetrahydropteroyltriglutamate methyltransferase are all in the chloroplasts (SHAH and COSSINS 1970b) but that the latter enzyme may also be present in the mitochondria (CLANDININ and COSSINS 1974). However, none of these studies has been carried out using appropriate techniques (e.g., see above) and they must be regarded with some reserve.

### 4.4.3 Regulation

Evidence for the regulation of the pathway in vivo has been obtained from the studies of OAKS (1965a) which showed that lysine and threonine each inhibits its own biosynthesis from [$^{14}$C]-labelled carbon precursors. Subsequently similar evidence has been obtained (OAKS et al. 1970, DUNHAM and BRYAN 1971, DAVIES and MIFLIN 1978, HENKE and WILSON 1974). It has also been shown that methionine synthesis is inhibited by the addition of exogenous methionine (BRIGHT S.W.J. unpubl.) and by lysine plus threonine by using [$^{35}$S]$O_4^{2-}$ feeding to germinating seedlings (BRIGHT et al. 1978b). GIOVANELLI et al. (1980) have also fed [$^{35}$S]$O_4$ to *Lemna* and obtained results that suggest that methionine (or a derivate thereof) controls de novo assimilation of sulphate into cystathionine and its products, suggesting that the regulatory steps in methionine formation are at reaction (*13*) (Fig. 4) or before.

Further evidence for complex regulatory patterns in this pathway can be deduced from the effects of exogenous amino acids on the growth of plant tissues. Whilst not all of the inhibitory effects of amino acids on plant tissues are due to their interference with amino acid metabolism (e.g., see FILNER 1969) it is possible to interpret the effects of certain combinations of amino acids in this way. Thus the inhibitory effects of lysine and threonine on plant growth can be relieved by methionine or methionine precursors (DUNHAM and BRYAN 1969, FURUHASHI and YATAZAWA 1970, GREEN and PHILLIPS 1974, BRIGHT et al. 1978a, HENKE et al. 1974) and can be interpreted in terms of feedback inhibition of the exogenous amino acids on early steps in the pathway. Lysine and threonine separately are not generally very inhibitory on their own (but see FILNER 1969) but together have a marked synergistic effect on growth (HENKE et al. 1974, BRIGHT et al. 1978a). This synergism could be explained by concerted inhibition of aspartate kinase by the two amino acids (for which there is no enzymic evidence), by additive inhibition of lysine and threonine isoenzymes (for which there is evidence see below), and/or cumulative effects on aspartate kinase and homoserine dehydrogense.

Studies on the regulation of the biosynthesis of lysine and threonine in isolated pea leaf chloroplasts have also shown that the pathway is subject to end-product inhibition (MILLS et al. 1980). Addition of exogenous lysine decreases total lysine synthesis and that of threonine, whereas threonine only inhibits its own synthesis, suggesting that the flux through the common part of the pathway (reactions *1* and *2*, Fig. 4) is predominantly regulated by lysine and that the main site of threonine action is after reaction *2* (MILLS et al. 1980). Since threonine also inhibits homoserine synthesis (LEA et al. 1979b) it suggests reaction (*10*) as the sensitive site.

Many studies on the regulation of isolated enzymes have been reported in the literature. Initial work on aspartate kinase suggested that the enzyme in maize (BRYAN et al. 1970), barley (AARNES 1977a), wheat (*Triticum aestivum*) (BRIGHT et al. 1978b), and sunflower (*Helianthus annuus*) (AARNES 1974) was lysine-sensitive, whereas in germinating pea seeds it was inhibited by threonine (AARNES and ROGNES 1974). Subsequent studies have changed this picture. Firstly, the presence of lysine- and threonine-sensitive isoenzymes is suggested by the additive inhibitions of these two amino acids reported for aspartate kinases from mustard (*Sinapis alba*) (AARNES 1974), carrot (*Daucus carota*) tissue cultures (DAVIES and MIFLIN 1977), developing pea and bean (*Vicia*) cotyledons (BRIGHT et al. 1980; ROGNES et al. 1980), young pea leaves (LEA et al. 1979b) and the suggestion substantiated by the physical separation of two forms of the enzyme from carrots (DAVIES and MIFLIN 1978, SAKANO and KOMAMINE 1978) and soyabean (MATTHEWS and WIDHOLM 1979a, c). The relative amounts of the two isoenzymes of aspartate kinase change as a function of the physiological state of the tissue; physiologically active dividing cells have proportionally more lysine-sensitive isoenzyme than older non-dividing tissues (DAVIES and MIFLIN 1978, SAKANO and KOMAMINE 1978, LEA et al. 1979b, BRIGHT et al. 1980). Secondly, lysine-sensitive asparte kinases have a second regulatory site. Studies by SHEWRY and MIFLIN (1977) showed that the barley enzyme was cooperatively inhibited by lysine and methionine at low levels of lysine. Similar, slight effects of methionine, in the presence of lysine, were also observed with the pea leaf enzyme (LEA et al. 1979b). This effect has been followed up by ROGNES et al. (1980) who have shown that the lysine-sensitive aspartate kinase from *Hordeum vulgare, Cucumis sativus, Pisum sativum, Zea mays* and *Vicia faba* is synergistically regulated by lysine and S-adenosylmethionine at low levels of both effectors. Thus the addition of 0.1 mM S-adenosylmethionine decreases the amount of lysine required for half-maximal inhibition of *H. vulgare* aspartate kinase from 340 to 48 μM. The previously reported effect of methionine was probably due to the presence of methionine adenosyltransferase (*16*) (Fig. 4) in the extracts and all the necessary reactants present in the assay tube for S-adenosylmethionine to be formed upon the addition of methionine. Thirdly, the lysine-sensitive activity of barley can be separated into at least two peaks by DEAE-cellulose chromatography and there is genetic evidence that these are separate lysine-sensitive iso-enzymes (BRIGHT et al. 1981 and unpublished).

Therefore it appears that most plants have a common mechanism of control for aspartate kinase operating on the lysine-sensitive isoenzyme(s) and the outstanding question is whether or not all higher plants have a threonine-sensitive aspartate kinase; recent work in our laboratory (S.E. ROGNES unpublished) suggests that this might be the case. Since the reports of concerted lysine + threonine inhibition of the *Triticum* (WONG and DENNIS 1973) and *Cucumis* enzymes (AARNES 1974) have not been confirmed by subsequent studies (BRIGHT et al. 1978b, ROGNES et al. 1980) it would appear that the controls at the aspartate kinase step are more universal among different species than was considered likely previously (e.g., see BRYAN 1976, 1980, MIFLIN 1977, MIFLIN et al. 1979).

The control of lysine synthesis may also be achieved at the first enzyme past the branch point, dihydrodipicolinate synthase (*4*), since its activity is inhibited by lysine in extracts of *Zea mays, Triticum aestivum, Pisum sativum* and *Spinacia oleracea* (CHESHIRE and MIFLIN 1975, MAZELIS et al. 1977, WALLSGROVE and MAZELIS 1981). The sensitivity of the enzyme to lysine is considerable, with 50% of the maximal inhibitory effect occurring at 11 μM lysine.

Homoserine dehydrogenase is only affected by threonine with no reported effects by methionine (BRYAN 1969, DICAMELLI and BRYAN 1975, 1980, AARNES and ROGNES 1974, AARNES 1977a, WALTER et al. 1979). When the two classes of the enzyme are considered it is found that only the larger chloroplastic enzyme is inhibited; the smaller, cytosolic enzyme is virtually insensitive to threonine (DICAMELLI and BRYAN 1975, 1980, WALTER et al. 1979, SAINIS et al. 1981). When cereal tissues of increasing ages are extracted, the degree of sensitivity of homoserine dehydrogenase seems to decrease (DICAMELLI and BRYAN 1975, MATTHEWS et al. 1975, SAINIS et al. 1981). This effect appears to depend on changes in the sensitive chloroplastic class and not on the relative amount of the two classes (DICAMELLI and BRYAN 1975, SAINIS et al. 1981). The exact physiological significance of this finding is not clear and it has not been found in certain plant species (BRYAN et al. 1979, SAINIS et al. 1981) or by all workers with one species (e.g., cf. AARNES 1977a and SAINIS et al. 1981). Homoserine dehydrogenase has also been reported to be inhibited by cysteine (BRYAN 1969, AARNES and ROGNES 1974, DI MARCO and GREGO 1975, AARNES 1977a, SAINIS et al. 1981). The lower molecular weight, cytosol enzyme appears to be much more sensitive, the pea and barley enzymes having $K_i$'s of 2 μM and 50 μM respectively (SAINIS et al. 1981). Again the physiological significance of the observation is not clear.

Regulatory controls may also be expected at the steps utilizing phosphohomoserine (*12* and *13*). So far no feedback inhibition of cystathione synthase has been reported (e.g., see MADISON and THOMPSON 1976). However, threonine synthase is subject to an unusual control in that it is strongly activated by S-adenosylmethionine (MADISON and THOMPSON 1976, AARNES 1978, THOEN et al. 1978a); for the *Beta vulgaris* enzyme there was barely detectable activity in the absence of S-adenosylmethionine, but it was stimulated 20-fold by its presence at a concentration of 0.5 mM. This activation is antagonized by cysteine (MADISON and THOMPSON 1976).

From these experiments a picture of how the pathway might be regulated in vivo can be built up. Synthesis of excess lysine alone would shut off the lysine branch at (*3*) but only partly affect the flux through (*1*), carbon would continue to flow to methionine and S-adenosylmethionine alone until sufficient of the latter accumulated to activate threonine synthase (*12*) and synergistically inhibit the lysine-sensitive aspartate kinase. The flux to threonine in the presence of a sufficiency of lysine and S-adenosylmethionine would be dependent on threonine-sensitive aspartate kinase, once sufficient threonine accumulated it would lead to cessation of flux through (*1*). A subsequent fall in the lysine content in the presence of sufficient threonine and S-adenosylmethionine could be accommodated as the inhibition of the lysine-sensitive aspartate kinase (at least the synergistic part) and of dihydrodipicolinate synthase would be removed; carbon would then be preferentially converted from aspartate semialdehyde to lysine because of the threonine block on the formation of homoserine.

Although changes in the amount as opposed to the activity of amino acid enzymes in the presence of end-products have not been easy to demonstrate (see BRYAN 1976, 1980, MIFLIN 1977, STEWART and RHODES 1977a) there are some recent suggestions that aspartate kinase may be subject to small repressive effects by lysine (SAKANO 1979, MATTHEWS and WIDHOLM 1979c) and cystathionine synthase by methionine (GIOVANELLI pers. comm.).

Recent work has also provided genetic evidence for the controls on this pathway. Several workers have used screening to select for lines of tissue culture or intact plants that can grow in the presence of lysine or methionine analogues or inhibitory concentrations of lysine plus threonine. WIDHOLM (1976) has selected lines resistant to ethionine (also selected in *Chlorella*, SLOGER and OWENS 1974), 5-hydroxylysine and aminoethylcysteine (respectively one methionine and two lysine analogues) which overproduce the corresponding amino acids but nothing is known of the enzymic mechanism involved. HIBBERD et al. (1980) have selected lines of maize callus resistant to lysine plus threonine and regenerated shoots from them. One of the lines (D33) had an altered aspartate kinase with a $K_i$ for lysine of 530 μM compared with a value of 62 μM for the parental line and the tissue cultures had enhanced levels of threonine and methionine. Unfortunately the line proved infertile and a genetic analysis was not possible. BRIGHT et al. (1979a, b, 1980, 1981) have selected mutated barley embryos resistant to either aminoethylcysteine (AEC) or lysine plus threonine. The AEC resistance of R.906 was due to the presence of a single recessive gene *aec* which led to an inability to take up lysine and AEC, but not leucine. Three other lines R2501, R3004, R3202 contained dominant mutations conferring resistance to lysine plus threonine. The most characterized line R2501 accumulated threonine in the soluble pool of amino acids in the seed, also appeared to overproduce methionine and one of its lysine-sensitive aspartate kineases was much less sensitive to lysine (BRIGHT et al. 1981). The general conclusion, therefore, is that the enzyme feedback inhibitions that have been described in vitro are relevant to the control of the pathway in the plant and that alterations in these controls lead to the enhanced production of certain amino acids and the relief of toxicity by various compounds.

## 4.5 Synthesis of Glycine, Serine and Cysteine

The carbon skeletons of glycine and serine are readily interconverted and these two compounds are also of importance in pathways outside those of amino acid and protein synthesis, particularly the photorespiratory pathway (see Sect. 5.1 and LEA and MIFLIN 1980). Cysteine also derives its carbon from serine and thus is logically included in this group. The biosynthetic routes important in the synthesis of these amino acids are summarized in Fig. 5. Reactions *1–5* plus *7–9* comprise the photorespiratory pathway which has been reviewed elsewhere (TOLBERT 1979, KEYS 1980); it and other reactions common to carbohydrate metabolism will not be dealt with in detail. A number of in vivo studies

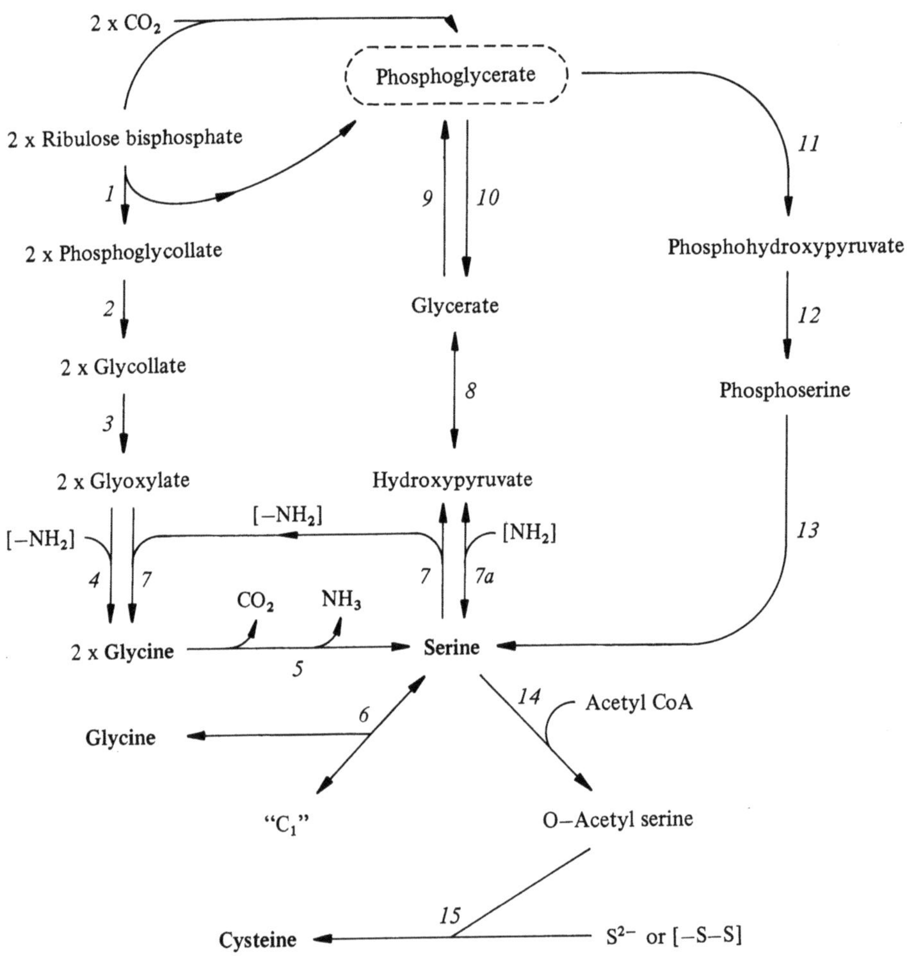


**Fig. 5.** The pathway of synthesis of glycine, serine and cysteine

have been carried out which are consistent with the reactions outlined in Fig. 5. Thus the feeding of specifically labelled glyoxylate, glycollate and glycine and determination of the ratio of isotope in the specific carbons of serine have verified the operation of reactions 3–5 (WANG and WAYGOOD 1962, RABSON et al. 1962, MIFLIN et al. 1966, MARKER and WHITTINGHAM 1967). The in vivo conversion of [$^{32}$P]-phosphoglycerate to [$^{32}$P]-phosphoserine demonstrates that the key reactions of the phosphorylated pathway to serine production can occur in vivo (HANFORD and DAVIES 1958). The nature of the distribution of the label in serine, glycine and phosphoglyceric acid, after short periods of photosynthesis, also indicates that serine can arise from $C_3$ compounds (CHANG and TOLBERT 1965) and phosphoserine has been shown to be an early labelled intermediate of [$^{14}$C]$O_2$ fixation (CHAPMAN and LEECH 1976, DALEY and BIDWELL 1977). Evidence of cysteine synthesis from serine comes from the feeding experiments of ELLIS (1963) in which [$^{14}$C]-serine was converted to [$^{14}$C]-cysteine and from the presence of O-acetyl serine in plants (SMITH 1977).

### 4.5.1 Enzymic Evidence

Glycine is formed by transamination from glyoxylate which may arise by the photorespiratory pathway or in the glyoxylate cycle. The glycine transaminases responsible for glycine formation in peroxisomes (*4*) (*7*) have been shown to use either glutamate, serine or alanine as amino donors; in non-green tissues transamination chiefly involves glutamate or alanine as amino donors. Purification of enzymes (BROCK et al. 1970, KING and WAYGOOD 1968) and mutant evidence (SOMERVILLE and OGREN 1980a) suggest that serine:glyoxylate aminotransferase EC 2.6.1.45 (*7*) and glutamate:glyoxylate aminotransferase EC 2.6.1.4 (*4*) are separate proteins; whether there is a separate alanine:glyoxylate aminotransferase is not clear at present.

Transamination to form serine can occur by means of both reaction (*7*) or by serine:pyruvate aminotransferase EC 2.6.1.51 (*7a*). The serine:glyoxylate aminotransferase is, however, considered to act in the direction of glycine formation during the photorespiratory pathway and, conversely, there is good evidence that pyruvate or oxoglutarate are unable to act as acceptors in this pathway (SOMERVILLE and OGREN 1980a). Thus only *7a* is probably relevant to serine synthesis.

The conversion of glycine to serine in the photorespiratory pathway is catalyzed by a system (*5*) that operates in intact mitochondria (see TOLBERT 1979, KEYS 1980). It is probably similar in operation to the same reaction catalyzed by rat liver mitochondria and may involve pyridoxal phosphate, lipoic acid, tetrahydrofolate and FAD as cofactors. In the process one molecule of glycine is oxidatively decarboxylated and deaminated to a $C_1$ fragment, $CO_2$ and $NH_3$ and yielding $2H^+ + 2e^-$. The second glycine then combines with the $C_1$ fragment to form serine. The $2H^+ + 2e^-$ are accepted by a flavoprotein and passed down the electron transport chain giving rise to ATP. The enzyme complex has not been broken down into its soluble parts. The reaction is thus dependent on the intactness of the mitochondria and appears to be irreversible. Besides its importance in photorespiration it has also been found in mitochondria from non-green tissues (KISAKI et al. 1971, BIRD et al. 1972b), although GARDESTROM et al. (1980) concluded that only mitochondria from green tissue were capable of oxidizing glycine.

Glycine may, however, be formed from serine by serine hydroxymethyltransferase EC 2.1.2.1 (*6*). This enzyme has been found in a number of plants, e.g., *Zea mays* seedlings (HAUSCHILD 1959) *Nicotiana tabacum* roots (PRATHER and SISLER 1966) cauliflower buds (MAZELIS and LIU 1967) and spinach leaves (KISAKI et al. 1971). The activity is dependent on 5,10-methylenetetrahydrafolate and pyridoxal phosphate. The equilibrium of the reaction is slightly in favour of glycine synthesis depending on the relative concentrations of the substrate. Although it may be a component of the glycine to serine system in mitochondria it is likely that the activity observed in most studies was due to a protein separate from that involved in the mitochondrial system.

Serine may arise from phosphoglyceric acid via the action of phosphoglycerate phosphatase EC 3.1.3.18 (*10*) and glycerate dehydrogenase EC 1.1.1.29 (*8*). The former enzyme is widely distributed, particularly in leaf tissue (RANDALL et al. 1971), and is specific in its choice of substrate; the latter enzyme is also present in leaf extracts and uses NADH as a cofactor. The equilibrium of the latter reaction, however, favours glycerate formation and it may be that the enzyme is more important in the flow of carbon away from serine rather than towards it. Serine can also be synthesized by a phosphorylated pathway from phosphoglycerate (reactions *11–13*). The first enzyme in this pathway is phosphoglycerate dehydrogenase EC 1.1.95 (*11*) which has been found in plant tissues (CHEUNG et al. 1968) and purified some 450-fold by SLAUGHTER and DAVIES (1968a). Phosphohydroxypyruvate is transaminated to phosphoserine by an aminotransferase EC 2.6.1.52 (*12*) a reaction shown to be catalyzed by plant extracts (CHEUNG et al. 1968, SLAUGHTER and DAVIES 1968a, LARSON and ALBERTSSON 1979). Serine is then produced by the action of phosphoserine phosphatase EC 3.1.3.3 (*13*), found in extracts of spinach leaves (LARSON and ALBERTSSON 1979) and pea epicotyls (SLAUGHTER and DAVIES 1969).

Serine is generally considered to be acetylated to O-acetyl serine by serine acetyltransferase EC 2.3.1.30 (*14*) prior to its sulphydration. Extracts capable of catalyzing this reaction have been obtained from plants (NGO and SHARGOOL 1974, SMITH and THOMPSON 1971 and see GIOVANELLI et al. 1980). Cysteine formation is then catalyzed by O-acetyl serine (thiol)-lyase (cysteine synthase) EC 4.2.99.8 (*15*) which has been purified from many species [e.g., rape (*Brassica napus*) leaves MASADA et al. 1975, *Phaseolus* sp. BERTAGNOLLI and WEDDING 1977, also see GIOVANELLI et al. 1980] and two forms of the enzyme have been reported in mungbean (*Phaseolus aureus*) (MASADA et al. 1975) and *Phaseolus* (BERTAGNOLLI and WEDDING 1977). There is some controversy as to whether or not free or bound sulphide is the natural substrate of the enzyme. Plants are capable of reducing both free and bound sulphite and the relative importance of the two pathways is not firmly resolved (see ANDERSON 1980), although mutant evidence from *Chlorella* (SCHMIDT et al. 1974) suggests that it is the "bound" pathway that is normally used in sulphate assimilation. Thus, although all the studies with cysteine synthase have been done with free sulphide, this may not be the normal physiological donor.

### 4.5.2 Sub-Cellular Localization

The photorespiratory pathway appears to be distributed between several organelles. Evidence is available that shows that reactions *3, 4, 7* and *8* in Fig. 5 are present in leaf peroxisomes (see TOLBERT 1979). Reactions *4, 7,* and *8* have also been shown to be present in microbodies from other plant tissues (see BEEVERS 1979). Reaction *5* has been described only in intact mitochondria (see KEYS 1980). In contrast to the glycine decarboxylation complex, serine hydroxymethyl transferase has been reported to be present in chloroplasts, although it was not possible to exclude its presence in the cytosol (BIRD et al. 1972a, SHAH and COSSINS 1970a, KISAKI et al. 1971). Recently LARSON and ALBERTSSON (1979) have studied the localization of the phosphorylated pathway and reported the presence of enzymes catalyzing reactions *11, 12,* and *13* in chloroplasts, although they consider that a parallel pathway may also exist outside the chloroplast.

Of the two important enzymes of cysteine synthesis, only the distribution of cysteine synthase has been studied in detail. Although some workers report that it is a soluble enzyme (e.g., ASCANO and NICHOLAS 1976, MASADA et al. 1975, SMITH 1972), this probably reflects the fact that the plastids were ruptured and there is now reasonably strong evidence that the enzyme is present in the chloroplasts (FANKHAUSER et al. 1976, NG and ANDERSON 1978, 1979). Intact, isolated chloroplasts also carry out the light-dependent synthesis of cysteine from *O*-acetylserine and sulphate (NG and ANDERSON 1979).

### 4.5.3 Regulation

The photorespiratory production of glycine and serine is primarily regulated by the relative competition of $O_2$ and $CO_2$ for ribulose bisphosphate carboxylase (see TOLBERT 1979) and not by the rates of production of glycine and serine. However, there is evidence that 3-phosphoglycerate dehydrogenase (*11*) is inhibited by serine (SLAUGHTER and DAVIES 1968b) and certain nucleotides (SLAUGHTER 1973) and activated by L-methionine (SLAUGHTER 1970). This last effect required relatively large concentrations of methionine; unfortunately, S-adenosylmethionine, which is now known to be the regulatory metabolite of methionine in the aspartate pathway, was not tested.

## 4.6 Synthesis of the Aromatic Amino Acids

This group consists of tryptophan, phenylalanine and tyrosine, all of which are synthesized by a pathway in which the initial steps are in common and involving the intermediate shikimic acid (Fig. 6); the pathway is thus often called the shikimate pathway. Besides leading to the formation of aromatic amino acids for protein synthesis the pathway also provides precursors for a number of secondary metabolites produced in higher plants. Early labelling studies in plants, suggesting the existence of the proposed pathway particularly to the formation of shikimic acid, have been reviewed by HASLAM (1974). Beyond this BELSER et al. (1971) have shown that shikimate is metabolized to tryptophan.

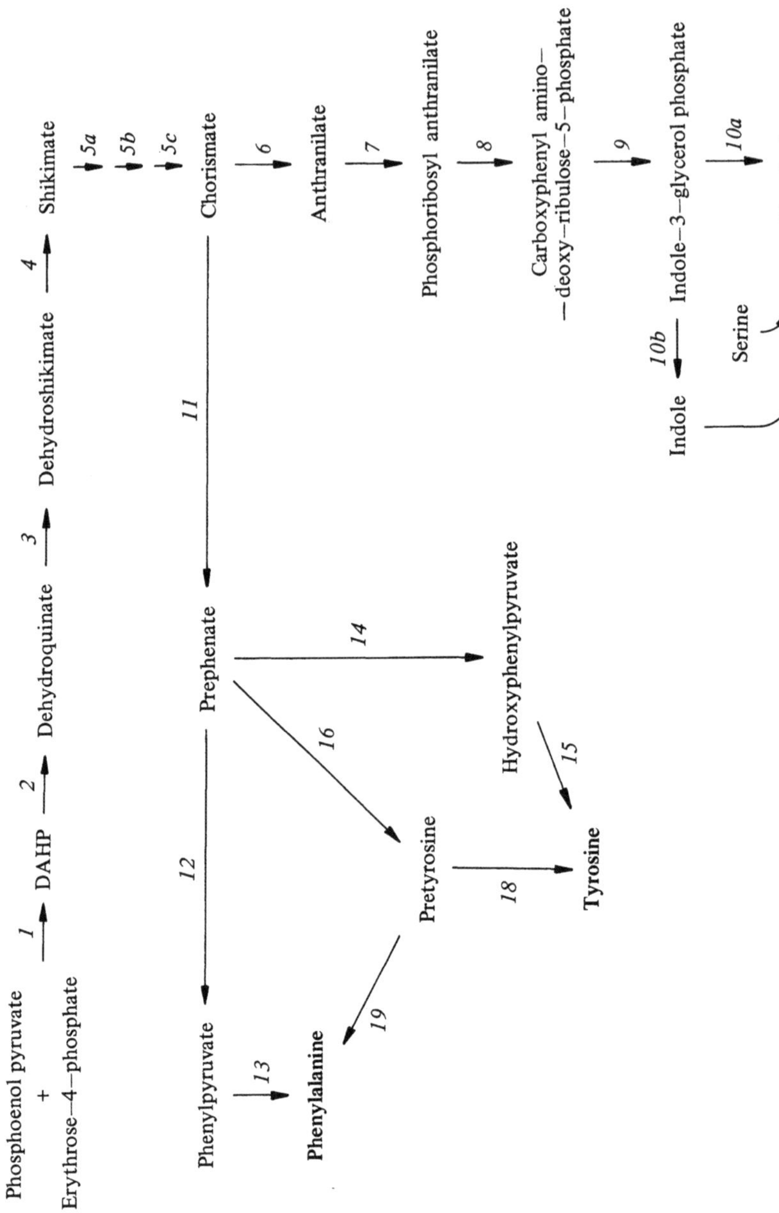


**Fig. 6.** The pathway of synthesis of the aromatic amino acids

### 4.6.1 Enzymic Evidence

The first enzyme in the pathway is 3-deoxy-D-arabino heptulosonate-7-phosphate (DAHP) synthase (*1*). Cell-free extracts demonstrating this activity have been derived from green algae (WEBER and BOCK 1968), pea seedlings (ROTHE et al. 1976), mung beans (NANDY and GANGULI 1961, MINAMIKAWA and URITANI 1967) and cauliflower (*Brassica oleracea*) florets (HUISMAN and KOSUGE 1974).

The enzyme is generally stimulated by divalent metal ions. In peas separate isoenzymes have been reported (ROTHE et al. 1976); whether these are in any way related to the multiple purposes of the pathway is not yet clear. Reaction *2* is catalyzed by 3-dehydroquinate synthase EC 4.6.1.3 which has been found in mung beans (YAMAMOTO 1977) and sorghum (*Sorghum vulgare*) seedlings (SAIJO and KOSUGE 1978). The enzyme uses NAD as a cofactor and is stimulated by $Cu^{2+}$ or $Co^{2+}$ but not by $Mn^{2+}$, $Mg^{2+}$ or $Ca^{2+}$. The next step (*3*) is catalyzed by 3-dehydroquinate dehydratase EC 4.2.1.10 which has been widely studied (e.g., see NANDY and GANGULI 1961, BALINSKY and DAVIES 1961b, c, MINAMIKAWA 1967, BOUDET and LECUSSAN 1974, BOUDET et al. 1977, KOSHIBA 1978a). Although the enzyme has been purified separately from the next one (*4*) in the pathway (BALINSKY and DAVIES 1961c) there is some evidence that it can exist aggregated with it (BOUDET et al. 1975) and that in certain species multiple forms, with different properties, may exist (BOUDET et al. 1977). Shikimate dehydrogenase EC 1.1.1.25 (*4*) leads to the formation of shikimate. Again the enzyme has been found in many plants (e.g., BALINSKY et al. 1971, SANDERSON 1966, BOUDET and LECUSSAN 1974). BALINSKY and DAVIES (1961a) have partially purified this activity and shown it to be NADP-specific. In many plants the enzyme is present in several forms (FEIERABEND and BRASSEL 1977, ROTHE 1974). Shikimate is then phosphorylated by shikimic kinase EC 2.7.1.71 (*5a*) which has been found in mung beans (KOSHIBA 1978b) and sorghum seedlings (BOWEN and KOSUGE 1979). The condensation of shikimate-5-phosphate with phosphoenolpyruvate is catalyzed by 3-enolpyruvolylshikimate-5-phosphate synthase EC 2.5.1.19 (*5b*) which has not been studied in detail although some activity has been reported in mung beans (KOSHIBA 1978b). The product of reaction *5b*, is converted to chorismate in bacteria by chorismate synthase EC 4.6.1.4 (*5c*) although we are not aware that the activity has been found in higher plants.

Chorismate lies at the branching of the pathway; in the formation of tryptophan the first reaction (*6*) is its conversion to anthranilate by anthranilate synthase (EC 4.1.3.27) which catalyzes the transfer of the amide-amino group of glutamine and the loss of the pyruvyl side chain of chorismate. The activity has been found in extracts of several species of plants (BELSER et al. 1971, WIDHOLM 1972a, 1972b, 1973). Anthranilate is converted to N-phosphoribosyl anthranilate by the action of anthranilate phosphoribosyl transferase EC 2.4.2.18 (*7*) which is in turn isomerized by phosphoribosyl anthranilate isomerase (*8*) and then converted to indole-3-glycerol-phosphate (IGP) by IGP synthase EC 4.1.1.48 (*9*). The only detailed report of these enzymes is in peas and maize by HANKINS et al. (1976). They showed that the activities were the result of separable enzymes and not of aggregates or single multifunctional proteins as in the enteric bacteria, some fungi and also *Euglena* (LARA and MILLS 1972). The final step in the pathway has been widely studied in bacteria and plants (e.g., see DELMER and MILLS 1968a, CHEN and BOLL 1971a, b, NAGAO and MOORE 1972, WIDHOLM 1973, HANKINS et al. 1976) and is catalyzed by tryptophan synthase EC 4.2.1.20 (*10*). The enzyme is complex and consists of two subunits A and B and shows three activities (see Fig. 6, reaction *10a*, *b* and *c*). For reaction *10a* both components are required whereas the A subunit catalyzes *10c* and the B, *10b*. Reactions *a* and *b* require pyridoxal phosphate and probably only *10a* is of any physiological significance. Mixing experiments using the *E. coli* B subunit and the tobacco A subunit show that the enzymes are partly homologous, since this combination can catalyze reaction *10a* (DELMER and MILLS 1968b).

The other part of the pathway to phenylalanine and tyrosine has recently been shown to occur in bacteria in two ways, depending upon species (PATEL et al. 1977) and, possibly, by both ways within one plant, e.g., *Vigna* (RUBIN and HENSEN 1979). The first enzyme chorismate mutase EC 5.4.99.5 (*11*), which leads to the production of prephenate, is common to both routes. The enzyme has been found in many plants (e.g., pea – COTTON and GIBSON 1969, algae – WEBER and BOCK 1969, 1970, ZURAWSKI and BROWN 1975, mung bean – GILCHRIST et al. 1972, oak – GADAL and BOUYSSOU 1973). It exists as a number of different isoenzymes (e.g., see WOODIN et al. 1978) whose occurrence may have regulatory significance (see below). In the first described pathway (reactions *12–15*) prephenate is oxidatively decarboxylated by prephenate dehydrogenase EC 1.3.1.12, which has been found in legumes (GAMBORG and KEELEY 1966, RUBIN and JENSEN 1979), to form p-hydroxyphenylpyruvate. This compound is then transaminated by a broad-specificity aromatic aminotransferase (GAMBORG 1965, REDKINA et al. 1969, FOREST and WIGHTMAN 1973) to form tyrosine. In the formation of phenylalanine, prephenate is converted to phenyl pyruvate by prephenate dehydratase EC 4.2.1.51 (*12*) which is also transaminated by the same transminase. The alternative pretyrosine pathway was originally described for cyanobacteria (see PATEL et al. 1977) and involves, first, a transamination of prephenate to form pretyrosine (*16*) and then its subsequent conversion, to either tyrosine or phenylalanine, by pretyrosine dehydrogenase (*18*) or pretyrosine dehydratase (*19*) respectively.

### 4.6.2 Sub-Cellular Localization

There is now considerable evidence to show that chloroplasts can carry out all of the reactions necessary to form the aromatic amino acids. The most convincing is that isolated, intact chloroplasts can synthesize phenylalanine, tyrosine and tryptophan from $[^{14}C]O_2$ in a light-dependent reaction (BICKEL et al. 1978, BUCHOLZ et al. 1979). Studies on enzyme distribution have also shown that all the enzymes from anthranilate synthase to tryptophan synthetase are present in etioplasts of pea (GROSSE 1976) and walnuts (GROSSE 1977). Little enzyme location work has been done with the phenylalanine and tyrosine enzymes but, presumably, they are also present in plastids. It is unlikely that the chloroplast is the only location in which all or part of the pathway exists and recent evidence on the distribution of isoenzymes of shikimate dehydrogenase suggest that they exist in the chloroplast (FEIERABEND and BRASSEL 1977) in the microbodies and the cytosol (ROTHE 1974).

### 4.6.3 Regulation

Evidence for feed-back regulation in vivo comes from studying the effects of externally applied tryptophan on the flow of carbon from shikimic acid (BELSER et al. 1971) or glucose (OAKS 1965a) to tryptophan and on the effect of tyrosine and phenylalanine on the growth of barley seedlings (MIFLIN 1969). Consistent with the in vivo effects, BELSER et al. (1971) found that anthranilate synthase was very sensitive to inhibition by tryptophan. Direct evidence of the importance of this feedback inhibition (at least for tryptophan synthesis) is that cell lines resistant to 5-methyltryptophan, and whose anthranilate synthetase is less sensitive to feed-back regulation, accumulate several times the normal level of free tryptophan in their cells (WIDHOLM 1972a, b). When detailed studies were done it was found that normal plants have two isoenzymes of anthranilate synthetase, one sensitive and one insensitive to feedback regulation, with the former greatly in excess of the latter, whereas, in contrast, the resistant cells have a predominance of the insensitive form (CARLSON and WIDHOLM 1978). On the other branch of the pathway it appears that chorismate mutase (CM) is the major regulatory enzyme and exists in multiple forms with different regulatory characteristics. In mung beans isozyme CM-1 is inhibited by phenylalanine and tyrosine, although not additively, and this inhibition was antagonized by equimolar amounts of tryptophan. The other isoenzyme CM-2 was unaffected by amino acids or several other products of the pathway (GILCHRIST and KOSUGE 1974, 1975). A similar control seems to exist for the oak enzyme (GADAL and BOUYSSOU 1973). The inhibition pattern is more complex in alfalfa (*Medicago sativa*) where there are three isoenzymes, two (CM-1 and CM-3) sensitive to controls by the aromatic amino acids whereas all CM-1 and CM-2 were inhibited by chlorogenoquinone and caffeic acid and CM-3 by ferulic acid (WOODIN and NISHIOKA 1973a).

Inhibition of the synthesis of phenylalanine, tyrosine and tryptophan from $[^{14}C]O_2$ by isolated spinach chloroplasts has been reported (BICKEL et al. 1978, BUCHOLZ et al. 1979). However, the effects of phenylalanine and tyrosine are seen only on their own synthesis and not on that of the other amino acid which is not entirely consistent with the isolated enzyme studies (albeit using different species) reported above. Tryptophan appeared to inhibit the synthesis of all three amino acids and the authors suggested that it may have a site of action between shikimate and chorismate.

In summary the controls at present are unclear, probably because the enzymes of the pathway often exist in multiple forms, possibly in different locations within the cell, producing different secondary metabolite end-products. Many further studies are required combining careful separation of the different enzyme forms, determination of their location and the particular end-products of that form's pathway.

## 4.7 Synthesis of Histidine

Very little information regarding histidine biosynthesis in plants is available. It is generally assumed that the route of synthesis is the same as in bacteria shown in Fig. 7 and DOUGALL and FULTON (1967a) showed that histidinol blocked the incorporation of $^{14}C$ from glucose into histidine. The only enzymic evidence in favour of this assertion is the demonstration of the presence of

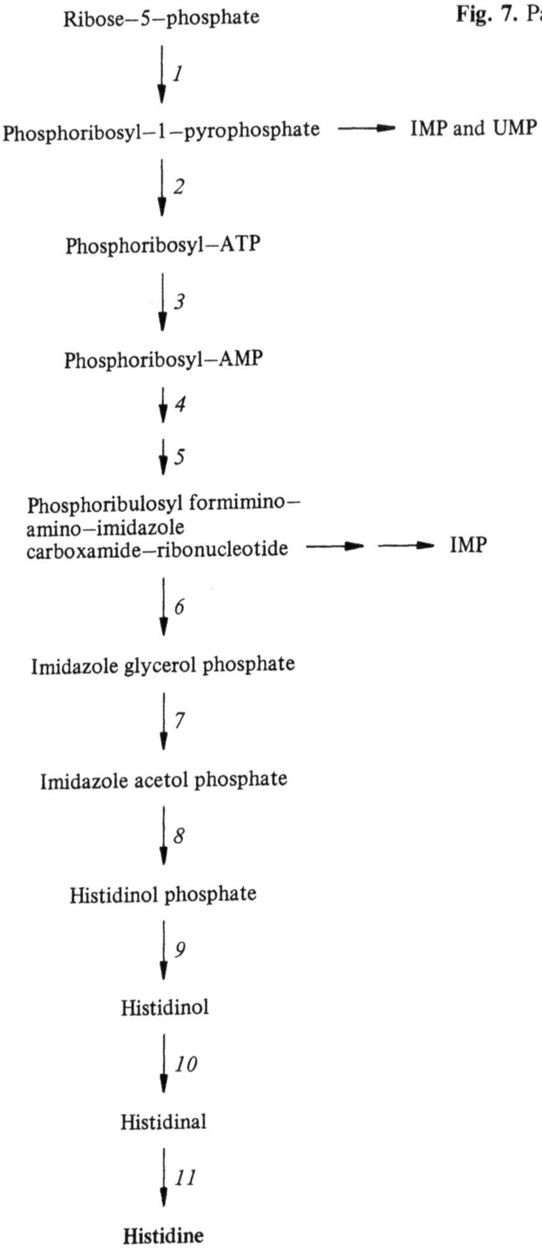


**Fig. 7.** Pathway for the synthesis of histidine

the enzymes ATP-phosphoribosyl transferase EC 2.4.2.17 (*2*) imidazol glycerol-phosphate dehydratase EC 4.2.1.19 (*7*), histidinol phosphatase EC 3.1.3.15 (*9*) in extracts of barley, oat and pea shoots (WIATER et al. 1971) and the unpublished observation of the presence of histidinol dehydrogenase EC 1.1.1.13 (*11*) in rose tissue culture cells (see DAVIES 1971). Circumstantial evidence in favour of the pathway and its feedback regulation has come from feeding studies using the herbicide aminotriazole, which inhibits reaction *7* and leads to the accumulation of imidazole glycerol (presumably formed from the corresponding phosphate), however this accumulation is abolished and the inhibition relieved by the addition of histidine (SIEGEL and GENTILE 1966, DAVIES 1971). There is also some suggestive evidence for the feedback control of plant ATP-phosphoribosyl transferase (*2*) by histidine (WIATER et al. 1971).

## 5 Amino Acid Catabolism

Whilst most attention has been devoted to amino acid biosynthesis it is essential to remember that in the plant certain amino acids are also continuously being broken down. We shall consider the catabolism of glycine and nitrogen transport compounds, which involve the production of large amounts of ammonia, and the metabolism of proline which has a specific role in the metabolism of plants undergoing different types of water stress (see also LEA and MIFLIN 1980, MAZELIS 1980, STEWART and LAHRER 1980).

### 5.1 Photorespiration

The photorespiratory pathway as shown in Fig. 5 arises because of the dual function of the ribulose-1,5-bisphosphate carboxylase-oxygenase enzyme in the chloroplast (TOLBERT 1979, KEYS 1980). At high $CO_2:O_2$ ratios 3-phosphoglycerate is the only product of the reaction but as the ratio decreases the synthesis of phosphoglycollate occurs which is hydrolyzed to glycollate in the chloroplast, oxidized to glyoxylate in the peroxisome and transaminated to yield glycine. The conversion of glycine to serine in the mitochondria is probably the most well-known reaction in the photorespiratory pathway as it involves the evolution of $CO_2$, and $NH_3$ (see Fig. 5, reaction (*5*)). The $CO_2$ evolution is frequently used as a measure of the rate of photorespiration. The ammonia, produced in stoichiometrically equal amounts, is not, however, released to the atmosphere in the same manner as the $CO_2$ (WOO et al. 1978, KEYS et al. 1978). ZELITCH (1979) has published a table collating a wide range of data for $C_3$ plants which suggest that the rate of photorespiration varies between 14% and 75% of the rate of photosynthetic $CO_2$ fixation. Thus in the leaf ammonia could be evolved in the mitochondria at a rate tenfold higher than that produced by nitrate reduction. If the ammonia was not rapidly reassimilated the plant would soon die due to a lack of nitrogen. The evolution of ammonia can be readily demonstrated by treating a plant with MSO; KEYS et al. (1978) initially showed that $[^{15}N]H_3$ was released from MSO-treated wheat leaves fed $[^{15}N]$-glycine. Studies with MSO-treated *Chlamydomonas* (CULLIMORE and SIMS 1980) have shown that ammonia evolution increased at high $O_2:CO_2$ ratios in the atmosphere. At high photorespiration rates in the presence of MSO, there is a rapid breakdown of protein in *Chlamydomonas* in order to supply the necessary amino groups for the glycine-serine interconversion.

Although mitochondria contain GDH, we were unable to show that the enzyme was involved in the reassimilation of photorespiratory evolved ammonia (KEYS et al. 1978, WALLSGROVE et al. 1980), a result later confirmed by HARTMANN and EHMKE (1980) (see also Sect. 2.5.2). The present hypothesis on ammonia recycling (WALLSGROVE et al. 1980) suggests that ammonia is rapidly assimilated by cytoplasmic GS (see Sect. 2.5.1), and the glutamine is converted to glutamate via glutamate synthase in the chloroplast. Such a system allows a rapid recycling of glutamate to act as a substrate for GS. However, it is not clear at the present time exactly what role glutamate released by protein breakdown may have in feeding into actively metabolizing pools in the peroxisome involved in glycine synthesis and in the cytoplasm involved in glutamine synthesis.

SOMERVILLE and OGREN (1980a, b) have recently isolated mutants of *Arabidopsis* that are unable to photorespire, and are killed when transferred to a normal atmosphere from high levels of $CO_2$. Some of the mutants were shown to evolve ammonia rapidly. Enzymological studies showed the plants lacked leaf ferredoxin-dependent glutamate synthase, although they contained the NADH-enzyme in the root, thus allowing them to synthesize sufficient amino acids for the whole plant. Clearly in *Arabidopsis* a major role of glutamate synthase in the leaf is to recycle the glutamine synthesized from the evolution of ammonia during photorespiration.

## 5.2 Nitrogen Transport Compounds

A detailed discussion of the metabolism of nitrogen transport compounds has recently been published by LEA and MIFLIN (1980). In plant organs, when rapid production of ammonia exceeds the supply of carbon skeletons required for amino acid synthesis, e.g., germinating seeds, nitrogen-fixing root nodules, senescing leaves, it is essential that the nitrogen is stored and transported in a non-toxic but economical form. Typical compounds that are used for this process are asparagine, arginine and the ureides, allantoin and allantoic acid.

### 5.2.1 Asparagine

BAUER et al. (1977) utilizing pea leaves and RHODES et al. (1980) with *Lemna* have demonstated rapid turnover of $^{15}N$ in asparagine pools. Their evidence suggests that ammonia is first incorporated into the amide position of glutamine but is then transferred to aspartate to yield asparagine. The synthesis of aspartate is probably via oxaloacetate from the tricarboxylic acid cycle, although in the root nodule oxaloacetate may be formed by the action of phosphenolpyruvate carboxylase (CHRISTELLER et al. 1977). Cyanide is rapidly converted to asparagine when fed to plants, but there is little evidence that this is a major pathway of asparagine biosynthesis (see LEA and MIFLIN 1980 for a full discussion). It is probable that this route is simply a means of detoxifying cyanide, although COONEY et al. (1980) have once again misguidedly resurrected the possibility that, in asparagus, the cyanide pathway is important in asparagine synthesis.

This was based on their inability to measure glutamine-dependent asparagine synthetase, due probably to incorrect isolation procedures.

Asparagine synthetase (AS) catalyzes the amidation of aspartate by glutamine in an ATP-dependent reaction:

$$\text{Aspartate} + \text{glutamine} + \text{ATP} \xrightarrow{Mg^{2+}} \text{asparagine} + \text{glutamate} + \text{AMP} + \text{PPi}$$

The enzyme has been isolated from the cotyledons of a number of germinating seeds (ROGNES 1970, 1975, STREETER 1973, LEA and FOWDEN 1975, DILWORTH and DURE 1978, KERN and CHRISPEELS 1978), root nodules (SCOTT et al. 1976) and maize roots (STULEN and OAKS 1977, STULEN et al. 1979, OAKS et al. 1980). The enzyme is notoriously unstable, and measurement of activity has been prevented by the presence of inhibitors, asparaginase and the lack of chloride ions which are necessary for optimum activity (ROGNES 1980). Irrespective of the sources, there is evidence that the enzyme is able to utilize ammonia in place of glutamine as a substrate of asparagine synthesis; STULEN et al. (1979) have claimed that in maize roots the $K_m$ for ammonia is low enough that direct incorporation into asparagine may occur under physiological conditions.

There are three possible routes of asparagine catabolism in higher plants.

*Asparaginase* (EC 3.5.1.1). Asparagine → aspartate + $NH_3$

The presence of the enzyme was suggested by the earlier workers in plant metabolism (see MCKEE 1962) and has been purified (500-fold) from maturing seeds of *Lupinus polyphyllus* (LEA et al. 1978). Attempts to isolate the enzyme from other lupin and legume seeds were initially unsuccessful, despite the ability of ATKINS et al. (1975) to detect the enzyme in crude extracts of the maturing seeds of *L. albus*. The problem was eventually solved by the demonstration in maturing pea seeds of an enzyme that had an absolute requirement for $K^+$ ions (SODEK et al. 1980). Maximum activity of the enzyme did not develop until 21 days after flowering, but was sufficient to provide the N required for the increase in protein content of the seeds during maturation. The presence of asparaginase in pea seeds explains the ability of isolated legume cotyledons to grow on asparagine as a sole N source in a defined medium (MILLERD et al. 1975, THOMPSON et al. 1977, LEA et al. 1979a). Attempts by a number of workers to detect asparaginase activity from leaves at levels comparable to those in maturing seeds have so far been unsuccessful.

*Asparagine Aminotransferase*. The 2-amino group of asparagine may be transaminated to yield 2-oxosuccinamic acid, which can convert chemically to a dimer (see LEA and MIFLIN 1980). The reaction has been detected in crude extracts (MEISTER et al. 1952, STREETER 1977, LLOYD and JOY 1978, COONEY et al. 1980), but the specificity of the oxoacid acceptor appears to vary with the plant material tested. As no purified preparation of an asparagine transaminase has yet been isolated, it is not clear whether it is a distinct new enzyme or is a result of the multiple specificity frequently displayed by aspartate aminotransferases (WIGHTMAN and FOREST 1978). STREETER (1977) suggested 2-oxosuccinamate could be broken down to oxaloacetate and ammonia, a result also confirmed by LLOYD and JOY (1978). However, in pea leaves the majority of 2-oxosuccinamate was reduced to 2-hydroxysuccinamate which accumulated (LLOYD and JOY 1978).

Thus in tissues that do not contain high asparaginase activity a split pathway is proposed:

$$\text{Asparagine} \rightarrow \text{2-oxosuccinamate} \begin{cases} \nearrow \text{2-hydroxysuccinamate} \rightarrow \text{malate} + NH_3 \\ \searrow \text{oxaloacetate} + NH_3 \end{cases}$$

*Asparagine Amidotransferase.* In 1974 both FOWLER et al., and DOUGALL suggested that NAD(P)-dependent glutamate synthase could use asparagine as a substrate amide donor, however, these results were later shown to be due to the presence of an artefact (MIFLIN and LEA 1975).

Thus it would appear that asparagine can be split by either the first or second routes into three parts: (1) an amide N group that is liberated as ammonia, (2) a readily transaminated amino group, (3) a four C dicarboxylic acid that is readily metabolized in the tricarboxylic acid cycle.

### 5.2.2 Ureides

Early work on ureide metabolism has been discussed by REINBOTHE and MOTHES (1962), and a more recent review on the transport of ureides has been presented by LEA and MIFLIN (1980). Two pathways of synthesis have been suggested, the first the reversal of allantoic acid breakdown by the condensation of urea and glyoxylic acid is backed by little experimental evidence; the second is via inosine 5′-monophosphate which is an intermediate in the synthesis of purine-derived nucleic acids. The pathway of inosine 5′-monophosphate synthesis required six ATP molecules, two activated $C_1$ fragments and the input of four nitrogen atoms derived from glutamine, glycine and aspartate:

Inosine 5′-monophosphate → Inosine → Hypoxanthine

$$\text{Hypoxanthine} \xrightarrow[\text{NAD} \ \ \text{NADH}]{} \text{Xanthine} \xrightarrow[\text{NAD} \ \ \text{NADH}]{(1)} \text{Uric acid} \xrightarrow{(2)} \text{Allantoin}$$

$$\text{Allantoin} \xrightarrow{(3)} \text{Allantoic acid}$$

(1) Xanthine dehydrogenase; (2) Uricase; (3) Allantoinase

A proposed pathway of the conversion of inosine-5′-monophosphate to allantoic acid is shown above.

The ureides allantoin and allantoic acid can account for up to 90% of the total N transported from the nodules of certain tropical legumes (e.g., *Glycine, Phaseolus* and *Vigna*). There is a strong correlation between nodulation and the production of allantoin (MATSUMOTO et al. 1977a), and $^{15}N_2$ and $^{14}CO_2$

feeding studies have shown direct incorporation into the ureides in nodules (MATSUMOTO et al. 1977b, HERRIDGE et al. 1978). The data of MATSUMOTO et al. (1977b) suggests that the purines were converted directly into the ureides and not incorporated into DNA or RNA primarily. Although xanthine oxidase EC 1.2.3.2 activity has been detected in the bacteroids of root nodules (FUJIHARA and YAMAGUCHI 1978), TRIPLETT et al. (1980) have proposed that uric acid synthesis takes place in the plant "cytosol" catalyzed by xanthine dehydrogenase EC 1.2.1.37. The conversion of hypoxanthine to uric acid by this method also generates two molecules of NADH, which greatly reduces the total cost of allantoin synthesis. Large amounts of uricase (EC 1.7.3.3) have been detected in root nodules and have been shown to be in either the bacteroid (TAJIMA et al. 1977), plant (HERRIDGE et al. 1978) or both fractions (TRIPLETT et al. 1980). On maturation the enzyme activity disappears in the roots but remains in the pods and leaves. Allantoinase (EC 3.5.2.5), which may be considered as involved in the synthesis of allantoic acid, or in the breakdown of allantoin is widely distributed throughout the plant. In the nodule it is apparently present in both the bacteroid and plant fractions (TAJIMA et al. 1977, HERRIDGE et al. 1978, TRIPLETT et al. 1980).

$$\text{Allantoin} \xrightarrow{(1)} \text{Allantoic acid} \xrightarrow{(2)} \text{Glyoxylate} + 2 \text{ molecules urea}$$

$$\text{Glyoxylate} + 2 \text{ molecules urea} \xrightarrow{H_2O\ (3)} 4NH_3 + 2CO_2$$

(1) Allantoinase; (2) Allantoicase; (3) Urease

Ureides are degraded to urea and glyoxylate as shown above. The key enzyme in the breakdown is allantoicase EC 3.5.3.4 which is present in shoots and pods (TAJIMA et al. 1977). Urease EC 3.5.1.5 carries out the final reaction which liberates all the nitrogen transported in the ureides as ammonia. The properties and distribution of the enzyme have recently been reviewed (THOMPSON 1980).

### 5.2.3 Arginine

Although arginine is frequently found as a nitrogen storage compound in plants, it is particularly predominant in conifers and fruit trees (BIDWELL and DURZAN 1975).

The four possible routes of arginine breakdown are set out in Fig. 8. Reaction (*1*) is the classical reaction involving the action of arginase (EC 3.5.3.1) to yield ornithine and urea. Arginase has been found in pumpkin (SPLITTSTOESSER 1969), germinating broad beans (*Vicia faba*) (KOLLOFFEL and VAN DIJKE 1975) and grape (*Vitus*) tissues (ROUBELAKIS and KLIEWER 1978a, b); the enzyme is also thought to be involved in the breakdown of canavanine (the oxygen analogue of arginine) in Jack bean (*Canavalia ensiformis*) tissues (ROSENTHAL 1970, WHITESIDE and THURMAN 1971). Ornithine contains two amino groups, either of which may be transaminated leading to the eventual formation of glutamate or proline (reactions *2–7* see Sect. 4.3). The synthesis of the polyamines agmatine, spermine and spermidine is apparently connected with $K^+$, $Mg^{2+}$

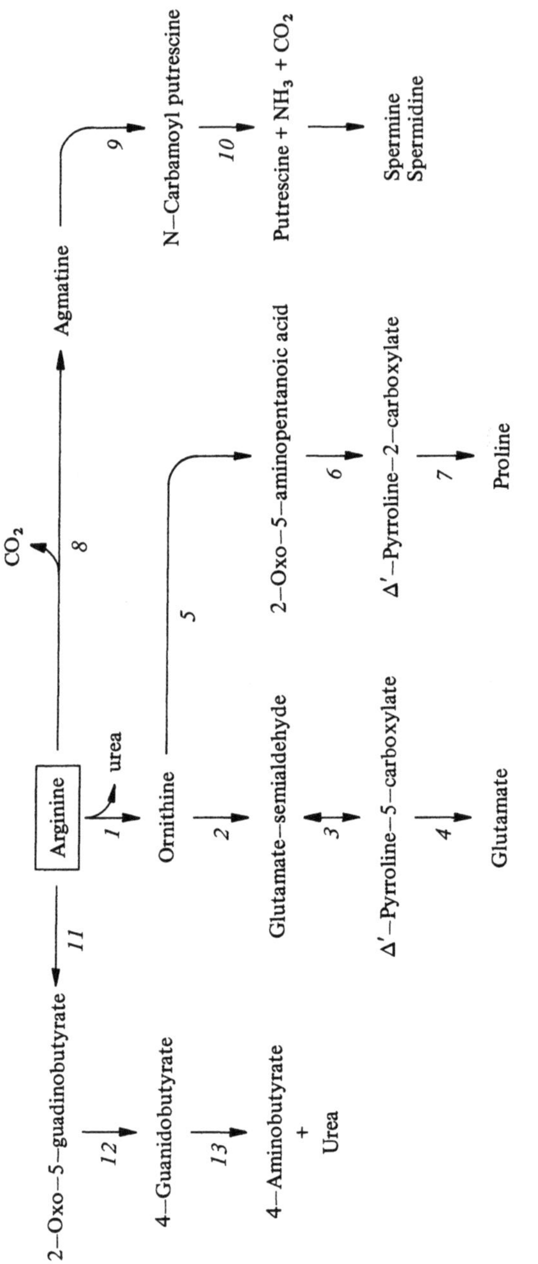


**Fig. 8.** Possible routes of arginine metabolism in higher plants

and $Ca^{2+}$ deficiencies in plants (SMITH TA 1977, SMITH et al. 1979). It is thought that polyamines can substitute for the metal ions particularly in protein synthesis reactions where the stabilization of tRNA is important. Arginine is first decarboxylated to agmatine by arginine decarboxylase EC 4.1.1.19 (*8*), followed by the hydrolysis of the imino nitrogen group to yield N-carbamyl putrescine

(*9*) which is further hydrolyzed to yield putrescine (*10*). Putrescine may be converted to spermine and spermidine by a series of methylation reactions involving *S*-adenosylmethionine. The conversion of arginine to a number of monosubstituted guanidines particularly in conifers has been discussed by BIDWELL and DURZAN (1975). A major pathway is the oxidation and transamination of arginine to yield 2-oxo-5-guadinobutyrate *11*, which may be decarboxylated to yield 4-guadinobutyrate *12*. A lyase is then able to split off urea to yield 4-aminobutyrate *13* which may be readily metabolized via succinate (STREETER and THOMPSON 1972).

*Acknowledgements.* We are indebted to Susan Wilson for all her efforts in preparing the manuscript and wish to record our gratitude.

## References

Aarnes H (1974) Aspartate kinase from some higher plants and algae. Physiol Plant 32:400–402

Aarnes H (1976) Homoserine kinase from barley seedlings. Plant Sci Lett 7:187–194

Aarnes H (1977a) A lysine-sensitive aspartate kinase and two molecular forms of homoserine dehydrogenase from barley seedlings. Plant Sci Lett 9:137–145

Aarnes H (1977b) Partial purification and characterization of methionine adenosyl transferase from pea seedlings. Plant Sci Lett 10:381–390

Aarnes H (1978) Regulation of threonine synthesis in barley seedlings (*Hordeum vulgare* L) Planta 140:185–192

Aarnes H, Rognes SE (1974) Threonine-sensitive aspartate kinase and homoserine dehydrogenase from *Pisum sativum*. Phytochemistry 13:2717–2724

Anderson JW (1980) Assimilation of inorganic sulphate into cysteine. In: Miflin BJ (ed) Amino acids and their derivatives. The biochemistry of plants, vol V. Academic Press, London New York, pp 203–233

Anderson JW, Done J (1977a) A polarographic study of glutamate synthase activity in isolated chloroplasts. Plant Physiol 60:354–359

Anderson JW, Done J (1977b) A polarographic study of ammonia assimilation by isolated chloroplasts. Plant Physiol 60:504–508

Anderson NW, Thompson JF (1979) Cysteine lyase: β-cystathionase from turnip roots. Phytochemistry 18:1953–1958

Arima Y (1978) Glutamate synthase in rice root extracts and the relationship among electron donors, nitrogen donors and its activity. Plant Cell Physiol 19:955–961

Arima Y, Kumazawa K (1977) Evidence of ammonium assimilation via glutamine synthetase – glutamate synthase system in rice seedling roots. Plant Cell Physiol 18:1121–1129

Ascano A, Nicholas DJD (1976) Purification and properties of O-acetyl-L-serine sulphydrylase from wheat leaves. Phytochemistry 16:889–893

Atkins CA, Pate JS, Sharkey PJ (1975) Asparagine metabolism – key to the nitrogen nutrition of developing legume seeds. Plant Physiol 56:807–812

Awonaike KO (1980) Studies on the development and properties of ammonia assimilating enzymes in *Phaseolus vulgaris* L. root nodules. Ph D Thesis, Univ London

Awonaike KO, Led PJ, Miflin BJ (1981) The location of the enzymes of ammonia assimilation in root moduls of *Phaseolus vulgaris*. Plant Sci Lett 23:189–195

Baker JE, Thompson JF (1961) Assimilation of ammonia by nitrogen-starved cells of *Chlorella vulgaris*. Plant Physiol 36:208–212

Balinsky D, Davies DD (1961a) Preparation and properties of dehydroshikimic reductase. Biochem J 80:292–296

Balinsky D, Davies DD (1961b) Mode of attachment of shikimic acid and dehydroshikimic acid to dehydroshimkimate reductase. Biochem J 80:296–300

Balinsky D, Davies DD (1961c) Aromatic biosynthesis in higher plants. Biochem J 80:300–304

Balinsky D, Dennis AW, Cleland WW (1971) Kinetic and isotope-exchange studies on shikimate dehydrogenase from *Pisum sativum*. Biochemistry 10:1947–1952

Barash I, Mor H, Sodon T (1975) Evidence for ammonium dependent *de novo* synthesis of glutamate dehydrogenase in oat leaves. Plant Physiol 56:856–858

Barratt DHP (1980) Method for the detection of glutamine synthetase activity on starch gels. Plant Sci Lett 18:249–255

Bassham JA, Kirk M (1964) Photosynthesis of amino acids. Biochim Biophys Acta 90:553–562

Bassham JA, Larsen PO, Lawyer AL, Cornwell KL (1981) Relationships between nitrogen metabolism and photosynthesis. In Bewley JD (ed) 'Nitrogen and Carbon Metabolism' Martinus Nijhoff. The Hague, pp 135–163

Bauer A, Urquart AA, Joy KW (1977) Amino acid metabolism of pea leaves: diurnal changes and amino acid synthesis from $^{15}$N-nitrate. Plant Physiol 59:915–919

Beevers H (1979) Microbodies in higher plants. Annu Rev Plant Physiol 30:159–193

Beevers L, Storey R (1976) Glutamate synthase in developing cotyledons of *Pisum sativum*. Plant Physiol 57:862–866

Belser W, Murphy JB, Delmer DP, Mills SE (1971) End product control of trytophan biosynthesis in extracts and intact cells of the higher plant *Nicotiana tabacum* var Wisconsin 38. Biochim Biophys Acta 237:1–10

Bertagnolli BL, Wedding RT (1977) Purification and initial kinetic characterization of different forms of O-acetylserine sulphydrylase from seedlings of two species of *Phaseolus*. Plant Physiol 60:115–121

Bertani A, Tonelli C, Gavazzi G (1980) Determination of $\Delta'$ pyrolline-5-carboxylic acid reductase in proline requiring mutants of *Zea mays* L. Maydica 25:17–24

Bickel H, Palme L, Schultz G (1978) Incorporation of shikimate and other precursors into aromatic amino acids and prenylquinones of isolated spinach chloroplasts. Phytochemistry 17:119–124

Bidwell RGS, Durzan (1975) Some recent aspects of nitrogen metabolism. In: Davies PJ (ed) Historical and current aspects of plant physiology. Cornell Univ Press, New York, pp 152–262

Bird IF, Cornelius MJ, Keys AJ, Whittingham CP (1972a) Oxidation and phosphorylation associated with the conversion of glycine to serine. Phytochemistry 11:1587–1594

Bird IF, Keys AJ, Whittingham CP (1972b) In: Forti G, Avron M, Melandri A (eds) Proc Int Congr Photosynthesis Res. The Hague, Junk NV, pp 2215–2224

Bleckman GI, Kagan ZS, Kretovich WL (1971) Purification and some properties of "biosynthetic" L-threonine dehydratase from subcellular structures of pea seedlings. Biokhimiya 36:1050–1060

Boggess SF, Aspinall D, Paleg LG (1976a) The significance of end-product inhibition of proline biosynthesis and of compartmentation in relation to stress induced proline accumulation. Aust J Plant Physiol 3:513–525

Boggess SF, Stewart CR, Aspinall D, Paleg LG (1976b) Effects of water stress on proline synthesis from radioactive precursors. Plant Physiol 58:398–401

Boland MJ (1979) Kinetic mechanism of NADH-dependent glutamate synthase from lupin nodules. Eur J Biochem 99:531–539

Boland MJ, Benny AG (1977) Enzymes of nitrogen metabolism in legume nodules, purification and properties of NADH-dependent glutamate synthase from lupin nodules. Eur J Biochem 79:335–362

Boland MJ, Fordyce AM, Greenwood RM (1978) Enzymes of nitrogen metabolism in legume nodules, a comparative study. Aust J Plant Physiol 5:553–561

Borstlap AC (1970) Antagonistic effects of branched chain amino acids on the growth of *Spirodela polyrhiza* L Schleiden. Acta Bot Neerl 19:211–215

Borstlap AC (1975) Metabolic fate of exogenously supplied branched-chain amino acids in *Spirodela polyrhiza*. Acta Bot Neerl 24:203–206

Boudet AM, Lecussan R (1974) Generalité de l'association 5-deshydroquinate hydro-lyase, shikimate: NADP$^+$ oxydoreductase chez les végétaux superieurs. Planta 119:71–79

Boudet AM, Lecussan R, Boudet A (1975) Characterization and properties of two dehydroquinate hydro-lyases in higher plants. Planta 124:67–75
Boudet AM, Boudet A, Bouyssou H (1977) Taxonomic distribution of isoenzymes of dehydroquinate hydrolase in the angiosperms. Phytochemistry 16:919–922
Bowen JR, Kosuge T (1979) In vitro activity, purification and characterisation of shikimate kinase from Sorghum. Plant Physiol 63:382–396
Bright SWJ, Wood EA, Miflin BJ (1978a) The effect of aspartate-derived amino acids (lysine, threonine, methionine) on the growth of excised embryos of wheat and barley. Planta 139:113–117
Bright SWJ, Shewry PR, Miflin BJ (1978b) Aspartate kinase and the synthesis of aspartate-derived amino acids in wheat. Planta 139:119–125
Bright SWJ, Norbury PB, Miflin BJ (1979a) Isolation of a recessive barley mutant resistant to S-2-aminoethylcysteine. Theor Appl Genet 55:1–4
Bright SWJ, Featherstone L, Miflin BJ (1979b) Lysine metabolism in a barley mutant resistant to S-2-aminoethylcysteine. Planta 146:629–633
Bright SWJ, Lea PJ, Miflin BJ (1980) The regulation of methionine biosynthesis and metabolism in plants and bacteria. In: Elliott K, Wheland J (eds) Ciba Foundation Symposium No 72. Sulphur in biology. Exerpta Medica, Amsterdam, pp 101–104
Bright SWJ, Rognes SE, Miflin BJ (1981) Threonine accumulation in the seeds of a barley mutant with an altered aspartate kinase. Biochem Genet. In press
Brock BL, Wilkinson DA, King J (1970) Glyoxylate aminotransferases from oat leaves. Can J Biochem 48:486–492
Brown CM, Dilworth MJ (1975) Ammonia assimilation by *Rhizobium* cultures and bacteroids. J Gen Microbiol 86:39–48
Brown CM, Macdonald-Brown DS, Meers JL (1974) Physiological aspects of microbial inorganic nitrogen metabolism. Adv Microbiol Physiol 11:1–52
Bryan JK (1969) Studies on the catalytic and regulatory properties of homoserine dehydrogenase of *Zea mays* roots. Biochem Biophys Acta 171:205–216
Bryan JK (1976) Amino acid biosynthesis and its regulation. In: Bonner J, Varner JF (eds) Plant Biochemistry, 3rd edn. Academic Press, London New York, pp 525–560
Bryan JK (1980) Synthesis of the aspartate family and branched-chain amino acids. In: Miflin BJ (ed) The biochemistry of plants, vol V. Amino acids and their derivatives. Academic Press, London New York, pp 403–452
Bryan PA, Cawley RD, Brunner CE, Bryan JK (1970) Isolation and characterisation of a lysine sensitive aspartokinase from a multicellular plant. Biochim Biophys Res Commun 41:1211–1217
Bryan JK, Lissik EA, DiCamelli C (1979) Changes in enzyme regulation during plant growth. In: Hewitt EJ, Cutting CV (eds) Nitrogen assimilation of plants. Academic Press, London New York, pp 423–430
Buchanan BB (1980) The role of light in the regulation of chloroplast enzymes. Annu Rev Plant Physiol 31:341–374
Bucholz D, Reupke B, Bickel H, Schultz G (1979) Reconstitution of amino acid synthesis by combining spinach chloroplasts with other leaf organelles. Phytochemistry 18:1109–1112
Burdge EL, Matthews BF, Mills WR, Widholm JM, Wilson KG, Carlson LR, DeBonte LR, Oaks A (1979) Association of six aspartate and aromatic amino acid pathway enzymes with pea root and leaf plastids. Plant Physiol 63:Suppl 26
Butler GW, Shen L (1963) Leucine biosynthesis in higher plants. Biochim Biophys Acta 71:456–458
Burton EG, Sakami W (1969) The formation of methionine from the monoglutamate form of methyltetrahydrofolate by higher plants. Biochim Biophys Res Commun 36:228–234
Canvin DT, Atkins CA (1974) Nitrate, nitrite and ammonia assimilation by leaves: Effect of light, carbon dioxide, and oxygen. Planta 116:207–224
Carlson JE, Widholm JM (1978) Separation of two forms of anthranilate synthase from 5-Me tryptophan-susceptible and -resistant cultured *Solanum tuberosum* cells. Physiol Plant 44:251–255
Chang W-H, Tolbert NE (1965) Distribution of $C^{14}$ in serine and glycine after $^{14}CO_2$ photosynthesis by isolated chloroplasts. Plant Physiol 40:1048–1052

Chapman DJ, Leech RM (1976) Phosphoserine as an early product of photosynthesis in isolated chloroplast and leaves of *Zea mays* seedlings. FEBS Lett 68:160–164
Chen J, Boll WG (1971a) Tryptophan synthetase: purification and some properties of an inhibitor from pea roots. Can J Bot 49:821–832
Chen J, Boll WG (1971b) Tryptophan synthetase a 2 component enzyme from pea plants. Can J Bot 49:1155–1163
Cheshire RM, Miflin BJ (1975) The control of lysine biosynthesis in maize. Phytochemistry 14:695–698
Cheung GP, Rosenblum IY, Sallach HJ (1968) Comparative studies of enzymes related to serine metabolism in higher plants. Plant Physiol 43:1813–1820
Chiu JY, Shargool PD (1979) Importance of glutamate synthase in glutamate synthesis by soybean cell suspension cultures. Plant Physiol 63:409–415
Ch'ng AL (1980) Biosynthesis of amino acids from inorganic nitrogen sources in a marine alga. PhD Thesis, Univ Melbourne
Chou KH, Splittstoesser WE (1972) Glutamate dehydrogenase from pumkin cotyledons. Characterisation and isoenzymes. Plant Physiol 49:550–554
Christeller JG, Laing WA, Sutton WD (1977) Carbon dioxide fixation by lupin root nodules. Plant Physiol 60:47–50
Clandinin MT, Cossins EA (1974) Methionine biosynthesis in isolated *Pisum sativum* mitochondria. Phytochemistry 13:585–591
Cocking EC, Yemm EW (1961) Synthesis of amino acids and proteins in barley seedlings. New Phytol 60:103–116
Cooney DA, Jayaram HN, Swengros SG, Alter SC, Levine M (1980) The metabolism of L-asparagine in *Asparagus officinalis*. Int J Biochem 11:69–83
Cotton RGH, Gibson F (1969) The biosynthesis of phenylalanine and tyrosine in the pea (*P. sativum*): chorismate mutase. Biochim Biophys Acta 156:187–189
Cullimore JV, Sims AP (1980) An association between photorespiration and protein catabolism: studies with *Chlamydomonas*. Planta 150:392–396
Cullimore JV, Sims AP (1981a) Occurrence of two forms of glutamate synthase in *Chlamydomonas reinhardii*. Phytochemistry 20:597–600
Cullimore JV, Sims AP (1981b) Pathway of ammonia assimilation in illuminated and darkened *Chlamydomonas reinhardii*. Phytochemistry 20:933–940
Daley LS, Bidwell RGS (1977) Phosphoserine and phosphohydroxypyruvic acid: evidence for their role in as early intermediates in photosynthesis. Plant Physiol 60:109–114
Datko AH, Giovanelli J, Mudd SH (1974) Homocysteine biosynthesis in green plants; O-phosphorylhomoserine as the physiological substrate for cystathionine synthase. J Biol Chem 249:1139–1155
Datko AH, Mudd SH, Giovanelli J (1977) Studies of the homocysteine-forming sulfhydrolase. J Biol Chem 252:3436–3445
Davies DD, Teixeira AW (1975) The synthesis of glutamate and the control of glutamate dehydrogenase in pea mitochondira. Phytochemistry 14:647–656
Davies HM, Miflin BJ (1977) Aspartate kinase from carrot cell suspension culture. Plant Sci Lett 9:323–332
Davies HM, Miflin BJ (1978) Regulatory isoenzymes of aspartate kinase and the control of lysine and threonine biosynthesis in carrot cell suspension culture. Plant Physiol 62:536–541
Davies ME (1971) Regulation of histidine biosynthesis in cultured plant cells: evidence from studies on amitrole toxicity. Phytochemistry 10:783–788
Davis RH, Mora J (1968) Mutants of *Neurospora crassa* deficient in ornithine-$\delta$-transaminase. J Bacteriol 96:383–388
Delmer DP, Mills SE (1968a) Tryptophan synthetase from *Nicotiana tabacum*. Biochim Biophys Acta 167:431–443
Delmer DP, Mills SE (1968b) Tryptophan biosynthesis in cell cultures of *Nicotiana tabacum*. Plant Physiol 43:81–87
DiCamelli CA, Bryan JK (1975) Changes in enzyme regulation during growth of maize. Plant Physiol 55:999–1005
DiCamelli CA, Bryan JK (1980) Comparison of sensitive and desensitized forms of maize homoserine dehydrogenase. Plant Physiol 65:176–183

Dilworth MF, Dure L (1978) Developmental biochemistry of cotton seed embryogenesis and germination. Plant Physiol 61:698–702

Di Marco G, Grego S (1975) Homoserine dehydrogenase in *Pisum sativum* and *Ricinus communis*. Phytochemistry 14:943–947

Dodd WA, Cossins EA (1970) Homocysteine-dependent transmethylase catalyzing the synthesis of methionine in germinating pea seed. Biochim Biophys Acta 201:461–470

Dougall DK (1970) Threonine deaminase from Pauls Scarlet Rose tissue cultures. Phytochemistry 9:959–964

Dougall DK (1974) Evidence for the presence of glutamate synthase in carrot cell cultures. Biochem Biophys Res Commun 58:639–646

Dougall DK, Bloch J (1976) A survey of the presence of glutamate synthase in plant cell suspension cultures. Can J Bot 54:2924–2927

Dougall DK, Fulton MM (1967a) Isotope competition experiments using glucose-U-$^{14}$C and potential intermediates. Plant Physiol 42:941–945

Dougall DK, Fulton MM (1967b) Biosynthesis of protein amino acids in plant tissue cultures III. Studies on biosynthesis of arginine. Plant Physiol 42:387–390

Duffus CM, Rosie R (1978) Metabolism of ammonium ion and glutamate in relation to nitorgen supply and utilisation during grain development in barley. Plant Physiol 61:750–574

Duke SH, Collins M, Soberalske RM (1980) Effects of Potassium fertilisation on nitrogen fixation and nodule enzymes of nitrogen metabolism in alfalfa. Crop Sci 20:213–218

Dunham VL, Bryan JK (1969) Synergistic effects of metabolically related amino acids on the growth of a multicellular plant. Plant Physiol 44:1601–1608

Dunham VL, Bryan JK (1971) Synergistic effects of metabolically related amino acids on the growth of a multicellular plant II. Studies of amino acid incorporation. Plant Physiol 47:91–97

Dunn SD, Klucas RV (1973) Studies on possible routes of ammonia assimilation in soybean root nodule bacteroids. Can J Microbiol 19:1493–1499

Edge PA, Ricketts TR (1978) Studies on ammonium assimilating enzymes of *Platymonas striata* Butcher (Prasmophyceae). Planta 138:123–125

Ehmke A, Hartmann T (1976) Properties of glutamate dehydrogenase from *Lemna minor*. Phytochemistry 15:1611–1617

Ehmke A, Hartmann T (1978) Control of glutamate dehydrogenase from *Lemna minor* by divalent metal ions. Phytochemistry 17:6347–641

Eid S, Waly Y, Abdelal AT (1974) Separation and properties of two ornithine carbamoyltransferases from *Pisum sativum* seedlings. Phytochemistry 13:99–102

Elliott WH (1953) Isolation of glutamine synthetase and glutaminotransferase from green peas. J Biol Chem 201:661–672

Ellis RJ (1963) Cysteine biosynthesis in beet discs. Phytochemistry 2:1291–1336

Emes MJ, Fowler MW (1978) The intracellular location of the enzymes of nitrate assimilation in the apices of seedling pea roots. Planta 144:249–253

Estigneeva ZG, Pushkin AV, Radyukina NA, Kretovich VL (1977) Induction of glutamine synthetase by ammonium and its localisation in chloroplasts and cytosol of pea leaves. Dokl Akad Nauk SSR 237:962–964

Fankhauser H, Brunold C, Erismann KH (1976) Subcellular localization of O-acetylserine sulphydralase in spinach leaves. Experientia 32:1494–1497

Farago A, Denes G (1967) Mechanism of arginine biosynthesis in *Chlamydomonas reinhardti:* II Purification and properties of N-acetylglutamate-5-phosphotransferase, the allosteric enzyme of the pathway. Biochim Biophys Acta 136:6–18

Feierabend J, Brassel D (1977) Subcellular location of shikimate dehydrogenase in higher plants. Z Pflanzenphysiol 82:334–346

Filner P (1969) Control of nutrient assimilation. A growth-regulatory mechanism in cultured plant cells. Dev Biol Suppl 3:206–226

Forest JC, Wightman F (1973) Kinetic studies with a multispecific aminotransferase from bush bean (*Phaseolus vulgaris* L). Can J Biochem 51:332–343

Fowden L, Lea PJ, Bell EA (1979) The nonprotein amino acids of plants. In: Meister A (ed) Advances in enzymology, vol 50. John Wiley, New York, pp 118–175

Fowler MW, Jessup W, Stephan-Saskissian G (1974) Glutamate synthase type activity in higher plants. FEBS Lett 46:340–342
Fujihara S, Yamaguchi M (1978) Effects of allopurinol on the metabolism of allantoin in soybean plants. Plant Physiol 62:134–138
Furuhashi K, Yatazawa M (1970) Methionine-lysine-threonine-isoleucine interrelationships in the amino acid nutrition of rice callus tissue. Plant Cell Physiol 11:569–578
Gadal P, Bouyssou H (1973) Allosteric properties of chorismate mutase from *Quercus pedunculata*. Physiol Plant 28:7–13
Gardeström P, Bergman A, Ericson I (1980) Oxidation of glycine via the respiratory chain in mitochondria prepared from different parts of spinach. Plant Physiol 65:389–391
Gamborg OL (1965) Transamination in plants. The specificity of an amino transferase from mung bean. Can J Biochem 43:723–730
Gamborg OL, Keeley FW (1966) Aromatic metabolism in plants I. A study of the prephenate dehydrogenase from bean plants. Biochim Biophys Acta 115:65–72
Garland WJ, Dennis DT (1977) Steady-state kinetics of glutamate dehydrogenase from *Pisum sativum* L. mitochondria. Arch Biochem Biophys 182:614–625
Gavazzi G, Racchi M, Tonelli C (1975) A mutation causing proline requirement in *Zea mays*. Theor Appl Genet 46(7):339–346
Gayler KR, Morgan WR (1976) An NADP-dependent glutamate dehydrogenase in chloroplasts from the marine green alga *Caulerpa simpliciuscula*. Plant Physiol 58:283–287
Gengenbach BG, Walter TJ, Hibberd KA, Green CE (1978) Feedback regulation of lysine, threonine and methionine biosynthetic enzymes in corn. Crop Sci 18:472–476
Gilchrist DG, Kosuge T (1974) Regulation of aromatic amino acid biosynthesis in higher plants. Arch Biochem Biophys 164:95–105
Gilchrist DG, Kosuge T (1975) Properties of an aromatic amino acid insensitive chorismate mutase from mung bean. Arch Biochem Biophys 171:36–42
Gilchrist DG, Kosuge T (1980) Aromatic amino acid biosynthesis and its regulation. In: Miflin BJ (ed) Amino acids and their derivatives. The biochemistry of plants, vol V. Academic Press, London New York, pp 507–531
Gilchrist DG, Woodin TS, Johnson MS, Kosuge T (1972) Regulation of aromatic amino acid biosynthesis in higher plants I. Evidence for a regulatory form of chorismate mutase in etiolated mung bean seedlings. Plant Physiol 49:52–57
Giovanelli J, Mudd SH (1967) Synthesis of homocysteine and cysteine by enzyme extracts of spinach. Biochem Biophys Res Commun 27:150–156
Giovanelli J, Mudd SH (1971) Transsulfuration in higher plants. Partial purification and properties of β-cystathionase of spinach. Biochim Biophys Acta 227:654–670
Giovanelli J, Mudd SH, Datko AH (1974) Homoserine esterification in green plants. Plant Physiol 54:725–736
Giovanelli J, Mudd SH, Datko AH (1978) Homocysteine biosynthesis in green plants; physiological importance of the transsulphuration pathway in *Chlorella sorokiniana* growing under steady state conditions with limiting sulphate. J Biol Chem 253:5665–5677
Giovanelli J, Mudd SH, Datko AH (1980) Sulphur amino acids in plants. In: Miflin BJ (ed) The biochemistry of plants, vol V. Amino acids and their derivatives. Academic Press, London New York, pp 454–505
Givan C (1980) Amino transferase in higher plants. In: Miflin BJ (ed) The biochemistry of plants, vol V. Amino acids and their derivatives. Academic Press, London New York, pp 329–357
Glenn E, Maretzki A (1977) Properties and subcellular distribution of two partially purified ornithine transcarbamoylases in cell suspension of sugarcane. Plant Physiol 60:122–126
Green CE, Phillips RL (1974) Potential selection system for mutants with increased lysine, threonine, and methionine in cereal crops. Crop Sci 14:827–830
Groat RG, Soulen JK (1977) Kinetic properties of L-glutamate dehydrogenase from pea root mitochondria. Plant Physiol 59:70S
Grosse W (1976) Enzymes of tryptophan biosynthesis in etioplasts of *Pisum sativum* L. Z Pflanzenphysiol 80:463–468
Grosse W (1977) Ammonia detoxification and tryptophan synthesis in leucoplasts of walnuts. Z Pflanzenphysiol 83:249–255

Guiz C, Hirel B, Shedlofsky G, Gadal P (1979) Occurrence and influence of light on the relative proportions of two glutamine synthetases in rice leaves. Plant Sci Lett 15:271–277

Hanford J, Davies DD (1958) Formation of phosphoserine form 3-phosphoglycerate in higher plants. Nature (London) 182:532

Hankins CN, Largen MT, Mills SF (1976) Some physical characteristics of the enzymes of L-tryptophan biosynthesis in higher plants. Plant Physiol 57:101–104

Hanson AD, Tully RE (1979) Light stimulation of proline synthesis in water-stressed barley leaves. Planta 145:45–51

Harel E, Lea PJ, Miflin BJ (1977) The localisation of enzymes of nitrogen assimilation in maize leaves and their activities during greening. Planta 134:195–200

Hartmann T, Ehmke A (1980) Role of mitochondrial glutamate dehydrogenase in the reassimilation of ammonia produced by glycine-serine transformation. Planta 149:207–208

Haslam E (1974) The shikimate pathway. Halsted, New York, Toronto

Hasse K, Ratych OT, Salkinow J (1967) Transamination and decarboxylation of ornithine in higher plants. Hoppe Seyler's Z Physiol Chem 348:843–851

Hauschild HH (1959) The interconversion of glycine and serine in *Zea mays*. Can J Biochem Physiol 37:887–894

Haystead A (1973) Glutamine synthetase in the chloroplast of *Vicia faba*. Planta 111:271–274

Heber U, Walker DA (1979) The chloroplast envelope-barrier or bridge? Trends Biochem Sci 4:252–256

Henke RR, Wilson KG (1974) In vitro evidence for metabolic control of amino acid and protein synthesis by exogenous lysine and threonine in *Mimulus cardinalis*. Planta 121:155–166

Henke RR, Wilson KG, McClure JW, Treick RW (1974) Lysine-methionine-threonine interactions in growth and development of *Mimulus cardinalis* seedlings. Planta 116:333–345

Herridge DF, Atkins CA, Pate JS, Rainbird RM (1978) Allantoin and allantoic acid in the nitrogen economy of the cow pea. Plant Physiol 62:495–498

Hibberd KA, Walter T, Green CE, Gengenbach BG (1980) Selection and characterisation of a feedback-insensitive tissue cultures of maize. Planta 148:183–187

Huisman OC, Kosuge TC (1974) Regulation of aromatic amino acid biosynthesis in higher plants II. 3-Deoxy-arabino-heptalosonic acid 7-phosphate synthetase from cauliflower. J Biol Chem 249:6842–6848

Ito O, Yoneyama T, Kumazawa K (1978) The effect of light on ammonium assimilation and glutamine metabolism in the cells isolated from spinach leaves. Plant Cell Physiol 19:1109–1119

Jackson C, Dench JE, Morris P, Lui SC, Hall DO, Moore AL (1979) Photorespiratory nitrogen cycling: evidence for a mitochondrial glutamine synthetase. Biochem Trans 7:1122–1124

Joy KW (1971) Glutamate dehydrogenase changes in *Lemna* not due to enzyme induction. Plant Physiol 47:445–446

Joy KW (1973) Control of glutamate dehydrogenase from *Pisum sativum* roots. Phytochemistry 12:1031–1040

Kagan ZS, Kretovich WL, Tscheuschner G (1963) Biosynthesis of isoleucine from its α, β-dihydroxy analogue in various plant seedlings. Plant Physiol 10:485–489

Kaiser JJ, Lewis OAM (1980) Nitrate nitrogen assimilation in the leaves of *Helianthus annuus* L. New Phytol 85:235–241

Kanamori M, Wixom RL (1963) Studies in valine biosynthesis V. Characterisation of the purified dihydroxyacid dehydrolase from spinach leaves. J Biol Chem 238:998–1105

Kanamori T, Matsumoto H (1972) Glutamine synthetase from rice plant roots. Arch Biochem Biophys 152:404–412

Kennedy IR (1966) Primary products of symbiotic nitrogen fixation II. Pulse labelling of seradella nodules with $^{15}N_2$. Biochim Biophys Acta 130:295–303

Kern R, Chrispeels MJ (1978) Influence of the axis on the enzymes of protein and amide metabolism in the cotyledons of mung beans seedlings. Plant Physiol 62:815–819

Keys AJ (1980) Synthesis and interconversion of glycine and serine. In: Miflin BJ (ed) The biochemistry of plants, vol V. Amino acids and their derivatives. Academic Press, London New York, pp 359–374

Keys AJ, Bird IF, Cornelius MJ, Lea PJ, Wallsgrove RM, Miflin BJ (1978) The photorespiratory nitrogen cycle. Nature (London) 275:741–743

Kindt R, Pahlich E, Asched I (1980) Glutamate dehydrogenase from peas: Isolation, quaternary structure, and influence of cations on activity. Eur J Biochem 112:533–540

King J, Waygood ER (1968) Glyoxylate amino transferases from wheat leaves. Can J Biochem 46:771–779

King J, Wu Y-FW (1971) Partial purification and kinetic properties of glutamic dehydrogenase from soybean cotyledons. Phytochemistry 10:915–928

Kingdon HS (1974) Feedback inhibition of glutamine synthetase from green pea seeds. Arch Biochem Biophys 163:429–431

Kiritani K, Wagner RP (1970) $\alpha$, $\beta$-dihydroxy acid dehydratase. In: Tabor H, Tabor CW (eds) Methods in enzymology, vol XVII A. Academic Press, London New York, pp 755–766

Kisaki T, Yoshida N, Imai A (1971) Glycine decarboxylase and serine formation in spinach leaf mitochondrial preparation with reference to photorespiration. Plant Cell Physiol 12:275–288

Kleczkowski K, Cohen P (1964) Purification of ornithine transcarbamylase from pea seedlings. Arch Biochem Biophys 107:271–278

Kolloffel C, van Dijke HD (1975) Mitochondrial arginase activity from cotyledons of developing and germinating seeds of Vicia faba. Plant Physiol 55:507–510

Koshiba T (1978a) Purification of two forms of the associated 3-dehydroquinate hydro-lyase and shikimate: $NADP^+$ oxidoreductase in *Phaseolus mungo* seedlings. Biochim Biophys Acta 522:10–18

Koshiba T (1978b) Shikimate kinase and 5 enolpyruvolshikimate-3-phosphate synthase in *Phaseolus mungo* seedlings. Z Pflanzenphysiol 88:353–355

Kretovich WL, Evstigneeva ZG (1974) Comparative study of plant organism glutamine synthetase. In: Dore K, Fox SW, Debronn A, Polovska TE (eds) The origin of life and evolutionary biochemistry. Plenum Press, New York, pp 245–262

Kretovich VL, Karyakin TI, Charakch'yan VV, Kaloshina GS, Sidelnikova LI, Shaposhnikovgl (1978) Biosynthesis of asparatic acid from oxaloacetic acid and ammonium in plants. Dokl Akad Nauk SSSR 243:793–796

Kurz WCW, Rokosh DA, La Rue TA (1975) Enzymes of ammonia assimilation in *Rhizobium* bacteroids. Can J Microbiol 21:1009–1012

Lara J, Mills SE (1972) Tryptophan synthetase in *Euglena gracilis* strain G. J Bacteriol 110:1100–1106

Larson C, Albertsson E (1979) Enzymes related to serine syntheses in spinach chloroplasts. Physiol Plant 45:7–10

Lea PJ, Fowden L (1975) The purification and properties of glutamine-dependent asparagine synthetase isolated from *Lupinus albus*. Proc R Soc London Ser B 192:13026

Lea PJ, Miflin BJ (1974) An alternative route for nitrogen assimilation in higher plants. Nature (London) 251:614–616

Lea PJ, Miflin BJ (1975) The occurrence of glutamate synthase in algae. Biochem Biophys Res Commun 64:856–862

Lea PJ, Miflin BJ (1979) Photosynthetic ammonia assimilation. In: Gibbs M, Latzko E (eds) Encyclopaedia of plant physiology, vol VI. New series. Springer, Berlin Heidelberg New York, pp 445–456

Lea PJ, Miflin BJ (1980) Transport and metabolism of asparagine and other nitrogen compounds within the plant. In: Miflin BJ (ed) The biochemistry of plants, vol V. Amino acids and their derivatives. Academic Press, London New York, pp 569 607

Lea PJ, Thurman DA (1972) Intracellular location and properties of plant L-glutamate dehydrogenase. J Exp Bot 23:440–449

Lea PJ, Fowden L, Miflin BJ (1978) The purification and properties of asparaginase from *Lupinus* species. Phytochemistry 17:217–222

Lea PJ, Hughes J, Miflin BJ (1979a) Glutamine and asparagine dependent protein synthesis in maturing cotyledons cultured in vitro. J Exp Bot 30:529–537
Lea PJ, Mills WR, Miflin BJ (1979b) The isolation of a lysine-sensitive aspartate kinase from pea leaves and its involvement in homoserine biosynthesis in isolated chloroplasts. FEBS Lett 98:165–168
Lea PJ, Mills WR, Wallsgrove RM, Miflin BJ (1981) The assimilation of nitrogen and the synthesis of amino acids in chloroplasts and blue-green bacteria. In: Schiff JA (ed) The origin of the chloroplasts. Elsevier, New York, in press
Lee DW (1973) Glutamate dehydrogenase isoenzyme in *Ricinus communis* seedlings. Phytochemistry 12:2631–2634
Lee JA, Stewart GR (1978) Ecological aspects of nitrogen metabolism. Adv Bot Res 6:1–43
Leech RM (1977) Subcellular fractionation techniques in enzyme distribution studies. In: Smith H (ed) Regulation of enzyme synthesis and activity in higher plants. Academic Press, London New York, pp 289–327
Leech RM, Kirk P (1968) An NADP-dependent L-gutamate dehydrogenase from chloroplasts of *Vicia faba*. Biochem Biophys Res Commun 32:685–690
Lewis OAM (1975) An $^{15}N$, $^{14}C$ study of the role of the leaf in the nitrogen nutrition of the seed of *Datura stramonium* L. J Exp Bot 26:361–372
Lewis OAM, Pate JS (1973) The significance of transpirationally derived nitrogen in protein synthesis in fruiting plants of pea (*Pisum sativum* L). J Exp Bot 24:596–606
Lewis OAM, Probyn TA (1978a) The effect of nitrate feeding levels on the pathway of nitrogen incorporation into photosynthesizing leaf metabolism of *Datura stramonium* L. Plant Soil 49:625–631
Lewis OAM, Probyn TA (1978b) $^{15}N$ incorporation and glutamine synthetase inhibition studies of nitrogen assimilation in beans of the nitrophile *Datura stramonium* L. New Phytol 81:519–526
Lloyd NDH, Joy KW (1978) 2-Hydroxysuccinamic acid: a product of asparagine metabolism in plants. Biochim Biophys Res Commun 81:186–192
Lu TS, Mazelis M (1975) L-Ornithine:2-oxoacid aminotransferase from Squash (*Cucurbita pepo* L.) cotyledons. Plant Physiol &&:502–506
MacHolan L, Zobac P, Hekelova J (1965) Activity, utilization and metabolism of 5-amino-2-oxovaleric acid ($\Delta'$-pyrolline-2-carboxylic acid) in pea seedlings and bakers yeast. Hoppe-Seyer's Z Physiol Chem 349:97–106
Madison JT, Thompson JF (1976) Threonine synthase from higher plants – stimulation by S-adenosylmethionine and inhibition by cysteine. Biochem Biophys Res Commun 71:684–691
Mann AF, Fentem PA, Stewart GR (1979) Identification of two forms of glutamine synthetase in barley (*Hordeum vulgare*). Biochem Biophys Res Commun 88:515–521
Mann AF, Fentem PA, Stewart GR (1980) Tissue localization of barley (*Hordeum vulgare*) glutamine synthetase isoenzymes. FEBS Lett 110:265–267
Marker AFH, Whittingham CP (1967) The site of synthesis of sucrose in green plant cells. J Exp Bot 18:732–739
Masada M, Fukushima K, Tamura G (1975) Cysteine synthase from rape leaves. J Biochem 77:1107–1115
Matoh T, Ida S, Takahashi E (1980) Isolation and characterization of NADH-glutamate synthase from pea (*Pisum sativum*). Plant and Cell Physiol 21:1461–1474
Matsumoto T, Yamatazawa M, Yamamoto Y (1977a) Incorporation of nitrogen-15 into allantoin in nodulated soyabean plants supplied with molecular nitrogen. Plant Cell Physiol 18:353–359
Matsumoto T, Yatazawa M, Yamamoto Y (1977b) Distribution and change in the contents of allantoin and allantoic acid in developing nodulated an non-nodulated soyabean plants. Plant Cell Physiol 18:459–462
Matthews BF, Widholm JM (1978) Regulation of lysine and threonine synthesis in carrot cell suspension cultures and whole carrot roots. Planta 141:315–321
Matthews BF, Widholm JM (1979a) Enzyme expression in soyabean cotyledon of callus and cell suspension culture. Can J Bot 57:299–304

Matthews BF, Widholm JM (1979b) Regulation of homoserine dehydrogenase in developing organs of soyabean seedlings. Phytochemistry 18:395–400
Matthews BF, Widholm JM (1979c) Expression of aspartokinase, dihydrodipicolinic acid synthase and homoserine dehydrogenase during growth of carrot cell suspension cultures on lysine- and threonine-supplemented media. Z Naturforsch 34c:1177–1185
Matthews BF, Gurman AW, Bryan JK (1975) Changes in enzyme regulation during growth of maize. Plant Physiol 55:991–998
Mazelis M (1980) Amino acid catabolism. In: Miflin BJ (ed) The biochemistry of plants, vol V. Amino acids and their derivatives. Academic Press, London New York, pp 541–567
Mazelis M, Creveling RK (1978) The enzymology of lysine biosynthesis in higher plants. Diaminopimelate decarboxylase from wheat germ. J Food Biochem 2:29–37
Mazelis M, Fowden L (1969) Conversion of ornithine into proline by enzymes from germinating peanut cotyledons. Phytochemistry 8:801–810
Mazelis M, Liu ES (1967) Serine transhydroxymethylase of cauliflower (*Brassica oleracea* var botrytis L.). Plant Physiol 42:1763–1768
Mazelis M, Miflin BJ, Pratt HM (1976) A chloroplast-localized diaminopimelate decarboxylase in higher plants. FEBS Lett 64:197–200
Mazelis M, Whatley FR, Whatley J (1977) The enzymology of lysine biosynthesis in higher plants. FEBS Lett 84:236–240
McKay G, Shargool PD (1977) The biosynthesis of ornithine from glutamate in higher plant tissue. Plant Sci Lett 9:189–193
McKee HS (1962) Nitrogen metabolism in plants. Clarendon Press, Oxford
McKenzie GH, Ch'ng AL, Gayler KR (1979) Glutamine synthethase/glutamine:2-oxoglutarate aminotransferase in chloroplasts from the marine alga *Caulerpa simpliciuscula*. Plant Physiol 63:578–582
McNamer AD, Stewart CR (1974) NAD-dependent proline dehydrogenase in *Chlorella*. Plant Physiol 53:440–444
McParland RH, Guevara JG, Becker RR, Evans HJ (1976) The purification and properties of glutamine synthetase from the cytosol of soybean root nodules. Biochem J 153:597–606
Meeks JC, Wolk CP, Schilling N, Shaffer PW, Avissar Y, Chien W-S (1978) Initial organic products of fixation of [$^{13}$N] dinitrogen by root nodules of soybean (*Glycine max*). Plant Physiol 61:980–983
Meister A, Soben HA, Trich SV, Fraser PE (1952) Transamination and associated deamidation of asparagine and glutamine. J Biol Chem 197:319–330
Meister A, Radhakrishnan AN, Buckley SD (1957) Enzymatic synthesis of L-pipecolic acid and L-proline. J Biol Chem 229:789–800
Mestichelli LJJ, Gupta RN, Spenser ID (1979) The biosynthetic route from ornithine to proline. J Biol Chem 254:640–647
Meulen PYF van der, Bassham JA (1959) Study of inhibition of azaserine and diazo-oxo-norleucine (DON) on the algae *Scenedesmus* and *Chlorella*. J Am Chem Soc 81:2233–2239
Miflin BJ (1969) The inhibitory effects of various amino acids on the growth of barley seedlings. J Exp Bot 20:810–819
Miflin BJ (1970) Studies on the subcellular location of particulate nitrate and nitrite reductase, glutamic dehydrogenase and other enzymes in barley roots. Planta 93:160–170
Miflin BJ (1971) Cooperative feedback control of barley acetohydroxy acid synthetase by leucine, isoleucine and valine. Arch Biochem Biophys 146:542–550
Miflin BJ (1974) The location of nitrite reductase and other enzymes related to amino acid biosynthesis in the plastids of root and leaves. Plant Physiol 54:550–555
Miflin BJ (1977) Modification controls in time and space. In: Smith H (ed) Regulation of enzyme synthesis and activity in higher plants. Academic Press, London New York, pp 23–40
Miflin BJ (ed) (1980a) Amino acids and their derivatives, vol V. The biochemistry of plants. Academic Press, London New York
Miflin BJ (1980b) Nitrogen metabolism and amino acid biosynthesis in crop plants. In:

Carlson PS (ed) The biology of crop productivity. Academic Press, London New York, pp 255–296

Miflin BJ, Cave PR (1972) The control of leucine, isoleucine and valine biosynthesis in a range of higher plants. J Exp Bot 23:511–516

Miflin BJ, Lea PJ (1975) Glutamine and asparagine as nitrogen donors for reductant-dependant glutamate synthesis in pea roots. Biochem J 149:403–409

Miflin BJ, Lea PJ (1976) The pathway of nitrogen assimilation in plants. Phytochemistry 15:873–885

Miflin BJ, Lea PJ (1977) Amino acid metabolism. Annu Rev Plant Physiol 28:299–329

Miflin BJ, Lea PJ (1980) Ammonia assimilation. In: Miflin BJ (ed) The biochemistry of plants, vol V. Amino acids and their derivatives. Academic Press, London New York, pp 169–202

Miflin BJ, Shewry PR (1979) The synthesis of proteins in normal and high lysine barley seeds. In: Laidman D, Wyn Jones R (eds). Academic Press, London New York, pp 239–273

Miflin BJ, Marker AFH, Whittingham CP (1966) The metabolism of glycine and glycolate by pea leaves in relation to photosynthesis. Biochim Biophys Acta 120:266–273

Miflin BJ, Bright SWJ, Davies HM, Shewry PR, Lea PJ (1979) Amino acids derived from asparatate: their biosynthesis and its regulation in plants. In: Hewitt EJ, Cutting CV (eds) Academic Press, London New York, pp 335–358

Mills WR, Wilson KG (1978) Amino acid biosynthesis in isolated chloroplast; metabolism of labelled aspartate and sulphate. FEBS Lett 92:129–132

Mills WR, Lea PJ, Miflin BJ (1980) Photosynthetic formation of the aspartate family of amino acids in isolated chloroplasts. Plant Physiol 65:1166–1172

Minamikawa T (1967) A study of DAHP synthetase in higher plants. Plant Cell Physiol 8:695–707

Minamikawa T, Uritani J (1967) Deoxyarabinoheptulosonate phosphate synthase in sweet potato roots. J Biochem (Tokyo) 61:367–372

Mitchell CA, Stocking CR (1975) Kinetics and energetics of light-driven chloroplast glutamine synthesis. Plant Physiol 55:59–63

Moller BC (1974) Lysine biosynthesis in barley (*Hordeum vulgare* L). Plant Physiol 54:638–643

Moore R, Black CC (1979) Nitrogen assimilation pathways in leaf mesophyll and bundle sheath cells of $C_4$ photosynthesis plants formulated from comparative studies with *Digitaria sanguinalis* (L.) Scop. Plant Physiol 64:309–313

Morris CJ, Thompson JF (1975) Acetyl coenzyme A-glutamate acetyltransferase and $N^2$-acetylornithine-glutamate acetyltransferase of chlorella. Plant Physiol 55:960–967

Morris CJ, Thompson JF (1977) Formation of N-acetylglutamate by extracts of higher plants. Plant Physiol 59:684–687

Morris CJ, Thompson JF, Johnson CM (1969) Metabolism of glutamic acid and N-acetyl glutamic acid in leaf discs and cell-free extracts of higher plants. Plant Physiol 44:1023–1026

Mudd SH (1960) S-Adenosylmethionine formation by barley extracts. Biochim Biophys Acta 38:354–355

Nagao RT, Moore TC (1972) Partial purification and properties of tryptophan synthetase of pea plants. Arch Biochem Biophys 149:402–413

Nagel M, Hartmann T (1980) Glutamate dehydrogenase from *Medicago sativa* L. Purification and comparative kinetic studies of the organ-specific multiple forms. Z Naturforsch 35C:406–415

Nandy M, Ganguli NC (1961) Biological synthesis of 5-dehydroshikimic acid by a plant extract. Biochim Biophys Acta 48:608–610

Nauen W, Hartmann T (1980) Glutamate dehydrogenase from *Pisum sativum*. Planta 148:7–16

Ng BH, Anderson JW (1978) Chloroplast cysteine synthases of *Trifolium repens* and *Pisum sativum*. Phytochemistry 17:879–886

Ng BH, Anderson JW (1979) Light-dependent incorporation of selenite and sulphite into selenocysteine and cysteine by isolated pea chloroplasts. Phytochemistry 18:573–580

Ngo TT, Shargool PD (1974) The enzymatic synthesis of L-cysteine in higher plant tissues. Can J Bot 52:6
Nicklish A, Geske W, Kohl JG (1976) Relevance of glutamate synthase and glutamate dehydrogenase to the nitrogen assimilation of primary leaves of wheat. Biochem Physiol Pflanz 170:85–90
Nishimura M, Douce R, Akazawa T (1980) Absence of glutamine synthetase in mitochondria isolated from spinach leaf protoplasts. Plant Physiol 65:(Suppl) 14
Noguchi M, Koiwai A, Tamaki E (1966) Studies on nitrogen metabolism in tobacco plants VII. $\Delta'$ pyrroline-5-carboxylate reductase from tobacco leaves. Agric Biol Chem 30:452–456
Noguchi M, Koiwai A, Yokoyama M, Tamaki E (1968) Studies on nitrogen metabolism in tobacco plants IX. Effect of various compounds on proline biosynthesis in green leaves. Plant Cell Physiol 9:35–47
Oaks A (1965a) The effect of leucine on the biosynthesis of leucine in maize root tips. Plant Physiol 40:149–155
Oaks A (1965b) The synthesis of leucine in maize embryos. Biochim Biophys Acta 111:79–89
Oaks A, Mitchell IJ, Barnard RA, Johnson FT (1970) The regulation of proline biosynthesis in maize roots. Can J Bot 48:2249–2258
Oaks A, Jones K, Misra S (1979) A comparison of glutamate synthase obtained from maize endosperms and roots. Plant Physiol 63:793–795
Oaks A, Stulen I, Jones K, Winspear MJ, Misra S, Boesel IL (1980) Enzymes of nitrogen assimilation in maize roots. Planta 148:477–484
Ohyama T, Kumazawa K (1978) Incorporation of $^{15}N$ into various nitrogenous compounds in intact soybean nodules after exposure to $^{15}N_2$ gas. Soil Sci Plant Nutr (Tokyo) 24:525–533
Ohyama T, Kumazawa K (1980) Nitrogen assimilation in soyabean nodules 1. The role of GS/GOGAT system in the assimilation of ammonia produced by $N_2$-fixation. Soil Sci Plant Nutr (Tokyo) 26:109–115
O'Neal TD, Joy KW (1973a) Localization of glutamine synthetase in chloroplasts. Nature New Biol (London) 246:61–62
O'Neal TD, Joy KW (1973b) Glutamine synthetase of pea leaves I. Purification, stabilization and pH optima. Arch Biochem Biophys 159:113–122
O'Neal TD, Joy KW (1974) Glutamine synthetase of pea leaves. Divalent cation effects, substrate specificity and other properties. Plant Physiol 54:773–779
O'Neal TD, Joy KW (1975) Pea leaf glutamine synthetase, regulatory properties. Plant Physiol 55:968–974
O'Neal TD, Naylor AW (1969) Partial purification and properties of carbamoyl phosphate synthetase of Alaska pea (*Pisum sativum* L cv. Alaska). Biochem J 113:271–279
O'Neal TD, Naylor AW (1976) Some regulatory properties of pea leaf carbamoyl phosphate synthase. Plant Physiol 57:23–28
Ong BJ, Jackson JF (1972) Enzymic aspects of the control of carbamoyl phosphate synthesis and utilization. Biochem J 129:583–593
Pahlich E (1972) Evidence that the multiple molecular forms of glutamate dehydrogenase from pea seedlings are conformers. Planta 104:78–88
Pahlich E, Gerlitz CHR (1980) Deviations from Michaelis-Menten behaviour of plant glutamate dehydrogenase. Phytochemistry 19:11–13
Pahlich E, Joy KW (1971) Glutamate dehydrogenase from pea roots: purification and properties of the enzyme. Can J Biochem 49:127–138
Pahlich E, Ott W, Schad B (1980) Immunochemical investigations with highly purified glutamate dehydrogenase from pea seeds by means of the Ouchtenlony test. J Exp Bot 31:419–423
Patel N, Pierson DL, Jensen RA (1977) Dual enzymatic routes to L-tyrosine and L-phenylalanine via pretyrosine in *Pseudomonas aeruginosa*. J Biol Chem 252:5839–5846
Planque K, de Vries GE, Kijne JW (1978) The relationship between nitrogenase and glutamine synthetase in bacteroids of *Rhizobium leguminosarum* of various ages. J Gen Microbiol 106:173–178

Prather CW, Sisler EC (1966) Purification and properties of serine hydroxymethyltransferase from *Nicotiana rusticana* L. Plant Cell Physiol 7:457–466

Probyn TA, Lewis OAM (1979) The route of nitrate-nitrogen assimilation in the root of *Datura stramonium*. J Exp Bot 30:299–305

Quail PH (1979) Plant cell fractionation. Annu Rev Plant Physiol 30:425–484

Rabson R, Tolbert NE, Kearney PC (1962) Formation of serine and glyceric acid by the glycolate pathway. Arch Biochem Biophys 98:154–163

Racchi ML, Gavazzi G, Monti D, Manitto P (1978) An analysis of the nutritional requirements of the *pro* mutant in *Zea mays*. Plant Sci Lett 13:347–364

Radyukina NA, Pushkin AV, Estigneeva ZG, Kretovich VL (1977) The assimilation of ammonia in *Glycine* nodules during the vegetation process. Dokl Biochem 234:193–195

Randall DD, Tolbert NE, Gremel D (1971) 3-Phosphoglycerate phosphatase in plants. Plant Physiol 48:480–487

Rasulov AS, Estigneeva ZG, Kretovich VL (1977) Physiochemical properties of *Chlorella* glutamine synthetase. Dokl Akad Nauk SSSR 233:726–729

Ratajczak L, Ratajczak W, Mazurowa H, Woznya A (1979) Localisation of glutamate dehydrogenase and glutamate synthase in roots and nodules of *Lupinus* seedlings. Biochem Physiol Pflanz 174:289–295

Rathnam CKM, Edwards GE (1976) Distribution of nitrate-assimilating enzymes between mesophyll protoplasts and bundle sheath cells in leaves of three groups of $C_4$ plants. Plant Physiol 57:881–885

Ray TB, Black CC (1979) The $C_4$ pathway and its regulation. In: Gibbs M, Latzko E (eds) Encyclopaedia of plant physiology, vol VI. Springer, Berlin Heidelberg New York, pp 77–101

Redkina TV, Uspenskaya ZV, Kretovich WL (1969) Purification and properties of L-glutamate phenylpyruvate aminotransferase of plant origin. Biokhimiya 34:247–250 (Engl edn)

Reinbothe H, Mothes K (1962) Urea, Ureides and guanidines in plants. Annu Rev Plant Physiol 13:129–150

Rena AB, Splittstoesser WE (1975) Proline dehydrogenase and pyrroline-5-carboxylate reductase from pumpkin seedlings. Phytochemistry 14:657–661

Rhodes D, Rendon GA, Stewart GR (1976) The regulation of ammonia assimilating enzymes in *Lemna minor*. Planta 129:203–210

Rhodes D, Sims AP, Stewart GP (1979) Glutamine synthetase and the control of nitrogen assimilation in *Lemna* minor L. In: Hewitt EJ, Cutting CV (eds) Nitrogen assimilation in plants. Academic Press, London New York, pp 501–520

Rhodes D, Sims AP, Folkes BF (1980) Pathway of ammonia assimilation in illuminated *Lemna*. Phytochemistry 19:357–365

Robertson JG, Farnden KJF, Warburton MP, Banks JM (1975a) Induction of glutamine synthetase during nodule development in lupin. Aust J Plant Physiol 2:265–272

Robertson JG, Warburton MP, Farnden KJF (1975b) Induction of glutamate synthase during nodule development in lupin. FEBS Lett 55:33–37

Rognes SE (1970) Glutamine-dependent asparagine synthetase from yellow lupin seedlings. FEBS Lett 10:62–66

Rognes SE (1975) Glutamine-dependent asparagine synthetase from *Lupinus luteus*. Phytochemistry 14:1977–1982

Rognes SE (1980) Anion regulation of lupin asparagine synthetase: chloride activation of the glutamine-utilizing reactions. Phytochemistry 19:2287–2294

Rognes SE, Lea PJ, Miflin BJ (1980) S-Adenosylmethionine, a novel regulator of aspartate kinase. Nature (London) 287:357–359

Rosenthal GA (1970) Canavanine utilisation in the developing plant. Plant Physiol 46:273–276

Rosenthal GA, Naylor AW (1969) Purification and general properties of arginosuccinate lyase from Jack Bean, *Canavalia ensiformis* L. DC, Biochem J 112:415

Rothe GM (1974) Intracellular compartmentation and regulation of two shikimate dehydrogenase isoenzymes in *Pisum sativum*. Z Pflanzenphysiol 74:152–159

Rothe GM, Maurer W, Mielke C (1976) Study of 3-deoxy-D-arabino-heptulosonic acid-7 phosphate synthase in higher plants. Ber Dtsch Bot Ges 89:163–177
Roubelakis KA, Kliewer WM (1978a) Ornithine carbamoyltransferase: isolation and some properties. Plant Physiol 62:337–339
Roubelakis KA, Kliewer WM (1978b) Enzymes of the Krebs:Henseleit cycle in *Vitis vinifera* L. II. Arginosuccinate synthetase and lyase. Plant Physiol 62:340–343
Rowell P, Stewart WDP (1976) Alanine dehydrogenase of the $N_2$-fixing blue-green alga *Anabaena cylindrica*. Arch Microbiol 107:115–124
Rubin JL, Jensen RA (1979) Enzymology of L-tryosine biosynthesis in mung bean *Vigna radiata* L. Plant Physiol 64:727–734
Sahulka J, Lisa L (1979) Regulation of glutamine synthetase level in isolated pea roots. Biochem Physiol Pflanz 174:646–652
Saijo R, Kosuge T (1978) The conversion of 3-deoxyarabino-heptulosonate-7-phosphate to 3-dehydroquinate by sorghum seedling preparations. Phytochemistry 17:223–225
Sainis J, Mayne RG, Wallsgrove RM, Lea PJ, Miflin BJ (1981) Localisation and characterisation of homoserine dehydrogenase isolated from barley and pea leaves. Planta 152:491–496
Sakano H (1979) Derepression and repression of lysine-sensitive aspartokinase during in vitro culture of carrot root tissue. Plant Physiol 63:583–585
Sakano K, Komamine A (1978) Changes in the proportion of two aspartokinases in carrot root tissue in response in vitro culture. Plant Physiol 61:115–118
Sanderson GW (1966) 5-Dehydroshikimate reductase in the tea plant (*Camellia sinensis* L.): Properties and distribution. Biochem J 98:248
Santarius KA, Stocking CR (1969) Intracellular localization of enzymes in leaves and chloroplast membranes, permeability to compounds involved in amino acid biosynthesis. Z Naturforsch B246:1170–1179
Sasaoka K (1961) Studies on homoserine dehydrogenase in pea seedlings. Plant Cell Physiol 2:231–242
Scheid H-W, Ehmke A, Hartmann T (1980) Plant NAD-dependent glutamate dehydrogenase. Purification, molecular properties and metal ion activation of the enzymes from *Lemna minor* and *Pisum sativum*. Z Naturforsch 35C:213–221
Scher WI, Vogel HJ (1957) Occurrence of ornithine $\delta$-transaminase: a dichotomy. Proc Natl Acad Sci USA 53:796–803
Schmidt A, Abrams WR, Schiff JA (1974) Reduction of adenosine-5′-phosphosulphate to cysteine in extracts from *Chlorella* and mutants blocked for sulphate reduction. Eur J Biochem 47:423–434
Schnyder J, Rottenberg M, Erismann KH (1975) The synthesis of threonine and thiothreonine from O-phosphohomoserine by extracts prepared from higher plants. Biochem Physiol Pflanz 167:605–608
Scott DB, Robertson J, Farnden KJF (1976) Ammonia assimilation in lupin nodules. Nature (London) 262:703–705
Sen D, Schulman HM (1980) Enzymes of ammonia assimilation in the growth of developing soybean root nodules. New Phytol 85:243–250
Seneviratne AS, Fowden L (1968) Diamino acid metabolism in plants with special reference to $\alpha$, $\beta$-diaminoproprionic acid. Phytochemistry 7:1047–1056
Shah SPJ, Cossins EA (1970a) The biosynthesis of glycine and serine by isolated chloroplasts. Phytochemistry 9:1545–1551
Shah SPJ, Cossins EA (1970b) Pteroylglutamate and methionine biosynthesis in isolated chloroplasts. FEBS Lett 7:267–270
Shargool PD (1971) Purification of arginosuccinate synthetase from cotyledons of germinating peas. Phytochemistry 10:2029–2032
Shargool PD (1973) The response of soybean arginosuccinate synthetase to different energy charge values. FEBS Lett 33:348–350
Shargool PD (1975) Degradation of argininosuccinate lyase by a protease synthesized in soybean cell suspension cultures. Plant Physiol 55:632–635
Shargool PD, Cossins EA (1968) Further studies of L-arginine biosynthesis in germinating pea seeds. Can J Biochem 46:393–399

Shargool PD, Steeves T, Weaver MG, Russell M (1978) The localization within plant cells of enzymes involved in arginine biosynthesis. Can J Biochem 56:273–279
Sharma RK, Mazumder R (1970) Purification properties and feedback control of L threonine dehydratase from spinach. J Biol Chem 245:3008–3025
Shatilov VR, Sofin AV, Zabrodina TM, Mutuskin AA, Pshenova KV, Kretovich VL (1978) Ferredoxin-dependent glutamate synthase of *Chlorella*. Biokhimiya 43:1492–1495
Shepard DV, Thurman DA (1973) Effect of nitrogen sources upon the activity of glutamate dehydrogenase of *Lemna gibba*. Phytochemistry 12:1937–1946
Shewry PR, Miflin BJ (1977) Properties and regulation of aspartate kinase from barley seedlings (*Hordeum vulgare* L). Plant Physiol 59:69–73
Shimura Y, Vogel HJ (1966) Diaminopimelate decarboxylase of *Lemna perspusilla*: partial purification and some properties. Biochim Biophys Acta 118:396–404
Siegel JN, Gentile AC (1966) Effect of 3-amino 1,2,4-triazole on histidine metabolism. Plant Physiol 41:670–672
Skokut TA, Wolk CP, Thomas J, Meeks JC, Shaffer PW (1978) Initial organic products of assimilation of [$^{13}$N] ammonium and [$^{13}$N] nitrate by tobacco cells cultured on different sources of nitrogen. Plant Physiol 62:299–304
Slaughter JC (1970) The activiation of 3PGA dehydrogenase by methionine. FEBS Lett 7:245–247
Slaughter JC (1973) Inhibition of 3-phosphoglycerate dehydrogenase by compounds other than L-serine. Phytochemistry 12:2627–2629
Slaughter JC, Davies DD (1968a) The isolation and characterisation of 3-phosphoglycerate dehydrogenase from peas. Biochem J 109:743–749
Slaughter JC, Davies DD (1968b) Inhibition of 3-phosphoglycerate dehydrogenase of L-serine. Biochem J 109:749–755
Slaughter JC, Davies DD (1969) The enzymes of serine biosynthesis. J Exp Bot 20:451–456
Sloger M, Owens LD (1974) Control of free methionine production in wild type and ethionine-resistant mutants of *Chlorella sorokiniana*. Plant Physiol 53:469–473
Smith IK (1972) Studies of L-cysteine biosynthesis enzymes in *Phaseolus vulgaris*. Plant Physiol 50:477–479
Smith IK (1977) Evidence for O-acetylserine in *Nicotiana tabacum*. Phytochemistry 165:1293–1294
Smith IK, Thompson (1971) Purification and characterization of L-serine transacetylase and O-acetyl-L-serine sulfhydrylase from kidney bean seedlings (*P. vulgaris*). Biochim Biophys Acta 227:288–295
Smith JE (1962) Mitochondrial transamination in sunflower hypocotyls. Biochim Biophys Acta 57:183–185
Smith TA (1977) Recent advances in the biochemistry of plant amines. In: Reinhold L, Harborne JB, Swain T (eds) Progress in phytochemistry, vol IV. Pergamon Press, Oxford, pp 27–82
Smith TA, Bagni N, Serafini Fracassini D (1979) In: Hewitt EJ, Cutting CV (eds) Nitrogen assimilation of plants. Academic Press, London New York, pp 557–570
Sodek L (1978) Partial purification and properties of diaminopimelate decarboxylase from maize endosperm. Rev Bras Bot 1:65–69
Sodek L, Da Silva WJ (1977) Glutamate synthase: A possible role in nitrogen metabolism of the developing maize endosperm. Plant Physiol 60:602–605
Sodek L, Lea PJ, Miflin BJ (1980) Distribution and properties of a potassium-dependent asparaginase isolated from developing seeds of *Pisum sativum* and other plants. Plant Physiol 65:22–26
Somerville CR, Ogren WL (1980a) Photorespiration mutants of *Arabidopsis thaliana* deficient in serine and glyxoylate aminotransferase activity. Proc Natl Acad Sci USA 77(5): 2684–2687
Somerville CR, Ogren WL (1980b) Photosynthesis is inhibited in mutants of *Arabidopsis* deficient in glutamate synthase (GOGAT) activity. Nature (London) 286:257–259
Spencer PW, Titus JS (1974) The occurrence and nature of ornithine carbamoyltransferase in senescing apple leaf tissue. Plant Physiol 54:382–385

Splittstoesser WE (1969) The appearance of arginine and arginase in pumpkin cotyledons, characterisation of arginase. Phytochemistry 8:753–758
Splittstoesser WE, Fowden LF (1973) Ornithine transaminase from *Cucurbita maxima* cotyledons. Phytochemistry 12:1565–1568
Stasiewicz S, Dunham VL (1979) Isolation and characterisation of two forms of glutamine synthetase from soybean hypocotyl. Biochem Biophys Res Commun 87:627–634
Stewart CR (1979) The effect of ammonium, glutamine, methionine sulphoximine and azaserine on asparagine synthesis in soybean leaves. Plant Sci Lett 14:269–273
Stewart GR, Lahrer F (1980) Accumulation of amino acids and related compounds in relation to environmental stress. In: Miflin BJ (ed) The biochemistry of plants, vol V. Amino acids and their derivatives. Academic Press, London New York, pp 609–635
Stewart GR, Rhodes D (1976) Evidence for the assimilation of ammonia via the glutamate pathway in nitrate-grown *Lemna minor* L. FEBS Lett 64:296–299
Stewart GR, Rhodes DA (1977a) Control of enzyme levels in the regulation of nitrogen assimilation. In: Smith H (ed) Regulation of enzyme synthesis and activity in higher plants. Academic Press, London New York pp 1–22
Stewart GR, Rhodes D (1977b) A comparison of the characteristics of glutamine synthetase and glutamate dehydrogenase from *Lemna minor* L. New Phytol 79:257–265
Stewart GR, Rhodes D (1978) Nitrogen metabolism of halophytes. III. Enzymes of ammonia assimilation. New Phytol 80:307–316
Stewart GR, Mann AF, Fentem PA (1980) Enzymes of glutamate formation: glutamate dehydrogenase, glutamine synthetase and glutamate synthase. In: Miflin BJ (ed) The biochemistry of plants, vol V. Amino acids and their derivatives. Academic Press, London New York, pp 271–327
Stewart WDP (1979) $N_2$-fixation and photosynthesis in microorganisms. In: Gibbs M, Latzko E (eds) Encylopaedia of plant physiology, vol VI. Springer, Berlin Heidelberg New York, pp 457–471
Stone SR, Copeland L, Kennedy IR (1979) Glutamate dehydrogenase of lupin nodules: Purification and properties. Phytochemistry 18:1273–1278
Stone SR, Copeland L, Heyde E (1980) Glutamate dehydrogenase of lupin nodules: Kinetics of the deamination reaction. Arch Biochem Biophys 199:550–559
Storey R, Beevers L (1978) Enzymology of glutamine metabolism related to senescence and seed development in the pea (*Pisum sativum* L). Plant Physiol 61:494–500
Storey R, Reporter M (1978) Amino acid metabolism in developing soyabean (*Glycine max*). Can J Bot 56:1349–1356
Streeter JG (1973) In vivo and in vitro studies on asparagine biosynthesis in soyabean seedlings. Arch Biochem Biophys 157:613–624
Streeter JG (1977) Asparaginase and asparagine transaminase in soybean leaves and root nodules. Plant Physiol 60:235–239
Streeter JG, Thompson JF (1972) In vivo and in vitro studies on $\gamma$-aminobutyrate metabolism with the radish plant. Plant Physiol 49:579–584
Stripf R, Werner D (1978) Enzymic activity in bacteroids and plant cytoplasm during development of nodules of *Glycine max*. Z Naturforsch 33c:375–281
Stulen I, Oaks A (1977) Asparagine synthetase in corn roots. Plant Physiol 60:680–683
Stulen I, Israelstam GF, Oaks A (1979) Enzymes of asparagine synthesis in maize roots. Planta 146:237–241
Suni J, Denes G (1967) Mechanism of arginine biosynthesis in *Chlamydomonas reinhardti* III. Purification and properties of ornithine transaminase. Acta Biochim Biophys Acad Sci Hung 2:291–302
Tajima S, Yatazawa M, Yamamoto Y (1977) Allontoin production and its utilisation in relation to nodule formation in soy beans. Soil Sci Plant Nutr (Tokyo) 23:225–235
Taylor AA, Stewart GR (1981) Tissue and subcellular localization of enzymes of arginine metabolism in *Pisum sativum*. Biochem Biophys Res Commun 101:1281–1289
Thoen A, Rognes SE, Aarnes H (1978a) Biosynthesis of threonine from homoserine in pea seedlings I: Homoserine kinase. Plant Sci Lett 13:103–112
Thoen A, Rognes SE, Aarnes H (1978b) Biosynthesis of threonine from homoserine in pea seedlings II. Threonine synthase. Plant Sci Lett 13:113–119
Thompson JF (1980) Arginine synthesis, proline synthesis and related processes. In: Miflin

BJ (ed) The biochemistry of plants, vol V. Amino acids and their derivatives. Academic Press, London New York, pp 375–402
Thompson JF, Madison JT, Madison AE (1977) In vitro culture of immature cotyledons of soya bean (*Glycine max* L. Merr) Ann Bot 41:29–39
Thurman DA, Palin C, Laycock MV (1965) Isoenzymatic nature of L-glutamic dehydrogenase of higher plants. Nature (London) 207:193–194
Tolbert NE (1979) Glycolate metabolism by higher plants and algae. In: Gibbs M, Latzko E (eds) Encyclopaedia of plant physiology, vol VI. Springer, Berlin Heidelberg New York, pp 338–352
Tripplet EW, Blevins DG, Randall DD (1980) Allantoic acid synthesis in soybean root nodule cytosol via xanthine dehydrogenase. Plant Physiol 65:1203–1206
Umbarger HE (1978) Amino acid biosynthesis and its regulation. Annu Rev Biochem 47:533–606
Upchurch RG, Elkan GH (1978) Ammonia assimilation in *Rhizobium japonicum*, colonial derivatives differing in nitrogen-fixing efficiency. J Gen Microbiol 104:219–225
Vansuyt G, Vallee JC, Prevost J (1979) Pyrroline-5-carboxylate reductase and proline dehydrogenase in *Nicotiana tabacum* (Xanthi n.c.) as a function of its development. Physiol Veg 17(1):95–105
Vennesland B, Guerrero MG (1979) Reduction of nitrate and nitrite. In: Gibbs M, Latzko E (eds) Encyclopaedia of plant physiology, vol VI. Springer, Berlin Heidelberg New York, pp 425–444
Vogel HJ (1959) On biochemical evolution: lysine formation in higher plants. Proc Natl Acad Sci USA 45:1717–1721
Vogel HJ, Hirovonen AP (1971) Diaminopimelate decarboxylase. In: Tabor H, Tabor CW (eds) Methods in enzymology, vol XVIIB. Academic Press, London New York, pp 146–150
Wahnbaeck-Spencer R, Henke RR, Mills WR, Burdge El, Wilson KG (1979) Intracellular location of $\beta$-aspartate kinase activity in spinach (*Spinacea oleracea*). Plant Physiol 63:(Suppl) 146
Walker JB, Myers J (1953) The formation of arginosuccinic acid from arginine and fumarate. J Biol Chem 203:143–152
Wallsgrove RM, Mazelis M (1980) The enzymology of lysine biosynthesis in higher plants. FEBS Lett 116:189–192
Wallsgrove RM, Mazelis M (1981) The enzymology of lysine biosynthesis in higher plants; partial purification and characterization of spinach leaf dihydrodipicolinate synthase. Phytochemistry 20:2651–2655
Wallsgrove RM, Harel E, Lea PJ, Miflin BJ (1977) Studies on glutamate synthase from the leaves of higher plants. J Exp Bot 28:588–596
Wallsgrove RM, Lea PJ, Miflin BJ (1979a) Distribution of the enzyme of nitrogen assimilation within the pea leaf cell. Plant Physiol 63:232–236
Wallsgrove RM, Lea PJ, Miflin BJ (1979b) The reduction of inorganic nitrogen and its assimilation into glutamine in pea chloroplasts. In: Hewitt EJ, Cutting CV (eds) Nitrogen assimilation in plants. Academic Press, London New York, pp 431–433
Wallsgrove RM, Lea PJ, Mills WR, Miflin BJ (1979c) The regulation and subcellular distribution of enzymes of the aspartate pathway in *Pisum sativum* leaves. Plant Physiol 63:(Suppl)26
Wallsgrove RM, Keys AJ, Bird IF, Cornelius MJ, Lea PJ, Miflin BJ (1980) The location of glutamine synthetase in leaf cells and its role in the reassimilation of ammonia released in photorespiration. J Exp Bot 31:1005–1017
Walter TJ, Connelly JA, Gengenbach BG, Wold F (1979) Isolation and characterization of two homoserine dehydrogenases from maize suspension cultures. J Biol Chem 254:1349–1355
Wang D, Waygood ER (1962) Carbon metabolism of $^{14}C$-labelled amino acids in wheat leaves. Plant Physiol 37:826–832
Washitani I, Sato S (1977a) Studies on the function of proplastids in the metabolism of in vitro cultured tobacco cells. I. Localization and distribution of NR and NADP-GDH. Plant Cell Physiol 18(1):117–126
Washitani I, Sato S (1977b) Studies on the function of proplastids in metabolism of

in vitro cultures tobacco cells II. Glutamine synthetase/glutamine synthase pathway. Plant Cell Physiol 18:505–512
Washitani I, Sato S (1978) Studies on the function of proplastids in the metabolism of in vitro cultured tobacco cells. V. Primary transamination. Plant Cell Physiol 19:43–50
Weber HL, Bock A (1968) Comparative studies on the regulation of deoxyarabinoheptulosonate-7-phosphate synthase activity in blue-green and green algae. Arch Mikrobiol 61:159–168
Weber HL, Bock A (1969) Regulation of the chorismic acid branch point in aromatic biosynthesis in blue-green and green algae. Arch Mikrobiol 66:250–258
Weber HL, Bock A (1970) Chorismate mutase from *E. gracilis*: purification and regulatory properties. Eur J Biochem 16:244–251
Webster G (1964) Enzymes of peptide and protein metabolism. In: Paech K, Tracey MV (eds) Modern methods of plant analysis, vol VII, Springer, Berlin Heidelberg New York, pp 392–430
Weissman GS (1976) Glutamine synthetase regulation by energy charge in sunflower roots. Plant Physiol 57:339–343
Whiteside JA, Thurman DA (1971) The degradation of canavanine by Jack bean cotyledons. Planta 98:279–284
Wiater A, Krajewska-Grynkiewicz K, Kloptowski T (1971) Histidine biosynthesis and its regulation in higher plants. Acta Biochim Pol 18:299–307
Widholm JM (1972a) Cultured *Nicotiana tabacum* cells with an altered anthranilate synthetase which is less sensitive to feedback inhibition. Biochim Biophys Acta 261:52–58
Widholm JM (1972b) Anthranilate synthetase from 5-methyltryptophan-susceptible and -resistant cultured *Daucus carota* cells. Biochim Biophys Acta 279:48–57
Widholm JM (1973) Measurement of the five enzymes which convert chorismate to tryptophan in cultured *Daucus carota* cell extracts. Biochim Biophys Acta 320:217–226
Widholm JM (1976) Selection and characterization of cultured carrot and tobacco cells resistant to lysine, methionine and proline analogs. Can J Bot 54:1523–1529
Wightman F, Forest JC (1978) Properties of plant aminotransferases. Phytochemistry 17:1455–1471
Wong KF, Dennis DT (1973) Aspartokinase from wheat germ: isolation, characterization and regulation. Plant Physiol 51:322–326
Woo KC, Canvin DT (1980) Effect of ammonia, nitrite, glutamate and inhibitors of N metabolism on photosynthetic carbon fixation in isolated spinach leaf cells. Can J Bot 58:511–516
Woo KC, Berry JA, Turner G (1978) Release and refixation of ammonia during photorespiration. Carnegie Inst Washington Yearb 77:240–245
Woo KC, Jakinen M, Canvin DT (1980) Nitrate reduction by a dicarboxylate shuttle in a reconstituted system from spinach leaves. Aust J Plant Physiol 7:123–130
Woodin TS, Nishioka L (1973a) Evidence for three isoenzymes of chorismate mutase in alfalfa. Biochim Biophys Acta 309:211–223
Woodin T, Nishioka L (1973b) Chorismate mutase isoenzyme patterns in three fungi. Biochim Biophys Acta 309:224–231
Woodin TS, Nishioka L, Hsu A (1978) Comparison of chorismate mutase isozyme patterns in selected plants. Plant Physiol 61:949–952
Yamamoto E (1977) Partial purification and some properties of 3-dehydroquinate synthase from *Phaseolus mungo* seedlings. Plant Cell Physiol 18:995–1007
Yoneyama T, Kumazawa K (1975) A kinetic study of the assimilation of $^{15}N$-labelled nitrate in rice seedlings. Plant Cell Physiol 16:21–26
Yue SB (1968) Isoenzymes of glutamate dehydrogenase in plants. Plant Physiol 44:453–457
Zelitch I (1979) Photorespiration: Studies with whole tissues. In: Gibbs M, Latzko E (eds) Encyclopaedia of plant physiology, vol VI. Springer, Berlin Heidelberg New York, pp 353–367
Zurawski G, Brown KD (1975) Chorismate mutase of *Chlamydomonas reinhardi*. Biochim Biophys Acta 377:473–481

# 2 Transfer RNA and Aminoacyl-tRNA Synthetases in Plants

J.H. WEIL and B. PARTHIER

## 1 Introduction

The first step in protein biosynthesis is the attachment of free amino acids to cognate transfer RNA's. This process can be separated into two reactions, which are both catalyzed by the same cognate aminoacyl-transfer RNA synthetase (L-amino acid:tRNA ligase (AMP forming), EC 6.1.1.x, also called activation enzyme), which specifically recognizes the corresponding amino acid, ATP and the corresponding tRNA (or isoaccepting tRNA's):

1. amino acid activation:
   amino $\text{acid}_1$ + ATP + $\text{enzyme}_1 \rightleftarrows$ ($\text{aminoacyl}_1$-AMP) $\text{enzyme}_1$ + $PP_i$
2. tRNA aminoacylation:
   ($\text{aminoacyl}_1$-AMP) $\text{enzyme}_1$ + $\text{tRNA}_1 \rightleftarrows$ $\text{aminoacyl}_1$-$\text{tRNA}_1$ + AMP + $\text{enzyme}_1$

This general process involves mechanisms common to all organisms. The very intriguing problem of the nature of the sites on the tRNA molecule which are recognized by the cognate aminoacyl-tRNA synthetase and are thus responsible for the necessary specificity of the tRNA aminoacylation process has not been resolved yet, although various studies point to some areas of the tRNA structure which are on the same side of the molecule when its three-dimensional structure is considered (SCHIMMEL and SÖLL 1979); the answer to this question will probably have to await the crystallization of a complex between a tRNA and its cognate aminoacyl-tRNA synthetase and the analysis of the X-ray diffraction pattern of the complex. Description of the molecular and functional properties of the two types of macromolecules, tRNA and synthetases, which guarantee the genetically correct insertion of an amino acid into its position in the nascent polypeptide chain, will be the main emphasis of this chapter. It also includes information on their interactions in vitro, their compartmentalization within the plant cell, their biosyntheses, and their possible involvement in plant physiological processes.

The later steps of the translation process, involving the transfer of the amino acid from the aminoacyl-tRNA to the growing polypeptide chain, and the coding properties of plant tRNA's, have been reviewed previously (WEIL 1979) and are discussed in Chapter 3. So far there is no information suggesting that in the plant kingdom the genetic code shows any deviation from the "universal" one, but it will be interesting to see if the exceptions recently observed in yeast and mammalian mitochondria (where UGA has been found to code for tryptophan instead of being a non-sense codon, and AUA for methionine instead of isoleucine) also apply to plant organelles. In the past, due to several technical difficulties, plant cells and organs have been much less used as sources for tRNA and synthetase preparations and for aminoacylation studies than bacteria or animal tissues. However, in recent years progress in the field of protein biosynthesis has occurred using green plants, and many interesting reports are available which are worthy of consideration and discussion in an Encyclopedia volume, since only a few pertinent reviews are available (LEA and NORRIS 1977, WEIL 1979, WOLLGIEHN and PARTHIER 1980).

## 2 Transfer RNA's (tRNA's)

### 2.1 Occurrence and Intracellular Localization of Plant tRNA's

Fractionation methods developed to separate tRNA's from a variety of organisms (see Sect. 2.2), especially column chromatography and more recently two-dimensional gel electrophoresis, have been applied to the tRNA's from a large number of plants (including green algae, such as *Euglena*). Isoaccepting tRNA's have thus been characterized for practically all 20 amino acids, and in many instances it has been shown that some of these isoacceptors are specifically located in one or another compartment of the plant cell, namely the cytoplasm, the chloroplasts or mitochondria (Table 1). However, as discussed previously (WEIL 1979), each peak in a chromatographic profile for instance, does not necessarily correspond to a distinct tRNA species, because there are several possibilities for artefacts to occur, including changes in the three-dimensional configuration of the tRNA molecule, aggregation (formation of dimers for example), partial degradation of the 3′ terminal CCA sequence, incomplete post-transcriptional modification, loss of a rare base (such as the Y base in the case of $tRNA^{Phe}$), etc.

Recent studies, using two-dimensional polyacrylamide gel electrophoresis, on the chloroplast tRNA populations from various plants have led to the characterization of 27 chloroplast tRNA species specific for 16 amino acids in the case of spinach (*Spinacia oleracea*), 27 chloroplast tRNA specific for 16 amino acids in the case of *Phaseolus,* 25 chloroplast tRNA species specific for 17 amino acids in the case of maize and 23 chloroplast tRNA species-specific for 18 amino acids in the case of *Euglena* (MUBUMBILA et al. 1980). It therefore appears that chloroplasts contain a complete set of tRNA's needed for protein synthesis inside the organelle. As discussed below (Sect. 2.4) these chloroplast-specific tRNA's are coded for by chloroplast DNA, which contains about 30 tRNA cistrons in green algae and higher plants. It is thus unlikely that chloroplasts require an import of tRNA's from the cytoplasm, as suggested by CHIU et al. (1975) for mitochondria although such an import has not been experimentally ruled out in the case of chloroplasts. MARTIN et al. (1977) have been able to show that purified yeast mitochondria contain only one $tRNA^{Lys}$ species which is not coded for by mitochondrial DNA, and which probably does not

**Table 1.** Multiplicity and intracellular organization of plant tRNA's

| tRNA's accepting | Plant and tissue | Number of iso-acceptors | Comments | Number of organellar isoacceptors | Comments | Reference |
|---|---|---|---|---|---|---|
| Ala | Cotton seedlings | 4 | | | | Merrick and Dure (1972) |
| | Tobacco tissue culture | 1–2 | | | | Cornelis et al. (1975) |
| | *Euglena gracilis* chloroplasts | | | | 1 sequenced | Orozco et al. (1980), Graf et al. (1980) |
| | Maize chloroplasts | | | | 1 sequenced | Koch et al. (1981) |
| Arg | Wheat embryo and seedlings | 3 | | | | Vold and Sypherd (1968) |
| | Cotton seedlings | 7 | | | | Merrick and Dure (1972) |
| | Tobacco tissue culture | 3 | | | | Cornelis et al. (1975) |
| Asn | Cotton seedlings | 1 | | | | Merrick and Dure (1972) |
| Asp | Cotton seedlings | 4 | | | | Merrick and Dure (1972) |
| Glu | Wheat embryos and seedlings | 3 | | | | Vold and Sypherd (1968) |
| | Tobacco tissue culture | 2 | | | | Cornelis et al. (1975) |
| | *Euglena gracilis* | 3 | | 1 chloro | | Barnett et al. (1969) |
| Gly | Wheat embryo and seedlings | 1 | | | | Vold and Sypherd (1968) |
| | Wheat germ | 2 | 1 sequenced | | | Marcu et al. (1977) |
| | Cotton seedlings | 3 | | | | Merrick and Dure (1972) |
| | Soyabean seedlings | 2 | Shifts on chilling | | | Yang and Brown (1974) |
| | Tobacco tissue culture | 2 | | | | Cornelis et al. (1975) |
| His | Wheat embryos and seedlings | 2 | | | | Vold and Sypherd (1968) |
| | Cotton seedlings | 2 | | 1 chloro | | Merrick and Dure (1972) |
| | Maize chloroplasts | | | | 1 sequenced | Schwarz et al. (1980) |

Table 1 (continued)

| tRNA's accepting | Plant and tissue | Number of iso-acceptors | Comments | Number of organellar isoacceptors | Comments | Reference |
|---|---|---|---|---|---|---|
| Ile | Lupin seeds and cotyledons | 4–5 | | | | LEGOCKI et al. (1968), AUGUSTYNIAK and PAWELKIEWICZ (1978a, 1979), KEDZIERSKI et al. (1980) |
| | *Euglena gracilis* | 3–5 | | 1 chloro<br>1 mito | 1 chloro sequenced | BARNETT et al. (1969), REGER et al. (1970), KISLEV et al. (1972), GOINS et al. (1973), SELSKY (1978), OROZCO et al. (1980), GRAF et al. 1980 |
| | Cotton seedlings | 4 | | 2 chloro | | MERRICK and DURE (1972) |
| | Tobacco tissue culture | 2 | | | | CORNELIS et al. (1975) |
| Leu | Lupin seeds and cotyledons | 3–5 | | | | LEGOCKI et al. (1968), KEDZIERSKI et al. (1980) |
| | Wheat embryos and seedlings | 2 | | | | VOLD and SYPHERD (1969) |
| | Wheat leaves | 6 | | 3 chloro | | KARWOWSKA et al. (1979), |
| | Soyabean cotyledons and seedlings | 6 | Shifts during development | | | ANDERSON and CHERRY (1969), CHERRY and OSBORNE (1970), BICK et al. (1970), BICK and STREHLER (1971), YANG and BROWN (1974), VENKATARAMAN and DELEO (1972), PILLAY and CHERRY (1974), LESTER and CHESTER (1979), LESTER et al. (1979) |
| | Pea cotyledons and leaves | 4 | Shifts during development | | | CHERRY and OSBORNE (1970), WRIGHT et al. (1972), PATEL and PILLAY (1976) |
| | Pea roots | 5–6 | | | | VANDERHOEF and KEY (1970), BABCOCK and MORRIS (1973) |
| | *Phaseolus* leaves | 9 | | 3 chloro<br>4 mito | 3 sequenced | BURKARD et al. (1970), GUILLEMAUT et al. (1975), WILLIAMS et al. (1973), OSORIO-ALMEIDA et al. (1980) |
| | Cotton seedlings | 6 | | 3 chloro | | MERRICK and DURE (1972) |

| | | | | | | |
|---|---|---|---|---|---|---|
| | Tobacco leaves | 6 | | 2 chloro<br>2 mito | | GUDERIAN et al. (1972) |
| | Tobacco tissue culture | 5 | | | | CORNELIS et al. (1975) |
| | *Euglena gracilis* | 6 | | 2 chloro | | PARTHIER (1977), PARTHIER and NEUMANN 1977 |
| | *Mercurialis annua* flowers | 6 | | | | BAZIN et al. (1975) |
| | Tomato fruits | 3 (+2) | Shifts on ripening | | | METTLER and ROMANI (1976) |
| | Apple and pear fruits | 3 | | | | ROMANI et al. (1975) |
| | Spinach leaves | | | | 1 chloro sequenced | CANADAY et al. (1980b) |
| | Maize chloroplasts | | | | 1 sequenced | SCHWARZ et al. (1980) |
| Lys | Wheat embryos and seedlings | 5 | Shifts during development | | | VOLD and SYPHERD (1968) |
| | Pea roots | 4 | | | | VANDERHOEF and KEY (1970) |
| | *Vigna sinensis* | 3 | | | | HAGUE and KOFOLD (1971) |
| | Cotton seedlings | 4 | | 1 chloro | | MERRICK and DURE (1972) |
| | Wheat grains | 3–4 | | | | NORRIS et al. (1975a) |
| | Tomato fruits | 5 | Shifts upon ripening | | | METTLER and ROMANI (1976) |
| | Tomato tissue culture | 3 | | | | CORNELIS et al. (1975) |
| | *Phaseolus* leaves | 4 | | 1 chloro<br>1 mito | | JEANNIN et al. (1976) |
| | Apple and pear fruits | 4 | | | | ROMANI et al. (1975) |
| | Lupin seeds and cotyledons | 5–6 | Shifts during devel-opment | 1 chloro<br>1 mito | | AUGUSTYNIAK and PAWELKIEWICZ (1978b, 1979), KEDZIERSKI et al. (1980) |
| Met | *Phaseolus* leaves | 8 | $tRNA_i^{Met}$ sequenced | 3 chloro<br>3 mito | $tRNA_F^{Met}$ sequenced | BURKARD et al. (1969), GUILLEMAUT et al. (1972), GUILLEMAUT et al. (1973), GUILLEMAUT and WEIL (1975), CANADAY et al. (1980a) |
| | Pea roots | 3 | | | | VANDERHOEF and KEY (1970) |
| | Wheat leaves | 5 | $tRNA_i^{Met}$ sequenced | 2 chloro | | LEIS and KELLER (1970), GHOSH et al. (1974, 1979) |
| | Cotton seedlings | 3 | | 2 chloro | | MERRICK and DURE (1972) |
| | *Euglena gracilis* | 3 | | 1 chloro | | GOINS et al. (1973) |
| | Tomato fruits | 2 | Shifts upon ripening | | | METTLER and ROMANI (1976) |

**Table 1** (continued)

| tRNA's accepting | Plant and tissue | Number of iso-acceptors | Comments | Number of organellar isoacceptors | Comments | Reference |
|---|---|---|---|---|---|---|
| | Tobacco tissue culture | 2 | | | | CORNELIS et al. (1975) |
| | *Scenedesmus obliquus* | 5 | $tRNA_i^{Met}$ sequenced | 2 chloro | $tRNA_F^{Met}$ and $tRNA_M^{Met}$ sequenced | JAY and JONES (1977), OLINS and JONES (1980), |
| | Lupin seeds | 2–3 | | | | GULEWICZ and TWARDOWSKI (1980) |
| | Spinach chloroplasts | | | | $tRNA_F^{Met}$ sequenced | CALAGAN et al. (1980) |
| Phe | Wheat embryos and seedlings | 1 | | | | VOLD and SYPHERD (1968) |
| | Wheat germ | 2 | 1 sequenced | | | DUDOCK et al. (1969) |
| | *Euglena gracilis* | 2 | 1 sequenced | 1 chloro | 1 sequenced | BARNETT et al. (1969), REGER et al. (1970), CHANG et al. (1976) |
| | Pea roots | 4 | 1 sequenced | | | VANDERHOEF and KEY (1970), EVERETT and MADISON (1976) |
| | Cotton seedlings | 4 | | 2 chloro | | MERRICK and DURE (1972) |
| | Barley seedlings | 9 | | 2 chloro | | HIATT and SNYDER (1973) |
| | Barley embryos | 2 | 2 sequenced | | | LABUDA et al. (1974), WOWER et al. (1979), JANOWICZ et al. (1979) |
| | Lupin seeds | 2 | 1 sequenced | | | AUGUSTYNIAK et al. (1974), RAFALSKI et al. (1977) |
| | *Phaseolus* leaves | 4 | | 2 chloro 1 mito | 1 chloro sequenced | GUILLEMAUT et al. (1976), GUILLEMAUT and KEITH (1977), JEANNIN et al. (1978) |
| | Spinach leaves | | | | 1 chloro sequenced | CANADAY et al. (1980b) |
| | Tobacco tissue culture | 1 | A new peak in crown-gall | | | CORNELIS et al. (1975) |
| | Soyabean leaves | 2 | | 1 chloro | | SWAMY and PILLAY (1980) |
| Pro | Wheat embryos and seedlings | 2 | Shifts during development | | | VOLD and SYPHERD (1968) |
| | Wheat grains | 3 | | | | NORRIS et al. (1975a) |

| | | | | | | |
|---|---|---|---|---|---|---|
| | Pea roots | 4 | | | | Vanderhoef and Key (1970) |
| | *Phaseolus* leaves | 5 | | 1 chloro<br>2 mito | | Jeannin et al. (1976) |
| Ser | Wheat embryos and seedlings | 2 | Shifts during development | | | Vold and Sypherd (1968) |
| | Pea roots | 3 | | | | Vanderhoef and Key (1970) |
| | Soyabean cotyledons | 3 | | | | Pillay and Cherry (1974) |
| | *Mercurialis annua* flowers | 6 | | | | Bazin et al. (1975) |
| Thr | Wheat embryos and seedlings | 2 | | | | Vold and Sypherd (1968) |
| | Pea roots | 3 | | | | Vanderhoef and Key (1970) |
| | *Euglena gracilis* | 2 | | 1 chloro | | Parthier and Neumann (1977) |
| | Spinach chloroplasts | | | 2–3 | 1 sequenced | Kashdan et al. (1980) |
| Trp | Cotton seedlings | 3 | | 2 chloro | | Merrick and Dure (1972) |
| | *Chlamydomonas reinhardii* | 2 | | 1 chloro | | Preddie et al. (1973) |
| | Soyabean leaves | 2 | | 1 chloro | | Swamy and Pillay (1980) |
| Tyr | Pea roots | 4 | Differences upon division | 1 mito | | Vanderhoef and Key (1970) |
| | Soyabean seedlings, cotyledons and leaves | 3–4 | Shifts during devel-opment | 2 chloro | | Bick et al. (1970), Pillay and Cherry (1974), Locy and Cherry (1978), Swamy and Pillay (1980) |
| | Tobacco tissue culture | 2 | | | | Cornelis et al. (1975) |
| | *Mercurialis annua* flowers | 5 | | | | Bazin et al. (1975) |
| | Tomato fruits | 2 | | | | Mettler and Romani (1976) |
| | Lupin cotyledons | 2 | | | | Kedzierski et al. (1980) |
| Val | Wheat embryos and seedlings | 3 | | | | Vold and Sypherd (1968) |
| | Lupin seeds and cotyledons | 2–4 | | | | Legocki et al. (1968), Kedzierski et al. (1980) |
| | Cotton seedlings | 5 | | 1 chloro | | Merrick and Dure (1972) |
| | *Phaseolus* leaves | 4 | | 1 chloro | | Burkard et al. (1970) |
| | Tobacco tissue culture | 3 | | | | Cornelis et al. (1975) |
| | *Mercurialis annua* flowers | 3 | | | | Bazin et al. (1975) |

function in mitochondrial protein synthesis since it is not aminoacylated by the mitochondrial lysyl-tRNA synthetase.

As far as transport of tRNA's in the other direction is concerned, it has been reported that *Euglena* chloroplasts selectively export a specific class of chloroplast DNA-coded tRNA's found on cytoplasmic polysomes (McCREA and HERSHBERGER 1978), but this report is in contradiction with many other observations and has been seriously questioned (SCHWARTZBACH et al. 1979).

## 2.2 Extraction, Fractionation and Purification of Plant tRNA's

### 2.2.1 Extraction of Plant tRNA's

As an example, the scheme for the extraction of total tRNA from *Phaseolus* hypocotyls is outlined on Fig. 1. Details of the extraction procedure and of various techniques which can be used to prevent the action of ribonucleases, eliminate contaminants such as carbohydrates, polyphosphates, DNA's and other RNA's (especially rRNA's) and free the tRNA's from esterified amino acids have been discussed previously (WEIL 1979). The determination of tRNA concentration by measuring the absorbance at 260 nm, the tRNA yields usually obtained from plant material, and the evaluation of tRNA purity (or of the relative concentration of a tRNA specific for a given amino acid) by measuring the attachment of a radioactive amino acid (aminoacylation reaction) have also been reviewed (WEIL 1979).

### 2.2.2 Fractionation and Purification of Plant tRNA's

A large number of chromatographic procedures have been used to fractionate and purify individual tRNA species; they are discussed in the same review article (WEIL 1979). Among those which have been used most often and have given the best results, are benzoylated DEAE-cellulose or BD-cellulose (KOTHARI and TAYLOR 1972), Sepharose 4B (HOLMES et al. 1975) and reverse phase chromatography or RPC (KOTHARI and TAYLOR 1973); RPC-5 especially (PEARSON et al. 1971, KELMERS and HEATHERLY 1971) have allowed a very good fractionation of tRNA's from many plants. For instance *Phaseolus vulgaris* cytoplasmic, chloroplastic and mitochondrial leucyl-tRNA's (GUILLEMAUT et al. 1975), lysyl-tRNA's, prolyl-tRNA's (JEANNIN et al. 1976), methionyl-tRNA's (GUILLEMAUT et al. 1973, GUILLEMAUT and WEIL 1975) and phenylalanyl-tRNA's (JEANNIN et al. 1978) have been successfully fractionated using RPC-5. Several fractionation methods are based on the difference between charged and uncharged tRNA's; some of them require chemical modification of either the aminoacyl-tRNA or the uncharged tRNA's. These and other techniques, including those based on the interaction between complementary anticodons, or on the interaction between an aminoacyl-tRNA, GTP and protein synthesis elongation factor EF-Tu, or on the specific interaction between an aminoacyl-tRNA and the cognate aminoacyl-tRNA synthetase bound to a modified cellulose matrix, have also been reviewed (WEIL 1979).

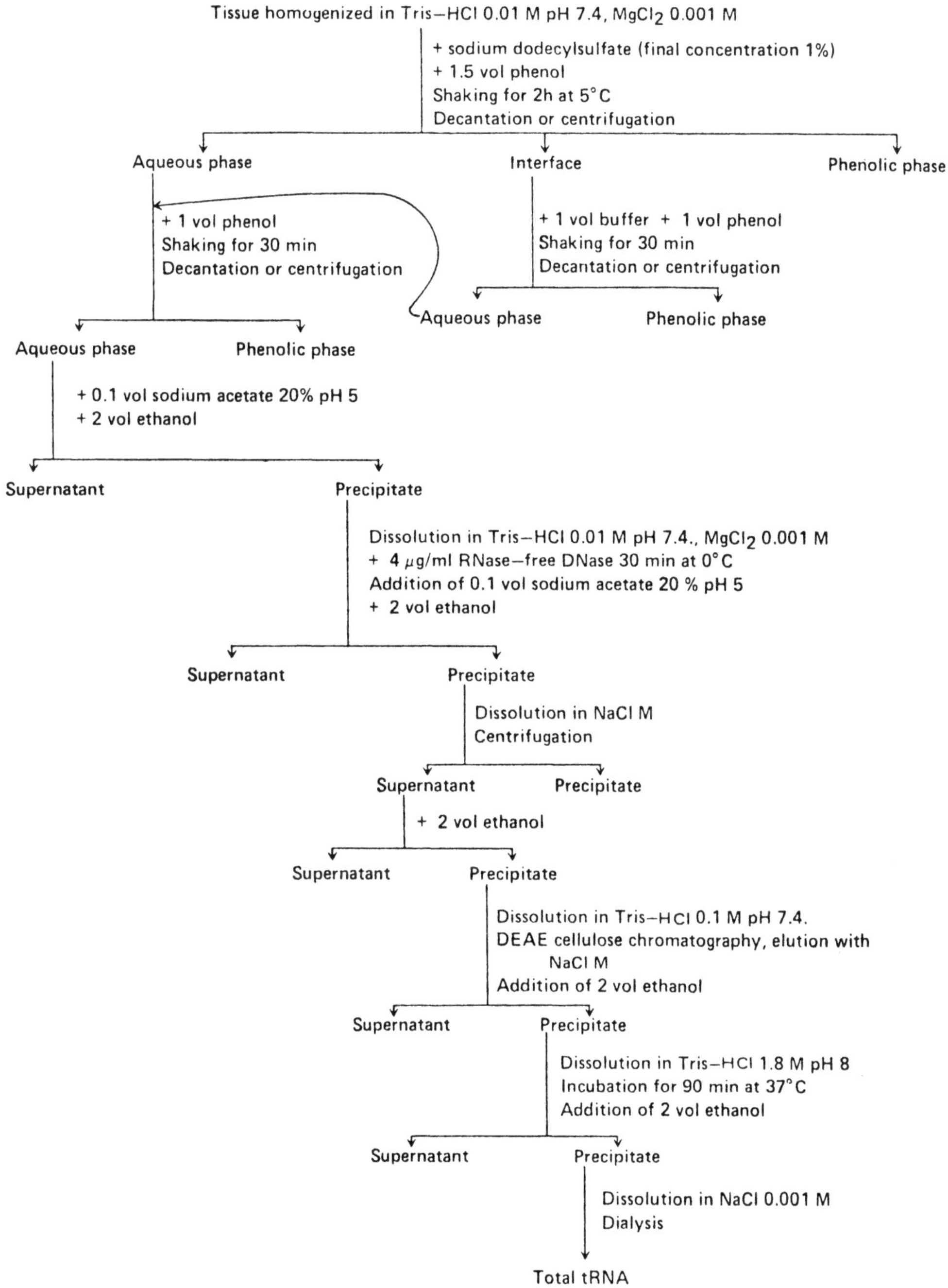


**Fig. 1.** General scheme for the extraction of total tRNA from *Phaseolus* hypocotyls

**Fig. 2.** Two-dimensional gel electrophoretic fractionation of spinach chloroplast tRNA's. (DRIESEL et al. 1979). The amino acid accepted by each identified tRNA spot is indicated (as revealed by aminoacylation of the tRNA extracted from the gel). Unidentified spots are designated by *capital letters*. Weak spots are indicated by *dotted circles*

More recently two-dimensional polyacrylamide gel electrophoresis has frequently been used, as it has the advantage of allowing the simultaneous purification of most (or all) individual tRNA's contained in a total tRNA preparation. The details of this very powerful technique and its applications for the fractionation and purification of chloroplast tRNA's have been described elsewhere (BURKARD et al. 1980). This technique has, for instance, allowed the fractionation of chloroplast tRNA's from spinach shown in Fig. 2 (DRIESEL et al. 1979), and from *Phaseolus, Zea mays* and *Euglena* (MUBUMBILA et al. 1980).

## 2.3 Structure of Plant tRNA's

### 2.3.1 New Methods for Sequence Determination

As plant tRNA's are difficult to label in vivo, nucleotide sequence determinations of plant cytoplasmic or chloroplastic tRNA's have only progressed recently when in vitro end-group labelling techniques have been developed (for a review see SILBERKLANG et al. 1979); these methods allow the determination of a tRNA sequence even when only small amounts (10–20 μg) of purified tRNA are available (as in the case of organellar tRNA's). These techniques include:

In vitro labelling at the 5′ end of the tRNA (which requires prior dephosphorylation) or of oligonucleotides (obtained by partial or total enzymatic hydrolysis, with $T_1$ ribonuclease for instance) using [$^{32}P_\gamma$]-ATP and $T_4$ polynucleotide kinase. In vitro labelling at the 3′ end can also be used, either by linking a [$^{32}$P]Np to the 3′ terminal OH of the tRNA using $T_4$ RNA ligase (ENGLAND and UHLENBECK 1978), or by adding a labelled adenylic residue at the 3′ end of the tRNA using [$^{32}P_\alpha$]-ATP and tRNA nucleotidyl transferase.

Mobility shift analysis by two-dimensional homochromatography: The 5′ (or 3′) labelled tRNA or oligonucleotide is hydrolyzed either with endonuclease $P_1$, or with snake venom phosphodiesterase, although using this exonuclease has the disadvantage that its action is slowed down by modified nucleotides. The resulting radioactive digestion products are fractionated in the first dimension primarily according to their charge (at pH 3.5), and in the second dimension according to their size (using a RNA hydrolysate called "homomixture"). The nucleotide sequence (up to 20 nucleotides) can be deduced from the characteristic mobility shifts between successive spots on the autoradiogram.

Read-off sequencing gels: The 5′ (or 3′) labelled tRNA is cleaved at G residues with RNAase $T_1$, at A residues with RNAase $U_2$, at C and U residues with RNAase A (other enzymes allow one to differentiate between C and U), and at all four residues using boiling water or alkali. The partial hydrolysis products obtained in the different digestions are separated electrophoretically according to their size by placing them side by side on the same denaturing polyacrylamide gel and the tRNA sequence can then be read off the autoradiogram of the gel, because the presence of a radioactive band in each slot indicates whether a G, an A, or a pyrimidine has allowed the hydrolysis to occur.

Another technique (STANLEY and VASSILENKO 1978) uses partial non-specific hydrolysis by hot water or formamide to produce less than one cleavage per molecule (on an average). The fragments are labelled at their 5′ end with $^{32}$P, separated by size on a polyacrylamide gel, recovered from the gel, hydrolyzed to mononucleotides (by alkali or nuclease $P_1$) and the 5′ $^{32}$P mononucleotides are identified (including the unusual ones) by thin-layer chromatography or paper electrophoresis. The 5′ $^{32}$P mononucleotide of the longest fragment corresponds to the second nucleotide from the 5′ end of the tRNA molecule (the first one has not been labelled as there was no prior dephosphorylation), and the 5′ $^{32}$P mononucleotides of the successive shorter fragments give the tRNA sequence from the 5′ end to the 3′ end.

Recently, as DNA sequencing techniques have been developed, it has become easier to determine the sequence of tRNA genes, especially when cloned fragments containing tRNA cistrons can be obtained, and to deduce that of the corresponding tRNA's from them. This method does not allow the characterization of modified nucleotides since they are the result of post-transcriptional events.

Since the first tRNA structure, that of yeast $tRNA^{Ala}$, was published by HOLLEY et al. (1965), the sequences of about 135 tRNA's have been determined and are given in the compilation of GAUSS and SPRINZL et al. (1981). The list includes tRNA's from a great variety of sources, such as mycoplasma, phages, bacteria, blue-green algae, fungi, animals and even man, but only a few plant tRNA's have so far been sequenced.

All tRNA sequences can be arranged to fit the cloverleaf model proposed by HOLLEY et al. (1965). The general features of this model, the invariant and semi-invariant nucleotides, and the special features of initiator tRNA's have been reviewed (WEIL 1979).

### 2.3.2 Structure of Plant Cytoplasmic tRNA's

The cytoplasmic tRNA's$^{Phe}$ from wheat (*Triticum esculentum*) (DUDOCK et al. 1969), pea (*Pisum sativum*) (EVERETT and MADISON 1976), lupin (*Lupinus luteus*) (RAFALSKI et al. 1977) and barley (*Hordeum vulgare*) (JANOWICZ et al. 1979, WOWER et al. 1979) have been sequenced and found to be identical, except for a G-C base pair present in wheat, pea and barley tRNA's$^{Phe}$ which is replaced by an A-U pair in 80% of the lupin tRNA$^{Phe}$ molecules. In barley embryos, 2 tRNA's$^{Phe}$ have been detected which differ only by the Y base (WOWER et al. 1979).

*Euglena* cytoplasmic tRNA$^{Phe}$ has also been sequenced, but it resembles mammalian (rather than plant) cytoplasmic tRNA's$^{Phe}$ (CHANG et al. 1978), whereas *Euglena* chloroplastic tRNA$^{Phe}$ is very similar to higher plant chloroplastic tRNA's$^{Phe}$ (see Sect. 2.3.3).

Wheat germ cytoplasmic tRNA$_1^{Gly}$ has been sequenced (MARCU et al. 1977) and found to have some unusual features: T is absent (and is replaced by U), there is a 2′-O-methyl cytidine in the acceptor stem (which is usually devoid of modified nucleotides) and the D stem has only one Watson-Crick base pair (A-U) in addition to three weaker base pairs (G-U, A-C and G-$\psi$).

Three plant cytoplasmic initiator tRNA's (tRNA$_i^{Met}$) have been sequenced so far, from wheat (GHOSH et al. 1974, 1979), *Phaseolus* (CANADAY et al. 1980a) and *Scenedesmus obliquus* (OLINS and JONES 1980). All three present structural features characteristic of eukaryotic initiators (1) a base-paired 5′ nucleotide, (2) a AUCG sequence in loop IV, (3) a pyrimidine 14-purine 24 base-pair in the D stem (Fig. 3). The sequence of *S. obliquus* tRNA$_i^{Met}$ has 87% homology with wheat germ tRNA$_i^{Met}$ and 87% with *Phaseolus* tRNA$_i^{Met}$.

### 2.3.3 Structure of Chloroplastic tRNA's

The first two chloroplastic tRNA's which have been sequenced are the tRNA's$^{Phe}$ from *Euglena* (CHANG et al. 1976) and *Phaseolus* (GUILLEMAUT and KEITH 1977). They were found to resemble prokaryotic tRNA$^{Phe}$ in that they contain the same hypermodified nucleotides, namely ms$^2$i$^6$A and acp$^3$U (but no Y), and in their overall structure, as for instance *Phaseolus* chloroplast tRNA$^{Phe}$ has 78.9% homology with *E. coli* tRNA$^{Phe}$ and 86.8% homology with blue-green alga tRNA$^{Phe}$ (CHANG et al. 1978). The fact that *Euglena* and *Phaseolus* chloroplast tRNA's$^{Phe}$ are so similar (neglecting differences in post-transcriptional modifications, they differ by five nucleotides only) should be interpreted with caution, as far as the evolution of chloroplast tRNA structures is concerned, because it may represent an exceptional situation. In fact, when cross-hybridization experiments were performed it was found that while spinach total chloroplast tRNA hybridized to *Phaseolus* chloroplast DNA practically as well as

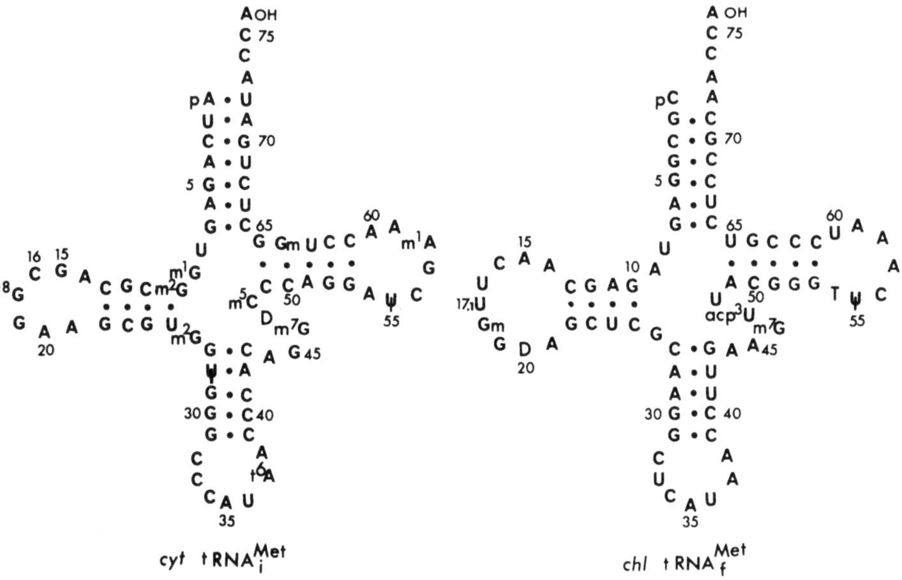


**Fig. 3.** Nucleotide sequences of *Phaseolus* chloroplast and cytoplasmic initiator tRNA's$^{Met}$ (CANADAY et al. 1980a)

*Phaseolus* total chloroplast tRNA, the hybridization of *Euglena* total chloroplast tRNA to *Phaseolus* chloroplast DNA was very poor; of the 14 individual *Euglena* chloroplast tRNA's tested, only *Euglena* chloroplast tRNA$^{Phe}$ hybridized to *Phaseolus* chloroplast DNA (MUBUMBILA et al. 1980). Chloroplast tRNA's of bean and spinach, two dicotyledonous plants, are very similar, as judged from the results of cross-hybridization experiments (see above) and from the very high degree of homology between the sequences of spinach and bean chloroplast tRNA's$^{Phe}$ (98.7%) and tRNA's$_3^{Leu}$ (98.8%) which were compared recently by CANADAY et al. (1980b), and between their tRNA's$_F^{Met}$ (see below).

*Phaseolus* chloroplast initiator tRNA (tRNA$_F^{Met}$) has been sequenced (CANADAY et al. 1980a) and found to have structural features characteristic of prokaryotic initiators: (1) an unpaired 5′ nucleotide, (2) a T$\psi$C sequence in loop IV, (3) a A$_{11}$-U$_{24}$ base-pair in the D stem (Fig. 3). All three features are also found in the chloroplast tRNA$_F^{Met}$ from spinach (CALAGAN et al. 1980) and from *S. obliquus* (MCCOY and JONES 1980). *S. obliquus* chloroplastic tRNA$_m^{Met}$ has also been recently sequenced (MCCOY and JONES 1980).

The three tRNA$^{Leu}$ isoacceptors from *Phaseolus* chloroplasts have been sequenced (OSORIO-ALMEIDA et al. 1980) and shown to differ in their length, nucleotide sequence and anticodon. *Phaseolus* chloroplast tRNA$_3^{Leu}$ has an unusual anticodon (UA m7G) containing a m7G in a position where it had never been found before. Interestingly, spinach chloroplast tRNA$_3^{Leu}$ has the same anticodon (CANADAY et al. 1980b). Spinach chloroplast tRNA$^{Thr}$ has been sequenced recently (KASHDAN et al. 1980).

As already mentioned, cloned chloroplast DNA fragments carrying tRNA genes have been obtained, making it possible to sequence tRNA genes, for instance those of *Euglena* or *Zea mays* chloroplast $tRNA^{Ala}$ and $tRNA^{Ile}$, which are found in the spacer between the 16S and the 23S rRNA genes and contain, in the case of maize, rather long introns of 806 and 950 nucleotides respectively (OROZCO et al. 1980, GRAF et al. 1980, KOCH et al. 1981). Maize chloroplast $tRNA^{His}$ and $tRNA^{Leu}$ genes have also been sequenced (SCHWARZ et al. 1980).

The three-dimensional conformation of tRNA's has been reviewed (WEIL 1979) and will not be discussed here, as no information is available on the three-dimensional conformation of plant tRNA's.

## 2.4 Organization and Expression of tRNA Genes in Nuclear and Organellar Genomes

The problem of the origin of tRNA's can be approached by DNA-tRNA hybridization techniques, using radioactive tRNA labelled either in vivo with $^{32}P$, in the case of unicellular algae such as *Euglena* or *Chlamydomonas*, for instance, or in vitro with $^{125}I$ (COMMERFORD 1971) or with $^{32}P$ using one of the above-mentioned post-labelling techniques (see Sect. 2.3.1). Hybridization is usually performed as described by GILLESPIE and SPIEGELMAN (1965) with denatured DNA, or DNA fragments obtained upon action of a restriction endonuclease, immobilized on nitrocellulose filters, and labelled total tRNA or individual tRNA purified by two-dimensional gel electrophoresis (see Sect. 2.2.2).

### 2.4.1 tRNA Genes in the Nuclear Genome

According to GRUOL and HASELKORN (1976), the *Euglena* cell has 800 nuclear genes for each cytoplasmic rRNA and approximately 740 nuclear genes for tRNA's. If these 740 genes code for 50 different RNA species, a reasonable assumption, the average reiteration frequency is only about 15, a very low figure compared with an average reiteration frequency of about 200 in *Xenopus laevis* which has about 7,800 tRNA genes (CLARKSON et al. 1973).

### 2.4.2 tRNA Genes in the Chloroplast Genome

GRUOL and HASELKORN (1976) found 10,000 chloroplast tRNA genes in the *Euglena* cell, a figure which corresponds to 25 tRNA cistrons on each chloroplast chromosome, taking into account that there are 400 chloroplast chromosomes per *Euglena* cell. MCCREA and HERSHBERGER (1976) and SCHWARTZBACH et al. (1976) also found that the *Euglena* chloroplast genome contains about 25 tRNA cistrons. The chloroplast genome of maize was found to contain about 26 tRNA cistrons (HAFF and BOGORAD 1976) and that of pea between 30 and 40 tRNA cistrons (TEWARI et al. 1977).

Considerable progress has recently been achieved in the localization of tRNA genes on the chloroplast genome, thanks to the construction of restriction endonuclease cleavage site map of chloroplast DNA on one hand, and to the purification of individual chloroplast tRNA's by two-dimensional gel electrophoresis (see Sect. 2.2.2) on the other. This allows hybridization experiments to be per-

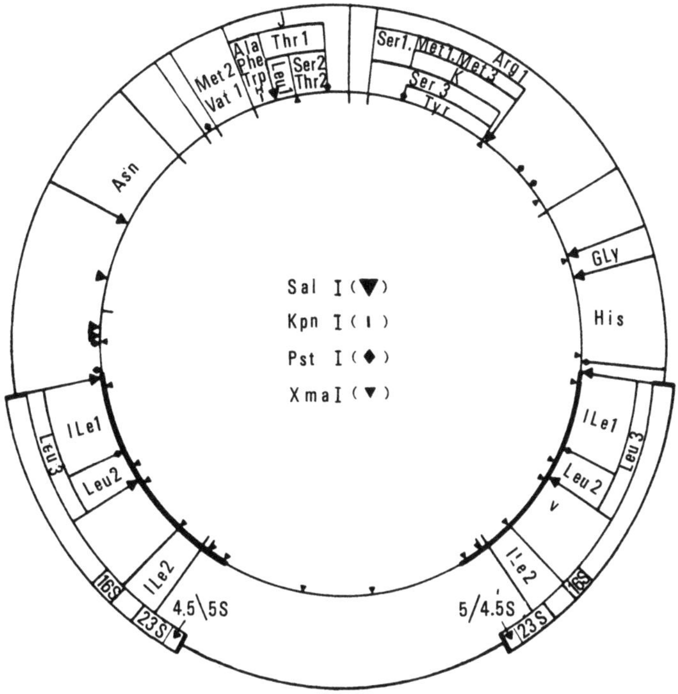


**Fig. 4.** Localisation of tRNA genes on the restriction endonuclease cleavage site map of spinach chloroplast DNA. The smallest DNA fragment to which hybridization was observed is indicated for each chloroplast tRNA tested. The *thick line* corresponds to the two copies of the inverted repeat region, containing the two polycistronic ribosomal DNA units; the arrangement of the rRNA genes in each unit is: 16S-spacer-23S-4,5/5S, with the transcription polarity in that order (as in *E. coli*)

formed between tRNA's and DNA fragments whose position on the physical map of the chloroplast genome is known. After it was observed that 4S RNA (total tRNA) hybridizes to DNA fragments which were not clustered but scattered throughout the chloroplast genome of *Euglena* (Hallick et al. 1978) and *Chlamydomonas* (Malnoe and Rochaix 1978), hybridization of individual spinach chloroplast tRNA's to chloroplast DNA fragments allowed the mapping of 21 tRNA genes on the spinach chloroplast genome, as shown in Fig. 4 (Driesel et al. 1979). Similar studies have been performed to map maize, *Phaseolus* and *Euglena* chloroplast tRNA genes on the corresponding chloroplast DNA.

These studies have clearly shown that chloroplast isoaccepting tRNA's are often coded for by genes which are not only different, but even located in different areas of the chromosome. A peculiar feature of chloroplast tRNA gene organization is the presence of tRNA genes, especially of $tRNA^{Ile}$ and $tRNA^{Ala}$, in the spacer between the 16S and the 23S rRNA genes, in the case of *Chlamydomonas* (Malnoe and Rochaix 1978), spinach (Bohnert et al. 1979), *Euglena* (Keller et al. 1980, Orozco et al. 1980, Graf et al. 1980) and *Zea mays* (Koch et al. 1981). It should be pointed out that in *Escherichia coli*

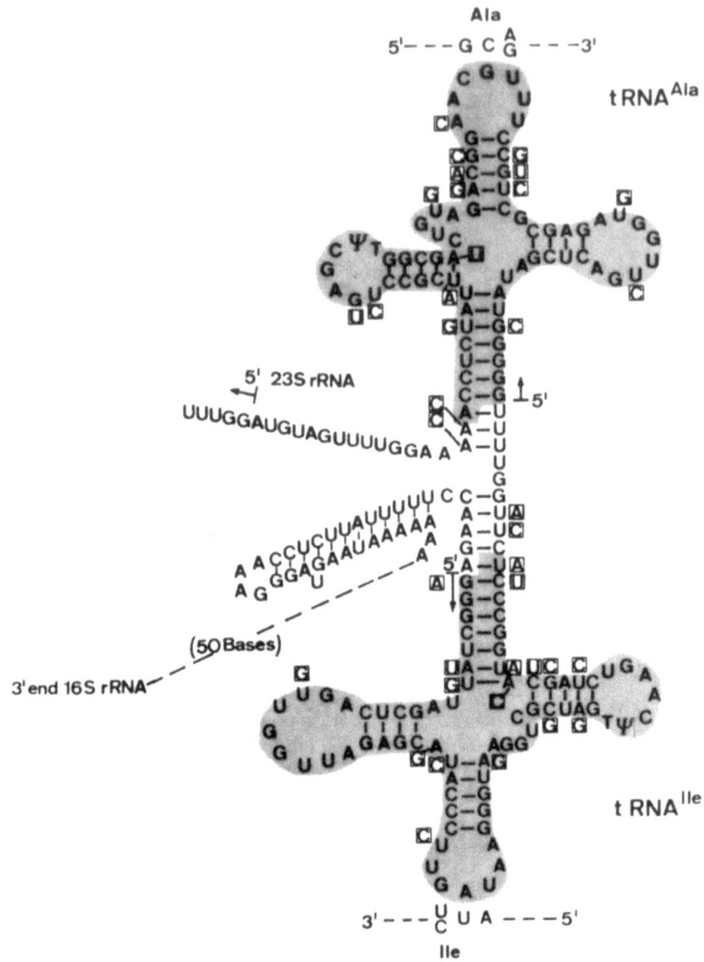


**Fig. 5.** Possible secondary structure of the dimeric $tRNA^{Ile,Ala}$ precursor region as predicted from the corresponding DNA sequence of *Euglena gracilis* chloroplast rDNA spacer (GRAF et al. 1980). The two tRNA cloverleaf structures (*shaded areas*) are shown without modifications of the bases, except for the Tψ sequence in the two T-arms. Bases which differ in the corresponding *E. coli* tRNA species (YOUNG et al. 1979, SEKIYA and NISHIMURA 1979) are indicated *within boxes*. It should be noted that the acceptor stems of the two tRNA species together with the short intergenic region can form an uninterrupted base stack of 24 base pairs. The 5′ ends of the mature tRNA species and of the 23S rRNA are indicated by 5′ → (with the *arrows* pointing in the 5′ to 3′ direction of the RNA chain). The tRNA precursors do not contain the CCA sequences found at the 3′ end of mature tRNA's

ribosomal operons, genes for $tRNA^{Ala}$ and $tRNA^{Ile}$ are also found in the spacer between the 16S and the 23S rRNA genes (YOUNG et al. 1979).

As already mentioned (Sect. 2.3.3) chloroplast $tRNA^{Ile}$ and $tRNA^{Ala}$ genes have been sequenced in *Euglena* (OROZCO et al. 1980, GRAF et al. 1980), see Fig. 5, and in maize (KOCH et al. 1981). In maize, the chloroplast $tRNA^{Ala}$

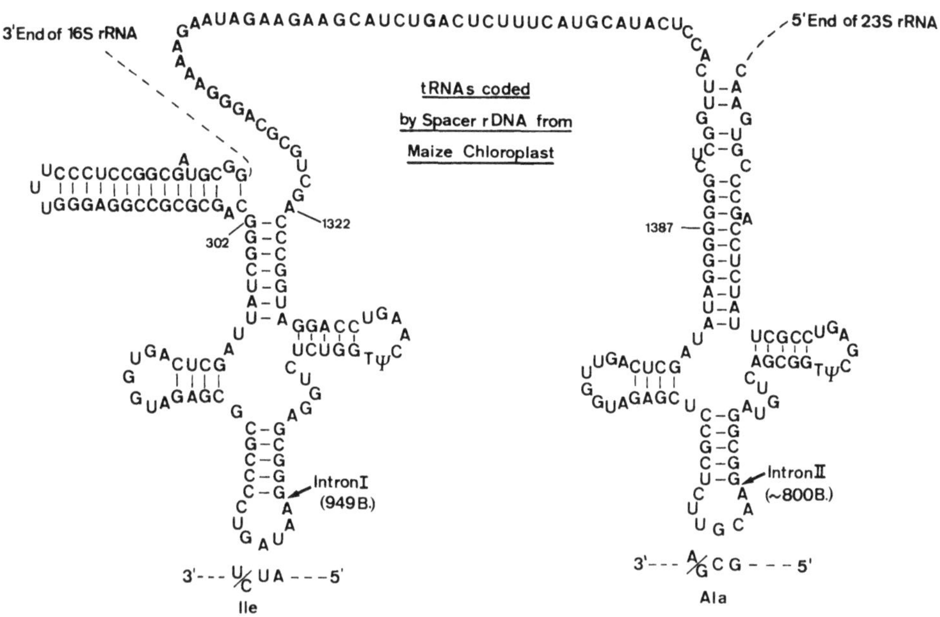


**Fig. 6.** Possible secondary structure of the dimeric tRNA$^{Ile, Ala}$ precursor region, as predicted from the corresponding DNA sequence of *Zea mays* chloroplast rDNA spacer (KOCH et al. 1981). The two cloverleaf structures are shown without modifications of the bases, except for the T$\psi$ sequence in the two T-arms. *Numbers* refer to distances from the 3′ end of the 16S rRNA gene (SCHWARZ and KÖSSEL 1980). The (tentative) positions of the two introns are indicated

gene has an intron of 806 base pairs and that of tRNA$^{Ile}$ an intron of 949 base pairs, both located between the second and third base distal to the anticodon (see Fig. 6), which is in agreement with the positions of introns found in yeast cytoplasmic tRNA genes (VALENZUELA et al. 1978). The gene order on maize chloroplast genome is: 16S–301 base pairs–tRNA$^{Ile}$ (with its 949 base pairs intron)–61 base pairs–tRNA$^{Ala}$ (with its 806 base pairs intron)–134 base pairs–23S (KOCH et al. 1981). Sequence studies on chloroplast tRNA genes in *Euglena* (OROZCO et al. 1980) and in maize (SCHWARZ et al. 1980) have shown that the 3′ CCA end is not coded for and is therefore added post-transcriptionally.

### 2.4.3 tRNA Genes in the Plant Mitochondrial Genome

Whereas mitochondrial tRNA's from other organisms, such as yeast or animal cells, have been extensively studied, little is known about plant mitochondrial tRNA's, except that plant mitochondria contain specific tRNA's and that some of them have been characterized (Table 1). While several yeast mitochondrial tRNA's have been sequenced (GAUSS and SPRINZL 1981) and while tRNA genes have been mapped on yeast mitochondrial DNA (WESOLOWSKI and FUKUHARA 1979), no plant mitochondrial tRNA has been sequenced so far, and the mapping of tRNA genes appears difficult because of the unexplained heterogeneity observed upon cleavage of plant mitochondrial DNA with restriction endonucleases (QUETIER and VEDEL 1977, BONEN et al. 1980). Wheat mitochondrial 4S RNA was shown to hybridize with many large restriction fragments of mitochondrial DNA, suggesting that the tRNA genes are broadly distributed throughout the mitochondrial genome, with some apparent clustering in regions containing the 18S and 5S rRNA genes (BONEN and GRAY 1980).

### 2.4.4 Biosynthesis of Plant tRNA's

Little is known about the expression of tRNA genes, or in other words about tRNA biosynthesis in plant nuclei or organelles. However, now that cloned chloroplast DNA fragments carrying tRNA genes have become available, it should be possible to study the in vitro transcription of tRNA genes using either a chloroplastic or a bacterial (*E. coli*) DNA-dependent RNA polymerase and to see whether tRNA genes are transcribed individually or in clusters. Furthermore it should be possible to identify the signals responsible for the initiation and the termination of tRNA gene transcription, and to study the mechanisms involved in the regulation of tRNA biosynthesis. As in other organisms, plant tRNA's genes are certainly transcribed to yield precursors which undergo maturation processes including cleavage by specific nucleases and enzymatic modifications such as methylation, isopentenylation, $\psi$ formation, etc. (for a review see WEIL 1979). Some plant tRNA methylases have been studied and organelle-specific tRNA methylases have been characterized (for a review see WEIL 1979). Among the more "exotic" base modifications found in tRNA's, the formation of the hypermodified nucleosides isopentenyl-adenosine ($i^6A$) and 2 methylthio-isopentenyl-adenosine ($ms^2i^6A$) is of special interest because of their possible involvement in codon recognition, and because these cytokinins are plant hormones which exert regulatory functions on plant growth and development (for a review see WEIL 1979).

## 2.5 Functions of tRNA's

### 2.5.1 Role of tRNA's in Protein Biosynthesis

The main function of a transfer RNA, as its name indicates, is to accept the corresponding amino acid and to transfer it to the growing polypeptide chain. To ensure fidelity in translation, it is essential that the correct amino acid is attached to each tRNA, and that for each codon of the mRNA the correct aminoacyl-tRNA is selected by a specific codon–anticodon interaction, thus allowing incorporation of the amino acids into the growing polypeptide chain in the correct order, as specified by the genetic information (cf. Chapter 3).

### 2.5.2 Other Biological Functions of tRNA's

tRNA's have been shown to have other roles, for instance in regulatory processes, or as a primer for reverse transcriptase, or in the transfer of an amino acid to a specific acceptor (such as a N-acetyl-muramide peptide, a phosphatidyl-glycerol, the N-terminus of a polypeptide). These functions have been reviewed (WEIL 1979) and will not be discussed here, as they do not specifically concern the plant kingdom, except in the case of the aminoacylation of plant viral RNA's which have been shown to have a tRNA-like structure at their 3′ end (cf. Vol. 14B, Chapter 10).

## 2.6 tRNA's and Plant Development

Changes in the tRNA populations, usually observed by comparing chromatographic fractionation profiles obtained under different physiological or pathological conditions, have been described in a large number of reports. These changes in the tRNA populations of microorganisms, animals and plants, have been reviewed by LITTAUER and INOUYE (1973), those observed more specifically in plants have been reviewed by LEA and NORRIS (1977) and by WEIL (1979), so that they will not be discussed here in detail. They include:

- Changes in the chromatographic profiles of seryl-, lysyl- and prolyl-tRNA's upon differentiation of wheat seedlings (VOLD and SYPHERD 1968).
- Variations in the relative levels of several tRNA's in soyabean cotyledons, following germination and during development and senescence (ANDERSON and CHERRY 1969, BICK et al. 1970, VENKATARAMAN and DELEO 1972, PILLAY and CHERRY 1974, SHRIDHAR and PILLAY 1976, PILLAY and GOWDA 1980).
- Differences in the relative levels of the isoaccepting $tRNA^{Tyr}$ species found in dividing and non-dividing cells of pea roots (VANDERHOEF and KEY 1970).
- Changes in the relative levels of the isoaccepting $tRNA^{Phe}$ species of barley following germination (HIATT and SNYDER 1973).
- Changes in the $tRNA^{Leu}$ species of germinating pea seedlings and developing pods (PATEL and PILLAY 1976).
- Differences in the distribution of polysome-bound $tRNA\text{'s}^{Ile}$ and $tRNA^{Lys}$ isoacceptors in lupin cotyledons, depending on the development stage (AUGUSTYNIAK and PAWELKIEWICZ 1979).
- Light-stimulated synthesis of chloroplast tRNA species in *Euglena* cells (BARNETT et al. 1969, PARTHIER 1977) and light-stimulated formation of minor (methylated) nucleotides in wheat germ $tRNA^{Phe}$ (RACZ et al. 1979).
- Shift in the ratio of two glycyl-tRNA's in soyabean seedlings hypocotyls upon chilling stress (YANG and BROWN 1974).
- Changes in the level of aminoacylation (from 0 to 81%) of two $tRNA\text{'s}^{Met}$ in dry and 2-day old lupin cotyledons (KEDZIERSKI et al. 1979).
- Cytokinin-induced variations in the relative levels of $tRNA\text{'s}^{Leu}$ in soyabean cotyledons (BICK et al. 1970, PILLAY and CHERRY 1974, WRIGHT et al. 1972, LESTER et al. 1979), and cytokinin-induced appearance of female flowers (containing a tRNA population different from that of male flowers) on genetically male *Mercurialis annua* (a dioecious species) plants (BAZIN et al. 1975).
- Changes in the relative proportions of several tRNA's during ethylene-induced ripening of tomato fruits (METTLER and ROMANI 1976).

- Increase in the $tRNA^{Gln}$ content of maize endosperm, where zein, the major maize seed's storage protein, which has a high glutamine content (about 20%), is synthesized (VIOTTI et al. 1978), an observation supporting the theory of functional adaptation of RNA's (GAREL 1974).
- Presence of an extra species of $tRNA^{Phe}$, lacking the Y base, in crown-gall tumour tissues as compared to tobacco-normal tissues (CORNELIS et al. 1975).

It has, however, been pointed out (GUSSECK 1977) that if a complete aminoacylation of all the isoacceptors present in a total tRNA preparation is not achieved, the observed changes may not reflect true variations in the relative proportions of isoaccepting tRNA's, but may result from differences in the aminoacylation rates of the various isoacceptors, and therefore from differences in their aminoacylation extent under conditions of incomplete acylation. But assuming that the observed changes in the tRNA population are real, their biological significance can still be questioned, since in many studies, especially the earlier ones, the existence (or appearance) of organellar tRNA's was not taken into account, thus making it difficult to interpret the variations observed.

However, studies performed in animal systems have recently suggested that protein synthesis may be regulated at the translational level by the availability of one (or several) isoaccepting tRNA species. For instance, a minor tRNA species could be a limiting factor if it is the only tRNA whose anticodon is able to read a codon present in mRNA molecules, which must be translated in a given tissue or at a given time in differentiation. Similar situations may also exist in plants, but more work is needed to determine more precisely the relationships between the observed variations in the tRNA population and the events occurring during plant growth and development.

# 3 Aminoacyl-tRNA Synthetases

## 3.1 Preparation, Fractionation and Purification of the Enzymes

Compared with the long list of synthetases (Aa-RS) purified to homogeneity from bacteria, yeast and animals (KISSELEV and FAVOROVA 1974, SCHIMMEL and SÖLL 1979), very few of these enzymes have been purified from plant material (see Table 3). This situation may be due to a higher degree of instability of many plant synthetases, the need for more rigorous homogenization methods, and the release of active proteolytic enzymes from lysosomes, harmful for enzyme activity during the lengthy procedures of many enzyme purifications. Therefore SH-reducing substances, such as mercaptoethanol, dithiotreitol, glutathione, should be routinely included in the homogenization buffer, and glycerol should be used for stabilization.

Suitable procedures are now available, which have been successfully developed for the purification of Aa-RS from microbial or animal cells. Some of the procedures used for plant Aa-RS purification include: affinity chromatography on Sepharose columns containing either the corresponding amino acid

**Table 2.** Purification steps of chloroplastic (E1) and cytoplasmic (E2) valine-tRNA synthetase from green *Euglena gracilis* (IMBAULT et al. 1979, SARANTOGLOU et al. 1980)

| Purification steps | | Yield (%) | | Spec. activity (units/mg prot.) | | Purification (fold) | |
|---|---|---|---|---|---|---|---|
| E1 | E2 | E1 | E2 | E1 | E2 | E1 | E2 |
| Homogenate (35,000 *g* supernatant) | | 100 | 100 | 1.1 | 0.5 | 1 | 1 |
| $(NH_4)_2SO_4$ precipitation | | 66 | | 0.9 | | 0.8 | |
| Hydroxyapatite chromatography | | 58 | 54 | 1.4 | 5 | 1.3 | 10 |
| | Phosphocellulose | | 30 | | 54 | | 108 |
| DEAE cellulose | DEAE cellulose | 34 | 15 | 15.8 | 80 | 14.3 | 160 |
| Blue Dextran Sepharose | Blue Dextran Sepharose | 19 | 5 | 500 | 380 | 454 | 760 |
| Sephadex G-200 | | 6 | | 1,090 | | 990 | |

or the cognate tRNA (BARTKOWIAK and PAWELKIEWICZ 1972, JAKUBOWSKI and PAWELKIEWICZ 1975, SWAMY and PILLAY 1979, 1980) or hydrophobic chromatography on Aminohexyl Sepharose (JAKUBOWSKI and PAWELKIEWICZ 1973, 1974, 1975). Since affinity chromatography with tRNA may also retain other proteins interacting with tRNA (methylases, nucleotidyl transferase), purification in a successive two-column step is recommended, in which the first column contains all the tRNA species minus the cognate tRNA and the second column has only the cognate tRNA. A very efficient purification was achieved by the use of Blue Dextran 2000 bound to Sepharose 4B (IMBAULT et al. 1979, SARANTOGLOU et al. 1980; cf. Table 2). For routine Aa-RS purification the following steps are employed: crude homogenate or post-ribosomal supernatant, $(NH_4)_2SO_4$ precipitation or partial fractionation, chromatography through DEAE-cellulose, DEAE-Sephadex, phosphocellulose, and gel filtration through Sephadex G-200. The procedure for the isolation for homogenous Aa-RS preparations from plant material is complicated by the fact that organelle synthetases can contaminate cytoplasmic Aa-RS, if both species have similar properties. However, in the case of several plant synthetases studied so far, they can be separated by hydroxyapatite chromatography and distinguished by tRNA charging specificity (cf. Sects. 3.2 and 3.4.1). Chloroplastic and cytoplasmic Val-RS from *Euglena gracilis,* both purified to homogeneity by IMBAULT et al. (1979) and SARANTOGLOU et al. (1980), exemplify the achievement of active Aa-RS isoenzyme preparation (Table 2). After common enrichment of the two enzymes by a series of classical steps, hydroxyapatite chromatography separates the two isoenzymes. Thus hydroxyapatite chromatography can be regarded as the method of choice for the separation of several organelle Aa-RS species from their cytoplasmic counterparts (KRAUSPE and PARTHIER 1973, 1974). Complete purification was achieved by affinity elution from a Blue Dextran Sepharose column using tRNA, namely $tRNA^{Val}$ from yeast in the case of cytoplasmic Val-RS (SARANTOGLOU et al. 1980).

### 3.2 Intracellular Localization and Enzyme Heterogeneity

The theoretically expected occurrence of three different synthetases (isoenzyme species), for each of the 20 protein amino acids in the cytoplasm, in the chloroplasts and in the mitochondria have been unequivocally verified only in a few cases. In *Phaseolus* three isoenzyme species have been separated for Met, Pro, Lys and Phe activation (GUILLEMAUT and WEIL 1975, JEANNIN et al. 1976, 1978). In the case of *Euglena* Ile-RS, Phe-RS and Glu-RS from organelles and cytoplasm have been distinguished by chromatography (REGER et al. 1970, KISLEV et al. 1972, HECKER et al. 1974, BARNETT et al. 1976). However, since obtaining pure mitochondrial preparations is questionable with these procedures, As-RS obtained from this type of organelle preparation might not be free of contaminations by Aa-RS from other cell compartments.

Aa-RS for most, if not all, of the 20 amino acids are located within chloroplasts, as has been demonstrated with isolated organelles after at least one purification step (ALIEV and PHILIPPOVICH 1968, LANZANI et al. 1969, BRANTNER and DURE 1975, PARTHIER et al. 1978, and the references cited earlier in this Section). These results have been confirmed, when chloroplast-specific Aa-RS species could not be detected in bleached *Euglena* mutants, such as $W_3$BUL or others (REGER et al. 1970, KRAUSPE and PARTHIER 1973, 1974, HECKER et al. 1974, KISLEV et al. 1972). However, the impact of these findings should not be over-estimated, since the mutants contain proplastid-like structures, which develop prolamellar bodies and increase organellar Phe-RS or Leu-RS activity upon illumination of dark-grown mutant cells (PARTHIER and NEUMANN 1977). In addition, at least some of the synthetase species in the two types of *Euglena* organelles are indiscernible from each other, either by tRNA specificity (cf. Sect. 3.4.1) or by chromatographic methods, as illustrated for Leu-RS species in Fig. 7.

Whereas the methods for obtaining purified chloroplasts in large amounts seems to be satisfactory, preparation of pure and intact proplastids, etioplasts, and mitochondria is still a matter of discussion as it is for other stroma or matrix enzymes. In particular mitochondrial preparations from higher green plants and *Euglena* (where mitochondria seem to exist as a reticular structure) are often contaminated with attached cytoplasmic constituents including synthetases (Fig. 7D); parallel determination of marker enzymes of the compartments is essential but often not done. Further it is very difficult to distinguish between the Aa-RS species of mitochondria and non-green plastids, e.g., in bleached or mutant *Euglena* cells, seeds or etiolated seedlings. Such separations are necessary for studies on chloroplast biogenesis. In all future work the use of immunological methods should prove of value.

Despite the application of several techniques (chromatographic enzyme separation, organelle isolation, light-induced biosynthesis, plastid mutants of tRNA and synthetases, inhibitor studies, and heat-stability of the enzyme), plastid Thr-RS could not be detected in *Euglena gracilis*. Only one Thr-RS species with typical characteristics of cytoplasmic synthetase was observed (PARTHIER and KRAUSPE 1973, KRAUSPE and PARTHIER 1974, PARTHIER and NEUMANN 1977, PARTHIER et al. 1978). The occurrence of an extremely labile chloroplast

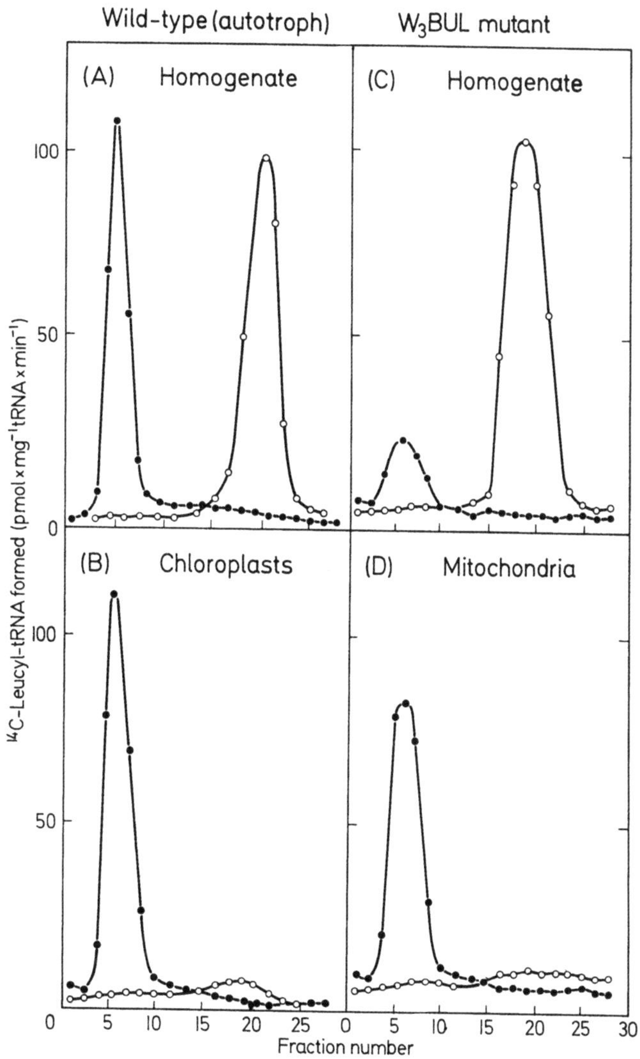


**Fig. 7 A–D.** Elution pattern from hydroxyapatite chromatography of organelle and cytoplasmic leucyl-tRNA synthetases of *Euglena gracilis,* wild-type, grown autotrophically and plastid mutant $W_3BUL$, grown heterotrophically (Krauspe and Parthier 1973, Parthier et al. 1978, Müller-Uri et al. 1981). **A** wild-type, homogenate; **B** isolated and purified chloroplasts therefrom; **C** $W_3BUL$ mutant, homogenate; **D** isolated and purified mitochondria therefrom. Cell homogenates or disrupted organelles were layered on hydroxyapatite columns and the proteins eluted with a linear 0.01 to 0.3 M potassium phosphate gradient. Fractions (5 ml) were collected and aliquots checked for aminoacylation activity with [$^{14}$C]-Leu towards (●) tRNA from *Anacystis nidulans,* chloroplasts or mitochondria from *Euglena,* and towards (○) *Euglena* cytoplasmic tRNA. Either *first peak* represents organellar and the *second one* cytoplasmic Leu-RS

Thr-RS as an explanation of this finding would be astonishing, since all other chloroplastic Aa-RS species in *Euglena* are more stable than their cytoplasmic counterparts.

Two Met-RS have been purified from wheat germ (ROSA and SIGLER 1977) and two Tyr-RS demonstrated in pea roots (COWLES and KEY 1972), both probably cytoplasmic, though the organellar origin of each one of them was not determined.

## 3.3 Functional and Molecular Properties

### 3.3.1 Assays of Activity

Three methods are available for the determination of aminoacyl-tRNA synthetase activity. According to the reaction scheme described in the Introduction, both the first (ATP-$PP_i$ exchange) and the second step (esterification of amino acid to tRNA) can serve separately as assay procedures, in addition to the overall reaction. The latter is the method mostly used, whereas in earlier papers pyrophosphate exchange assay and hydroxamate assay have been frequently reported.

*Hydroxamate Assay.* Amino acid in the presence of ATP and enzyme is activated by forming the aminoacyl-AMP-enzyme complex + $PP_i$. Addition of $NH_2OH$ results in aminoacylhydroxamate + AMP + enzyme. Thus, hydroxylamine acts as an acceptor for the amino acid instead of tRNA. A disadvantage of this method is the lack of specificity, when crude enzyme extracts are used, since other amino acids using these enzymes can interfere. Secondly, rather high amounts of amino acids are required, and the end products, especially $PP_i$, inhibit the reaction.

*Pyrophosphate Exchange Assay.* The aminoacyl-AMP-enzyme complex is obtained as above, however, addition of $^{32}PP_i$ permits measurement of the amount of [$^{32}P$]-ATP formed in the reverse reaction. Adenylate binding to the enzyme is the rate-limiting step. After stopping the reaction by the addition of acid, labelled ATP may be separated from labelled $PP_i$ by adsorption to and elution from charcoal or other methods. This assay procedure is about twice as sensitive as the hydroxamate assay. However, certain acid-activating enzymes other than Aa-RS can interfere in crude enzyme preparations and in addition the amino acid liberation by proteases may give rise to a high background activity. Pyrophosphate exchange is often stimulated by addition of cognate tRNA (MEHLER 1970, KEDZIERSKI and PAWELKIEWICZ 1970, JAKUBOWSKI 1978b, SCHIMMEL 1979).

*Aminoacyl-tRNA Esterification Assay.* Besides the fact that binding of amino acids to tRNA is the physiological type of reaction catalyzed by synthetases, this step is of much higher specificity also in in vitro assays (in spite of many exceptions, cf. Sect. 3.3.1). Semi-automated assay procedures for the simultaneous determination of Aa-RS activities have been described (McCune et al. 1977, Smith and Santi 1979) and techniques to measure the degree of in vivo aminoacylation which involves the use of periodate oxidation have been developed (e.g., Ehresmann et al. 1974) which now replace indirect methods (Hall and Tao 1970).

The esterification assay will be the basis for the data discussed throughout the following sections, in spite of the criticisms against its suitability for quantitative assays of total Aa-RS levels, in which various isoacceptor tRNA's or heterologous tRNA sources have been used (Gusseck 1977, cf. also Sect. 2.6). Discrepancies between the pyrophosphate exchange and esterification assays may be accounted for, inter alia, by the fact that those enzymes which aminoacylate tRNA less efficiently often form tightly bound aminoacyl adenylate complexes with the enzyme (Mehler 1970).

### 3.3.2 Kinetic Parameters

Michaelis-Menten constants ($K_m$) derived from Lineweaver-Burk plots of the aminoacylation reaction are common characteristics of enzyme activities and in many cases reported for the three substrates involved in Aa-RS catalysis. The $K_m$ values found are in the same orders of magnitude for all plant Aa-RS studied so far, i.e., $10^{-6}$ to $10^{-5}$ M for the amino acid, $10^{-5}$ to $10^{-3}$ M for ATP, $10^{-8}$ to $10^{-6}$ M for cognate (isoacceptor) and total tRNA, respectively. Hydroxyapatite-fractionated chloroplast and cytoplasmic Leu-RS from *Euglena gracilis* are reported to have very similar $K_m$ values (Krauspe and Parthier 1974). Highly purified Val-RS isoenzymes from the same organism (Table 2) show the following $K_m$ parameters (in mol): Chloroplastic Val-RS $1.5 \times 10^{-5}$ (Val), $5 \times 10^{-5}$ (ATP), $6 \times 10^{-8}$ ($tRNA^{Val}_{Eugl}$) and $8 \times 10^{-8}$ ($tRNA^{Val}_{Coli}$) (Imbault et al. 1979); cytoplasmic Val-RS $5 \times 10^{-5}$ (Val), $7 \times 10^{-5}$ (ATP), $5 \times 10^{-8}$ ($tRNA^{Val}_{yeast}$). Correspondingly, chloroplast and cytoplasmic Tyr-RS from soyabean cotyledons (Locy and Cherry 1978) indicate the following apparent $K_m$ values: chloroplastic Tyr-RS $4.9 \times 10^{-6}$ (Tyr), $2.14 \times 10^{-4}$ (ATP), $2.2 \times 10^{-8}$ ($tRNA^{Tyr}_{soyabean}$), fractionated by BD-cellulose chromatography); cytoplasmic Tyr-RS $6.8 \times 10^{-6}$ (Tyr), $4.9 \times 10^{-5}$ (ATP), $8.9 \times 10^{-8}$ (total tRNA from soyabean cotyledons).

$K_m$ values estimated in the pyrophosphate exchange reaction are in the order of $10^{-5}$ to $10^{-3}$ M for the amino acid and $10^{-4}$ M for ATP (Burkard et al. 1970, Norris and Fowden 1973, Krauspe and Parthier 1975, Burnell and Shrift 1977), $10^{-3}$ M for $PP_i$ (Cys-RS, Burnell and Shrift 1977). For further references for kinetic data see Lea and Norris (1977).

Aa-RS activities can be markedly altered based upon changes in pH or ionic strength in the assay milieu. As a rule, pH optima are found between pH 7 and 8.5, but exceptional high pH optima (below 6 and about 9) are likewise reported (Tao and Hall 1971, Jakubowski and Pawelkiewicz 1975). Optimal Mg:ATP ratios can vary between 10 and 1 and differ widely for various Aa-RS, even for the same enzyme obtained from different species and also between chloroplastic and cytoplasmic isoenzymes from the same species (e.g., Burkard et al. 1970, Tao and Hall 1971, Krauspe and Parthier 1974).

**Table 3.** Aminoacyl-tRNA synthetases from plant sources: Purification and molecular weights

| Enzyme species (EC) | Plant species | Organ, compartment | Purification (fold) | Subunit structure | MW ($10^3$) | Reference |
|---|---|---|---|---|---|---|
| Arg-RS (6.1.1.19) | *Triticum* sp. | Embryos | | $\alpha$ | 73 | CARIAS et al. (1978) |
| | *Lupinus luteus* | Seeds | | $\alpha_2$ | 140 $(70)_2$ | BARCISZEWSKI et al. (1979) |
| Asp-RS (6.1.1.12) | *Phaseolus aureus* | Seeds | 170 | | | LEA and FOWDEN (1973) |
| Asn-RS (6.1.1.13) | 3 Legume species | Seeds | 150 | | | LEA and FOWDEN (1973) |
| Cys-RS (6.1.1.16) | *Phaseolus aureus* | Seeds | 300 | $\alpha$ | 60 | SHRIFT et al. (1976), BURNELL and SHRIFT (1977) |
| | *Astragalus* sp. | Seeds | 49–109 | | | BURNELL and SHRIFT (1979) |
| Glu-RS (6.1.1.17) | *Phaseolus aureus* | Seeds | 50 | | | LEA and FOWDEN (1972) |
| Leu-RS (6.1.1.4) | *Lupinus luteus* | Seeds | | | 170 | LEGOCKI and PAWELKIEWICZ (1967) |
| | *Aesculus hippocast.* | Cotyledons | 100 | | | ANDERSON and FOWDEN (1970b) |
| | *Gossypium hirsutum* | Cotyledons | 23 | | | BRANTNER and DURE (1975) |
| | *Triticum* sp. | Embryos | | $\alpha_2$ | 110 $(63)_2$ | CARIAS et al. (1978) |
| | *Euglena gracilis* | Cytoplasm | 150 | $\alpha$ | 110 | KRAUSPE and PARTHIER (1974), |
| | *Euglena gracilis* | Chloroplast | 150 | $\alpha$ | 100 | KRAUSPE and PARTHIER (1974) |
| | *Euglena gracilis* | Cytoplasm | 1030 | $\alpha$ | 116 | SARANTOGLOU et al. (1981) |
| | *Euglena gracilis* | Chloroplast | 990 | $\alpha$ | 100 | IMBAULT et al. (1981) |
| Met-RS (6.1.1.10) | *Triticum* sp. | Embryos | | $\alpha_2$ | 165 $(74)_2$ | CHAZAL et al. (1975) |
| | *Triticum* sp. | Embryos | | $\alpha$; $\alpha$ | 105; 70 | ROSA and SIGLER (1977) |
| | *Lupinus luteus* | Seeds | 545 | $\alpha_2$ | 170 $(85)_2$ | JOACHIMIAK et al. (1978) |

| | | | | | | |
|---|---|---|---|---|---|---|
| Phe-RS (6.1.1.20) | *Triticum* sp. | Chloroplasts | 10 | | | Lanzani et al. (1969) |
| | *Aesculus hippocast.* | Cotyledons | 800 | | | Anderson and Fowden (1970a) |
| | *Triticum* sp. | Embryos | | $\alpha_2\,\beta_2$ | 250 $(50, 80)_2$ | Carias and Julien (1976) |
| | *Lupinus luteus* | Seeds | | $\alpha_2\,\beta_2$ | 260 (59, 75) | Barciszewski et al. (1979) |
| | *Glycine max* | Cotyledons | 180 | $\alpha$ | 65 | Swamy and Pillay (1979) |
| | *Glycine max* | Cotyledons | 2,228 | $\alpha$ | 80 | Swamy and Pillay (1980) |
| Pro-RS (6.1.1.15) | *Phaseolus aureus* | Seeds | 250 | | | Peterson and Fowden (1965) |
| | *Phaseolus aureus* | Seeds | 350 | | | Norris and Fowden (1972) |
| | *Delonix regia* | | 75 | | | Norris and Fowden (1972) |
| Ser-RS (6.1.1.11) | *Lupinus luteus* | Seeds | 1,500 | $\alpha_2$ | 110 $(55)_2$ | Jakubowski and Pawelkiewicz (1975) |
| Thr-RS (6.1.1.3) | *Aesculus hippocast.* | Cotyledons | 30 | | | Anderson and Fowden (1970b) |
| Trp-RS (6.1.1.2) | *Lupinus luteus* | Seeds | 125 | $\alpha_4$ | 200 $(37)_4$ (?) | Jakubowski and Pawelkiewicz (1975) |
| Tyr-RS (6.1.1.1) | *Glycine max* | Cytoplasm | 374 | $\alpha_2$ | 126 $(61)_2$ | Locy and Cherry (1978) |
| | *Glycine max* | Chloroplasts | | $\alpha_2$ | 98 $(43)_2$ | Locy and Cherry (1978) |
| Val-RS (6.1.1.9) | *Aesculus hippocast.* | Cotyledons | 25 | | | Anderson and Fowden (1970b) |
| | *Lupinus luteus* | Seeds | 600–1,000 | $\alpha$ | 125 | Jakubowski and Pawelkiewicz (1974, 1975) |
| | *Euglena gracilis* | Chloroplasts | 990 | $\alpha$ | 126 | Imbault et al. (1979) |
| | *Euglena gracilis* | Cytoplasm | 760 | $\alpha$ | 126 | Sarantoglou et al. (1980) |

$Mg^{2+}$ optima ranging from 2.5 to 20 mM are rather broad and varying. Spermidine, $Ca^{2+}$ and even $Co^{2+}$ can replace $Mg^{2+}$ at least partially; stabilization effects of these ions for the enzyme molecules may also play a role. Monovalent cations such as $K^+$ and $NH_4^+$ are needed for optimal activities unless supplied in higher concentrations than 100–150 mM, where they exert inhibitory effects (FOWDEN and FRANKTON 1968, ANDERSON and FOWDEN 1970b, BURKARD et al. 1970, LEA and FOWDEN 1973, KRAUSPE and PARTHIER, 1973). Phosphate ions also depress synthetase activities in higher concentrations. This should be kept in mind for experiments in which the activities of non-dialyzed fractions eluted from hydroxyapatite columns are quantitatively compared.

### 3.3.3 Molecular Structure and Stability

Sizes of Aa-RS expressed in molecular weights are very variable. Even from a single organism such as *E. coli*, from which all 20 Aa-RS have been purified, synthetases ranging in molecular weight from 60,000 to 230,000 (SCHIMMEL and SÖLL 1979) have been reported. In addition four different types of subunit structures: $\alpha$, $\alpha_2$, $\alpha_4$, and $\alpha_2$, $\beta_2$ were also determined. This variability seems to be valid also for plant Aa-RS, as far as can be seen from the synthetase species isolated and purified so far (Table 3). For example, agreement exists in $\alpha_2\,\beta_2$ subunit structure and approximate molecular weight between Phe-RS from lupin seeds and wheat germ (as well as for Phe-RS of yeast, animal tissue and bacteria, cf. KISSELEV and FAVOROVA 1974, SCHIMMEL and SÖLL 1979). In contrast, Phe-RS in soyabean cotyledons seems to consist of a monomer (SWAMY and PILLAY 1979, 1980). Arg-RS can occur as monomer or dimer of about 70,000 molecular weight (CARIAS et al. 1978, BARCISZEWSKI et al. 1979). In wheat germ, ROSA and SIGLER (1977) observed two species of monomeric Met-RS. It is unclear whether one of them is from organelles.

Molecular weights and subunit structures of chloroplastic and cytoplasmic Val-RS are identical in *Euglena gracilis* (IMBAULT et al. 1979, SARANTOGLOU et al. 1980), whereas both isoenzymes of Tyr-RS in soyabean cotyledons differ in molecular weight (LOCY and CHERRY 1978). The two Leu-RS isoenzymes from *Euglena* showed similar but not identical molecular weight when estimated by Sephadex G-200 gel filtration (KRAUSPE and PARTHIER 1974). Only one protein band of 56,000 molecular weight was observed on SDS gel electrophoresis for the chloroplastic and mitochondrial Leu-RS, suggesting $\alpha_2$ subunit structure if not produced by proteolytic cleavage (unpublished data). Aminoacyl-tRNA synthetases in eukaryotic cells can occur as high molecular weight complexes ($10^5$ to more than $10^6$). There is evidence that these complexes exist in vivo but are destroyed during rigorous homogenization. They may also contain proteins other than synthetases, tRNA and translation factors. After such synthetase complexes had been characterized from animal cells (e.g., VENNEGOOR and BLOEMENDAL 1972, KELLERMANN et al. 1979, DANG and YANG 1979, further refs. cf. SCHIMMEL and SÖLL 1979), they were found in wheat germ with sizes of 18–20S, and even 40–60S (QUINTARD et al. 1978). Synthetase–synthetase as well as synthetase–ribosome interactions are suggested as means of stabilization of these enzymes in lupin seedlings (JAKUBOWSKI 1979), since purified synthetase became partially or totally inactive in the absence of other proteins. Under

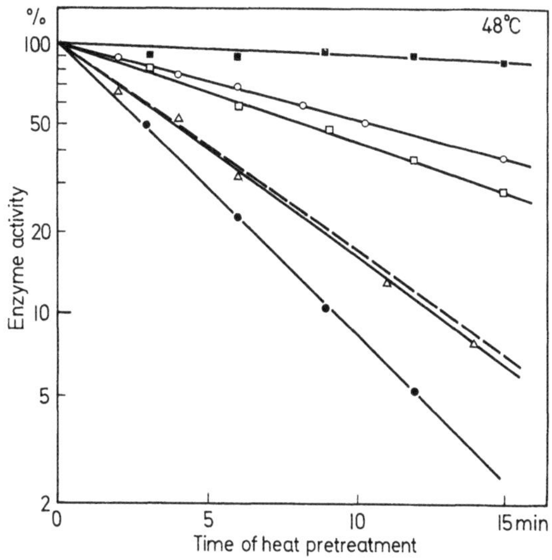


**Fig. 8.** Effect of L-leucine, ATP or other nucleoside triphosphates on the heat stability of cytoplasmic Leu-RS from *Euglena gracilis*. (KRAUSPE and PARTHIER 1975). The enzyme was used after elution from hydroxyapatite (Fig. 6, A) and pretreated for 0–15 min at 48 °C without (----) or in the presence of: □ 0.285 mM L-leucine; ○ 5 mM ATP; ● 5 mM ITP or GTP; △ 5 mM CTP or UTP; ■ 5 mM ATP+0.285 mM L-leucine. After cooling to 0° the mixtures were made complete with the necessary reagents for aminoacylation assay for 7 min at 25 °C

assay conditions protective interaction can be simulated by addition of spermine or spermidine (JAKUBOWSKI 1980).

Substrates of Aa-RS are able to stabilize the enzymes during incubation at assay temperatures (25 °C) or higher temperatures (33–58 °C), e.g., ATP (NORRIS and FOWDEN 1973, KRAUSPE and PARTHIER 1974, 1975, SWAMY and PILLAY 1980), cognate amino acid (NORRIS and FOWDEN 1973, PARFAIT 1973) or cognate tRNA (KEDZIERSKI and PAWELKIEWICZ 1970). Stabilizing effects of $Mg^{2+}$ may be indispensable for tRNA–synthetase complex formation. In general, synthetases from chloroplasts are more stable than from cytoplasm (NORRIS and FOWDEN 1973, PARTHIER and KRAUSPE 1973, KRAUSPE and PARTHIER 1974, 1975). A quantitative study performed with Leu-RS isoenzymes from *Euglena* chloroplasts and cytoplasm revealed that the latter synthetase was heat-inactivated after 15 min at a half-life temperature ($T_{1/2}$=50% inactivation) of 44 °C or with a half-life time ($t_{1/2}$) at 48 °C of 4 min (Fig. 8). The corresponding values for chloroplastic Leu-RS were 51 °C and 30 min, respectively. ATP (5 mM) added before preincubation of the cytoplasmic enzyme was able to delay the inactivation process by 6 °C ($T_{1/2}$) and 4 min ($t_{1/2}$), respectively. The protection constant $\pi$ (ATP)$=1.6\times10^{-5}$ M (25 °C) for cytoplasmic Leu-RS is more than one order of magnitude smaller than the $K_m$ (ATP) for this enzyme (KRAUSPE and PARTHIER 1975). Other purine nucleoside triphosphates such as GTP, ITP and XTP accelerated heat inactivation markedly, pyrimidine nucleoside triphosphates were without effect (cf. Fig. 8).

### 3.4 Substrate Specificities

#### 3.4.1 Transfer RNA

A remarkable feature of aminoacylating enzymes is their high affinity towards cognate tRNA from homologous sources. The preciseness of intermolecular recognition is based on a general structural adaptation between the two macromolecules and additional but highly specific interactions between certain elements of the polypeptide and tRNA chains (Schimmel and Söll 1979). Nevertheless, certain Aa-RS are able to acylate heterologous tRNA's (review: Jacobson 1971). Transfer RNA substrate specificity in a heterologous cell-free aminoacylation assay involves two aspects: (1) Specificity towards non-cognate tRNA's from the same material including various isoacceptor species; (2) specificity towards cognate tRNA but of different biological origins. The former aspect has great importance for assays using total tRNA fractions, crude enzyme fractions, or both. Knowledge on the latter aspect is necessary for compartmentation studies, including mutual contaminations during cell fractionation, and may be of evolutionary interest.

Heterologous aminoacylation has been studied with prokaryote-type tRNA's and eukaryote-type synthetases, and vice versa. Using different plant material this class-specificity in aminoacylation has been confirmed many times for organelle and cytoplasmic synthetases and tRNA's (Vanderhoef et al. 1972, Barnett et al. 1978, Weil 1979, cf. Sect. 3.2). Apparently, all plant species containing more than one isoacceptor tRNA species in their tRNA pattern (cf. Table 1) possess at least two different species of Aa-RS (isoenzymes) in a cell homogenate.

Most of these isoenzymes show more or less strict specificity towards cognate isoacceptor tRNA species separable from each other by reverse phase chromatography (Sect. 2.1). Chloroplastic and cytoplasmic Leu-RS from higher plants (Cherry and Osborne 1970, Kanabus and Cherry 1971, Wright et al. 1974, Guillemaut et al. 1975, Brantner and Dure 1975, Klyachko and Parthier 1980) as well as from *Euglena gracilis* (Parthier et al. 1972, Krauspe and Parthier 1973, 1974) are particularly good examples of a high specificity in recognizing tRNA's from the two homologous cell compartments. Similar specificities have been observed for Pro-RS and Lys-RS (Jeannin et al. 1976) and Phe-RS (Jeannin et al. 1978) from *Phaseolus* chloroplasts and cytoplasm and for Asp-, Val-, Lys-, Ile-, Ser-, Tyr-RS from *Euglena* (Parthier et al. 1978). This variation of organelle-specificity for Aa-RS in *Euglena* was not confirmed, when synthetases from greening pumpkin cotyledons were separated by hydroxyapatite chromatography (Fig. 9): Prokaryotic and eukaryotic tRNA's were ambiguously aminoacylated by most of synthetase fractions. Non-specificity concerns both chloroplastic (E1) and cytoplasmic (E2) isoenzymes and could be due to either unspecific tRNA recognition ability of one synthetase species, or homologous tRNA recognition by two isoenzymes not separable by this method. In *Euglena*, the second case seems to be true for Ile-RS isoenzymes, but chloroplastic Phe-RS acylates either homologous or heterologous tRNA's (Parthier and Krauspe 1973, Parthier et al. 1978). Cytoplasmic Thr-RS charges tRNA equally well from cytoplasm, chloroplasts, mitochondria, *E. coli* and the blue-green alga, *Anacystis nidulans* (Parthier and Krauspe 1973, Krauspe and Parthier 1974).

A comparison, in charging specificity, between tRNA's and synthetases from chloroplasts and blue-green algae is of special interest, since speculations have been made about a possible phylogenetic origin of chloroplasts from ancestral blue-green algae. Early data reported by Parthier (1971, 1972) indicated that

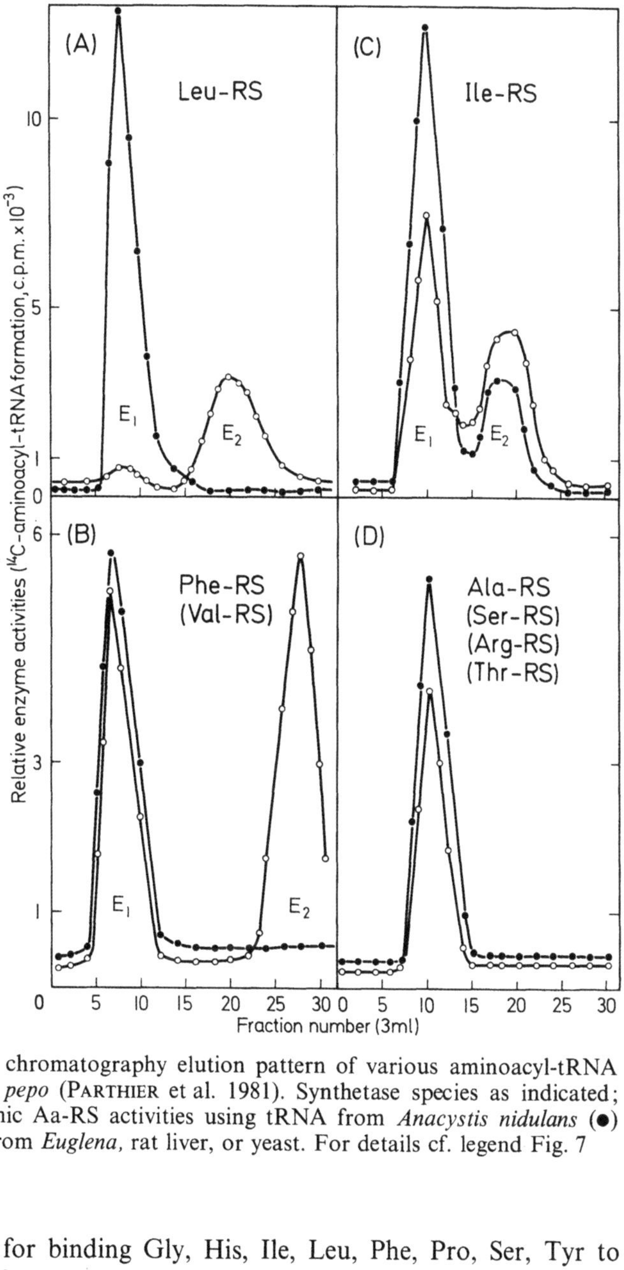


**Fig. 9 A–D.** Hydroxyapatite chromatography elution pattern of various aminoacyl-tRNA synthetases from *Cucurbita pepo* (PARTHIER et al. 1981). Synthetase species as indicated; *E1* organelle; *E2* cytoplasmic Aa-RS activities using tRNA from *Anacystis nidulans* (●) or cytoplasmic tRNA (○) from *Euglena*, rat liver, or yeast. For details cf. legend Fig. 7

several Aa-RS (specific for binding Gly, His, Ile, Leu, Phe, Pro, Ser, Tyr to tRNA) in crude 100,000 *g* supernatants from *Anacystis nidulans* acylated *E. coli* tRNA to about the same extent as homologous *Anacystis* tRNA and to a much less extent total tRNA from green *Euglena* cells. Total tRNA from plastid-mutant *Euglena* cells or from rat liver cytoplasm was not aminoacylated. As an exception, Arg-RS acylated heterologous tRNA from various sources severalfold better than homologous tRNA. It was later shown (PARTHIER and KRAUSPE 1974) that Leu-RS from *Anacystis* acylated BD-cellulose-separated chloroplastic $tRNA^{Leu}$ from *Euglena,* and vice versa, chloroplastic Leu-RS acylated *Anacystis* $tRNA^{Leu}$ to similar extents. Cytoplasmic Leu-RS was unable to charge *Anacystis* tRNA, and *Anacystis* Leu-RS failed to aminoacylate cytoplasmic tRNA separated from total *Euglena* tRNA by BD-cellulose chromatography.

BEAUCHEMIN et al. (1973) using reversed phase chromatography for isoacceptor-tRNA separation, obtained identical aminoacylation pattern with *Anacystis* tRNA for Met, Leu and Ser with both *E. coli* and *Anacystis* synthetases. A more detailed communication by SELSKY (1978) described Ile-tRNA formation in heterologous systems from *Nostoc,* a filamentous blue-green alga, and green and bleached *Euglena.* He found that the two isoacceptor $tRNA^{Ile}$ species from *Nostoc* were aminoacylated by enzymes from green as well as plastid-free *Euglena* cells. However, the *Nostoc* enzyme was unable to acylate any $tRNA^{Ile}$ isoacceptor species prepared from bleached *Euglena* cells, indicating aminoacylating specificity of the enzyme from blue-green algae for heterologous tRNA's of chloroplast origin only. Moreover, in the heterologous system with *Nostoc* tRNA and *Euglena* enzymes 5% of mischarging [$^{14}$C]-isoleucine to $tRNA^{Val}$ was observed, which underlines the necessity of specificity estimations prior to experiments with heterologous components.

Heterologous aminoacylation between tRNA and Aa-RS from chloroplasts versus mitochondria or mitochondria versus cytoplasm has been studied less extensively. *Euglena* is a favourite organism because of the occurrence of plastid mutants which can be easily used for mitochondria preparation (although most of these bleached mutants are not free of proplastid-like structures, cf. PARTHIER and NEUMANN 1977). Mitochondrial or cytoplasmic $tRNA^{Ile}$ can be aminoacylated by either mitochondrial or cytoplasmic synthetaes, but not by the chloroplast enzyme (KISLEV et al. 1972). In higher plants Leu-RS from mitochondria aminoacylates not only the homologous isoacceptor tRNA species but also cytoplasmic species (GUDERIAN et al. 1972, GUILLEMAUT et al. 1975, SINCLAIR and PILLAY 1980). Cytoplasmic and mitochondrial synthetases failed to be separated from each other by hydroxyapatite chromatography.

This type of organelle specificity cannot be generalized, since in heterologous systems from *Phaseolus* tissues chloroplastic, mitochondrial and cytoplasmic Met-RS (GUILLEMAUT and WEIL 1975) and Pro-RS as well as Lys-RS (JEANNIN et al. 1976) were able to aminoacylate the cognate tRNA isoacceptor species from both types of organelles, whereas cytoplasmic tRNA's were only charged by the respective cytoplasmic Aa-RS. Mitochondrial Phe-RS from *Phaseolus vulgaris* aminoacylates $tRNA^{Phe}$ from bean mitochondria and chloroplasts, but not from the cytoplasm (JEANNIN et al. 1978). *Euglena* mitochondrial Leu-RS shows the same chromatographic elution characteristics as does chloroplastic Leu-RS (Fig. 7) and aminoacylates tRNA from mitochondria and blue-green algae to the same extent, but not cytoplasmic tRNA. In conclusion, strict specificity exists between the macromolecular components of aminoacyl-tRNA formation from chloroplasts and cytoplasm (with few exceptions). Mitochondrial tRNA and synthetases behave differently; cytoplasmic components from various eukaryotic organisms seem to be very similar in their structures and are exchangeable on a broad scale.

### 3.4.2 Amino Acids

The activation of amino acid analogues in place of protein amino acids has been extensively studied in L. Fowden's laboratory, often with synthetases from both analogue-free and analogue-producing plants. Because of the structural similarities between amino acid and analogue, the latter mimics the correct substrate in the catalytic centre of the enzyme. Consequently, the analogue can be incorporated into the polypeptide chain, provided it has been attached to tRNA and did not sterically inhibit substrate aminoacyl-AMP-enzyme complex formation or esterification to the cognate tRNA. Whether a possible misacylation process has any importance for protein biosynthesis in the intact cell is not only a question of different intracellular compartmentation of enzymes and analogues (as secondary plant products they are probably accumulated in high amounts within the vacuole) but also could depend on the proof-reading efficiency of a synthetase, a property which helps to assure the precise incorporation of the correct amino acid during the ribosome-mediated polypeptide formation (YARUS 1979).

Leu-RS from *Aesculus hippocastanum* activated several amino acids structurally related to leucine, such as L-hypoglycin, azaleucine, norleucine and isoleucine (ANDERSON and FOWDEN 1970b); all decreased leucing activation competitively in the ATP-$PP_i$ exchange reaction. The carbon skeleton of these compounds is not branched before the $\gamma$-carbon atom whereas $\beta$-carbon branched valine was not activated by Leu-RS. Glu-RS prepared from *Phaseolus aureus* seeds activated several $\gamma$-substituted glutamic acids, which occur in certain plant species, e.g., erythro-$\gamma$-methyl-L-glutamic acid in *Caesalpinia bonduc* seeds or threo-$\gamma$-hydroxy-L-glutamic acid in *Hemerocallis fulva* leaves. When Glu-RS from the two plant species were prepared and compared for substrate specificity by ATP-$PP_i$ exchange activities, it became clear (LEA and FOWDEN 1972) that each of them did not activate its own natural product. Both accepted each other's analogue, and *P. aureus* Glu-RS activated all analogues tested. Thus it was suggested that the two analogue-containing plant species have developed discriminatory mechanisms based on enzyme alterations that prevents the plants from incorporating their own, possibly harmful product into polypeptides (LEA and FOWDEN 1972). Likewise, Pro-RS from *Delonix regia, Convallaria* and *Polygonatum,* plant species containing azetidine-2-carboxylic acid, the lower homologue of Pro, activated this analogue to a much lower extent than did Pro-RS from *Beta vulgaris, Phaseolus aureus* or *Asparagus officinalis,* i.e., species not producing this substance (NORRIS and FOWDEN 1972).

The Arg-activating enzyme from *Canavalia ensiforme,* a plant species accumulating canavanine, a highly toxic Arg analogue, in considerable amounts in their seeds, rejected canavanine as a substrate in pyrophosphate exchange assays (FOWDEN and FRANKTON 1968). Lys-RS from the same species misacylated arginine and homoarginine but did not accept canavanine. In this connection it is interesting that the larvae of a bruchid beetle, *Caryedes brasiliensis,* subsist solely on the seeds of *Dioclea megacarpa,* which contains more than 8% L-canavanine by dry weight. ROSENTHAL et al. (1976) found that Arg-RS prepared from the larvae discriminated between Arg and canavanine, so that no canavanyl-protein was synthesized. Finally, both chloroplast and cytoplasmic Arg-RS from *Phaseolus vulgaris* used canavanine as a substrate; but $PP_i$ exchange activity of the cytoplasmic synthetase was inhibited at higher concentrations of the analogue, whereas chloroplastic Arg-RS activity appeared to have been increased (BURKARD et al. 1970).

Phe-RS from four *Aesculus* species, among them *A. californica* containing 2-amino-4-methylhex-4-enoic acid, was able to activate this Phe analogue with approximately the same maximum velocity as phenylalanine. Since the analogue was not found in the proteins of *A. californica,* discrimination against this amino acid probably occurs in the transfer step to tRNA (ANDERSON and FOWDEN 1970a), though steric hindrance at the site of Phe binding to the synthetase or subcellular compartmentation of analogue and substrate or enzyme, respectively, cannot be excluded. Besides other Phe analogues, 1-amino-2-phenylethane 1-phosphoric acid was reported to be a true competitive inhibitor for Phe-RS from *A. hippocastanum,* since the calculated inhibitor constant ($K_i$) of $1.7 \times 10^{-5}$ M was identical with the $K_m$ (Phe) of $1.6 \times 10^{-5}$ M (ANDERSON and FOWDEN 1970c). On the other hand, 3-hydroxy-methyl-Phe, a natural product of *Caesalpinia tinctoria* proved to be a suitable substrate for Tyr-RS from *Phaseolus aureus* (NORRIS et al. 1975b). Fluor-phe was activated by Phe-RS and to a lesser extent also by Tyr-RS (SMITH and FOWDEN 1968). Both L-selenocysteine and α-aminobutyric acid served as substrates of purified Cys-RS from *Phaseolus aureus,* a species not accumulating selenium, as well as Se-accumulating species of *Astragalus* (BURNELL and SHRIFT 1977, 1979).

*Phaseolus aureus* contains an Asp-RS which was able to utilize α-aminomalonate and threo-β-hydroxyaspartate as substrates; Asn-RS from *Phaseolus aureus* and *Vicia sativa,* species producing β-cyanoalanine, accepted this asparagine analogue (LEA and FOWDEN 1973). Albizziine was shown to inhibit glutamine activation by Gln-RS, irrespective of whether the enzyme was prepared from *P. aureus* or from the albizziine-producing species, *Albizzia julibrissin* (LEA and FOWDEN 1973). In these examples transfer of significant amounts of the analogues into polypeptide chains seemed unlikely, but it was not shown whether the analogues acted as competitive inhibitors in the amino acid-dependent pyrophosphate exchange reaction or were indeed esterified to tRNA.

To summarize, there is good evidence that amino acid analogues do interfere in the aminoacylation process in vitro and in vivo, as was shown by analogues found in proteins. As a rule, however, aminoacylation represents the discriminatory step where the plant "decides" whether the correct amino acid or its analogue is selected to become incorporated into proteins. It has been suggested (LEA and FOWDEN 1972, NORRIS and FOWDEN 1973, LEA and NORRIS 1977) that amino acid analogues in many higher plants might have evolutionary importance in the sense of constraining certain Aa-RS to discriminate the normal amino acid from the non-proteinaceous counterpart and thus preventing possible toxic effects as a result of their incorporation into cell proteins.

### 3.4.3 ATP

ATP is an essential substrate for Aa-RS catalysis as a supply of energy and as an intermediate. None of the other nucleoside triphosphates can replace it, although deoxy-ATP was found to be utilized to a certain degree (PETERSON and FOWDEN 1965, JAKUBOWSKI and PAWELKIEWICZ 1975). Apart from the observations that ATP can protect synthetases against thermal inactivation (cf. Sect. 3.3.3), the enzymes bind modified substrates and inhibitors such as adenosine and adenosine analogues at their ATP sites in the $PP_i$ exchange reaction (LAWRENCE et al. 1974). Substitution in position 8 of the purine ring yielded highly effective inhibitors.

Little is known in this respect on ATP-site properties of plant synthetases. After KRAUSPE and PARTHIER (1975) had demonstrated differences in the effects of ATP and other purine nucleoside triphosphates on acitivity and stability of *Euglena* chloroplastic and cytoplasmic Leu-RS, KRAUSPE et al. (1978) prepared ATP analogues containing alkylated or phosphorylated groups. Most of them irreversibly inhibited cytoplasmic Leu-RS but had no effect on chloroplast Leu-RS. The effect of adenosine-5′-chloro-ethylphosphate was directed to the ATP site, whereas adenosine-5′-mesylcarbonyldiphosphate interacted with the binding site for the 3′terminal adenosine of the tRNA.

### 3.5 Biosynthesis of Synthetases

When dark-grown *Euglena* cells were illuminated and aminoacylation was measured with tRNA from green cells or chloroplasts during the greening process, increases in Aa-RS activities were first observed by REGER et al. (1970); these findings were confirmed by several authors e.g., PARTHIER et al. (1972), HECKER et al. (1974), BARNETT et al. (1976). The rates of increase for individual Aa-RS differed considerably (PARTHIER et al. 1972, PARTHIER 1973, KRAUSPE and PARTHIER 1973, 1974). It was suggested that at least the majority of Aa-RS species showing severalfold increased activity were of chloroplastic origin, since aminoacylation of cytoplasmic tRNA was not enhanced. Light induction experiments using hydroxyapatite separation of crude enzyme preparations into organelle and cytoplasmic isoenzymes supported this assumption as well as studies with plastid mutants of *Euglena gracilis,* indicating nuclear coding sites for synthetases (REGER et al. 1970, HECKER et al. 1974, PARTHIER and KRAUSPE 1975, PARTHIER and NEUMANN 1977, however, cf. Sect. 3.4.1).

Addition of chloramphenicol or nalidixic acid, an inhibitor of prokaryotic DNA replication, resulted in about 50% inhibition of light-induced aminoacylation activity for almost all chloroplastic Aa-RS investigated so far, after 48 h of illumination. Since low concentrations of cycloheximide stimulated Aa-tRNA formation above the values with non-treated cells, the results were first misinterpreted to mean that chloroplasts might be the sites of synthesis for plastidic Aa-RS (PARTHIER et al. 1972). Again using inhibitors of gene expression, PARTHIER (1973) and HECKER et al. (1974) independently demonstrated that organelle Aa-RS are synthesized in the cytoplasmic compartment. Short-term treatments with cycloheximide at any time during the greening phase resulted in the instantaneous block of plastid Leu-RS increase (Fig. 10), whereas the partial inhibition by chloramphenicol at a later stage suggests indirect or secondary effects of this drug.

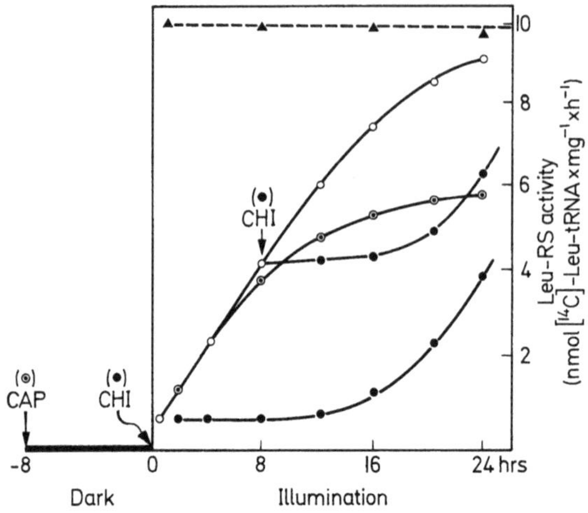


**Fig. 10.** Effects of chloramphenicol and cycloheximide on the level of plastid-specific leucyl-tRNA synthetase during light-induced chloroplast formation in *Euglena gracilis*. (PARTHIER 1973, PARTHIER and KRAUSPE 1975, modified and combined). The inhibitors were applicated at the times indicated by *arrows*. Leu-RS activities were determined in the 100,000 *g* supernatants of the homogenized cells after sonication using *Anacystis* tRNA or *Euglena* cytoplasmic tRNA. ○ control (without inhibitor); ● 8 μg/ml cycloheximide; ⊙ 1 mg/ml D-*threo*-chloramphenicol; ▲ control, cytoplasmic Leu-RS

It was also shown that cycloheximide in low doses lost its action in *Euglena* cells under lengthy experimental conditions (PARTHIER 1973, SPARE et al. 1978). Since cell division is completely interrupted by low amounts of cycloheximide, but synthesis of cytoplasmic proteins connected with chloroplast formation is inhibited at higher concentrations and thus recovers earlier from inhibition, a larger number of chloroplasts per cell was observed concomitantly with a higher level of chloroplastic Aa-RS activity, in comparison with non-treated controls (NEUMANN and PARTHIER 1973). Similarly, chloroplastic Aa-RS activity levels in illuminated *Euglena* increased much more when the cells were grown in carbohydrate-free media ("resting media") in place of full growth media. The glucose component of the latter seemed to play a repressive regulatory role in the formation of chloroplast constitutents produced in the cytoplasm (PARTHIER 1981). Plastid transcription products may likewise control synthetase biosynthesis in *Euglena*, as was deduced from experiments with various inhibitors of gene expression (LESIEWICZ and HERSON 1975), although the authors doubted that available data would permit the unequivocal designation of the sites of transcription and translation of chloroplast Phe-RS (SPARE et al. 1978).

In order to circumvent critical objections against inhibitor experiments and against equalization of enzyme activity and biosynthesis, incorporation experiments either in vivo or in cell-free systems are required. De novo synthesis and negligible, if any, turnover of chloroplast Leu-RS in greening *Euglena* cells have been obtained with $D_2O$ density labelling and CsCl gradient centrifugation (NOVER 1976). With the same method GORE and WRAY (1978) demonstrated that obviously cytoplasmic Leu-RS in tobacco cell cultures was synthesized de novo. They also observed a balance between synthesis and degradation of the enzyme. This is in accordance with unaltered activities of cytoplasmic Leu-RS in light-induced *Euglena* (KRAUSPE and PARTHIER 1974) and differs from the chloroplast Leu-RS which showed no turnover in the light period (NOVER 1976). The latter enzyme seems to be metabolically stable even in darkened *Euglena* cells, where its activity is diluted out together with the redistribution of the plastids in the progeny cells (PARTHIER 1981).

In summary, there is good evidence that the Aa-RS isoenzymes of the plastids are encoded in the nucleus, synthesized on cytoplasmic polysomes and post-translationally compartmentalized within the chloroplasts. The mechanism of import is as unclear as the control of transcription or translation by light. Chloroplast synthetases seem to be metabolically more stable than their cytoplasmic counterparts, which may be synthesized within the physiological regime and by the control of a balanced turnover.

### 3.6 Synthetases and Developmental Processes

Alterations in isoacceptor tRNA species during the development of plant organs or cells (cf. Sect. 2.6, LEA and NORRIS 1977) need not necessarily involve changes in the amount or activity of respective synthetases. Nevertheless, a number of papers report increases of Aa-RS levels during developmental processes, e.g., seed germination and seedling growth, chloroplast differentiation, leaf senescence. In no reliable experiment has the appearance or disappearance of a certain Aa-RS species been observed, i.e. alterations are of quantitative nature. This raises two problems which are not always rigorously excluded: (1) Aa-RS increase may be part of a general activation of the whole gene expression apparatus upon internal or external induction; (2) rates of activity changes may reflect alteration of "reference units" during physiological processes.

*Seed Germination and Seedling Growth.* Specific activity increases of Aa-RS in crude enzyme extracts from germinating seeds and growing seedlings have been reported (HENSHALL and GOODWIN 1964, ANDERSON and FOWDEN 1969, NORRIS et al. 1973, TAO and KHAN 1974, KEDZIERSKI et al. 1979). These increases could be due to the activity of the plumule and radicle within the embryo, while those of cotyledons or endosperm remained constant or decreased during development (NORRIS et al. 1973). In developing pea roots the activities of four synthetase species were compared in zones of cell division, elongation and maturation (COWLES and KEY 1973). Though variations in the individual enzyme activities were observed, they were as expected highest in the first and lowest in the third zone. In tobacco (*Nicotiana tabacum*) cell cultures WRAY et al. (1974) measured an increased activity of most Aa-RS up to the mid-exponential phase of growth but then decreases occurred.

Aminoacylation enzyme and nucleotidyltransferase activities during early lupin seed germination may be limited by ATP deficiency (DZIEGIELEWSKI et al. 1979), since no inhibitor of the enzyme reaction could be detected. On the other hand, unidentified substances from seedlings inhibitory in aminoacylation have been described. A heat-stable, low molecular weight substance was extracted from etiolated pea hypocotyls or cotyledons, but lacking in the plumule. The inhibitor showed a UV maximum at 262 nm (pH 1–2) which shifted to 286 nm at pH 9–10. It interacted with Phe-tRNA formation competitively, quite independent of the source of enzyme. The inhibitor disappeared during growth and greening of the seedlings (PARTHIER 1968). HECKER and MÜLLER (1974) isolated a sugar-containing inhibitor of 2,000–4,000 molecular weight in germinating *Agrostemma* seeds, which blocked non-competitively pyrophosphate exchange and tRNA aminoacylation tested with several synthetases. In lupin (*Lupinus*) seeds a factor which stimulates Aa-tRNA formation with homologous synthetase and tRNA from *Lupinus* causes inhibition in a heterologous system using components from yeast or bacteria (BARTKOWIAK and RADLOWSKI 1977).

*Chloroplast Development.* Details of light-induced chloroplast formation in *Euglena* have been described in Section 3.5. During cotton seed development plastid-specific Aa-RS increased 4–16-fold (BRANTNER and DURE 1975), although this increase could not be a light-dependent phenomenon as it was observed in both etiolated and greening cotyledons. Similar results were obtained with chloroplastic Leu-RS in illuminated *Cucurbita* cotyledons. Little stimulation of the enzyme activity was observed during the greening process (KLYACHKO and PARTHIER 1980), but there was a remarkable enhancement of activity upon cytokinin addition.

*Hormonal Effects.* Several reports exist which demonstrate changes in Aa-RS activities in plant tissues or organs treated with hormones. Others were unable to find a correlation between hormone addition and response in aminoacylation. Leu-RS activity in senescing soyabean cotyledons did not change upon cytokinin treatment of the material, although two of the six isoacceptor $tRNAs^{Leu}$ were aminoacylated to a higher degree than the others (ANDERSON and CHERRY 1969, BICK et al. 1970), particularly those species presumed to represent cytoplasmic $tRNA^{Leu}$ (BICK and STREHLER 1971). Kinetin counteracts the decrease of total Aa-RS activities in ageing tobacco leaves (ANDERSON and ROWAN 1966). The marked decline in the specific activity of Leu-RS in detached tobacco leaves was not associated with a loss of acylation of any of the $tRNA^{Leu}$ isoacceptor species (NATHAN and RICHMOND 1974). In detached, but developing and greening pumpkin cotyledons, 6-benzyladenine caused a marked stimulation of both cytoplasmic and plastid Leu-RS in darkness and in light (KLYACHKO and PARTHIER 1980, PARTHIER et al. 1981). Thus, cytokinins might stimulate both plastid and cytoplasmic protein synthesis apparatus via gene expression (promotion of the nucleo-cytoplasmic compartment). However, one should keep in mind that these responses of synthetases or other constituents can be likewise explained in terms of secondary hormone effects, which are embedded in a sequence of biochemical and physiological events following primary hormone action.

Interestingly, BAZIN et al. (1975) observed sex-dependent differences in tRNA isoacceptors and corresponding synthetases for Leu, Val, Ser, Tyr. These enzymes, extracted from female flowers of *Mercurialis annua,* were more active than and showed characteristic differences to those of male flowers. Since cytokinin converted male flower primordia to female ones, the endogenous level of this hormone group might regulate the pattern of Aa-RS activities and isoacceptor tRNA's during the development of sexual organs in *Mercurialis*. This example is a first indication that phytohormones could influence the aminoacylation process qualitatively.

*Acknowledgement.* The authors wish to thank Dr. D.T.N. PILLAY (Department of Biology, University of Windsor, Canada) for his help in editing the manuscript.

# References


Aliev KA, Philippovich II (1968) Differences of tRNA and aminoacyl-tRNA synthetases of chloroplasts and cytoplasm from pea seedlings (russ) Mol Biol (USSR) 2:364–373

Anderson JW, Fowden L (1969) A study of the aminoacyl-sRNA synthetases of *Phaseolus vulgaris* in relation to germination. Plant Physiol 44:60–68

Anderson JW, Fowden L (1970a) Properties and substrate specificities of the phenylalanyl-tRNA synthetases of *Aesculus* species. Biochem J 119:677–690

Anderson JW, Fowden L (1970b) Properties and substrate specificity of the leucyl-, the threonyl- and the valyl-tRNA synthetases from *Aesculus* species. Biochem J 119:691–697

Anderson JW, Fowden L (1970c) 1-amino-2-phenylethane-1-phosphonic acid: a specific competitive inhibitor of phe-tRNA synthetase. Chem Biol Interact 2:53–55

Anderson JW, Rowan KS (1966) Activity of aminoacyl-ribonucleic acid synthetases in tobacco-leaf tissue in relation to senescence and to the action of 6-furfurylaminopurine. Biochem J 101:15–18

Anderson MB, Cherry JH (1969) Differences in leucyl-tRNAs and synthetases in soybean seedlings. Proc Natl Acad Sci USA 62:202–209

Augustyniak H, Pawelkiewicz J (1978a) Isolation and properties of the main isoleucine tRNAs from *Lupinus luteus* seeds. Acta Biochim Pol 25:81–89

Augustyniak H, Pawelkiewicz J (1978b) Lysyl-tRNAs from *Lupinus luteus* seeds. Phytochemistry 17:15–18

Augustyniak H, Pawelkiewicz J (1979) Preferential binding of isoaccepting species of $tRNA^{Lys}$ and $tRNA^{Ile}$ from lupin cotyledons to polyribosomes. Biochim Biophys Acta 565:148–153

Augustyniak H, Barciszewski H, Rafalski A, Zawielak J, Szyfter K (1974) Phenylalanine tRNA of *Lupinus luteus* seeds. Phytochemistry 13:2679–2684

Babcock DF, Morris RO (1973) Specific degradation of a plant leucyl-tRNA by a factor in the homologous synthetase preparation. Plant Physiol 52:292–297

Barciszewski J, Joachimiak A, Rafalski A, Barciszewska A, Twardowski T, Wiewiorowski M (1979) Conservation of the structure of plant tRNAs and aminoacyl-tRNA synthetases. FEBS Lett 102:194–197

Barnett WE, Pennington CJ, Fairfield SA (1969) Induction of *Euglena* transfer RNAs by light. Proc Natl Acad Sci USA 63:1261–1268

Barnett WE, Schwartzbach SD, Farelly JG, Schiff JA, Hecker LI (1976) Comments on the translational and transcriptional origin of *Euglena* chloroplastic aminoacyl-tRNA synthetases. Arch Microbiol 109:201–203

Barnett WE, Schwartzbach SD, Hecker LI (1978) The transfer RNAs of eukaryotic organelles. Progr Nucleic Acid Res Mol Biol 21:143–179

Barrell BG, Bankier AT, Drouin J (1979) A different genetic code in human mitochondria. Nature (London) 282:189–194

Bartkowiak S, Pawelkiewicz J (1972) The purification of aminoacyl-tRNA synthetases by affinity chromatography. Biochim Biophys Acta 272:137–140

Bartkowiak S, Radlowski M (1977) A factor affecting stimulation of aminoacylation in plants. Biochim Biophys Acta 474:619–628

Bazin M, Chabin A, Durand R (1975) Comparison between four isoaccepting transfer ribonucleic acids and corresponding synthetases in male and female flowers of the dioecious species *Mercurialis annua*. Dev Biol 44:288–297

Beauchemin N, Larne B, Cedergren RJ (1973) The characterization of the tRNAs and aminoacyl-tRNA synthetases of the blue-green alga, *Anacystis nidulans*. Arch Biochem Biophys 156:17–25

Bick MD, Strehler B (1971) Leucyl-transfer RNA synthetase changes during soybean cotyledon senescence. Proc Natl Acad Sci USA 68:224–228

Bick MD, Liebke H, Cherry JH, Strehler B (1970) Changes in leucyl- and tyrosyl-tRNA of soybean cotyledons during plant growth. Biochim Biophys Acta 204:175–182

Bohnert HJ, Driesel AJ, Crouse EJ, Gordon K, Herrmann RG, Steinmetz A, Mubumbila M, Keller M, Burkard G, Weil JH (1979) Presence of a tRNA gene in the spacer sequence between the 16S and 23S rRNA genes of spinach chloroplast DNA. FEBS Lett 103:52–56

Bonen L, Gray MW (1980) Organization and expression of the mitochondrial genome of plants. I. The genes for wheat mitochondrial ribosomal and transfer RNAs: evidence for an unusual arrangement. Nucleic Acids Res 8:319–335

Bonen L, Huh TY, Gray MW (1980) Can partial methylation explain the complex fragment patterns observed when plant mitochondrial DNA is cleaved with restriction endonucleases? FEBS Lett 111:340–346

Brantner JH, Dure LS (1975) The developmental biochemistry of cotton seed embryogenesis and germination VI. Levels of cytosol and chloroplast aminoacyl-tRNA synthetases during cotyledon development. Biochim Biophys Acta 414:99–114

Burkard G, Eclancher B, Weil JH (1969) Presence of N-formyl-methionyl-transfer RNA in bean chloroplasts. FEBS Lett 4:285–287

Burkard G, Guillemaut P, Weil JH (1970) Comparative studies of the tRNAs and the aminoacyl-tRNA synthetases from the cytoplasm and chloroplasts of *Phaseolus vulgaris*. Biochim Biophys Acta 224:184–198

Burkard G, Steinmetz A, Keller M, Mubumbila M, Crouse E, Weil JH (1980) Resolution of chloroplast tRNAs by two-dimensional gel electrophoresis. In: Edelman M, Hallick RB, Chua NH (eds) Methods in chloroplast molecular biology. Elsevier/North-Holland, (Amsterdam New York, in press

Burnell JN, Shrift A (1977) Cystein-tRNA synthetase from *Phaseolus aureus*. Purification and properties. Plant Physiol 60:670–674

Burnell JN, Shrift A (1979) Cysteinyl-tRNA synthetase from *Astragalus* species. Plant Physiol 63:1095–1097

Calagan JL, Pirtle RM, Pirtle IL, Kashdan MA, Vreman HJ, Dudock BS (1980) Homology between a chloroplast and a prokaryotic initiator tRNAs: Nucleotide sequence of spinach chloroplast methionine initiator tRNA. J Biol Chem 255:9981–9984

Canaday J, Guillemaut P, Weil JH (1980a) The nucleotide sequences of initiator transfer RNAs from bean cytoplasm and chloroplasts. Nucleic Acids Res 8:999–1008

Canaday J, Guillemaut P, Gloeckler R, Weil JH (1980b) Comparison of the nucleotide sequences of chloroplast $tRNAs^{Phe}$ and $tRNAs^{Leu}$ from spinach and bean. Plant Sci Lett 20:57–62

Carias JR, Julien R (1976) Phénylalanyl-tRNA synthétase des embryons de blé. Purification, masse moléculaire, structure, propriétés. Biochimie 58:253–259

Carias JR, Mouricout M, Quintard B, Thomes JC Julien R (1978) Leucyl-tRNA and arginyl-tRNA synthetases of wheat germ. Inactivation and ribosome effects. Eur J Biochem 87:583–590

Chang SH, Hecker LI, Silberklang M, Brum CK, RajBhandary UL, Barnett WE (1976) Nucleotide sequence of phenylalanine transfer RNA from the chloroplasts of *Euglena gracilis*. Cell 9:717–724

Chang SH, Brum CK, Schnabel JJ, Heckman JI, RajBhandary UL, Barnett WE (1978) Similarities in nucleotide sequence between *Euglena* gracilis and mammalian cytoplasmic phenylalanine tRNAs. Fed Proc 37:1768

Chang SH, Lin FK, Hecker LI, Heckman JE, RajBhandary UL, Barnett WE, (1979) Nucleotide sequence of blue-green algae phenylalanine tRNA. In: Schimmel P, Söll D, Abelson J (eds) Transfer RNA: Biological aspects, Cold Spring Harbor meeting on tRNA, abstracts p 45

Chazal P, Thomes JC, Julien R (1975) Methionine-tRNA-ligase from wheat germ: Purification and properties. FEBS Lett 56:268–272

Chazal P, Thomes JC, Julien R (1977) Methionyl-tRNA synthétase des embryons de blé: dissociation en sous-unités. Eur J Biochem 73:607–615

Cherry JH, Osborne DJ (1970) Specificity of leucyl-tRNA and synthetase in plants. Biochem Biophys Res Commun 40:763–769
Chiu N, Chiu A, Suyama Y (1975) Native and imported tRNA in mitochondria. J Mol Biol 99:37–50
Clarkson SG, Birnstiel ML, Serra V (1973) Reitereated transfer RNA genes of *Xenopus laevis*. J Mol Biol 79:391–410
Commerford SL (1971) Iodination of nucleic acids in vitro. Biochemistry 10:1993–1999
Cornelis P, Claessen E, Claessen J (1975) Reversed phase chromatography of isoaccepting tRNAs from healthy and crown-gall tissues from *Nicotiana tabacum*. Nucleic Acids Res 2:1153–1161
Cowles JR, Key JL (1972) Demonstration of two tyrosyl-tRNA synthetases of pea roots. Biochim Biophys Acta 281:33–44
Cowles JR, Key JL (1973) Changes in certain aminoacyl transfer ribonucleic acid synthetase activities in developing pea roots. Plant Physiol 51:22–25
Crick FHC (1966) Codon-Anticodon pairing: The wobble hypothesis. J Mol Biol 19:548–555
Dang CV, Yang DCH (1979) Disassembly and gross structure of particulate aminoacyl-tRNA synthetases from rat liver. J Biol Chem 254:5350–5356
Driesel AJ, Crouse EJ, Gordon K, Bohnert HJ, Herrmann RG, Steinmetz A, Mubumbila M, Keller M, Burkard G, Weil JH (1979) Fractionation and identification of spinach chloroplast tRNAs and mapping of their genes on the restriction map of chloroplast DNA. Gene 6:285–306
Dudock BS, Katz G, Taylor EK, Holley RW (1969) Primary structure of wheat germ phenylalanine transfer RNA. Proc Natl Acad Sci USA 62:941–945
Dziegielewski T, Kedzierski W, Pawelkiewiez J (1979) Levels of aminoacyl-tRNA synthetases, tRNA nucleotidyltransferase and ATP in germinating lupin seeds. Biochim Biophys Acta 564:37–42
Ehresmann B, Imbault P, Weil JH (1974) Determination of the degree of in vivo tRNA aminoacylation in yeast cells. Anal Biochem 61:548–556
England TE, Uhlenbeck OC (1978) 3′-terminal labelling of RNA with $T_4$ RNA ligase. Nature (London) 275:560–561
Everett GA, Madison JT (1976) Nucleotide sequence of phenylalanine transfer ribonucleic acid from pea (*Pisum sativum*, Alaska). Biochemistry 15:1016–1021
Fowden L, Frankton JB (1968) The specificity of aminoacyl-sRNA synthetases with special reference to arginine activation. Phytochemistry 7:1077–1086
Garel JP (1974) Functional adaptation of tRNA population. J Theor Biol 43:211–228
Gauss DH, Sprinzl M (1981) Compilation of tRNA sequences. Nucl Acids Res 9:r1–r23
Ghosh K, Ghosh HP, Simsek M, RajBhandary UL (1974) Initiator methionine transfer ribonucleic acid from wheat embryo. Purification, properties and partial nucleotide sequence. J Biol Chem 249:4720–2729
Ghosh RP, Ghosh K, Simsek M, RajBhandary UL (1978) Primary sequence of wheat germ initiator $tRNA_i^{Met}$. Cold Spring Harbor meeting on tRNA, abstracts p 6
Gillespie D, Spiegelman S (1965) A quantitative assay for DNA-RNA hybrids with DNA immobilized on a membrane. J Mol Biol 12:829–842
Goins DJ, Reynolds RJ, Schiff JA, Barnett WE (1973) A cytoplasmic regulatory mutant of *Euglena*. Constitutivity for the light inducible chloroplast transfer RNAs. Proc Natl Acad Sci USA 70:1749–1752
Gore NR, Wray JL (1978) Leucine-tRNA ligase from cultured cells of *Nicotiana tabacum* var. Xanthi. Plant Physiol 61:20–24
Graf L, Kössel H, Stutz E (1980) Sequencing of 16S–23S spacer in a ribosomal RNA operon of *Euglena gracilis* chloroplast reveals two tRNA genes. Nature (London) 286:908–910
Gruol DJ, Haselkorn R (1976) Counting the genes for stable RNA in the nucleus and chloroplasts of *Euglena*. Biochim Biophys Acta 447:82–95
Guderian RH, Pulliam RL, Gordon MP (1972) Characterization and fractionation of tobacco leaf transfer RNA. Biochim Biophys Acta 262:50–65

Guillemaut P, Keith G (1977) Primary structure of bean chloroplast $tRNA^{Phe}$: comparison with *Euglena* chloroplast $tRNA^{Phe}$. FEBS Lett 84:351–356
Guillemaut P, Weil JH (1975) Aminoacylation of *Phaseolus vulgaris* cytoplasmic, chloroplastic and mitochondrial $tRNAs^{Met}$ and E. coli $tRNAs^{Met}$ by homologous and heterologous enzymes. Biochim Biophys Acta 407:240–248
Guillemaut P, Burkard G, Weil JH (1972) Characterization of N-formyl-methionyl-tRNA in bean mitochondria and etioplasts. Phytochemistry 11:2217–2219
Guillemaut P, Burkard G, Steinmetz A, Weil JH (1973) Comparative studies on the $tRNAs^{Met}$ from the cytoplasm, chloroplasts and mitochondria of *Phaseolus vulgaris*. Plant Sci Lett 1:141–149
Guillemaut P, Steinmetz A, Burkard G, Weil JH (1975) Aminoacylation of $tRNA^{Leu}$ species from *E. coli* and from the cytoplasm, chloroplasts and mitochondria of *Phaseolus vulgaris* by homologous and heterologous enzymes. Biochim Biophys Acta 378:64–72
Guillemaut P, Martin R, Weil JH (1976) Purification and base composition of a chloroplastic $tRNA^{Phe}$ from *Phaseolus vulgaris*. FEBS Lett 63:273–277
Gulewicz K, Twardowski T (1980) Purification of $tRNA_2^{Met}$ from yellow lupin seeds and characterization by high pressure liquid chromatography. Bull Acad Pol Sci (in press)
Gusseck DJ (1977) On dealing with anomalies in the transfer RNA aminoacylation reaction in partially purified systems. Arch Biochem Biophys 182:533–539
Haff HA, Bogorad L (1976) Hybridization of maize chloroplast DNA with transfer ribonucleic acids. Biochemistry 15:4105–4109
Hague DR, Kofold EC (1971) The coding properties of lysine accepting transfer ribonucleic acids from black-eyed peas. Plant Physiol 48:305–311
Hall TC, Tao KL (1970) Rates of aminoacyl-transfer ribonucleic acid synthesis in vivo and in vitro by bean leaves. Biochem J 177:853–859
Hallick RB, Gray PW, Chelm BK, Rushlow KE, Orozco EM (1978) *Euglena gracilis* chloroplast DNA structure, gene mapping and RNA transcription, In: Akoyunoglou G, Argyroudi-Akoyunoglou JH (eds) Chloroplast development. Elsevier/North-Holland, Amsterdam New York pp 619–622
Hecker LI, Egan J, Reynolds RJ, Nix CE, Schiff JA, Barnett WE (1974) The sites of transcription and translation for *Euglena* chloroplastic aminoacyl-tRNA synthetases. Proc Natl Acad Sci USA 71:1910–1914
Hecker M, Müller H (1974) Untersuchungen über das Verhalten eines während der Keimung von *Agrostemma*-Samen gebildeten Inhibitors der Aminoacyl-tRNS-Synthetase. Biochem Physiol Pflanz 165:419–428
Henshall JD, Goodwin TW (1964) Amino acid-activating enzymes in germinating pea seedlings. Phytochemistry 3:677–691
Hiatt VS, Snyder LA (1973) Phenylalanine transfer RNA species in early development of barley. Biochim Biophys Acta 324:57–68
Holley R, Apgar J, Everett GA, Madison JT, Marquisee M, Merrill SH, Penswick J, Zamir A (1965) Structure of a ribonucleic acid. Science 147:1462–1465
Holmes WM, Hurd RE, Reid BR, Rimerman RA, Hatfield GW (1975) Separation of transfer ribonucleic acid by Sepharose chromatography using reverse salt gradient. Proc Natl Acad Sci USA 72:1068–1071
Imbault P, Sarantoglou V, Weil JH (1979) Purification of the chloroplastic valyl-tRNA synthetase from *Euglena gracilis*. Biochem Biophys Res Commun 88:75–84
Imbault P, Colas B, Sarantoglou V, Boulanger Y, Weil JH (1981) Chloroplast leucyl-tRNA synthetase from *Euglena gracilis*. Purification, kinetic analysis and structural characterization. Biochemistry 20:5855–5859
Jacobson KB (1971) Reaction of aminoacyl-tRNA synthetases with heterologous tRNA's. Prog Nucleic Acid Res Mol Biol 11:461–488
Jakubowski H (1978a) Yellow lupin (*Lupinus luteus*) aminoacyl-tRNA synthetases. Isolation and some properties of enzyme-bound valyl adenylate and seryl adenylate. Biochim Biophys Acta 521:584–596
Jakubowski H (1978b) Valyl-tRNA synthetase from yellow lupin seeds. Instability of enzyme-bound noncognate adenylases versus cognate adenylate. FEBS Lett 95:235–238
Jakubowski H (1979) A role for protein-protein interactions in the maintenance of active forms of aminoacyl-tRNA synthetases. FEBS Lett 103:71–76

Jakubowski H (1980) Polyamines and yellow lupin aminoacyl-tRNA synthetases. FEBS Lett 109:63–66

Jakubowski H, Pawelkiewicz J (1973) Chromatography of plant aminoacyl-tRNA synthetases on ω-aminoalkyl Sepharose columns. FEBS Lett 34:150–154

Jakubowski H, Pawelkiewicz J (1974) Valyl-tRNA synthetase of yellow lupin seeds. Purification and some properties. Acta Biochim Pol 21:271–282

Jakubowski H, Pawelkiewicz J (1975) The plant aminoacyl-tRNA synthetases. Purification and characterization of valyl-tRNA, tryptophanyl-tRNA and seryl-tRNA synthetases from yellow lupin seeds. Eur J Biochem 52:301–310

Janowicz Z, Wower JM, Augustyniak J (1979) Primary structure of barley embryo $tRNA^{Phe}$ and its identity with wheat germ $tRNA^{Phe}$. Plant Sci Lett 14:177–183

Jay FT, Jones DS (1977) Transfer ribonucleic acid from *Scenedesmus obliquus*: Purification of the major formylatable methionine-accepting species. Phytochemistry 16:1329–1332

Jeannin G, Burkard G, Weil JH (1976) Aminoacylation of *Phaseolus vulgaris* cytoplasmic, chloroplastic and mitochondrial $tRNAs^{Pro}$ and $tRNAs^{Lys}$ by homologous and heterologous enzymes. Biochim Biophys Acta 442:24–31

Jeannin G, Burkard G, Weil JH (1978) Characterization of *Phaseolus vulgaris* cytoplasmic, chloroplastic and mitochondrial $tRNAs^{Phe}$. Aminoacylation by homologous and heterologous enzymes. Plant Sci Lett 13:75–81

Joachimiak A, Barciszewski J, Twardowski T, Barciszewska M, Wiewiorowski M (1978) Purification and properties of methionyl-tRNA synthetase from yellow lupin seeds. FEBS Lett 93:51–54

Kanabus J, Cherry JH (1971) Isolation of an organ-specific leucyl-tRNA synthetase from soybean seedlings. Proc Natl Acad Sci USA 68:873–876

Karwowska U, Gozdzicka-Josefiak A, Augustyniak J (1979) Chloroplast-specific leucine tRNAs from wheat. Acta Biochim Pol 26:319–326

Kashdan MA, Pirtle RM, Pirtle IL, Calagan JL, Vreman HJ, Dudock BS (1980) Nucleotide sequence of a spinach chloroplast threonine tRNA. J Biol Chem 255:8831–8835

Kedzierski W, Pawelkiewicz J (1970) Stabilization of isoleucyl-tRNA synthetase from yellow lupin seeds by transfer RNA. Acta Biochim Pol 17:41–51

Kedzierski W, Sulewski T, Pawelkiewicz J (1979) Levels of aminoacylation of methionine tRNAs in germinating lupin cotyledons. Plant Sci Lett 14:373–380

Kedzierski W, Augustyniak H, Pawelkiewicz J (1980) Aminoacylation of four tRNA species in lupin cotyledons. Planta 147:439–443

Keller M, Burkard G, Bohnert HJ, Mubumbila M, Gordon K, Steinmetz A, Heiser D, Crouse E, Weil JH (1980) Transfer RNA genes associated with the 16S and 23S rRNA genes of *Euglena* chloroplast DNA. Biochem Biophys Res Commun 95:47–54

Kellermann O, Brevet A, Tonetti H, Waller J-P (1979) Macromolecular complexes of aminoacyl-tRNA synthetases from eukaryotes. Eur J Biochem 99:541–550

Kelmers AD, Heatherly DE (1971) Columns for rapid chromatographic separation of small amounts of tracer-labeled transfer ribonucleic acids. Anal Biochem 44:486–495

Kislev N, Selsky MI, Norton C, Eisenstadt JM (1972) tRNA and aminoacyl-tRNA synthetases of chloroplasts, mitochondria and cytoplasm from *Euglena gracilis*. Biochim Biophys Acta 287:256–269

Kisselev LL, Favorova OO (1974) Aminoacyl-tRNA synthetases, some recent results and achievements. Adv Enzymol 40:141–225

Klyachko NL, Parthier B (1980) Cytokinin control of aminoacyl-tRNA synthetases and ribulose-bisphosphate carboxylase in developing and greening excised *Cucurbita* cotyledons. Biochem Physiol Pflanz 175:333–345

Koch W, Edwards K, Kössel H (1981) Sequencing of the 16S–23S spacer in a ribosomal RNA operon of *Zea mays* chloroplast DNA reveals two split tRNA genes. Cell 25:203–213

Kothari RM, Taylor MW (1972) RNA fractionation on modified celluloses. III. BD-cellulose. J Chromatogr 73:479

Kothari RM, Taylor MW (1973) RNA fractionation on reversed phase columns. J Chromatogr 86:289–324

Krauspe R, Parthier B (1973) Chloroplast- and cytoplasmic-specific aminoacyl-transfer ribonucleic acid synthetases of *Euglena gracilis*: Separation, characterization and site of synthesis. Biochem Soc Symp 38:111–135

Krauspe R, Parthier B (1974) Chloroplast and cytoplasmic aminoacyl-tRNA synthetases of *Euglena gracilis*. Biochem Physiol Pflanz 165:18–36

Krauspe R, Parthier B (1975) Influence of purine nuclectides on the heat-stability of aminoacyl-tRNA synthetases. Biochem Physiol Pflanz 168:257–266

Krauspe R, Kovaleva GK, Gulyaev NN, Baranova LA, Agalova MB, Severin ES, Sokolova NI, Shabarova ZA, Kisselev LL (1978) Inhibition of leucyl-tRNA synthetases by modifying ATP analogs. Biokhimiya 43:656–661

Labuda D, Janowicz Z, Haertle T, Augustyniak J (1974) Isolation and chromatographic behaviour of phenylalanine tRNA from barley embryos. Nucleic Acids Res 1:1703–1712

Lagerkvist U (1978) "Two out of three": An alternative method for codon reading. Proc Natl Acad Sci USA 75:1759–1762

Lanzani GA, Manzocchi A, Galante E, Menegus F (1969) Some properties of the phenylalanyl tRNA synthetase activity from wheat seedling chloroplasts. Enzymologia 37:97–110

Lawrence F, Shire D, Waller J-P (1974) The effect of adenosine analogues on the ATP-pyrophosphate exchange reaction catalysed by methionyl-tRNA synthetase. Eur J Biochem 41:73–81

Lea PJ, Fowden L (1972) Stereospecificity of glutamyl-tRNA synthetase isolated from higher plants. Phytochemistry 11:2129–2139

Lea PJ, Fowden L (1973) Amino acid substrate specificity of asparaginyl-, aspartyl- and glutamyl-tRNA synthetase isolated from higher plants. Phytochemistry 12:1903–1916

Lea PJ, Norris RD (1976) The use of amino acid analogues in study on plant metabolism. Phytochemistry 15:585–595

Lea PJ, Norris RD (1977) tRNA and aminoacyl-tRNA synthetases from higher plants. Prog Phytochem 4:121–167

Legocki AB, Pawelkiewicz J (1967) Amino acid-activating enzymes in yellow lupin seeds, and purification of leucyl-sRNA synthetase. Acta Biochim Pol 14:313–322

Legocki AB, Szymkowiak A, Hierowski M, Pawelkiewicz J (1968) Heterogeneity of transfer ribonucleic acids from yellow lupin seeds. Acta Biochim Pol 15:197–203

Leis JP, Keller EB (1970) Protein chain-initiating methionine tRNAs in chloroplasts and cytoplasm of wheat leaves. Proc Natl Acad Sci USA 67:1593–1599

Lesiewicz JL, Herson DS (1975) A reinvestigation of the sites of transcription and translation of *Euglena* chloroplastic phenylalanyl-tRNA synthetase. Arch Microbiol 105:117–121

Lester BR, Cherry JH (1979) Purification of leucine tRNA isoaccepting species from soybean cotyledons. I. Benzoylated diethylaminocellulose fractionation, N-hydroxysuccinimide modification and characterization of product. Plant Physiol 63:79–86

Lester BR, Morris RO, Cherry JH (1979) Purification of leucine tRNA isoaccepting species from soybean cotyledons. II. RPC-II purification, ribosome binding and cytokinin content. Plant Physiol 63:87–92

Littauer UZ, Inouye H (1973) Regulation of tRNA. Annu Rev Biochem 42:439–470

Locy RO, Cherry JH (1978) Purification and characterization of two tyrosyl-tRNA synthetase activities from soybean cotyledons. Phytochemistry 17:19–27

Macino G, Coruzzi G, Nobrega FG, Li M, Tzagoloff A (1979) Use of the UGA terminator as a tryptophan codon in yeast mitochondria. Proc Natl Acad Sci USA 76:3784–3785

Malnoë P, Rochaix JD (1978) Localization of 4S RNA genes on the chloroplast genome of *Chlamydomonas reinhardii*. Mol Gen Genet 166:269–275

Marcu KB, Mignery RE, Dudock BS (1977) Complete nucleotide sequence and properties of the major species of glycine tRNA from wheat germ. Biochemistry 16:797–806

Martin RP, Schneller JM, Stahl AJC, Dirheimer G (1977) Studies of yeast mitochondrial tRNAs by two-dimensional polyacrylamide gel electrophoresis: characterization of isoaccepting species and search for imported cytoplasmic tRNAs. Nucleic Acids Res 4:3497–3510

McCoy JG, Jones DS (1980) The nucleotide sequence of *Scenedesmus obliquus* chloroplast $tRNA_f^{Met}$. Nucl Acids Res 8:5089–5093

McCrea JM, Hershberger CL (1976) Chloroplast DNA codes for transfer RNA. Nucleic Acids Res 3:2005–2018

McCrea JM, Hershberger CL (1978) Chloroplast DNA codes for tRNA from cytoplasmic polyribosomes. Nature (London) 274:717–719

McCune SA, Yu PL, Nance WE (1977) A semiautomated assay procedure for the determination of aminoacyl-tRNA synthetase activity. Anal Biochem 79:618–622
Mehler AH (1970) Induced activation of amino acid activating enzymes by amino acids and tRNA. Prog Nucleic Acids Res Mol Biol 10:1–23
Merrick WC, Dure LS (1972) The developmental biochemistry of cotton seed embryogenesis and germination. IV. Levels of cytoplasmic and chloroplastic transfer ribonucleic acid species. J Biol Chem 247:7988–7999
Mettler IJ, Romani RJ (1976) Quantitative changes in tRNA during ethylene-induced ripening (ageing) of tomato fruits. Phytochemistry 15:25–28
Mubumbila M, Burkard G, Keller M, Steinmetz A, Crouse E, Weil JH (1980) Hybridization of bean, spinach, maize and *Euglena* chloroplast tRNAs with homologous and various species. Biochim Biophys Acta 609:31–39
Müller-Uri F, Krauspe R, Parthier B (1981) Mitochondrial leucyl-tRNA-synthetase of *Euglena gracilis*. Biochem Physiol Pflanz 176:841–851
Nathan I, Richmond A (1974) Leucyl transfer RNA synthetase in senescing tobacco leaves. Biochem J 140:169–173
Neumann D, Parthier B (1973) Effects of nalidixic acid, chloramphenicol, cycloheximide, and anisomycin on structure and development of plastids and mitochondria in greening *Euglena* cells. Exp Cell Res 81:255–268
Norris RD, Fowden L (1972) Substrate discrimination by prolyl-tRNA synthetase from various higher plants. Phytochemistry 11:2921–2935
Norris RD, Fowden L (1973) Substrate protection during selective heat inactivation of aminoacyl-tRNA synthetases and its use in enzyme studies. Biochim Biophys Acta 312:695–707
Norris RD, Fowden L (1974) Cold-lability of prelyl-tRNA synthetase from higher plants. Phytochemistry 13:1677–1687
Norris RD, Lea PJ, Fowden L (1973) Aminoacyl-tRNA synthetases in *Triticum aestivum* L. during seed development and germination. J Exp Bot 24:615–625
Norris RD, Lea PJ, Fowden L (1975a) tRNA species in the developing grain of *Triticum aestivum*. Phytochemistry 14:1683–1686
Norris RD, Watson R, Fowden L (1975b) The activation of amino acid analogues by phenylalanyl- and tyrosyl-tRNA synthetases from plants. Phytochemistry 14:393–396
Nover L (1976) Density labeling of chloroplast-specific leucyl-tRNA synthetase in greening cells of *Euglena gracilis*. Plant Sci Lett 7:403–407
Olins PO, Jones DS (1980) Nucleotide sequence of *Scenedesmus obliquus* cytoplasmic initiator tRNA. Nucleic Acid Res 8:715–730
Orozco EM, Rushlow KE, Dodd JR, Hallick RB (1980) *Euglena gracilis* chloroplast ribosomal RNA transcription units. II. Nucleotide sequence homology between the 16S–23S ribosomal RNA spacer and the 16S ribosomal leader regions. J Biol Chem 255:10997–11001
Osorio-Almeida ML, Guillemaut P, Keith G, Canaday J, Weil JH (1980) Primary structure of three leucine transfer RNAs from bean chloroplast. Biochem Biophys Res Commun 92:102–108
Parfait R (1973) Arginyl-tRNA synthetase from *Bacillus stearothermophilus*. Heat inactivation and substrate induced protection. FEBS Lett 29:323–325
Parthier B (1968) Spezifischer Inhibitor der Aminoacyl-Transfer-RNS-Synthese aus Erbsenkeimlingen. Naturwissenschaften 55:653
Parthier B (1971) Species-specific reactions of cell-free polypeptide synthesis. Biochem Physiol Pflanz 162:45–59
Parthier B (1972) Sites of synthesis of chloroplast proteins. Symp Biol Hung 13:235–248
Parthier B (1973) Cytoplasmic site of synthesis of chloroplast aminoacyl-tRNA synthetase in *Euglena gracilis*. FEBS Lett 38:70–74
Parthier B (1977) Light-induced chloroplast differentiation in *Euglena gracilis* In: Nover L, Mothes K (eds) Cell differentiation in microorganisms, higher plants and animals. Fischer, Jena and Elsevier, Amsterdam, pp 602–624
Parthier B (1981) Chloroplast development in *Euglena*: Regulatory aspects. In: Levandow-

sky M, Hutner SH (eds) Biochemistry and physiology of protozoa, vol IV. Academic Press, London New York, pp 261–300
Parthier B, Krauspe R (1973) Assignment to chloroplast and cytoplasm of three *Euglena gracilis* aminoacyl-tRNA synthetases with ambiguous specificity for transfer RNA. Plant Sci Lett 1:221–227
Parthier B, Krauspe R (1974) Chloroplast and cytoplasmic transfer RNA of *Euglena gracilis*. Transfer $RNA^{Leu}$ of blue-green algae as a substitute for chloroplast $tRNA^{Leu}$. Biochem Physiol Pflanz 165:1–17
Parthier B, Krauspe R (1975) Specificity and synthesis of plastid-specific aminoacyl-tRNA synthetase in *Euglena gracilis*. Colloq Int CNRS 240:233–239
Parthier B, Neumann D, (1977) Structural and functional analysis of some plastid mutants of *Euglena gracilis*. Biochem Physiol Pflanz 171:547–560
Parthier B, Krauspe R, Samtleben S (1972) Light-stimulated synthesis of aminoacyl-tRNA synthetases in greening *Euglena gracilis*. Biochim Biophys Acta 277:335–341
Parthier B, Mueller-Uri F, Krauspe R (1978) The aminoacyl-tRNA synthetases of *Euglena* chloroplasts. In: Akoyunoglou G, Argyroudi-Akoyunoglou JH (eds) Chloroplast development. Elsevier North-Holland Biomed Press, Amsterdam New York, pp 687–693
Parthier B, Lerbs S, Klyachko NL (1981) Plastogenesis and cytokinin action. In: Péaud-Lenoël C, Guern J (eds) Metabolism and molecular activities of cytokinins. Springer, Berlin Heidelberg New York, 275–286
Patel, HV, Pillay DTN (1976) Leucine specific transfer ribonucleic acids and synthetases in the cotyledons of mature and germinating pea seeds. Phytochemistry 15:401–405
Pearson RL, Weiss JF, Kelmers AD (1971) Improved separation of transfer RNAs on polychlorotrifluorethylene-supported reversed phase chromatography column. Biochim Biophys Acta 228:770–774
Peterson PJ, Fowden L (1965) Purification, properties and comparative specificities of the enzyme prolyl-transfer ribonucleic acid synthetase from *Phaseolus aureus* and *Polygonatum multiflorum*. Biochem J 97:112–124
Pillay DTN, Cherry JH (1974) Changes in leucyl-, seryl-, and tyrosyl-tRNAs in ageing soybean cotyledons. Can J Bot 52:2499–2504
Pillay DTN, Gowda S (1980) Gerontology (in press)
Preddie DL, Preddie EC, Guerrini AM, Cremona T (1973) Two isoaccepting species of tryptophan-tRNA from *Chlamydomonas reinhardi*. Can J Bot 51:951
Quetier F, Vedel F (1977) Heterogeneous population of mitochondrial DNA molecules in higher plants. Nature (London) 268:365–368
Quintard B, Monricout M, Carias JF, Julien R (1978) Occurrence of aminoacyl-tRNA synthetase complexes in quiescent wheat germ. Biochem Biophys Res Commun 85:999–1006
Racz I, Juhasz A, Kiraly I, Lasztity D (1979) The effect of hight on the nucleotide composition of $tRNA^{Phe}$ of wheat germ. Plant Sci Lett 15:57–61
Rafalski A, Barciszewski J, Gulewicz K, Twardowski T, Keith G (1977) Nucleotide sequence of $tRNA^{Phe}$ from the seeds of lupin. Comparison of the major species with wheat germ $tRNA^{Phe}$. Acta Biochim Pol 24:301–318
Reger BJ, Fairfield SA, Epler JL, Barnett WE (1970) Identification and origin of some chloroplast aminoacyl-tRNA synthetases and tRNAs. Proc Natl Acad Sci USA 67:1207–1213
Romani RJ, Sprole BV, Mettler IJ, Tuskes ES (1975) Extraction and purification of tRNA from fruit tissues. Phytochemistry 14:2563–2567
Rosa MP, Sigler PB (1977) Isolation and characterization of two methionine: tRNA ligases from wheat germ. Eur J Biochem 78:141–151
Rosenthal GA, Dahlmann DL, Janzen DH (1976) A novel means dealing with L-canavanine, a toxic metabolite. Science 192:256–257
Sarantoglou V, Imbault P, Weil JH (1980) The use of affinity elution from blue dextran sepharose by yeast $tRNA_2{}^{Val}$ in the complete purification of the cytoplasmic valyl-tRNA synthetase from *Euglena gracilis*. Biochem Biophys Res Commun 93:134–140
Sarantoglou V, Imbault P, Weil JH (1981) Purification of *Euglena gracilis* cytoplasmic leucyl-tRNA synthetase. Plant Sci Lett 22:291–297

Schimmel PR (1979) Understanding the recognition of transfer RNAs by aminoacyl transfer RNA synthetases. Adv Enzymol 49:187–221

Schimmel PR, Söll D (1979) Aminoacyl-tRNA synthetases: General features and recognition of tRNAs. Annu Rev Biochem 48:601–648

Schwartzbach SD, Hecker LI, Barnett WE (1976) Transcriptional origin of *Euglena* chloroplast tRNA. Proc Natl Acad Sci USA 73:1984–1988

Schwartzbach SD, Barnett WE, Hecker LI (1979) Evidence that *Euglena* chloroplasts do not export tRNAs. Nature (London) 280:86–87

Schwarz Z, Steinmetz A, Bogorad L (1980) Personal communication

Schwarz Z, Kössel H (1980) The primary structure of 16S rDNA from *Zea mays* chloroplast is homologous to *E. coli* 16S rRNA. Nature (London) 283:739–742

Sekiya T, Nishimura S (1979) Sequence of the gene for isoleucine $tRNA_1$ and the surrounding region in a ribosomal RNA operon of *E. coli.* Nucleic Acids Res 6:575–592

Selsky MI (1978) Reverse-phase chromatographic analysis of *Nostoc* and *Euglena* isoleucyl-tRNAs aminoacylated *in vitro* in homologous and heterologous systems. Biochim Biophys Acta 520:555–567

Shridhar V, Pillay DTN (1976) Changes in leucyl-tRNAs and aminoacyl-tRNA synthetases in developing and ageing soybean cotyledons. Phytochemistry 15:1809–1812

Shrift A, Bechard D, Harcup C, Fowden L (1976) Utilization of selenocysteine by a cysteinyl-tRNA synthetase from *Phaseolus aureus.* Plant Physiol 58:248–252

Silberklang M, Gillum AM, RajBhandary UL (1979) Use of *in vitro* $^{32}P$-labeling in the sequence analysis of non-radioactive tRNAs. In: Moldave K, Grossman L (eds) Methods in enzymology, vol LIX. Academic Press, London New York, pp 53–109

Sinclair DG, Pillay DTN (1980) Localization of tRNAs and aminoacyl-tRNA synthetases in cytoplasm, chloroplasts and mitochondria of *glycine max* L. Z Pflanzenphysiol (in press)

Smith IK, Fowden L (1968) Studies on the specificities of the phenylalanyl- and tyrosyl-sRNA synthetases from plants. Phytochemistry 7:1065–1075

Smith RA, Santi DV (1979) Simultaneous measurement of charging levels of multiple aminoacyl-tRNAs. Anal Biochem 99:372–378

Spare W, Lesiewicz JL, Herson DS (1978) The effect of cycloheximide on *Euglena gracilis* phenylalanyl-tRNA synthetases. Arch Microbiol 118:289–292

Stanley J, Vassilenko S (1978) A different approach to RNA sequencing. Nature (London) 274:87–89

Swamy GS, Pillay DTN (1979) Purification of phenylalanine transfer ribonucleic acid synthetase from soybean (*Glycine max*) cotyledon by affinity chromatography. Z Pflanzenphysiol 93:403–410

Swamy GS, Pillay DTH (1980) Purification and some properties of phenylalanyl-tRNA synthetase from soybean (*Glycine max* L.). Plant Sci Lett 20:99–107

Tao HL, Hall TC (1971) Factors controlling aminoacyl-transfer ribonucleic acid synthesis in vitro by a plant system. Biochem J 121:495–501

Tao KL, Khan AA (1974) Increase in activities of aminoacyl-tRNA synthetases during cold-treatment of dormant pear embryos. Biochem Biophys Res Commun 59:764–770

Tewari KK, Kolodner R, Chu NM, Meeker RM (1977) Structure of chloroplast DNA. In: Bogorad L, Weil JH (eds) Nucleic acids and protein synthesis in plants. Plenum Press, New York London, pp 15–36

Valenzuela P, Venegas A, Weinberg F, Bishop R, Rutter WJ (1978) Structure of yeast phenylalanine tRNA genes: an intervening DNA segment within the region coding for the tRNA. Proc Natl Sci USA 75:190–194

Vanderhoef LN, Key JL (1970) The fractionation of transfer ribonucleic acid from roots of pea seedlings. Plant Physiol 46:294–298

Vanderhoef LN, Travis RL, Murray MG, Key JL (1972) Interspecies aminoacylation of transfer ribonucleic acid from several higher plants, *Neurospora*, yeast and *E. coli.* Biochim Biophys Acta 269:413–418

Venkataraman R, Deleo P (1972) Changes in leucyl-tRNA species during ageing of detached soybean cotyledons. Phytochemistry 11:923–927

Vennegoor C, Bloemendal H (1972) Occurrence and particle character of aminoacyl-tRNA synthetases in the post-microsomal fraction from rat liver. Eur J Biochem 26:462–473

Viotti A, Balducci C, Weil JH (1978) Adaptation of the tRNA population of maize endosperm for zein synthesis. Biochem Biophys Acta 517:125–132

Vold BS, Sypherd PS (1968) Modification in transfer RNA during the differentiation of wheat seedlings. Proc Natl Acad Sci USA 59:453–458

Weil JH (1979) Cytoplasmic and organellar tRNAs in plants. In: Hall TC, Davies JW (eds) Nucleic acids in plants. CRC Press, Boca Raton, pp 143–192

Wesolowski M, Fukuhara H (1979) The genetic map of tRNA genes of yeast mitochondria: Correction and extension. Mol Gen Genet 170:261–275

Williams CR, Williams A, George SH (1973) Hybridization of leucyl transfer ribonucleic acid isoacceptors from green leaves with nuclear and chloroplastic deoxyribonucleic acid. Proc Natl Acad Sci USA 70:3498–3501

Wollgiehn R, Parthier B (1980) RNA and protein synthesis in plastid differentiation. In: Reinert J (ed) Results and problems in cell differentiation, vol X. Springer, Berlin Heidelberg New York, pp 97–145

Wower JM, Janowicz ZA, Augustyniak J (1979) Determination of the nucleotide sequence of phenylalanine tRNA from barley embryo. Acta Biochim Pol 26:369–381

Wray JL, Brice EB, Fowden L (1974) Development of aminoacyl-tRNA synthetases in cultured *Nicotiana tabacum* cells. Phytochemistry 13:697–701

Wright RD, Pillay DTN, Cherry JH (1972) Changes in leucyl-tRNA species of pea leaves during senescence and after zeatin treatment. Mech Ageing Dev 1:403–412

Wright RD, Kanabus J, Cherry JH (1974) Multiple leucyl-tRNA synthetases in pea seedlings. Plant Sci Lett 2:347–355

Yang JS, Brown GN (1974) Isoaccepting transfer ribonucleic acids during chilling stress in soybean seedling hypocotyls. Plant Physiol 53:694–698

Yarus M (1979) The accuracy of translation. Prog Nucleic Acid Res Mol Biol 23:195–225

Young RA, Macklis R, Steitz JA (1979) Sequence of the 16S–23S spacer region in two ribosomal RNA operons of *E. coli*. J Biol Chem 254:3264–3271

# 3 Ribosomes, Polysomes and the Translation Process

A. MARCUS

## 1 Introduction

The molecular constituents most directly determining the structure and function of cells are the proteins, the end products of the expression of the cell's genetic information. Synthesis of all proteins probably occurs by a process in which ribosomes catalyze the polymerization of amino acids in an alignment determined by the sequence of nucleotides in a nucleic acid template. The initial polypeptide produced is often modified by removal of a peptide fragment, methylation, acetylation, glycosylation, and/or phosphorylation to attain its final functional state. In situations where a protein is destined to pass across a membrane, the protein is often synthesized with a hydrophobic N-terminal region that is discarded after the protein has traversed the membrane. This chapter is concerned with the synthesis of the primary peptide product and focuses on the mechanisms used by the cell to accomplish this synthesis. The processes by which the production of different protein species might be controlled is also briefly considered (see Chapter 4 for post-translational modifications).

## 2 Ribosomes

Ribosomes are generally described as 80S (those found in the cytoplasm of eukaryotes) or 70S (those found in prokaryotes or organelles of eukaryotes), depending upon their sedimentation characteristics. When the $Mg^{2+}$ concentration of a "solution" of ribosomes is lowered sufficiently, dissociation occurs, with the 80S ribosome yielding 40S and 60S subunits, and the 70S ribosome giving 30S and 50S subunits. The 60S subunit contains three RNA's: 25S RNA, 5.8S RNA, and 5S RNA, having molecular weights of $1.3 \times 10^6$, 5.4 and $3.75 \times 10^3$, respectively. The 5.8S RNA is hydrogen-bonded to 25S RNA and is released only when these bonds are disrupted. The 5S RNA is released directly when ribosome preparations are deproteinized. The 40S subunit has one RNA, an 18S species having a molecular weight of $0.7 \times 10^6$. In addition to these RNA's, the 80S ribosome contains 70–80 different proteins. At each step in the synthesis of a protein, a number of the ribosomal components interact with appropriate non-ribosomal protein factors, with guanosine and adenosine di- and triphosphates, with mRNA, with the nascent peptidyl tRNA, and with the entering aminoacyl-tRNA. The spatial relationships between the ribosomal components and the three-dimensional structures of the ribosome are important in allowing these reactions to proceed in an ordered manner. Reviews of current knowledge

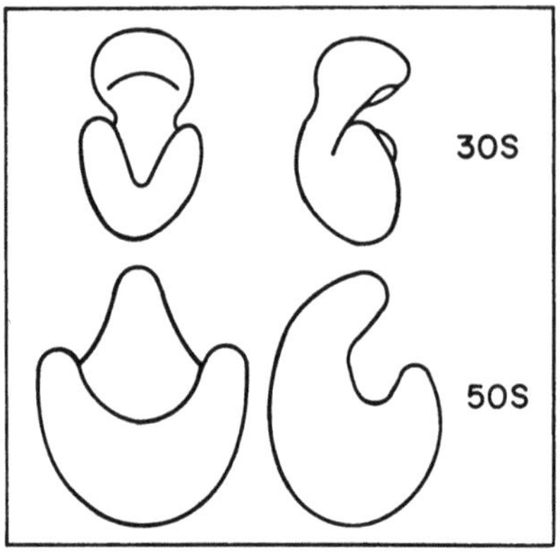


**Fig. 1.** Three-dimensional models of the 30S and 50S subunits. The two forms shown are related to each other by rotation through a 90° angle along the long axis. (STOFFLER and WITTMANN 1977)

of ribosome structure are presented by STOFFLER and WITTMANN (1977) and BRIMACOMBE et al. (1978). Several general points are worthwhile stressing. Many of the ribosomal proteins are elongated, coming through to the ribosomal surface at several points. In the 70S ribosomes, the shape of the smaller subunit resembles that of an embryo (see Fig. 1), i.e., a particle with two lobes of unequal size protruding from the main body, the lobes being separated by a cleft. Affinity labeling studies combined with electronmicroscopy have established that both the 3′ end of the 16S RNA of the small subunit and the decoding region of the ribosome (the area containing the mRNA codon hydrogen-bonded to the cognate tRNA anticodon) are uniquely localized in the cleft region of the upper part of the subunits (KEREN-ZUR et al. 1979, OLSON and GLITZ 1979). The shape of the larger subunit is that of an armchair with an assymetric notch. In the 70S ribosome, the small subunit sits horizontally transverse with respect to the 50S subunit with a channel between the subunits (see Fig. 2). Details of the secondary and tertiary interactions between the ribosomal proteins, the rRNA's, and the components of the protein synthesizing system have been obtained by in situ immunoelectronmicroscopy, by cross-linking experiments, and by specific chemical interaction with reactive groups exposed at the ribosome surface. Examples of this latter approach are the kethoxal attack on a nonpaired guanine residue, and the modification by N-ethylmaleimide of amino acid residues in a protein. Interactions primarily between regions of RNA have also been implicated in specific steps of the protein synthetic process. For example, the binding of tRNA to the 50S subunit involves the interaction of the TΨCG loop of the tRNA (see Fig. 6) to a complementary region on 5S RNA, while mRNA recognition includes an interaction with the RNA of the small subunit (see Sect. 6.1). Unraveling the details of the overall process still remains a formidable challenge, particularly in eukaryotic systems.

## 3 Translation

The overall process whereby ribosomes convert the information present in messenger RNA (mRNA) into proteins is referred to as "translation". A general

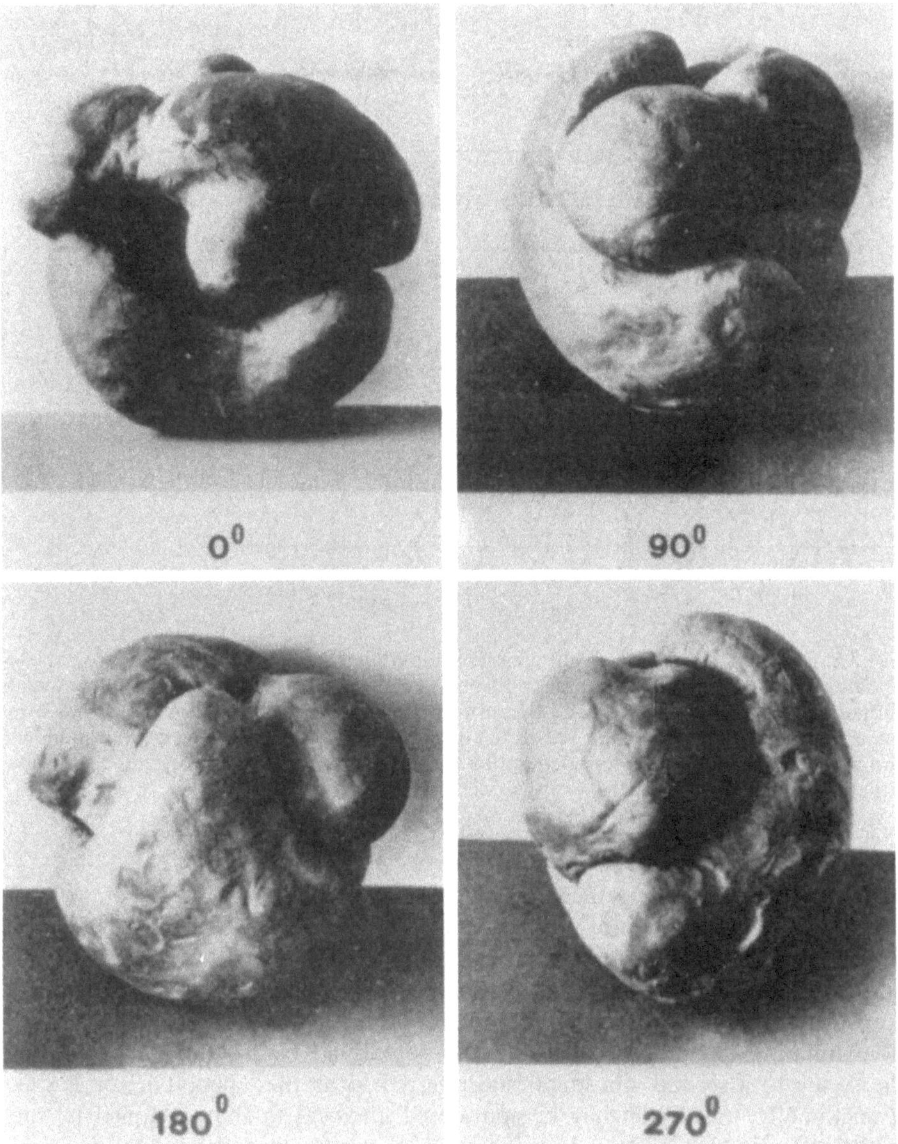


**Fig. 2.** Plasticine model of the 70S monomeric ribosome of *E. coli*. (STOFFLER and WITTMANN 1977)

scheme depicting the process is shown in Fig. 3. A ribosome attaches to an mRNA in such manner as to align an initiating trinucleotide sequence (AUG or GUG) on the mRNA with the initiating species of methionyl tRNA. A second aminoacyl-tRNA directed by the nucleotide triplet at the 3′ side of the initiating mRNA sequence is then attached to a "decoding" ribosome site. In all subsequent transitions, this site serves as the acceptor for incoming

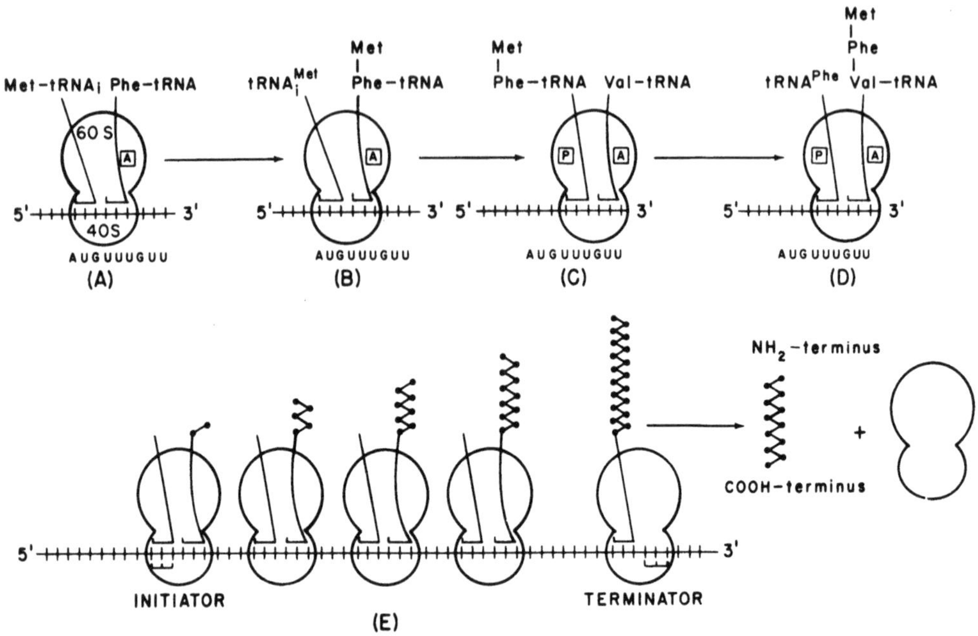


**Fig. 3A–E.** Schematic model of protein chain biosynthesis. Binding of Met-tRNA$_i$ at the initiator site and Phe-tRNA at the ribosomal *A* site (**A**) is followed by peptide bond formation (**B**). Ejection of tRNA$_i^{Met}$, translocation of the peptidyl tRNA from the *A* to the *P* site, and the binding of Val-tRNA then occurs (**C**), and a new peptide bond is formed (**D**). This sequence is repeated until a termination codon is reached. The steady-state situation results in the formation of the polyribosome (**E**)

aminoacyl-tRNA, and is therefore referred to as the "acceptor" or A site. Following the attachment of both aminoacyl-tRNA's, a peptide bond is formed in which the carboxyl group of the initiating aminoacyl-tRNA attaches to the amino group of the first aminoacyl-tRNA. The ribosomes then move a distance equivalent to 3 nucleotides in the 5′ to 3′ direction relative to the mRNA. The initiating tRNA is ejected and the peptidyl tRNA is translocated from the A site to a second ribosomal site referred to as the "peptidyl" or P site. [Functionally, the P site may be considered identical to that occupied by the initiating aminoacyl-tRNA. It is possible, however, that the actual physical site occupied by the two species may differ (THACH and THACH 1971, SEAL and MARCUS 1972).] A new aminoacyl-tRNA, coded for by the incoming mRNA triplet, now attaches to the vacant A site, and the process of peptide formation is repeated. In this manner the mRNA is translated in the 5′ to 3′ direction until a termination triplet is reached, whereupon the ribosome and the completed polypeptide are released. The functioning complex in which several ribosomes are moving along the mRNA, each elongating a peptide chain, is referred to as a polyribosome. Such structures have been identified in vivo in electronmicrographs (Fig. 4) and in vitro by sucrose density gradient analysis of isolated ribosomes (Fig. 5).

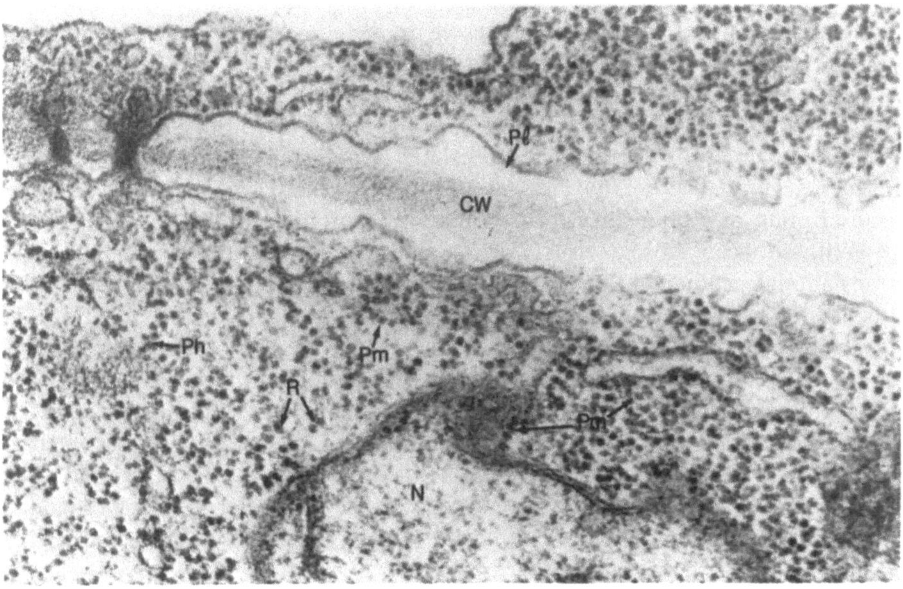


**Fig. 4.** Electronmicrograph of a thin section of parts of two adjacent cells in a shoot apex of pea showing ribosomes and polyribosomes. *CW* cell wall; *Pl* plasmalemma; *N* nucleus; *Pm* membrane-bound polysomes; *R* ribosome; *Ph* polysome helix. (By courtesy of A.D. GREENWOOD, Department of Botany and Plant Technology, Imperial College of Science and Technology, London)

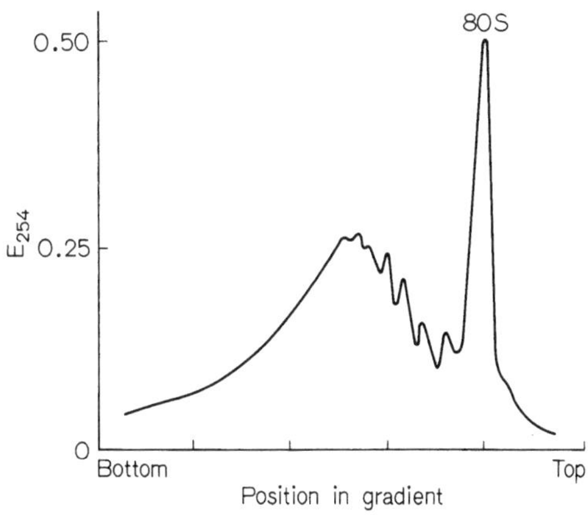


**Fig. 5.** Sucrose density profiles of ribosomes and polysomes from *Pisum sativum*. Ribosomes recovered after a 6-h centrifugation through a 1 *M* sucrose cushion. $E_{254}$ absorbance at 254 nm. (LEAVER and DYER 1974)

Most eukaryotic mRNA's are monocistronic, so that the translational process is complete once the termination codon is reached. [A possible further requirement may be a special mRNA release reaction, a process thus far described only in *Escherichia coli* (HIRASHIMA and KAJI 1973).] A number of polycistronic mRNA's have, however, been reported both in animal and plant viruses. Generally these mRNA's lack termination signals between cistrons, so that one translation product is made and specific proteolytic cleavages then provide the discrete cellular proteins. The protease that does the processing may be itself coded for by the viral RNA, and in one case the protease itself is synthesized as a precursor polypeptide that is processed by a cellular enzyme in the presence of ATP (PELHAM 1979a). With several polycistronic viral RNA's (SHIH and KAESBERG 1976, GOULD and SYMONS 1978, PELHAM 1978, SALVATO and FRAENKEL-CONRAT 1977), only the cistron nearest the 5′ end is read from the full-length RNA. The internal cistrons are translated from smaller sub-genomic RNA species generated either by cleavage of the full-length RNA or by selective transcription of a replicative form of the viral genome.

With tobacco mosaic virus (TMV) and tobacco rattle virus (TRV), two translational products have been obtained in vitro, one of these being a read-through of a leaky termination codon (PELHAM 1978, 1979b). The read-through product of TMV-RNA translation has been found in vivo. In another in vitro study, the number of initiation sites for translation of poliovirus RNA was found to be affected by the presence of a ribosomal extract from polio-infected cells (BROWN and EHRENFELD 1979) suggesting the existence of specific factors that allow internal initiation. Finally, one of the cowpea mosaic virus (CPMV) RNA's, and carnation mottle virus RNA appear to be functionally polycistronic, i.e., several initiation sites function in the intact RNA (SALOMON et al. 1978, PELHAM 1979a).

The transport of proteins across an intracellular membrane, although formally distinct from translation, is nevertheless linked mechanistically. A sequence of hydrophobic amino acids encoded in the 5′-region of the mRNA (the "signal" sequence), is initially synthesized on membrane-free polyribosomes. The nascent chain then attaches to a membrane, giving rise to rough endoplasmic reticulum (RER), i.e., membrane-bound polyribosomes. Depending upon the type of protein (secretory, integral to the membrane, peripheral to the membrane), the growing chain either passes through the membrane or becomes embedded in the correct transmembrane orientation. If the protein passes through the membrane, the signal sequence is removed during the vectorial cotranslational transport (LINGAPPA et al. 1979a, see also LODISH and ROTHMAN 1979). A variation on this scheme occurs with several mitochondrial and chloroplast proteins. In these cases synthesis of a precursor protein is completed on membrane-free polyribosomes and the precursor is subsequently transported to the organelle, presumably by attaching to receptor on the target membrane. Proteolytic cleavage to the mature product may occur within the organelle so that transport would become irreversible (DOBBERSTEIN et al. 1977, MACCECCHINI et al. 1979). A recent study of ovalbumin synthesis and secretion has shown that the signal sequence may be internal to the protein chain and that it need not be cleaved to achieve vectorial transport (LINGAPPA et al. 1979b). The irreversibility of the translocation seems to be a property of a receptor system within the membrane.

## 4 The Genetic Code and Messenger RNA

The designation of specific amino acids by trinucleotide sequences in a mRNA is accomplished by antiparallel Watson-Crick hydrogen bonding to an anticodon

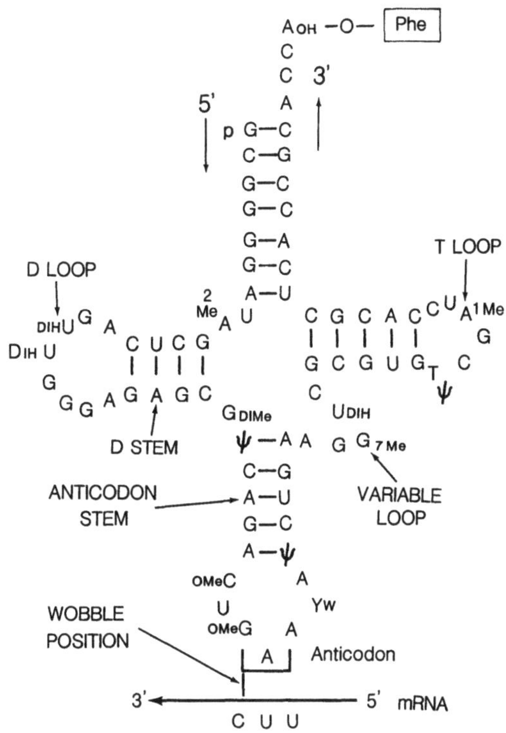


**Fig. 6.** Structure of wheat germ tRNA$^{Phe}$ in the clover leaf model (DUDOCK and KATZ 1969). A theoretical mRNA is included to show the codon–anticodon interaction as well as the "wobble" position

region in a transfer-RNA (tRNA). The tRNA thus serves as the decoding device with its 3′ end covalently attached to a specific amino acid. A typical illustration is shown in Fig. 6 for wheat germ phenylalanyl-tRNA (DUDOCK and KATZ 1969).

The initial evidence for mRNA recognition by aminoacyl-tRNA was the classical demonstration that an in vitro system from *E. coli* programmed with an mRNA consisting entirely of U residues [poly U], synthesized polyphenylalanine from phenylalanyl-tRNA (NIRENBERG and MATTHEI 1961). Subsequent experiments utilized more sophisticated synthetic oligonucleotides and analyzed the polypeptides synthesized (see Table 1; KHORANA et al. 1966). The genetic code clearly utilizes nonoverlapping triplets and contains no punctuation between these triplets. Finally, a ribosome-binding assay was developed (NIRENBERG and LEDER 1964) which allowed a direct correlation of specific aminoacyl-tRNA's with nucleotide triplets of known sequence. Cumulatively, these studies allowed the assignment of each of the triplet codons to a particular amino acid, i.e., the genetic code (Table 2). Verification of the code for plants has come from the analysis of amino acid replacements in the coat protein of tobacco mosaic virus (TMV) point mutations (WITTMANN and WITMANN-LIEBOLD 1966), and in vitro ribosome-binding studies with synthetic triplets (HATFIELD and RICE 1978). The recent finding that mitochondria recognize the UGA codon as tryptophan and the CUA codon as threonine (MACINO et al. 1979) provides the first exceptions to the universality of the code.

**Table 1.** Peptide synthesis with synthetic oligonucleotides. (After KHORANA et al. 1966)

| Polymer | mRNA<br>5′ end → 3′ end | Peptide product<br>$NH_2$ → COOH |
|---|---|---|
| $(UC)_n$ | UCU CUC UCU CUC | Ser·Leu·Ser·Leu |
| $(AG)_n$ | AGA GAG AGA GAG | Arg·Glu·Arg·Glu |
| $(UG)_n$ | UGU GUG UGU GUG | Val·Cys·Val·Cys |
| $(AC)_n$ | ACA CAC ACA CAC | Thr·His·Thr·His |
| $(UAUC)_n$ | UAU CUA UCU AUC UAU CUA | Thr·Leu·Ser·Ile·Thr·Leu |
| $(UUAC)_n$ | UUA CUU ACU UAC UUA CUU | Leu·Leu·Thr·Tyr·Leu·Leu |

**Table 2.** The genetic code

| 1st | 2nd | | | | 3rd |
|---|---|---|---|---|---|
| | U | C | A | G | |
| U | Phe | Ser | Tyr | Cys | U |
| | Phe | Ser | Tyr | Cys | C |
| | Leu | Ser | TERM | TERM | A |
| | Leu | Ser | TERM | Trp | G |
| C | Leu | Pro | His | Arg | U |
| | Leu | Pro | His | Arg | C |
| | Leu | Pro | Gln | Arg | A |
| | Leu | Pro | Gln | Arg | G |
| A | Ile | Thr | Asn | Ser | U |
| | Ile | Thr | Asn | Ser | C |
| | Ile | Thr | Lys | Arg | A |
| | Met | Thr | Lys | Arg | G |
| G | Val | Ala | Asp | Gly | U |
| | Val | Ala | Asp | Gly | C |
| | Val | Ala | Glu | Gly | A |
| | Val | Ala | Glu | Gly | G |

The table shows the amino acid requirement of all the possible 64 trinucleotide codons. (TERM refers to terminator.) The triplets coding for initiation are AUG and GUG

There are 61 codons utilized by 20 amino acids. The existence of several isoacceptor tRNA's that differ in their anticodon sequence (and therefore respond to different codons) provides a partial explanation for the quantitative imbalance. A further observation is that in many cases the same aminoacyl tRNA recognizes more than one codon. This was explained by a "wobble" hypothesis (CRICK 1966), stating that the nucleotide at the 5′ end of the anticodon was not as spatially confined as the other two anticodon bases, and as such could form hydrogen bonds with one of several bases (specifically I with either A, C, or U, G with U or C, and U with G or A). The basic idea of variance at the 5′ end of the anticodon has stood experimental test, but the extent of the variance has proved to be greater than originally postulated (see GOLDMAN et al. 1979). Indeed, it appears that reading "two out of three" codons may be sufficient to attach an amino acid, albeit with reduced affinity (SAMUELSSON et al. 1980). Finally, the utilization of the different codons for a specific amino acid varies considerably with different mRNA's (see, for example, codon usage in the synthesis of turnip yellow mosaic virus coat protein; GUILLEY and BRIAND 1978). In specialized cells, the codon frequency in the proteins synthesized is often mirrored in the distribution of the aminoacyl tRNA's (see, for example, HEINDELL et al. 1978). The mechanism linking these two phenomena is unknown.

## 5 Synthesis of Aminoacyl-tRNA

The "charging" reaction whereby an amino acid is attached to its cognate tRNA occurs in two steps, as shown in Fig. 7. The synthetase enzyme catalyzes a reaction between ATP and the specific amino acid, forming enzyme-bound aminoacyl adenylate. In the presence of a specific tRNA, the same synthetase enzyme transfers the amino acid to the terminal adenylic acid of the tRNA. This sequence of reactions converts the amino acid into a complex that functions to specifically direct the amino acid to an alignment against the mRNA as described in Section 3.

The specificity of the synthetase for both the amino acid and the tRNA requires that the synthetase has at least two different recognition sites. The tRNA in turn must have at least five areas of recognition: the anticodon site for interacting with mRNA, sites for interaction with the synthetase and for attachment to ribosomes (in addition to the codon-anticodon binding), sites for recognition by enzymes whose function is to modify specific bases within the tRNA sequence (see Fig. 6) and a site for tRNA pyrophosphorylase (an enzyme that reversibly removes the 3′ CCA terminus).

An additional site of specificity in tRNA is indicated by the observation in prokaryotes that aminoacylated tRNA often functions as a repressor of

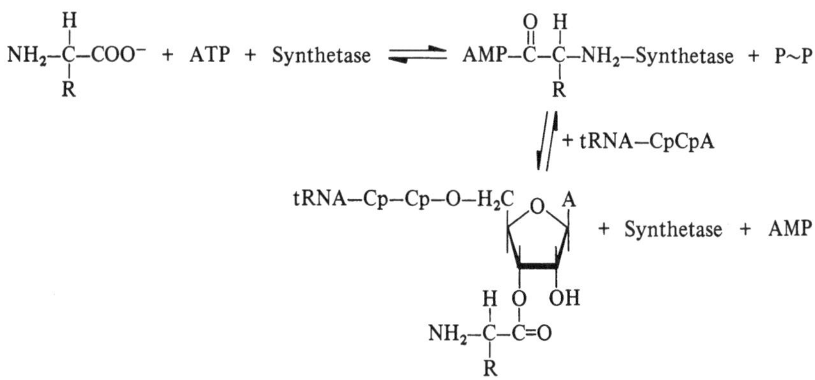


**Fig. 7.** Activation of an amino acid and its attachment to tRNA. Enzyme-bound aminoacyl adenylate formed by the reaction with ATP transfers the aminoacyl group to the adenylic acid at the 3′ terminus of tRNA

the transcription of genes leading to the biosynthesis of the particular amino acid. The regulatory function of one such tRNA has been associated with a specific modified nucleotide, pseudouracil (see Turnbough et al. 1979). Specificity within the tRNA molecule is undoubtedly generated by the remarkable system of stems and loops within its structure (Rich and Kim 1978). It should be noted, however, that a large number of base modifications in tRNA have little or no effect on the reactivity of either the tRNA with synthetase or of the aminoacylated modified tRNA in the various partial reactions of protein synthesis (Ofengand 1977, see however Ramberg et al. 1978).

The specificity of a synthetase for its cognate amino acid is not absolute. Since the tRNA anticodon (by alignment with the mRNA) determines the positioning of the amino acid within the peptide chain, such "misactivation" would result in the replacement of a correct amino acid by an incorrect one. Since protein synthesis is known to occur with extraordinary fidelity, i.e., error insertions occurring less than once per $10^4$ amino acid residues (Yarus 1979), it seems certain that an "editing" process eliminates potential errors. Two basic types of proofreading steps have been proposed. In one scheme, the aminoacyl tRNA synthetase itself acts as a hydrolytic enzyme towards the noncognate aminoacyl tRNA (Yarus 1979), while in the second scheme, the energy of GTP is utilized to maintain fidelity (Thompson and Stone 1977).

A somewhat different situation involving amino acid specificity occurs with the proline-tRNA synthetase of mungbean. The enzyme is able to "activate" azetidine-2-carboxylic acid and transfer it to $tRNA^{Pro}$ (Peterson and Fowden 1963). If supplied in vivo the analog is incorporated into protein and causes cell toxicity. In *Polygonatum multiflorum* where azetidine-carboxylic acid is a normal constituent, survival is attained by the inability of the proline synthetase to activate the analog (see Chapter 2 for details of aminoacylation).

## 6 Synthesis of the Protein Chain

### 6.1 Initiation

Before going into the details of the eukaryotic process, it is appropriate to consider prokaryotic protein chain initiation. In both the prokaryotic and eukaryotic systems an important role is played by an initiator Met-tRNA. The first amino acid residue incorporated into peptide linkage is methionine. Furthermore, translation of synthetic polynucleotides in prokaryotic systems can be primed by Met-tRNA. It is necessary, however, that the Met-tRNA be N-substituted by a formyl group, and the tRNA be the specific initiator species, $tRNA^{fMet}$. A second major observation in prokaryotic systems is that translation in vitro of a phage mRNA requires, in addition to elongation factors (see Sect. 6.2), three protein components that function at the level of initiation; IF1 (a basic protein with a molecular weight of 9,500), IF2 (molecular weight about 100,000) and IF3 (a thermostable protein of molecular weight 22,000). Table 3 describes the most recent consensus of the reaction in which these

**Table 3.** Intermediates in prokaryotic protein chain initiation

(1) 70S $^{IF1}$ 30S + 50S $^{+IF3}$ 30·IF3
(2) 30S·IF3 + mRNA ⇌ 30S·IF3·mRNA
(3) 30S·IF3·mRNA + GTP + IF2·fMet-tRNA ⇌ 30S·mRNA·GTP·IF2·fMet-tRNA + IF3
(4) 30S initiation complex + 50S ⇌ 70S·mRNA·fMet-tRNA + GDP + IF2 + $P_i$

The scheme is based on studies of VOORMA, BOSCH and coworkers. (See BOSCH and VAN DER HOFSTADT 1979)

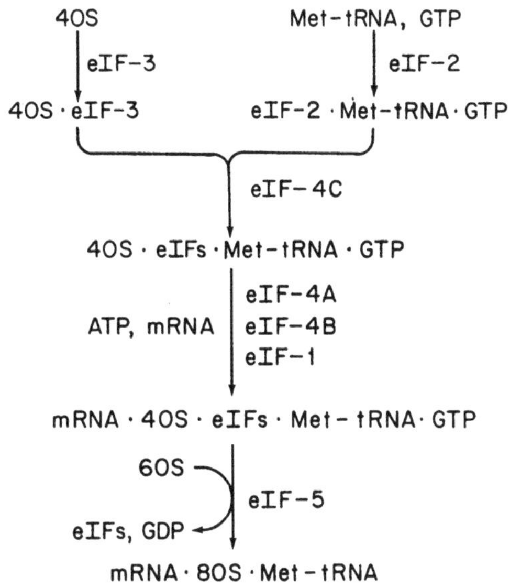


**Fig. 8.** A working model of eukaryotic protein chain initiation. (Modified from BENNE and HERSHEY 1978)

initiation factors interact with the ribosome and the initiator tRNA (VAN DER HOFSTAD et al. 1977, 1978a). IF1 catalyzes the dissociation of the 70S ribosome allowing the 30S subunit to stochiometrically bind IF3, thereby shifting the equilibrium in favor of the free subunits. The 30S·complex then binds mRNA. Initiator tRNA enters as a binary complex with IF2, and IF3 is displaced. GTP, or its nonhydrolyzable analog, facilitates this reaction. Finally, the 50S subunit attaches concomitant with the release of IF2 and the hydrolysis of GTP. Though not shown in this scheme, IF1 probably remains attached to the ribosome until the final stage of 70S initiation complex formation (VAN DER HOFSTAD et al. 1978b). Two points in the scheme are particularly significant: the role of the binary complex in attaching the initiator tRNA, with GTP serving only in an auxillary capacity, and the binding of mRNA to the 30S subunit prior to the attachment of initiator tRNA. The latter conclusion has been contested and evidence has been obtained for an alternative scheme in which an IF2·fMet-tRNA·30S subunit (34S complex) is formed initially and then interacts with mRNA and IF3 to form a 46S mRNA·fMet-tRNA·30S subunit complex (KAEMPFER and JAY 1979).

The sequence of events considered to occur in eukaryotic protein chain initiation (Fig. 8) consists of four steps: (1a) formation of native 40S ribosomal

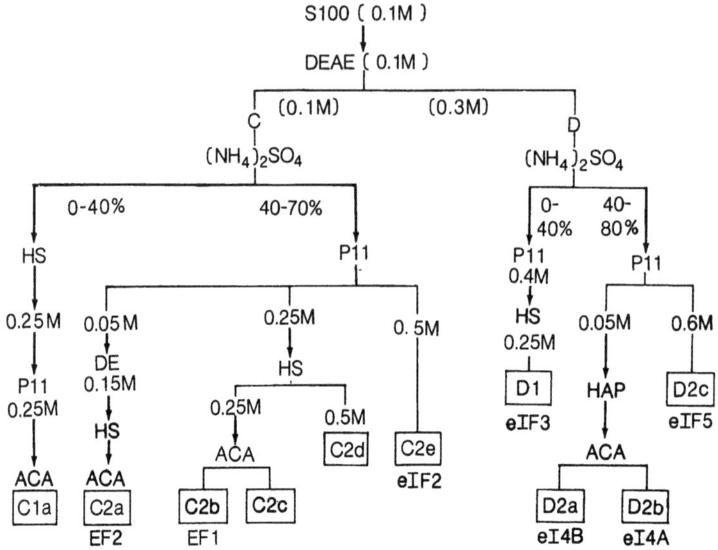


**Fig. 9.** Fractionation of wheat germ S100. The flow diagram describes the preparation of the different factors. *HS* heparin sepharose; *P11* phosphocellulose; *HAP* hydroxyapatite. *Numbers in parentheses* are the KCl concentration used to elute the factor. For *HAP*, 0.05 M K phosphate pH 7.6 is the eluting solution. *ACA* is acrylamide-agarose 44 (run in 0.18 M KCl). The *boxed components* are those referred to in the text and in Tables 4 and 5. The presumed equivalent nomenclature in the standardized system (ANDERSON et al. 1977) is indicated below each of the components

subunits (40S·eIF3 complexes), and (1 b) forming a ternary complex containing Met-tRNA binding protein, and GTP; (2) attachment of the ternary complex to the 40S ribosomal subunit forming preinitiation complex I; (3) attachment of mRNA at the correct initiation sequence to yield preinitiation complex II; (4) joining of the 60S ribosomal subunit resulting in an 80S initiation complex. The observations leading to this scheme are as follows. In vitro amino acid polymerization with a system extracted from rabbit reticulocyte ribosomes requires, in addition to elongation factors, five different protein components (SCHREIER et al. 1977, BENNE and HERSHEY 1978). In extracts of wheat germ, amino acid polymerization with TMV-RNA as a template requires at least six such factors (see Fig. 9). Tables 4 and 5 compare the requirements of the two systems. Where possible the different factors have been identified by a standardized terminology (ANDERSON et al. 1977). In the reticulocyte system, the protein synthesis assay measures polymerization of free amino acids with a pH 5 supernatant added as a source of aminoacyl tRNA synthetase. In the wheat germ system, precharged aminoacyl tRNA's are used in the assay (see legend to Table 4). Because of this difference, the extent of requirement for the different factors is more stringent in the wheat germ system. Another consequence of the difference in the assays is that a requirement for a free amino acid cannot be determined in the reticulocyte assay. The wheat germ system demonstrates exactly such a phenomenon with the observation that amino acid

**Table 4.** Initiation factor requirements for mRNA-dependent amino acid polymerization

| Conditions | Reticulocyte system Incorporation (pmol) | | Wheat germ system Incorporation (pmol) |
|---|---|---|---|
| | 2.5 µl pH 5 | 4 µl pH 5 | |
| Complete | 4.2 | 6.4 | 9.5 |
| No factors | 0.6 | 0.6 | 0.5 |
| eIF1 Omitted | 3.2 | 4.5 | – |
| eIF2 Omitted | 1.0 | 2.7 | – |
| eIF3 Omitted | 0.7 | 1.5 | 1.0 |
| eIF4A Omitted | 0.9 | 1.9 | 0.7 |
| eIF4B Omitted | 1.1 | 2.1 | 1.0 |
| eIF4C Omitted | 2.5 | 4.1 | – |
| eIF5 Omitted | 1.1 | 1.9 | 0.7 |
| EF2 Omitted | – | – | 2.7 |

The data for the reticulocyte system are calculated from BENNE and HERSHEY (1978). (Substantially greater activity is reported by SCHREIER et al. (1977), with approximately the same level of requirement for the different factors.) The data for the wheat germ system (SEAL, S.N. and MARCUS, A., unpublished observations) are taken from an assay in a vol of 0.2 ml containing 2.2 mM dithiothreitol, 20 µM GTP, 1 mM ATP, 20 mM Hepes KOH pH 7.6, 2.0 mM $MgAc_2$, 70 µM spermine, 50 mM KCl, 8 mM creatine phosphate, 8 µg creatine phosphokinase, 125 µg ribosomes (washed with 0.6 M KCl), 5 µg TMV-RNA, 9.5 µg wheat tRNA containing 35 pmol (specific activity 450 cpm/pmol) of five [$^{14}$C]-aminoacyl-tRNA's (leucine, valine, serine, lysine, and phenylalanine) prepared with an excess of the other 15 unlabeled amino acids, 30 µM [$^{12}$C]amino acids corresponding to the five [$^{14}$C]-aminoacyl tRNA's, 30 µM Asn, 30 µM Ala, 30 µM Thr, 250 µM glutamate, and the following levels of factors – 4 µg C1a ($P_{11}$ step), 6 µg C2b+c (HS step), 5.5 µg D1 (eIF3), 8 µg D2a (eIF4B), 3 µg D2b (eIF4A), 2.6 µg D2c (eIF5), 1 µg C2a (EF2). The C and D notations correspond to the fractionation shown in Fig. 9, while the eIF (eukaryotic initiation factor) and EF (elongation factor) designations are according to ANDERSON et al. (1977)

incorporation is stimulated about fourfold by free glutamic acid (Table 5), and to a lesser extent by free asparagine and threonine. When the polymerization reaction is carried out with globin mRNA or AMV-RNA 4 as templates, only glutamate is required, suggesting that the stimulation by asparagine and threonine may be due to a deficiency of charged Asn- and Thr-tRNA's. Such a deficiency is of course of concern in using precharged aminoacyl tRNA's for the assay of polymerizing activity.

The specific functions ascribed to the particular components of the polymerizing system, as noted in Fig. 8, are tentative. Component eIF2 from rabbit reticulocyte ribosomal wash is a protein of molecular weight 120,000 containing three subunits of 32,000($\alpha$), 35,000($\beta$) and 55,000($\gamma$) (LLOYD et al. 1980). An equivalent factor preparation from calf liver is reported to have a molecular weight of 95,000 with only two subunits of 38 and 48,000 (STRINGER et al. 1979). Factor eIF2 can bind Met-tRNA and GTP to form a ternary complex that is retained by nitrocellulose filters at low $Mg^{2+}$. The addition of GTP is generally necessary for the formation of the eIF2·Met-tRNA complex (see, however, SAFER et al. 1975 and RANU and WOOL 1976 for less stringent GTP requirement),

**Table 5.** Requirement of purified wheat germ factors for amino acid incorporation

| Conditions | Incorporation (pmol) |
|---|---|
| 1. Complete | 10.6 |
| No factors | 0.5 |
| Cla Omitted | 1.0 |
| C2b (EF1) Omitted | 1.3 |
| C2c Omitted | 6.1 |
| 2. Complete | 9.4 |
| Glutamate omitted | 2.4 |

The assay system was similar to that of Table 4 except that the ribosomes were washed with 0.1 M KCl and factors D1 (eIF3), D2a (eIF4B), D2b (eIF4A) and D2c (eIF5) were replaced by 12.5 µl starting fraction D in experiment 1 while in experiment 2, the D complement included factors D1 and D2c and 15 µg of D2a+b (HAP step). Other additions were 0.2 µg Cla, 4 µg C2b (EF1) and 0.8 µg C2c in experiment 1; 4 µg factor Cla ($P_{11}$ step) and 6 µg C2b+c (HS step) in experiment 2. Requirement for factors corresponding to Cla and C2c have as yet not been reported in mammalian systems

and the reaction is fairly specific for the initiator species of Met-tRNA. The ternary complex can be bound to a 40S ribosomal subunit, although whether or not GTP remains stably bound in the 40S ribosome complex is not clear (see TRACHSEL and STAEHLIN 1978 and PETERSON et al. 1979). What seems certain is that in the overall reaction leading to the 80S ribosome initiation complex, the same GTP that is initially involved in the attachment of Met-tRNA functions later in the subunit joining reaction where it is cleaved to GDP and Pi (MERRICK 1979). Factor eIF3 (a protein of molecular weight 450,000–700,000 with 8–12 polypeptides; BENNE and HERSHEY 1978) stimulates the formation of 40S ribosome·eIF2·Met-tRNA complexes, particularly when the level of eIF2 is not in great excess, presumably by stabilizing the 40S·ribosome complexes (PETERSON et al. 1979). The primary function of factor eIF3 may be to modify the 40S subunit so as to allow mRNA attachment (see below). Both in vivo and in vitro it functions as an anti-association factor, resulting in the conversion of the 40S subunit to a particle that sediments at 46S and has a buoyant density of 1.41 in cesium chloride (THOMPSON et al. 1977, SUNDQUIST and STAEHLIN 1975). Radiolabeled eIF3 attaches to the 40S subunit (BENNE and HERSHEY 1978, TRACHSEL and STAEHLIN 1978) and the attachment is greatly stabilized when the ternary eIF2·Met-tRNA·GTP complex is also present (PETERSON et al. 1979). Overall, the data obtained with reticulocyte factors provide convincing evidence for the formation of a 40S ribosome eIF2·eIF3·Met-tRNA complex, i.e., preinitiation complex I. The allocation of factor eIF4C to the formation of this complex is related to its partial stabilization in vitro of the various Met-tRNA·40S ribosome complexes (BENNE and HERSHEY 1978).

In the wheat germ system the role of eIF2 is unclear. Purified factors with eIF2 activity have been obtained (SPREMULLI et al. 1979, BENNE et al. 1980, see also Fig. 9). However, the polymerization reaction is stimulated by these fractions to only a limited extent. Furthermore, direct assay of the components of the polymerizing system for ability to catalyze the binding of Met-tRNA to ribosomes in the presence of fractions D1 (eIF3) and D2C (eIF5) indicate that there is little eIF2 activity in any of the required fractions. One possible explanation is that the activity of an exceedingly low level of eIF2 is stimulated by the presence of another fraction to the point that, despite its low level, eIF2 would not be rate-limiting for polymerization. Such eIF2-stimulating activity has been reported by RANU et al. (1978) and DE HARO and OCHOA (1978) in reticulocyte extracts, and is present in wheat germ supernatant fraction C2c (see Fig. 9). As noted in Table 5, this component significantly stimulates the polymerization reaction. A further consideration with regard to eIF2 is that the ability to remove most of this activity and yet maintain the polymerization reaction makes it certain that the level of this factor is not rate-limiting to translation. In this regard, the plant system appears to differ from the mammalian systems, where regulation of protein synthesis is considered to occur at the eIF2 step (see AMESZ et al. 1979).

A 48S ribosome complex containing mRNA, i.e., preinitiation complex II, has been shown to be formed in vivo in reticulocyte lysates either endogenously (BUHL et al. 1980) or in the presence of inhibitors that stop the adjunction of the 60S ribosome (GROSS 1979, SAFER et al. 1979). Allocation of the functions of factor eIF3 and the eIF4 group to this reaction is based on in vitro studies with reticulocyte factors in which binding of labeled globin mRNA to 40S ribosome subunits was shown to be stimulated by the addition of these components (TRACHSEL et al. 1977, BENNE and HERSHEY 1978). Radiolabeled factors, however, could not be detected in the 40S ribosome complex, indicating that these factors either function catalytically or that their association is unstable. Finally, ascribing eIF5 to the role of subunit joining is based on in vitro studies with both the reticulocyte and the wheat germ system in which addition of this factor converts 40S ribosome complexes (labeled either with Met-tRNA or radioactive mRNA) to 80S ribosome complexes. Subunit joining requires the hydrolysis of GTP since 40S ribosome complexes made with either GMP-PCP or GMP-PNP do not attach 60S subunits despite the presence of eIF5.

The functioning of a number of factors at the level of mRNA attachment raises the possibility that differences in affinity of a particular mRNA for one of these components could result in selective translation. In the reticulocyte system, factor eIF4A is required in unusually large amounts for translation of EMC-RNA (TRACHSEL et al. 1977), and factors eIF4B and eIF2 have been reported to shift the translation of a mixture of globin mRNA's from the $\beta$-mRNA to the $\alpha$-mRNA (KABAT and CHAPPELL 1977, DI SEGNI et al. 1979). SONENBERG et al. (1979) have isolated a cap-binding protein and have suggested that this protein serves specifically for the attachment of capped mRNA's. Recently TRACHSEL et al. (1980) purified a protein that could restore the ability to translate a capped mRNA to extracts of poliovirus-infected cells, and have suggested that the protein is the cap-binding protein. In the wheat germ system, a striking competitive translation has been demonstrated (HERSON et al. 1979). When a saturating level of a noncapped mRNA, STNV-RNA, is translated together with increasing levels (as much as three times the saturating level) of globin mRNA, globin mRNA translation is repressed, while translation of STNV-RNA is unaffected. When translated alone, globin mRNA is more active (pmol amino acid per min per pmol mRNA) than STNV-RNA. Similar experiments with a series of mRNA's (all of which are capped) showed a hierarchy of competitive translational ability. Further studies showed that supplementing with one factor (Cla, see Fig. 9) shifts the translation in the direction of the "weaker" mRNA. The interpretation of these results is that when the level of mRNA's is such that all of factor Cla is in the form of an mRNA·Cla complex, then the relative affinity of different mRNA's for this factor will determine which mRNA is translated. In this manner, factor Cla is functioning as an mRNA-discriminating component. Finally, it seems not unreasonable that the structural compexity of factor eIF3 may allow it to play some role in mRNA selection.

The 5′ caps of most viral and cellular mRNA's strongly facilitates their translation (SHATKIN 1976). This observation, together with the fact that most eukaryotic mRNA's are monocistronic (see Sect. 3), prompted the suggestion that the 40S ribosomal subunit must initially attach at the 5′ end of an mRNA and then migrate to the first AUG codon (KOZAK 1978). Such a scheme in effect prevents ribosome attachment at internal AUG codons. The fact that there is considerable translational competition between different mRNA's despite an equivalent cap structure requires, however, that at some point before formation of the final initiation complex, the functional attachment site on the mRNA must be such as to allow an equilibrium reaction. Presumably the affinity of this site on the mRNA for a factor attached to the ribosome determines the translational advantage of a particular mRNA (see also KOZAK 1980). Finally, there is evidence that an 80S ribosome can attach in a stable manner to an mRNA at a position 5′ to the first AUG (AHLQUIST et al. 1979, FILIPOWICZ and HAENNI 1979).

In prokaryotes, mRNA attachment to ribosomes has been ascribed in part to nucleotide homology between a CCUCC tract in the 3′ terminal region of 16S rRNA and a G-rich region about 10 bases 5′ to the AUG initiator codon in the mRNA (ECKHARDT and LURHMANN 1979). This pyrimidine tract is not found in eukaryotic 18S rRNA's, although there is a substantial homology in the 3′ terminal 220 nucleotides of the small subunit rRNA's of eukaryotes and prokaryotes (SAMOLS et al. 1979). One additional requirement for the attachment of an mRNA to ribosomes in eukaryotic systems is ATP (GIESEN et al. 1976, SCHREIER et al. 1977). The mechanism of action of the ATP is unknown, except that it must be hydrolyzed (APP-imidophosphate is inactive). Finally, it is worth noting that the 40S ribosome complexes generally are not sufficiently stable to withstand centrifugation through gradients. Treatments with glutaraldehyde can be used to fix the complexes (GIESEN et al. 1976), and more recently the inclusion of GMP-PNP (GPP-imidophosphate) in the gradient has been found to increase their stability (PETERSON et al. 1979).

## 6.2 Elongation

As described in Section 3 (see also Fig. 3), the elongation reactions begins with the codon-directed GTP-dependent attachment of aminoacyl tRNA to the ribosomal acceptor, or "A" site. The factor catalyzing this reaction, EF1, has been isolated from wheat germ generally in a high molecular weight aggregate, referred to as $EF1_H$. In the fractionated mRNA-dependent system (Fig. 9 and Table 5), EF1 is in the monomer form with a molecular weight of 52,000 ($EF1_L$). In the presence of GTP or GMP-PNP, EF1 binds all species of aminoacyl tRNA except Met-$tRNA_i^{Met}$. In vivo, it appears that $EF1_L$, complexed with GTP, binds to the 80S ribosome immediately following the initiation reaction and remains firmly attached until synthesis of the protein chain is complete (GRASMUK et al. 1977). The bound $EF1_L$·GTP complex attaches the aminoacyl tRNA appropriate to the mRNA codon at the "A" site and hydrolysis of the GTP then modifies the bound aminoacyl tRNA so that it can participate in peptide bond formation (see Fig. 3). This latter reaction is carried out by peptidyl transferase, a ribosome-bound enzyme. An additional factor (EF-Ts, molecular weight 30,000) seems to catalyze an exchange of new GTP for the ribosome-bound GDP, thus recycling $EF1_L$ for the next round of aminoacyl tRNA binding (BOLLINI et al. 1974, SLOBIN 1979). The next step is the attachment to the ribosome of elongation factor 2 (EF2) in a complex with GTP. There now ensue three reactions (see Fig. 3): ejection of the deacylated tRNA from the "P" site, movement of the peptidyl tRNA from the "A" to the "P" site, and the movement of the mRNA a distance of three nucleotides. Energy from the hydrolysis of GTP participates in these reactions, but specific details of the process are not known. The preparation of EF2 from extracts of wheat germ is described in Fig. 9. The molecular weight of the factor is approximately 80,000.

The process of elongation continues with a repetition of the cycle, i.e., aminoacyl tRNA attaches to the ribosome-bound $EF1_L$ and a peptide bond is formed, etc., until the protein chain is completed.

## 6.3 Termination

Termination of the peptide chain occurs when a terminator codon (UAA, UAG, or GUA, see Fig. 3) reaches the ribosomal "A" site. In prokaryotes a specific termination factor(s) recognizes this codon in a GTP-requiring reaction and in some manner activates the ribosomal peptidyl transferase to hydrolyze the peptidyl tRNA. How were terminator codons found? Initially it was observed that certain types of genetic errors ("nonsense mutants") could be reversed if the mutant genes were crossed into specific cell lines. (The cell lines capable of the reversal are referred to as suppressors.) The mutants were found to be of two types, based on the ability of the different suppressor strains to reverse the mutation. Analysis of the amino acids substituted into the "suppressed" products, together with a consideration of the codons for these amino acids, established two of the mutations as the codons UAG (amber) and UAA (ochre) (WEIGERT et al. 1967). In a similar manner, a third nonsense mutation was established to be UGA. That these codons function as terminators was confirmed by in vitro translation of a bacteriophage mRNA having a UGA at a known position and the demonstration by analysis of the product that translation had terminated at this codon (CAPECCHI 1967). Suppressor systems have aminoacyl tRNA's recognizing the terminator codon, and are thus able to insert an amino acid. Generally, this involves a change in the anticodon of the tRNA so as to allow it to base pair with the terminator codon. A more subtle change of a base in a region other than the anticodon may also allow a particular aminoacyl tRNA to suppress a terminator codon [see CASKEY (1977) for a detailed discussion]. Finally, direct evidence of the function of these codons in normal termination has been obtained by sequencing specific mRNA's. The terminator codons follow directly after the codon for the C-terminal amino acid, and in many cases two such codons are found in tandem arrangement.

The mechanism of the termination reaction is only partially clarified (see CASKEY 1977). The assays for the reaction are usually carried out with model systems (CASKEY et al. 1974, MENNINGER and WALKER 1974). In bacteria two factors have been isolated that recognize the terminator codons and a third factor is stimulatory. In both reticulocytes and extracts of brine shrimp there seems to be only one termination factor and this factor recognizes all three of the codons (REDDINGTON and TATE 1979). In view of the consideration that overall protein synthesis could be regulated at the termination step (BERGMANN and LODISH 1979) it would appear desirable to develop a more natural assay system, i.e., one that would measure the release of peptidyl-tRNA from polysomes.

## 7 Regulation of Protein Synthesis

During the development of an organism, changes occur in both the overall rate of protein synthesis and in the specific complement of proteins that are synthesized. Modulation of the overall rate of synthesis can, in theory, be brought about by a change in the level of the component that is rate-limiting. Assuming that one can maintain the in vivo regulation with in vitro preparations, such a possibility should be experimentally testable.

With regard to specific proteins, there are a number of steps at which regulation could occur. The primary control in eukaryotes, as in prokaryotes, is undoubtedly at the level of transcription. However, specific molecular mechanisms to activate a particular gene have not been elucidated thus far. Other potential sites for regulation of the overall rate of protein synthesis are the availability of RNA polymerases, modulation of premature termination of transcripts (SALDITT-GEORGIEFF et al. 1980), the rate of processing of nuclear precursors of mRNA, and the transport of mRNA from the nucleus. In certain situations, preformed mRNA's have been found to be stored as mRNP's and released at a later period in the physiological development of the organism (see Chap. 15, this Vol. for a more detailed discussion). Regulation of the synthesis of specific proteins, if it occurs at the level of translation, will presumably involve competition between mRNA's. A reasonable mechanism would be the differences in affinity of the mRNA's for one of the initiation factors (see Sect. 6.1). Finally, the turnover rates of both the mRNA and the protein product allow further possibilities for regulating the levels of particular proteins.

# 8 Epilogue

The general outlines of the process of protein synthesis are known but the details of the various reactions are poorly understood. Although much information on the process in eukaryotes has come from the reticulocyte system, the greater activity of the fractionated wheat germ system and the better resolution obtained make this system attractive for future studies. The questions that are of most immediate interest are: the molecular nature of eIF2 and the ability of modified forms of this factor to interact in the reaction leading to the formation of an 80S initiation complex, and the mechanism of interaction of the eIF3 and eIF4 factors with ATP in attaching mRNA to the 40S subunit. There are also indications of an interaction between protein synthesis and RNA polymerization. Bacteriophage Qβ is known to utilize components of the host protein-synthesizing system in its RNA replicase and recent studies have suggested that prokaryotic RNA polymerase can be regulated by protein synthesis components (Debenham et al. 1980).

With regard to the regulation of protein synthesis, while information on potential mechanisms is available (see Sect. 7), experimental data analyzing these possibilities is scanty. Thus both the biochemistry and the biology of protein synthesis provide a challenge for future research.

*Acknowledgements.* Supported by grant PCM79-00268 from the National Science Foundation; by grants GM-15122, CA-06927, and RR-05539 from the National Institutes of Health; and by an appropriation from the Commonwealth of Pennsylvania.

# References

Ahlquist P, Dasgupta R, Shih DS, Zimmern D, Kaesberg P (1979) Two step binding of eukaryotic ribosomes to brome mosaic virus RNA3. Nature (London) 281:277–282

Amesz H, Goumans H, Haubrich-Morree T, Voorma HO, Benne R (1979) Purification and characterization of a protein factor that reverses the inhibition of protein synthesis by the heme-regulated translational inhibitor in rabbit reticulocyte lysates. Eur J Biochem 98:513–520

Anderson WF, Bosch L, Cohn WE, Lodish H, Merrick WC, Weissbach H, Wittman HG, Wool IG (1977) International symposium on protein synthesis. FEBS Lett 76:1–10

Benne R, Hershey JWB (1978) The mechanism of action of protein synthesis initiation factors from rabbit reticulocytes. J Biol Chem 253:3078–3087

Benne R, Kasperaitis M, Voorma HO, Ceglarz E, Legocki AB (1980) Initiation factor eIF-2 from wheat germ; purification, functional comparison to eIF-2 from rabbit reticulocytes and phosphorylation of its subunits. Eur J Biochem 104:109–117

Bergmann JE, Lodish HF (1979) A kinetic model of protein synthesis; application to hemoglobin synthesis and translational control. J Biol Chem 254:11927–11937
Bollini R, Soffientini AN, Bertani A, Lanzani GA (1974) Some molecular properties of the elongation factor EF1 from wheat embryos. Biochemistry 13:5421–5425
Bosch L, van der Hofstadt GAJM (1979) Initiation of protein synthesis in prokaryotes. Methods Enzymol 60:11–15
Brimacombe R, Stoffler G, Wittman HG (1978) Ribosome structure. Annu Rev Biochem 47:217–249
Brown BA, Ehrenfeld E (1979) Translation of poliovirus RNA in vitro: changes in cleavage pattern and initiation sites by ribosomal wash. Virology 97:396–405
Buhl W, Sarre TF, Hilse K (1980) Characterization of a native mRNA containing preinitiation complex from rabbit reticulocytes: RNA and protein constituents. Biochem Biophys Res Commun 93:979–987
Capecchi MR (1967) Polypeptide chain termination in vitro. Isolation of a release factor. Proc Natl Acad Sci USA 58:1144–1151
Caskey CT (1977) Peptide chain termination. In: Weissbach H, Pestka S (eds) Molecular mechanisms of protein biosynthesis. Academic Press, London New York, pp 443–465
Caskey CT, Beaudet AL, Pate WP (1974) Mammalian release factor; in vitro assay and purification. Methods Enzymol 30:293–303
Crick FHC (1966) Codon-anticodon pairing: the wobble hypothesis. J Mol Biol 19:548–555
Debenham PG, Pongs O, Travers AA (1980) Formylmethionine-tRNA alters RNA polymerase specificity. Proc Natl Acad Sci USA 77:870–874
Di Segni G, Rosen H, Kaempfer R (1979) Competition between α- and β-globin messenger ribonucleic acids for eucaryotic initiation factor 2. Biochemistry 18:2847–2854
Dobberstein B, Blobel G, Chua NH (1977) In vitro synthesis and processing of a putative precursor for the small subunit of ribulose-1,5 bisphosphate carboxylase of *Chlamydomonas reinhardtii*. Proc Natl Acad Sci USA 74:1082–1085
Dudock BS, Katz G (1969) Large oligonucleotide sequences in wheat germ phenylalanine transfer ribonucleic acid. J Biol Chem 244:3069–3074
Eckhardt H, Luhrmann R (1979) Blocking of the initiation of protein biosynthesis by a pentanucleotide complementary to the 3′ end of *Escherichia coli* 16S rRNA. J Biol Chem 254:11,185–11,188
Filipowicz W, Haenni A-L (1979) Binding of ribosomes to 5′-terminal leader sequences of eukaryotic messenger RNAs. Proc Natl Acad Sci USA 76:3111–3115
Giesen M, Roman R, Seal SN, Marcus A (1976) Formation of an 80S methionyl-tRNA initiation complex with soluble factors from wheat germ. J Biol Chem 251:6075–6081
Goldman E, Holmes WM, Hatfield GW (1979) Specificity of codon recognition by *Escherichia coli* $tRNA^{Leu}$-isoaccepting species as determined by protein synthesis in vitro directed by phage RNA. J Mol Biol 129:567–585
Gould AR, Symons RH (1978) Alfalfa mosaic virus RNA. Determination of the sequence homology between the four RNA species and a comparison with the four RNA species of cucumber mosaic virus. Eur J Biochem 91:269–278
Grasmuk H, Noland RD, Drews J (1977) Further evidence that elongation factor 1 remains bound to ribosomes during peptide chain elongation. Eur J Biochem 79:93–102
Gross M (1979) Control of protein synthesis by hemin. Evidence that the hemin-controlled translational repressor inhibits formation of 80S initiation complexes from 48S intermediate initiation complexes. J Biol Chem 254:2370–2377
Guilley H, Briand JP (1978) Nucleotide sequence of turnip yellow mosaic virus coat protein mRNA. Cell 15:113–122
Haro de C, Ochoa S (1978) Mode of action of the hemin-controlled inhibitor of protein synthesis: studies with factors from rabbit reticulocytes. Proc Natl Acad Sci USA 75:2713–2716
Hatfield D, Rice M (1978) Patterns of codon recognition by isoacceptor aminoacyl tRNAs from wheat germ. Nucleic Acids Res 5:3491–3502
Heindell HC, Liu A, Paddock GV, Studnicka GM, Salser WA (1978) The primary sequence of rabbit α-globin mRNA. Cell 15:43–54

Herson D, Schmidt A, Seal SN, Marcus A, van Vloten-Doting L (1979) Competitive mRNA translation in an in vitro system from wheat germ. J Biol Chem 254:8245–8249
Hirashima A, Kaji A (1973) Role of elongation factor G and a protein factor on the release of ribosomes from messenger ribonucleic acid. J Biol Chem 248:7580–7587
Hofstad van der GAJM, Foekens JA, Bosch L, Voorma HO (1977) The involvement of a complex between formylmethionyl-tRNA and initiation factor IF-2 in prokaryotic initiation. Eur J Biochem 77:69–75
Hofstad van der GAJM, Buitenhek A, Bosch L, Voorma HO (1978a) Initiation factor IF-3 and the binary complex between initiation factor IF-2 and formylmethionyl-tRNA are mutually exclusive on the 30-S ribosomal subunit. Eur J Biochem 89:213–220
Hofstad van der GAJM, Buitenhek A, van den Elsen PJ, Voorma HO, Bosch L (1978b) Binding of labelled initiation factor IF-1 to ribosomal particles and the relationship to the mode of IF-1 action in ribosome dissociation. Eur J Biochem 89:221–228
Kabat D, Chappell MR (1977) Competition between globin messenger ribonucleic acids for a discriminating initiation factor. J Biol Chem 252:2685–2690
Kaempfer R, Jay G (1979) Binding of messenger RNA in initiation of prokaryotic translation. Methods Enzymol 60:332–343
Keren-Zur M, Boublik M, Ofengand J (1979) Localization of the decoding region on the 30S *Escherichia coli* ribosomal subunit by affinity immunoelectron microscopy. Proc Natl Acad Sci USA 76:1054–1058
Khorana HG, Buchi H, Ghosh H, Gupta N, Jacob TM, Kossel H, Morgan R, Narang SA, Ontsuka E, Wells HD (1966) Polynucleotide synthesis and the genetic code. Cold Spring Harbor Symp Quant Biol 31:39–49
Kozak M (1978) How do eucaryotic ribosomes select initiation regions in messenger RNA? Cell 15:1109–1123
Kozak M (1980) Influence of mRNA secondary structure on binding and migration of 40S ribosomal subunits. Cell 19:79–90
Leaver C, Dyer JA (1974) Caution in the interpretation of plant ribosomes studies. Biochem J 144:165–167
Lingappa VR, Cunningham BA, Jazurinski SM, Hopp TP, Blobel G, Edelman GM (1979a) Cell-free synthesis and segregation of $B_2$-microglobulin. Proc Natl Acad Sci USA 76:3651–3655
Lingappa VR, Lingappa JR, Blobel G (1979b) Chicken ovalbumin contains an internal signal sequence. Nature (London) 281:117–121
Lloyd MA, Osborne JC, Safer B, Powell GM, Merrick WC (1980) Characteristics of eukaryotic initiation factor 2 and its subunits. J Biol Chem 255:1189–1193
Lodish HF, Rothman JE (1979) The assembly of cell mebranes. Sci Am 240:48–63
Maccecchini ML, Rudin Y, Blobel G, Schatz G (1979) Import of proteins into mitochondria: precursor forms of the extramitochondrially made F1-ATPase subunits in yeast. Proc Natl Acad Sci USA 76:343–347
Macino G, Coruzzi G, Nobrega FG, Li M, Tzagaloff A (1979) Use of the UGA terminator as a tryptophan codon in yeast mitochondria. Proc Natl Acad Sci USA 76:3784–3785
Menninger JR, Walker C (1974) An assay for protein chain termination using peptidyl-tRNA. In: Moldave L, Grossman L (eds). Methods Enzymol 30:303–310
Merrick WC (1979) Evidence that a single GTP is used in the formation of 80S initiation complexes. J Biol Chem 254:3708–3711
Nirenberg M, Leder P (1964) RNA codeword and protein synthesis. Science 145:1399–1407
Nirenberg M, Matthei J (1961) Dependence of cell-free protein synthesis in *E. coli* upon naturally occurring or synthetic polyribonucleotides. Proc Natl Acad Sci USA 47:1588–1602
Ofengand J (1977) tRNA and aminoacyl-tRNA synthetases. In: Weissbach H, Pestka S (eds) Molecular mechanisms of protein biosynthesis, Academic Press, London New York, pp 7–79
Olson HM, Glitz DG (1979) Ribosome structure: localization of 3′ end of RNA in small subunit by immunoelectronmicroscopy. Proc Natl Acad Sci USA 76:3769–3773
Pelham HRB (1978) Leaky UAG termination condon in tobacco mosaic virus RNA. Nature (London) 272:469–471

Pelham HRB (1979a) Synthesis and protoeolytic processing of cowpea mosaic virus proteins in reticulocyte lysates. Virology 96:463–477
Pelham HRB (1979b) Translation of tobacco rattle virus RNAs in vitro: four proteins from three RNAs. Virology 97:256–265
Peterson DT, Merrick WC, Safer B (1979) Binding and release of radiolabeled eukaryotic initiation factors 2 and 3 during 80S initiation complex formation. J Biol Chem 254:2509–2516
Peterson PJ, Fowden L (1963) Different specificities of proline activating enzymes from some plant species. Nature (London) 200:148–151
Ramberg ES, Ishaq M, Rulf S, Mueller B, Horowitz J (1978) Inhibition of transfer RNA function by replacement of uridine and uridine-derived nucleosides with 5-fluorouridine. Biochemistry 17:3978–3985
Ranu RS, Wool IG (1976) Preparation and characterization of eukaryotic initiation factor EIF-3. J Biol Chem 251:1926–1935
Ranu RS, London IM, Das A, Dasgupta A, Majumder A, Ralston R, Roy R, Gupta NK (1978) Regulation of protein synthesis in rabbit reticulocyte lysates by the heme-regulated protein kinase: inhibition of interaction of Met-tRNA$_f^{Met}$ binding factor with another initiation factor in formation of Met-tRNA$_f^{Met}$·40S ribosomal subunit complexes. Proc Natl Acad Sci USA 75:745–749
Reddington MA, Tate WP (1979) A polypeptide chain release factor from the undeveloped cyst of the Brine shrimp, *Artemia Salina.* FEBS Lett 97:335–338
Rich A, Kim SH (1978) The three-dimensional structure of transfer RNA. Sci Am 238:52–62
Safer B, Adams SL, Anderson FW, Merrick WC (1975) Binding of Met-tRNA$_f$ and GTP to homogeneous initiation factor MP. J Biol Chem 250:9076–9082
Safer B, Kemper W, Jagus R (1979) The use of [$^{14}$C]eukaryotic initiation factor 2 to measure the endogenous pool size of eukaryotic initiation factor 2 in rabbit reticulocyte lysate. J Biol Chem 254:8091–8094
Salditt-Georgieff M, Harpold M, Chen-Kiang S, Darnell JE Jr (1980) The addition of 5′ cap structure occurs early in hnRNA synthesis and prematurely terminated molecules are capped. Cell 19:69–78
Salomon R, Bar-Joseph M, Soreq H, Gozes I, Lettauer UZ (1978) Translation in vitro of carnation mottle virus RNA. Virology 90:288–298
Salvato MS, Fraenkel-Conrat H (1977) Translation of tobacco necrosis virus and its satellite in a cell-free wheat germ system. Proc Natl Acad Sci USA 74:2288–2292
Samols DR, Hagenbuchle O, Gage LP (1979) Homology of the 3′ terminal sequences of the 18S rRNA of *Bombyx mori* and the 16S rRNA of *Escherichia coli.* Nucleic Acids Res 7:1109–1118
Samuelsson T, Elias P, Lustig F, Axberg T, Folsch G, Akesson B, Lagerkvist U (1980) Aberrations of the classic codon reading scheme during protein synthesis in vitro. J Biol Chem 255:4583–4588
Schreier MH, Erni B, Staehlin T (1977) Initiation of mammalian protein synthesis. I. Purification and characterization of seven initiation factors. J Mol Biol 116:727–753
Seal SN, Marcus A (1972) Reactivity of ribosomally bound methionyl-tRNA with puromycin and the locus of pactamycin inhibition of chain initiation. Biochem Biophys Res Comm 46:1895–1902
Shatkin AJ (1976) Capping of eucaryotic mRNAs. Cell 9:645–653
Shih DS, Kaesberg P (1976) Translation of the RNAs of Brome mosaic virus: the monocistronic nature of RNA 1 and RNA 2. J Mol Biol 103:77–88
Slobin LI (1979) Eucaryotic elongation factor Ts is an integral component of rabbit reticulocyte elongation factor 1. Eur J Biochem 96:287–293
Sonenberg N, Rupprecht KM, Hecht SM, Shatkin AJ (1979) Eukaryotic mRNA cap binding protein purification by affinity chromatography on Sepharose-coupled m$^7$ GDP. Proc Natl Acad Sci USA 76:4345–4349
Spremulli LL, Walthall BJ, Lax SR, Ravel JM (1979) Partial purification of the factors required for the initiation of protein synthesis in wheat germ. J Biol Chem 254:143–148
Stoffler G, Wittmann HG (1977) Primary structure and three dimensional arrangement of proteins within the *Escherichia coli* ribosome. In: Weisbach H, Pestka S (eds) Molecu-

lar mechanisms of protein biosynthesis. Academic Press, London New York, pp 117–202
Stringer EA, Chaudhuri A, Maitra U (1979) Purified eukaryotic initiation factor 2 from calf liver consits of two polypeptide chains of 48,000 and 38,000 daltons. J Biol Chem 254:6845–6848
Sundquist IC, Staehlin T (1975) Structure and function of free 40S ribosomal subunits: characterization of initiation factors. J Mol Biol 99:401–418
Thompson HA, Sadnik I, Scheinbuks J, Moldave K (1977) Studies on native ribosomal subunits from rat liver. Purification and characterization of a ribosomal dissociation factor. Biochemistry 16:2221–2230
Thompson R, Stone R (1977) Proofreading of the codon-anticodon interaction of ribosomes. Proc Natl Acad Sci USA 74:198–202
Thach SS, Thach RE (1971) Translocation of messenger RNA and "accommodation" of fMet-tRNA. Proc Natl Acad Sci USA 68:1791–1795
Trachsel H, Staehlin T (1978) Binding and release of eukaryotic initiation factor eIF-2 and GTP during protein synthesis initiation. Proc Natl Acad Sci USA 75:204–208
Trachsel H, Erni B, Schreier MH, Staehlin T (1977) Initiation of mammalian protein synthesis. II The assembly of the initiation complex with purified initiation factors. J Mol Biol 116:755–767
Trachsel H, Sonenberg N, Shatkin AJ, Rose JK, Leong K, Bergmann JE, Gordon J, Baltimore D (1980) Purification of a factor that restores translation of vesicular stomatitis virus mRNA in extracts from poliovirus infected HeLa cells. Proc Natl Acad Sci USA 77:770–774
Turnbough CL Jr, Neil RJ, Landsberg R, Ames BN (1979) Pseudouridylation of tRNAs and its role in regulation in *Salmonella typhimurium*. J Biol Chem 254:5111–5119
Weigert MG, Lanka E, Garen A (1967) Amino acid substitutions resulting from suppression of nonsense mutations. III Tyrosine insertion by the SO-4 gene. J Mol Biol 23:401–404
Wittmann HG, Witmann-Liebold B (1966) Protein chemical studies of two RNA viruses and their mutants. Cold Spring Harbor Symp Quant Biol 31:163–172
Yarus M (1979) Relationship of the accuracy of aminoacyl-tRNA synthesis to that of translation. In: Schimmel PR, Soll D, Abelson JN (eds) Transfer RNA. Cold Spring Harbor Monogr, pp 501–515

# 4 Post-Translational Modifications

L. BEEVERS

## 1 Introduction

During the past three decades there have been major improvements in the techniques for isolating and characterizing proteins. As a consequence of the application of these improved procedures it has become apparent that proteins are not simple linear polymers of amino acids but have three-dimensional structure. Many proteins consist of several component peptide chains and they may contain in addition to amino acids other associated components such as metals, carbohydrates, various prosthetic groups as well as substituted amino acids. Proteins are often segregated into specific cellular organelles.

The developments in molecular biology in the 1960's and early 1970's provided an explanation of the mechanism by which the protein amino acids become arranged in linear sequence. However, it is only in more recent years that attention has been focused on the process by which the nascent polypeptides, produced during translation on the ribosome, become post-translationally associated with specific organelles or are modified by folding or addition of such constituents as metals, carbohydrates, prosthetic groups etc. and it is these post-translational modifications which are discussed in this chapter.

## 2 Cleavage of N-Terminal Amino Acids

Initiation of protein synthesis in the cytoplasm of higher plants requires the participation of proteins isolated from cytoplasmic ribosomes by salt washing (WELLS and BEEVERS 1975) or recovered from the post-ribosomal supernatant (BENNE et al. 1980). Chloroplast protein synthesis is initiated by other specific protein factors apparently associated with chloroplast ribosomes. In all these instances the proteinaceous factors are responsible for inserting tRNA-met at the initial acceptor site in the ribosome. This initially incorporated tRNA-met is specific for the initiation process and can be separated from the tRNA-met involved in insertion of internal methionyl residues of the peptide chain (LEIS and KELLER 1971). The chloroplast initiator methionyl-tRNA incorporates N-formyl methionine, whereas cytoplasmic initiator methionyl-tRNA incorporates unsubstituted methionine. A result of these initiation sequences is that all peptides synthesized in the chloroplasts initially contain an N-terminal formyl methionine residue and all peptides synthesized in the cytoplasm have N-terminal methionine. However, reference to amino acid sequences of purified proteins

(DAYHOFF 1972) indicates that any of the 20 protein amino acids can occur in the N-terminal position. Thus an early event of post-translational modification is usually the cleavage of N-terminal methionyl or formyl methionyl residues. So far peptidases responsible for removal of these components have not been characterized from plant sources.

## 3 Secondary and Tertiary Structure

Although the conformation of only a few plant proteins has been determined, it is apparent that, rather than consisting of linear arrays of amino acids they contain such secondary structures as α-helices and tertiary structure. (See Chap. 8, this Vol.) Formation of disulphide-bridges between peptidyl cysteine residues is commonly found, but it is not completely established how the folding and formation of these disulphide-bridges is achieved. The observations that not all of the cysteine residues of a protein are involved in disulphide-bond formation and that pairing always occurs between specific cysteinyl residues suggest that the formation of correct disulphide-linkages may be under cellular control. However it has also been proposed (EPSTEIN et al. 1963) that the gain in thermodynamic stability upon folding provides a driving force for correct in vitro folding and disulphide-bond formation. Suggestions for this thermodynamic control of folding have originated from experiments with ribonuclease A. When reduced, unfolded RNAse A is reoxidized in 8 M urea a scrambled set of disulphide-bonds is formed, however removal of the urea and addition of sulphydryl reagent results in the reshuffling of disulphide-bonds and leads to the formation of a folded protein with correct disulphide-bonds and native structure of RNAse A. The recovery of the native RNAse A involves a disulphide interchange reaction. In vivo disulphide interchange may be mediated by one or a series of enzymes (FREEDMAN 1979) and enzymatic activity catalyzing disulphide interchange has been characterized from wheat (*Triticum vulgare*) embryos and demonstrated in extracts from other plant tissues (GRYNBERG et al. 1978). It appears that reduced glutathione serves as cofactor in the interchange reaction which is measured as the capacity to restore enzyme activity to randomly crosslinked ribonuclease.

### 3.1 Quaternary Structure

Single polypeptide chains may assemble into complexes that range in complexity from simple oligomeric proteins with molecular weights of less than 100,000 to supramolecular complexes with molecular weights in the millions. The largest complexes may become part of the cellular architecture as for example microtubules, intermediate-sized assemblies occur in multienzyme complexes and ribosomal subunits, and smaller multisubunit proteins may be enzymes. No comprehensive list is available of the subunit composition of proteins derived from plants although the review by KLOTZ et al. (1975) includes subunit composition of some proteins of plant origin. The dissociation of proteins into subunits is usually achieved by denaturing agents such as urea, guanidine hydrochloride or sodium dodecyl sulphate. Under these conditions by far the majority of multimeric proteins dissociate into two or four monomeric subunits. The component subunits of many multisubunit proteins associated with organelles may be assembled at different cellular locations. It is possible that the association of a subunit produced in the cytoplasm with a subunit translated by organellar ribosomes provides a mechanism for assuring confinement and sequestration of a particular protein within an organelle. The mechanism of subunit association is not understood, however it may involve non-covalent interaction between the amino acid residues of component subunits in proteins and/or the formation of interchain disulphide bonds perhaps mediated by protein disulphide isomerase. The observations that conformational changes occur following addition of various extrinsic factors such as substrate, metal ions, and other ligands to enzyme solutions indicates that quaternary structure is capable of some degree of rearrangement. This conformational rearrangement provides a mechanism of regulation of activity of multisubunit enzymes. It has been proposed that the light regulation of enzyme activity may involve structural modifications of the enzyme brought about by changes in the level of effectors such as NADPH and ATP or by alteration in the enzyme sulphydryl status (BUCHANAN 1980).

## 4 Modification of Protein Amino Acids

The polypeptide chains of proteins are assembled from the 20 amino acids specified in the genetic code. However, over 140 amino acids or their derivatives have been identified as constituents of proteins of different organisms (UY and WOLD 1977). This diversity of amino acids is brought about by post-translational modifications of peptidyl amino acids by such processes as methylation, phosphorylation, hydroxylation, acetylation etc.

### 4.1 Methylation

In proteins from animal systems the amino acids lysine, arginine, histidine, glutamic acid and aspartic acid may be methylated (PAIK and KIM 1975). Methylation of glutamic acid and aspartic acid produces methyl esters which are susceptible to acid hydrolysis and this renders their detection in proteins difficult. While histones have been found to contain methylated aspartate and glutamate, evidence for the occurrence of other proteins containing these derivatives is based for the most part on detection of protein methylases which can transfer methyl residues from S-adenosyl methionine to various substrate proteins (PAIK and KIM 1975). The occurrence of these enzymes in plants has not been described.

3-N-methylhistidine has so far been reported only in proteins of animal origin and there appears to be considerable interspecies variability in the methylation of histidine residues of specific proteins.

Data compiled by Matsuoka cited by PAIK and KIM (1975) demonstrate the occurrence of ε-N-dimethyllysine, ε-trimethyllysine, $N^G$-monomethylarginine, $N^G$, $N^G$-dimethylarginine and $N^G$, $N'^G$-dimethylarginine in proteins of plants. It has been demonstrated that methylated lysine residues occur in ribosomal proteins of yeast (KRUISWIJK et al. 1978b), histone III from peas (PATTHY et al. 1973), and cytochrome c from higher plants and some fungi (RAMSHAW et al. 1974). Although several proteins containing methylated arginine have been identified from animal systems (PAIK and KIM 1975) apart from the reported association of methyl arginine residues with ribosomal proteins from yeast (KRUISWIJK et al. 1978b) the nature of the proteins containing this substituted amino acid in plants has not been established.

In those proteins containing methyllysine it is found that only certain lysine residues are methylated and mono-, di- and trimethyllysine derivatives are encountered. In histone III where lysine 9 and 27 are methylated, the residues are adjacent to arginine and serine residues which it is suggested may provide the recognition sites for the protein methylase. A protein methylase has been characterized from plants; the enzyme showed activity towards histone III and IIb but not histone IV (PATTERSON and DAVIES 1967). Although the timing of histone methylation has not been established in plants, it has been found that in animal cell tissue cultures and rat livers the amino acid modification occurs in the $G_2$ phase of the cell cycle after DNA and histone synthesis. The function of methylation in histones is not known, however, the process, unlike that of acetylation and phosphorylation, appears to be irreversible (ISENBERG 1979).

Although the location of methylated lysine residues in the ribosomal proteins from yeast has not been established, it has been shown that only specific proteins contain the modified amino acid. Proteins $S_{31}$ and $S_{32}$ of the small ribosomal subunit and $L_{15}$ and $L_4$ of the large subunit are methylated in vivo (KRUISWIJK et al. 1978b). These proteins belong to a group that associates with ribosomal particles at a relatively late stage, and it is suggested that methylation of the proteins may have a functional rather than a structural role.

In higher plant cytochrome c ε-N-trimethyllysine residues occur at residues 80 and 94, whereas in fungal cytochrome c only one trimethyllysine is usually encountered at residue 80 (RAMSHAW et al. 1974; POLASTRO et al. 1977). Protein methylases capable of transferring methyl residues from *S*-adenosyl methionine to specific lysine residues in horse heart cytochrome c have been prepared from *Neurospora* and yeast (NOCHUMSON et al. 1977, DI MARIA et al. 1979). The enzymes which appear to be located in the cytosol showed no activity with histones and produced monomethyl-, dimethyl- and trimethyllysines. Apocytochrome c is a better substrate than the holocytochrome c. In vivo studies with yeast suggest that protein synthesis is required for methylation (FAROOGUI et al. 1980), however, SCOTT and MITCHELL (1969) have demonstrated a temporal separation of methylated and unmethylated cytochrome c synthesis in *Neurospora*. It is suggested that methylation of the apocytochromes occurs in the cytosol and the trimethylation of the lysine residues enhances association of the enzyme with the mitochondria (POLASTRO et al. 1978).

### 4.2 Phosphorylation

Although many phosphorylated proteins have been characterized from animals there are only a few detailed descriptions of the phosphorylation of specific plant proteins. It is nevertheless apparent that groups of proteins phosphorylated in animals are similarly modified in plants. Thus histones, nucleoproteins, ribosomal proteins and enzymes which have been shown to be phosphorylated from animal sources (RUBIN and ROSEN 1975) are also capable of being phosphorylated in plants (TREWAVAS 1976).

Although the extent of phosphorylation of H1 histone in plants does not appear to have been determined, a protein kinase with specificity to it has been purified from soyabean (*Glycine max*) hypocotyls (LIN and KEY 1976). Extensive phosphorylation of chromatin-associated proteins can be detected in vivo (VAN LOON et al. 1975), the phosphate being primarily associated with serine residues. Phosphorylation of the chromatin-associated proteins is apparently mediated by protein kinase. The kinase involved in modification of the non-histone proteins is separable and distinct from that involved in histone phosphorylation (LIN and KEY 1976, MURRAY et al. 1978a, MURRAY et al. 1978b).

The extent of phosphorylation of ribosomal proteins is subject to debate. If ribosomes are isolated from plants which have been incubated with [$^{32}$P] orthophosphate, radioactivity is associated with a phosphoprotein of molecular weight 42,000 associated with the small ribosomal subunit (TREWAVAS 1973). The phosphorylation occurs principally on serine residues. In contrast, if isolated ribosomes are incubated with ATP in the presence of protein kinase, several phosphorylated proteins can be recovered from both the large and small subunit (SIKORSKI et al. 1979). In yeast two proteins of the small subunit and four proteins of the large subunit are phosphorylated from $^{32}$P orthophosphate in vivo (KRUISWIJK et al. 1978a). The finding that more ribosomal proteins became phosphorylated in vitro than in vivo in plant systems is similar to the situation encountered in animal systems where the consensus is that only the proteins of one small subunit are normally phosphorylated (WOOL 1979). A protein kinase presumably responsible for the phosphorylation of the ribosomal proteins has been demonstrated in ribosomes from plants (TREWAVAS and STRATTON 1977).

Several thylakoid polypeptides of the pea chloroplast incorporate [$^{32}$P]-orthophosphate in the light. The labelling of the polypeptides, which appear to be located on the outer surface of the thylakoid membrane, is apparently mediated by a light-stimulated protein kinase (BENNETT 1979, 1980).

KREBS and BEAVO (1979) list over 21 enzymes that undergo phosphorylation in animals, however only detailed studies of the phosphorylation of the pyruvate dehydrogenase complex have been reported from plants (RANDALL and RUBIN 1977, RUBIN and RANDALL 1977, RAO and RANDALL 1980, RALPH and WOJCIK

1978). It has been demonstrated that protein synthesis in animals can be regulated by the phosphorylation of a component involved in the assembly of an initiation complex (RANA 1980). Unfortunately to date the proteins involved in assembly of the initiation complex in plant cytoplasm have not been extensively characterized (BENNE et al. 1980), however a preliminary report indicates that while the proteinaceous cofactor $EIF_2$ from wheat germ can undergo phosphorylation, this modification has no effect on enzyme activity.

The kinases so far characterized from plants fit into category 5 of the classification of KREBS and BEAVO (1979), i.e., non-specified or messenger-independent protein kinases. To date there are no reports from plant sources of the cyclic AMP, cyclic GMP, $Ca^{2+}$, or double-stranded RNA-dependent kinases found in animal tissues. However, there is an increasing number of reports of the presence of cyclic AMP and of the enzymes involved in its metabolism from plant sources (BROWN et al. 1979a, b).

It is generally considered that phosphorylation of proteins represents a mechanism for modulating their activity. Thus it has been conjectured that phosphorylation of histones may regulate cell division, phosphorylation of nuclear acidic proteins may modulate gene expression, and phosphorylation of enzymes may control their activity (TREWAVAS 1976). Given the capacity to control the functioning of proteins depending on the phosphorylation status, it is apparent that for fine control in addition to mechanisms for phosphorylating proteins there is a need for dephosphorylation. This dephosphorylation appears to be mediated by protein phosphatases (KREBS and BEAVO 1979). Although dephosphorylation has been reported to activate the plant pyruvate dehydrogenase complex (RAO and RANDALL 1980), little information on the phosphatases involved is available. BENNETT (1980) has indicated that chloroplast thylakoids contain a bound phosphoprotein phosphatase which, like the phosphatases from animals, is stimulated by $Mg^{2+}$ and inhibited by NaF.

## 4.3 ADP-Ribosylation

Plant cells contain enzymes which catalyze the transfer of the ADP moiety of NAD (ADP ribosyl nicotinamide) to macromolecular acceptors with the concomitant release of nicotinamide and a proton (PAYNE and BAL 1976). If only one ADP residue is transformed, an ADP ribosylated macromolecule is produced. It is also possible to transfer sequential ADP residues to produce poly(ADP-ribose) macromolecules as indicated:

$$\text{n(ADPR-N)} + \text{X} \rightarrow \text{(ADPR)n} - \text{X} + \text{nN} + \text{nH}^{+}.$$

The macromolecular acceptor may be histones or other chromatin-associated proteins (WHITBY et al. 1979). When nuclei are incubated with NAD poly (ADP-ribose) accumulates within the histones $H_1$, $H_2$ and $H_2B$. Studies with animals (HAYAISHI and UEDA 1977) indicate that ADP ribosylation involves ester linkage to the carboxyl groups of aspartate and glutamate.

Although the biological significance of poly(ADP-ribose) has not been established, it has been implicated in modulation of DNA replication and transcription. It is speculated that fluctuations in the degree of poly ADP ribosylation could determine regulatory function of histones or other nuclear proteins. The extent of ADP ribosylation could be controlled by activities of the poly (ADP-ribose) synthetase and degradative enzymes which have been demonstrated in plant nuclei (WHITBY and WHISH 1978).

## 4.4 Hydroxylation

Many plant glycoproteins contain carbohydrate residues covalently linked to hydroxyproline. This hydroxyproline is formed by a post-translational modification of proline residues in the polypeptide chain by the enzyme peptidyl proline hydroxylase (CHRISPEELS 1970). The proline hydroxylase utilizes molecular oxygen and requires ascorbate, $\alpha$-ketoglutarate and ferrous ($Fe^{2+}$) ions (SADAVA and CHRISPEELS 1971a, SADAVA et al. 1973). Hydroxylation of proline residues occurs after release of the nascent peptide from the polysomes, and according to early reports this process occurs in the cytosol (SADAVA and CHRISPEELS 1971b). In animals prolyl hydroxylase has been located in the cisternal space of the endoplasmic reticulum (PETERKOFSKY and ASSAD 1976). It would appear worthwhile to reinvestigate the location of the peptidyl proline hydroxylase in plants, using newer techniques of cell fractionation.

Not all proline residues are hydroxylated (GLEESON and CLARKE 1979). However, since the primary structure of hydroxyproline containing proteins has not been fully established, it has not been determined what characteristics of the peptide chain dictate which peptidyl proline residues will be hydroxylated.

## 4.5 Acetylation

In studies with animals (primarily calf thymus) it has been established that all of the histones may be acetylated. Acetyl serine is the amino terminal residue of the major classes $H_1$, $H_2$ and $H_4$. In addition, amino acid sequence studies have established that a limited and specific set of internal peptidyl lysine residues is subject to acetylation to form $\varepsilon$-N-acetyl lysine (ALLFREY 1977, ISENBERG 1979).

In the limited studies of amino acid sequences and substitutions of plant histones amino terminal acetyl serine and internal acetyl lysine residues have been detected in histone $H_4$ (DELANGE et al. 1969). However, in contrast to the $H_3$ from calf thymus, no internal acetyl lysine residues were found in this histone from peas (PATTHY et al. 1973). The degree of acetylation of histones I, 2A, 2B from plants does not appear to have been investigated.

The degree of acetylation of lysine residues in the histones from animals is variable and it has been suggested that the acetylation of histones may regulate chromatin structure (ALLFREY 1977). Thus the failure to detect acetyl lysine in histone $H_3$ from peas may merely reflect the developmental stage, rather than demonstrating an evolutionary inability of plants to carry out this process.

The mechanism of acetylation of histone seryl and lysyl residues has not been worked out. Amino terminal serine acetylation could proceed by acetylation of a terminal serine, exposed by cleavage of precursor methionine, by an enzymatically catalyzed transfer of acetyl residues from acetyl CoA. An alternative mechanism proposes that seryl-tRNA is acetylated with the production of acetyl seryl-tRNA. It is speculated that the acetyl serine replaces the methionine cleaved from the histone precursor (ALLFREY 1977). The modification of amino terminal serine appears to occur in the cytoplasm and the reaction is essentially irreversible.

In contrast, acetylation of lysine residues in histones is apparently rapidly reversible and occurs predominantly in the nucleus. The acetylation of lysine residues is mediated by acetyl transferases (different acetyl transferases show specificity for particular histones) which utilize acetyl CoA as the acetyl donor. Removal of acetyl residues from peptidyl lysine in histones is catalyzed by deacetylases (ALLFREY 1977).

### 4.6 Non-Protein Amino Acids

Hydrolysates of certain arabinogalacto proteins indicate the presence of a basic amino acid tentatively identified as ornithine (GLEESON and CLARKE 1979, JERMYN and YEOW 1975, ALLEN et al. 1978). If subsequent analyses confirm the presence of this amino acid as a component of the peptide chain of these glycoproteins, mechanisms for its formation will need to be established. Ornithine is synthesized from glutamate and is usually encountered in plants as an intermediate in the biosynthesis of arginine or citrulline. Alternatively, ornithine can be produced during the catabolism of arginine by the enzyme arginase and it would appear simpler to use the latter mechanism.

## 5 Conjugated Proteins

Various prosthetic groups can associate with proteins to form conjugated proteins. The polypeptide to which the prosthetic group is associated is called the apoprotein and if the conjugate possesses catalytic activity it is termed the holoenzyme.

### 5.1 Haemoproteins

In these conjugated proteins, which include cytochromes, peroxidases and catalases, an iron porphyrin (haem) is associated with a protein unit. The linkage of the haem to the protein in most instances involves direct iron binding to the imidazole ring of peptidyl histidine and electrostatic interactions between carboxylate groups of the porphyrins and positively charged protein residues, or thioether links between cysteine residues of the protein and vinyl groups of the porphyrins. Catalase differs from other haemoproteins in that the haem iron appears to be ligated by side chain carboxyl groups. As discussed in Section 6, addition of the prosthetic group appears to occur as a post-translational event.

## 5.2 Porphyroproteins

It has been established by the use of detergents and polyacrylamide gel electrophoresis that chlorophyll (magnesium-containing tetraporphyrin) exists as a chlorophyll–protein complex. Originally it appeared that the chlorophyll was associated with two major complexes termed P-700–chlorophyll a protein complex and the light-harvesting chlorophyll a/b complex. However, recent studies (MARKWELL et al. 1979) indicate that different detergents produce complexes of varying composition and electrophoretic mobility. There is no information on the peptidyl or porphyrin residues involved in the assembly of the chlorophyll protein complexes.

Linear as opposed to the cyclic tetrapyrroles of haem and chlorophylls occur in phytochrome of higher plants and the phycoerythrins and phycocyanin of blue-green algae. The association of these linear tetrapyrroles to the apoprotein has been shown to involve an ester linkage between the OH group of peptidyl serine and propionic residues of the inner pyrrole, probably ring C, in addition to a thioether bond between cysteine and a 2-carbon side chain of pyrrole ring A in phycoerythrins and phycocyanins (KILLILEA et al. 1980). Although the structure of the tetrapyrrole component of phytochrome has been established, the nature of the association with the apoprotein has not been determined.

## 5.3 Flavoproteins

The nature and mechanism of attachment of the flavin to the apoprotein in flavoprotein has not been established. On the basis that mercurial and sulphydryl reagents impede the interaction of apoproteins with flavin nucleotides it appears that peptidyl sulphydryl groups may be involved. Modification of peptidyl tyrosine residues also inhibits flavin binding, thus implicating the involvement of this amino acid in formation of the conjugated protein.

## 5.4 Metalloproteins

Complexes of metals and proteins have been classified as metalloproteins or metal–protein complexes. Metal atoms of metalloproteins are bound so firmly that they are not removed from the protein by the isolation procedure, hence the highly purified protein ultimately contains stoichiometric quantities of metal. When the metalloprotein is catalytically active the stable association of a particular metal with an apoenzyme demonstrates a specific biological role for the metal in this system. When the metal atom is loosely bound to the protein, the association is both chemically and biologically more tenuous, hence the term metal–protein complex is used (VALLEE and WACKER 1970).

Metalloproteins which do not serve an enzymatic function include ferritin, metallothioneins, calmodulin and lectins. Ferritin is a multimeric protein with which iron becomes associated and probably functions in the storage of the metal (HYDE et al. 1963). In metallothioneins, metal ions such as zinc or copper are bound by three cysteinyl residues (VASAK et al. 1980). It has been suggested (CURVETTO and RAUSER 1979) that metallothioneins can function to complex potentially toxic metal ions in plants, however it is also possible that they operate in metal homeostasis and provide metals for the activation of apoenzymes (UDOM and BRADY 1980). A great deal of interest is being shown currently in calmodulin (GRAND et al. 1980; MEANS and DEDMAN 1980). This is a low molecular weight protein with up to four unidentified calcium-binding sites which modulates the activity of other cellular enzymes. Various metal ions are associated with many plant lectins and they frequently require two different metals for complete activity in the binding of carbohydrates, however the nature of association has not been established in most instances (LIENER 1976).

## 6 Metalloenzymes

The review by CLARKSON and HANSON (1980) documents the association of calcium, manganese, copper, iron, zinc and molybdenum in metalloenzymes.

The best characterized calcium-containing metalloenzyme in plants is α-amylase, however nothing is known of the mechanism of Ca linkage to the protein. In other calcium-binding proteins which have been characterized, the ion appears to be coordinated with carboxyls of glutamate and aspartate (KRETSINGER 1976).

Although manganese can stimulate the activity of a number of isolated enzymes, the nature of the association of the metal and the protein is not established.

Copper-containing metalloenzymes participate in cellular oxidation reductions. In plastocyanins the copper is linked to peptidyl methionine, cysteine and two histidine residues, whereas in the blue copper protein stellacyanin from *Rhus vernicifera* two histidines and two cysteines complex the copper (HILL and LEE 1979). The mechanism by which copper is introduced into the metalloprotein complex is unknown.

Although VALLEE (1977) indicates that more than 70 zinc-containing enzymes have been identified from animal sources, the occurrence of the metal in the same enzymes from plants has been confirmed in relatively few instances. Zinc has, however, been demonstrated in alcohol dehydrogenase, carbonic anhydrase and superoxide dismutase (CLARKSON and HANSON 1980) and more recently in RNA polymerase II (PETRANY et al. 1977).

Within the zinc-containing enzymes the primary structure of the enzyme protein probably dictates the relative position of those amino acid side chains destined to be ligated when the apoprotein combines with the metal ion. Evidence suggests that the metal ion is not incorporated into the growing polypeptide chain until the protein is fully formed. It appears that cysteinyl, histidyl, tyrosyl residues and the carboxyl groups of aspartic acid and glutamic acid are ligated with zinc (VALLEE and WACKER 1970).

Iron metalloproteins participate in redox reactions and the iron may be associated with sulphur groups in the *iron sulphur proteins* or associated with haem residues in *haemoproteins*. The iron sulphur proteins are implicated in the oxidation of NADH, succinate and xanthine, and in addition appear to be involved in photosynthetic electron transport and nitrite reduction in plants. In the iron sulphur proteins so far characterized, i.e., mainly lower molecular weight redox proteins, the iron or $Fe_2S_2$ or $Fe_4S_4$ is complexed to peptidyl cysteine residues.

Apoferredoxin, the iron sulphide-free protein, may be converted to native ferredoxin non-enzymatically in a reaction mixture containing ferrous or ferric salts and sodium sulphide in the presence of aqueous mercaptoethanol (HONG and RABINOWITZ 1967). It is not known whether ferredoxin is formed in an analogous manner in vivo. Iron occurs in cells and also sulphide is present in plants as a result of sulphite reductase activity, an enzyme in the normal pathway of reduction of assimilated sulphate. However, it is unlikely that Fe, S or $Fe_2S_3$ are utilized directly in the conversion of apoenzymes to holoenzymes since the iron sulphur salts are extremely insoluble. An apparent enzymatic reconstitution of parsley ferredoxin from apoferredoxin in the presence of $FeCl_3$, thiosulphate dithiothreitol and rhodanese (thiosulphate:cyanide sulphurtransferase) has been reported (FINAZZI-AGRO et al. 1971, TOMATI et al. 1974). However, it has been indicated (BRODRICK and RABINOWITZ 1977) that this may actually have been a chemical rather than enzymatic reconstitution.

The most fully characterized haemoproteins are the cytochromes. Although the nature of the association between the prosthetic haem group with the apoprotein has not been established in plants, it has been shown that in cytochrome c from horse heart the haem residue is bound to the polypeptide chain through covalent sulphur linkages to specific peptidyl cysteine residues. In addition axial iron ligands occur between the imidazole residue of peptidyl histidine and a methionyl sulphur.

In *Neurospora* apocytochrome c is synthesized on cytoplasmic ribosomes and released. Apparently the apocytochrome c then migrates to the inner mitochondrial membrane, where the haem group is covalently linked to the apoprotein. The change in conformation following haem addition leads to a trapping of the holocytochrome c in the membrane (KORB and NEUPERT 1978). It is not established whether the addition of the haem group to the apoenzyme is enzymatic or due to a chemical association with the appropriately aligned configuration of amino acid residues produced during folding of the apoprotein.

Molybdenum is a component of nitrate reductase (HEWITT 1974) and xanthine dehydrogenase (MENDEL and MULLER 1976). From studies with mutants of the fungi *Neurospora* and *Aspergillus* deficient in nitrate reductase it has been established that molybdenum is associated with a molybdenum-containing cofactor. Since the molybdenum-containing cofactor prepared from a variety of enzyme sources can, by complementation in vitro, restore nitrate reductase activity in mutant strains (LEE et al. 1974), it appears that there is a common molybdenum cofactor for a variety of molybdoproteins. A similarity of the molybdenum-containing cofactor for nitrate reductase and xanthine dehydrogenase in higher plants is indicated by the observation that certain mutants contain aberrant activity of both enzymes (MENDEL and MULLER 1976). Recent studies of the molybdenum cofactors from a variety of enzyme sources have confirmed their similarity and in addition have demonstrated the presence of pteridine residues in the cofactor (JOHNSON et al. 1980). It is speculated that the pteridine residues, in addition to being involved in metal complexing and participating in oxido-reductions reactions during enzyme catalysis, may also participate in the strong non-covalent interactions which anchor the cofactor to various apoproteins. A non-covalent association of a molybdenum-containing cofactor of nitrate reductase from higher plants is demonstrated by the ability to remove the cofactor from the enzyme by acid treatment or by passage through AMP-Sepharose. The molybdenum-containing cofactor removed by these procedures was able to activate the apoprotein obtained from molybdenum-deficient plants (NOTTON and HEWITT 1979).

## 7 Glycoproteins

During the past 15 years it has become well established that glycoproteins are normal constituents and of widespread occurrence in plants (See also Chap. 8, this Vol.). It has been demonstrated that the glycosyl moiety of the glycoproteins can contain hexoses (mannose, glucose, galactose), deoxyhexose (fucose), hexosamines (N-acetylglucosamine) and pentoses (xylose and arabinose) (BROWN and KIMMINS 1977, SHARON and LIS 1979). Of these carbohydrates

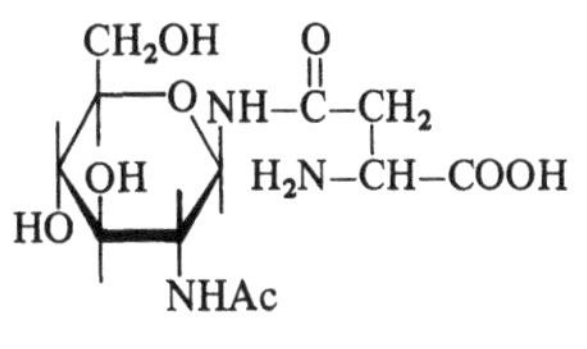

β–N–Acetylglucosaminyl–L–asparagine

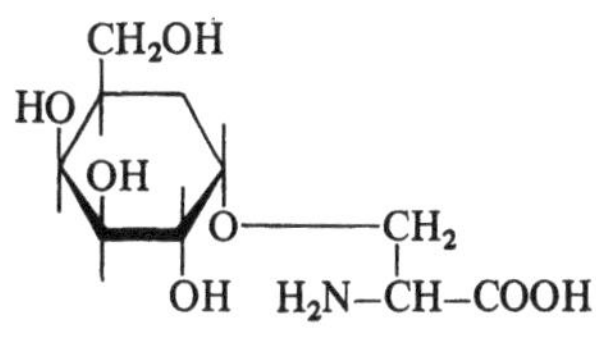

Galactosyl–α–L–serine

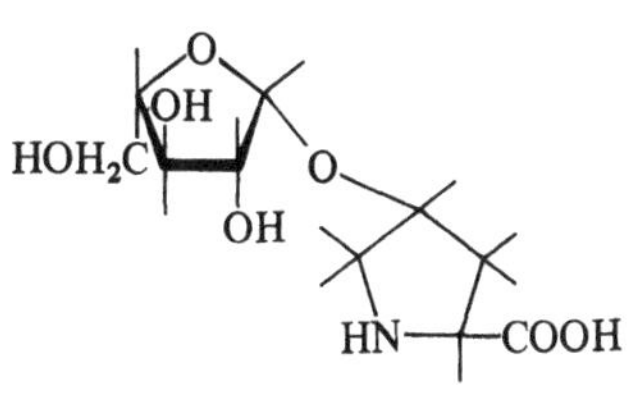

L–Arabinosyl–β–4–hydroxy–L–proline

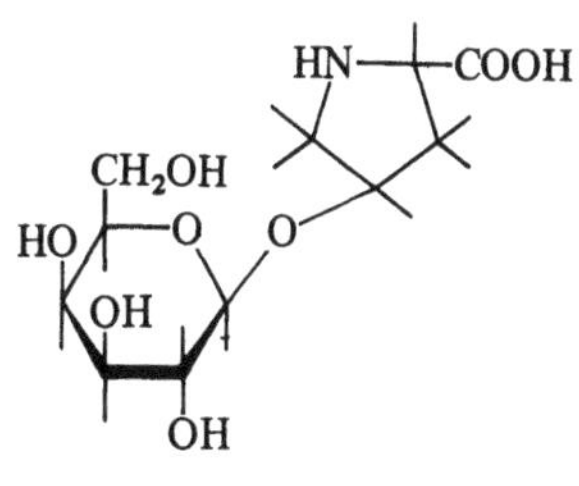

Galactosyl–α–4–hydroxy–L–proline

**Fig. 1.** Carbohydrate-peptide linkages in plant glycoproteins

mannose, galactose, N-acetylglucosamine and arabinose may be involved in carbohydrate peptide linkages. Basically two types of carbohydrate peptide linkage have been characterized: the N-glycosidic and the O-glycosidic (Fig. 1). While both types of glycosidic linkage have been reported in glycoproteins from animals it appears that apart from yeast mannoproteins the glycoproteins from plants are specifically N- or O-glycosidic. However, there can be a variety of O-glycosidic linkages within the same glycoproteins.

### 7.1 O-Glycosidic Linkages

The best characterized glycoproteins containing this glycopeptide are extensin, potato lectin, and the arabinogalacto-protein from the style of *Gladiolus*.

In extensin, arabinose residues in the furanose (*f*) configuration are linked to peptidyl hydroxyproline (Hyp) in the following sequences (AKIYAMA and KATO 1977):

[*L*-Ara*f*β(1-2)-L-Ara*f*β(1-2)-L-Ara*f*β(1-5)Hyp] or
[*L*-Ara*f*β(1-3)-L-Ara*f*β(1-2)-L-Ara*f*β(1-2)-L-Ara*f*β(1-5)Hyp].

The arabinosyl-hydroxyproline linkages in extensin are alkali stable, but are hydrolyzed by mild acid treatment. The hydroxyproline-rich polypeptide released by mild acid hydrolysis is more susceptible to hydrolysis by trypsin than the native glycoprotein. The trypsin digests of the acid-treated hydroxyproline contain the pentapeptide–Ser-Hyp-Hyp-Hyp-Hyp in which the serine residues may be galactosylated (LAMPORT et al. 1973). Although most of the characterization of extensin was performed on the glycoprotein extracted from cells grown in tissue culture, similar hydroxyproline-rich glycoproteins (HPRG) have been detected in intact plants (ESQUERRE-TUGAYE and LAMPORT 1979). It is interesting that the extent of glycosylation of hydroxyproline is higher in plants infected with fungi than in healthy plants, and that the extent of glycosylation of serine does not vary significantly upon infection but declines with increasing age. Apparently the processes involved in forming the two types of carbohydrate to protein linkage are under independent control (ESQUERRE-TUGAYE and LAMPORT 1979). This varying degree of glycosylation may account for some of the heterogeneity reported in putative precursors of HPRG's (TANAKA and UCHIDA 1979).

In addition to extensin and HPRG, other hydroxyproline-poor glycoproteins have also been isolated from the cell walls of plants (SELVENDRAN 1975, BROWN and KIMMINS 1978). It appears that there may be two of these which are differentially produced in the plant by wounding or virus infection. Thus if leaves of *Phaseolus* are wounded a glycoprotein enriched in xylose is produced; following viral infection glucose is the major carbohydrate constituent of the glycoproteins. Peptidyl serine and hydroxyproline participate in the glycopeptide linkage. Most of the substituted hydroxyproline residues contain arabinose, galactose and glucose but some have arabinose only. Serine residues contain arabinose, galactose and glucose (BROWN and KIMMINS 1978, 1979).

### 7.1.1 Arabinogalactan-Proteins

Various glycoproteins containing arabinose and galactose as principal carbohydrate components have been reported in plants. POPE (1977) found soluble glycoproteins (80%–95% carbohydrate) which were enriched in hydroxyproline in the medium of plant tissue cultures. About 50% of hydroxyproline residues were substituted with a mixed arabinogalactan polysaccharide which contained equal proportions of arabinose and galactose in addition to xylose and glucose. It was suggested that the glycopeptide was composed of hydroxyproline linked to glucose and galactose. Similar hydroxyproline galactose linkages have been reported in the arabinogalactan peptide from wheat endosperm and a cell wall glycoprotein of a green alga (MILLER et al. 1972).

GLEESON and CLARKE (1979) have isolated an arabinogalactan-protein from gladiolus styles by affinity chromatography in tridacnin (a galactose-binding protein from the clam *Tridacna maxima*) or by precipitation with $\beta$-glucosyl artificial carbohydrate antigen. This glycoprotein is similar to the extracellular glycoproteins characterized by POPE (1977) in that it is 90% carbohydrate, however the galactose/arabinose ratio is 6:1. The protein moiety has a high serine content and it is proposed that a $1 \rightarrow 3$-$\beta$ polygalactose backbone with a $1 \rightarrow 6$-$\beta$ galactose side chain and a terminal $\alpha$-L arabinofuranosyl residue is linked to the peptidyl serine. The gladiolus stylar arabinogalactan is low in hydroxyproline and thus differs from other $\beta$-lectins which are also precipitated by $\beta$-glucosyl artificial carbohydrate antigen (JERMYN and YEOW 1975, ANDERSON et al. 1977).

Potato lectin binds specifically to $\beta$-1–4 linked oligosaccharides of N acetylglucosamine. The purified protein contains 47% by weight arabinose and 3% galactose. The protein moiety is characterized by a high hydroxyproline content, 11% (ALLEN et al. 1978). All of these hydroxyproline residues are substituted with oligoarabinose with an arabinose/hydroxyproline ratio of 3.4:1 demonstrating the occurrence of L$(Ara)_3$-Hyp and L$(Ara)_4$-Hyp constituents similar to those encountered in extensin. In the lectin and extensin, arabinose is linked to hydroxyproline in the $\beta$-arabinofuranoside configuration, and thus differs from the proteoglycan of rice bran in which $\alpha$-arabinofuranoside linkages are reported (YAMAGISHI et al. 1976). Nine or ten serine residues in the potato lectin are substituted with $\alpha$-galactopyranoside (ALLEN et al. 1978).

## 7.2 Yeast Mannan

Yeasts synthesize a number of mannoproteins which possess partly O- and N-glycosidically linked carbohydrate moieties. The O-glycosidically linked components consist of manno-mono, or manno-oligosaccharides linked to peptidylseryl or threonyl residues (BALLOU 1976).

## 7.3 N-Glycosidic Linkages

The N-glycosidic linkage usually involves an association between peptidyl asparagine and N-acetylglucosamine. This linkage which was initially described in plants by LIS et al. (1969) has now been reported in horseradish peroxidase (CLARKE and SHANNON 1976), stem bromelain (YASUDA et al. 1970), lima bean lectin (MISAKI and GOLDSTEIN 1977), 7S protein from soyabean (YAMAUCHI et al. 1976), vicilin from mung bean (ERICSON and CHRISPEELS 1976), a legumin preparation from peas (BROWDER and BEEVERS 1978) and a lectin from sainfoin (*Onobrychis viciifolia*) seed (NAMEN and HAPNER 1979). The N-acetylglucosaminyl asparagine linkage also occurs in many fungal glycoproteins (BALLOU 1974, 1976; ONISHI et al. 1979). In addition a novel N-glycosidic linkage has been reported in cutinase, an extracellular glycoprotein from the fungus *Fusarium solani* f. *pisi*. In this glycoprotein the N-terminal glycine is in amide linkage with glucuronic acid (LIN and KOLATTUKUDY 1977).

Structural analyses of the glycosyl moiety of soyabean agglutinin (LIS and SHARON 1978), lima bean agglutinin (MISAKI and GOLDSTEIN 1977), stem bromelain (ISHIHARA et al. 1979) and sainfoin lectin (NAMEN and HAPNER 1979) indicate that all of these glycoproteins contain mannosyl residues covalently linked to two N-acetylglucosaminyl components in the characteristic manner depicted below.

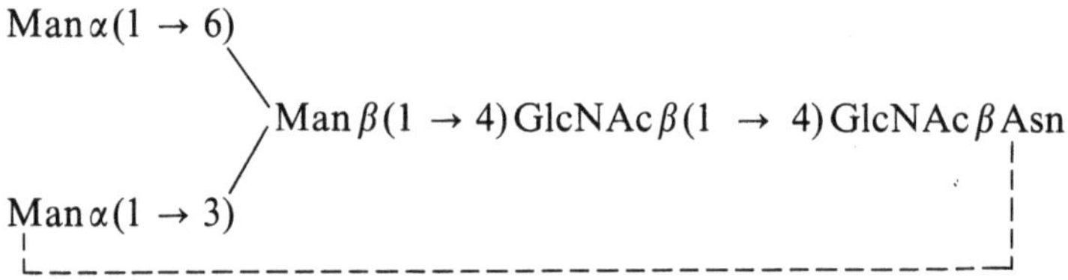


Core oligosaccharide

This core oligosaccharide is encountered in mannose-rich or oligomannosidic glycoproteins from yeast (BALLOU 1974, 1976), and animals (KORNFELD and KORNFELD 1976). Additional carbohydrates, mannose, xylose and fucose in plants are added to the core oligosaccharide, resulting in heterogeneity of the glycosyl moiety of the glycoproteins (LIS and SHARON 1978, ISHIHARA et al. 1979).

Not all of the peptidyl asparagines undergo substitution. MARSHALL (1974), on the basis of studies with animal tissues, has suggested that N-glycosylation of asparagine requires that the amide be located in an amino acid "sequon" consisting of Asn-X-Ser or Asn-X-Thr. Carbohydrate linkage to asparagine residues in sequons of this type have been reported in bromelain (GOTO et al. 1976), horseradish peroxidase (WELINDER 1976), vicilin from mung bean (ERICSON and CHRISPEELS 1976), 7S protein of soyabean (YAMAUCHI et al. 1976) and sainfoin lectin (NAMEN and HAPNER 1979). Although it was originally proposed that X could be any amino acid, recent studies by LEHLE (personal communication) indicate that if X is proline the acceptor properties of the sequon are lost. It is also found that not all asparagine residues associated in appropriate amino acid sequons are glycosylated; presumably tertiary structures render certain of these inaccessible.

## 7.4 Miscellaneous Glycoproteins

Exogenously supplied mannose and glucosamine become incorporated into proteins associated with various cell membranes in plants (NAGAHASHI et al. 1980,

ROBERTS and POLLARD 1975). Glycopeptides containing varying levels of arabinose, fucose, xylose, mannose, galactose, glucose, N-acetylgalactosamine, N-acetylglucosamine and mannosamine have been detected in matrix proteins and membranes derived from the endoplasmic reticulum, glyoxysomes and mitochondria from castor bean endosperm (MELLOR et al. 1980). Although none of the glycoproteins or the nature of glycopeptide linkages in these membrane systems has been characterized, these studies nevertheless demonstrate the extensive occurrence of glycoproteins in plants. Additionally the observation that lectins can be used to purify glycosidases (NEELY and BEEVERS 1980), coupled with the reports of the glycoprotein nature of fructosidase (FAYE and BERJONNEAU 1979) and α-amylase (RODAWAY 1978) and other enzymes cited by BROWN and KIMMINS (1977), provide further evidence for the glycosylation of many plant proteins.

### 7.5 Formation of the O-Glycopeptide Bonds

There is very little information available on the mechanism of formation of the variety of O-glycosidic linkages encountered in glycoproteins from plants. KARR (1972) reported the in vitro glycosylation of hydroxyproline by a particulate system from a suspension of sycamore cells using UDP-L arabinose as substrate. The acceptor hydroxyproline-rich protein was present in the particulate fraction. Alkaline hydrolysis of the product glycosylated in vitro released Hyp-$Ara_4$, Hyp-$Ara_3$, Hyp-$Ara_2$ and Hyp-Ara, suggesting that assembly of the side chain proceeded by the sequential transfer of monosaccharides of arabinose. The mechanism of glycosyl transfer and the number of enzymes involved in assembly of the Hyp-oligoarabinosides has not been determined. UDP-arabinose arabinosyl transferase(s) responsible for attachment of arabinose residues to peptidyl hydroxyproline has been located in the Golgi apparatus (GARDNER and CHRISPEELS 1975). The finding of arabinose-containing oligosaccharide lipids in the endoplasmic reticulum of castor bean endosperm (MELLOR et al. 1980) suggests that lipid intermediates may be involved in glycosylation of hydroxyproline, however this possibility has not been investigated thus far.

The mechanism of biosynthesis of O-glycosidic linkages between galactose and hydroxyproline or between galactose and serine has not been elucidated. MELLOR and LORD (1978) have demonstrated the incorporation of galactose into glycoproteins present in the ER and glyoxysomes of castor bean, and have subsequently reported the involvement of lipid-linked intermediates in the transfer of galactose from UDP galactose to exogenous proteins by preparations from castor bean endosperm (MELLOR and LORD 1979b). However, the nature of the linkage of galactose to the glycosylated protein was not established and it was not determined whether galactose residues were being added to a core oligosaccharide or being incorporated into glycopeptide linkages with peptidyl serine or hydroxyproline.

The synthesis of the O-mannosidic bond in fungal cells involves the participation of two enzymes and lipid intermediates according to the following sequence:

GDP-Man + Dolicholphosphate → Dolichol-phospho-Man + GDP
Dolichol-phospho-Man + (Acceptor protein)-Ser/Thr →
→ (Acceptor protein)-Ser/Thr-Man + Dolichol phosphate

The enzymes catalyzing the reaction sequence have been solubilized and partially purified and characterized from the yeast *Saccharomyces cerevisiae* (BABCZINSKI et al. 1980). A similar system for introduction of O-glycosidically linked mannose into proteins has been demonstrated in *Fusarium solani* f. *pisi* (SOLIDAY and KOLATTUKUDY 1979).

## 7.6 Formation of N-Glycopeptide Bonds

On the basis of studies with yeasts, animals, and higher plants it has become apparent that formation of the N-glycosidic linkage between peptidyl asparagine and N-acetyl glucosamine occurs by attachment of the core oligosaccharides. Thus studies aimed at following formation of N-glycosidic linkages also involve investigation of mechanism of production of core oligosaccharides. The proposed sequence of reactions is outlined in Fig. 2.

The initial reaction involves a transfer of N-acetylglucosamine-1-phosphate from UDP-GlcNAC to a lipid monophosphate to form GlcNAc-P-P lipid. A further N-acetyl glucosamine is transferred from UDP-GlcNAc to form $(GlcNAc)_2$-P-P lipid or N-acetylchitibiosyl-P-P lipid.

The $\beta$-linked mannosyl residue of the core oligosaccharide is transferred from GDP-Man to the $(GlcNAc)_2$-P-P lipid. Subsequent mannosyl residues of the core oligosaccharide that are $\alpha$-linked are transferred from dolichol-P-mannose. This intermediate is formed by transfer of mannosyl residues from GDP-Man to dolichol phosphate. The assembled oligosaccharide is transferred en bloc from the (Man)-$(GlcNAc)_2$-P-P lipid to peptidyl asparagine. Although the complete enzymatic sequence and all proposed intermediates have not been fully characterized, studies with particulate preparations from cotton fibres, beans (*Phaseolus aureus* and *Phaseolus vulgaris*) and peas reviewed by ELBEIN (1979) and SHARON and LIS (1979) are consistent with assembly of N-glycosidic linkage in this fashion in plants.

In animals and yeast the same lipid dolichol has been shown to be associated with N-acetylglucosamine and mannose. The lipid component involved in mannolipid assembly in plants has also been shown to be dolichol phosphate (DELMER et al. 1979). It has been reported that the lipid attached to N-acetylglucosamine in plants is also dolichol (LEHLE et al. 1976, BRETT and LELOIR 1977), however ERICSON et al. (1978) indicate that dolichol will not function in N-acetylglucosamine-lipid formation and suggest that a lipid other than dolichol may be required.

Studies with animal (PARODI and LELOIR 1979) and yeast (PARODI 1979) systems have indicated that two or three glucose residues are added to the core oligosaccharide-P-P lipid from UDP-glucose via Glc-P dolichol prior to transfer of the oligosaccharide to the acceptor protein. These glucosyl residues and some mannosyl units are trimmed from the glycoprotein before addition of further carbohydrates from sugar nucleotides.

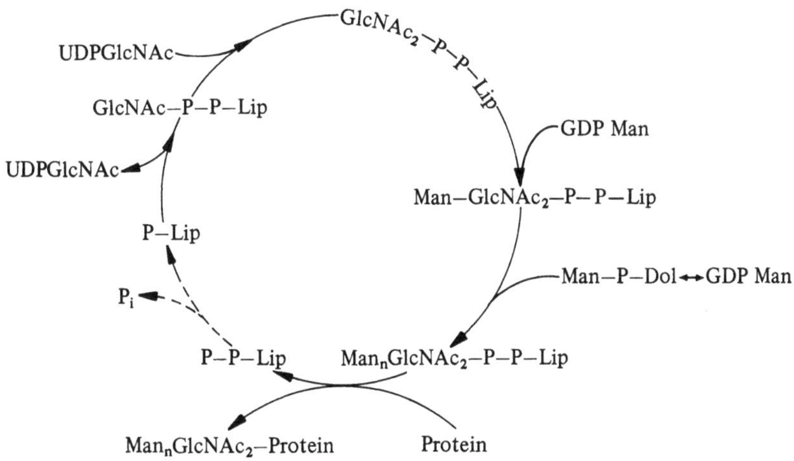


**Fig. 2.** Proposed sequence for glycopeptide assembly involving lipid linked intermediates

Glucosylation of polyprenol phosphates and formation of lipid-linked oligosaccharides, consisting of glucose only, which can become incorporated into a lectin, have been reported to occur in peas (HOPP et al. 1979). However, there is so far no evidence in higher plant systems for glucosylation and trimming of core oligosaccharide involved in the assembly of glycoproteins containing the N-glycosidic linkage.

The enzymes involved in the formation of mannosyl lipid have been located in the endoplasmic reticulum (NAGAHASHI and BEEVERS 1978, LEHLE et al. 1977, LEHLE et al. 1978, MELLOR and LORD 1979a) and formation of N-acetylglucosaminyl lipid and oligosaccharide lipid also occurs in the endoplasmic reticulum (NAGAHASHI and BEEVERS 1978). Treatment of particulate fractions from yeast with detergents solubilizes transferases involved in the formation of dolichol diphosphate-(N-acetyl glucosamine)$_{1-2}$ and production of dolichol monophosphate mannose. Enzymes involved in the transfer of oligosaccharide or mannosyl units from lipid intermediates to acceptor exogenous synthetic peptides can also be released from membranes by detergent treatment (PALAMARCZYK et al. 1980).

Glucosylation of polyprenol phosphates has been shown to occur in the endoplasmic reticulum (HOPP et al. 1979) and plasma membrane preparations (CHADWICK and NORTHCOTE 1980). It has been proposed that biosynthesis of cellulose may require the participation of dolichol monophosphate glucose and dolichol pyrophosphate oligoglucan (HOPP et al. 1978). Thus the conflicting reports on the cellular location of the polyprenyl phosphate UDP-glucose glucosyl transferase may represent differential localization of transferases involved either in glycoprotein biosynthesis (endoplasmic reticulum) or cellulose biosynthesis (plasma membrane).

Other glycosyl transferases have been located in the Golgi apparatus (BRETT and NORTHCOTE 1975, NAGAHASHI and BEEVERS 1978) however, these preparations do not produce the acidic lipid intermediates involved in core oligosaccharide synthesis. It is possible that the Golgi apparatus-associated transferases may be responsible for the addition of such glycoside residues as xylose, fucose and mannose to the core oligosaccharides. The addition of terminal N-acetyl glucosamine residues to vicilin (DAVIES and DELMER 1979) and horseradish peroxidase (CLARKE and SHANNON 1976) may be catalyzed by a Golgi apparatus-associated UDP-GlcNAc transferase which shows a preference for manganese rather than magnesium for optimal activity (NAGAHASHI et al. 1978).

## 8 Compartmentalization and Organelle Biogenesis

Many proteins are compartmentalized in the cell or associated with specific organelles, and protein segregation is achieved by various mechanisms. Some organelles, e.g., chloroplasts and mitochondria, have their own protein-synthesizing machinery and are capable of producing a limited number of their own proteins. The mitochondrial system manufactures about 10% of the organellar protein represented by about a dozen hydrophobic polypeptides of the mitochondrial membrane. The remaining mitochondrial polypeptides are synthesized on cytoplasmic ribosomes and are post-translationally transported into the mitochondrion (SCHATZ and MASON 1974). In chloroplasts only 3 of more than 30 polypeptides of the inner and outer envelope are made by intact chloroplasts; the remainder are made in the cytoplasm. Twenty-four out of a total of 33 major thylakoid membrane polypeptides are also manufactured outside the chloroplast (CHUA and GILLHAM 1977). The major component of the soluble stromal protein is the enzyme ribulose-1,5-bisphosphate carboxylase. This enzyme is made up of eight copies each of large (molecular weight 55,000) and small (molecular weight 11,000) subunits and the large subunit is synthesized inside the chloroplast, whereas the small subunit is made on cytoplasmic ribosomes. In addition to the many membrane polypeptides and the small subunit of ribulose-1,5-bisphosphate carboxylase, most other stromal proteins and chloroplast ribosomal proteins are also synthesized on cytoplasmic ribosomes (CHUA and SCHMIDT 1979).

The subcellular localization of the cytoplasmic ribosomes engaged in the synthesis of compartmentalized protein appears to determine the mechanism by which the proteins are transported into the organelles. Cytoplasmic ribosomes may be bound to the outer membrane of an organelle, bound to the endoplasmic reticulum membrane, or free in the cytoplasm.

### 8.1 Mitochondria

Very little information is available concerning the location of cytoplasmic ribosomes involved in the synthesis of mitochondrial proteins in plants. There are reports of 80S cytoplasmic ribosomes, functional in in vitro amino acid incorporation, being associated with outer mitochondrial membranes in yeast (KELLEMS et al. 1974). Some of the nascent peptides synthesized in this system are not released by puromycin and are insensitive to protease digestion, thus indicating that the newly produced proteins are discharged internally into the mitochondria. If proteins are incorporated into mitochondria as they are synthesized, then inhibition of protein synthesis should terminate transport and there should be no pool of newly synthesized mitochondrial protein outside the organelle. Pulse labelling and inhibitor studies have demonstrated a tight coupling of protein synthesis and transport in some yeasts (see CHUA and SCHMIDT 1979). However, in studies of *Neurospora crassa* (HALLERMAYER et al. 1977) and *Saccharomyces cereviscae* (SCHATZ 1979), it appears that some mitochondrial proteins or their precursors accumulate in the cytoplasm and are transported into the organelle by a post-translational mechanism. For several mitochondrial proteins (see Table 1) it appears that a nascent peptide with an N-terminal extension sequence of 20–60 amino acids is synthesized by free cytoplasmic ribosomes and released into the cytoplasm. The extension sequence apparently associates with the mitochondrial surface and there is a post-translational energy-

**Table 1.** Organellar proteins which undergo post-translational shortening and transport and mechanisms which may maintain directionality of transport

| Organelle | Protein | Modification |
|---|---|---|
| Mitochondria | F1 ATPase α subunit | Combination of subunits |
| | F1 ATPase β subunit | into oligomeric protein |
| | F1 ATPase γ subunit | |
| | Cytochrome bc, complex subunit V | Attachment to membrane |
| | Cytochrome c | Attachment of haem residue |
| | Cytochrome c-peroxidase | |
| Chloroplast | Ribulose 1-5 bisphosphate carboxylase small subunit | Association with large subunit |
| | Ferredoxin | Association of prosthetic group |
| | Light-harvesting chlorophyll a/b protein | Addition of chlorophyll association with thylakoid membrane |
| Glyoxysome | Malate dehydrogenase | |
| | Catalase | Addition of haem residue |
| | Malate synthetase | Glycosylation |
| | Isocitrate lyase (transported) | Methylation |
| Protein bodies | Hordein | Combination of subunits |
| | Zein | Combination of subunits |
| | Vicilin (transported) | Glycosylation, combination of subunits |
| | Legumin (transported) | Conversion of subunits, cleavage |
| Vacuolar proteins | Protease inhibitors | |
| Secreted proteins | α-Amylase | Glycosylation |

dependent migration of the polypeptide into the mitochondria (NELSON and SCHATZ 1979). Unidirectional transport is maintained by proteolytic cleavage of the precursor, modification of -SH linkages or covalent attachment of prosthetic groups as occurs for example during the addition of haem group to the apoprotein of cytochrome c (KORB and NEUPERT 1978).

## 8.2 Chloroplasts

There is so far no convincing evidence for the biosynthesis of chloroplast proteins by cytoplasmic ribosomes bound to the endoplasmic reticulum; rather it appears that the cytoplasmically synthesized proteins of this organelle are made on free ribosomes as precursors and then transported post-translationally into the chloroplast (Table 1). The precursors are converted to the size of native protein by proteinases present in the post-ribosomal supernatant in *C. reinhardtii* (DOBBERSTEIN et al. 1977) or in the chloroplasts of peas (HIGHFIELD and ELLIS 1978). Since the processed precursor becomes resistant to proteolytic digestion, it appears that post-translational shortening is accompanied by or coincides with transport. In the case of the ribulose-1,5-bisphosphate carboxylase small subunit directionality of transport is probably maintained by immediate combination with the large subunit and resultant production of the holoenzyme.

Membrane proteins synthesized on chloroplast ribosomes are also subject to post-translational modification by the addition of prosthetic groups as indicated by the in vitro synthesis of cytochrome b-559 and P-700 chlorophyll a protein complex by isolated chloroplasts. In addition, some membrane proteins synthesized by chloroplast ribosomes may be formed initially as higher molecular weight precursors which are then subjected to post-translational shortening; however, these observations are preliminary (ZIELINSKI and PRICE 1980; GREBANIER et al. 1979).

## 8.3 Glyoxysomes

LORD and BOWDEN (1978) have postulated that all of the glyoxysomal enzymes are synthesized on membrane-bound polysomes, segregated into the endoplasmic reticulum and subsequently packaged into glyoxysomes which are budded off from the endoplasmic reticulum. Synthesis of glyoxysomal proteins by ribosomes bound to the endoplasmic reticulum (ER) and their sequestration into the cisternae of the ER involves translocation of the polypeptide chain across the membrane during translation. It has been proposed that all proteins secreted into the cisternae of the ER are initially synthesized as precursors which contain at their N-termini short-chain extensions of hydrophobic amino acids termed "the signal peptide" as shown in Fig. 3. It is postulated that the signal peptide "emerges" from the ribosome and binds to specific receptors in the ER, resulting in the attachment of polysomes to the ER membranes. The proteins synthesized on these attached polysomes are co-translationally transported into the ER cisternae during which time the signal sequence is removed. Accordingly, pro-

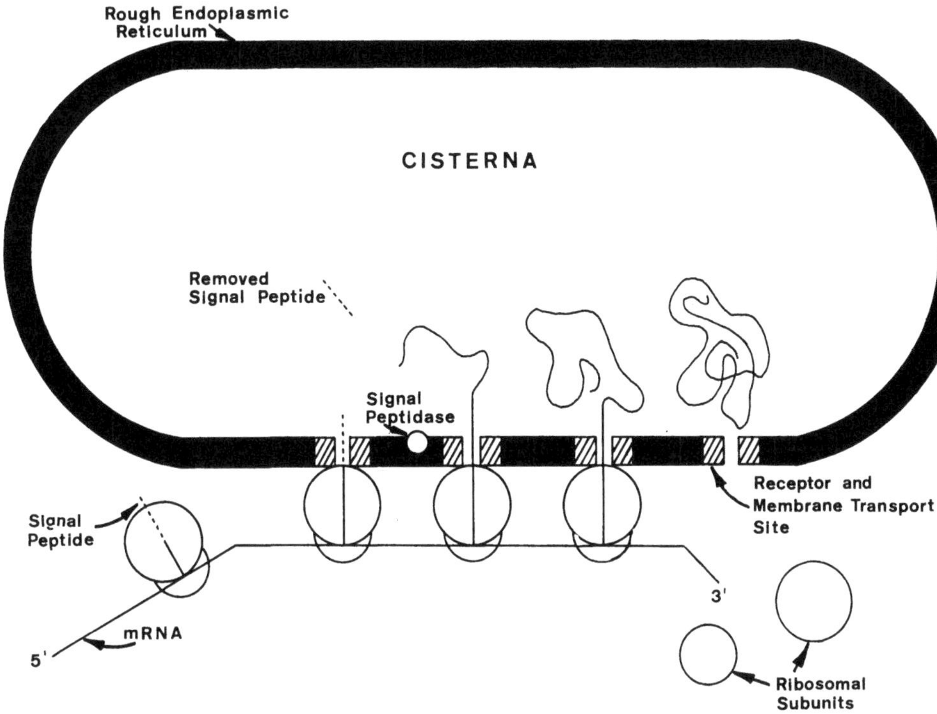


**Fig. 3.** Schematic illustration of the signal hypothesis. (Modified from BLOBEL 1977)

teins enter the cisternae only during translation, and translation of mRNA for proteins accumulated in the cisternae, in a cell-free system, should result in the synthesis of a precursor protein 16–30 amino acid residues ($\sim$2000 molecular weight) larger than the in vivo protein. In contrast, peptides of the same size as the in vivo proteins should be produced if endoplasmic reticulum membranes are included in the cell-free incorporating system. If rough endoplasmic reticulum, as opposed to mRNA or polysomes released from membranes, is used to direct in vitro protein synthesis, the peptides produced are resistant to proteolysis by exogenous proteases, indicating that they are sequestered into the cisternal space of the RER vesicles.

RIEZMAN et al. (1980) have studied the in vitro translation of four glyoxysomal enzymes with a view to determining their site and mode of synthesis. Monospecific antibodies against isocitrate lyase, catalase, malate synthase and malate dehydrogenase were used to immunoprecipitate these enzymes or their precursors from the in vitro translation products in a wheat germ system programmed with mRNA from cucumber cotyledons. The immunoprecipitates of malate dehydrogenase and catalase produced in vitro had a higher molecular weight than the in vivo enzymes. These findings indicate that these enzymes are synthesized as precursors which are post-translationally processed to a lower molecular weight native enzyme. WALK and HOCK (1978) have also reported on the post-translational chain shortening of malate dehydrogenase. Malate synthetase immunoprecipitated from in vitro cell-free systems or from cellular extracts had a molecular weight of 57,000 (RIEZMAN et al. 1980). Since the native malate synthetase is a glycoprotein and the in vitro product cannot be glycosylated (due to the absence of membranes containing glycosyltransferases), it is reasoned that the polypeptide of the native enzyme must be shorter than the in vitro product. Isocitrate lyase produced in vitro had a lower molecular weight than the purified enzyme. Since RIEZMAN et al. (1980) could not demonstrate that this enzyme was a glycoprotein (in contrast to the reports of FREVERT and KINDL 1978), it was suggested that the processing of the in vitro product must involve a gain in molecular weight perhaps by methylation or acetylation. Although these data on chain shortening are partially consistent with the proposal of LORD and BOWDEN (1978) on the synthesis of glyoxysomal proteins by ribosomes of the ER, RIEZMAN et al. (1980) were unable to demonstrate differences in the in vitro production of the peptides of the glyoxysomal enzymes by membrane or free polsyomes. No information is currently available concerning the sequestration of the newly synthesized glyoxysomal enzymes.

## 8.4 Protein Bodies – Cereal Endosperm

Evidence for post-translational chain shortening and sequestration of the product has been provided from studies of the synthesis of prolamins (alcohol soluble proteins) from cereal endosperms. These prolamins are accumulated in protein bodies. Fragments of RER of barley endosperm can be used in a cell-free protein-synthesizing system from wheat germ to direct the synthesis of hordein polypeptides soluble in isoproponal (CAMERON-MILLS and INGVERSEN 1978, CAMERON-MILLS et al. 1978a). Polyribosomes detached from the membranes of the endoplasmic reticulum also direct the synthesis of hordein polypeptides, in contrast to free cytoplasmic polyribosomes which have no significant capacity to make alcohol-soluble proteins. Hordein polypeptides produced by RER fragments were inaccessible to proteolysis by exogenous proteases and had a higher molecular weight than the native polypeptides. Thus the hordein polypeptides produced by ribosomes associated with ER were sequestered. Since the proteases involved in chain shortening are considered to be associated with ER membranes, it is not clear why in this study the polypeptides produced by the RER vesicles were not processed to the same molecular weight as the native polypeptides.

Extracted mRNA from corn (*Zea mays*) endosperm directs the synthesis of ethanol-soluble zein polypeptides in a wheat germ cell-free system. The in vitro products have molecular weights about 2000 larger than native zein. When RER vesicles were placed in the wheat germ system polypeptides of the same molecular weight as those of native zein were synthesized (LARKINS and HURKMAN 1978). Electron microscopy and sucrose density gradient analysis indicated that dilated ER produced the protein bodies which contain zein.

### 8.4.1 Protein Bodies – Legume Seeds

The reserve proteins of legume seeds accumulate in protein bodies which apparently originate from the vacuole. Polypeptides precipitable with antibodies against the reserve protein vicilin are synthesized in a wheat germ cell-free system fortified with membrane-bound polysomes from *Vigna radiata* (formerly *Phaseolus aureus*) cotyledons (BOLLINI and CHRISPEELS 1979). Free polysomes did not direct the synthesis of vicilin-like polypeptides. The polypeptides produced in the in vitro system had a lower molecular weight than those of in vivo vicilin, and it was reasoned that the discrepancy in molecular weights was due to the fact that the in vitro product was not glycosylated. SUN et al. (1979) have also suggested that the difference in molecular weights of in vitro and in vivo products was due to lack of post-translational glycosylation in the in vitro cell free system.

In peas the post-translational modifications of the reserve proteins, legumin and vicilin, appear to be complex. Three polypeptides immunoprecipitable with anti-vicilin antibodies can be isolated from the translation products of polyribosomes or poly(A)+-RNA in a wheat germ or rabbit reticulocyte lysate system. Two of these polypeptides synthesized in vitro correspond in molecular weight (70,000 and 50,000) within the resolution of the gel system ($\pm 1{,}000$) to those of native vicilin. A 47,000 molecular weight peptide synthesized in vitro did not correspond to any polypeptide from vicilin prepared from mature seed, but was present in vicilin from developing seeds (CROY et al. 1980b). A further report by HIGGINS and SPENCER (1980) concludes "at least some of the pea storage protein polypeptides such as the 75,000 and 50,000 molecular weight component of vicilin are synthesized as precursors which can be modified by the addition of membranes from dog pancreas during translation." Unpublished work by BOULTER and coworkers using more highly resolving gels and pea seed membranes confirm these results. Since vicilin is a glycoprotein it is possible that glycosylation of subunit polypeptide may mask to some extent any alteration in molecular weight produced by removal of a signal sequence. It is also possible that the signal sequences occur internally in the polypeptide and are not post-translationally cleaved; such a situation has been reported in ovalbumin in animal systems (LINGAPPA et al. 1979).

Legumin isolated from developing and mature peas can be dissociated into two subunits of molecular weights of about 40,000 and 20,000. When anti-legumin antibodies are used to precipitate products from a cell-free in vitro translation system directed by polyribosomes or poly (A)+-RNA from developing pea, a single polypeptide of molecular weight of 60,000 is detected (CROY et al. 1980a). Similar results are obtained with polyribosomes from *Vicia faba* (CROY et al. 1980a). It appears that legumin is synthesized as a 60,000 molecular weight polypeptide which is post-translationally cleaved to the 40,000 and 20,000 molecular weight subunits. This proteolytic cleavage is distinct from the removal of the signal sequence (described earlier) which occurs cotranslationally as proteins are transported across the ER membranes. Since total polysome preparations were utilized in the above studies of legumin and vicilin synthesis, it is not clear where the reserve globulins of peas are synthesized. However, work in the author's laboratory has shown that RER vesicles can be translated in a wheat germ cell-free system to produce polypeptides of molecular weight similar to those of the subunits of legumin and vicilin. The peptides are inaccessible to proteolytic digestion so it appears that reserve proteins of peas are synthesized and sequestered by the endoplasmic reticulum. There is as yet no firm information on the pathway from the site of synthesis to the site of deposition.

### 8.5 Vacuolar Proteins

Proteinase inhibitors accumulate in the vacuoles of leaf cells of tomato in response to wounding. Translation of mRNA from wounded plants in a rabbit reticulocyte cell-free system produced peptides that could be precipitated with antibodies prepared against the proteinase inhibitors. Electrophoresis of the immunoprecipitates indicates that the in vitro products were 2000 molecular weight larger than the inhibitor prepared from leaves (NELSON and RYAN 1980). Thus the inhibitors are synthesized as precursors which are post-translationally converted to the native protein. It is not established where in the cell the inhibitors are synthesized, thus the peptide cleaved during processing may be a signal sequence directing the translational products into the ER. Alternatively the sequence may be directly involved in the process of inhibitor transport into the vacuolar compartment of the cell.

### 8.6 Secreted Proteins

Aleurone cells of cereal grains produce a variety of hydrolytic enzymes which are secreted into the endosperm. On the basis of ultrastructural studies which demonstrate the production of extensive RER during the time of hydrolytic enzyme production, it appears that the enzymes are synthesized by ribosomes associated with the endoplasmic reticulum. Translation of mRNA or polysomes from wheat aleurones in a reticulocyte lysate system indicated that one of the secreted hydrolytic enzymes, α-amylase, is synthesized as a precursor 1500 larger in the molecular weight than the native enzyme. Attempts to localize the protease involved in the processing with the membranes of the endoplasmic reticulum have been unsuccessful. Moreover there was no major difference in the in vitro products of free or membrane-bound polysomes (OKITA et al. 1979). Thus although there is evidence for post-translational modification of this secreted protein, the site of its synthesis and processing require further investigation.

## 9 Conclusions

Although to a great extent the tremendous diversity of proteins results from variations in the primary sequence of the 20 amino acids coded for by the genetic code, further complexity of proteins is achieved by modifications of the peptidyl amino acids.

It is also apparent that modifications of peptidyl amino acids are frequently necessary to confer catalytic activity upon otherwise inert polypeptides and to assure that appropriate proteins become localized at their correct subcellular sites.

This review chapter has shown that while many of the post-translational modifications have been described in general terms in plant systems there is need for much more detailed studies of the specifics of modification. The nature of subunit interaction in polymeric proteins and the identity of amino acids ligated to prosthetic groups are largely unresolved. Although recent studies have characterized the sequence of synthesis of some glycoproteins, the mechanism by which metals and other prosthetic groups become associated with specific peptidyl amino acids are not well resolved.

*Acknowledgement.* Supported in part by grant PCM 7728273 from the National Science Foundation.

## References

Akiyama Y, Kato K (1977) Structure of hydroxy proline arabinoside from tobacco cells. Agric Biol Chem 41:79–81

Allen AK, Desai NN, Neuberger A, Creeth JM (1978) Properties of potato lectin and the nature of its glycoprotein linkages. Biochem J 171:665–674

Allfrey VG (1977) Post-synthetic modifications of histone structure. A mechanism for the control of chromosome structure by the modulation of histone-DNA interactions. In: Li HJ, Eckhardt R (eds) Chromatin and chromosome structure. Academic Press, New York, pp 167–192

Anderson RL, Clarke AE, Jermyn MA, Knox RB, Stone BA (1977) A carbohydrate-binding arabinogalactan-protein from liquid suspension cultures of endosperm from *Lolium multiflorum*. Aust J Plant Physiol 4:143–158

Babczinski P, Haelbeck A, Tanner W (1980) Yeast mannosyl transferases requiring dolichyl phosphate and dolichyl phosphate mannose as substrate. Eur J Biochem 105:509–515

Ballou CE (1974) Some aspects of the structure, immunochemistry and genetic control of yeast mannans. Adv Enzymol 40:239–270

Ballou CE (1976) Structure and biosynthesis of the mannan component of the yeast cell envelope. Adv Microb Physiol 14:93–158

Benne R, Kasperaites M, Voorma HO, Ceglarz E, Legocki AB (1980) Initiation factor $EIF_2$ from wheat germ, purification functional comparisons with $EIF_2$ from rabbit reticulocytes and phosphorylation of its subunit. Eur J Biochem 104:109–117

Bennet J (1979) Chloroplast phosphoproteins. The protein kinase of thylakoid membranes is light dependent. FEBS Lett 103:342–344

Bennet J (1980) Chloroplast phosphoproteins: Evidence for a thylakoid bound phosphoprotein phosphatase. Eur J Biochem 104:85–89

Blobel G (1977) Synthesis and segregation of secretory proteins: The signal hypothesis. In: Brinkley BR, Porter KR (eds) International cell biology, Rockefeller Univ Press, New York, pp 318–325

Bollini R, Chrispeels MJ (1979) The rough endoplasmic reticulum is the site of reserve protein synthesis in developing *Phaseolus vulgaris* cotyledons. Planta 146:487–501

Brett CT, Leloir LF (1977) Dolichyl monophosphate and its sugar derivatives in plants. Biochem J 161:93–101

Brett CT, Northcote DH (1975) The formation of oligoglucans linked to lipid during synthesis of $\beta$-glucan by characterized membrane factions isolated from peas. Biochem J 148:107–117

Brodrick JW, Rabinowitz JC (1977) Biosynthesis of iron sulfur proteins. In: Lovenberg W (ed) Iron sulfur proteins III. Academic Press, New York, pp 101–119

Browder SK, Beevers L (1978) Characterization of the glycopeptide bond in legumin from *Pisum sativum* L. FEBS Lett 89:145–148
Brown EC, Edwards MJ, Newton RP, Smith CJ (1979) Plurality of cyclic nucleotide phosphodiesterases in *Spinacea oleracea*: subcellular distribution, partial purification and properties. Phytochemistry 18:1943–1948
Brown EG, Najafi AL, Newton RP (1979) Adenosine 3′5′-cyclic monophosphate, adenylate cyclose and a cyclic AMP binding protein in *Phaeolus vulgaris*. Phytochemistry 18:9–14
Brown RG, Kimmins WC (1977) Glycoproteins. In: Northcote DH (ed) Int Rev of Biochem Plant Biochemistry II Vol 13. Univ Park Press, pp 183–209
Brown RG, Kimmins WC (1978) Protein polysaccharide linkages in glycoproteins from *Phaseolus vulgaris*. Phytochemistry 17:29–33
Brown RG, Kimmins WC (1979) Linkage analysis of hydroxyproline poor glycoprotein from *Phaseolus vulgaris*. Plant Physiol 63:557–561
Buchanan BB (1980) Role of light in the regulation of chloroplast enzymes. Ann Rev Plant Physiol 31:341–374
Cameron-Mills V, Ingversen J (1978) *In vitro* synthesis and transport of barley endosperm proteins: reconstitution of functional rough microsomes from polyribosomes and stripped microsomes. Carlsberg Res Commun 43:471–489
Cameron-Mills V, Ingversen J, Brandt A (1978) Transfer of *in vitro* synthesized barley endosperm proteins into the lumen of the endoplasmic reticulum. Carlsberg Res Commun 43:91–102
Chadwick CM, Northcote DH (1980) Glucosylation of phosphorylpolysoprenol at the plasma membrane of soya bean (*Glycine max*) protoplasts. Biochem J 186:411–421
Chrispeels MJ (1970) Synthesis and secretion of hydroxyproline containing macromolecules in carrots. Plant Physiol 45:223–227
Chua N-N, Gillham NW (1977) The sites of synthesis of the principal thylakoid membrane polypeptides in *Chlamydomonas reinhardtii*. J Cell Biol 74:441–452
Chua N-N, Schmidt GW (1979) Transport of proteins into mitochondria and chloroplasts. J Cell Biol 81:461–483
Clarke J, Shannon LM (1976) The isolation and characterization of the glycopeptide from horseradish peroxidase isoenzyme C. Biochim Biophys Acta 427:428–442
Clarkson DT, Hanson JB (1980) The mineral nutrition of higher plants. Annu Rev Plant Physiol 31:239–293
Croy RRD, Gatehouse JA, Evans IM, Boulter D (1980a) Characterization of the storage protein subunits synthesized *in vitro* by polyribosomes and RNA from developing pea (*Pisum sativum* L) I Legumin. Planta 148:49–56
Croy RRD, Gatehouse JA, Evans IM, Boulter D (1980b) Characterization of the storage protein subunits synthesized *in vitro* by polyribosomes and RNA from developing pea (*Pisum sativum* L). II Vicilin. Planta 148:57–63
Curvetto NR, Rauser WE (1979) Isolation and characterization of copper binding proteins from roots of *Agrostis gigantea* tolerant to excess copper. Plant Physiol 63:Suppl 326
Davies HM, Delmer DP (1979) Seed reserve-protein glycosylation in an *in vitro* preparation from developing cotyledons of *Phaseolus vulgaris*. Planta 146:513–520
Dayhoff MO (1972) Atlas of protein sequence and structure. Nat Biomed Res Fdt, Silver Spring, Maryland
Delange RJ, Fambrough DM, Smith EL, Bonner J (1969) Calf and pea histone IV III. Complete amino acid sequence of pea seedling Histone IV; comparison with the homologous calf thymus histone. J Biol Chem 244:5669–5679
Delmer DP, Kulow C, Ericson MC (1979) Glycoprotein synthesis in plants II. Study of the mannolipid intermediate. Plant Physiol 61:25–29
DiMaria P, Polastro E, Delange RJ, Kim S, Paik WK (1979) Studies on cytochrome *c* methylation in yeast. J Biol Chem 254:4645–4652
Dobberstein B, Blobel G, Chuna N-H (1977) *In vitro* synthesis and processing of a putative precursor for the small subunit of ribulose 1-5-bisphosphate carboxylase of *Chlamydomonas reinhardtii*. Proc Natl Acad Sci USA 74:1082–1085
Elbein AD (1979) The role of lipid-linked saccharides in the biosynthesis of complex carbohydrates. Annu Rev Plant Physiol 30:239–272

Epstein CJ, Goldberger RF, Anfinsen CB (1963) The genetic control of tertiary protein structure: Studies with model systems. Cold Spring Harbor Symp Quant Biol 28:439–449
Ericson MC, Chrispeels MJ (1976) The carbohydrate moiety of mung bean vicillin. Aust J Plant Physiol 3:763–769
Ericson MC, Gafford JT, Elbein AD (1978) Evidence that the lipid carrier for N-acetyl-glucosamine is different from that for mannose in mung bean and cotton fibers. Plant Physiol 61:274–277
Esquerre-Tugaye M-T, Lamport DTA (1979) Cell surface in plant microorganism interactions I. A structural investigation of cell wall hydroxyproline rich glycoproteins. Plant Physiol 64:314–319
Farooqui J, Kim S, Paik WK (1980) *In vitro* studies on yeast cytochrome *c* methylation in relation to protein synthesis. J Biol Chem 255:4468–4473
Faye L, Berjonneau C (1979) Evidence for the glycoprotein nature of radish $\beta$-fructosidase. Biochemie 61:51–59
Finazzi-Agro A, Cannella C, Graziani MT, Cavallini D (1971) A possible role for rhodanese: The formation of 'labile' sulfur from thiosulfate. FEBS Lett 16:172–174
Freedman RB (1979) How many distinct enzymes are responsible for the several cellular processes involving thiol-protein disulphide interchange. FEBS Lett 97:201–210
Frevert J, Kindl H (1978) Plant microbody protein. V. Purification and glycoprotein nature of glyoxysomol isocitrate lyase from cucumber cotyledons. Eur J Biochem 92:95–110
Gardner M, Chrispeels MJ (1975) Involvement of the Golgi apparatus in the synthesis and secretion of hydroxyproline rich cell wall glycoproteins. Plant Physiol 55:536–541
Gleeson PA, Clarke AE (1979) Structural studies on the major component of *Gladiolus* style mucilage an arabinogalacton protein. Biochem J 181:607–621
Goto K, Murachi T, Takahasi N (1976) Structural studies on stem bromelain isolation, characterization and alignment of the cyonogen bromide fragments. FEBS Lett 62:93–95
Grand RJA, Nairn AC, Perrys V (1980) The preparation of calmodulins from barley (*Hordeum* sp) and basidiomycete fungi. Biochem J 185:755–760
Grebanier AE, Steinback KE, Bogorad L (1979) Comparison of the molecular weights of proteins synthesized by isolated chloroplasts with those which appear during greening in *Zea mays*. Plant Physiol 63:436–439
Grynberg A, Nicolas J, Drapron R (1978) Some characteristic of protein disulfide isomerase (E.C. 5.3.4.1) from wheat (*Triticum vulgare*) embryo. Biochemie 60:547–551
Hallermayer G, Zimmerman R, Neupert W (1977) Kinetic studies of the transport of cytoplasmically synthesized proteins into the mitochondria in intact cell of *Neurospora crassa*. Eur J Biochem 81:523–532
Hayaishi O, Ueda K (1977) Poly (ADP-ribose) and ADP-ribosylation of proteins. Annu Rev Biochem 46:95–116
Hewitt EJ (1974) Aspects of trace element requirements in plants and microorganisms: the metalloenzymes of nitrate and nitrite reduction. In: MTP international review of science biochemistry series Vol 11. Butterworths, London, pp 199–246
Higgins JV, Spencer D (1980) Biosynthesis of pea seed proteins evidence for precursor forms from *in vitro* and *in vivo* studies. In: Leaver CJ (ed) Genome organisation and expression in plants. Plenum Press, New York, pp 245–258
Highfield PE, Ellis RS (1978) Synthesis and transport of the small subunit of chloroplast ribulose bisphosphate carboxylase. Nature 271:420–424
Hill HAO, Lee WK (1979) Investigations of the structure of the blue copper protein from *Rhus vernicifera* stellacyanin by nuclear magnetic spectroscopy. J Inorg Biochem 11:101–113
Hong J-S, Rabinowitz JC (1967) Preparation and properties of clostridial apoferredoxins. Biochem Biophys Res Commun 29:246–252
Hopp HE, Romero P, Daleo GR, Pont-Lezica R (1978) Synthesis of cellulose precursors. The involvement of lipid-linked sugars. Eur J Biochem 84:561–571
Hopp HE, Romero P, Pont-Lezica R (1979) Subcellular localization of glycosyl transferases involved in lectin glucosylation in *Pisum sativum* seedlings. Plant Cell Physiol 20:1063–1069

Hyde BB, Hodge AJ, Kahn A, Birmstiel M (1963) Studies on phytoferritin I, Identification and localization. J Ultrastruct Res 9:248–258
Isenberg I (1979) Histones. Annu Rev Biochem 48:159–161
Ishihara H, Takhashi N, Oguri S, Tejima S (1979) Complete structure of the carbohydrate moiety of stem bromelain. An application of the almond glycopeptidase for structural studies of glycopeptides. J Biol Chem 254:10715–10719
Jermyn MA, Yeow M (1975) A class of lectins present in the tissues of seed plants. Aust J Plant Physiol 2:501–531
Johnson JL, Hainline BE, Rajagopalan KV (1980) Characterization of the molybdenum cofactor of sulfite oxidase, xanthine oxidase and nitrate reductase. Identification of a pteridine as a structural component. J Biol Chem 255:1783–1786
Karr AL (1972) Isolation of an enzyme system which will catalyze the glycosylation of extensin. Plant Physiol 50:275–282
Kellems RE, Allison VF, Butow RA (1974) Cytoplasmic type 80s ribosomes associated with yeast mitochondria II. Evidence for the association of cytoplasmic ribosomes with the outer mitochondrial membrane *in situ*. J Biol Chem 249:3297–3303
Killilea SD, O'Carra P, Murphy RF (1980) Structure and apoprotein linkages of phycoerythrobilin and phycocyanobilin. Biochem J 187:311–320
Klotz IM, Darnall DW, Langerman NR (1975) In: Neurath H, Hill RL (eds) The proteins Vol I. Academic Press, New York, pp 293–411
Korb H, Neupert W (1978) Biogenesis of cytochrome *c* in *Neurospora crassa*. Synthesis of apocytochrome $c_1$ transfer to mitochondria and conversion to holocytochrome *c*. Eur J Biochem 91:609–620
Kornfeld R, Kornfeld S (1976) Comparative aspects of glycoprotein structure. Annu Rev Biochem 45:215–237
Krebs EG, Beavo JA (1979) Phosphorylation and dephosphorylation of enzymes. Annu Rev Biochem 48:923–959
Kretsinger RH (1976) Evolution and function of calcium-binding proteins. Int Rev Cytol 46:323–393
Kruiswijk T, Dettey JT, Planta RJ (1978a) Modification of yeast ribosomal proteins. Phosphorylation. Biochem J 175:213–219
Kruiswijk T, Kurst A, Planta RJ, Mayer WH (1978b) Modification of yeast ribosomal proteins. Methylation. Biochem J 175:221–225
Larkins BA, Hurkman WJ (1978) Synthesis and deposition of zein in protein bodies of maize endosperm. Plant Physiol 62:256–263
Lamport DTA, Katona L, Roerig S (1973) Galactosylserine in extension. Biochem J 133:125–131
Lee K-Y, Pan S-S, Erickson R, Nason A (1974) Involvement of molybdenum and iron in the *in vitro* assembly of assimilatory nitrate reductase utilizing *Neurospora* mutant nit-1. J Biol Chem 249:3941–3952
Lehle L, Fartaczek F, Tanner W, Kauss H (1976) Formation of polyprenol-linked mono- and oligosaccharides in *Phaseolus aureus*. Arch Biochem Biophys 175:419–426
Lehle L, Bauer F, Tanner W (1977) The formation of glycosidic bonds in yeast glycoproteins. Intracellular localization of the reactions. Arch Microbiol 114:77–81
Lehle L, Bowles DJ, Tanner W (1978) Subcellular site of mannosyl transfer to dolichyl phosphate in *Phaseolus auerus*. Plant Sci Letts 11:27–34
Leis JP, Keller EB (1971) N-formylmethionyl t-$RNA_f$ of wheat chloroplasts. Its synthesis by a wheat transformylase. Biochemistry 10:889–894
Liener IE (1976) Phytochemagglutinins (phytolectins). Annu Rev Plant Physiol 27:291–319
Lin PPC, Key JL (1976) Lysine rich $H_1$ kinase from soybean hypocotyls. Biochem Biophys Res Commun 73:396–407
Lin TS, Kolattukudy PE (1977) Glucuronyl glycine, a novel N-terminus in a glycoprotein. Biochem Biophys Res Commun 75:87–93
Lingappa VR, Lingappa JR, Blobel G (1979) Chicken ovalbumin contains an internal signal sequence. Nature 281:117–121
Lis H, Sharon N (1978) Soybean agglutinin – A plant glycoprotein structure of the carbohydrate unit. J Biol Chem 253:3468–3476

Lis H, Sharon N, Katchalski E (1969) Identification of the carbohydrate-protein linking group in soybean hemagglutinin. Biochim Biophys Acta 192:364–366

Van Loon LC, Trewavas A, Chapman KSR (1975) Phosphorylation of chromatin associated proteins in *Lemna* and *Hordeum*. Plant Physiol 55:288–292

Lord JM, Bowden L (1978) Evidence that glyoxysomal malate synthase is segregated by the endoplasmic reticulum. Plant Physiol 61:266–270

Markwell JP, Thornber JP, Boggs RT (1979) Higher plant chloroplast. Evidence that all the chlorophyll exists as chlorophyll protein complexes. Proc Natl Acad Sci USA 76:1233–1235

Marshall RD (1974) The nature and metabolism of the carbohydrate-peptide linkages of glycoproteins. Biochem Soc Symp 40:17–26

Means AR, Dedman JR (1980) Calmodulin – an intracellular calcium receptor. Nature 285:73–77

Mellor RB, Lord JM (1978) Incorporation of D-($^{14}$C) galactose into organelle glycoprotein in Castor bean endosperm. Planta 141:329–332

Mellor RB, Lord JM (1979a) Subcellular localization of mannosyl transferase and glycoprotein biosynthesis in castor bean endosperm. Planta 146:147–153

Mellor RB, Lord JM (1979b) Involvement of a lipid intermediate in the transfer of galactase from UDP-($^{14}$C) galactase to exogenous protein in castor bean endosperm homogenates. Planta 147:89–96

Mellor RB, Krusius T, Lord JM (1980) Analysis of glycoconjugate saccharides in organelles isolated from castor bean endosperm. Plant Physiol 65:1073–1075

Mendel RR, Muller AJ (1976) A common genetic determinant of Xanthine dehydrogenase and nitrate reductase in *Nicotiana tabacum*. Biochem Physiol Pflanz 170:538–541

Miller DH, Lamport DTA, Miller N (1972) Hydroxyproline heterooligosaccharides in *Chlamydomonas*. Science 176:918–920

Misaki A, Goldstein IJ (1977) Glycosyl moiety of the lima bean lectin. J Biol Chem 252:6995–6999

Murray MC, Guilfoyle TJL, Key JL (1978a) Isolation and characterization of a chromatin associated protein kinase from soybean. Plant Physiol 61:1023–1030

Murray MC, Guilfoyle TJ, Key JL (1978b) Isolation and preliminary characterization of a casein kinase from cauliflower nuclei. Plant Physiol 62:434–437

Nagahashi J, Beevers L (1978) Subcellular localization of glycosyl transferases involved in glycoprotein biosynthesis in the cotyledons of *Pisum sativum*. Plant Physiol 61:451–459

Nagahashi J, Mense RM, Beevers L (1978) Membrane associated glycosyl transferase in cotyledons of *Pisum sativum*. Differential effects of magnesium and manganese ions. Plant Physiol 62:766–772

Nagahashi J, Browder SK, Beevers L (1980) Glycosylation of pea cotyledon membranes. Plant Physiol 65:648–657

Namen AE, Hapner KD (1979) The glycosyl moiety of lectin from sainfoin (*Onobrychis viciifolia*). Biochim Biophys Acta 580:198–209

Neely RS, Beevers L (1980) Glycosidases from cotyledons of *Pisum sativum* L. J Exp Bot 31:299–312

Nelson CE, Ryan CA (1980) *In vitro* synthesis of preproteins of vacuolar compartmented proteinase inhibitors that accumulate in wounded tomato plants. Proc Natl Acad Sci USA 77:1975–1979

Nelson N, Schatz G (1979) Energy dependent processing of cytoplasmically made precursors to mitochondrial proteins. Proc Natl Acad Sci USA 76:4365–4369

Nochumson S, Durban E, Kim S, Paik WK (1977) Cytochrome *c* specific protein methylase III from *Neurospora*. Biochem J 165:11–18

Notton BA, Hewitt EJ (1979) Structure and properties of higher plant nitrate reductase, especially *Spinacea oleracea*. In: Hewitt EJ, Cutting CV (eds) Nitrogen assimilation of plants: proceedings of a symposium held at Long Ashton Research Station, Univ Bristol, Academic Press, New York, pp 227–244

Okita TW, De Caleya R, Rappaport L (1979) Synthesis of a possible precursor of α-amylase in wheat aleurone cells. Plant Physiol 63:195–200

Onishi HR, Tkacz JS, Lampen JO (1979) Glycoprotein nature of yeast alkaline phosphatase formations of active enzyme in the presence of tunicamycin. J Biol Chem 254:11943–11952
Paik WK, Kim S (1975) Protein methylation: Chemical enzymological and biological significance. Adv Enzymol 42:227–286
Palamarczyk G, Lehle L, Mankowski T, Chojnacki T, Tanner W (1980) Specificity of solubilized yeast glycosyl transferase for polyprenyl derivatives. Eur J Biochem 105:517–523
Parodi AJ (1979) Biosynthesis of yeast glycoproteins. Processing of the oligosaccharides transferred from dolichol derivatives. J Biol Chem 254:10051–10060
Parodi AJ, Leloir LF (1979) The role of lipid intermediates in the glycosylation of proteins in the eucaryotic cell. Biochim Biophys Acta 559:1–37
Patterson BD, Davies DD (1967) Specificity of the enzymatic methylation of pea histone. Biochem Biophys Res Commun 34:791–794
Patthy L, Smith EL, Johnson JD (1973) Histone III. V The amino acid sequence of pea embryo histone III. J Biol Chem 248:6834–6840
Payne JF, Bal AK (1976) Cytological detection of poly (ADP-ribose) polymerase. Exp Cell Res 99:428–432
Peterkofsky B, Assad R (1976) Submicrosomal localization of prolyl hydroxylase from chick embryo limb bone. J Biol Chem 251:4770–4777
Petrany P, Jendrisak JJ, Burgess RR (1977) RNA polymerase II from wheat germ contains tightly bound zinc. Biochem Biophys Res Commun 74:1031–1038
Polastro E, Looze Y, Leonis J (1977) Biological significance of methylation of cytochromes from ascomycetes and plants. Phytochemistry 16:639–641
Polastro E, Deconinck MM, Devogel MR, Mailier EL, Looze Y, Schneck AG, Leonis J (1978) Evidence that trimethylatioin of iso-l-cytochrome *c* from *Saccharomyces cerevisiae* affects interactions with the mitochondrion. FEBS Lett 86:17–20
Pope DG (1977) Relationship between hydroxyproline containing proteins secreted into the cell wall and medium by suspension cultured *Acer pseudoplatanus* cells. Plant Physiol 59:894–900
Ralph RK, Wojcik SJ (1978) Pyruvate metabolism and cytokinin action. Plant Sci Lett 12:227–232
Ramshaw JAM, Peacock D, Meatyard BT, Boulter D (1974) Phylogenetic implications of the amino acid sequence of cytochrome *c* from *Enteromorpha intestinalis*. Phytochemistry 13:2783–2789
Rana S (1980) Regulation of protein synthesis in eukaryotes by the protein kinase that phosphorylates initiation factor $elF_2$. Evidence for a common mechanism of inhibition of protein synthesis. FEBS Lett 112:211–215
Randall DD, Rubin PM (1977) Plant pyruvate dehydrogenase complex. II ATP dependent inactivation and phosphorylation. Plant Physiol 59:1–3
Rao KP, Randall DD (1980) Plant pyruvate dehydrogenase complex: Inactivation and reactivation by phosphorylation and dephosphorylation. Arch Biochem Biophys 200:461–466
Riezman H, Weir EM, Leaver CJ, Titus DE, Becker WM (1980) Regulation of glyoxysomal enzymes during germination of cucumber 3. *In vitro* translation and characterization of four glyoxysomal enzymes. Plant Physiol 65:40–46
Roberts RM, Pollard WE (1975) The incorporation of D-glucosamine into glycolipids and glycoproteins of membrane preparations from *Phaseolus aureus*. Plant Physiol 55:431–436
Rodaway SJ (1978) Composition of α-amylase secreted by aleurone layers of Himalaya barley. Phytochemistry 17:385–390
Rubin CS, Rosen OM (1975) Protein phosphorylation. Annu Rev Biochem 44:831–887
Rubin PM, Randall DD (1977) Regulation of plant pyruvate dehydrogenase complex by phosphorylation. Plant Physiol 60:34–39
Sadava D, Chrispeels MJ (1971a) Peptidyl proline hydroxylase from carrot disks. Biochim Biophys Acta 227:278–287

Sadava D, Chrispeels MJ (1971b) Intracellular site of proline hydroxylation in plant cells. Biochemistry 10:4290–4294

Sadava D, Walker F, Chrispeels MJ (1973) Hydroxyproline rich cell wall protein (extensin). Biosynthesis and accumulation in growing pea epicotyls. Dev Biol 30:42–48

Schatz G (1979) How mitochondria import proteins from the cytoplasm. FEBS Lett 103:201–211

Schatz G, Mason TL (1974) The biosynthesis of mitochondrial proteins. Annu Rev Biochem 43:51–87

Scott WA, Mitchell HK (1969) Secondary modification of cytochrome *c* by *Neurospora crassa*. Biochemistry 8:4282–4289

Selvendran RR (1975) Cell wall glycoproteins and polysaccharides of parenchyma of *Phaseolus coccineus*. Phytochemistry 14:2175–2180

Sharon N, Lis H (1979) Comparative biochemistry of plant glycoproteins. Biochem Soc Trans 7:783–799

Sikorski MM, Przybl D, Legocki AB, Gasior E, Zalac J, Borkowski T (1979) The ribosomal proteins of wheat germ distribution and phosphorylation *in vitro*. Plant Sci Letts 15:387–397

Soliday CL, Kolattukudy PE (1979) Introduction of O-glycosidally linked mannose into proteins via mannosyl phosphoryl dolichol by microsomal fractions from *Fusarium solani* f *pisi*. Arch Biochem Biophys 197:367–378

Sun SM, Ma Y, Buchbinder BLU, Hall TC (1979) Comparison of polypeptides synthesized *in vitro* and *in vivo* in the presence and absence of a glycosylation inhibitor. Plant Physiol 63:Supp 95

Tanaka M, Uchida T (1979) Heterogeneity of hydroxyproline containing glycoproteins in protoplasts from a *Vinca rosea* suspension culture. Plant Cell Physiol 20:1295–1306

Tomati U, Matarese R, Federici G (1974) Ferredoxin activation by rhodanese. Phytochemistry 13:1703–1706

Trewavas A (1973) The phosphorylation of ribosomal protein in *Lemna minor*. Plant Physiol 51:760–767

Trewavas A (1976) Post-translational modification of proteins by phosphorylation. Annu Rev Plant Physiol 27:349–374

Trewavas A, Statton BR (1977) The control of plant growth by protein kinases. In: Nucleic acid and protein synthesis in plants. Plenum Press, New York, pp 309–319

Udom AO, Brady FO (1980) Reactivation *in vitro* of zinc-requiring apoenzymes by rat liver zinc thionein. Biochem J 187:329–335

Uy R, Wold F (1977) Post-translational covalent modifications of proteins. Science 198:891–896

Vallee BL (1977) Recent advances in zinc biochemistry. In: Addison AW, Cullen WR, Dolphin D, James BR (eds) Biological aspects of inorganic chemistry. Wiley-Interscience, New York

Vallee BL, Wacker WEC (1970) Metalloproteins In: Neurath H (ed) The Proteins Vol 5 2nd edn. Academic Press, New York, pp 1–92

Vasak M, Galdes A, Allen H, Hill O, Kagi JHR, Brenmer I, Young BW (1980) Investigations of the structure of metallothioneins by proton nuclear magnetic resonance spectroscopy. Biochemistry 19:416–425

Walk RA, Hock B (1978) Cell free synthesis of glyoxysomal malate dehydrogenase. Biochem Biophys Res Commun 81:636–643

Welinder KG (1976) Covalent structure of the glycoprotein horseradish peroxidase (EC1.11.1.7). FEBS Lett 72:19–23

Wells GN, Beevers L (1975) Protein synthesis in the cotyledons of *Pisum sativum* L. Messenger RNA-independent formation of a methionyl-tRNA initiation complex. Arch Biochem Biophys 170:384–391

Whitby AJ, Whish WJD (1978) Poly (adenosine diphosphate ribose) glycohydrolase in germinating wheat embryos. Biochem Soc Trans 6:619–620

Whitby AJ, Stone PR, Whish JD (1979) Effect of polyamines and magnesium on poly-(ADP

ribose) synthesis and ADP-ribosylation of histones in wheat (*Triticum aestivum*). Biochem Biophys Res Commun 90:1295–1304
Wool IG (1979) The structure and function of eukaryotic ribosomes. Annu Rev Biochem 48:719–754
Yamagishi T, Matsuda K, Watanabe T (1976) Characterization of the fragments obtained by enzymic and alkaline degradation of rice bran proteoglycans. Carbohydrate Res 50:63–74
Yamauchi F, Thanh VH, Kawase M, Shibasaki K (1976) Separation of the glycopeptides from soybean 7S protein: Their amino acid sequences. Agric Biol Chem 40:691–696
Yasuda Y, Takahashi N, Murachi T (1970) The composition and structure of carbohydrate moiety of stem bromelain. Biochemistry 9:25–32
Zielinski RE, Price CA (1980) Synthesis of thylakoid membrane proteins by chloroplasts isolated from spinach. Cyt b559 and P700 chlorophyll a proteins. J Cell Biol 85:433–445

# 5 Protein Degradation

PH. MATILE

## 1 Introduction

In the past two decades the general excitement over the discovery of the genetic code and the elucidation of the mechanism of protein synthesis has largely drawn attention away from the phenomena of protein degradation. At the present time there is, however, an increasing interest in cellular digestion of endogeneous protein and the vital importance of this process is becoming more and more recognized. Research workers dealing with plants have of course always been aware of the importance of proteolysis in development because it is very conspicuous in germinating seeds and in senescent leaves. Yet, like their colleagues dealing with animals, they are largely ignorant as to the organization of proteolysis in living cells.

At first glance, protein degradation appears to be simply a matter of interaction between proteins and proteolytic enzymes. From a biochemical point of view it appears to be important to study the proteases capable of degrading, for example, seed proteins (Sects. 6.2, 6.3) and leaf proteins (Sect. 6.4) to amino acids which, in the case of seed germination, are a nutritional prerequisite for the developing plant or, in the case of senescence, are withdrawn from the leaves and deposited in one way or another for re-use in the next period of development. Yet proteolysis is much more fascinating than a mere interaction between substrate and enzymes. In living cells the proteases, although specific with regard to the peptide bonds linking certain amino acids, are unspecific with regard to the protein species attacked and are therefore a potential danger to metabolism which depends on enzyme proteins. The unspecificity of endoeptidases is illustrated by the fact that their activities are normally measured with substrates such as haemoglobin, casein, gelatine, serum albumin, hide powder and with a variety of synthetic substrates. Few studies use relevant natural substrates such as the storage protein vicilin (BAUMGARTNER and CHRISPEELS 1977), the major protein of chloroplasts, ribulose-1,5-bisphosphate carboxylase (WITTENBACH 1978) or phytochrome (PIKE and BRIGGS 1972) to measure protease activity. Indeed, the unspecificity of proteases with regard to protein species implies that in living cells protein degradation must be organized in such a way that the cytoplasm, the truly living entity, is not impaired. A corresponding concept has been presented in *The Lytic Compartment of Plant Cells* (MATILE 1975) and evidence is being accumulated which suggests that proteolytic enzymes are located in plant vacuoles (lysosomes), and are thus separated by membranes from the cytoplasm. It will be seen that in the case of degradation of storage proteins in germinating seeds, the lysosomal function of a distinct type of vacuole, the protein body, is now well documented. In other cases, such as in senescent leaves, the role and the details of subcellular compartmentation in the regulation of proteolysis are not yet clear. Although progress in the understanding of the subcellular distribution of proteases has been made, important pieces of the puzzle are still missing.

Protein degradation, however, marks not only the beginning and the end of plant development – seed germination and senescence – it is a continuous process representing the catabolic wing of protein turnover (see Sect. 5.6) and appears to be selective with regard to individual species of protein as demonstrated by their half-life. Researchers dealing with animal systems disagree on whether and how lysosomes are involved in protein turnover (KOLATA 1977) and what the mechanism of selective and organized protein degradation is. As far as plants are concerned, almost nothing is known about the subcellular organization of protein degradation as related to turnover. Therefore, this chapter is largely devoted to those fields of current research which lend themselves to a discussion of the problems of why and how proteolysis takes place in plant cells. The viewpoint of subcellular compartmentation will be emphasized.

## 2 Proteolysis in Germinating Seeds

### 2.1 The Role of Protein Bodies

Protein metabolism associated with seed development and germination will be dealt with in Chaps. 7 and 14, this Volume; the mobilization of storage proteins of seeds has recently been reviewed by ASHTON (1976). It is well established for seeds of a large variety of species that storage proteins are deposited in a distinct organelle, the protein body (reviews PERNOLLET 1978, WEBER and NEUMANN 1980). The boundary of protein bodies is formed by a single membrane which undoubtedly is homologous to a vacuolar membrane. In fact, in developing seeds, protein bodies may develop from preexisting vacuoles and in germinating seeds, when the storage proteins are gradually mobilized and the opaque contents disappear, the term vacuole (the optically empty space) is appropriate. Upon germination the small aleurone vacuoles normally fuse together and eventually form a single large central vacuole. There is little doubt that the breakdown of storage proteins takes place within the protein bodies. This has recently been demonstrated in a convincing fashion (NISHIMURA and BEEVERS 1979), employing vacuoles isolated from the endosperm of germinated castor beans. Originally the lysosomal nature of protein bodies was deduced from the presence of protease activity in the organelles isolated from cotyledons of germinating pea seeds (MATILE 1968). This hypothesis is now supported by numerous direct and indirect observations, and the protein body has become a unique model system for the study of the biochemical and structural prerequisites of digestive processes in plant cells.

## 2.2 Localization of Proteases in Protein Bodies

The presence of protease activities in protease bodies isolated from various sources has been reported by MATILE (1968), YATSU and JACKS (1968), ORY and HENNINGSEN (1969), ST. ANGELO et al. (1969), MORRIS et al. (1970), SCHNARRENBERGER et al. (1972), ABDEL-GAWARD and ASHTON (1973), ADAMS and NOVELLIE (1975), HARRIS and CHRISPEELS (1975), CHRISPEELS et al. (1976), KONOPSKA and SAKOWSKI (1978), NISHIMURA and BEEVERS (1978), SHUTOV et al. (1978) and others. The isolation of intact protein bodies which retain their soluble constituents, including the hydrolases, may be a difficult task. Therefore the analysis of distributions of proteases in storage cells requires appropriate techniques. YATSU and JACKS (1968) employed a non-aqueous medium for isolating protein bodies from dry cotton seeds and thereby avoided the release of water-soluble material even from damaged organelles. The protein body fraction contained all of the acid protease activity, whilst the complete absence of a cytosolic marker enzyme demonstrated the purity of the isolated material. In contrast, an attempt to localize an endopeptidase in protein bodies of cotyledonary cells of germinated mung beans was only partially successful; although a considerable proportion of the activity was present in the supernatant fraction (CHRISPEELS et al. 1976). Evidently, in germinated seeds, the protein bodies become fragile as the matrix of storage proteins disappears, and they are easily ruptured when the tissue is homogenized. In fact, the endopeptidase of mung bean cotyledons was later definitely localized in the protein bodies by means of monospecific antibodies of the protease and immunofluorescence microscopy (BAUMGARTNER et al. 1978). NISHIMURA and BEEVERS (1978) chose protoplasts prepared from the endosperm of germinated castor beans to avoid drastic mechanical means for liberating intact "aleurone vacuoles"; acid protease and other hydrolases "were present in the isolated vacuoles in amounts, indicating a primarily vacuolar localization in vivo".

Work on the subcellular localization of the numerous proteolytic enzyme activities which occur in germinating seeds is still incomplete. It appears that protein bodies typically contain acid endopeptidase(s). In addition, carboxypeptidase has been localized in protein bodies (HARRIS and CHRISPEELS 1975, BAUMGARTNER and CHRISPEELS 1976, NISHIMURA and BEEVERS 1978). A neutral protease activity was found to be absent from protein bodies isolated from pea seeds (KONOPSKA and SAKOWSKI 1978). In the case of mung beans it has been demonstrated that acid endopeptidase, in cooperation with carboxypeptidase, is capable of completely digesting vicilin, the major reserve protein, to amino acids (BAUMGARTNER and CHRISPEELS 1977).

Some endopeptidases of seeds have been purified and characterized thoroughly. Only the endopeptidase of mung beans (*Phaseolus aureus*) has been shown to function in the mobilization of storage protein by using its natural substrate, vicilin, and to be localized in protein bodies. It was accordingly named "vicilin peptidohydrolase" (BAUMGARTNER and CHRISPEELS 1977). It accounts for practically all of the endopeptidase activity extractable from germinated mung beans. Vicilin peptidohydrolase is a sulfhydryl protease; it is not inhibited by phenylmethylsulphonyl-fluoride, a potent inhibitor of plant carboxypeptidases and other serine proteases. It has a molecular weight of 23,000, a pH optimum of 5.1, and an isoelectric point of 3.75. Its specificity has been tested with esters of amino acids and p-nitrophenol coupled to N-carbobenzoxy and N-tert-butoxycarbonyl residues and the esters of asparagine and glutamine are hydrolyzed most readily.

Mung bean cotyledons also contain a specific inhibitor protein of vicilin peptidohydrolase, which, however, is not located in the protein bodies and therefore, in vivo, has most probably no effect on the hydrolysis of vacuolar reserve proteins (BAUMGARTNER and CHRISPEELS 1976). The protection of the cytoplasm against endopeptidases released from accidentally broken protein bodies has been proposed as a possible function of this inhibitor, but this is difficult to demonstrate. Incidently, the presence of protease inhibitors may lead to erroneous results when protease activities are determined in crude homogenates. The presence of inhibitors of endogeneous proteases in Scots pine (*Pinus sylvestris*) seeds demonstrates that they may be widespread (SALMIA et al. 1978, SALMIA and MIKOLA 1980, see also Chap. 8, this Vol.).

### 2.3 Regulation of Proteolysis in Protein Bodies

The disappearance of storage proteins in germinating seeds is generally associated with a more or less marked increase in proteolytic activities (see CHRISPEELS and BOULTER 1975, HARRIS and CHRISPEELS 1975, CHRISPEELS et al. 1976, TULLY and BEEVERS 1978, FELLER 1979, see also BEEVERS 1976). In the case of germinating cow peas (*Vigna unguiculata*), it has been demonstrated that protein mobilization is closely related not only to enhanced protease activity but also to the location and timing of protein breakdown within the cotyledons (HARRIS et al. 1975).

In the case of mung beans (*Vigna radiata*) the major endoeptidase is completely absent in the ungerminated seeds; the antigen of monospecific vicilin peptidohydrolase antibodies does not appear in the extracts before the third day of germination. Moreover, density labelling experiments have provided further support to the view that the enzyme is synthesized de novo in the course of germination (CHRISPEELS et al. 1976). Hence, the mobilization of storage protein in mung beans appears to be regulated primarily by the synthesis of the crucial hydrolase. From the viewpoint of subcellular compartmentation of proteolysis it is important to consider the results of endopeptidase localizations by immunofluorescence microscopy (BAUMGARTNER et al. 1978): the enzyme is located initially in small "spots" within the cytoplasm, and later appears in the protein bodies. These observations are consistent with ultrastructural features of germination in *Vigna* cotyledons: there is little doubt that the distinct foci in the cytoplasm are identical with ER-derived vesicles observed by HARRIS et al. (1975) and CHRISPEELS et al. (1976). These vesicles, which are primary lysosomes, eventually fuse with protein bodies, turning them into secondary lysosomes (digestive vacuoles). The endopeptidase thus appears to be synthesized in the ER and transported in vesicular form to the compartment in which proteolysis takes place. This strict compartmentation of an "unspecific" digestive enzyme is in agreement with the concept of the "lytic compartment of plant cells" (MATILE 1975).

Compartmentation has not always been considered in the context of regulation of proteolysis in seeds. An example of an alternative concept concerns trypsin inhibitors which in *Vigna unguiculata* apparently inhibit certain endogeneous proteolytic activities; ROYER et al. (1974) proposed a model of regulation which is based on interactions between inhibitors and proteases in vitro. However, since it is known from studies of BAUMGARTNER and CHRISPEELS (1976) and CHRISPEELS and BAUMGARTNER (1978) that such inhibitors are located in the cytoplasm, they most probably have no regulatory function in vivo in the mobilization of storage proteins taking place in the protein bodies.

### 2.4 Autophagic Function of Protein Bodies

In germinating mung beans the trypsin inhibitor (CHRISPEELS and BAUMGARTNER 1978), as well as the inhibitor of vicilin peptidohydrolase (BAUMGARTNER and CHRISPEELS 1976), gradually disappears. Proteolysis evidently not only concerns the storage proteins present in the protein bodies but cytoplasmic proteins as well.

It is important to note that isolated protein bodies, or vacuoles derived from them, not only contain proteolytic enzymes, but also a variety of other hydrolases (MATILE 1968, NISHIMURA and BEEVERS 1978, ADAMS and NOVELLIE 1975). The spectrum of hydrolases contained in protein bodies isolated from mung bean cotyledons – α-mannosidase, N-acetyl-β-glucosaminidase, RNase, acid phosphatase, phosphodiesterase and phospholipase D – suggests that these organelles are not only involved in protein degradation but in the intracellular digestion of cytoplasmic constituents generally (VAN DER WILDEN et al. 1980). One of the unsolved problems in this context concerns the mechanism by which cytoplasmic material is transported into the digestive compartment. Recently VAN DER WILDEN et al. (1980) presented ultrastructural evidence for the internalization of cytoplasmic material including ribosomes and mitochondria in protein bodies. This process appears to be initiated surprisingly early in germination, and it may be speculated that it continues as germination proceeds and is, in fact, the background of autophagic activity which in the senescent storage organs and tissues results in an extensive mobilization of cytoplasm. Protein bodies may not only represent a model system with regard to proteolysis of storage proteins but may also be used for the study of degradation of cytoplasmic proteins and other constituents taking place upon senescence.

## 3 Protein Degradation in Leaves

### 3.1 Leaf Proteases

Knowledge about leaf proteases is scarce compared with the abundance of data available on seed proteases (see RYAN 1973). Recent attempts to characterize the proteolytic system of leaves of several species have yielded evidence for the existence of a variety of distinct enzymes including both endo- and exopeptidases (e.g., SOPANEN and LAURIÈRE 1976, SOPANEN and CARFANTAN 1976, DRIVDAHL and THIMANN 1978, FRITH et al. 1978b, SANTARIUS and BELITZ 1978, THOMAS 1978, FELLER 1979, RAGSTER and CHRISPEELS 1979). Some endopeptidases have been characterized in detail. Several acid proteinases isolated from wheat leaves (FRITH et al. 1978a, b) hydrolyze haemoglobin as well as ribulose-1,5-bisphosphate carboxylase, a possible natural substrate (PEOPLES et al. 1979). One of these enzymes has been characterized as a pepsin-like endopeptidase having carboxyl functions as well as serine in its active centre (FRITH et al. 1978b, c). DRIVDAHL and THIMANN (1978) isolated a similar acid serine protease from oat leaves. In contrast to the insensitivity of this enzyme to sulphhydryl reagents, another acid protease activity of wheat leaves was clearly due to a sulphhydryl-type of endopeptidase (WITTENBACH 1979). FELLER (1979) observed that in bean leaves the main caseinolytic activity is optimal in the range of pH 7 and is slightly sensitive to reducing agents. Another neutral protease of oat leaves described by DRIVDAHL and THIMANN (1978) is inhibited by the serine reagent phenylmethylsulphonyl-fluoride as well as by sulphhydryl reagents and is possibly an exopeptidase. Two alkaline endopeptidases which digest Azocoll have been isolated from soyabean (*Glycine max*) leaves (RAGSTER and CHRISPEELS 1979). One of them is a metalloenzyme (inhibition by EDTA), the other is a sulphhydrylprotease, and neither of them has serine in the active centre. A final example is orange leaves, which contain an acid carboxypeptidase, possibly a metalloenzyme containing zinc in the active site; this exopeptidase has a broad substrate specificity and belongs to the group of serine proteases (ZUBER 1976).

Further work on leaf proteases will undoubtedly result in the discovery of additional enzymes. Yet it seems that the elucidation of protein degradation in leaves does not primarily depend on the comprehensiveness of knowledge about the proteolytic system. Nor does it seem that the knowledge about the catalytic and physical properties can contribute decisively to solving the problem of how proteolysis in leaves is regulated. Rather it will be necessary to investigate the subcellular locations of proteases in order to find the keys to the functions of individual enzymes in the proteolytic processes which take place in the course of leaf development.

### 3.2 Proteases and the Degradation of Leaf Protein

Proteins are most probably degraded throughout the development of leaves. Upon the initiation of senescence, protein turnover becomes unbalanced with increased degradation and declining synthesis resulting in a progressive loss of protein. The continuation of protein synthesis in senescing leaves and several other features of ageing (WOOLHOUSE 1967, SIMON 1967) suggest that the structural organization within living cells is a prerequisite of protein mobilization, metabolism of breakdown products and withdrawal of nitrogen-containing compounds from the leaves. Since proteolysis in senescent leaves is such a conspicuous phenomenon, most of the work has been done on the proteases involved in the final phase of leaf development. About 50%–70% of the total protein is normally mobilized in leaves whether these are attached to the plant or detached in order to experimentally induce senescence. In senescent green leaves the mobilization of protein proceeds over a period of several days or even weeks, whereas in the ephemeral petals of morning glory it is completed within about a day (see MATILE 1975).

Researchers have certainly expected to find correlations between the rate of protein loss in senescing leaves and changes of proteolytic activities in order to draw conclusions about the functional significance of certain proteases. Indeed, correlations have been discovered in a number of cases, yet in others, protein losses in senescent leaves were not found to be related to the development of proteolytic activities.

In senescing leaves of gramineous plants the rapid loss of protein appears to be generally associated with a concomitant increase in protease activity (MARTIN and THIMANN 1972, PETERSON and HUFFAKER 1975, FELLER et al. 1977, FELLER and ERISMANN 1978, THOMAS 1978, WITTENBACH 1978, 1979). In leaf sections of *Lolium temulentum* acid protease activity increased fourfold during the first 3 to 4 days of senescence; subsequently this activity was gradually replaced by proteolytic activity with a neutral pH optimum; since the activities of the two forms of proteases appeared to be reciprocally related, THOMAS (1978) speculated that the "neutral protease activity may be the result of limited self-hydrolysis or ageing of the acid form."

In senescing leaves of leguminous plants, protein degradation was found to be associated with a decline of endopeptidase activities (STOREY and BEEVERS 1977, FELLER 1979, RAGSTER and CHRISPEELS 1979). In tobacco (ANDERSON and ROWAN 1965) and in *Tropaeolum* (BEEVERS 1968) a clear relationship between protein mobilization and acid protease activity was not detectable. In the senescent corolla of morning glory acid protease activity (measured with haemoglobin as substrate) was practically unchanged during the period of most rapid loss of protein (MATILE and WINKENBACH 1971); in contrast, the activity measured with hide powder as substrate was increased two- to threefold, and in leaf discs this increase was completely inhibited in the presence of cycloheximide (A. LÜSCHER, unpublished results). The significance of this result, however, is doubtful as cycloheximide has no effect on the rate of protein degradation in morning glory (MATILE and WINKENBACH 1971). Again, the most encouraging results in this respect have been obtained with leaves of a cereal, wheat: PETERSON and HUFFAKER (1975) demonstrated that both the loss of protein (ribulose-1,5-bisphosphate carboxylase) and the increase in proteolytic activity were inhibited in the presence of cycloheximide. Hence, attempts to demonstrate a functional significance of proteases in protein degradation associated with the ageing of leaves, appear to depend largely on the plant species selected. The negative results may be important, however, as they may draw attention to the possible involvement of factors other than the activity of proteases. Such a factor may be the importance of subcellular compartmentation of protein degradation. Despite the possibility that the presence of endogenous inhibitors may falsify the assessment of protease activities in crude homogenates, the breakdown of protein in the living cells may be a matter of contact between proteins and proteases rather than of levels of protease activities. The example of breakdown of proteins located in chloroplasts is perhaps suitable for illustrating the possible role of subcellular compartmentation.

### 3.3 Degradation of Chloroplastic Protein

Chloroplasts are the site of the most abundant of leaf proteins, ribulose-1,5-bisphosphate carboxylase (RBPC). This enzyme complex, which accounts for most of the so-called fraction I protein, is the main protein component lost in senescing leaves (review by HUFFAKER and PETERSON 1974). WITTENBACH (1977) induced the typical changes associated with senescence such as losses of chlorophyll, soluble protein, and photosynthetic capacity, by placing wheat seedlings in the dark. Under these conditions the loss of RBPC measured immunochemically accounted for as much as 80% of the total loss of soluble protein (WITTENBACH 1978). In the flag leaf of field-grown wheat 40%–45% of the total soluble protein consists of RBPC; in the course of senescence this decreases to about 20%, indicating that RBPC is degraded much faster than other soluble leaf proteins (WITTENBACH 1979). Corresponding results were presented by PEOPLES and DALLING (1978). In senescing leaves of the labiate *Perilla frutescens* activities of several key enzymes of carbon dioxide assimilation change in a strikingly different fashion; RBPC was found to belong to a group of enzymes characterized by a rapid decline at the onset of senescence, whilst other enzyme activities declined much later (BATT and WOOLHOUSE 1975).

These findings suggest that the breakdown of chloroplasts in senescing leaves is a sequential and organized process. It is very interesting that during rapid degradation of RBPC, labelled amino acids are still incorporated into this protein complex (PETERSON and HUFFAKER 1975), the synthesis of which is known to be partly mediated by 70 S chloroplastic ribosomes. WITTENBACH (1978) showed that senescence in primary leaves of wheat is completely reversible during the initial period of rapid decline of RBPC. These observations suggest that the degradation of chloroplastic proteins is not identical with a general deterioration of the organelle. At least during the initial period of protein loss, the organization within senescent chloroplasts appears to be maintained.

In detached primary barley leaves, as already mentioned, the rapid decline of RBCP protein is negatively correlated with the increase in proteolytic activity (PETERSON and HUFFAKER 1975). The involvement of this activity (measured with azocasein as substrate) in the degradation of RBPC was apparent from the inhibitory effects of cycloheximide and kinetin on both loss of RBPC protein and increase of protease activity. In contrast, WITTENBACH (1978) showed that in primary wheat leaves the initial reversible stage of protein loss is not associated with an increased protease activity. Hence, on the basis of measurements of activities, it is difficult to draw conclusions about specific functions of proteases. WITTENBACH (1978) discovered an interesting feature of the major endopeptidase of wheat leaves. This acid thiol-protease has a much higher affinity for RBPC protein than for casein. WITTENBACH (1978) argued that if casein is a good representative for an average soluble protein, this property of the endopeptidase could provide an explanation for the preferential degradation of RBPC during the initial phase of senescence in wheat leaves. This argument, however, would only be valid if protein degradation takes place within the senescing chloroplasts. Hence, a key to the understanding of how the breakdown of chloroplast proteins is organized lies in the subcellular location of the relevant proteases.

At least as far as the origin of protease is concerned, the chloroplasts are not autonomous with regard to the degradation of their major protein since studies with inhibitors have shown that the synthesis of protease is on 80S cytoplasmic ribosomes (PETERSON and HUFFAKER 1975). This is the case with a number of other proteins involved in the metabolism of plastids and it is possible that the protease is incorporated into senescent chloroplasts and eventually exercises its function directly within the organelle. This kind of subcellular organization of proteolysis would, however, require a very short-lived and highly specific protease; otherwise the continuity of protein synthesis and the reversibility of chloroplast breakdown would be difficult to understand.

Results obtained so far on the subcellular distribution of endopeptidases in leaf cells are not fully conclusive. Attempts to localize proteases in isolated chloroplasts, have been more or less unsuccessful (see WITTENBACH 1978, RAGSTER and CHRISPEELS 1979, PEOPLES and DALLING 1979). Chloroplasts isolated from protoplasts of barley mesophyll contained only 4%–5% of the total acid protease activity (HECK et al. 1981) and the preparations were contaminated with a similar percentage of established extra-chloroplastic marker enzyme activities. Moreover, in density gradients, the distribution of acid protease did not provide any indication of coincidence with the chloroplast bands. These negative results were predictable since BOLLER and KENDE (1979 succeeded in localizing protease activity in vacuoles isolated from pineapple leaf protoplasts. Indeed, the bulk of acid protease activity present in barley mesophyll protoplasts is associated with vacuoles (HECK et al. 1981) and there is no indication that the subcellular distribution of protease is changed in senescent barley leaves. Therefore, the degradation of chloroplastic proteins probably takes place in vacuoles rather than in the chloroplasts.

The situation continues, however, to be unsatisfactory, as it is difficult to understand how proteins of chloroplasts such as RBPC contact the vacuolar digestive enzymes during senescence. As the breakdown of chloroplasts is probably a sequential process, it must be assumed that selected portions of chloroplast constituents are externalized on the one hand and eventually internalized in vacuoles on the other. SCHÖTZ and DIERS (1965) have presented ultrastructural evidence for the "emission of plastid parts into the cytoplasm" in leaves of *Oenothera*. Similar observations were made in other species and indications as to the incorporation into vacuoles of material originating from plastids have been reported (e.g., GIFFORD and STEWART 1968, VILLIERS 1971, OLIVEIRA and BISALPUTRA 1977).

Furthermore, on the basis of available data it cannot be excluded that chloroplasts do contain proteases. HOCHKEPPEL (1973) isolated an endopeptidase from senescent tobacco leaves which preferentially hydrolyzes peptide bonds between hydrophobic amino acids. The corresponding antigen was detectable in green leaves although the enzyme activity was absent. Apparently this protease, which readily digests the hydrophobic proteins of thylakoid membranes, is activated upon the ageing of tobacco leaves. It is feasible that such a protease, which is subjected to senescence-specific activation, is located in chloroplasts and is responsible for the deterioration of the photosynthetic membrane system. In this context it is interesting that protein degradation has been observed in isolated chloroplasts (CHOE and THIMANN 1975), and that this process is influenced by light (PANIGRAHI and BISWAL 1979), an environmental factor known to play an important role in the senescence of leaves. At least in part, the protein losses in isolated chloroplasts could be caused by endogeneous proteases of these organelles.

# 4 Yeast

## 4.1 Proteolysis in Bakers' Yeast

With regard to the fundamental problems of protein degradation, bakers' yeast *Saccharomyces cerevisiae* is coming of age as an almost ideal model organism. Protein degradation associated with cell differentiation can be induced experimentally and the changes in enzyme composition that occur when cells are subjected to certain environmental conditions have been studied in considerable detail. Moreover, the various components of the proteolytic system have been characterized and the subcellular location of proteases is largely known. Last but not least, the recent isolation of mutants lacking individual proteases has opened up new possibilities in the approach to the problems of how proteins are degraded in fungal cells.

In yeast cells protein is continuously turned over. In culture conditions which permit exponential growth, the overall rate of protein degradation is low, but protein turnover is enhanced when the yeast cells are starved (HALVORSON 1958, BAKALKIN et al. 1976, BETZ 1976).

In stationary yeast cells specific proteins may be formed in the absence of net protein synthesis. For example, $\alpha$-glucosidase induced in the presence of maltose is synthesized at the expense of degraded proteins (see HALVORSON 1960). Extensive degradation and simultaneous synthesis of protein take place during ascospore formation (HOPPER et al. 1974). As much as 60%–70% of the original vegetative protein is broken down and largely replaced by newly synthesized protein which is incorporated into the ascospores. Not only the pre-existing vegetative protein but also proteins synthesized at different times during sporulation are degraded, turnover eventually resulting in a marked change in enzyme composition (BETZ and WEISER 1976a, b). The importance of protein degradation in the biochemical differentiation of cells is clear also from the phenomenon of di-auxic growth. In the presence of glucose, the synthesis of a number of enzymes involved in oxidative metabolism is selectively repressed. Moreover, upon the addition of glucose to cells adapted to oxidative metabolism, pre-existing enzymes of this group are inactivated and degraded.

This phenomenon, which has been termed catabolite inactivation (HOLZER 1976) or glucose-dependent catabolite degradation (NEEF et al. 1978), is ideally suited for the study of proteolysis as related to the adaptation of metabolism to changing environmental (nutritional) conditions.

Another fascinating proteolytic event in yeast metabolism is the activation of chitin synthetase. This enzyme is responsible for the formation of chitin in the primary septum, and thus is vitally important for budding. It is uniformly distributed in the plasmalemma as an inactive zymogen which in vitro can be transformed proteolytically into the active enzyme (CABIB and FARKAS 1971, CABIB 1976, DURAN et al. 1979). This example is particularly interesting because on the one hand it demonstrates that proteolysis may be highly organized in space and time, and on the other, it represents a class of processes involving limited proteolysis, the importance of which is now generally appreciated. Many proteins are synthesized with sequences that contain the information for the insertion of the protein into membranes, or for its translocation across specific membranes. These signal sequences are eventually removed (see BLOBEL 1979). In yeast, an example of the processing of a nascent protein is given by cytochrome c oxidase: the four subunits of this constituent of the inner mitochondrial membrane are synthesized as one polypeptide chain on cytoplasmic ribosomes; after entry into the mitochondrion it is fragmented proteolytically into the functional proteins (see WOLF et al. 1979).

Although the activation of chitin synthetase or the selective degradation of cytoplasmic proteins associated with catabolite inactivation at first glance seem to require the action of highly specific proteases, which are able to recognize distinct polypeptides, it is puzzling to learn that for one reason or another these proteolytic processes have been attributed to a few constituents of the proteolytic system of yeast.

## 4.2 The Proteolytic System of Yeast

The yeast *Saccharomyces cerevisiae* is an example illustrating the presence of a complete proteolytic machinery in a single cell. The system comprises two endopeptidases, the proteases A and B, two carboxypeptidases Y (yeast) and S (*Saccharomyces*), at least three aminopeptidases and a dipeptidase (see WOLF et al. 1979). Protease A is an acid endopeptidase and B is a neutral serine-protease. Carboxypeptidase Y also possesses a serine residue in its active centre, whereas S is a metalloenzyme ($Zn^{2+}$), as are the aminopeptidases. The determination of protease activities in crude homogenates may yield erroneous results caused by the presence of specific inhibitor proteins of proteases A, B and carboxypeptidase Y. Pre-incubations are necessary therefore to degrade and inactivate these inhibitors, either through the action of endogenous proteases (SAHEKI and HOLZER 1975) or by added thermolysin as demonstrated by WIEMKEN et al. (1979).

The role of yeast proteases in turnover and cellular differentiation processes is apparent from a number of studies in which enhanced proteolytic activities were reported under conditions known to increase protein degradation (WIEMKEN

1969, LENNEY et al. 1974, MATILE et al. 1971, BAKALKIN et al. 1976, FREY and RÖHM 1978). In particular, cell differentiation taking place during sporulation is associated with enhanced protease activities (KLAR and HALVORSON 1975, BETZ and WEISER 1976a). BAKALKIN et al. (1976) showed that in the presence of inhibitors of serine proteases (protease B and carboxypeptidase Y) and in conjunction with pepstatin, an inhibitor of protease A, the degradation of protein was markedly depressed. The irreversible inactivation and elimination of NADP-dependent glutamate dehydrogenase which take place in glucose-starved yeast cells can be inhibited by treatment with phenylmethylsulphonyl-fluoride, a specific inhibitor of serine proteases (MAZON 1978), pointing to the possible involvement of protease B. It is interesting that this particular enzyme has been rediscovered twice, as the factor causing the inactivation of tryptophan synthase in vitro (HOLZER et al. 1973, SAHEKI and HOLZER 1974) and, also in vitro, the activation of chitin synthetase (CABIB and ULANE 1973, ULANE and CABIB 1974, HASILIK and HOLZER 1973). Of the two yeast endopeptidases only protease B catalyzed the conversion of the (inactive?) precursor of carboxypeptidase Y to a protein identical in size to the active form (HASILIK and TANNER 1978).

Although proteinase B appears to be a multifunctional enzyme, at least in vitro, there are also examples of proteolytic inactivation caused by protease A. Studies on the proteolytic inactivation of a variety of enzymes of the intermediary metabolism by the action of the proteases A and B, and carboxypeptidase Y have yielded evidence favouring a selective action for the individual enzymes (AFTING et al. 1976, JUSIC et al. 1976, NEEF et al. 1978, HOLZER and SAHEKI 1976). Yet recent studies with a mutant lacking protease B or carboxypeptidase Y suggest that the involvement of individual proteases in distinct proteolytic processes can hardly be deduced from in vitro studies (review WOLF et al. 1979).

WOLF and FINK (1975) have isolated a carboxypeptidase Y mutant. This mutation has no effect on growth and sporulation, demonstrating that the lack of one exopeptidase can be made good by other enzymes of this type. Carboxypeptidase S, which was discovered through this mutation (WOLF and WEISER 1977), has been found to be not required specifically for proteolysis associated with sporulation (WOLF and EHMANN 1978a). Since protease B, in vitro, catalyzes a number of significant proteolytic processes, the characteristics of a corresponding mutant (WOLF and EHMANN 1978b) could be particularly exciting. It is somewhat disappointing to learn that this mutation has no effect on budding and chitin deposition in the primary septum even though protease B is known to activate chitin synthetase in vitro. Hence, this activation, which must take place locally in the region of the primary septum towards the end of the budding phase, is perhaps catalyzed by a distinct, so far undetected, protease. Proteolytic inactivation of several enzymes proceeds in the normal fashion in the protease B mutant (WOLF and EHMANN 1979). In would appear that the gap in the proteolytic system can be filled by other proteases and that apparent specificities demonstrated in vitro may have a limited significance with regard to proteolysis in vivo. However, the deficiency in protease B is not without consequence; protein degradation in starved cells and during sporulation proceeds at a reduced rate, the appearance of ascospores is delayed and the frequence of sporulation is reduced (WOLF and EHMANN 1979). Mutants, which eventually may be constructed lacking all of the principal proteases, would probably not be viable. It appears to be an advantage of a diversified system of proteases, that the lack of an individual enzyme does not seriously impair the vitality of yeast cells.

### 4.3 Compartmentation of Proteolysis

The subcellular distribution of most of the constituents of the proteolytic system of yeast has been investigated. WIEMKEN et al. (1979) have recently presented an impressive quantitative analysis which demonstrates that the following proteases are predominantly, perhaps exclusively, located in the large vacuoles present in the protoplasts: proteases A and B, carboxypeptidase Y and one of the amino-peptidases. This analysis is distinguished by the employment of an almost ideal method for lyzing protoplasts gently and releasing intact vacuoles as well as by the use of a reliable vacuolar marker substance, polyphosphate, which allowed the determination of yields of isolated and purified vacuoles. In earlier attempts to localize proteases only the enrichment of these enzymes in preparations of isolated vacuoles was demonstrated (MATILE and WIEMKEN 1967, LENNEY et al. 1974, MATILE et al. 1971, HASILIK et al. 1974, FREY and RÖHM 1978, CABIB et al. 1973). Proteases that are located outside the vacuoles comprise two amino-peptidases and the dipeptidase (WIEMKEN et al. 1979, FREY and RÖHM 1978).

With regard to the protein degradations that have been attributed to the vacuolar proteases, WIEMKEN et al. (1979) have shown that the proteases are located in the vacuolar compartment they are not in direct contact with their potential substrates which are located outside the vacuoles. Incidentally, in vivo, the vacuolar proteases A and B, and carboxypeptidase Y are fully active, since the corresponding inhibitor proteins are known to be located in the cytosol (MATERN et al. 1974, LENNEY et al. 1974).

So far, however, it has not been demonstrated that protein digestion takes place within vacuoles. It cannot be excluded that proteolysis is brought about by proteases released from vacuoles, yet the presence of potent inhibitor proteins in the cytosol suggests that this is rather unlikely. The function of these inhibitors is completely unknown. From the viewpoint of the potential danger associated with lysosomes and their digestive machinery, the cytoplasmic protease inhibitors may be interpreted as safety devices against proteases that are released from accidentally damaged vacuoles.

Should it turn out that the vacuolar compartment is, in fact, the site of proteolysis, an explanation for the transport of cytoplasmic constituents into the vacuoles will have to be found. A corresponding process involving the invagination of the vacuolar membrane which results in the internalization into vacuoles of portions of cytoplasmic material has been deduced from electron micrographs (WIEMKEN and NURSE 1973, reviews MATILE 1975, MATILE and WIEMKEN 1976). However, the significance of these morphological events in the intracellular digestion of protein remains to be elucidated.

It should also be borne in mind that the vacuolar proteases are possibly not the only hydrolases that are involved in protein degradation. Apart from proteases responsible for limited proteolysis in the processing of nascent proteins, distinct proteases may cause protein degradation in mitochondria. Using inhibitors of cytoplasmic and mitochondrial protein synthesis as well as protease inhibitors, KAL'NOV et al. (1978, 1979) were able to demonstrate the degradation of products of mitochondrial protein synthesis by virtue of a protease or of proteases, most probably not identical with any of the vacuolar proteases. Hence, it may turn out that in yeast the vacuoles are not the exclusive site of protein degradation.

## 5 Concepts of Protein Degradation

Work with the unicellular eukaryotic organism, yeast, has produced overwhelming evidence for the concurrence of protein synthesis and protein degradation within the same cell, but the problem of *how* intracellular digestion of proteins is organized and regulated is unsolved. On the basis of available data, it is justified to assume that the digestive enzymes have a limited access to the cytoplasmic constituents. The concept of the lysosome or of the lytic compartment relies on the spatial separation of the degradative enzymes from the centres of metabolism in the cytoplasm (see MATILE 1975). This concept is supported by the fact that proteases and a variety of other hydrolases are located in plant cell vacuoles as well as by morphological observations of the presence of decaying cytoplasmic material in these compartments. Yet this concept has recently been questioned by LEIGH (1979), who argued that the hydrolases present in vacuoles may merely be remains of autophagic activity associated with vacuolation in expanding meristematic cells (see MARTY 1978), or else with the digestion of storage proteins in aleurone vacuoles (see NISHIMURA and BEEVERS 1978, MATILE 1975). The involvement of (vacuolar?) proteases in protein mobilization that occurs in senescing leaves suggests, however, that mature vacuoles may have important lysosomal functions and may even be supplemented with proteases when an enhanced proteolysis is induced. Nevertheless, the criticism is fully justified because decisive demonstations are still lacking. The localization of breakdown products of intracellular protein degradation in vacuoles has only been shown in the case of aleurone vacuoles (NISHIMURA and BEEVERS 1979) and the degradative processes associated with the engulfment of cytoplasmic components by invaginating vacuolar membranes have never been clearly demonstrated. Last but not least, it is an open question whether or not several mechanisms, lysosomal and non-lysosomal, are responsible for the various processes of protein degradation.

It may be comforting to take note that the fundamental problems of intracellular protein degradation are also unsolved in the case of animal cells. Although the role of lysosomes in the turnover of protein is firmly established, it is not yet decided whether non-lysosomal pathways are likewise involved (see

DEAN 1979). In animal systems it has been recognized that turnover rates depend markedly on the physical properties of proteins; hydrophobic proteins and proteins of high molecular weight are degraded preferentially (see BOHLEY et al. 1979). A concept of selective protein degradation, which takes into account the changeable conformation of proteins was presented by HOLZER (1976) and was developed with regard to (glucose-dependent) catabolite inactivation in yeast, but it could easily be employed to explain other phenomena of selective protein degradation. The principal features of this concept are allosteric effects and interconversions produced in enzymes by catabolites. These structural changes are thought to convert the enzymes into "vulnerable" proteins which are recognized and eventually internalized in lysosomes (vacuoles). This concept is characterized by the suggested degradation in vacuoles of those enzymes which are inactivated and no longer integrated in metabolism. The specificity of proteolysis, according to HOLZER (1976), is provided by the interactions between proteins and effective catabolites. A similar concept which also takes into account the possibility of non-lysosomal proteolysis has been presented by BOHLEY et al. 1979. In fact, since plant cells possess two classes of organelles, the mitochondria and the plastids, which have many features in common with prokaryotes, protein breakdown outside the vacuoles is not unlikely to occur, and concepts of regulation of proteolysis developed for bacteria (see SCHIMKE and KATUNUMA 1975, GOLDBERG and JOHN 1976, SWITZER 1977) should perhaps also be considered in future attempts to elucidate protein degradation in plant cells.

*Acknowledgement.* The author is indebted to Sonia Türler and to Dorli Furrer for their help with the manuscript.

## References

Abdel-Gaward HA, Ashton FM (1973) Protein hydrolysis in protein bodies isolated from squash seed cotyledons. Plant Physiol 51:289

Adams CA, Novellie L (1975) Acid hydrolases and autolases and autolytic properties of protein bodies and spherosomes isolated from ungerminated seeds of *Sorghum bicolor* (Linn.) Moench. Plant Physiol 55:7–11

Afting EG, Lynen A, Hinze H, Holzer H (1976) Effects of yeast proteinase A, proteinase B and carboxypeptidase Y on yeast phosphofructokinase. Z Physiol Chem 357:1771–1778

Anderson JW, Rowan KS (1965) Activity of peptidases in tobacco leaf tissue in relation to senescence. Biochem J 97:741–746

Ashton FM (1976) Mobilization of storage proteins of seeds. Annu Rev Plant Physiol 27:95–117

Bakalkin GY, Kal'nov SL, Zubatov AS, Luzikov VN (1976) Degradation of total cell protein at different stages of *Saccharomyces cerevisiae* yeast growth. FEBS Lett 63:218–224

Batt T, Woolhouse HW (1975) Changing activities during senescence and sites of synthesis of photosynthetic enzymes in leaves of labiate, *Perilla frutescens* (L) Britt. J Exp Bot 26:569–579

Baumgartner B, Chrispeels MJ (1976) Partial characterization of a protease inhibitor which inhibits the major endopeptidase present in the cotyledons of mung beans. Plant Physiol 58:1–6

Baumgartner B, Chrispeels MJ (1977) Purification and characterization of vicilin peptidohydrolase, the major endopeptidase in the cotyledons of mung bean seedlings. Eur J Biochem 77:223–233

Baumgartner B, Tokuyasu KT, Chrispeels MJ (1978) Localization of vicilin peptidohydrolase in the cotyledons of mung bean seedlings by immunofluorescence microscopy. J Cell Biol 79:10–19

Beevers L (1968) Growth regulator control of senescence in leaf discs of nasturtium (*Tropaeolum majus*). In: Wightman F, Setterfield G (eds) Biochemistry and physiology of plant growth substances. Runge Press, Ottawa, pp 1417–1435

Beevers L (1976) Nitrogen metabolism in plants. Edward Arnold, London

Betz H (1976) Inhibition of protein synthesis stimulates intracellular protein degradation in growing yeast cells. Biochem Biophys Res Commun 72:121–130

Betz H, Weiser U (1976a) Protein degradation and proteinases during yeast sporulation. Eur J Biochem 62:65–76

Betz H, Weiser U (1976b) Protein degradation during yeast sporulation, enzyme and cytochrome patterns. Eur J Biochem 70:385–396

Blobel G (1979) Determinants in protein topology. In: Holzer H, Tschesche H (eds) Biological functions of proteinases. Springer, Berlin Heidelberg New York, pp 102–109

Bohley P, Kirschke H, Langner J, Miehe M, Riemann S, Salama Z, Schön E, Wiederanders B, Ansorge S (1979) Intracellular protein turnover. In: Holzer H, Tschesche H (eds) Biological functions of proteinases. Springer, Berlin Heidelberg New York, pp 17–34

Boller TH, Kende H (1979) Hydrolytic enzymes in the central vacuole of plant cells. Plant Physiol 63:1123–1132

Cabib E (1976) The yeast primary septum: a journey into three-dimensional biochemistry. Trends Biochem Sci 1:275–277

Cabib E, Farkas V (1971) The control of morphogenesis: an enzymatic mechanism for initiation of septum formation in yeast. Proc Natl Acad Sci USA 68:2052–2056

Cabib E, Ulane R (1973) Chitin synthetase activating factor from yeast, a protease. Biochem Biophys Res Commun 50:186–191

Cabib E, Ulane R, Bowers B (1973) Yeast chitin synthetase: separation of the zymogen from its activating factor and recovery of the latter in the vacuole fraction. J Biol Chem 248:1451–1458

Choe HT, Thimann KV (1975) The metabolism of oat leaves during senescence III. The senescence of isolated chloroplasts. Plant Physiol 55:828–834

Chrispeels MJ, Baumgartner B (1978) Trypsin inhibitor in mung bean cotyledons, purification, characteristics, subcellular localization and metabolism. Plant Physiol 61:617–623

Chrispeels MJ, Boulter D (1975) Control of storage protein metabolism in the cotyledons of germinating mung beans: role of endopeptidase. Plant Physiol 55:1031–1037

Chrispeels MJ, Baumgartner B, Harris N (1976) Regulation of reserve protein metabolism in the cotyledons of mung bean seedlings. Proc Natl Acad Sci USA 73:3168–3172

Dean RT (1979) Lysosomes and intracellular proteolysis. In: Holzer H, Tschesche H (eds) Biological functions of proteinases. Springer, Berlin Heidelberg New York, pp 49–54

Drivdahl RH, Thimann KV (1978) Proteases of senescing oat leaves. II. Reaction to substrates and inhibitors. Plant Physiol 61:501–505

Duran A, Cabib E, Bowers B (1979) Chitin synthetase distribution on the yeast plasma membrane. Science 203:363–365

Feller U (1979) Nitrogen mobilization and proteolytic activities in germinating and maturing bush beans (*Phaseolus vulgaris* L.). Z Pflanzenphysiol 95:413–422

Feller U, Erismann KH (1978) Changes in gas exchange and in activities of proteolytic enzymes during senescence of wheat leaves (*Triticum aestivum* L.). Z Pflanzenphysiol 90:235–244

Feller U, Soong TST, Hageman RH (1977) Leaf proteolytic activities and senescence during grain development of field grown corn (*Zea mays* L.). Plant Physiol 59:290–294

Frey J, Röhm KH (1978) Subcellular localization and levels of aminopeptidases and dipeptidase in *Saccharomyces cerevisiae* (BBA 68601). Biochim Biophys Acta 527:31–41

Frith GJT, Gordon KHJ, Dalling MJ (1978a) Proteolytic enzymes in green wheat leaves.

I. Isolation on DEAE-cellulose of several proteinases with acid pH optima. Plant Cell Physiol 19:491–500

Frith GJT, Swinden LB, Dalling MJ (1978b) Proteolytic enzymes in green wheat leaves. II. Purification by affinity chromatography and some properties of proteinases with acid pH optima. Plant Cell Physiol 19:1029–1042

Frith GJT, Peoples MB, Dalling MJ (1978c) Proteolytic enzymes in green wheat leaves. III. Inactivation of acid proteinase II by diazoacetyl-DL-norleucine methyl ester and 1,2-epoxy-3(p-nitrophenoxy)-propane. Plant Cell Physiol 19:819–824

Gifford EM, Stewart KD (1968) Inclusions of the proplastids and vacuoles in the shoot apices of *Bryophyllum* and *Kalanchoe*. Am J Bot 55:269–279

Goldberg AL, John ACST (1976) Intracellular protein degradation in mammalian and bacterial cells, part 2. Annu Rev Biochem 45:748–803

Halvorson HO (1958) Intracellular protein and nucleic acid turnover in resting yeast cells. Biochem Biophys Acta 27:255–266

Halvorson HO (1960) The induced synthesis of proteins. Adv Enzymol 22:99–156

Harris N, Chrispeels MJ (1975) Histochemical and biochemical observations on storage protein metabolism and protein body autolysis in cotyledons of germinating mung beans. Plant Physiol 56:292–299

Harris N, Chrispeels MJ, Boulter D (1975) Biochemical and histochemical studies on protease activity and reserve protein metabolism in cotyledons of germinating cowpeas (*Vigna unguiculata*). J Exp Bot 26:544–554

Hasilik A, Holzer H (1973) Participation of the tryptophan synthase inactivating system from yeast in the activation of chitin synthase. Biochem Biophys Res Commun 53:552–559

Hasilik A, Tanner W (1978) Biosynthesis of the vacuolar yeast glycoprotein carboxypeptidase Y. Eur J Biochem 85:599–608

Hasilik A, Müller H, Holzer H (1974) Compartmentation of the tryptophan-synthase-proteolyzing system in *Saccharomyces cerevisiae*. Eur J Biochem 48:111)117

Heck U, Martinoia E, Matile PH (1981) Subcellular localization of acid proteinase in barley mesophyll protoplasts. Planta 151:198–200

Hochkeppel HK (1973) Isolierung einer Endopeptidase aus alternden Tabakblättern und ihre Beziehung zum Vergilben. Z Pflanzenphysiol 69:329–343

Holzer H (1976) Catabolite inactivation in yeast. Trends in Biochem Sci 1:178–181

Holzer H, Saheki T (1976) Mechanisms of regulation of proteolytic processes in yeast. Tokaj J Exp Clin Med 1:115–125

Holzer H, Katsunuma T, Schött EG, Ferguson AR, Hasilik A, Betz H (1973) Studies on a tryptophan synthase inactivating system from yeast. Adv Enzyme Regul 11:53–60

Hopper AK, Magee PT, Welch SK, Friedman M, Hall BD (1974) Macromolecule synthesis and breakdown in relation to sporulation and meiosis in yeast. J Bacteriol 119:619–628

Huffaker RC, Peterson LW (1974) Protein turnover in plants and possible means of its control. Annu Rev Plant Physiol 25:363–392

Jusic M, Hinze H, Holzer H (1976) Inactivation of yeast enzymes by proteinase A and B and carboxypeptidase Y from yeast. Z Physiol Chem 357:735–740

Kal'nov SL, Serebryakova NV, Zubatov AS, Luzikov VN (1978) Proteolysis of products of mitochondrial protein synthesis in isolated mitochondria of *Saccharomyces cerevisiae*. Biokhimiya 43:526–531

Kal'nov SL, Novikova LA, Zubatov AS, Luzikov VN (1979) Participation of a mitochondrial proteinase in the breakdown of mitochondrial translation products in yeast. FEBS Lett 101:355–358

Klar AJS, Halvorson HO (1975) Proteinase activities of *Saccharomyces cerevisiae* during sporulation. J Bacteriol 124:863–869

Kolata GB (1977) Protein degradation: putting the research together. Science 198:596–598

Konopska L, Sakowski R (1978) Enzyme activities in aleurone grains of pea seeds during the early stages of germination. Biochem Physiol Pflanz 173:536–540

Leigh RA (1979) Do plant vacuoles degrade cytoplasmic components? Trends Biochem Sci 4:N37–38

Lenney JF, Matile PH, Wiemken A, Schellenberg M, Meyer J (1974) Activities and cellular

localization of yeast proteases and their inhibitors. Biochem Biophys Res Commun 60:1378–1383
Martin C, Thimann KV (1972) The role of protein synthesis in the senescence of leaves. I. The formation of protease. Plant Physiol 49:64–71
Marty F (1978) Cytochemical studies on gerl, provacuoles and vacuoles in root meristematic cells of Euphorbia. Proc Natl Acad Sci USA 75:852–856
Matern H, Betz H, Holzer H (1974) Compartmentation of inhibitors of proteinases A and B and carbohypeptidase Y in yeast. Biochem Biophys Res Commun 60:1051–1057
Matile PH (1968) Aleurone vacuoles as lysosomes. Z Pflanzenphysiol 58:365–368
Matile PH (1975) The lytic compartment of plant cells. Cell biology monographs. Vol 1. Springer, Wien New York
Matile PH, Wiemken A (1967) The vacuole as the lysosome of the yeast cell. Arch Mikrobiol 56:148–155
Matile PH, Wiemken A (1976) Interaction between cytoplasm and vacuole. In: Stocking CR, Heber U (eds) Transport in Plants III. Springer, Berlin Heidelberg New York, pp 255–287
Matile PH, Winkenbach F (1971) Function of lysosomes and lysosomal enzymes in the senescing corolla of the morning glory (*Ipomoea purpurea*). J Exp Bot 22:759–771
Matile PH, Wiemken A, Guyer W (1971) A lysosomal aminopeptidase isozyme in differentiating yeast cells and protoplasts. Planta 96:43–53
Mazon MJ (1978) Effect of glucose starvation on the nicotine-amide dinucleotide phosphate-dependent glutamate dehydrogenase of yeast. J Bacteriol 133:780–785
Morris GFI, Thurman DA, Boulter D (1970) The extraction and chemical composition of aleurone grains (protein bodies) isolated from seeds of *Vicia faba*. Phytochemistry 9:1707–1714
Neef J, Hägele E, Neuhaus J, Heer U, Mecke D (1978) Evidence for catabolite degradation in the glucose dependent inactivation of yeast cytoplasmic malate dehydrogenase. Eur J Biochem 87:489–495
Nishimura M, Beevers H (1978) Hydrolases in vacuoles from castor bean endosperm. Plant Physiol 62:44–48
Nishimura M, Beevers H (1979) Hydrolysis of protein in vacuoles isolated from a higher plant tissue. Nature 277:412–413
Oliveira L, Bisalputra T (1977) Ultrastructural studies in the brown alga *Ectocarpus* in culture: ageing. New Phytol 78:131–138
Ory RL, Henningsen KW (1969) Enzymes associated with protein bodies isolated from ungerminated barley seeds. Plant Physiol 44:1488–1498
Panigrahi PK, Biswal UC (1979) Ageing of chloroplasts in vitro I. Quantitative analysis of the degradation of pigments, proteins and nucleic acids. Plant Cell Physiol 20:775–779
Peoples MB, Dalling MJ (1978) Degradation of ribulose-1,5-bisphosphate carboxylase by protelytic enzymes from crude extracts of wheat leaves. Planta 138:153–160
Peoples MB, Dalling MJ (1979) Intracellular localization of acid peptide hydrolases in wheat leaves. Plant Physiol Suppl 63/5 Abstract Nr 882, p 159
Peoples MB, Frith GJT, Dalling MJ (1979) Proteolytic enzymes in green wheat leaves. IV. Degradation of ribulose 1.5-bisphosphate carboxylase by acid proteinases isolated on DEAE-cellulose. Plant Cell Physiol 20:253–258
Pernollet JC (1978) Protein bodies of seeds: ultrastructure, biochemistry, biosynthesis and degradation. Phytochemistry 17:1473–1480
Peterson LW, Huffaker RC (1975) Loss of ribulose 1.5-diphosphate carboxylase and increase in proteolytic activity during senescence of detached primary barley leaves. Plant Physiol 55:1009–1015
Pike CS, Briggs WR (1972) Partial purification and characterization of a phytochrome degrading neutral protease from etiolated oat shoots. Plant Physiol 49:521–530
Ragster L, Chrispeels MJ (1979) Azocoll-digesting proteinases in soybean leaves. Characteristics and changes during leaf maturation and senescence. Plant Physiol 64:857–862
Royer A, Miège MN, Grange A, Miège J, Mascherpa KM (1974) Inhibiteurs anti-trypsine et activités protéolytiques des albumines de graine de *Vigna unguïculata*. Planta 119:1–16

Ryan CA (1973) Proteolytic enzymes and their inhibitors in plants. Annu Rev Plant Physiol 24:173–196

Saheki T, Holzer H (1974) Comparisons of the tryptophan synthase inactivating enzymes with proteinases from yeast. Eur J Biochem 42:621–626

Saheki T, Holzer H (1975) Proteolytic activities in yeast. Biochim Biophys Acta 384:203–214

Salmia MA, Mikola JJ (1980) Inhibitors of endogenous proteinases in the seeds of Scots pine, *Pinus silvestris*. Physiol Plant 48:126–130

Salmia MA, Nyman SA, Mikola JJ (1978) Characterization of the proteinases present in germinating seeds of Scots pine, *Pinus silvestris*. Physiol Plant 42:252–256

Santarius K, Belitz HD (1978) Proteinase activity in potato plants. Planta 141:145–153

Schimke RT, Katunuma N (1975) Intracellular protein turnover. Academic Press, New York

Schnarrenberger C, Oeser A, Tolbert NE (1972) Isolation of protein bodies on sucrose gradients. Planta 104:185–194

Schötz F, Diers L (1965) Elektronenmikroskopische Untersuchungen über die Abgabe von Plastidenteilen ins Plasma. Planta 66:269–292

Shutov AD, Korolova TN, Vaintraub IA (1978) Participation of dormant vetch seed proteinases in the degradation of reserve proteins during germination (in Russian). Fiziol Rast 25:735–742

Simon EW (1967) Types of leaf senescence. Symp Soc Exp Biol 21:215–230

Sopanen T, Carfantan N (1976) Activities of various peptidases in the senescing petals of tulip. Physiol Plant 36:247–250

Sopanen T, Laurière CH (1976) Activities of various peptidases in the first leaf of wheat. Physiol Plant 36:251–254

St. Angelo AJ, Ory RL, Hansen HJ (1969) Localization of an acid proteinase in hempseed. Phytochemistry 8:1135–1138

Storey R, Beevers L (1977) Proteolytic activity in relationship to senescence and cotyledonary development in *Pisum sativum*. Planta 137:34–44

Switzer RL (1977) In vivo inactivation of microbial enzymes. Annu Rev Microbiol 31:135–157

Thomas H (1978) Enzymes of nitrogen mobilization in detached leaves of *Lolium temulentum* during senescence. Planta 142:161–169

Tully RE, Beevers H (1978) Proteases and peptidases of castor bean endosperm-enzyme characterization and changes during germination. Plant Physiol 62:746–750

Ulane RE, Cabib E (1974) The activating system of Chitin synthetase from *Sacharomyces cerevisiae*. Purification and properties of an inhibitor of the activating factor. J Biol Chem 249:3418–3422

Villiers TA (1971) Lysosomal activities of the vacuole in damaged and recovering plant cells. Nature 233:57–58

Weber E, Neumann D (1980) Protein bodies, the storage organelles in plant seeds. Biochem Physiol Pflanz 175:279–306

Wiemken A (1969) Eigenschaften der Hefevakuole. Doctoral Thesis Nr. 4340, Swiss Federal Institute of Technology, Zürich

Wiemken A, Nurse P (1973) The vacuole as a compartment of amino acid pools in yeast. Proc Third Int Specialized Symp on Yeast Otaniemi/Helsinki. Part II, pp 331–347

Wiemken A, Schellenberg M, Urech K (1979) Vacuoles: The sole compartments of digestive enzymes in yeast (*Saccharomyces cerevisiae*)? Arch Microbiol 123:23–25

van der Wilden W, Herman EM, Chrispeels MJ (1980) Protein bodies of mung bean cotyledons as autophagic organelles. Proc Natl Acad Sci USA 77:428–432

Wittenbach VA (1977) Induced senescence of intact wheat seedlings and its reversibility. Plant Physiol 59:1039–1042

Wittenbach VA (1978) Breakdown of ribulose bisphosphate carboxylase and change in proteolytic activity during dark-induced senescence of wheat seedlings. Plant Physiol 62:604–608

Wittenbach VA (1979) Ribulose bisphosphate carboxylase and proteolytic activity in wheat leaves from anthesis through senescence. Plant Physiol 64:884–887

Wolf DH, Ehmann C (1978a) Carboxypeptidase S from yeast: regulation of its activity during vegetative growth and differentiation. FEBS Lett 91:59–62
Wolf DH, Ehmann C (1978b) Isolation of yeast mutants lacking proteinase B activity. FEBS Lett 92:121–124
Wolf DH, Ehmann C (1979) Studies on a proteinase B mutant of yeast. Eur J Biochem 98:375–424
Wolf DH, Fink GR (1975) Proteinase C (carboxypeptidase Y) mutant of yeast. J Bacteriol 123:1150–1156
Wolf DH, Weiser U (1977) Studies on a carboxypeptidase Y mutant of yeast and evidence for a second carboxy-peptidase activity. Eur J Biochem 73:553–556
Wolf DH, Ehmann C, Beck I (1979) Genetic and biochemical analysis of intracellular proteolysis in yeast. In: Holzer H, Tschesche H (eds) Biological functions of proteinases. Springer, Berlin Heidelberg New York, pp 55–72
Woolhouse HW (1967) The nature of senescence in plants. Symp Soc Exp Biol 21:179–213
Yatsu LY, Jacks TJ (1968) Association of lysosomal activity with aleurone grains in plant seeds. Arch Biochem Biophys 124:466–471
Zuber H (1976) Carboxypeptidase C. Methods Enzymol 45:561–568

# 6 Physiological Aspects of Protein Turnover

D.D. DAVIES

## 1 Introduction

Towards the end of the 19th century, plant physiologists considered the possibility that proteins undergo continuous breakdown and resynthesis. Despite the objections of Pfeffer, the idea of protein turnover was revived in the 1920's and a great deal of work on the nitrogen balance of detached leaves was interpreted in terms of the temporal separation of degradation and synthesis. MOTHES (1933) however, proposed that in leaves there was simultaneous synthesis and degradation of protein involving separate systems for synthesis and degradation. PAECH (1935), on the other hand, argued that control of the nitrogen balance was exercized through mass action. GREGORY and SEN (1931) noted that in leaves of potassium-starved barley there was an accumulation of amino acids, which, according to the mass action proposals of Paech, should have led to enhanced protein synthesis. However, the reverse happened and the continued accumulation of amino nitrogen suggested to them that the routes of protein synthesis and degradation were separable and that the rates of these processes might vary independently. They proposed a "PROTEIN CYCLE" according to which proteins undergo a cycle of synthesis and degradation (i.e., protein turnover), but also included the proposition that protein degradation is linked to the production of $CO_2$. The extent to which protein turnover contributes to respiration is discussed in Section 3, but it is not central to the concept of turnover itself. The first direct evidence for protein turnover in plants was provided independently by HEVESEY et al. (1940) and VICKERY et al. (1940), who reported the assimilation of $[^{15}N]H_3$ into leaf protein, despite a net loss of protein from leaves. Some of the assumptions involved in the experiments of HEVESEY et al. (1940) are open to doubt because they used leaves of different ages and the net protein content of these leaves was not measured before and after the period of labelling; consequently, the $^{15}N$ incorporation could be due to gross protein synthesis rather than to protein replacement. This uncertainty had earlier been identified by MOTHES (1931) who found that the extent of protein degradation depended on the age of the leaf. In young leaves protein increased in light and dark, in medium-aged leaves protein only increased in the light, and in old leaves protein decreased in both light and dark. This uncertainty about the reality of protein degradation in actively growing tissue is still not completely resolved. Methods of measuring protein turnover do not distinguish between populations of cells which may have different rates of protein turnover. Thus, for example, AMENTA and SARGUS (1979) have concluded that L-cells exist in at least two states with respect to protein degradation:

a sub-population that is actively replicating and does not degrade cellular proteins and a sub-population which degrades most of its labelled proteins but is not capable of replication.

This uncertainty about the spatial and temporal coincidence of protein synthesis and degradation has bedevilled research on protein turnover. With the advent of radioactive amino acids, it became a routine experiment to measure the apparent rates of protein synthesis and degradation. Nevertheless, interest in the process of protein *degradation* lapsed because a number of workers reported the absence of protein degradation in exponentially growing *Escherichia coli*. KOCH and Levy (1955) and MANDELSTAM (1958a, b) concluded that whilst protein degradation occurs at 2%–5% per h in non-growing *E. coli*, it was very slow or absent in exponentially growing *E. coli*. These findings were taken to mean that protein turnover has little physiological significance during active growth. Workers with plant tissue cultures (BIDWELL et al. 1964) came to the opposite conclusion – namely that protein degradation is slow in non-growing cultures and rapid in fast-growing cultures. Little work on protein degradation in plants was undertaken at this time, partly due to the difficulties involved in interpreting the kinetics, but perhaps of more significance was the rapid development of molecular biology which was much more concerned with protein synthesis than with degradation.

However, reports from a number of laboratories (WILLETTS 1967) established that protein degradation does take place in exponentially growing *E. coli* and that in HeLa cells, protein degradation occurs at a constant rate irrespective of growth rate (EAGLE et al. 1959). This renewed interest led to the development of new techniques for measuring degradation and to the first major review of protein turnover in plants (HUFFAKER and PETERSON 1974).

## 2 The Measurement of Protein Turnover

Protein turnover implies simultaneous synthesis and degradation:

net synthesis = gross synthesis – degradation.

Where gross synthesis equals degradation, the definition of turnover as "the process of renewal of a given substance" by ZILVERSMIT et al. (1943) can be applied. When gross synthesis and degradation are not equal, turnover may be defined as "the rate of synthesis *or* degradation, whichever is the smaller" (REINER 1953) or alternatively as "the flux of amino acids through proteins" (HUFFAKER and PETERSEN 1974). However, as KOCH (1962) pointed out, for most biochemists, the term turnover connotes the degradative process more than the synthetic process, whichever is the larger. Thus, GOLDBERG and DICE (1974) define protein turnover as the "hydrolysis of intracellular proteins to their component amino acids". However, protein turnover can only occur when synthesis and degradation occur simultaneously, and it was this point which led SWICK et al. (1956) to define turnover rates for biosynthesis and degradation and to calculate the corresponding turnover rate constants. In this review, I will be mainly concerned with the degradative aspects of turnover.

## 2.1 The Measurement of Gross Protein Synthesis

The most frequently used method involves supplying the tissue with radioactive amino acids and using the rate of incorporation to evaluate the rate of protein synthesis. The major difficulties in this method stem from compartments in the cell which contain amino acids which may equilibrate with the labelled amino acid at different rates, producing a number of pools of amino acids with different specific activities. When the cell is extracted, the pools mix to give a specific activity for the amino acid which may be very different from the specific activity at the site of protein synthesis.

Workers with plants have concluded that in addition to a vacuolar pool there is at least one other amino acid pool, and amino acids must pass through this pool before entry into protein. For example, HOLLEMAN and KEYS (1967) observed that when [$^{14}$C]-leucine is supplied to soyabean (*Glycine max*) hypocotyls, there is a lag phase before the incorporation into protein is linear (Fig. 1). The existence of this lag enables the size of the leucine pool to be determined. The lag time, obtained by extrapolating the linear part of the incorporation curve to meet the time axis, gives the time when the amount of leucine entering the pool equals the amount of leucine in the pool, i.e., the pool size = rate of leucine entering protein × $t_{lag}$. It should be noted that the rate of incorporation of labelled leucine into protein becomes linear when the protein precursor pool becomes saturated – the acid soluble pool of amino acids, which includes the vacuolar pool, reaches saturation *after* the protein labelling becomes linear.

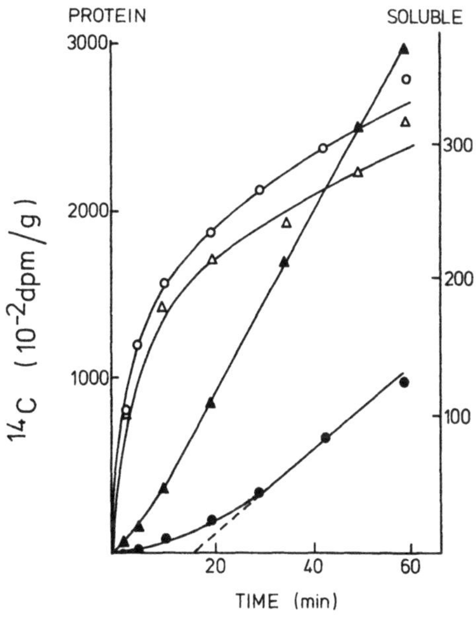


**Fig. 1.** Accumulation and incorporation of [$^{14}$C]-leucine in soyabean hypocotyls after a 30-min and a 4-h pre-incubation period. Incorporation into protein ● after 30 min; ▲ after 4 h. Into soluble amino acids; ○ after 30 min; △ after 4 h. (After HOLLEMAN and KEYS 1967)

Quantitative information about the size of the precursor pool makes it possible to check COWIE's (1962) suggestion that the precursor pool is composed of protein-forming templates. HOLLEMAN and KEYS (1967) have estimated that the size of the leucine and valine precursor pool exceeds the total RNA of soyabean tissue by a factor of 12 and concluded that the precursor pool is not composed of aminoacyl-tRNA. However, if it is assumed that the labelled amino acid passes through a single precursor pool of aminoacyl-tRNA before entering protein, the specific activity of the aminoacyl-tRNA gives the specific activity of the amino acid entering protein.

TREWAVAS (1972) measured the specific activity of methionyl tRNA in a growing population of *Lemna* to make possible the calculation of the gross rate of protein synthesis.

The incorporation of radioactivity into protein is described by the equation used by REINER (1953)

$$\frac{dP^*}{dt} = (F^* - P^*)\frac{Vs}{P} \tag{1}$$

where $F^*$ = Specific activity of methionyl-tRNA

$P^*$ = Specific activity of protein
$Vs$ = rate of protein synthesis
and $P$ = amount of protein.

If exponential growth is maintained, Vs and P will be directly dependent on the size of the culture, so

$$\frac{Vs}{P} = k_s \tag{2}$$

where $k_s$ is the rate constant of synthesis, and

$$\frac{dP^*}{dt} = (F^* - P^*)k_s \tag{3}$$

Under conditions in which the specific activity of methionyl-tRNA is constant, integration of (3) gives

$$P^* = F^*(1 - e^{-k_s t}) \tag{4}$$

so that $k_s$ can be evaluated from a plot of log $P^*$ against time.

## 2.2 The Measurement of Protein Degradation

In the chase period following a labelling period, the degradation of protein can, in principle, be measured by determining the loss of label from the protein.

However, if the labelled amino acid can recycle back into protein, it will reduce the apparent rate of degradation. Consider a simple model of protein turnover and amino acid recycling

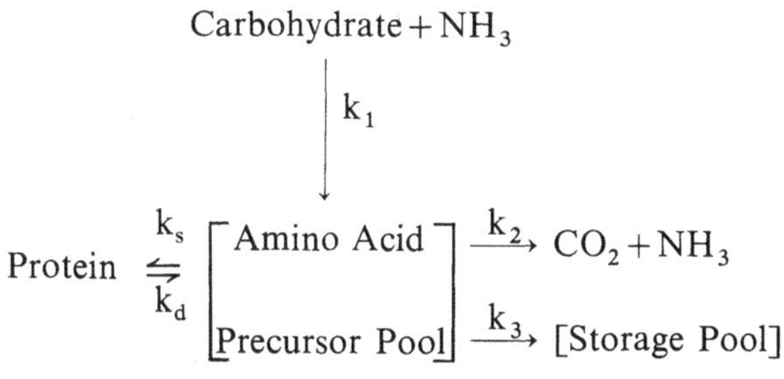


Scheme I

If recycling of amino acids is very small, i.e., $k_2+k_3 \gg k_5$, and assuming a steady state, the loss of radioactivity from protein during the chase period is

$$\frac{dP^*}{dt} = -k_d P^* \tag{5}$$

where P* is the specific activity of the labelled protein. The integrated form of Eq. (5) is

$$P^*_{(t)} = P^*_{(0)} e^{-k_d t} \tag{6}$$

The plot of log P* against time will be linear and the slope will give the rate constant of degradation $k_d$.

If recycling is significant, then the labelled amino acid released from the protein will equilibrate in the precursor pool and re-enter protein, so that the loss of radioactivity from the protein will be apparently less

$$\frac{dP^*}{dt} = k_d(F^* - P^*) \tag{7}$$

where F* is the specific activity of the precursor amino acid.

The use of this equation to evaluate protein degradation in liver has been discussed by Poole (1971). However, in plants where there may be large amino acid pools in the vacuole, the specific activity of the acid-soluble amino acid pool will differ from that of the precursor pool and the methods developed by Poole will not, in general, be applicable.

An alternative approach is to select amino acids which are metabolically active (i.e., $k_2 \gg k_s$). This approach has been assessed in *Lemna* by measuring the extent of recycling for a number of amino acids (Davies and Humphrey 1978). The results in Table 1 support the view that glutamate and alanine might be suitable precursors for studies of protein turnover.

**Table 1.** Recycling of amino acids in *Lemna minor* growing in a complete medium. (After DAVIES and HUMPHREY 1978)

| Method | Amino acid | Recycling (%) |
|---|---|---|
| Density labelling | Leucine | 43 |
| [$^3$H]+[$^{14}$C]-labelling | Leucine | 50 |
| [$^3$H]+[$^{14}$C]-labelling | Glutamate | 29 |
| [$^3$H]+[$^{14}$C]-labelling | Arginine | 48 |
| [$^3$H]+[$^{14}$C]-labelling | Lysine | 45 |
| [$^3$H]+[$^{14}$C]-labelling | Isoleucine | 50 |

Because of the difficulties inherent in these methods, alternative approaches have been developed and a few of these are briefly mentioned here.

### 2.2.1 Density Labelling

The use of density labelling to distinguish between newly synthesized and pre-existing DNA was introduced by MESELSON and STAHL (1958) and led to a similar method for the separation of newly synthesized from pre-existing proteins (HU et al. 1962). This method has been adapted to measure the rate of protein degradation (BOUDET et al. 1975; DAVIES and HUMPHREY 1976).

Plants are first grown in a medium containing $^2H_2O$ and then in the same solution lacking $^2H_2O$. The heavy and light proteins respectively produced can be separated by isopycnic centrifugation. If a [$^{14}$C]-labelled amino acid is given as a pulse at the end of the period in $^2H_2O$, [$^{14}$C]-label appears in heavy protein, and, if protein degradation occurs when the plants are transferred to $H_2O$, [$^{14}$C]-label is lost from the heavy protein and hence from the peak obtained on isopycnic centrifugation. If recycling of the amino acids occurs, label appears in the light protein peak. If there is no protein turnover, label will stay in the heavy protein peak. Protein degradation can therefore be measured by following the decrease in radioactivity associated with the heavy protein peak. An experiment to measure the rate constant of protein degradation is illustrated in Fig. 2.

The logarithm of the normalized peak areas of heavy protein gave a linear relationship with time from which the half-life of *Lemna*-soluble protein appears to be 101 h.

However, $^2H_2O$ puts the plant under a stress, in which there is a reduction in protein synthesis and a large increase in protein degradation (COOKE et al. 1979a). The enhanced degradation appears to be due to damage to the tonoplast allowing vacuolar proteolytic enzymes to pass into the cytoplasm and increase protein breakdown (COOKE et al. 1980a). Hence, estimates of protein degradation obtained by density labelling should be considered with caution.

### 2.2.2 The Use of Tritiated Water ($^3H_2O$)

The use of $^3H_2O$ to measure protein degradation was independently developed by HUMPHREY and DAVIES (1975) and PINE and SCHIMKE (1978). The method

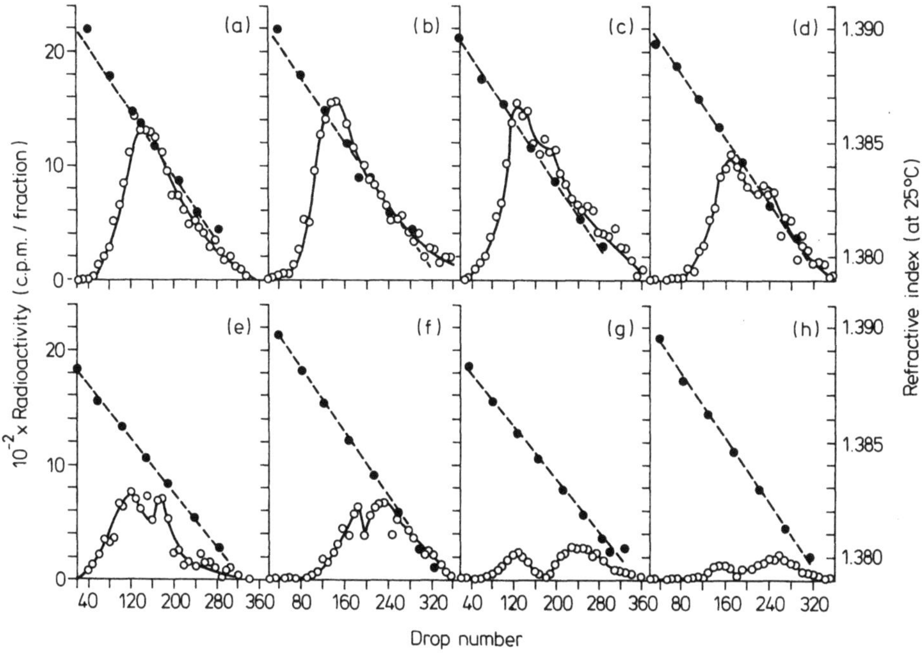


**Fig. 2a–h.** Measurement of protein turnover by using an 8 h exposure to [$^{14}$C]-leucine to label protein. *Lemma* fronds were density-labelled for 7 days with 50% $^2H_2O$, and on day 8 L-[U-$^{14}$C]-leucine (5 μCi/200 ml) was added. After 8 h, plants were transferred to a medium lacking [$^{14}$C]-leucine. After a further 40 h, plants were transferred to a medium lacking both $^2H_2O$ and [$^{14}$C]-leucine. Samples were removed at intervals, protein was extracted, low-molecular-weight substances were removed by passing through Sephadex G-100 and the heavy and light fractions separated by isopycnic centrifugation in KBr. ○ L-[U-$^{14}$C]-leucine in protein; ● refractive index at 25° C. **a** immediately after transfer from $^2H_2O$ medium; **b** 8 h after transfer; **c** 24 h after transfer; **d** 48 h after transfer; **e** 72 h after transfer; **f** 96 h after transfer; **g** 120 h after transfer; **h** 144 h after transfer. Drop number is the number of drops collected from the density gradient. (After BOUDET et al. 1975)

is based on the assumption that when cells are incubated with $^3H_2O$, $^3H$ rapidly equilibrates with H on the α-carbon atoms of most amino acids due to the exchange reaction catalyzed by transaminases (HILTON et al. 1954).

When cells are transferred back to $H_2O$, the transaminases ensure that the reverse sequence occurs and H replaces $^3H$ on the α-carbon of amino acids. However, if an amino acid labelled with $^3H$ on the α-carbon atom is incorporated into protein, the $^3H$ is no longer exchangeable and remains in the protein when the cells are transferred back to water.

However, if the protein undergoes hydrolysis to amino acids, the transaminases ensure that $^3H$ is lost from the amino acids. Hence, exchange reactions catalyzed by transaminases largely eliminate the problems of pools and recycling so that to measure protein degradation, it is only necessary to give cells a brief exposure to $^3H_2O$, then transfer the cells back to $H_2O$ and measure the amount of $^3H$ in protein as a function of time.

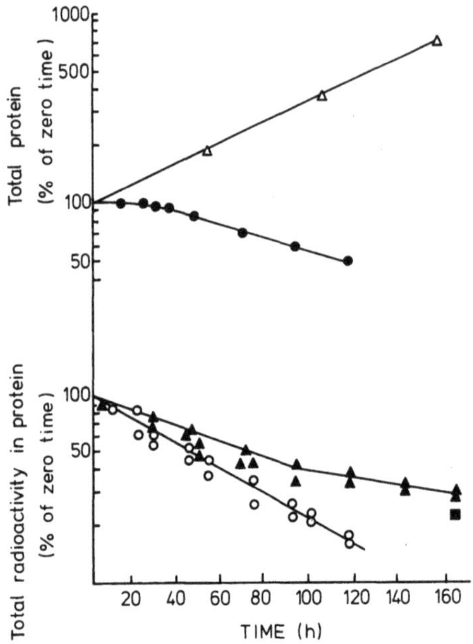


**Fig. 3.** Comparison of protein degradation in *Lemna* growing in complete culture medium and in distilled water. Fronds were labelled for 20 min with $^3H_2O$ before being chased for varying periods of time, before extracting protein. ▲ Total radioactivity in *Lemna* protein grown on complete medium expressed as % of radioactivity at zero time; ○ total radioactivity in *Lemna* protein grown on water; △ total protein in *Lemna* grown on complete medium; ● total protein in *Lemna* grown on water. (After HUMPHREY and DAVIES 1975)

An experiment using this method is illustrated in Fig. 3. There are two disadvantages. (1) Proteins with short half-lives are labelled to a greater extent than proteins with long half-lives, so that the method tends to give an *apparent* average half-life which is shorter than the *true* average half-life of protein. (2) The amount of $^3H$ which is incorporated is limited by the duration of exposure and so restricts the sensitivity of the method.

Increasing the time of exposure to $^3H_2O$ leads to $^3H$ entering non-exchangeable positions in the amino acids, thereby eliminating some of the advantages of the method. To increase the sensivity of the method and preserve its special advantage, it is necessary to measure the amount of $^3H$ on C-2 of protein amino acids without interference from $^3H$ in other positions. This can be achieved by hydrolyzing protein to its constituent amino acids, taking the amino acids to dryness and then treating with acetic anhydride and acetic acid to acetylate and racemize them. The azlactone re-arrangement releases $^3H$ into acetic acid which can be distilled and counted (HUMPHREY and DAVIES 1976).

In a typical experiment, *Lemna* fronds were grown on $^3H_2O$-labelled growth medium for 48 h then transferred to either [$^3H$]-free medium or water and grown for 7 days. Samples were taken at intervals and the total $^3H$ and 2-$^3H$ contents of the protein measured. When the former was measured the specific radioactivity showed an increase during the initial stage of the chase period (Fig. 4), suggesting that [$^3H$]-labelled amino acids continue to enter protein after removal of the fronds from the $^3H_2O$ containing growth medium. The initial increase in $^3H$ was not observed when the 2-$^3H$ content of the protein was determined. Instead there was a steady exponential decline from which the half-life of protein was estimated as 7.1 days in complete growth medium and 2 days in water.

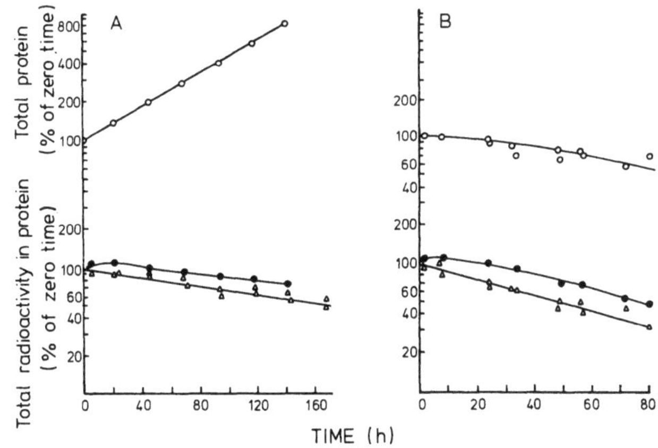


**Fig. 4***A*,*B*. Comparison of protein degradation in *Lemna minor* growing in complete culture medium (*A*) and in water (*B*). *Lemna minor* was cultured on [$^3$H]-labelled medium for 48 h, then removed, washed and transferred to either [$^3$H]-free medium or water. Samples were taken at intervals, the protein was isolated and hydrolysed, and both the $^3$H on the C-2 and the total [$^3$H]-content of the protein hydrolysates determined. The change in the [$^3$H]-content of the protein was used to determine rates of degradation. The *lines through the points* were determined by regression analysis. o total protein in *Lemna minor*; ● total [$^3$H]-radioactivity in *Lemna minor* protein expressed as a percentage of radioactivity at zero time; △ total [2-$^3$H]-radioactivity in *Lemna minor* protein expressed as a percentage of radioactivity at zero time. (After HUMPHREY and DAVIES 1976)

### 2.2.3 The Double Isotope Method for Measuring Relative Rates of Protein Degradation

In this method, two isotopic forms of an amino acid (e.g., [$^{14}$C]- or [$^3$H]-valine) are supplied at different times to establish two time points on the protein decay curve. The authors, ARIAS et al. (1969), specified the conditions under which the method is valid. [$^{14}$C]-valine is first supplied and labelled protein allowed to degrade for a given time. [$^3$H]-valine is then given and shortly afterwards the tissue is extracted and the ratio $^{14}$C/$^3$H determined in various protein fractions. A high $^{14}$C/$^3$H ratio indicates a fraction with a long half-life. The limitations of the method have been stressed by POOLE (1971) and by ZAK et al. (1977). The effect of recycling is to remove the linear relationship between the isotope ratio and the half-life. Despite these limitations, the method has proved useful and can be easily modified for special cases.

## 3 The Contribution of Protein Turnover to Respiration

In animals there is considerable evidence that proteins can be used as a metabolic fuel. Thus, during even short periods of starvation, there is significant breakdown of muscle proteins (MILLWARD et al. 1976). Since much energy is required for protein synthesis, protein turnover will constitute a significant proportion of the basal metabolic rate. The quantitative assessment is difficult, but estimates suggest that about 15% of the energy expenditure in man is due to protein turnover without net gain of protein.

Protein also contributes to production of carbon dioxide via the reactions of the alanine cycle (CAHILL 1970). The amino acids produced by protein degradation in muscle transaminate with pyruvate, produced from glucose by the reactions of glycolysis, to form keto acids and alanine. Some workers (GOLDSTEIN and NEWSHOLME 1976, GARBER et al. 1976) have suggested that most of the alanine was derived not from glucose but from other amino acids. More recently CHANG and GOLDBERG (1978) have estimated that 97% of the carbons of alanine, pyruvate and lactate released by muscle were derived from glucose. The alanine is transported to the liver, where it is deaminated, and the pyruvate released passes through the reactions of gluconeogenesis to form glucose, which will eventually once again undergo the reactions of glycolysis in the muscle. These reactions constitute a special case of protein metabolism and the extent to which proteins and amino acids can serve directly as sources of energy is uncertain (KREBS 1972), although in rat spleen, protein may contribute up to 30% of the substrates for respiration (SUTER and WEIDEMAN 1976).

The contribution of protein to respiration in plants has been reviewed by AP REES (1980). A number of amino acids, especially those with carbon skeletons closely related to respiratory intermediates, are readily oxidized to $CO_2$ by plant tissues, but release of $CO_2$ from the aromatic amino acids appears to be small (NOZZOLILLO et al. 1971, BURRELL and AP REES 1974). Amino acids that are metabolized to respiratory intermediates are not used solely as a source of energy. Thus, in fat-bearing seeds, the carbon skeletons of some amino acids participate in gluconeogenesis (STEWARD and BEEVERS 1967). The large differences in amino acid composition between seed storage proteins and the proteins derived from them, and also between the amino acid composition of leaves and the seed storage proteins derived from them (FOLKES and YEMM 1956, CHOU and SPLITTSTOESSER 1972) necessitates considerable interconversion of amino acids.

Quantitative or even semiquantitative information on the contribution of proteins to respiration is scanty. In excised leaves YEMM (1950) estimated that protein accounted for 20%–40% of the respired $CO_2$. BIDWELL et al. (1964) estimated the contribution of protein to respiratory $CO_2$ by carrot plants as 27% in fast-growing cultures and 7% in slowly growing cultures (no coconut milk). The theoretical basis of these calculations has been questioned by DAVIES and HUMPHREY (1976) since they appear to involve a misunderstanding of the kinetics of label-chase experiments. BIDWELL et al. (1964) analyze their experiments on the *incorrect* assumption that in a chase experiment the specific activity of a product *cannot exceed* the specific activity of the precursor. A similar misunderstanding is to be found in the analysis of chase experiments by OAKS and BIDWELL (1970). Since this question has been recently discussed (DAVIES 1980), the argument will not be repeated here, except to cite the work of ZILVERSMIT et al. (1943), which provides the mathematical basis for the correct interpretation of label-chase experiments.

The extent to which proteins and acids contribute to $CO_2$ production requires a knowledge of the rate constants shown in Scheme I. A method of evaluating

$$\frac{k_2+k_3}{k_s+k_2+k_3}$$

has been described by DAVIES and HUMPHREY (1978). This method makes it possible to estimate the % fraction of amino acids, derived from protein, which are metabolized rather than recycled into protein. When applied to *Lemna*, this analysis produces the "not unexpected" results that the % glutamate which is metabolized (71%) is greater than the % of leucine which is metabolized (50%). It would be difficult to contest the conclusion of AP REES (1980) that "there are not enough data available to permit assessment of the contribution of amino acids and protein to plant respiration. The available data fail to establish that protein is a major respiratory substrate except possibly when the plant is in extremis."

## 4 Protein Turnover During Seed Germination

There is a massive degradation of storage protein during germination, but since there is evidence for synthesis of enzymes in the cells which are hydrolyzing the storage proteins, the overall process is considered as a case of turnover in which degradation of storage protein greatly exceeds synthesis of enzymes.

### 4.1 Protein Degradation

A considerable body of evidence (RYAN 1973, ASHTON 1976) indicates that during seed germination there is usually a short delay, followed by an increase in endopeptidase activity. Three explanations for the increase in endopeptidase activity have been offered.

a) A decrease in the concentration of proteinase inhibitors.
b) The release of active proteases from inactive zymogens.
c) De novo synthesis of endopeptidases.

#### 4.1.1 Proteinase Inhibitors

Since their discovery in cereals (READ and HAAS 1938) these compounds have excited interest in relation to human nutrition and the physical chemistry of protein–protein interactions (FRITZ and TSCHESCHE 1971, FRITZ et al. 1974), but in recent years there has been much interest in relation to plant physiology (RYAN 1973, RICHARDSON 1977). The amino acid sequences of several of these inhibitors have been determined and many regions of structural homology are apparent (see Chap. 6, this Vol.). This suggests that a large number of plant proteins with the capacity to inhibit animal proteinases have evolved from a small number of ancestral molecules. This conclusion leads to the proposition that the inhibitors confer some selective advantage on the plants that contain them.

The possibility that proteinase inhibitors may play a role in controlling protein hydrolysis in seeds has until recently been generally discounted. Thus, the concentration of the trypsin inhibitor may *increase* during germination (PUSZTAI 1972, KIRSI 1974) and there is no correlation between endogenous protease activity and the concentration of the trypsin inhibitor. However, since these trypsin inhibitors do not inhibit the endogenous proteases (SOEDIGDO and GRUBER 1960, BURGER and SIEGELMAN 1966, HOBDAY et al. 1973), this is neither surprising nor meaningful. On the other hand, a number of endogenous inhibitors, active against endogenous proteases, have been demonstrated in the following seeds: rye (*Secale cereale*) (POLANOWSKI 1967), lettuce (*Lactuca sativa*) (SHAIN and MAYER 1965, 1968a), barley (MIKOLA and ENARI 1970), rice (*Oryza sativa*) (HORIGUCHI and KITAGISHI 1971), *Vigna unguiculata* (ROYER et al. 1974, GENNIS and CANTOR 1976), mung beans (*Phaseolus aureus*) (BAUMGARTNER and CHRISPEELS 1976). These observations have been interpreted by some (SHAIN and MAYER 1968b, ROYER et al. 1974) to mean that the inhibitors control the activity of proteases hydrolyzing seed storage proteins. This plausible model is unlikely to be applicable to all seeds.

In the case of the black-eyed pea protease inhibitor, GENNIS and CANTOR (1976) noted that the inhibitor stabilized the proteinase and suggested that on germination the seed protease is released as a result of the destruction of the disulphide-rich inhibitor by protein disulphide reductase. In the case of the mung bean protease inhibitor, BAUMGARTNER and CHRISPEELS (1976) noted that during germination the protease activity increased and the inhibitor decreased. However, the kinetics do not suggest a causal relationship because the inhibitor declines much faster than protease activity increases. These authors rejected the intuitive role for the protease inhibitor and proposed that the inhibitor may function in protecting the cytoplasm from accidental rupturing of the protease-containing protein bodies where storage protein hydrolysis occurs.

### 4.1.2 Activation of Zymogens

Despite the large-scale degradation of protein, there is surprisingly little information about the proteolytic systems involved. Early reports (YOUNG and VARNER 1959, HENSHALL and GOODWIN 1964) indicated only small increases in proteolytic activity during germination. These results may be attributed to the use of an unsatisfactory assay based on absorbance at 280 nm. Subsequently, it has become clear that there is an increase in proteolytic activity associated with germination. The possibility that this increased activity is due to the activation of a zymogen or proenzyme would parallel the activation of the well-known zymogens (pepsinogen, trypsinogen, prorennin and procarboxypeptidase) of the digestive tract.

Evidence consistent with this proposal is that several enzymes are released during germination from a precursor form, e.g., isocitritase and acid phosphatase in several seeds (PRESLEY and FOWDEN 1965), the R-enzyme in peas (SHAIN and MAYER 1968a) and ornithine-$\delta$-transaminase in peanuts (CAMERON and

MAZELIS 1971). Evidence for a trypsinogen-trypsin-like situation in seed germination was obtained by SHAIN and MAYER (1968b) who observed a large increase in proteolytic activity during the germination of lettuce seeds and purified the trypsin-like enzyme 450-fold. The increase in activity of this enzyme was not prevented by inhibitors of protein synthesis and there was no incorporation of $^{35}S$ from $[^{35}S]O_4$ into the enzyme. The authors concluded that the trypsin-like enzyme was released from a precursor form and suggested two possible mechanisms; either an autocatalytic process analogous to the action of pepsin on pepsinogen or a specific limited proteolysis catalyzed by either the carboxypeptidase or aminopeptidase which are known to be present in lettuce seeds.

#### 4.1.3 De Novo Synthesis of Endopeptidases

Density-labelling studies have shown that the gibberellic acid induced increase in the protease of barley aleurone cells represents de novo synthesis from amino acids released by proteolysis from storage proteins (JACOBSEN and VARNER 1967). BAUMGARTNER and CHRISPEELS (1977) have characterized the major endopeptidase of germinating mung beans, which is absent from dry seeds. Evidence (CHRISPEELS et al. 1976) that the increased activity was due to de novo synthesis was obtained by density labelling and by showing that when $[^{35}S]O_4$ was supplied to the seedling, $^{35}S$ was incorporated into the protease, purified by affinity chromatography and ion-exchange chromatography. Ultrastructural studies suggest that vesicles from the rough endoplasmic reticulum may mediate the transport of the proteinase from its site of synthesis to the protein bodies.

It seems likely that seeds employ different methods for controlling the degradation of their storage proteins. In some cases, initial proteolytic activity in the dry seed may be adequate for protein degradation. In other cases, removal of inhibitors, activation or de novo synthesis may contribute in varying degrees. In the case of the mung bean, the situation seems clear; the control of storage protein degradation has nothing to do with inhibitors or with zymogens, but is a straight-forward case of enzyme induction in which the vicilin peptidohydrolase, synthesized in the cytoplasm, is transported to the protein bodies (BAUMGARTNER et al. 1978). The function of the inhibitor of the vicilin peptidohydrolase would seem to be to protect the proteins of the cytoplasm during the time taken for the peptidohydrolase to enter the protein bodies.

### 4.2 Protein Synthesis

The density-labelling method has been used to show that there is de novo synthesis of a number of enzymes during the early stages of germination (FILNER et al. 1969). By allowing in vivo proteolysis to occur in $H_2{}^{18}O$, $[^{18}O]$-labelled amino acids are formed

$$\text{Reserve protein} + H_2{}^{18}O \rightarrow RCHNH_2\,C^{16}O^{18}OH.$$

The synthesis of new proteins such as protease, ribonuclease and α-amylase involves the incorporation of 50% of the $^{18}O$ in amino acids into protein and gives a sufficient increase in buoyant density to allow de novo synthesis to be demonstrated by isopycnic centrifugation. The introduction of $^2H_2O$ for density labelling by LONGO (1968) has enabled the de novo synthesis of other enzymes to be established. However, the presence of specific inhibitors complicates the situation in some cases. For example, maize seeds contain an inhibitor of alcohol dehydrogenase which increases on germination, concomitant with the decline in activity of alcohol dehydrogenase. This suggests that the activity of alcohol dehydrogenase during seed germination is controlled by the level of the inhibitor (HO and SCANDALIOS 1975). Further support for this view was the lack of a density shift in alcohol dehydrogenase when the seeds were germinated in the presence of $^2H_2O$ and $[^{15}N]H_4Cl$. However, when the seeds are germinated under anaerobic conditions, alcohol dehydrogenase activity increases (SCANDALIOS 1977, SACHS and FREELING 1978) but the inhibitor does not decrease. Evidence that de novo synthesis is involved has been obtained by the incorporation of $[^{14}C]$-leucine (SACHS and FREELING 1978) and by density labelling (LAI and SCANDALIOS 1977). The density shift observed was extremely small but the band width was significantly increased and, as discussed in Section 6.2, this observation is consistent with turnover of alcohol dehydrogenase.

# 5 Protein Turnover During Active Growth

The turnover of total protein during growth can be measured by a variety of methods and a comparison of some of these methods is shown in Table 2.

The measurement of turnover of individual proteins or enzymes presents a much more difficult problem, because what is usually measured is a change in enzyme *activity*. A loss of enzyme activity could be due to protein degradation but it could be due to inhibition or inactivation. The problem is clearly seen in the diurnal fluctuations in activity of a number of enzymes of *Kalanchoë blossfeldiana* (QUEIROZ 1974). Recognizing that changes in enzyme *activity* could be due to various regulatory processes such as feedback inhibition, QUEIROZ (1978) stresses that what is measured is enzyme *capacity* and it is these fluctuations which are recorded in Fig. 5. The weakness in these data is that the

**Table 2.** Comparison of of half-lives of *Lemna minor* soluble protein obtained by various methods

| Method | $t_{1/2}$ (days) |
|---|---|
| HELLEBUST and BIDWELL (1963) | 0.88 |
| KEMP and SUTTON (1971) | 12.25 |
| TREWAVAS (1972) | 7 |
| HUMPHREY and DAVIES (1975) | Fraction A 3 days<br>Fraction B 7 days |
| BOUDET et al. (1975) | 4.2 |
| HUMPHREY and DAVIES (1976) | 7.1 |

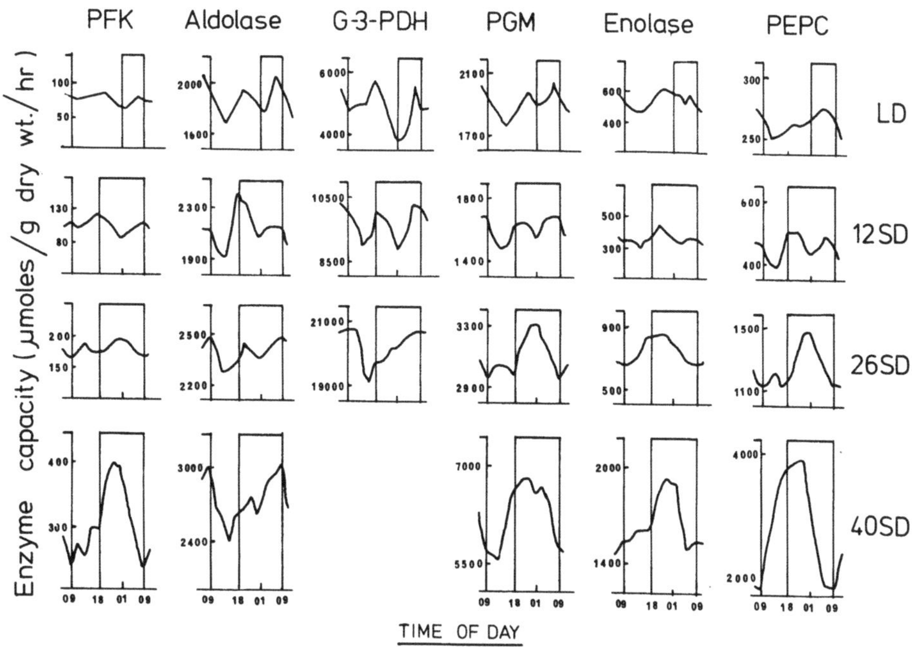


**Fig. 5.** Day/night rhythms of capacity for the enzymes of glycolysis and PEP carboxylase in long days (*LD*) and during the 12th, 26th and 40th short day (*SD*). Enzyme capacity is expressed as μMol product formed/g dry wt/h. (After PIERRE and QUEIROZ 1979)

authors did not carry out the necessary enzymological tests to eliminate some of the pitfalls in enzyme assay. The authors used polyethylene glycol to remove tannins, but they did not carry out molecular sieving to remove small molecules, they did not carry out dilution tests to check the presence of dissociable inhibitors or activators and they did not carry out mixing experiments to check the recovery of the enzymes (AP REES 1974). However, if we accept the changes in enzyme capacity shown in Fig. 5, it is necessary to explain them. QUEIROZ (1974) has discussed photoactivation and enzyme induction where the control of enzyme synthesis is assumed to operate via the phytochrome system. If only photoactivation is involved, and evidence for the light activation of a number of enzymes of *Kalanchoë* has been provided by GUPTA and ANDERSON (1978), then this falls outside the area of this review. However, light modulation of enzyme activity in the case of pyruvate phosphate dikinase (SUGIYAMA 1974) and NADP malate dehydrogenase (KAGAWA and HATCH 1977) appears to involve an enzyme-catalyzed disulphide exchange which BALLARD et al. (1974) have proposed as a necessary step before proteolysis of mammalian phosphoenol pyruvate carboxykinase can occur. A role for protein turnover in controlling these enzyme fluctuations cannot be ruled out, but other possibilities should be examined first. Thus the inactivation of phosphoenolpyruvate carboxylase in *Mesembryanthemum crystallinum* requires a heat-labile fraction whose properties are consistent with it being a specific inactivating protein and possibly a protease (WINTER and GREENWAY 1978).

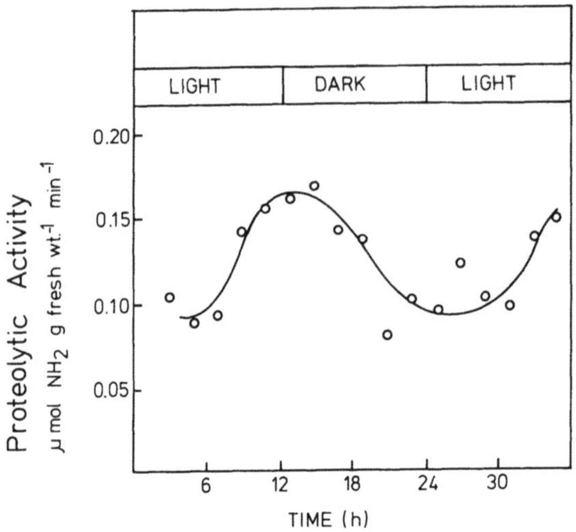


**Fig. 6.** Diurnal variation of a proteinase acting on ribuose bisphosphate carboxylase in wheat leaves. (After PEOPLES and DALLING 1978)

Circadian rhythms in enzyme activity or capacity are not restricted to CAM plants. For example, oscillations have been reported for both NAD and NADP-linked glyceraldehyde-3-phosphate dehydrogenase in *Chenopodium rubrum* (FROSCH et al. 1973) for phenylalanine ammonia lyase in *Lemna* (GORDON and KOUKKARI 1978) for several dehydrogenases in tobacco (*Nicotiana tabacum*) leaves (DE JONG and WOODLIEF 1979) and for ribulose bisphosphate carboxylase in wheat (DI MARCO et al. 1979). Pronounced oscillations have also been reported for enzymes of carbohydrate metabolism when sycamore cell cultures underwent a transition from one steady state to another (FOWLER and CLIFTON 1975). Protein turnover could be involved in all or none of these cases. Evidence bearing on this is the observation that the fluctuations in ribulose bisphosphate carboxylase activity were closely paralleled by the amount of fraction I protein measured by gel electrophoresis (DI MARCO et al. 1979) whilst fluctuations of a proteinase (possibly specific for ribulose bisphosphate carboxylase) showed a pattern complementary to that obtained for the carboxylase, with the maximum proteolytic activity at the end of the light period as seen in Fig. 6 (PEOPLES and DALLING 1978). However, this evidence must be balanced against reports that ribulose bisphosphate carboxylase is light-activated in vivo and that its activity is decreased in the dark (PEDERSEN et al. 1966). It is necessary to distinguish between factors which affect the activation state of the enzyme and those which affect the actual rate of catalysis, and in the case of ribulose bisphosphate carboxylase some factors such as $CO_2$ and pH can affect both processes (JENSEN and BAHR 1977). Thus the extent to which this enzyme undergoes light activation and dark inactivation is not clear (ROBINSON et al. 1979). Furthermore, there is uncertainty about the extent to which the enzyme undergoes turnover. PETERSON et al. (1973) concluded that ribulose bisphosphate carboxylase does not undergo simultaneous synthesis and degradation in light, whereas the remaining soluble protein was turned over rapidly. After prolonged darkness (in excess of 24 h) degradation of ribulose bisphosphate carboxylase was observed. On

**Table 3.** The half-life ($t_{1/2}$) and rate constant (kd) of degradation of total soluble protein and ribulose bisphosphate carboxylase protein in 13-day-old second leaves of *Zea mays* under continuous light (L) or 14 h light/10 h dark (L/D). Two methods of determing $t_{1/2}$ and kd were used, the $^3H_2O$ method of HUMPHREY and DAVIES (1976) and the [$^3$H] acetic anhydride method of Simpson and Varner (personal communication). The values shown are the means of three determinations (except where indicated) ±S.E.

| Conditions | Method | Total protein | | Ribulose bisphosphate carboxylase protein | |
|---|---|---|---|---|---|
| | | $t_{1/2}$ (days) | kd (day$^{-1}$) | $t_{1/2}$ (days) | kd (day$^{-1}$) |
| L | $^3H_2O$ | 7.54±0.67 | −0.095±0.009 | 7.79±0.92 | −0.01 ±0.011 |
| L | $^3$H-Ac.Anh. | 6.74±0.30 | −0.103±0.006 | 6.30±0.71 [a] | −0.112±0.012 [a] |
| L/D | $^3H_2O$ | 6.85±0.44 | −0.102±0.006 | 6.21±0.75 | −0.117±0.013 |
| L/D | $^3$H-Ac.Anh. | 6.17±0.90 | −0.112±0.016 | 5.77 [b] | −0.120 [b] |

[a] Only 2 determinations
[b] Only 1 determination

the other hand, collaborative experiments between Prof. Varner's laboratory and that of the author indicate that the half-life of ribulose bisphosphate carboxylase of maize leaves is close to that of the total protein (Table 3).

## 6 The Measurement of Enzyme Turnover

A number of methods are available for measuring the turnover of specific proteins but all involve assumptions which must be kept in mind when assessing experimental results.

### 6.1 Determination of the Rate Constant of Degradation from Changes in Enzyme Activity

The concentration of enzyme in a tissue is determined by the balance between synthesis and degradation.

In general

$$\frac{dE}{dt} = k_s - k_d E$$

where $k_s$ is the zero order rate constant of synthesis, $k_d$ is the first-order rate constant of degradation and E is the concentration of enzyme.

In the steady state

$$E_0 = \frac{k_s}{k_d}$$

If the plant passes from one steady state to another in which

$$E_t = \frac{k_s^1}{k_d^1}$$

where $k_s^1$ and $k_d^1$ are the new rate constants of synthesis and degradation, the transition from one state to another can be fitted to an equation derived by BERLIN and SCHIMKE (1965). For an increase in enzyme activity

$$E_t - E^1 = (E_t - E_0)\,e^{-k_d^1 t}$$

where $E^1$ is the concentration of enzyme at time t.

Hence, $\ln(E_t - E^1) = \ln(E_t - E_0) - k_d^1 t$, and the plot of $\ln(E_t - E^1)$ against time is linear, the slope gives the rate constant of enzyme degradation $k_d^1$ and the time taken for 50% increase in enzyme activity gives the half-life of enzyme degradation. Whilst it is clear that the concentration of enzyme in the steady state is determined by the rate constants of synthesis *and* degradation, it is not intuitively obvious that the time taken for the enzyme concentration to *increase* from $E_0$ to $E_t$ is determined solely by the rate constant of *degradation* $k_d^1$.

If the concentration of enzyme subsequently falls to the original concentration, the equation for the decline in enzyme activity may be written

$$E^1 = \frac{k_s}{k_d} + \left(E_t - \frac{k_s}{k_d}\right)e^{-k_d t}$$

$$\ln(E^1 - E_0) = \ln(E_t - E_0) - k_d t$$

The plot of $\ln(E^1 - E_0)$ against time is linear and the slope gives the rate constant of degradation $k_d$. It is essential to note that when applying this method, the equations used are derived from a simple model of the transition from one steady state to another in which the rate constants are assumed to change instantly and the model does not allow gradual changes in $k_s$ or $k_d$. Thus, any change in the rate constants must be rapid relative to the half-life of the enzyme. However, above all it must be remembered that what has been measured is the rate constant for the *loss of enzyme activity* which may or may not reflect enzyme degradation.

If the transition from one steady state to another involves the reduction of enzyme activity to virtually zero, i.e., $k_s = 0$, then

$$-\frac{dE}{dt} = k_d E$$

and $\ln E^1 = \ln E_t - k_d t$, so that the plot of $\ln E^1$ against time is linear. It should be noted that this plot can only be justified when $k_s = 0$ and this restriction has not always been kept in mind (see discussion in DAVIES 1980).

## 6.2 Determination of the Rate of Enzyme Degradation by Density Labelling

The main theoretical difficulty in the use of density labelling for measuring rates of enzyme synthesis or degradation is that it is only the density of *catalytically active* molecules which is determined. Thus density labelling measures the half-life of catalytically active molecules rather than the degradative half-life.

Consider a sample of cells in which an enzyme is maintained at a constant level. Initially the cells will have a population of enzyme molecules of density $d_1$ which can be isolated and banded by isopycnic centrifugation. If the cells are transferred to $^2H_2O$, $^2H$ will rapidly equilibrate with amino acids so that new enzyme molecules will be heavy with a density $d_2$. Thus after a given time, a certain number of "light" molecules of density $d_1$ will degrade and be replaced by an equivalent number of "heavy" molecules of density $d_2$. However, since light and heavy molecules have the same catalytic activity, isopycnic centrifugation will give a single band of density $d_3$ (Fig. 7). When all the light molecules have been replaced by heavy molecules, a single peak of density $d_2$ will be observed. In addition to measuring density changes, workers in this field also measure the bandwidth at half-peak height. Initially the population of light enzymes will have a band width $w_1$, which will increase as heavy molecules enter. The bandwidth will increase to a maximum ($w_2$) when there are equal numbers of heavy and light molecules and then decrease to the original bandwidth ($w_1$) when all the molecules are heavy. If there is no net change in the concentration of enzyme, i.e., enzyme synthesis equals enzyme degradation, the time taken to attain maximum bandwidth will give the half-life of the enzyme. However, it should be noted that this relationship does *not* apply if there is a net gain of enzyme. Under these conditions the maximum bandwidth will be attained in *less* than the half-life of enzyme degradation. This limitation has not always been recognized by workers in this field.

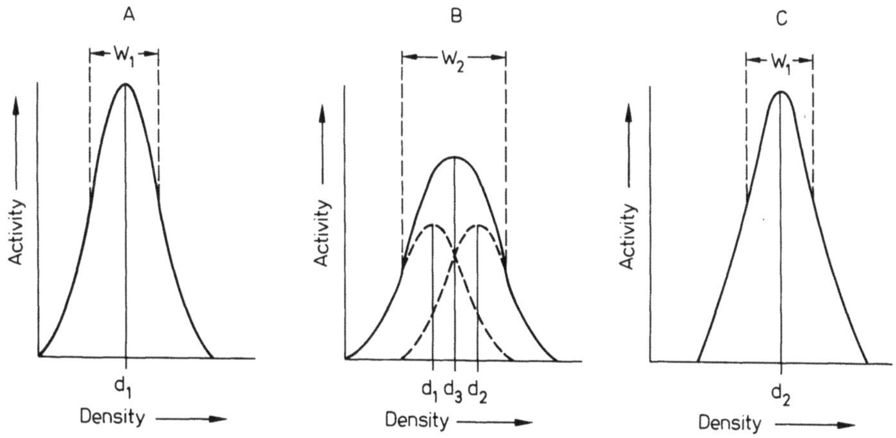


**Fig. 7A–C.** Predicted changes in density of an enzyme after transfer to a heavy labelled medium. (A steady state is assumed.) **A** Distribution of unlabelled enzyme (density $d_1$) in a density gradient (bandwidth $W_1$). **B** Distribution of equal numbers of heavy ($d_2$) and light ($d_1$) enzyme molecules in a density gradient ($d_3$ the measured density; $W_2$ bandwidth). **C** Distribution of labelled enzyme (density $d_2$; bandwidth $W_1$)

If there is a net gain of enzyme during the experiment, the measured profile of enzyme activity against density will be the arithmetic sum of two populations of enzyme molecules of different density. Assuming that the activity-density profile is Gaussian, the two sub-populations can be determined by a curve-fitting analysis described by AITCHISON et al. (1976) and LAMB et al. (1979).

## 6.3 Determination of the Rate of Enzyme Degradation by Immunology

A general model for enzyme degradation may be written

Active enzyme
↓ ↑
Modified, inactive or inhibited enzyme
↓
Denatured protein
↓
Amino acids

Immunological methods enable small quantities of an enzyme to be precipitated with a specific antibody. In a pulse-chase experiment the immunoprecipitate measures the radioactivity remaining in the active enzyme *plus* any derived protein which cross-reacts with the antibody. In cases where a decrease in enzyme is concomitant with a decreased immunoprecipitate, it is usual to conclude that enzyme degradation has occurred. Examples include nitrate reductase (SORGER et al. 1974), phenylalanine ammonia lyase (TANAKA and URITANI 1977, BETZ et al. 1978), ribulose bisphosphate carboxylase (WITTENBACH 1978) and isocitrate lyase (KAHN et al. 1979). An example of the sort of evidence obtained is shown in Fig. 8. When a given amount of antiserum was reacted with the

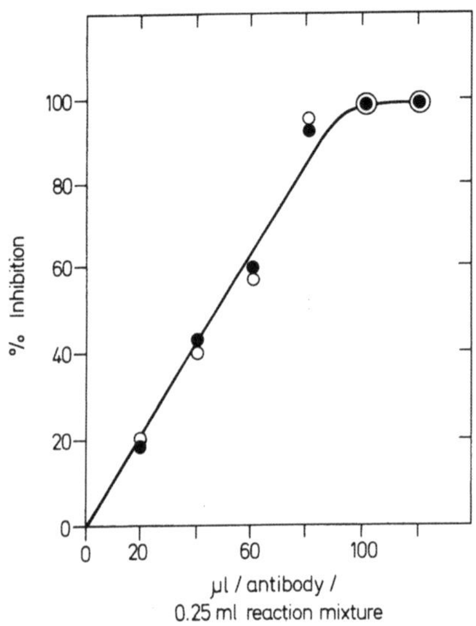


**Fig. 8.** Precipitation of isocitrate lyase activity by a varying amount of antibody in homogenates of cotyledons from flax seeds germinated either for 48 (0) or 96 h (●). Enzyme preparations of isocitrate lyase from cotyledons at these two stages were prepared as described in the text. In each experiment the volumes of homogenates were adjusted to contain the same initial activity. Antibody was added to each homogenate containing 0.2 unit of isocitrate lyase in a total volume of 0.25 ml of 50 mM TEMD buffer, pH 7.6. After a 12-h incubation at 4° C, the preparations were centrifuged at 12,000 *g* for 10 min at 4° C and each supernatant assayed for residual enzyme activity. (After KAHN et al. 1979)

same fixed number of units of isocitrate lyase, derived from seedlings germinated for 48 h or 96 h, the same inhibition was observed over a sixfold range of antibody concentrations. After 48 h of germination, isocitrate lyase activity is about 90% of maximum and after 96 h, the activity has declined to about 40%. The experiment establishes that during the period when isocitrate lyase activity was declining, there was no accumulation of material which could cross-react with the specific antibody. The possibility that loss of enzyme activity is associated with the accumulation of a modified protein, which does not react with antibody, cannot be excluded. The experiment does not establish that the enzyme has been degraded to amino acids and in the particular case of flax isocitrate lyase, the in vitro loss of activity appears to be due to an endopeptidase which does not release amino acids.

In the case of the flavanone synthase of parsley (*Petroselinum crispum*) cells, there was a large discrepancy between the rates of disappearance of catalytic activity and immunoprecipitability (SCHRÖDER et al. 1979). The discrepancy suggests that inactivation of flavanone synthetase occurs at a much faster rate than proteolytic degradation.

Immunoprecipitation utilizes the surface characteristics of the enzyme and if these characteristics are modified, there may be a loss of precipitability. Affinity chromatography utilizes the properties of the active site of the enzyme and therefore could be used jointly with immunology in studies of protein turnover, but even then a modified protein may escape detection and gel electrophoresis should be considered. In principle, gel electrophoresis is capable of providing direct evidence on the problem of degradation versus modification. If degradation has occurred, the protein band corresponding to the enzyme should disappear completely. If inactivation rather than degradation is involved, the modified protein should be retained somewhere on the gel; the technical problem is to recognize and separate this component from the other proteins present.

# 7 Protein Turnover During Senescence

Senescence is a complex physiological process which may be initiated by specific senescence signals (LINDOO and NOODEN 1977, MISHRA and GAUR 1977), and involves a range of hormonal interactions. Although this process is reviewed in Chapter 16, this Volume, specific points relating to protein turnover are discussed here.

## 7.1 Protein Synthesis in Senescing Leaves

Detached leaves and leaf discs undergo premature senescence in which there is a large decrease in the total protein. The loss of protein is correlated with an increase in protease activity in a number of species – oats (*Avena sativa*)

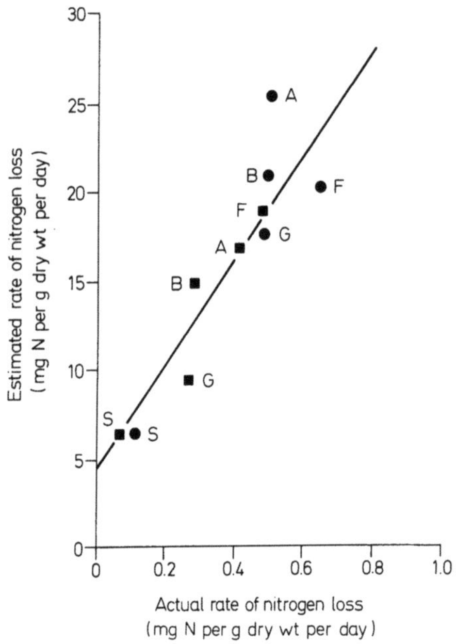


**Fig. 9.** Relation between the actual rate of nitrogen loss from senescing tissues and the rate estimated from protease activity values for tissues of the wheat varieties Argentine 1X (●) and Insigma (■). $y=4.38+29.82\times$, $\gamma=0.893$; $P<0.01$. *F* flag leaf; *A* leaf A; *B* leaf B; *G* glumes; *S* stem. (After DALLING et al. 1976)

(MARTIN and THIMANN 1972) barley (*Hordeum vulgare*) (PETERSON and HUFFAKER 1975), wheat (*Triticum esculentum*) (DALLING et al. 1976), maize (*Zea mays*) (FELLER et al. 1977), betel (*Piper betle*) (MISHRA and GAUR 1977) and *Lolium* (THOMAS 1978). The correlation between protease activity and in vivo rates of protein loss which has been shown to apply to leaves, glumes and stem of wheat (Fig. 9) may have economic significance, since leaf protease activity appears to be higher in wheats with high grain protein-N than in low protein cultivars (RAO and CROY 1972), and rice varieties giving high yields of grain protein contain higher levels of leaf protease than varieties with a lower capacity for yielding grain protein (PEREZ et al. 1973). However, in some species, the increased rate of protein degradation which occurs in senescence does not appear to be correlated with increased protease activity – tobacco (ANDERSON and ROWAN 1965), *Perilla* (WOOLHOUSE 1967), nasturtium (*Tropaeolum majus*) (BEEVERS 1968), apple (*Malus*) (SPENCER and TITUS 1972) and peas (STOREY and BEEVERS 1977).

The differences between species have been observed using detached leaves, attached leaves and leaf discs, and the different treatments complicate the interpretation. For example, MARTIN and THIMANN (1972) observed an increase in protease activity when detached oat leaves were placed horizontally over moist filter papers, and suggested that cytokinins may delay senescence by suppressing protease formation. VAN LOON et al. (1978) found that when detached oat leaves were placed horizontally in wet filter papers in darkness, they accumulated free amino acids and yielded increases in proteases active at pH 4.5 and pH 7.5. However, when the leaves were held vertically with their bases in water, protease activities and free amino acids decreased in parallel as in attached leaves (Table 4).

**Table 4.** Changes occurring in detached oat leaves, held vertically or horizontally for 4 days in the dark. (After VAN LOON et al. 1978)

| | Vertically on water (%) | Horizontally on wet filter paper (%) |
|---|---|---|
| Protease (pH 4.5) | − 5 | +65 |
| Protease (pH 7.5) | −27 | +16 |
| Protein | −80 | −57 |
| Chlorophyll | −90 | −91 |
| Free amino acids | | |
| (a) in leaf | +157 | +878 |
| (b) in solution | +559 | +153 |

## 7.2 Protein Synthesis in Ripening Fruit

An enhanced rate of amino acid incorporation into protein was observed during the early period of ripening of apples (HULME et al. 1968, 1971), pears (*Pirus*) (FRENKEL et al. 1968) and bananas (*Musa*) (BRADY et al. 1970). BRADY and O'CONNELL (1976) found the amino acid incorporation was into a wide range of banana proteins rather than a few specific ones, and concluded that the amino acid incorporation represented turnover. They calculated that the rate of turnover was sufficient to replace 25%–50% of the proteins in 24 h and concluded that the enhanced rate of protein synthesis was related to enhanced turnover rather than to the production of enzymes to control specific ripening processes.

On the other hand, there have been many reports of increases in the activity of certain enzymes during ripening (DILLEY 1970, IKI et al. 1978). The synthesis of enzymes for the removal of malic acid during the ripening of apples has been discussed by HULME et al. (1963), and FRENKEL et al. (1968) have reported that in ripening pears there is an incorporation of amino acids into NADP malic enzyme, which they suggest plays an important role in malate metabolism. These views are not mutually exclusive, since the production of some specific enzymes might underlie the general turnover of proteins and evidence from inhibitors of protein synthesis such as cycloheximide and fluorophenylalanine suggest that some protein synthesis may be essential for ripening (MCGLASSON et al. 1971).

## 7.3 Protein Degradation During Senescence

A general picture of the enzymes involved in protein degradation during senescence is beginning to emerge (see Chaps. 10 and 16, this Vol.). DRIVDAHL and THIMANN (1977, 1978) have identified three of the proteases which appear to be involved in the senescence of oat leaves; one, with a pH optimum of 3.8, appears to be quantitatively of little importance. An endopeptidase, which is an acid protease of the serine type, active at pH 4.2, is considered to initiate the increased proteolysis of senescence and to interact with a neutral SH requiring exopeptidase, active at pH 6.6. The acid protease from oats is similar to the acid proteinase of wheat active against ribulose bisphosphate carboxylase (PEOPLES and DALLING 1978).

In addition to the enzymological aspects of proteolysis, there are a number of perplexing physiological problems. Thus, proteolytic enzymes are present in leaves before the onset of senescence, yet remain relatively inactive. One possible explanation based on the presence of protease inhibitors has little experimental evidence (see Chap. 10, this Vol.). However, MATILE (1978) has persuasively argued that the vacuole of higher plants is effectively lysosomal and direct evidence for proteolysis in the vacuoles of castor bean endosperm has been obtained by NISHIMURA and BEEVERS (1979). Thus, at least part of the proteolytic system appears physically separated from the substrates, and with the onset of senescence, it is necessary to explain how this separation is overcome. The problem is best illustrated by reference to ribulose bisphosphate carboxylase which is located in the chloroplasts and accounts for up to 50% of the total leaf protein. There is little evidence that chloroplasts contain significant amounts of proteases. WITTENBACH (1978) reported that only 3% of the proteinase activity is associated with chloroplasts, and this seems to be the result of a general association with membranes. PEOPLES and DALLING (1979), on the other hand, report pH 4.5 haemoglobin-hydrolyzing activity co-sedimenting with wheat leaf chloroplasts purified by sucrose gradient centrifugation. Furthermore, the rate of protein degradation by isolated chloroplasts is very low (CHOE and THIMANN 1975). Thus, the report (PETERSON et al. 1973) that ribulose bisphosphate carboxylase does not turn over in barley leaves under conditions where other soluble proteins turned over rapidly is consistent with the compartmentation of the carboxylase in the chloroplast and the protease in the vacuole. HUFFAKER and MILLER (1978) have proposed that the chloroplast envelope is a barrier between the carboxylase in the stroma and the endopeptidase in the cytoplasm. During senescence there is a preferential degradation of the carboxylase (PETERSON and HUFFAKER 1975, WOOLHOUSE and BATT 1976, WITTENBACH 1978). It is possible that the carboxylase is transported to the vacuole, or that the protease enters the chloroplast. In this later case, it should be noted that the proteases which are assumed to degrade the carboxylase of wheat leaves is active at a pH of 5, whereas the pH of the stroma of the chloroplast is 7 in darkness and 8 in light. However, the pH within the thylakoid space may be as low as 4.5 (SCHULDINER et al. 1972) and although the major protein of the stroma is carboxylase (LYTTLETON and TSO 1958), some of the carboxylase may be associated with the thylakoid membrane (ARNTZEN and BRIANTAIS 1975). Furthermore, there are marked changes in the ultrastructure of chloroplasts during senescence (BUTLER and SIMON 1970, THOMAS 1977) but the way in which these structural changes relate to protein degradation is not clear.

The mechanism of transporting the proteins is not clear. The possibility that ATP is involved is suggested by the widespread observation that inhibitors of energy metabolism which severely reduce ATP levels also inhibit protein degradation (CHUA et al. 1979). The requirement for ATP has been interpreted to mean that the transport of proteins into the lysosome is energy-requiring (GOLDBERG and ST. JOHN 1976). Anaerobiosis retards senescence (JAMES 1953), suggesting that proteolysis is an energy-requiring process. However, it should be remembered that bacteria which do not have lysosomes also show an energy requirement for ATP (GOLDBERG et al. 1975), and ETLINGER and GOLDBERG (1977) have shown that reticulocytes contain a soluble ATP-dependent proteolytic system. On the other hand, attempts to demonstrate ATP dependence of the soluble proteases from pea leaves have been negative (STOREY and BEEVERS 1977).

# 8 The Specificity of Protein Degradation

GOLDBERG and ST. JOHN (1976) noted that the 16 mammalian proteins with the shortest reported half-lives were important metabolic control points, and argued that enzymes whose activities limit the flow of substrates through biochemical pathways have evolved especially short half-lives. This argument states the physiological significance of protein degradation and not the underlying factors which determine the rate of degradation.

Specificity may reside in the proteolytic system, in the extreme case every enzyme having its own protease. An example of this is the nitrate reductase-inactivating protein of maize roots (WALLACE 1973, 1974) which has been confirmed as a protease by YAMAYA et al. (1980). Nitrate reductase-inactivating proteins have also been isolated from rice cells (YAMAYA and OHIRA 1977) and soyabean leaves (JOLLY and TOLBERT 1978), but in both cases the purified preparations were devoid of proteolytic activity. The maize root nitrate reductase inactivating protein after 900-fold purification, has about 16% of the activity of trypsin against azocasein (YAMAYA et al. 1980). Consequently it is most unlikely that the inactivating enzyme is a protease specific for nitrate reductase. On the contrary, it is the inherent susceptibility of nitrate reductase to proteolytic attack which has created the illusion of a specific protease. This is an example of a general principle that specificity of proteolytic attack lies in the proteins themselves. In the extreme case a single protease could attack all enzymes and the rate of degradation would be determined by the physical properties of the enzymes. Thus the half-life of an enzyme is genetically determined and specified by its size and amino acid sequence. To establish the physical properties of proteins which define the susceptibility of those proteins to proteolytic attack, it is necessary to establish correlations between particular properties and the rate of protein degradation.

## 8.1 Correlation with Size

In general, larger proteins from rat liver are degraded more rapidly than smaller ones (DEHLINGER and SCHIMKE 1971), and it may be the size of the subunits rather than the overall protein size which determines the rate of degradation (DICE et al. 1973). DICE and GOLDBERG (1975a) found a correlation between the log of the subunit molecular weight and the $t_{\frac{1}{2}}$ of 33 rat liver proteins; the coefficient of correlation being −0.6. A correlation between size and $t_{\frac{1}{2}}$ was obtained for a group of membrane or organelle-associated proteins that were insoluble in phosphate-buffered saline but soluble in 1% Triton X-100 (DICE et al. 1979). Evidence for a correlation between size and $t_{\frac{1}{2}}$ for plant proteins has been obtained by DICE et al. 1973, ACTON and GUPTA 1979, COOKE and DAVIES 1980.

If protein degradation was a random process in which the proteolytic enzymes attacked one molecule, then transferred to another molecule, limited proteolysis could produce extensive loss of enzyme activity. The principle of parsimony requires that once an enzyme is attacked, proteolysis should go to completion. The generally accepted model of LINDERSTROM-LANG (1949) proposes that the first peptide bond cleavage is rate-limiting and the remainder of the protein then undergoes unfolding and rapid degradation. This model has been supported by KOMINAMI et al. (1972), who have shown that intermediate degradation products are much more rapidly degraded by trypsin than are the native proteins. On the basis of this model, the size correlation is easily explained by the target size theory.

Results which at least appear to be inconsistent with the size correlation have been obtained by KEMSHEAD and HIPKISS (1976) who found that extracts of *E. coli* could degrade proteins of molecular weight less than 29,000, but not larger proteins. These authors concluded that the size correlation with degradation is not valid for *E. coli* growing exponentially. No correlation between size and degradation was observed in any submitochondrial fractions obtained from rat liver by WALKER et al. (1978). BRINSTER et al. (1979) examined 56 proteins in mouse blastocysts and found that whilst there is a tendency for large proteins to be degraded more rapidly than small proteins, the regression of size on half-life was not significant. Similarly, HORST and ROBERTS (1979) examined 60 proteins in plasma membranes and found no significant correlation between half-life and size.

## 8.2 Correlation with Charge

DICE and GOLDBERG (1975b) have obtained a relationship between the isoelectric point and $t_{1/2}$ of 22 rat proteins. The straight line plot of the isoelectric point (sometimes estimated from the homologous proteins of other animals) against $t_{1/2}$ gave a correlation coefficient of −0.82. Subsequently, Dice et al. (1979) obtained correlations between isoelectric point and $t_{1/2}$ for proteins from an extended number of tissues and organisms. Evidence for a similar correlation with soluble plant proteins has been presented by ACTON and GUPTA (1979) and COOKE and DAVIES (1980), permitting the generalization that soluble acidic proteins degrade more rapidly than neutral or basic proteins. The correlation between charge and $t_{1/2}$ does not apply to membrane-bound proteins (DICE et al. 1979, HORST and ROBERTS 1979).

### 8.3 Correlation with "Abnormality"

Both growing and starving *E. coli*, degrade proteins with abnormal conformations more rapidly than normal proteins. Such abnormal proteins may result from mutations (GOLDSCHMIDT 1970, PLATT et al. 1970), errors in gene translation, premature release of immature polypeptides produced by puromycin or the incorporation of amino acid analogues (GOLDBERG 1972a, BRADLEY et al. 1976). PROUTY et al. (1975) have shown that within 1 h after exposing cells to canavanine, 50% of the analogue-containing proteins were degraded to acid-soluble forms. The abnormal proteins accumulate in a particulate fraction and the authors suggest that the presence of analogues makes them less soluble and causes them to aggregate. It is not clear that aggregation is an essential feature in the degradation of abnormal proteins, since certain mutations in the *lac* operon result in proteins that are rapidly hydrolyzed even though they are soluble (GOLDSCHMIDT 1970). A review of the degradation of abnormal proteins in mammalian systems has been published (GOLDBERG et al. 1978). Experiments which appear to be at variance with the correlation have been reported. JOHNSON and KENNY (1973) supplied hepatoma cells with fluorinated derivatives of tryptophane and isolated tyrosine aminotransferase containing fluorotryptophane. The catalytic activity of this abnormal protein was reduced about a third, but the half-life was not affected. A number of studies have reported the presence of faulty enzymes with reduced catalytic activity in aged cells. The accumulation of these faulty enzymes implies that their degradation rate is low. It is possible that the degradative machinery declines with age and so allows the accumulation of errors. Direct evidence for this proposition has been obtained (ZEELON et al. 1973). However, in senescence leading to death, protein degradation is enhanced in animals (BRADLEY et al. 1975) as well as in plants.

Approximately 80% of the intracellular proteins of mammalian cells are α-N-acetylated (BROWN and ROBERTS 1976), and JÖRNVALL (1975) has suggested that α-N-acetyl groups may protect proteins from proteolytic degradation. However, a direct test of this proposal has shown that α-N-acetylation does not protect proteins from degradation (BROWN 1979).

Since plants accumulate a great variety of non-protein amino acids structurally related to those found in proteins, the avoidance of abnormal proteins requires the evolution of mechanisms to exclude these analogues from incorporation. Discrimination seems to depend upon species differences in the substrate specificity of the amino acyl-tRNA synthetases. Thus for example, plants which contain azetidine-2-carboxylic acid, have evolved a prolyl-tRNA synthetase which fails to activate the analogue (LEA and NORRIS 1972, NORRIS and FOWDEN 1972).

### 8.4 Correlation with Amide Concentration

ROBINSON et al. (1970) proposed that many of the essential processes of living organisms are timed by deamidation of glutaminyl and asparaginyl residues. ROBINSON (1974) has presented evidence that variation in the sequence of amino acid residues around glutaminyl and asparaginyl residues can change their rates of deamidation. Robinson noted a correlation between the amide content of proteins and their turnover rates and a general discussion of the proposition that glutaminyl and asparaginyl residues function as the molecular clocks of proteins has been presented by ROBINSON and RUDD (1974). It could be that the molecular clock is the content of amides and the process of deamidation is irrelevant. Alternatively, the rate of deamidation may be a determinant. Rabbit muscle aldolase undergoes deamidation of an asparagine residue near the C-terminus of the subunits with a half-life of 8 days (in vivo). The sequence Asp-His is resistant to carboxypeptidase whereas Asn-His is sensitive (LAI et al. 1970). The in vitro deamidation of aspartate aminotransferase has been demonstrated by WILLIAMS and JOHN (1979) with a half-life at pH 7.5 and 25° C of 20 days.

There is little work on this aspect in plants, but it may be worth noting that the seed proteins which undergo rapid hydrolysis during germination are particularly rich in amides – thus the prolamin and glutelin of barley seed contain 23 and 10.3 of amide N/100 g N respectively (FOLKES and YEMM 1956).

## 8.5 Correlation with Disulphide Content

In response to wounding, tomato plants produce a hormone-like substance which is translocated through the vascular system (GREEN and RYAN 1973). When detached tomato leaves are supplied with this compound, they exhibit a specificity of protein turnover directed towards the accumulation of heat-stable proteins having disulphide cross linkages. GUSTAFSON and RYAN (1976) observed a marked stability of disulphide proteins against in vivo degradation and concluded that the state of oxidation of protein-bound half-cystine residues may be a principle factor influencing the susceptibility of leaf proteins to degradation. Other workers (VARANDANI and SHROYER 1973) have suggested that the initial step in protein degradation may involve the reduction of disulphide bonds.

The sulphhydryl-disulphide hypothesis of LEVITT (1962) relates the disulphide (-SS-) content of proteins to the environmental stress. Since environmental stress stimulates protein degradation (COOKE et al. 1979b), the proteins remaining after the stress would be expected to be rich in disulphide if the presence of sulphydryl (SH) confers instability on proteins. TOMATI and GALLI (1979) have shown that when maize plants are subjected to a water stress there is a marked decrease in the ratio SH/-SS-.

## 8.6 Correlation with Thermodynamic Properties

Many workers have suggested that enzymes exist in two forms, one of which is more susceptible to proteolysis (GOLDBERG 1972b). The equilibrium between these forms is affected by the binding of ligands which have been shown to affect degradative rates (LITWACK and ROSENFELD 1973). Others have suggested that the initial step in degradation may be denaturation (SEGAL 1976). HOPGOOD and BALLARD (1974) measured the rate of enzyme inactivation of nine enzymes present in rat liver whose half-lives are known. The rate of inactivation was correlated with the rate of degradation for seven of the nine enzymes.

The energy difference between native (fully folded) and fully unfolded forms is usually quite low (5–15 Kcal/mol) (TANFORD 1970, PACE 1975) and such forms are rapidly interconvertible (MCLENDON and SMITH 1978). Since an unfolded protein is potentially a very reactive substrate for proteolytic enzymes, the equilibrium concentration of the unfolded form which is determined by the unfolding equilibrium constant (Ku) might determine the overall rate of degradation. MCLENDON and RADANY (1978) measured the melting curves of nine enzymes (whose rates of degradation were known) to obtain the transition temperature (Tm) which is related to Ku, but cannot rigorously be used to measure Ku. They observed a linear correlation between the thermal unfolding temperature and $t_{1/2}$ for eight of the nine enzymes. It is of interest that the exception was 3-phosphoglyceraldehyde dehydrogenase which was also anomalous in the correlation between inactivation rate and $t_{1/2}$ reported by HOPGOOD and BALLARD (1974).

### 8.7 Correlation with Glycosylation

Glycoproteins can be readily isolated by affinity chromatography using plant lectins bound to agarose. Using this method, it can be shown that glycoproteins are present in the cytosolic fraction of tissues and that they are degraded more rapidly than non-glycosylated proteins (KALISH et al. 1979). MATHEWS et al. (1976) reported that two glycosylated proteins present in the plasmamembranes of neuroblastoma cells turnover more rapidly than the non-glycosylated proteins. Although the biochemical basis for degradation of glycoproteins is not known, it is probably not based on the susceptibility of glycoproteins to proteolytic attack. A possible basis may be that the carbohydrate group allows for "special handling" at the lysosome. PRICER and ASHWELL (1976) have reported that an asialoglycoprotein-binding protein exists in a variety of organelles including the lysosome. Increased degradation of soluble glycoproteins compared with non-glycosylated proteins from plants has also been observed (COOKE and DAVIES 1980a).

### 8.8 Correlation with Hydrophobicity

HILDEBRAND (1979) has taken exception to the use of the term "hydrophobic effects" by biologists. He argues that there is no hydrophobia between water and alkanes; there only is not enough hydrophilia to prize apart the hydrogen bonds of water so that alkanes can go into solution without assistance from attached polar groups. Apart from this semantic problem, there is growing evidence that relative rates of protein degradation are related to protein hydrophobicity. BOHLEY et al. (1977) partitioned proteins in a two-phase system of water and organic solvent; proteins with a short half-life accumulate preferentially at the interphase and to some extent in the organic phase. Moreover, short-lived proteins are always enriched in the floating fat-rich layer of the cytosol after high speed centrifugation. MANN and SHAH (1979) have shown that lysosome aminotransferase ($t_{1/2}=2$ h) and serine dehydratase ($t_{1/2}=4$ h) are tightly bound to detergent affinity columns, whereas arginase ($t_{1/2}=96$ h) malate dehydrogenase ($t_{1/2}=96$ h) and lactate dehydrogenase ($t_{1/2}=144$ h) are hardly retained on the column.

### 8.9 Interdependence or Independence of Correlates

DICE and GOLDBERG (1975b) compared the isoelectric points and molecular weights of 22 enzymes with their half-lives in rat liver. Finding no correlation between isoelectric points and molecular weights, the authors concluded that size and charge are independent factors influencing protein half-life. However, as DUNCAN et al. (1980) have pointed out, the data for the comparison were partly obtained from mammalian livers other than rat and from tissues other than liver, so that the correlation and for that matter the lack of correlation must be treated with caution. In a re-investigation they concluded that the correlates between size, charge and degradative rates are interdependent, since soluble acidic liver proteins are generally larger than basic proteins. The interdependence of the correlates has been recognized by MOMANY et al. (1976), who developed a parameter which relates the amino acid composition and subunit size of a protein to the in vivo degradative rate. The parameter was calculated for 11 rat liver proteins and the plot against $t_{1/2}$ was linear with a coefficient of correlation of $-0.96$.

## 9 Protein Degradation Under Stress

When *Lemna* fronds are placed under stress, there is a reduction in growth and protein synthesis and an increase in protein degradation (COOKE et al. 1979a, b). The interaction of benzyl adenine with stress conditions is shown in Table 5. The increased protein degradation does not appear to be due to an increase in activity of soluble proteolytic enzymes. Biochemical and ultrastructural evidence (COOKE et al. 1980a, b) suggests that stress, perhaps acting via hormones, affects the permeability of certain membranes, particularly the tonoplast, and allows the vacuolar proteolytic enzymes to interact with cytoplasmic proteins.

In animals the enhanced protein degradation associated with diabetes and starvation is fundamentally different from normal protein catabolism. Thus the correlations between size, isoelectric point and glycosylation, on the one hand, and protein half-life on the other is absent or markedly reduced in liver and muscle of diabetic and starved rats (DICE et al. 1978). The simplest explanation for these results invokes at least two components in the protein degradative system, as suggested by KNOWLES and BALLARD (1976).

**Table 5.** Effect of various stress conditions on the rate constants of growth, protein synthesis and protein degradation in *Lemna minor*. (After COOKE et al. 1979b)

| Growth condition | Frond growth $k_f$ day$^{-1}$ | Protein synthesis $k_s$ day$^{-1}$ | Protein degradation | |
|---|---|---|---|---|
| | | | $k_d$ day$^{-1}$ | $t_{1/2}$ day$^{-1}$ |
| Complete growth medium | 0.35 | 0.43 | 0.1 | 7.1 |
| $H_2O$ | 0.12 | 0.15 | 0.35 | 2.1 |
| $-NO_3$ | 0.18 | 0.21 | 0.25 | 2.8 |
| +Mannitol (0.5 M) | 0.03 | 0.15 | 0.28 | 2.5 |
| Complete+BA[a] (2 μM) | 0.37 | 0.51 | 0.09 | 7.7 |
| $H_2O$+BA (2 μM) | 0.16 | 0.17 | 0.18 | 3.8 |
| $-NO_3$+BA (2 μM) | 0.19 | 0.24 | 0.14 | 5.0 |
| Mannitol+BA (2 μM) | 0.03 | 0.17 | 0.12 | 5.8 |

[a] BA is benzyladenine

The correlations between size, isoelectric point, glycosylation and protein half-life have been investigated for *Lemna* growing under normal and stressed conditions (COOKE and DAVIES 1980). Under normal growth conditions, large proteins, acidic proteins and non-glycoproteins degrade relatively more rapidly than small, basic or glycoproteins. Thus, with the exception of glycoproteins, plants show the same correlatives as animals. However, when plants are stressed the correlations are unaffected, suggesting that basal protein degradation and enhanced degradation do not differ fundamentally. Thus in plants we can consider protein degradation in terms of a number of proteolytic enzymes sequestered in the vacuole. The proteolytic enzymes may leave the vacuole to degrade cytosolic proteins or the proteins may enter the vacuole. Under conditions of stress, the enhanced degradation is to be attributed to changes in the permeability of vacuolar and organelle membranes. This very simple model is unlikely to explain fully the complex nature of protein degradation; nevertheless, it is consistent with our present knowledge of protein turnover in plants and provides a basis for further experiments.

## References


Acton GJ, Gupta S (1979) A relationship between protein degradation rates *in vivo*, isoelectric points and molecular weights obtained by using density labelling. Biochem J 184:367–377

Aitchison PA, Aitchison JM, Yeoman MM (1976) Synthesis and loss of activity of glucose-6-phosphate dehydrogenase in dividing plant cells. Biochim Biophys Acta 451:393–407

Amenta JS, Sargus MJ (1979) Mechanism of protein degradation in growing and non-growing L-cell cultures. Biochem J 182:847–859

Anderson JW, Rowan KS (1965) Activity of peptidase in tobacco-leaf tissue in relation to senescence. Biochem J 97:741–746

Arias IM, Doyle D, Schimke RT (1969) Studies on the synthesis and degradation of the proteins of the endoplasmic reticulum of rat liver. J Biol Chem 244:3303–3315

Arntzen CJ, Briantais JM (1975) Chloroplast structure and function. In: Govindji (ed) Bioenergetics of photosynthesis. Academic Press, London New York, pp 51–107

Ashton FM (1976) Mobilization of storage proteins of seeds. Annu Rev Plant Physiol 27:95–117
Ballard FJ, Hopgood MF, Reshef L, Hanson RW (1974) Degradation of phosphoenolpyruvate carboxykinase (guanosine triphosphate) *in vivo* and *in vitro*. Biochem J 140:531–538
Baumgartner B, Chrispeels MJ (1976) Partial characterization of a protease inhibitor which inhibits the major endopeptidase present in the cotyledons of mung beans. Plant Physiol 58:1–6
Baumgartner B, Chrispeels MJ (1977) Purification and characterization of vicilin-peptohydrolase, the major endopeptidase in the cotyledons of mung bean seedlings. Eur J Biochem 77:223–233
Baumgartner B, Tokuyasu KT, Chrispeels MJ (1978) Localization of vicilin peptide hydrolase in the cotyledons of mung bean seedlings by immunofluorescence microscopy. J Cell Biol 79:10–17
Beevers L (1968) Growth regulator control of senescence in leaf discs of Nasturtium (*Topaeolum majus*). In: Wightman F, Setterfield G (eds) Biochemistry and physiology of plant growth substances. Runge Press, Ottawa, pp 1417–1435
Berlin CM, Schimke RT (1965) Influence of turnover rates on the responses of enzymes to cortisone. Mol Pharmacol 1:149–156
Betz B, Schäfer E, Hahlbrock K (1978) Light-induced phenylalanine ammonia lyase in cell suspension cultures of *Petroselenium hortense*. Arch Biochem Biophys 190:126–135
Bidwell RGS, Barr R, Steward FC (1964) Protein synthesis and turnover in cultured plant tissue: sources of carbon for synthesis and the fact fate of protein breakdown products. Nature (London) 203:367–373
Bohley P, Kirschke H, Langner J, Wideranders B, Ansorge S, Hanson H (1977) Primary reaction of intracellular protein catabolism. In: Turk V, Marks N (eds) Intracellular protein catabolism II. 2nd Int Symp J Stephan Inst, Ljubljana, Yugoslavia. Plenum Publ Co, New York, pp 108–110
Boudet A, Humphrey TJ, Davies DD (1975) The measurement of protein turnover by density labelling. Biochem J 152:409–416
Bradley MO, Dice JF, Hayflick L, Schimke RT (1975) Protein alterations in ageing W138 cells as determined by proteolytic susceptibility. Exp Cell Res 96:103–112
Bradley MO, Hayflick L, Schimke RT (1976) Protein degradation in human fibroblasts (W1-38). Effects of ageing viral transformation, and amino acid analogs. J Biol Chem 251:3521–3529
Brady CJ, O'Connell PBH (1976) On the significance of increased protein synthesis in ripening banana fruits. Aust J Plant Physiol 3:301–310
Brady CJ, Palmer JK, O'Connell PBH, Smillie RM (1970) An increase in protein synthesis during ripening of the banana fruit. Phytochemistry 9:1037–1047
Brinster RL, Brunner S, Joseph X, Levey IL, (1978) Protein degradation in the mouse blastocyst. J Biol Chem 254:1927–1931
Brown JL (1979) A comparison of the turnover of α-N-acetylated and nonacetylated mouse L-cell proteins. J Biol Chem 254:1447–1449
Brown JL, Roberts WK (1976) Evidence that approximately 80% of the soluble proteins from Ehrlich Ascites cells are $N^{\alpha}$-acetylated. J Biol Chem 251:1009–1014
Burger WC, Siegelman HW (1966) Location of a protease and its inhibitor in the barley kernel. Physiol Plant 19:1089–1093
Burrell MM, ap Rees T (1974) Metabolism of phenylalanine and tyrosine by rice leaves infected by *Piricularia oryzae*. Physiol Plant Pathol 4:497–508
Butler RD, Simon EW (1970) Ultrastructural aspects of senescence in plants. Adv Gerontol Res 3:73–129
Cameron EC, Mazelis M (1971) A nonproteolytic 'trypsin-like' enzyme. Plant Physiol 48:278–281
Cahill GF Jr (1970) Starvation in man. New Engl J Med 282:668–675
Chang TW, Goldberg AL (1978) The origin of alanine produced in skeletal muscle. J Biol Chem 253:3677–3684
Choe HT, Thimann KV (1975) The metabolism of oat leaves during senescence III. The senescence of isolated chloroplasts. Plant Physiol 55:838–834

Chou K-H, Splittstoesser WE (1972) Changes in amino acid content and the metabolism of α-aminobutyrate in *Cucurbita moschata* seedlings. Physiol Plant 26:110–114
Chrispeels MJ, Baumgartner B, Harris N (1976) Regulation of reserve protein metabolism in the cotyledons of mung bean seedlings. Proc Natl Acad Sci USA 73:3168–3172
Chua B, Kao RL, Rannels DE, Morgan HE (1979) Inhibition of protein degradation by anoxia and ischemia in perfused rat hearts. J Biol Chem 254:6617–6623
Cooke RJ, Davies DD (1980a) General characteristics of normal and stress enhanced protein degradation in *Lemna minor*. Biochem J 192:499–566
Cooke RJ, Grego S, Oliver J, Davies DD (1979a) The effect of deuterium oxide on protein turnover in *Lemna minor*. Planta 146:229–236
Cooke RJ, Oliver J, Davies DD (1979b) Stress and protein turnover in *Lemna minor*. Plant Physiol 64:1109–1113
Cooke RJ, Grego S, Roberts K, Davies DD (1980a) The mechanism of deuterium oxide-induced protein degradation in *Lemna minor*. Planta 148:374–380
Cooke RJ, Roberts K, Davies DD (1980b) The mechanism of stress-induced protein degradation in *Lemna minor*. Plant Physiol 66:1119–1122
Cowie DB (1962) In: Holden (ed) Amino acid pools. Elsevier, Amsterdam New York, pp 633–646
Dalling MJ, Boland G, Wilson JH (1976) Relation between acid proteinase activity and redistribution of nitrogen during grain development in wheat. Aust J Plant Physiol 3:721–730
Davies DD (1980) The measurement of protein turnover in plants. In: Woolhouse H (ed) Adv Bot Res 8:66–126
Davies DD, Humphrey TJ (1976) Protein turnover in plants. In: Sunderland N (ed) Perspectives in experimental biology, vol II. Botany. Pergamon Press, Oxford, pp 313–324
Davies DD, Humphrey TJ (1978) Amino acid recyling in relation to protein turnover. Plant Physiol 61:54–58
Dehlinger PJ, Schimke RT (1971) Size distribution of membrane proteins of rat liver and their relative rates of degradation. J Biol Chem 246:2574–2583
Dice JF, Goldberg AL (1975a) A statistical analysis of the relationship between degradative rates and molecular weights of proteins. Arch Biochem Biophys 170:213–219
Dice JF, Goldberg AF (1975b) Relationship between *in vivo* degradative rates and isoelectric points of proteins. Proc Natl Acad Sci USA 72:3893–3897
Dice JF, Dehlinger PJ, Schimke RT (1973) Studies on the correlation between size and relative degradation rate of soluble proteins. J Biol Chem 248:4220–4228
Dice JF, Walker CD, Byrne B, Cardiel A (1978) General characteristics of protein degradation in diabetes and starvation. Proc Natl Acad Sci USA 75:2093–2097
Dice JF, Hess EJ, Goldberg AL (1979) Studies on the relationship between the degradative rates of proteins *in vivo* and their isoelectric points. Biochem J 178:305–312
Dilley DR (1970) In: Hulme AC (ed) The biochemistry of fruit and their products vol I, Academic Press, London New York, pp 179–207
Drivdahl RH, Thimann KV (1977) Proteases of senescing oat leaves. 1. Purification and general properties. Plant Physiol 59:1059–1063
Drivdahl RH, Thimann KV (1978) Proteases of senescing oat leaves. II. Reaction to substrates and inhibitors. Plant Physiol 61:501–505
Duncan WE, Offerman MK, Bond JS (1980) Intracellular turnover of stable and labile soluble liver proteins. Arch Biochem Biophys 199:331–341
Eagle H, Piez KA, Fleischman R, Oyama VI (1959) Protein turnover in mammalian cell cultures. J Biol Chem 234:592–597
Etlinger JD, Goldberg AL (1977) A soluble ATP-dependent proteolytic system responsible for the degradation of abnormal protein in reticulocytes. Proc Natl Acad Sci USA 74:54–58
Feller UK, Soong T-ST, Hageman RH (1977) Proteolytic activities and leaf senescence during grain development of field-grown corn (*Zea mays* L). Plant Physiol 59:290–294
Filner PJ, Wray L, Varner JE (1969) Enzyme induction in higher plants. Science 165:358–367
Folkes BF, Yemm EW (1956) The amino acid content of the proteins of barley grains. Biochem J 62:4–11

Fowler MW, Clifton A (1975) Rhythmic oscillations in carbohydrate metabolism during growth of sycamore (*Acer pseudoplatanus* L.) cells in continuous (chemostat) culture. Biochem Soc Trans 3:395–398
Frenkel C, Klein I, Dilley DR (1968) Protein synthesis in relation to ripening of pome fruits. Plant Physiol 43:1146–1153
Fritz H, Tschesche H (1971) Proc 1st Int Res Conf Proteinase Inhibitors. Walter de Gruyter, Berlin
Fritz H, Tschesche H, Green LJ, Truscheit E (1974) Proteinase inhibitors. Bazer Symp, vol V. Springer, Berlin Heidelberg New York
Frosch S, Wagner E, Cumming BG (1973) Endogenous rhythmicity and energy transduction. 1 Rhythmicity in adenylate kinase, NAD- and NADP-linked glyceraldehyde-3-phosphate dehydrogenase in *Chenopodium rubrum*. Can J Bot 51:1355–1367
Garber AJ, Karl IE, Kipnis DM (1976) Alanine and glutamine synthesis and release from skeletal muscle. I. Glycolysis and amino acid release. J Biol Chem 251:826–835
Gennis LS, Cantor CR (1976) Double-headed protease inhibitors from black-eyed peas. I. Purification of two new protease inhibitors and the endogenous protease by affinity chromatography. J Biol Chem 251:734–740
Goldberg AL (1972a) Degradation of abnormal proteins in *Escherichia coli*. Proc Natl Acad Sci USA 69:422–426
Goldberg AL (1972b) Correlation between rates of degradation of bacterial proteins *in vivo* and their sensitivity to proteases. Proc Natl Acad Sci USA 69:2640–2644
Goldberg AL, Dice JF (1974) Intracellular protein degradation in mammalian and bacterial cells. Annu Rev Biochem 43:835–869
Goldberg AL, St. John AC (1976) Intracellular protein degradation in mammalian and bacterial cells, Part 2. Annu Rev Biochem 45:747–803
Goldberg AL, Olden K, Prouty WF (1975) In: Schimke RT, Katunuma N (eds) Intracellular protein turnover. Academic Press, London New York, pp 17–57
Goldberg AL, Kowit J, Etlinger J, Klemes Y (1978) Selective degradation of abnormal proteins in animal and bacterial cells. In: Segal HL, Doyle DJ (eds) Protein turnover and lysosomal function. Academic Press, London New York, pp 171–196
Goldschmidt R (1970) *In vivo* degradation of nonsense fragments in *E. coli*. Nature (London) 228:1151–1154
Goldstein L, Newsholme EA (1976) The formation of alanine from amino acids in diaphragm muscle of the rat. Biochem J 154:555–558
Gordon WR, Koukkari WL (1978) Circadian rhythmicity in the activities of phenylalanine ammonia lyase from *Lemna perpusilla* and *Spirodela polyrhiza*. Plant Physiol 62:612–615
Green TR, Ryan CA (1973) Wound-induced proteinase inhibitor in tomato leaves. Some effects of light and temperature on the wound response. Plant Physiol 51:19–21
Gregory FG, Sen GK (1937) Physiological studies in plant nutrition. VI. The regulation of respiration rate to the carbohydrate and nitrogen metabolism of the barley leaf as determined by nitrogen and potassium deficiency. Ann Bot (London) 1:521–561
Gupta VK, Anderson LE (1978) Light modulation of the activity of carbon metabolism enzymes in the crassulacean acid metabolism plant *Kalanchoë*. Plant Physiol 61:469–471
Gustafson G, Ryan CA (1976) Specificity of protein turnover in tomato leaves. Accumulation of proteinase inhibitors induced with the wound hormone PIIF. J Biol Chem 251:7004–7110
Hellebust JA, Bidwell RGS (1963) Protein turnover in wheat and snapdragon leaves. Can J Bot 41:969–983
Henshall JD, Goodwin TW (1964) The effect of red and far red light on carotenoid and chlorophyll formation in pea seedlings. Photochem Photobiol 3:243–247
Heversey G, Linderstrom-Lang K, Keston AS, Olsen C (1940) Exchange of N atoms in the leaves of the sunflower. CR Trav Lab Carlsberg Ser Chim 23:213–218
Hildebrand JH (1979) Is there a "hydrophobic effect"? Proc Natl Acad Sci USA 76:194
Hilton MA, Barnes FW, Henry SS, Enns T (1954) Mechanisms in enzymatic transamination. Rate of exchange of the hydrogen of aspartate. J Biol Chem 209:743–754
Ho DT, Scandalios JG (1975) Regulation of alcohol dehydrogenase in maize scutellum during germination. Plant Physiol 56:56–59

Hobday SM, Thurman DA, Barber DJ (1973) Proteolytic and trypsin inhibitory activites in extracts of germinating *Pisum sativum* seeds. Phytochemistry 12:1041–1046
Holleman JM, Keys JL (1967) Inactive and protein precursor pools of amino acids in the soybean hypocotyl. Plant Physiol 42:29–36
Hopgood MF, Ballard FJ (1974) The relative stability of liver cytosol enzymes incubated *in vitro*. Biochem J 144:371–376
Horiguchi T, Kiragishi K (1971) Studies on rice seed protease: V Protease inhibitor in rice seed. Plant Cell Physiol 12:902–915
Horst MN, Roberts RM (1979) Analysis of polypeptide turnover rates in Chinese hamster ovary cell plasma membranes using two-dimensional electrophoresis. J Biol Chem 254:5000–5007
Hu ASL, Brock RM, Halvorson HO (1962) Separation of labelled from unlabelled proteins by equilibrium density gradient sedimentation. Anal Biochem 4:489–504
Huffaker RC, Miller BL (1978) Reutilization of ribulose bisphosphate carboxylase. In: Siegelman HW, Hind G (eds) Photosynthetic carbon assimilation. Plenum Press, New York, pp 139–152
Huffaker RC, Peterson LW (1974) Protein turnover in plants and possible means of its regulation. Annu Rev Plant Physiol 25:363–392
Hulme AC, Jones JD, Wooltorton LSC (1963) The respiration climacteric in apple fruits. Proc Soc London Ser B 158:514–535
Hulme AC, Rhodes MJC, Galliard T, Wooltorton LSC (1968) Metabolic changes in excised fruit tissue IV. Changes occurring in discs of apple peel during the development of the respiration climacteric. Plant Physiol 43:1154–1161
Hulme AC, Rhodes MJC, Wooltorton LSC (1971) The relationship between ethylene and the synthesis of RNA and protein in ripening apples. Phytochemistry 10:749–756
Humphrey TJ, Davies DD (1975) A new method for the measurement of protein turnover. Biochem J 148:119–127
Humphrey TJ, Davies DD (1976) A sensitive method for measuring protein turnover based on the measurement of 2-$^3$H-labelled amino acids in protein. Biochem J 156:561–568
Iki K, Sekiguchi K, Kurata K, Tada T, Nakagawa H, Ogura N, Takehana H (1978) Immunological properties of *β*-fructofuranosidase from ripening tomato fruit. Phytochemistry 17:311–312
Jacobsen JV, Varner JE (1967) Gibberellic acid-induced synthesis of protease by isolated aleurone layers of barley. Plant Physiol 42:1596–1600
James WO (1953) Plant respiration. Oxford Univ Press
Jensen RG, Bahr JT (1977) Ribulose 1,5-bisphosphate carboxylase-oxygenase. Annu Rev Plant Physiol 28:379–400
Jörnvall H (1975) Acetylation of protein N-terminal amino groups structural observations on α-amino acetylated proteins. J Theor Biol 55:1–12
Johnson RW, Kenny FT (1973) Regulation of tyrosine aminotransferase in rat liver. XI. Studies on the relationship of enzyme stability to enzyme turnover in cultured hepatoma cells. J Biol Chem 248:4528–4531
Jolly SO, Tolbert NE (1978) NADH-nitrate reductase inhibitor from soybean leaves. Plant Physiol 62:197–203
Jong De DW, Woodlief WG (1979) Fluctuations in activity of dehydrogenases in leaves of tobacco genotypes as a function of time of day and leaf maturity. Plant Sci Lett 14:297–301
Kagawa T, Hatch MD (1977) Regulation of $C_4$ photosynthesis: characterization of a protein factor mediating the activation and inactivation of NADP-malate dehydrogenase. Arch Biochem Biophys 184:290–297
Kahn FR, Saleemuddin M, Siddiqi M, McFadden BA (1979) The appearance and decline of isocitrate lyase in flax seedlings. J Biol Chem 254:6938–6944
Kalish F, Chovick N, Dice JF (1979) Rapid *in vivo* degradation of glycoprotein isolated from cytosol. J Biol Chem 254:4475–4483
Kemp JD, Sutton DW (1971) Protein metabolism in cultured plant tissues. Calculation of an absolute rate of protein synthesis, accumulation and degradation in tobacco callus in vivo. Biochemistry 10:81–88

Kemshead JT, Hipkiss AR (1976) Degradation of abnormal proteins in *Escherichia coli*. Differential proteolysis *in vitro* of *E. coli* alkaline phosphatase cyanogen-bromide cleavage products. Eur J Biochem 77:185–192
Kirsi M (1974) Proteinase inhibitors in germination barley embryos. Physiol Plant 32:89–93
Knowles SE, Balard FJ (1976) Selective control of the degradation of normal and aberrant proteins in Reuber H35 hepatoma cells. Biochem J 156:609–617
Koch AL (1962) The evaluation of the rates of biological processes from tracer kinetic data. J Theor Biol 3:283–303
Koch AL, Levy HR (1955) Protein turnover in growing cultures of *Escherichia coli*. J Biol Chem 217:947–957
Kominami E, Kobayashi K, Kominami S, Katunuma N (1972) Properties of a specific protease for pyridoxal enzymes and its biological role. J Biol Chem 247:6848–6855
Krebs HA (1972) Some aspects of the regulation of fuel supply in omnivorous animals. Adv Enzyme Regul 10:397–420
Lai CY, Chen C, Horecker BL (1970) Primary structure of two COOH-terminal hexapeptides from rabbit muscle aldolase. A difference in the structure of the α and β subunits. Biochem Biophys Res Commun 40:461–468
Lai KY, Scandalios JG (1977) Differential expression of alcohol dehydrogenase and its regulation by an endogenous ADH-specific inhibitor during maize development. Differentiation 9:111–118
Lamb CM, Merritt TK, Butt VS (1979) Synthesis and removal of phenylalanine ammonia-lyase: activity in illuminated discs of potato tuber parenchyme. Biochim Biophys Acta 582:196–212
Lea PJ, Norris RD (1972) tRNA and aminoacyl-tRNA synthetases from plants. Phytochemistry 11:2897–2920
Levitt J (1962) A sulfhydryl-disulfide hypothesis of frost injury and resistance in plants. J Theor Biol 3:355–391
Linderstrom-Lang K (1949) Structure and enzymatic break-down of proteins. Cold Spring Harbor Symp Quant Biol 14:117–125
Lindoo SJ, Noodén LD (1977) Studies on the behaviour of the senescence signal in Anoka soybeans. Plant Physiol 59:1136–1140
Litwack G, Rosenfeld S (1973) Coenzyme dissociation, a possible determinant of short half-life of inducible enzymes in mammalian liver. Biochem Biophys Res Commun 52:181–188
Longo CP (1968) Evidence for *de novo* synthesis of isocitritase and malate synthetase in germinating peanut cotyledons. Plant Physiol 43:660–664
Loon Van JC, Haverkort AJ, Lokhorst GJ (1978) Changes in protease activity during leaf growth and senescence. Abstr Fed Euro Soc Plant Physiol Inaugural Meet 9–14 July
Lyttleton JW, Tso POP (1958) The localization of fraction. 1. Protein of green leaves in the chloroplasts. Arch Biochem Biophys 73:120–126
Mandelstam J (1958a) The free amino acids in growing and non-growing populations of *Escherichia coli*. Biochem J 69:103–110
Mandelstam J (1958b) Turnover of proteins in growing and non-growing populations of *Escherichia coli*. Biochem J 69:110–119
Mann DF, Shah K (1979) Correlation of rat liver enzyme half-life with retention in detergent affinity columns. Fed Proc Fed Am Soc Exp Biol 38:414, Abs 979
Marco Di G, Grego S, Tricoli D (1979) RuBP carboxylase-oxygenase in field-grown wheat. J Exp Bot 118:851–861
Martin C, Thimann KV (1972) The role of protein synthesis in the senescence of leaves. 1. The formation of protease. Plant Physiol 49:64–71
Mathews RA, Johnson TC, Hudson JE (1976) Synthesis and turnover of plasma membrane proteins and glycoproteins in a neuroblastoma cell line. Biochem J 154:57–64
Matile P (1978) Biochemistry and function of vacuoles. Annu Rev Plant Physiol 29:193–213
Meselson M, Stahl FW (1958) The replication of DNA in *E. coli*. Proc Natl Acad Sci USA 44:671–682
McGlasson WB, Palmer JK, Kendrell M, Brady CJ (1971) Metabolic studies with banana

fruit slices. II. Effects of inhibitors on respiration, ethylene production and ripening. Aust J Biol Sci 24:1103–1114

McLendon G, Radany E (1978) Is protein turnover thermodynamically controlled? J Biol Chem 253:6335–6337

McLendon G, Smith M (1978) Equilibrium and kinetic studies of unfolding of homologous cytochromes c. J Biol Chem 253:4004–4008

Mikola J, Enari TM (1970) Changes in the contents of barley proteolytic inhibitors during malting and mashing. J Inst Brew (London) 76:182–188

Millward DJ, Garlick PJ, Nanyelugo DO, Waterlow JC (1976) The relative importance of muscle protein synthesis and breakdown in the regulation of muscle mass. Biochem J 156:185–188

Mishra SD, Gaur BK (1977) Role of petiole in protein metabolism of senescing betel. Plant Physiol 59:915–919

Momany FA, Aguanno JJ, Larrabee, AR (1976) Correlation of degradative rates of proteins with a parameter calculated from amino acid composition and subunit size. Proc. Natl Acad Sci USA 73:3093–3097

Mothes K (1931) Nitrogen metabolism in higher plants. III. Effect of age and water content of leaf. Planta 686–731

Mothes K (1933) Investigations on the assimilation of ammonia. Planta 19:117–138

Nishimura M, Beevers H (1979) Hydrolysis of protein in vacuoles isolated from higher plant tissue. Nature (London) 277:412–413

Norris RD, Fowden L (1972) Substrate discrimination by prolyl-tRNA synthetase from various higher plants. Phytochemistry 11:2921–2935

Nozzolillo C, Paul KB, Godin C (1971) The fate of L-phenylalanine fed to germinating pea seeds, *Pisum sativum* (L) var. Alaska, during imbibition. Plant Physiol 47:119–123

Oaks A, Bidwell RGS (1970) Compartmentation of intermediary metabolites. Annu Rev Plant Physiol 21:43–66

Pace CN (1975) The stability of globular proteins. CRC Crit Rev Biochem 3:1–90

Paech K (1935) Regulation of protein metabolism and the condition of proteolytic enzymes in plants. Planta 24:78–129

Pedersen TA, Kirk M, Bassham JA (1966) Light-dark transients in levels of intermediate compounds during photosynthesis in air-adapted *Chlorella*. Physiol Plant 19:219–231

Peoples MB, Dalling MJ (1978) Degradation of ribulose-1,5-bisphosphate carboxylase by proteolytic enzymes from crude extracts of wheat leaves. Planta 138:153–160

Peoples MB, Dalling MJ (1979) Intracellular localization of acid peptide hydrolases in wheat leaves. Plant Physiol Suppl 63:159

Perez CM, Cagampang GB, Esmama BV, Monserrate RV, Juliano BO (1973) Protein metabolism in leaves and developing grains of rice differing in grain grotein content. Plant Physiol 51:537–542

Peterson LW, Huffaker RC (1975) Loss of ribulose-1,5-diphosphate carboxylase and increase in proteolytic activity during senescence of detached primary barley leaves. Plant Physiol 55:1009–1015

Peterson LW, Kleinkopf GE, Huffaker RC (1973) Evidence for lack of turnover of ribulose-1,5-diphosphate carboxylase in barley leaves. Plant Physiol 51:1042–1045

Pierre JN, Queiroz O (1979) Regulation of glycolysis and level of the crassulacean acid metabolism. Planta 144:143–151

Pine MJ, Schimke RT (1978) Measurement of protein synthesis and turnover in animal cells with tritiated water. In: Segal HL, Doyle DJ (eds) Protein turnover and lysosome function. Academic Press, London New York, pp 273–286

Platt T, Miller JH, Weber K (1970) *In vivo* degradation of mutant *Lac* repressor. Nature (London) 228:1154–1156

Polanowski A (1967) Trypsin inhibitor from rye seeds. Acta Biochim Pol 14:389–394

Poole B (1971) The kinetics of disappearance of labelled leucine from the free leucine pool of rat liver and its effect on the apparent turnover of catalase and other hepatic proteins. J Biol Chem 246:6587–6591

Presley HJ, Fowden L (1965) Acid phosphatase and isocitritase production during seed germination. Phytochemistry 4:169–176

Pricer WE, Ashwell G (1976) Subcellular distribution of a mammalian hepatic binding protein specific for asialoglycoproteins. J Biol Chem 251:7539–7544
Prouty WF, Karovsky MJ, Goldberg AL (1975) Degradation of abnormal proteins in *Escherichia coli*. Formation of protein inclusions in cells exposed to amino acid analogs. J Biol Chem 250:1112–1122
Pusztai A (1972) Metabolism of trypsin inhibitory proteins in the germinating seeds of kidney bean. Planta 107:121–129
Queiroz O (1974) Circadian rhythms and metabolic patterns. Annu Rev Plant Physiol 25:115–134
Queiroz O (1978) CAM: Rhythms of enzyme capacity and activity as adaptive mechanisms. In: Gibbs M, Latzko E (eds) Encycl Plant Physiol (New Ser). Photosynthesis, vol II. Springer, Berlin Heidelberg New York
Rao SC, Croy LI (1972) Protease and nitrate reductase seasonal patterns and their relation to grain production of high versus low protein wheat varieties. J Agric Food Chem 20:1138–1141
Read JW, Haas LW (1938) The baking quality of flour as affected by certain enzyme actions. Cereal Chem 15:59–68
Rees ap T (1974) In: Kornberg and Phillips (eds) MTP. International review of science, biochemistry series, vol VII. Butterworth, London, pp 84–107
Rees ap T (1980) Assessment of the contributions of metabolic pathways to plant respiration. In: Davies DD (ed) The biochemistry of plants, vol II. Metabolism and respiration. Academic Press, London New York, pp 1–29
Reiner J (1953) The study of metabolic turnover rates by means of isotopic tracers. Arch Biochem Biophys 46:53–79
Richardson M (1977) The proteinase inhibitors of plants and microorganisms. Phytochemistry 16:159–169
Robinson AB (1974) Evolution and the distribution of glutaminyl and asparaginyl residues in proteins. Proc Natl Acad Sci USA 71:885–888
Robinson AB, Rudd CJ (1974) Deamidation of glutaminyl and asparaginyl residues in peptides and proteins. Curr Top Cell Regul 8:247–295
Robinson AB, McKerrow JH, Cary P (1970) Controlled deamidation of peptides and proteins. An experimental hazard and a possible biological timer. Proc Natl Acad Sci USA 66:753–757
Robinson SP, McNeil PH, Walker DA (1979) Ribulose bisphosphate carboxylase – lack of dark inactivation of the enzyme in experiments with protoplasts. FEBS Lett 97:296–300
Royer A, Miege MN, Grange A, Miege J, Mascherpa JM (1974) Inhibiteurs anti-trypsine et activités protéolytiques des albumines de graine de *Vigna unguiculata*. Planta 119: 1–16
Ryan CA (1973) Proteolytic enzymes and their inhibitors in plants. Annu Rev Plant Physiol 24:173–196
Sachs MM, Freeling M (1978) Selective synthesis of alcohol dehydrogenase during anaerobic treatment of maize. Mol Gen Genet 161:111–115
Scandalios JG (1977) Isozymes: genetic and biochemical regulation of alcohol dehydrogenase. Phytochem Soc Symp 14:129–153
Schröder J, Kreuzaler F, Schäfer E, Hahlbrock K (1979) Concomitant induction of phenylalanine ammonia-lyase and flavanone synthase in RNAs in irradiated plant cells. J Biol Chem 254:57–65
Schuldiner S, Rottenberg H, Avron M (1972) Determination of ΔpH in chloroplasts. 2. Fluorescent amines as a probe for the determination of ΔpH in chloroplast. Eur J Biochem 25:64–70
Segal HL (1976) Mechanism and regulation of protein turnover in animal cells. Curr Top Cell Regul 11:183–201
Shain Y, Mayer AM (1965) Proteolytic enzymes and endogenous trypsin inhibitor in germinating lettuce seeds. Physiol Plant 18:853–859
Shain Y, Mayer AM (1968a) Activation of enzymes during germination – trypsin-like enzyme in lettuce. Phytochemistry 7:1491–1498

Shain Y, Mayer AM (1968b) Activation of enzymes during germination: Amylopectin-1,6-glucosidase in peas. Physiol Plant 21:765–776
Soedigdo R, Gruber M (1960) Purification and some properties of a protease from pea seeds *Pisum sativum* L., ssp *arvense* A and G. Biochim Biophys Acta 44:315–323
Sorger GJ, Debanne MT, Davies J (1974) Effect of nitrate on the synthesis and decay of nitrate reductase of *Neurospora*. Biochem J 140:395–403
Spencer PW, Titus JS (1972) Biochemical and enzymatic changes in apple leaf tissue during autumnal senescence. Plant Physiol 49:746–756
Steward CR, Beevers H (1967) Gluconeogenesis from amino acids in germinating castor bean endosperm and its role in transport to the embryo. Plant Physiol 42:1587–1595
Storey R, Beevers L (1977) Proteolytic activity in relationship to senescence and cotyledonary development in *Pisum sativum* L. Planta 137:37–44
Sugiyama T (1974) Proteinaceous factor reactivating an inactive form of pyruvate $P_i$ dikinase isolated from dark-treated maize leaves. Plant Cell Physiol 15:723–726
Suter D, Weidemann MJ (1976) Regulation of carbohydrate metabolism in lymphoid tissue. Biochem J 156:119–127
Swick RW, Koch AL, Handa DT (1956) The measurement of nucleic acid turnover in rat liver. Arch Biochem Biophys 63:226–233
Tanaka Y, Uritani I (1977) Synthesis and turnover of phenylalanine ammonia-lyase in root tissue of sweet potato injured by cutting. Eur J Biochem 73:255–260
Tanford C (1970) Protein denaturation. Adv Protein Chem 24:1–108
Thomas H (1977) Ultrastructure, polypeptide composition and photochemical activity of chloroplasts during foliar senescence of a non-yellowing mutant genotype of *Festuca pratensis* Huds. Planta 137:53–60
Thomas H (1978) Enzymes of nitrogen mobilization in detached leaves of *Lolium temulentum* during senescence. Planta 141:161–169
Tomati V, Galli E (1979) Water stress and SH dependent physiological activities in young maize plants. J Exp Bot 30:557–563
Trewavas A (1972) Determination of the rates of protein synthesis and degradation in *Lemna minor*. Plant Physiol 49:40–46
Varandani PT, Shroyer LA (1973) Insulin degradation. IV. Sequential degradation of insulin by rat kidney, heart and skeletal muscle homogenates. Biochim Biophys Acta 295:630–636
Vickery HB, Pucher GW, Schoenheimer R, Rittenberg D (1940) The assimilation of ammonia nitrogen by the tobacco plant: a preliminary study with isotopic nitrogen. J Biol Chem 135:531–539
Walker JH, Burgess RJ, Mayer RJ (1978) Relative rates of turnover of subunits of mitochondrial proteins. Biochem J 176:927–932
Wallace W (1973) A nitrate reductase inactivating enzyme form the maize root. Plant Physiol 52:197–201
Wallace W (1974) Purification and properties of a nitrate reductase inactivating enzyme. Biochim Biophys Acta 341:265–276
Willetts NS (1967) Intracellular protein breakdown in non-growing cells of *Escherichia coli*. Biochem J 103:462–466
Williams JA, John RA (1979) Generation of aspartate amino transferase multiple forms by deamidation. Biochem J 177:121–127
Winter K, Greenway H (1978) Phosphoenolpyruvate carboxylase from *Mesembryanthemum crystallinum*: its isolation and inactivation *in vitro*. J Exp Bot 29:539–546
Wittenbach VA (1978) Breakdown of ribulose bisphosphate carboxylase and change in proteolytic activity during dark-induced senescence of wheat seedlings. Plant Physiol 62:604–608
Woolhouse HW (1967) The nature of senescence in plants. Symp Soc Exp Biol 21:179–213
Woolhouse HW, Batt T (1976) The nature and regulation of senescence in plants. In: Sunderland N (ed) Perspectives in experimental biology, vol II. Botany. Pergamon Press, Oxford, pp 163–175
Yamaya T, Ohira K (1977) Purification and properties of a nitrate reductase inactivating factor from rice cells in suspension culture. Plant Cell Physiol 18:915–925

Yamaya T, Oaks A, Boesel IL (1980) Characteristics of nitrate reductase inactivating proteins obtained from corn roots and rice cell cultures. Plant Physiol 65:141–145

Yemm EW (1950) Respiration of barley plants. IV. Protein catabolism and the formation of amides in starving leaves. Proc R Soc London Ser B 136:632–647

Young JL, Varner JE (1959) Enzyme synthesis in the cotyledons of germinating seeds. Arch Biochem Biophys 84:71–77

Zak R, Martin AF, Prior G, Rabinowitz M (1977) Comparison of turnover of several myofibrillar proteins and critical evaluation of double isotope method. J Biol Chem 252:3430–3435

Zeelon P, Gershon H, Gershon D (1973) Inactive enzyme molecules in ageing organisms. Nematode fructose-1,6-diphosphate aldolase. Biochemistry 12:1743–1750

Zilversmit DB, Entenmay C, Fishler MC (1943) On the calculation of turnover time and "turnover rate" from experiments involving the use of labelling agents. J Gen Physiol 26:325–331

# 7 Structures of Plant Proteins

J.A.M. Ramshaw

## 1 Introduction

Studies on the amino acid sequences and three-dimensional structures of plant proteins have always lagged far behind those of proteins from other sources. Thus in 1969, only six complete sequences for plant proteins had been established compared to over 230 from other sources (see Dayhoff 1972). At that time the tertiary structure of only one plant protein had been established with a reasonable degree of certainty (Drenth et al. 1968). The major reason for this apparent lack of interest in plant proteins probably lies in the relative difficulty in obtaining sufficient quantities of material for study from plants when compared to other sources. This is due to both the lower intrinsic yields of the plant proteins and to specific difficulties in their preparation. Thus even a decade later plant enzymes have still been barely studied because the conservative nature of many of the biochemical pathways means that similar enzymes are more readily available elsewhere. Even when there are enzymes unique to plants, such as photosynthetic enzymes, these have not been examined.

One of the first major stimuli to sequence studies on plant proteins was the suggestion by Boulter that phylogenetic trees constructed using protein sequence data may prove a useful tool in examining plant affinities (see Boulter et al. 1970). These studies by Boulter and his colleagues have led to the collection of a substantial amount of data on the small electron transfer proteins, particularly cytochrome c and plastocyanin.

More recently a wider variety of plant proteins have been studied because of the unusual or valuable properties which they possess, e.g., sweetness, toxicity, inhibitor action. Thus by 1977 there had been a noticeable increase in the data available, 60 complete sequences from 15 types of protein (see Boulter 1977). However, this has increased to 107 complete sequences from 28 types of protein by 1979, in addition to the now substantial amount of partial sequence data. Despite all this sequence data, little information has emerged on the mechanism of action of the majority of these proteins. This must await further studies on their three-dimensional structures, as at present the structures of only eight have been reported.

It is hoped that this section will give a fairly complete bibliography of the structural data available for plant proteins (to mid-1980) which will enable the reader to locate easily the original papers on topics of interest. While listings of data can be of great use, for example the major work of Dayhoff (1972), they do not as a rule contain any discussion of the biological and structural importance of the data. Further, as stressed by Ambler (1976) in his discussion of standards and accuracy in amino acid sequence determination, typographical errors are a major source of the mistakes which occur in reports of sequence data. Thus relying on other than the original source may only lead to additional errors.

## 2 Enzymes

### 2.1 Ribulose 1,5-Bisphosphate Carboxylase

The enzyme ribulose 1,5-bisphosphate carboxylase (EC 4.1.1.39), sometimes called Fraction I protein, is probably the world's most abundant protein as it is the major soluble protein present (up to 50%) in the leaves of plants which fix carbon via the Calvin ($C_3$) pathway. Not only is the enzyme required for the photosynthetic fixation of $CO_2$, but also the enzyme has inherent oxygenase activity and can thus account for photorespiration, a metabolic process which decreases net $CO_2$ fixation (BOWES et al. 1971).

The enzyme from higher plants has a molecular weight of ~560,000 and contains eight protomeric units, each composed of a large subunit of ~56,000 and a small subunit of molecular weight 12,000–15,000 (see WILDMAN 1979). The inheritance of these two subunits is determined by two separate genetic systems, the small subunits exhibiting classical Mendelian segregation in reciprocal crossing experiments, whilst the large subunit is controlled by factors inherited solely via the maternal side in a reciprocal cross (KAWASHIMA and WILDMAN 1972, CHAN and WILDMAN 1972).

The structural gene for the large subunit in *Zea mays* has been shown to be contained within a 2,500 base pair sequence of chloroplast DNA and one copy of this DNA is present in each circular maize chloroplast DNA molecule (BEDBROOK et al. 1979). Since each chloroplast may contain 15–30 copies of the circular DNA genome and leaf cells may contain several hundred chloroplasts, the large subunit may be reiterated several thousandfold per cell. In contrast, the small subunit may have only one or very few copies per haploid genome (see CASHMORE 1979).

The complete sequence of the small subunit from spinach, *Spinacia oleracea*, has been determined (MARTIN 1979) and shown to consist of a single chain of 120 residues (see Fig. 1). Heterogeneity was present at a single position, 91, where both Tyr and Pro were observed. Partial data, mainly from the N-terminal, has been obtained for several other species (GIBBONS et al. 1975, STROBAEK et al. 1976; POULSEN et al. 1976, HASLETT et al. 1976, SCHMIDT et al. 1979). All show close structural homology, except for one peptide derived from barley, *Hordeum vulgare* (POULSEN et al. 1976) which bears no apparent relationship to any part of the spinach small subunit sequence. However, this subunit is synthesized in the cytoplasm as a larger molecule with perhaps some 50 extra residues which are subsequently lost during passage through the membrane into the chloroplast where the enzyme is assembled (CASHMORE et al. 1978, HIGHFIELD and ELLIS 1978).

A precursor to the small subunit has been isolated from *Chlamydomonas reinhardtii* which is converted to the mature form of the subunit by an endopeptidase present in the cell extracts (SCHMIDT et al. 1979). This precursor contains an N-terminal extension of 44 amino acid residues (see Fig. 1) and shows heterogeneity at 2 positions. The 16 residue peptide from barley which did not show homology with the spinach sequence, does not show homology to the precursor sequence from *Chlamydomonas*. The nature of the N-terminal of the protein is not completely clear. HASLETT et al. (1976) reported that in the sequences which they examined the N-terminal amino acid was Met, but that in 50% of the molecules the N-terminal was "frayed". In spinach, the observed N-terminal was Gln, frequently cyclized to pyrrolidone carboxylic acid, although the amino acid analysis data did suggest that a Met may be present (MARTIN 1979). Initial studies on the N-terminal sequence of the small subunit from tobacco, *Nicotiana tabacum*, showed heterogeneity at position 7 (GIBBONS et al. 1975). *N. tabacum* is believed to be an amphidiploid derived from the putative parents *N. sylvestris* and *N. tomentosiformis*. Sequence studies showed that residues 7 and 8 were Ile-Asn in *N. sylvestris* and were Tyr-Gly in *N. tomentosiformis*, while in *N. tabacum* both Ile and Tyr and Gly and Asn were found in these positions (STROBAEK et al. 1976). Thus, it was concluded that the amphidiploid *N. tabacum* had inherited two alleles for the small subunit, one from each parent, and that both the alleles continue to be retained and expressed.

The complete DNA sequence of the gene for the large subunit of the *Zea mays* enzyme has been determined and from this the amino acid sequence has been deduced (MCINTOSH et al. 1980). This deduced amino acid sequence, which comprises 475 amino acid residues is shown in Fig. 1, where it is compared with the sequences of some CNBr fragments obtained from the large subunit from barley (POULSEN et al. 1979). Although only about half of the barley sequence is available from these fragments, very few substitutions are found between the two sequences. Reactive affinity labels have been used to specifically identify the peptides which contain the essential Cys and Lys residues in the spinach enzyme (STRINGER and HARTMAN 1978, SCHLOSS et al. 1978). These residues belong to the large subunit, and the three putative catalytic site peptides are widely separated along the polypeptide chains (MCINTOSH et al. 1980). The barley peptide which did not show homology to any other small subunit peptides is homologous with residues 118–133 of the large subunit.

Ribulose 1,5-bisphosphate carboxylase has been readily crystallized in various forms from several species (see BAKER et al. 1977, JOHAL et al. 1980) but, in part because of its large size, no crystal structure analysis at atomic resolution has yet been reported. The observed molecular symmetry at low resolution, however, is consistent with the proposed octomeric structure of the enzyme (BAKER et al. 1977).

## 2.2 Proteases

Proteolytic activity in plants was reported over 180 years ago, but had clearly been recognized much earlier by inhabitants of tropical lands who used latex

```
A.

                                                 1    V           10
Chlamydomonas reinhardtii (Precursor peptide)   M A S I A K S S V S

11                20                  30                40    44
A A V A R P A R S S V R P M A A L K P A V K A A P V A A P A E A N D
                                                    V

                                  1                 10                  20
a. Spinacia oleracea            (M) Q  V  W  P  P  L  G  L  K  K  F  E  T  L  S  Y  L  P  P
b. Hordeum vulgare               M  Q  V  W  P  I  E  G  I  K  K  F  E  T  L  S  Y  L  P  P
c. Pisum satinum                 M  Q  V  W  P  P  I  G  K  K  K  F  E  T  L  S  W  L  P  P
d. Vicia faba                    M  Q  V  W  P  P  I  G  K  K  K  F  E  T  L  S  Y  L  P  P
e. Nicotiana tabacum             M  Q  V  W  P  P |I  N| K  K  K  Y  E  T  L  S  Y  L  P  D
                                                  |Y  G|
f. Chlamydomonas reinhardtii     M  M  V  W  T  P  V  N  N  K  M  F ........

 21                      30                      40                      50
a. L  T  T  E  Q  L  L  A  E  V  N  Y  L  L  V  K  G  W  I  P  P  L  E  F  E  V  K  D  G  F
b. L  S  T  E  A ................
c. L  T  P  D  Q ................
d. L  T  Q  D  Q ................
e. L  ?  Q  Q  Q ................

 51                      60                      70                      80
a. V  Y  R  E  H  D  L  S  P  G  Y  Y  D  G  R  Y  W  K  L  P  M  F  G  G  T  D  P  A  Q  V
b.                       ...............Y  W  T  L  W  K  L  P  M  F  G  I  P  B  A .......
c.                                      ..................M  F  G  P  P  D  A  S......

 81                      90                      100                     110
a. V  N  E  V  E  E  V  K  K  A  Y  P  D  A  F  V  R  F  I  G  F  B  B  K  R  E  V  Q  C  I

 111                     120
a. S  F  I  A  Y  K  P  A  G  Y


B.

                                  1                 10                  20
a. Zea mays                      M  S  P  Q  T  E  T  K  A  S  V  G  F  K  A  G  V  K  D  Y
b. Hordeum vulgare                                            ...........A  G  V  K  D  Y
c. Spinacia oleracea

 21                      30                      40                      50
a. K  L  T  Y  Y  T  P  E  Y  E  T  K  D  T  D  I  L  A  A  F  R  V  T  P  Q  L  G  V  P  P
b. K  L  T  Y  Y  T  P  E  Y  E  T  K  D  T  D  F  L  A  A  F  R  V  S  P  Q  P  G  V  P  P
c.

 51                      60                      70                      80
a. E  E  A  G  A  A  V  A  A  E  S  S  A  A  G  T  W  T  T  V  W  T  D  G  L  T  S  L  D  R
b. E  E  A  G  A  A  V  A  A  E..............
c.

 81                      90                      100                     110
a. Y  K  G  R  C  Y  H  I  E  P  V  P  G  D  P  D  Q  Y  I  C  Y  V  A  Y  P  L  D  L  F  E
b.
c.

 111                     120                     130                     140
a. E  G  S  V  T  N  M  F  T  S  I  V  G  N  V  F  G  F  K  A  L  R  A  L  R  L  E  D  L  R
b.       ..............F  T  S  I  V  G  N  V  F  G  F  K  A  L  -  A............
c.
```

**Fig. 1 A, B.** Sequences of ribulose-1,5-bisphosphate carboxylase. **A** Small subunit: *a Spinacia oleracea* (MARTIN 1979); *b Hordeum vulgare* (POULSEN et al. 1976); *c Pisum sativum; d Vicia faba* (HASLETT et al. 1976); *e Nicotiana tabacum* (STROBACK et al. 1976); *f Chlamydomonas reinhardtii* (SCHMIDT et al. 1979). Numbering is based on the mature form of the subunit. **B** Large subunit: *a Zea mays* (MCINTOSH et al. 1980); *b Hordeum vulgare* (POULSEN et al. 1979); *c Spinacia oleracea* (STRINGER and HARTMAN 1978, SCHLOSS et al. 1978), *mark putative catalytic site residues

```
 141                                        150                                              160                                          170
a. I  P  P  A  Y  S  K  T  F  Q  G  P  P  R  G  M  Q  V  E  R  D  K  L  N  K  Y  G  R  P  L
b.                                                             ...............Y  G  R  P  L
c.                                                             ...............Y  G  R  P  L

 171                                        180                                              190                                          200
a. L  G  C  T  I  K  P  K  L  G  L  S  A  K  N  Y  G  R  A  C  Y  E  C  L  R  G  G  L  D  F
b. L  G  C  T  I  K  P  K............
c. L  G  C* T  I  K* P  K.... ...........

 201                                        210                                              220                                          230
a. T  K  D  D  E  N  V  N  S  Q  P  F  M  R  W  R  D  R  F  V  F  C  A  E  A  I  Y  K  S  Q
b.                         ...............R  I  R  D  R  F  V  F  C  A  E  A  I  Y  K  S  Q
c.

 231                                        240                                              250                                          260
a. A  E  T  G  E  I  K  G  H  Y  L  N  A  T  A  G  T  C  D  E  M  I  K  G  A  V  F  A  R  Q
b. A  E  T  G  E  I  K  G  H  Y  L  N  A..........  .............I  K  G  A  V  F  A  R  Q
c.

 261                                        270                                              280                                          290
a. L  G  V  P  I  V  M  H  D  Y  L  T  G  G  F  T  A  N  T  T  L  S  H  Y  C  R  D  N  G  L
b. L  G  V  P...........H  D  Y  L  T  G  G  F  T  A  N  T  T  L  A  H  Y  C  R  D........
c.

 291                                        300                                              310                                          320
a. L  H  I  H  R  A  M  H  A  V  I  D  R  Q  K  N  H  G  M  H  F  R  V  L  A  K  A  L  R  M
b.                ...........A  V  I  D  R  Q.........
c.                                                                   .....................L

 321                                        330                                              340                                          350
a. S  G  G  D  H  I  H  S  G  T  V  V  G  K  L  E  G  E  R  E  I  T  L  G  F  V  D  L  L  R
b. S  G  G  D  H  I  H  S  G  T  V  V  G  K  L  E  G  E  R  E..............
c. S  E  G  D  H  I  H  S  G  T  V  V  G  K* L  E  G  E  R..............

 351                                        360                                              370                                          380
a. D  D  F  I  E  K  D  R  S  R  G  I  F  F  T  Q  D  W  V  S  M  P  G  V  I  P  V  A  S  G
b.                                                  ..............P  G  V  I  P  V  A  S  G
c.

 381                                        390                                              400                                          410
a. G  I  H  V  W  H  M  P  A  L  T  E  I  L  G  D  D  S  V  L  Q  F  G  G  G  T  L  G  H  P
b. G  I  H  V  W  H  M  P  A  L  T  E  I  F  G  D  D  S  V  L  Q  F  G  G  G  T  L  G  H  P
c.

 411                                        420                                              430                                          440
a. W  G  N  A  H  G  A  A  A  N  R  V  A  L  E  A  C  V  Q  A  R  N  E  G  R  D  L  A  R  E
b. W  G  N  A  P  G  A  A  A  N  R  V  A  L  E  A  C  V  Q  A  R  N  E  G  R  D  L  A  R  E  ...
c.

 441                                        450                                              460                                          470
a. V  Q  I  I  K  A  A  C  K  W  S  A  E  L  A  A  A  C  E  I  W  K  E  I  K  F  D  G  F  K
b.                ...........W  S  A  E  L  A  A  A  C  E  V  W  K...........
c.             ..............W  S  P  E  L  A  A  A  C* E  V  W  K...........

 471
a. A  M  D  T  I
b.
c.
```

**Fig. 1** (continued)

and leaves of papaya as meat tenderizers. Subsequently a variety of proteases with diverse properties have been isolated from a wide range of plant species and tissue types, although it is only recently that their role in cellular metabolism has been examined (see the review by RYAN 1973). Of these various enzymes, only one group, the plant sulphydryl proteases, has been the subject of any extensive structural investigations.

The latex from *Carica papaya* contains at least four distinct sulphydryl proteases, namely papain (EC 3.4.22.2) chymopapain (EC 3.4.22.6), and papaya peptidases A and B (LYNN 1979). In addition, similar proteases have been well characterized from pineapple, *Ananas comosus* (Bromelain, EC 3.4.22.4), fig, *Ficus glabrata* (Ficin) and the Chinese gooseberry, *Actinidia chinensis* (Actinidin). While all these enzymes show many similarities, including sizes, specificities and mechanism, there are nevertheless many differences between them. For example, all show different kinetic parameters; bromelain is a glycoprotein whereas papain, chymopapain, ficin and actinidin are not; actinidin has an acidic pI of 3.1, whereas the others have basic pI values betwen 8.7 and 9.6 (see, e.g., CARNE and MOORE 1978).

The complete amino acid sequences have been determined for both papain (MITCHEL et al. 1970) and actinidin (CARNE and MOORE 1978). A major part of the sequence of stem bromelain has also been determined (GOTO et al. 1976, 1980) and N-terminal and/or active site sequences are available for papaya peptidases A and B (LYNN and YAGUCHI 1979), ficin (HUSAIN and LOWE 1970) and chymopapain (TSUNODA and YASUNOBU 1966, LYNN and YAGUCHI 1979). This sequence data has been aligned (and numbered with respect to the actinidin sequence) and is shown in Fig. 2. Further, the three-dimensional structures of both papain (DRENTH et al. 1968) and actinidin (BAKER 1977, 1980; BAKER and DODSON 1980) have been reported (see Figs. 3, 4).

Examination of the aligned sequence data (Fig. 2) shows a close homology between all these enzymes, particularly around the active site Cys(25) and His(162) residues. Distinct differences exist, however, between the four enzymes obtained from papaya, indicating that they are distinct enzymes, closely related by common ancestry, rather than precursors of a single common structural type. The only glycoprotein is bromelain, which has a carbohydrate unit attached through Asn (121) in an Asn-Glu-Ser sequence; the complete structure of this carbohydrate moiety has been established (ISHIHARA et al. 1979). An homologous Asn-X-Ser sequence is not present in either papain or actinidin. The complete primary structures of actinidin and papain are very similar, having 220 and 212 residues each respectively and both having three disulphide bridges and one free (active site) sulphydryl group. The extra residues in the actinidin sequence compared to papain are readily accommodated at positions 59–60, 82, 171–174 and the C-terminal while one residue deleted is needed at position 224 (for numbering see Fig. 2).

A comparison of the 2·8 Å structures of actinidin (BAKER 1977) and papain (DRENTH et al. 1968) shows them to be essentially identical (see Fig. 3). While some small differences in conformation are apparent, the overall similarity between the two molecules, in which about 50% of the amino acid sequence residues are different, is most striking. The additional residues in actinidin

```
                    1                 10                  20                  30
a. Actinidin        L P S Y V D W R S A G A V V D I K S Q G E C G G C W A F S A
b. Papain           I P E Y V D W R Q K G A V T P V K N Q G S C G S C W A F S A
c. Bromelain        V P Q S I D W R D Y G A V T S V K N Q N P C G A C W G F A A
d. Chymopapain      Y P E S I D W R A K G A V T P V K R V P D S G E C Y......
e. Papaya pep-A     L P E N V D W R K K G A V T P V R............
f. Papaya pep-B     L P E S V D W R A K G A V T P V K............
g. Ficin            .............P I R Q Q G G C G S C W......

                    31                40                  50                  60
a.                  I A T V E G I N K I T S G S L I S L S E Q E L L D C G R T Q
b.                  V V T I E G I I K I R T G N L N Q Y S E Q E L L D C D R - -
c.                  I A T V E S V A S I Y K G I L Q P L S Q Q Q V D D C A K....

                    61                70                  80                  90
a.                  N T R G C D G G Y I T D G F Q F I I N D G G I N T D E N Y P
b.                  R S Y G C N G G Y P W S A L Q L V A Q V G - I H Y R N T P Y
c.                  .....

                    91                100                 110                120
a.                  Y T A Q D G D C D V A L Q D Q K Y V T I D T Y D N V P Y N N
b.                  Y E G V Q R Y C R S R E K G P Y A A K T D G V R Q V Q P Y N
c.                                                .............K A R V P R N

                    121               130                 140                150
a.                  E W A L Q T A V T Y Q P V S V A L D A A G D A F K Q Y A S G
b.                  Q G A L L Y S I A N Q P V S V V L Q A A G K D F Q L Y R G G
c.                  N E S S M Y A V S K Q P I T VAV V S A A - - N F Z L Y K S G
                    |
                    carb
                    151               160                 170                180
a.                  I F T G P C G T A V D H A I V I V G Y G T E G G V D Y W I V
b.                  I F V G P C G N K V D H A V A A V G Y N - - - - P G Y I L I
c.                  V G D G Y C K D K L N H A V T A I G Y N - - - - K A E F G D

g.                ......T G P C G T S L D H A V A L ...........

                    181               190                 200                210
a.                  K N S W D T T W G E E G Y M R I L R N V G G A - G T C G I A
b.                  K N S W G T G W G E N G Y I R I K R G T G N S Y G V C G L Y
c.                  G S G K K A R W G E A G Y I R M A R D V S S S S G I C G I A

                    211               220
a.                  T M P S Y P V K Y N N
b.                  T S S F Y P V K N
c.                  I D P L Y P T E G
```

**Fig. 2.** Sequences of the plant sulphydryl proteases. *a* Actinidin (CARNE and MOORE 1978); *b* Papain (MITCHEL et al. 1970); *c* Bromelain (GOTO et al. 1976, 1980); *d* Chymopapain (TSUNODA and YASUNOBU 1966, LYNN and YAGUCHI 1979) *e, f* Papaya peptidases A and B (LYNN and YAGUCHI 1979); *g* Ficin (HUSAIN and LOWE 1970). The numbering is based on the actinidin sequence. The active site residues are Cys (25) and His (162). Carbohydrate binding in bromelain is indicated as *-carb*. Disulphide bridges link residues 22–65, 56–98 and 156–207 in actinidin and papain

compared with papain all occur on the outside of the molecule, and none appears to affect significantly the overall conformation. Apart from these insertions, the only substantial difference between the two main chain conformations appears to be in the region 99–104 where both show α-helical turns but with directions at right angles to each other. There may be other smaller differences but crystallographic refinement at higher resolution will be needed to identify such differences with confidence.

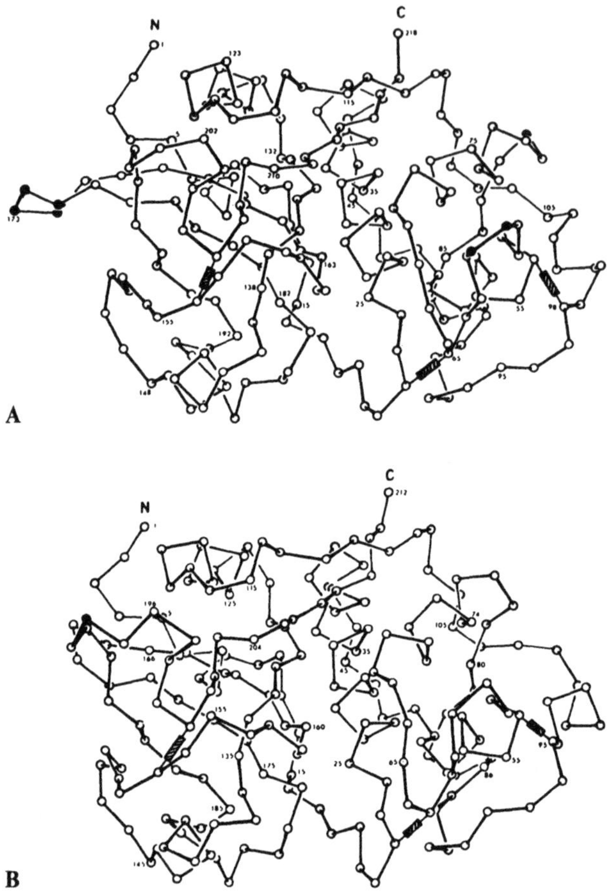

**Fig. 3.** α-Carbon plots of **A** actinidin and **B** papain plotted to give views from similar directions. Disulphide bridges are shown as -▨-, and additional residues in comparisons between the two structures are shown by *solid circles*. (BAKER 1977)

A detailed look at the actinidin structure (Fig. 4) shows it to be composed of two major domains. Domain 1 can be said to consist of residues 19–115 and 214–218 (the last C-terminal residues were not well defined) and domain 2 of residues 1–18 and 116–213. The N-terminal and C-terminal sections bind across domains and so appear as "straps" helping to hold the two halves of the molecule together. Domain 1 has no β-structure, but has several pieces of α-helix, most notably the five turns formed by residues 25–42. Domain 2, on the other hand, contains only one region of α-helix, formed by residues 120–129. The remainder of this domain is made up of an extensive although irregular piece of antiparallel β-sheet.

The catalytic sites of all the plant sulphydryl proteases have been shown to involve a single sulphydryl group (Cys 25) and an imidazole group (His 162) and in both actinidin and papain the arrangement of these residues is extremely similar. DRENTH et al. (1976) probed the mechanism of hydrolysis

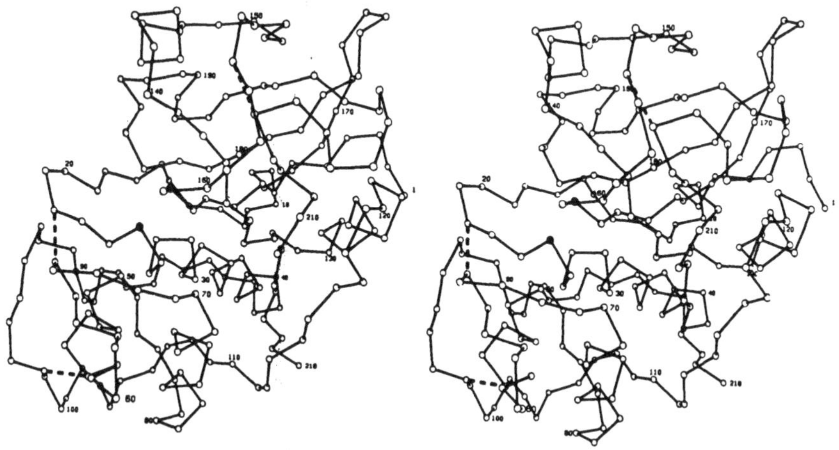

**Fig. 4.** A stereo view showing the polypeptide chain of actinidin. The catalytically important residues Cys (25) and His (162) are shown as *solid circles*, and the disulphide bridges as *dotted lines*. (BAKER 1977)

by papain by examining the structures of derivatives in which chloromethyl ketone substrate analogues were covalently bound to the enzyme through the active site Cys residue. Assuming the generally agreed upon scheme for hydrolysis via a covalent acyl-enzyme intermediate,

$$\text{acylation:} \quad E + S \rightleftharpoons ES \rightleftharpoons ES^{+}_{tetra} \rightarrow ES^{*} + P_1$$

$$\text{deacylation:} \quad ES^{*} + H_2O \rightleftharpoons ES^{+}_{tetra} \rightarrow E + P_2$$

their results indicated that the carbonyl oxygen of the $P_1$ residue is near two potential hydrogen bond donating groups, the backbone NH of Cys(25) and $NH_2$ of Gln(19), and is stabilized by formation of hydrogen bonds in the tetrahedral intermediate form. The nitrogen bond of the scissile peptide bond appeared close to the imidazole ring of the active site histidine residue, which indicated a role for this ring in protonating the N atom of the leaving group. This proton transfer would be facilitated by a rotation of the ring about the $C_\beta$–$C_\gamma$ bond; this rotation seemed possible from the observed data. The results also suggested that valine residues 133 and 151 (papain numbering 136 and 160 in Fig. 2) determined the substrate specificity of the enzyme.

### 2.3 Peroxidase

Peroxidase isoenzymes (donor: $H_2O_2$ oxidoreductase, EC 1.11.1.7) have been examined from a wide variety of plants. The isoenzymes obtained from horseradish, *Armoracia lapathifolia*, have been those most extensively studied. The crude extract from the horseradish plant contains seven major peroxidase isoenzymes and up to 13 additional minor isoenzymes. The physical properties of the purified

```
                                         carb
                              1          10   |          20              30
a. Amoracia lapathifolia   Pca L T P T F Y D N S C P N V S N I V R D T I V N E L R S D P
b. Brassica napus          Pca L T T N F Y S T S C P N L L S T V K S G V K S A V S S Q P

                                                                     carb
                               31          40                  50       |  60
                               R I A A S I L R L H F H D C F V N G C D A S I L L D N T T S
                               R M G A S I L R L F F H D C F V N G C D G S I L L D D T S S

                               61                70                  80                  90
                               F R T E K D A F G N A N S A R G F P V I D R M K A A V E S A
                               F T G E Q N A G P N R N S A R G F T V I N D I K S A V E K A

                               91                100                 110                120
                               C P R T V S C A D L L T I A A Q Q S V T L A G G P S W R V P
                               C P G V V S C A D I L A I A A R D S V V Q L G G P N W N V K

                               121               130                 140                150
                               L G R R D S L Q A F L D L A N A N L P A P F F T L P Q L K D
                               V G R R D A K T A S Q A A A N S N I P A P S M S L S Q L I S

                                             carb
                               151             | 160                 170                180
                               S F R N V G L N R S S D L V A L S G G H T F G K N Q C R F I
                               S F S A V G L S - T R D M V A L S G A H T I G Q S R C V N F

                                         carb                    carb
                               181         |     190               | 220                210
                               M D R L Y N F S N T G L P D P T L N T T Y L Q T L R G L C P
                               R A R V Y N - - - - - - - E T N I N A A F - A T L R QRS C P
                                         |
                                  carb   carb
                               211  |                220                 230                240
                               L N G N L S A L V D F D L R T P T I F D N K Y Y V N L E E Q
                               R AAG S G D ANL A P L D I N S A T S F D N S Y F K N L M A Q

                                                           carb                    carb
                               241               250         |       260             | 270
                               K G L I Q S D Q E L F S S P N A T D T I P L V R S F A N S T
                               R G L L H S D Q V L F - N G G S T D S I - - V R G Y S N S P

                               271               280                 290                300
                               Q T F F N A F V E A M D R M G N I T P L T G T Q G Q I R L N
                               S S F N S D F A A A M I K M G D I S P L T G S S G E I R K V

                               301          308
                               C R V V N S N S
                               C G K T N - - -
```

**Fig. 5.** The amino acid sequence of peroxidase. *a* Horseradish, *Amoracia lapathifolia,* Peroxidase isozyme C (WELINDER 1979) *b* Turnip, *Brassica napus,* Peroxidase isozyme 7 (MAZZA and WELINDER 1980). Sites of carbohydrate attachment are marked *-carb.* Disulphide bonds join residues 11–91, 44–49, 97–301 and 177–209

forms of all seven major isoenzymes have been studied by Shannon and colleagues (see CLARKE and SHANNON 1976). Of these, the most abundant, isoenzyme C, which accounts for about 50% of total peroxidase activity in horseradish roots, has been studied in greatest detail. It consists of a hemin prosthetic group, $2Ca^{2+}$ and 308 amino acid residues, including four disulfide bridges, in a single polypeptide chain that carries eight neutral carbohydrate side chains (WELINDER 1976, 1979). The complete sequence, showing the location of the carbohydrate and disulphides, is shown in Fig. 5. The eight carbohydrate units, which range in molecular weight from 1,600 to 3,000 (CLARKE and SHANNON 1976) and account for 20% of the molecular weight, are all attached to the polypeptide chain through Asn residues in the sequence Asn-X-Thr/Ser; every such sequence in the molecule except for Asn(286) had attached carbohydrate

(WELINDER 1979). The predominant carbohydrate residues found were mannose and glucosamine, with lesser amounts of fucose, xylose and arabinose (CLARKE and SHANNON 1976). The N-terminal of the protein is blocked as pyrrolidone carboxylic acid while the C-terminal is serine; however, some of the preparations lacked this residue, presumably as a result of the lability of the terminal Asn-Ser bond. No genetically determined heterogeneity was observed during the sequence determination (WELINDER 1979).

The complete amino acid sequence of the principle peroxidase of turnip, *Brassica napus*, has also been determined (MAZZA and WELINDER 1980). This enzyme, peroxidase 7, consists of a single chain of 296 amino acid residues (see Fig. 5) and one hemin group. It has only one neutral carbohydrate group compared with the eight in horseradish peroxidase C. The two proteins have the same arrangement of the four disulphide bonds and are clearly homologous. Nevertheless, the sequences are only 49% identical, indicating that their common evolutionary origin is distant. In addition, this turnip peroxidase shows many differences in its physicochemical and enzymatic properties and in its seasonal concentration dependence when compared with horseradish peroxidase C and other turnip peroxidases which suggests separate biological functions for the enzymes.

The peroxidases have many chemical and physical characteristics in common with globin monomers. Comparisons of the sequences show no obvious homologies, the globins having ~150 residues and peroxidase ~300 residues, there being no evidence of gene duplication. Nevertheless, there are only two histidine sequences in peroxidase and these are reminiscent of the proximal and distal histidine sequences of the globins and are capable of similar haem contacts. Further, peptide mapping of five isoperoxidases and subsequent sequence analysis of the most conserved peptides showed these peptides to be the histidine-containing peptides (WELINDER and MAZZA 1977). Final verification of such studies must await X-ray crystallographic analysis. However, although crystals of the enzyme have been obtained, their size and form have not been suitable for further analysis (BRAITHWAITE 1976).

## 2.4 ATP Synthase

The energy transducing ATPase complex appears to be universal in both eukaryotic and prokaryotic cells. The complex has various components of which one, the membrane factor $F_0$, consists of several subunits. The major one of the subunits is the proteolipid fraction which has a molecular weight of about 8,000, and is present in the complex probably as a hexamer. This proteolipid subunit alone constitutes the proton channel of the ATPase complex (see SEBALD et al. 1979). The sequence of the proteolipid fraction from spinach, *Spinacia oleracea*, chloroplasts has been determined (SEBALD et al. 1979) and is shown in Fig. 6. The sequence consists of a single polypeptide chain of 81 residues which has an N-formyl methionine as the N-terminal. The protein has very few polar amino acid residues, and those which are present are clustered in positions 41 to 52 between two extended regions of hydrophobic residues.

```
1                  10                  20                  30                  40
M N P L I A A A S V I A A G L A V G L A S I G P G V G Q G T A A G Q A V E G I A
41                 50                  60                  70                  80
R Q P E A E G K I R G T L L L S L A F M E A L T I Y G L V V A L A L L F A N P F V
```

**Fig. 6.** The amino acid sequence of ATP synthase, lipid-binding fragment, from spinach chloroplasts

Of the various inhibitors which are effective against the complex, it was shown that DCCD acts through residue Glu(61) of the proteolipid fraction. Also, the residues which may be substituted for oligomycin resistance are all found in the C-terminal hydrophobic region of this protein (see SEBALD et al. 1979).

## 2.5 Phosphorylase

Phosphorylase (EC 2.4.1.1) is an enzyme which catalyzes the reversible phosphorolysis of α-glucan. The enzymes from diverse sources have many common structural features including a subunit molecular weight of about 100,000 and one pyridoxal 5′-phosphate binding site per subunit. However, the regulatory properties are quite different depending on the source of the enzyme. For plant phosphorylase a regulatory mechanism has not been established and the allosteric transitions or chemical modification mechanisms used by other phosphorylases are absent.

The complete sequence of a plant phosphorylase has not yet been established, but the sequences of various segments from the potato enzyme have been reported (NAKANO et al. 1978, 1980a, 1980b, SCHILTZ et al. 1980). The sequences of a 57-residue fragment which encompassed the pyridoxal 5′-phosphate binding site (NAKANO et al. 1978) and of 10 Cys-containing peptides (NAKANO et al. 1980a) were all highly conserved when compared with the rabbit enzyme (see NAKANO et al. 1980b). However, initial studies on the N-terminal of the protein showed that no homology existed for the N-terminal 15 residues (SCHILTZ et al. 1980) and it was suggested that an N-terminal extension may exist for the potato enzyme. However, further studies on the N-terminal 81 residues (NAKANO et al. 1980b) have shown that the potato enzyme has only two additional residues at its N-terminal when compared with the rabbit enzyme. No homology exists between these enzymes for the N-terminal 35 residues whereas beyond this point extensive homology exists (NAKANO et al. 1980b).

## 2.6 Glycolate Oxidase

One of the steps in the photorespiratory process in plants is the oxidation of glycolate to glyoxylate, and this reaction is catalyzed by the enzyme glycolate oxidase. This enzyme contains FMN and has a subunit molecular weight of 37,000. The crystal structure of the enzyme at low resolution, 5.5 Å, has been reported (LINDQVIST and BRANDEN 1980). While the full course of the polypeptide chain is not visible at this resolution, the arrangement of the secondary structure of two-thirds of the molecule was shown and is similar to that of triose phosphate isomerase. The active site of the enzyme was also shown by determination of the binding site of the inhibitor thioglycolate (LINDQVIST and BRANDEN 1980).

# 3 Electron Transfer Proteins

Most of the proteins which are involved in the mitochondrial electron transfer process are common to all eukaryotic organisms. Plants, however, possess in addition the electron transfer proteins of the photosynthetic electron transfer pathway. A substantial amount of data for all these proteins has been collected by Boulter and his colleagues, principally on cytochrome c and plastocyanin, as part of an investigation into higher plant affinities using trees derived from protein sequence data. While these trees have provided several important new insights, it has been repeatedly stressed by Boulter (BOULTER et al. 1972, 1978, 1979, BOULTER 1973, 1974, 1977) that the affinity information derived from protein data should only be interpreted *in conjunction* with *all* the other biological and palaeontological data which exists.

## 3.1 Cytochrome c

Ever since the "rediscovery" of cytochrome c by KEILIN in 1925, following its initial description by MACMUNN in 1884 (see KEILIN 1966), cytochrome c has been one of the most widely studied of all proteins (for reviews see, MARGOLIASH and SCHEJTER 1966, DICKERSON and TIMKOVICH 1975). Cytochrome c is an electron carrier between two macromolecular complexes in the inner membrane of the mitochondrion, and is found in the mitochondria of all eukaryotes. Even though the three-dimensional structure of the molecule has been determined, the actual pathway of the electron flow through cytochrome c is still a matter of controversy (see DICKERSON and TIMKOVICH 1975).

The protein consists of a single polypeptide chain of 103–113 residues without any disulphide bridges and has a covalently bound haem group, linked to the polypeptide chain by thioether bridges to two Cys residues. Cytochrome c from plants has been extensively studied, principally by Boulter and his colleagues, and the complete sequences from 26 plants have been determined, 19 members of the Dicotyledonae, 5 members of the Monocotyledonae, 1 gymnosperm and 1 from a green alga. In addition the N-terminal sequences of cytochrome c from two red algae have been determined (see Fig. 7 for both the sequence data and the bibliography). This compares with around 70 sequences which are available from all species (see DICKERSON and TIMKOVICH 1975).

In general plant cytochromes c are similar to those from all other eukaryote sources. While a few positions have residues which are characteristic to the

| | |
|---|---|
| | 1 10 20 30 40 50 |
| a. *Phaseolus aureus* | Ac A S F B Z A P P G B S K S G E K I F K T K C A Q C H T V D K G A G H K Q G P N L N G L F G R Q S G T |
| b. *Helianthus annuus* | Ac A S F A E A P A G D P T T G A K I F K T K C A Q C H T V E K G A G H K Q G P N L N G L F G R Q S G T |
| c. *Sesamum indicum* | Ac A S F B Z A P P G B V K S G E K I F K T K C A Q C H T V D K G A G H K Q G P N L N G L F G R Q S G T |
| d. *Ricinus communis* | Ac A S F B Z A P P G B V K A G E K I F K T K C A Q C H T V E K G A G H K Q G P N L N G L F G R Q S G T |
| e. *Cucurbita maxima* | Ac A S F B E A P P G B S K A G E K I F K T K C A Q C H T V D K G A G H K Q G P N L N G L F G R Q S G T |
| f. *Fagopyrum esculentum* | Ac A S F S E A P P G N I K S G E K I F K T K C A Q C H T V E K G A G H K Q G P N L N G L F G R Q S G T |
| g. *Brassica oleracea and B. napus* | Ac A S F D E A P P G N S K E G E K I F K T K C A Q C H T V D K G A G H K Q G P N L N G L F G R Q S G T |
| h. *Abutilon threophrasti* | Ac A S F Q E A P P G N A K A G E K I F K T K C A Q C H T V E K G A G H K Q G P N L N G L F G R Q S G T |
| i. *Gossypium barbadense* | Ac A S F Q E A P P G N A K A G E K I F K I K C A Q C H T V D K G A G H K Q G P N L N G L F G R Q S G T |
| j. *Lycopersicon esculentum* | Ac A S F N E A P P G N P K A G E K I F K T K C A Q C H T V E K G A G H K E G P N L N G L F G R Q S G T |
| k. *Cannabis sativa* | Ac A S F B Z A P P G B S K A G E K I F K T K C A E C H T V G R G A G H K Q G P N L N G L F G R Q S G T |
| l. *Spinacia oleracea* | Ac A T F S E A P P G N K D V G A K I F K T K C A Q C H T V D L G A G H K Q G P N L N G L F G R Q S G T |
| m. *Nigella damascena* | Ac A S F B Z A P A G B S A S G F K I F K T K C A Z C H T V B Z G A G H K Z G P N L H G L F G R Q S G T |
| n. *Tropaeolum majus* | Ac A S F A E A P A G D N K A G D K I F K N K C A Q C H T V D K G A G H K Q G P N L N G L F G R Q S G T |
| o. *Acer negundo* | Ac A S F A E A P P G N P A A G E K I F K T K C A Q C H T V D K G A G H K Q G P N L N G L F G R Q S G T |
| p. *Sambucus nigra* | Ac A S F A E A P P G N P K A G E K I F K T K C N Q C H T V D K G A G H K Q G P N L N G L F G R Q S G T |
| q. *Pastinaca sativa* | Ac A S F A E A P P G D K D V G G K I F K T K C A Z C H T V Z L G A G H K Q G P N L N G L F G R Q S G T |
| r. *Solanum tuberosum* | Ac A S F G E A P P G N P K A G E K I F K T K C A Q C H T V D K G A G H K E G P N L N G L F G R Q S G T |
| s. *Guizotia abyssinica* | Ac A S F A E A P A G D A K A G E K I F K T K C A Z C H T V Z K G A G H K Q G P N L N G L F G R Q S G T |
| t. *Triticum aestivum* | Ac A S F S E A P P G N P D A G A K I F K T K C A Q C H T V D A G A G H K Q G P N L H G L F G R Q S G T |
| u. *Allium porrum* | Ac A T F S Z A P P G B Z K A G Q K I F K L K C A Q C H T V E K G A G H K Q G P N L N G L F G R Q S G T |
| v. *Oryza sativa* | Ac A S F S E A P P G N P K A G E K I F K T K C A Q C H T V D K G A G H K Q G P N L N G L F G R Q S G T |
| w. *Arum maculatum* | Ac A S F A E A P P G N P K A G E K I F K T K C A Q C H T V E K G A G H K Q G P N L N G L F G R Q S G T |
| x. *Zea maize* | Ac A S F S E A P P G N P K A G E K I F K T K C A Q C H T V E K G A G H K Q G P N L N G L F G R Q S G T |
| y. *Ginkgo biloba* | Ac A T F S E A P P G D P K A G E K I F K T K C A Z C H T V Z K G A G H K Q G P N L H G L F G R Q S G T |
| z. *Enteromorpha intestinales* | Ac S T F A B A P P G B P A K G A K I F K A K C A Z C H T V B A G A G H K Q G P N L N G A F G R T S G T |
| α. *Rhodymenia palmata* | A P A A A Y A D L K G N P T K G A K I F K.......... |
| β. *Porphyra umbilicalis* | A G N E Y K G A K I F K.......... |

```
      51        60                  70                  80                  90                  100                 110
a.    T A G Y S Y S T A N K N M A V I W E E K T L Y D Y L L N P K*K Y I P G T K M V F P G L K*K P Q D R A D L I A Y L K E S T A
b.    T A G Y S Y S A A N K N M A V I W E E N T L Y D Y L L N P K*K Y I P G T K M V F P G L K*K P Q E R A D L I A Y L K T S T A
c.    T P G Y S Y S A A N K N M A V I W G E N T L Y D Y L L N P K*K Y I P G T K M V F P G L K*K P Q E R A D L I A Y L K Z A T A
d.    T A G Y S Y S A A N K N M A V Q W G E N T L Y D Y L L N P K*K Y I P G T K M V F P G L K*K P Q D R A D L I A Y L K Z A T A
e.    T P G Y S Y S A A N K N R A V I W E E K T L Y D Y L L N P K*K Y I P G T K M V F P G L K*K P Q D R A D L I A Y L K E A T A
f.    T A G Y S Y S A A N K N K A V T W G E D T L Y E Y L L N P K*K Y I P G T K M V F P G L K*K P Q E R A D L I A Y L K B S T Z
g.    T A G Y S Y S A A N K N K A V E W E E K T L Y D Y L L N P K*K Y I P G T K M V F P G L K*K P Q D R A D L I A Y L K E A T A
h.    T P G Y S Y S A A N K N M A V N W G E N T L Y D Y L L N P K*K Y I P G T K M V F P G L K*K P Q D R A D L I A Y L K E S T A
i.    T A G Y S Y S A A N K N M A V Q W G E N T L Y D Y L L N P K*K Y I P G T K M V F P G L K*K P Q D R A D L I A Y L K E S T A
j.    T A G Y S Y S A A N K N M A V N W G E N T L Y D Y L L N P K*K Y I P G T K M V F P G L K*K P Q E R A D L I A Y L K E A T A
k.    T A G Y S Y S A A N K N M A V T W Z Z K T L Y D Y L L N P K*K Y I P G T K M V F P G L K*K P Z B R A D L I A Y L K E S T A
l.    A A S Y S Y S A A N K N K A V I W S E D T L Y E Y L L N P K*K Y I P G T K M V F P G L K*K P Q D R A D L I A Y L K D S T Q
m.    V A G Y S Y S A A N K N K A V N W E E K T L Y D Y L L N P K*K Y I P G T K M V F P G L K*K P Z Z R A B L L A Y L K E S T A
n.    T A G Y S Y S A A N K N K A V L W Z Z A T L Y D Y L L N P K*K Y I P G T K M V F P G L K*K P Q D R A D L I A Y L K E S T A
o.    T A G Y S Y S A A N K N M A V N W G Y N T L Y D Y L L N P K*K Y I P G T K M V F P G L K*K P Q D R A D L I A Y L K Q S T A A
p.    T A G Y S Y S A A N K N M A V N W E E K T L Y D Y L L N P K*K Y I P G T K M V F P G L K*K P Q D R A D L I A Y L K Q S T A
q.    T A G Y S Y S A A N K N K A V L W A B B T L Y D Y L L N P K*K Y I P G T K M V F P G L K*K P Q D R A D L I A Y L K H A T A
r.    T A G Y S Y S N A N K N M A V T W G E N T L Y D Y L L N P K*K Y I P G T K M V F P G L K*K P Q D R A D L I A Y L K E A T A
s.    T A G Y S Y S A A N K N K A V A W Z Z B S L Y D Y L L N P K*K Y I P G T K M V F P G L K*K P Z Z R A D L I A Y L K A S T A

t.    T A G Y S Y S A A N K N K A V E W E E N T L Y D Y L L N P K*K Y I P G T K M V F P G L K*K P Q D R A D L I A Y L K K A T S S
u.    A A G Y S Y S A A N K N M A V V W Z Z B T L Y D Y L L N P K*K Y I P G T K M V F P G L K*K P Q D R A D L I A Y L K E S T A
v.    T P G Y S Y S T A N K N M A V I W E E N T L Y D Y L L N P K*K Y I P G L K M V F P G L K*K P Q E R A D L I S Y L K E A T S
w.    T A G Y S Y S A A N K N M A V I W E E S T L Y D Y L L N P K*K Y I P G T K M V F P G L K*K P Q E R A D L I A Y L K E S T A
x.    T A G Y S Y S A A N K N K A V V W E E N T L Y D Y L L N P K*K Y I P G T K M V F P G L K*K P Q E R A D L I A Y L K E A T A

y.    T A G Y S Y S T G N K N K A V N W G Z Z T L Y E Y L L N P K*K Y I P G T K M V F P G L K*K P Z Z R A D L I S Y L K Q A T S Q E

z.    A A G F S Y S A A B K B K T A D W B Z B T L Y D Y L L N P K*K Y I P G T K M V F A G L K K P Z B R A D L I A F L K D A T A
```

**Fig. 7.** The amino acid sequences of cytochrome c from plants. *a* Mung Bean, *Phaseolus aureus* (THOMPSON et al. 1970a); *b* Sunflower, *Helianthus annuus* (RAMSHAW et al. 1970); *c* Sesame, *Sesamum indicum; d* Castor, *Ricinus communis* (THOMPSON et al. 1970b); *e* Pumpkin, *Cucurbita maxima* (THOMPSON et al. 1971b); *f* Buckwheat, *Fagopyrum esculentum; g* Cauliflower, *Brassica oleracea* and Black Rape, *B. napus* (THOMPSON et al. 1971b; RICHARDSON et al. 1971); *h Abutilon theophrasti; i* Cotton, *Gossypium barbadense* (THOMPSON et al. 1971a); *j* Tomato, *Lycopersicon esculentum* (SCOGIN et al. 1972; BROWN and BOULTER 1975); *k* Hemp, *Cannabis sativa* (WALLACE et al. 1973); *l* Spinach, *Spinacia oleracea* (BROWN et al. 1973); *m* Love-in-a-mist, *Nigella damascena* (BROWN and BOULTER 1973b); *n* Nasturtium, *Tropaeoleum majus; o* Box-elder, *Acer negundo; p* Elder, *Sambucus nigra; q* Parsnip, *Pastinaca sativa* (BROWN and BOULTER 1974); *r* Potato, *Solanum tuberosum* (MARTINEZ and ROCHAT 1974); *s* Niger, *Guizotia abyssinicia* (RAMSHAW and BOULTER 1975); *t* Wheat, *Triticum aestivum* (STEVENS et al. 1967); *u* Leek, *Allium porrum* (BROWN and BOULTER 1973a); *v* Rice, *Oryza sativa* (MORI and MORITA 1978, 1980); *w Arum maculatum; x* Maize, *Zea mays* (unpublished results, RICHARDSON DL and BOULTER D, cited in DICKERSON and TIMKOVICH 1975); *y Ginkgo biloba* (RAMSHAW et al. 1971); *z Enteromorpha intestinalis* (MEATYARD and BOULTER 1974) *α Rhodymenia palmata, β Porphyra umbilicalis* (MEATYARD et al. 1975). Trimethyllysine residues are indicated by *asterisks*

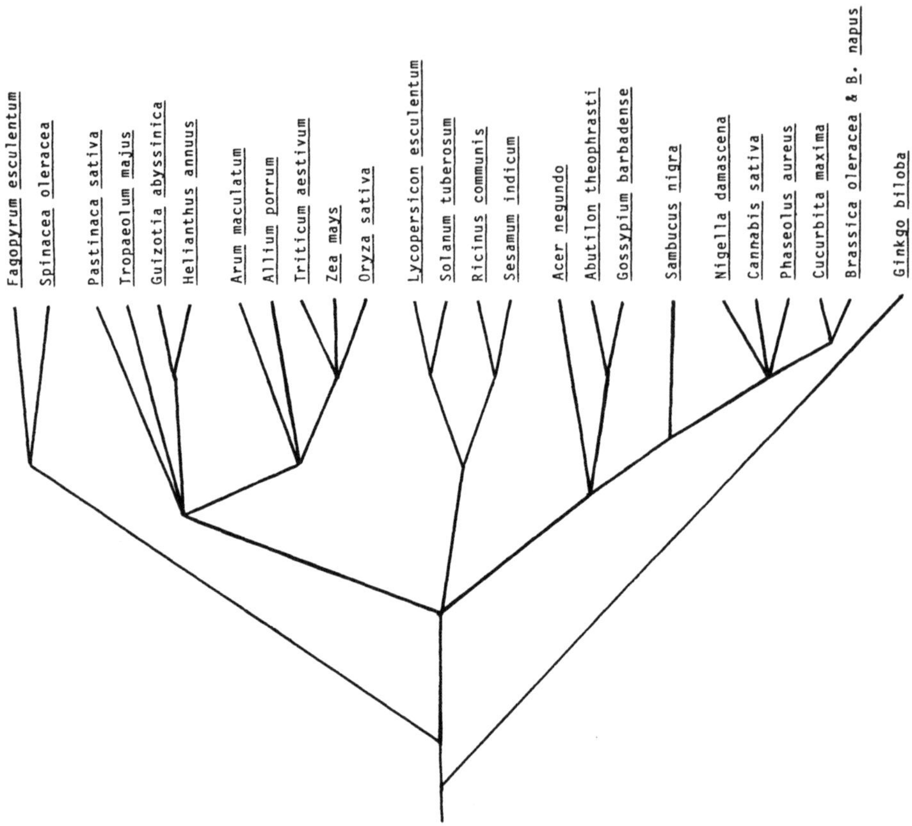


**Fig. 8.** A phylogenetic tree derived from the amino acid sequences of higher plant cytochromes c as given in Fig. 7, based on BOULTER (1973), and constructed using the ancestral sequence method (DAYHOFF 1972)

plant sequences, there are few distinctive features to the plant proteins, reflecting the slow rate of evolutionary change characteristic to the protein. The plant sequences are typically longer at the N-terminal than the animal cytochromes by eight residues, and the N-terminal is acetylated. In plant cytochromes two lysine residues are present as ε-N-trimethyllysine (except in *Enteromorpha* where only one is found); methylated lysine residues are not found in animal cytochromes c but some are found in fungal cytochromes c. When all the cytochromes in Fig. 7 are compared, 62 residues are invariant, and of these 25 are invariant in all eukaryotic cytochromes c. Little heterogeneity was unambiguously observed during the determination of the sequences of the plant cytochromes. Whether this reflects the monomorphic nature of the protein or results from the difficulties in obtaining data from the small quantities of pure protein available is uncertain. However, in the case of pumpkin (THOMPSON et al. 1971 b) heterogeneity was definitively established as it led to peptides which carried different charges.

Crystalline cytochrome c has been obtained and characterized from both rice (MORITA and IDA 1972) and spinach (MORITA et al. 1973), but as yet no detailed crystallographic analysis is available. However, as pointed out by BOULTER and RAMSHAW (1972), the primary structure similarities of plant cytochromes c to those from animal sources are so great that it is unlikely that any major differences in three-dimensional structure will exist. Therefore most of the conclusions drawn from the completed tertiary structures of horse, tuna and bonito cytochromes c (reviewed by DICKERSON and TIMKOVICH 1975) will be equally applicable to the plant proteins. Thus for example, Met(88) and His(26) will be the iron ligands derived from the polypeptide chain, and many of the invariant hydrophobic residues will be those creating the "haem pocket".

On the basis of the initial animal cytochrome c data, it was shown that a "molecular tree" could be constructed solely from protein data which was in excellent agreement with the phylogeny derived from classical fossil evidence (FITCH and MARGOLIASH 1967, DAYHOFF 1972). Further, from these data it could be deduced that cytochrome c was "evolving" very slowly, with approximately one substitution being fixed on average every 20 million years. Boulter therefore suggested (BOULTER et al. 1970) that protein sequence data, and in particular that from cytochrome c, would be extremely useful in examining affinities of higher plants for which no adequate fossil record exists. As a result of their extensive work in collecting cytochrome c sequence data, Boulter and his colleagues were able to establish successively larger affinity trees (BOULTER et al. 1970, 1972, BOULTER 1974) such that a tree relating all the sequences (Fig. 7) was produced (Fig. 8). The problems and assumptions needed in constructing such trees have been discussed in detail (BOULTER et al. 1972), together with the biological relevance of the various topologies (BOULTER et al. 1972, BOULTER 1973, 1974).

### 3.2 Cytochrome $c_6$

Cytochrome $c_6$ (or c-552) is a soluble c-type cytochrome which is found only in certain algae. It has a molecular weight of about 10,000 and has been shown to be associated with photosynthetic electron transfer. Initially it was called cytochrome f by analogy with the cytochrome found in higher plants which occupied a similar, though not necessarily identical place in the electron transfer chain. However, higher plant cytochrome f has a molecular weight of at least 33,000 and is tightly lipid-bound in most cases (see GRAY 1978). Recently WOOD (1978) has demonstrated that a large lipid-bound cytochrome f is also present in green alga, and that the cytochrome $c_6$ is a separate component which is interchangeable with plastocyanin in the electron transfer chain.

The sequences of cytochrome $c_6$ from a chrysophytan alga *Monochrysis lutheri* (LAYCOCK 1972), a red alga, *Porphyra tenera* (AMBLER and BARTSCH 1975) and a brown alga, *Alaria esculenta* (LAYCOCK 1975) have been described (Fig. 9). Also the sequences of this protein from *Euglena* and several blue-green bacteria have been described (see AITKEN 1979). The sequences (Fig. 9) have been aligned using the Cys residues, 14 and 17, through which the haem group

```
                           1                 10                  20                  30
a. Monochrys is lutheri    G D I A N G E Q V F T G D C A A C H S V Z Z Z K*T L E L S S
b. Porphyra tenare         A D L D N G E K V F S A N C A A C H A G G N N A I M P D K T
c. Alaria esculenta        I D I N N G E N I F T A N C S A C H A G G N N V I M P E K T

                           31                40                  50                  60
a.                         L W K - - A K S Y L A N F N G D E S A I V Y Q V T N G K N A
b.                         L K K - D V - - L E A N S M N T I D A I T Y Q V Q N G K N A
c.                         L K K - D A - - L A D N K M V S V N A I T Y Q V T N G K N A

                           61                70                  80
a.                         M P A F G G R L E D D E I A N V A S Y V L S K A G
b.                         M P A F G G R L V D E D I E D A A N Y V L S Q S E K G W
c.                         M P A F G S R L A E T D I E D V A N F V L T Q S D K G W D
```

**Fig. 9.** The amino acid sequences of cytochromes $c_6$. *a Monochrysis lutheri* (LAYCOCK 1972); *b Porphyra tenera* (AMBLER and BARTSCH 1975); *c Alaria esculenta* (LAYCOCK 1975). One residue of trimethyllysine is found in *M. lutheri* and is indicated by an *asterisk*

is attached. There are several regions which show a good match, particularly around this haem attachment site and around Met(61) which is likely to be the sixth iron ligand. Although the sequences are much shorter than the mitochondrial cytochromes c (see Fig. 7), there are several conserved features, suggesting that the folding patterns of the proteins may be similar. Comparison of the algal cytochrome $c_6$ sequences with those from the blue-green bacteria shows extensive homology, and the evidence drawn from these comparisons strongly favours the endosymbiotic hypothesis on chloroplast evolution (see AITKEN 1979).

### 3.3 Ferredoxin

Ferredoxins have been isolated and characterized from a wide variety of bacteria, algae and plants. They are small non-haem iron proteins with molecular weights of between 6,000 to 11,000 depending on the source. Those from higher plants, eukaryotic algae and the blue-green bacteria possess the 2Fe–2S active centre and have molecular weights around 11,000. The protein acts as an electron carrier in a variety of biological processes, including photosynthetic electron transport and nitrogen fixation. The evolutionary relationships between the plant and the other types of ferredoxins, which include gene duplications, have been discussed by DAYHOFF (1972).

The amino acid sequences of ferredoxins from a very wide range of plants have been reported. The data are shown in Fig. 10, together with the bibliographical data. All the sequences have between 93 and 98 residues. When compared there are 26 residues which are invariant; of these, four are the Cys residues, 39, 44, 47 and 77 which are required for the binding of the 2Fe atoms involved in the catalytic site of the protein. Although the three-dimensional structures of several types of ferredoxin have been determined (see ADMAN 1979) the tertiary structure of a plant ferredoxin has not yet been determined. However, the structure of a chloroplast type ferredoxin from *Spirulina platensis* has been reported at 2.5 Å resolution (FUKUYAMA et al. 1980). From this structure, it is expected that the plant ferredoxins will also have a tetrahedral arrangement

| | 1–50 (positions 1, 10, 20, 30, 40, 50) |
|---|---|
| a. *Colocasia esculenta* | A T Y K V K L V T - P S G - Q Q E F Q C P D D V Y I L D Q A E E V G I D L P Y S C R A G S C S S C A G K |
| b. *Leucaena glauca* | - A F K V KL/VL T - P D G -P/AK E F E C P D D V Y I L D Q A E E L G ID/EL P Y S C R A G S C S S C A G K |
| c. *Spinacia oleracea* | A A Y K V T L V T - P T G - N V E F Q C P D D V Y I L D A A E E E G I D L P Y S C R A G S C S S C A G K |
| d. *Medicago sativa* | A S Y K V K L V T - P E G - T Q E F E C P D D V Y I L D H A E E E G I V L P Y S C R A G S C S S C A G K |
| e. *Phytolacca americana I* | A T Y K V T L V T - P S G - T N T I D C P D D T Y V L D A A E E A G L D L P Y S C R A G S C S S C T G K |
| f. *Phytolacca americana II* | A A S Y K V T F V T - P S G - T N T I T C P A D T Y V L D A A E E S G L D L P Y S C R A G A C S S C A G K |
| g. *Phytolacca esculenta I* | A T Y K V T L V T - P S G - T Q T I D C P D D T Y V L D A A E E A G L D L P Y S C R A G S C S S CT/AG K |
| h. *Phytolacca esculenta II* | A A S Y K V T F V T - P S G - T K T I T C P A D T Y V L D A A E D T G L D L P Y S C R A G A C S S C A G K |
| i. *Brassica napus* | A T Y K V K F I T - P E G - E Q E V E C D D D V Y V L D A A E E A G I D L P Y S C R A G S C S S C A G K |
| j. *Pisum sativum I* | A S Y K V K I V T - P D G - T Q E F E C P S D V Y I L D H A E E V G L D L P Y......... |
| k. *Pisum sativum II* | A T Y N I K L I T - P E G - T K E I T C S D S E Y I L D A A E E K G L D L P Y......... |
| l. *Sambucus nigra* | A S Y K V K L I T - P D G - P Q E F E C P D D V Y I L E H A E E L G I D I P Y S C R A G S C S S C A G K |
| m. *Triticum aestivum* | A T Y K V K L V T - P E G - E V E L E V P D D V Y I L D Q A E E E G I D L P Y S C R A G S C S S C A G K |
| n. *Equisetum telmateia I* | - A Y K T V L K T - P S G - E F T L D V P E G T T I L D A A E E A G Y D L P F S C R A G A C S S C L G K |
| o. *Equisetum telmateia II* | - A Y K V T L K T - P D G - D I T F D V E P G E R L I D I A S E K A - D L P L S C Q A G A C S T C L G K |
| p. *Equisetum arvense I* | - A Y K T V L K T - P S G - E F T L D V P E G T T I L D A A E E A G Y D L P F S C R A G A C S S C L G K |
| q. *Equisetum arvense II* | - A Y K V T L K T - P D G - D I T F D V E P G E R L I D I G S E K A - D L P L S C Q A G A C S T C L G K |
| r. *Scenedesmus quadricauda* | A T Y K V T L K T - P S G - D Q T I E C P D D T Y I L D A A E E A G L D L P Y S C R A G A C S S C A G K |
| s. *Porphyra umbilicalis* | A D Y K I H L V S K E E G I D V T F D C S E A T Y I L D A A E E E G I E L P Y S C R A G A C S T C A G K |
| t. *Cyanidium caldarium* | A S Y K I H L V N K D Q G I D E T I E C P D D Q Y I L D A A E E Q G L D L P Y S C R A G A C S T C A G K |

| | 51–97 (positions 51, 60, 70, 80, 90, 97) |
|---|---|
| a. | V K V G D V D Q S D G S F L D D E Q I G E G W V L T C V A Y P V S D G T I E T H K E E E L T A |
| b. | L V E G D L D Q S D Q S F L D D E Q I E E G W V L T C A A Y P R S D V V I E T H K E E E L TG/A |
| c. | L K T G S L N Q D D Q S F L D D D Q I D E G W V L T C A A Y P V S D V T I E T H K E E E L T A |
| d. | V A A G E V N Q S D G S F L D D D Q I E E G W V L T C V A Y A K S D V T I E T H K E E E L T A |
| e. | V T A G T V D Q E D Q S F L D D D Q I E A G F V L T C V A F P K G D V T I E T H K E E D I V |
| f. | V T A G A V N Q E D G S F L E E E Q M E A G W V L T C V A Y P T S D V T I E T H K E E D L T A |
| g. | V T A G T V D Q E D Q S F L D D D Q I E A G F V L T C V A Y P K G D V T I E T H K E E D IV/A |
| h. | V T A G S V N Q E D G S F L D E E Q M E A G W V L T C V A Y P T S D V T I E T H K E E D L S A |
| i. | V V S G F V D Q S D E S F L D D D Q I A E G F V L T C A A Y P T S D V T I E T H K E E E L V |
| j. | |
| k. | |
| l. | L V A G S V D Q S D Q S F L D D E Q I E E G W V L T C V A Y P K S D V T I E T H K E E E L T A |
| m. | L V S G E I D Q S D Q S F L D D D Q M E A G W V L T C H A Y P K S D I V I E T H K E E E L T A |
| n. | V V S G S V D Q S E G S F L D D G Q M E E G F V L T C I A I P E S D L V I E T H K E E E L F |
| o. | I V S G T V D Q S E G S F L D D E Q I E Q G Y V L T C I A I P E S D V V I E T H K E D E L |
| p. | V V S G S V D E S E G S F L D D G Q M E E G F V L T C I A I P E S D L V I E T H K E E E L F |
| q. | I V S G T V D Q S E G S F L D D E Q I E Q G Y V L T C I A I P E S D V V I E T H K E D E L |
| r. | V E A G T V D Q S D Q S F L D D S Q M D G G F V L T C V A Y P T S D C T I A T H K E E D L F |
| s. | V T E G T V D Q S D Q S F L D D E Q M L K G Y V L T C I A Y P E S D C T I L T H V E Q E L Y |
| t. | L L E G E V D Q S D Q S F L D D D Q V K A G F V L T C V A Y P T S N A T I L T H Q E E S L Y |

**Fig. 10.** The amino acid sequences of plant ferredoxins. *a* Taro, *Colocasia esculenta* (RAO and MATSUBARA 1970); *b Leucaena glauca* (BENSON and YASUNOBU 1969a, 1969b); *c* Spinach, *Spinacia oleracea* (MATSUBARA and SASAKI 1968); *d* Alfalfa, *Medicago sativa* (KERESZTES-NAGY 1969); *e–h* Pokeweeds, *Phytolacca americana* (WAKABAYASHI et al. 1978), *P. esculenta* (WAKABAYASHI et al. 1980); *i* Rape, *Brassica napus* (TAKRURI and BOULTER 1980); *j* and *k* Pea, *Pisum sativum* (DUTTON et al. 1980); *l* Elder, *Sambucus nigra* (TAKRURI and BOULTER 1979a); *m* Wheat, *Triticum aestivum* (TAKRURI and BOULTER 1979b); *n–q* Horsetails, *Equisetum telmateia* (HASE et al. 1977a), *E. arvense* (HASE et al. 1977b); *r Scenedesmus* (SUGENO et al. 1969); *s Porphyra* (TAKRURI et al. 1978); *t Cyanidium* (HASE et al. 1978)

of the iron-sulphur clusters and that the arrangement of the polypeptide chain will be similar to that found for this ferredoxin.

When the sequence of ferredoxin from a mixed population of *Leucaena glauca* was examined, heterogeneity was observed at four positions (BENSON and YASUNOBU 1969a). Using ferredoxin isolated from individual trees, BENSON and YASUNOBU (1969b) showed the heterogeneity was probably due to either nonallelic nuclear genes or allelic genes of the chloroplast (position 7 and 14), as two distinct protein types were found. However, by using peptide mapping and amino acid compositions, SHIN et al. (1979) were unable to find any variation within or between the structures of the ferredoxins from three *Triticum* and two *Aegilops* species.

Ferredoxin seems well suited for analysis of higher plant affinities through trees constructed from protein sequence data (BOULTER et al. 1970) and this accounts for the recent increase in interest in this protein. However, it is also of use in establishing affinities between primitive members of the plant kingdom, particularly the algae. For example, structural studies on a ferredoxin from *Cyanidium caldarium*, and acido-thermal alga of uncertain classification, strongly suggest a close relationship to the red algal type (HASE et al. 1978).

Ferredoxin also illustrates the need for care when making comparisons between different sequences. In several species two different ferredoxins have been characterized, so it is important that only the homologous sequences between species are used. Complete sequence data on these duplicated ferredoxins from *Equisetum* and *Phytolacca* clearly show that the sequences from within a family are more closely related to each other than to any other sequence (see Fig. 10). This indicates that these gene duplications have been independent genetic events on the different phyletic lines. However, the duplication along the *Equisetum* line of descent occurred prior to the divergence of the two species which have been examined (see WAKABAYASHI et al. 1978, 1980).

### 3.4 Plastocyanin

Plastocyanin occurs in many algae and in higher plants, and functions as an electron transfer protein in photosynthesis. It is monomeric with a low molecular weight, 11,000, has no bound carbohydrate and contains one copper atom per molecule. Its sequence has been studied from an extremely wide variety of higher plants as part of an investigation of plant affinities using trees constructed from protein sequence data (see BOULTER 1974). Plastocyanin has several characteristics which ideally suit it to such studies. These include its small size, its easy extraction in good yield and the absence of any blocked N-terminal making it suitable for automatic sequencing methods. Plastocyanin has also been the subject of many chemical and spectroscopic studies, because of the unusual properties of the copper atom. Most notably the copper exhibits an intense blue colour, with an absorbance almost 1,000 times greater on a molar basis than the absorbance of simple copper amino acid complexes; this cannot be readily mimicked by "model" complexes. The chemical, spectroscopic and biological properties of plastocyanin (BOULTER et al. 1977) and of other similar

```
                                 1                10                  20                  30
a. Chlorella fusca             D V T V K L G A D S G A L V F E P S S V T I K A G E T V T W V
b. Phaseolus vulgaris            L E V L L G S G D G S L V F V P S E F S V P S G E K I V F K
c. Solanum tuberosum             L D V L L G G D D G S L A F I P G N F S V S A G E K I T F K
d. Sambucus nigra                V E I L L G G E D G S L A F I P S N F S V P S G E K I T F K
e. Cucurbita pepo                I E V L L G G D D G S L A F I P N D F S V A A G E K I V F K
f. Vicia faba                    V E V L L G A S D G G L A F V P N S F E V S A G D T I V F K
g. Spinacia oleracea             V E V L L G G G D G S L A F L P G D F S V A S G E E I V F K
h. Lactuca sativa                A E V L L G S S D G G L V F E P S T F S V A S G E K I V F K
i. Mercurialis perennis          L D V L L G S D D G G L A F V P N N F S I A A G E K I V F K
j. Capsella bursa-pastoris       I E V L L G G G D G S L A F V P N D F S I A K G E K I V F K
k. Rumex obtusifolius            I E I K L G G D D G A L A F V P G S F T V A A G E K I V F K
l. Solanum crispum               I E V L L G S D D G G L A F V P G N F S I S A G E K I T F K
m. Populus nigra                 I D V L L G A D D G S L A F V P S E F S I S P G E K I V F K

   31                40                  50                  60                  70
a.  N N A G F P H A I V F D E D E V P S G A N A E A L S - H - E D Y L N A P G E S Y
b.  N N A G F P H N V V F D E D E I P A G V D A V K I S M P E E E L L N A P G E T Y
c.  N N A G F P H N V V F D E D E I P A G V D A S K I S M A E E D L L N A A G E T Y
d.  N N A G F P H N V V F D E D E V P S G V D S A K I S M S E D D L L N A P G E T Y
e.  N N A G F P H N V V F D E D E I P S G V D A G K I S M N E E D L L N A P G E V Y
f.  N N A G F P H N V V F D E D E I P S G V D A A K I S M P E E D L L N A P G E T Y
g.  N N A G F P H N V V F D E D E I P G S V D A A K I S M S E E D L L N A P G E T Y
h.  N N A G F P H N V V F D E D E I P A G V D A S K I S M S E E D L L N A P G E T Y
i.  N N A G F P H N V V F D E D E I P S G V D A S K I S M D E A D L L N A P G E T Y
j.  N N A G F P H N V V F D E D E I P S G V D A S K I S M D E N D L L N A A G E T Y
k.  N N A G F P H N I V F D E D E V P A G V D A A K I S M S E E D L L N A P G E T Y
l.  N N A G F P H N V V F D E D E I P A G V D A S K I S M P E E D L L N A P G E T Y
m.  N N A G F P H N I V F D E D S I P S G V D A S K I S M S Z Z B L L N A K G E T F

   71                80                    91              99
a.  S A K F D T A G T Y G Y F C E P H Q G A G M K G T I T V Q
b.  V V T L D T K G T Y S F Y C S P H Q G A G M V G K V T V N
c.  S V T L S E K G T Y T F Y C A P H Q G A G M V G K V T V N
d.  S V T L T E S G T Y K F Y C S P H Q G A G M V G K V T V N
e.  K V N L T E K G S Y S F Y C S P H Q G A G M V G K V T V N
f.  S V K L D A K G T Y K F Y C S P H Q G A G M V G Q V T V N
g.  K V T L T E K G T Y K F Y C S P H Q G A G M V G K V T V N
h.  A V T L T E K G T Y S F Y C A P H Q G A G M V G K V T V N
i.  A V T L T E K G S Y S F Y C S P H Q G A G M V G K V T V N
j.  E V A L T E A G T Y S F Y C A P H Q G A G M V G K V T V N
k.  A V T L S E K G T Y S F Y C S P H Q G A G M V G K V T V Q
l.  S V T L S E K G T Y S F Y C S P H Q G A G M V G K V T V N
m.  E V A L S N K G E Y S F Y C S P H Q G A G M V G K V T V N
```

**Fig. 11.** The complete amino acid sequences of plastocyanin. *a Chlorella fusca* (Kelly and Ambler 1973); *b* French Bean, *Phaseolus vulgaris* (Milne et al. 1974); *c* Potato, *Solanum tuberosum* (Ramshaw et al. 1974a); *d* Elder, *Sambucus nigra* (Scawen et al. 1974); *e* Marrow, *Cucurbita pepo* (Scawen and Boulter 1974); *f* Broad Bean, *Vicia faba* (Ramshaw et al. 1974b); *g* Spinach, *Spinacia oleracea* (Scawen et al. 1975); *h* Lettuce, *Lactuca sativa* (Ramshaw et al. 1976); *i* Dog's Mercury, *Mercurialis perennis* and *j* Shepherd's Purse, *Capsella bursa-pastoris* (Scawen et al. 1978); *k* Dock, *Rumex obtusifolius* (Haslett et al. 1978a); *l Solanum crispum* (Haslett et al. 1978b); *m* Poplar, *Populus nigra* (unpublished results of Ambler RP cited by Freeman 1979)

"blue," Type I, copper proteins have been subject to review (Fee 1975, Solomon et al. 1980).

The complete sequence of plastocyanin from 13 plant species has been determined (Fig. 11), and in addition, the N-terminal sequences of plastocyanin from a further 50 higher plants have been reported (Fig. 12). (The references to the sequence data are given in the legends to Figs. 11 and 12.) When all the

| | |
|---|---|
| **Apiaceae** | |
| *Pastinaca sativa* | AEVKLGGDDGGLVFSPNSFTVAAGEKITFKNNAGFPHNIV... |
| *Anthriscus sylvestris* | AEVKLGGDDGSLAFVP$^{S}_{N}$SITVASGETITFKNNAGFPHNIV... |
| *Heracleum sphondylium* | AEVKLGGDDGGLVFSPNSFTVAAGEKITFKNNAGFPHNIV.... |
| *Heracleum mantegazzianum* | AEVKLGGDDGGLVFSPNSFTVAAGEKITFKNNAGFPHNIV... |
| *Aegopodium podagraria* | $^{A}_{V}$ELKLGADDGGLVFSPSSFTVAAGEKITFKNN..... |
| **Asteraceae** | |
| *Achillea millefolium* | $^{AD}_{IE}$VLLGANDGGLAF$^{E}_{V}$PAT$^{L}_{F}$SVPAGEKIVFKNNDGFPHNVV... |
| *Tanacetum vulgare* | IDVLLGANDGGLAFEPATFSVPAGEKIVFKNNSGFPHNVV... |
| *Ursinia anethoides* | $^{A}_{M}$EVLLGD$^{N}_{E}$DGGL$^{N}_{A}$F$^{E}_{V}$P$^{A}_{K}$T$^{L}_{F}$T$^{V}_{I}$PAGEKIVFKNNS$^{A}_{G}$FPHNVV.... |
| *Bellis perennis* | $^{I}_{A}$EVLLGDNNG$^{A}_{G}$LVF$^{E}_{V}$P$^{K}_{S}$T$^{L}_{F}$SVASGEKIVFKNNIGFPHNVV... |
| *Solidago altissima* | IEVLLGDNNGALVFEPATFSVAAGEEIVFKNNIGFPHNVV.... |
| *Calendula officinalis* | $^{F}_{I}$EVLLGD$^{E}_{N}$DGGLAF$^{E}_{V}$P$^{SNL}_{NTF}$SV$^{P}_{A}$SGEKIVFKNNSGFPHNVV.. |
| *Hieracium* sp. | VEVLLGDNDGGLVFEPSTLSVASGEKIVFKNNSGFPHNVV.... |
| *Sonchus oleraceus* | VEVLLGSSDGGLVFEPSTFSVASGEKIVFKNNAGFPHNVV.. |
| *Taraxacum officinale* | VEVLLGDNDGGLVFEPSTFSVPAGEKIVFKNNSGFPHNVV.... |
| *Tragopogon porrifolium* | VEVLLGDNDGSLVFEPSTFSVASGEKIVFKNNSGFPHNVV.. |
| *Arctium lappa* | IEVLLGANDGGLVFEPSTFSVASGEKIVFKNNSGFPHNVV... |
| *Centaurea nigra* | VDVLLGGDDGGLVFEPSTFSVASGEKIVFKNNAGFPHNVV.... |
| *Cirsium vulgare* | $^{V}_{I}$EVLLGASDGGLVFEPS$^{T}_{N}$FTVASGEKIVFKNNAGFPHNVV.. |
| *Guizotia abyssinica* | $^{F}_{L}$$^{D}_{E}$VLLGDNDGALAFEPSTFSVPSGEKIVFKNNSGFPHNVV... |
| *Helianthus annuus* | $^{F}_{I}$$^{D}_{E}$VLLGDNDGGLAFEPSTFSV$^{P}_{A}$AGEKIVFKNNSGFPHNVV.... |
| *Rudbeckia* sp. | L$^{F}_{D}$$^{I}_{V}$LLG$^{G}_{D}$N$^{L}_{D}$GGL$^{V}_{A}$FEPA$^{N}_{T}$FSV$^{A}_{P}$AGEKIVFKNNSGFPHNVV... |
| *Inula magnifica* | IEIKLGDDNAALVFEPATFSVAAGEKIVFKNNAGFPHNVV... |
| *Petasites hybridus* | $^{I}_{V}$EVLLGD$^{N}_{E}$DGGLAF$^{E}_{V}$PSTFSVASGEKIVFKNNSGFPHNVV.. |
| *Senecio jacobaea* | IEVLLGSSDGGLAFEPNAFSV$^{A}_{P}$PGEKIVFKNNAGFPHNVV.... |
| *Senecio vulgaris* | IEVLLGDNDG$^{G}_{A}$LAF$^{E}_{V}$PS$^{N}_{T}$FSVAAGEKIVFKNNSGFPHNVV.. |
| *Tussilago farfara* | $^{I}_{V}$EVLLGD$^{N}_{E}$DGGL$^{N}_{A}$F$^{E}_{V}$PSTFTVAPGEKIV$^{F}_{L}$KNNS$^{A}_{G}$FPHNVV..... |
| **Brassicaceae** | |
| *Capsella bursa-pastoris* | IEVLLGGGDGSLAFVPNDFSIAKGEKIVFKNNAGFPHNVV... |
| *Brassica oleracea* (2 var) | IDVLLGSGDGALAFVPNEFTIAKGEKIVFKNNAGFPHNVV.... |
| **Caprifoliaceae** | |
| *Viburnum tinus* | IE$^{I}_{V}$LLGGDDGSLAFVPGNFSV$^{P}_{A}$SGEKIVFKNNAGFPHN... |
| *Sambucus nigra* | VEILLGGEDGSLAF$^{V}_{I}$P$^{G}_{S}$NFSVPSGEKITFKNNAGFPHNVV... |
| *Lonicera periclymenum* | IEVLLGGDDGSLAFVPGSFTVPSGEKIIFKNNAGFPHNVV... |

**Fig. 12.** N-terminal sequence data for plastocyanin where complete sequences are not available. Data is taken from HASLETT and BOULTER (1976), HASLETT et al. (1977), BOULTER et al. (1978) and HASLETT et al. (1979)

**Fabaceae**

| | |
|---|---|
| *Cytisus battendieri* | VEVLLGSDDGGLAFVPDNFSVSAGEKIVFKNNAGFPHNVV... |
| *Lupinus* sp. | VEVLLGSDDGGLAFVPNDFSVNPGEKIVFKNNAGFPHNVV... |
| *Robinia pseudoacacia* | VEVLLGGDDGSLAFVPNDFSVSPGEKIVFKNNAGFPHNVV... |
| *Daviesia latifolia* | IEVLLGASDGSLAFVPNSFSVSPGEKITFKNNAGFPHNVV... |
| *Pisum sativum* | VEVLLGASDGGLAFVPSSLEVSAGETIVFKNNAGFP... |
| *Trifolium medium* | VEVLLGASDGGLAFVPNNFTVSAGDTIVFKNNAGFPHNVV... |
| *Vigna radiata* | LEVLLGAGDGSLVFVPSDFSVASGEEIVFKNNAGF... |

**Magnoliaceae**

| | |
|---|---|
| *Magnolia soulangeana* | IEVLLGGSDGTLAFVPKEFSVSPGEKIVFKNNAGFPHNVV... |
| *Liriodendron tulipifera* | IEVLLGDSDGNLVFVPKEFSVAPGEKIVFKNNAGFPHNIV... |

**Plantaginaceae**

| | |
|---|---|
| *Plantago major* | MDVLLGGDDGSLAFIPGSFEVAAGEKITFKNN... |

**Rosaceae**

| | |
|---|---|
| *Prunus serrulata splendens* | IEVLLGGDDGSLAFVPNSFSISPGEKIVFKNNAGFPHNIV... |
| *Crataegus monogyna* | IEVLLGSDDGGLAFVP$^{S}_{N}$SFSVAPGEKIVFKNNAGFPHNII... |

**Scrophulariaceae**

| | |
|---|---|
| *Antirrhinum majus* | LDVLLGGDDGSLAF$^{I}_{V}$PGTFEVAAGEKIVFKNNAGFP.... |
| *Digitalis purpurea* | LDVLLGGDDGSLAFIPGSFEVAAGEKITFKNNAGFPHNVV... |
| *Verbascum thapsus* | IEVTLGGDDGSLAFIPQNFEVAAGEKIVFKNNAGFPHNVV.... |

**Solanaceae**

| | |
|---|---|
| *Lycopersicon esculentum* | LEVLLGGDDGSLAFIPGNFSVSAGEKITFKNNAGPF..... |
| *Nicotiana tabacum* | IEVLLGSDDGGLAFVPGNFSVSAGEKITFKNNAGFPHNVV... |
| *Capsicum frutescens* | IEVLLGGDDGSLAFVPGTFSV?SGETITFKNNAGFPHNVV... |

**Fig. 12** (continued)

sequences are compared, 32 residues are invariant. Two particular regions where the sequences are either invariant or highly conserved are residues 31–45 and 82–94; these regions contain both the His residues and the single Cys residue, all of which had been implicated in the copper-binding site by chemical and spectroscopic methods (see BOULTER et al. 1977). These regions of the sequence are also clearly homologous to regions of azurin, another "blue", Type I copper protein (RYDEN and LUNDGREN 1979). Boulter has started to examine this enormous data set to produce trees relating the species (BOULTER et al. 1978, 1979), but progess on such work will be slow because of all the topological arrangements which are possible. Comparisons have shown that plastocyanin is evolving at least twice as fast as cytochrome c. While this makes determining inter-family relationships more difficult, it allows intra-family affinities to be examined, and means that sufficient variation exists in the N-terminal sequence alone for mean-

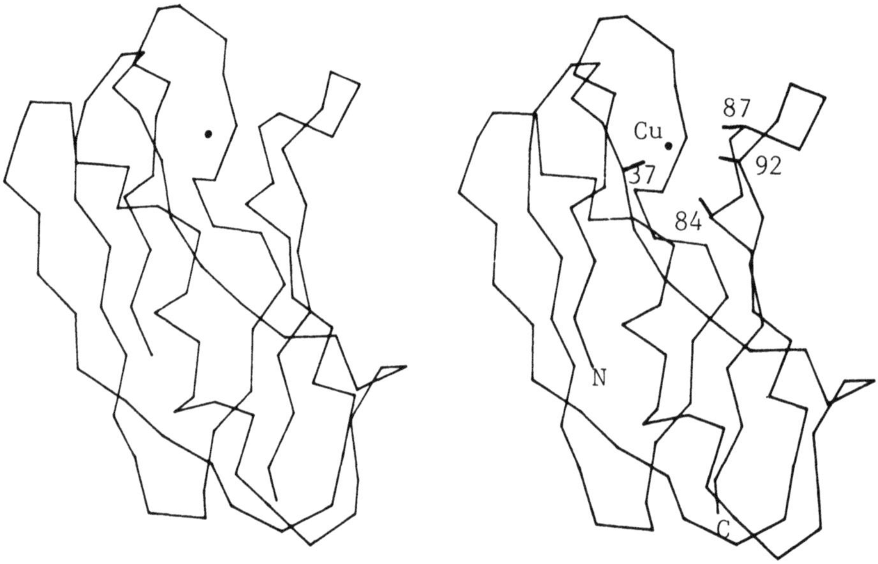


**Fig. 13.** Stereo drawing of the α-carbon positions of the polypeptide chain of plastocyanin. (COLMAN et al. 1978)

ingful comparisons to be made. For example, the affinities of 22 members from eight tribes of the Compositae have been examined in detail (BOULTER et al. 1978). However, as noted by BOULTER et al. (1979), the increased amount of parallel and back mutations which are observed in the plastocyanin data requires more sophisticated methods of data analysis to be developed, and means that, of necessity, the protein data must be examined in conjunction with other biological data.

Crystals of plastocyanin suitable for X-ray diffraction studies have been obtained from poplar (CHAPMAN et al. 1977), maize (*Zea mays*) and pea (*Pisum sativum*) (CHIRGADZE et al. 1977) and a structural analysis to 2.7 Å resolution for poplar plastocyanin in the oxidized, Cu(II) form, has been reported (COLMAN et al. 1978). This has enabled many of the hypotheses regarding the nature of the copper site to be resolved, although further refinement of the structure to the diffraction limit of 1.8 Å will be needed to obtain accurate bond angles and distances for the copper ligands. The identity of the ligands has, however, been unambiguously established. Two are the δ nitrogen atoms from the imidazole rings of histidine residues 37 and 87, and one the S atom from the single Cys, residue 84. The fourth ligand is the thioether sulphur of Met(92). This places the copper site at one end of the molecule (see Fig. 13) about 6 Å from solvent. In the oxidized Cu(II) form, the copper environment is not regular but is significantly distorted from tetrahedral (COLMAN et al. 1978). The folding of the polypeptide chain produces an elongated cylinder shape made from eight strands of chain, seven of which appear to be involved in extensive stretches of β structure (Fig. 13). The core of the molecule is entirely hydrophobic, notably by the clustering of six of the seven Phe residues. On the outside of the molecule the charged residues are not evenly distributed, and there is one large hydrophobic patch; it is possible that this is involved in interactions related to the protein's function (COLMAN et al. 1978). NMR studies on a wide variety of plastocyanins have suggested that this structure is highly conserved between plastocyanins, and that there will not be any major changes in the polypeptide folding between the oxidized and reduced forms of the protein (FREEMAN et al. 1978).

The sequence homologies which had been noted between plastocyanin and the bacterial protein azurin (see BOULTER et al. 1977, RYDEN and LUNDGREN 1979) could be readily fitted to the plastocyanin topology (COLMAN et al. 1978), and completion of the three-dimensional structure of azurin confirmed this structural homology of the two proteins, both in the copper site and the polypeptide folding (see ADMAN 1979). As with many other photosynthetic proteins, a homologous plastocyanin has been obtained from a blue-green bacterium, and its sequence shows great homology to the eukaryotic plastocyanins (AITKEN 1975).

## 3.5 Stellacyanin

Stellacyanin is, like plastocyanin (see Sect. 3.4), a small "blue", Type I, copper protein which contains one Cu atom per molecule. It is obtained from the Japanese lacquer tree, *Rhus vernicifera*, but as yet its specific function is unknown, although it is assumed to be involved in electron transfer. When compared to plastocyanin, there are several major spectral differences and stellacyanin is also a glycoprotein, containing about 30% carbohydrate (see FEE 1975).

The sequence of stellacyanin has been determined (Fig. 14) and shown to be a single chain of 107 residues (BERGMAN et al. 1977; the complete C-terminal is given in RYDEN and LUNDGREN 1979). The carbohydrate moieties are attached to the chain at three different positions, all having the characteristic Asn-X-Thr sequence. The polypeptide chain shows homology to the plastocyanin sequence, particularly around two of the histidine residues, 46 and 92 (RYDEN and LUNDGREN 1979), which may therefore be copper ligands as in plastocyanin. However,

```
1                10                  20                  30                  40
T V Y T V G D S A G W K V P F F G D V D Y D W K W A S N K T F H I G D V L V F K
                                                      |
                                                      carb

41               50                  60                  70                  80
Y D R R F H N V D K V T Q K N Y Q S C N D T T P I A S Y N T G B B R I N L K T V
                                      |
                                      carb

81               90                  100           107
G Q K Y Y I C G V P K H C D L G Q K V H I N V T V R S
                                          |
                                          carb
```

**Fig. 14.** The amino acid sequence of stellacyanin (BERGMAN et al. 1977; the complete C-terminal is given in RYDEN and LUNDGREN 1979). Carbohydrate binding is indicated by *-carb*

stellacyanin contains no Met residues and the single free Cys residue is not homologous to the Cys residue in either plastocyanin or azurin, so the nature of the Cu site is uncertain. Stellacyanin does, however, contain one disulphide bridge, where the individual Cys locations are homologous with the Cys and Met residues in plastocyanin; whether this bridge is present in the native protein, or is formed concomitantly with loss of copper is not clear. This gives an explanation for the Cu binding site, which can account for the spectroscopic properties of the protein and for the two-step binding of copper to apoprotein.

# 4 Toxic Proteins

In grouping the proteins in the following section as toxins, it must be emphasized that there is nothing absolute about toxicity, and that the word "poisonous" frequently means no more than "poisonous to man and/or his domestic animals" (see BELL 1978). Thus in each of the following groups of proteins, some, but not necessarily all, of each group show some level of toxicity to man or his animals, while others which may be very closely related may show no toxicity at all. However, in view of such close relationships these non-toxic proteins have nevertheless been included within this section.

## 4.1 Phytohaemagglutinins (Lectins)

The phytohaemagglutinins are a somewhat heterogeneous collection of proteins which have been grouped together because of their ability to cause the clumping of red blood cells in vitro. Some, but by no means all, are toxic to higher animals, but this toxicity is not necessarily related to their ability to agglutinate red blood cells. The term lectin, which is now frequently used, refers only to the selective character of many of these proteins in their binding specificities, which while of importance in other disciplines, is not essential to their toxic properties. The in vivo function of these proteins is not clear (see RYAN 1973), but it has been suggested that they may have enzymatic properties within the plant cell. For example, the major phytohaemagglutinin from mung bean seeds has been show to possess a strong α-galactosidase activity (HANKINS and SHANNON 1978). However, as a class they do not have a single function.

### 4.1.1 Concanavalin A

Concanavalin A (Con A) is a lectin extracted from the jack bean, *Canavalia ensiformis*, and is the most extensively studied of the plant lectins. The interest in the protein stems from its unusual effects on animal cells. Thus the protein will agglutinate cells transformed by oncogenic viruses, it inhibits growth of malignant cells in experimental animals, and it exhibits mitogenic activity (see WANG et al. 1975). These biological effects are related to its sugar-binding properties. Carbohydrates having the D-arabinopyranoside configuration at the C-3, C-4 and C-6 positions possess the minimum structural characteristics for binding to Con A. In addition to possessing a sugar-binding site, Con A also has two metal-binding sites per monomer. One site, termed S1, binds a variety of divalent transition metal ions, including $Mn^{2+}$, $Ni^{2+}$, $Co^{2+}$, $Cd^{2+}$, $Zn^{2+}$, $Fe^{2+}$ and $Cu^{2+}$. The other, termed S2, binds $Ca^{2+}$ and $Cd^{2+}$ (SHOHAM et al. 1973). The usual metal ions found are $Mn^{2+}$ and $Ca^{2+}$. Normally, S2 is not occupied until S1 is occupied, and occupation of both sites is needed for saccharide binding. However, under limiting conditions other binding effects have been observed (see SHERRY et al. 1978).

Con A exists in two pH-dependent forms, each composed of identical subunits of molecular weight 25,500. The form found at below pH 5.5 is predominantly dimeric, while at pH 7 and above it is tetrameric. Each subunit consists of a single polypeptide chain of 237 amino acids (see Fig. 15), although in some preparations a "nicked" form of the enzyme consisting of an association of two smaller fragments is found (WANG et al. 1975). The sequence (WANG et al. 1975; CUNNINGHAM et al. 1975) shows an unusual distribution of certain amino acid residues. The distribution of charged amino acids is more dense in the N-terminal region of the molecule, and while six of seven Tyr residues are in the N-terminal half of the molecule, all 11 Phe residues follow position 111.

The location of the metal- and saccharide-binding sites has been established by X-ray diffraction studies on Con A, which have provided a structure for the molecule at 2.0 Å resolution (EDELMAN et al. 1972, BECKER et al. 1975, REEKE et al. 1975). The α-carbon backbone of the protein is illustrated in Fig. 16. This shows that the predominant structure element is an extended polypeptide chain arranged in two antiparallel β-sheets, one of these sheets being involved in monomer–dimer association. Residues not involved in the β-structure are in regions of random coil. The $Mn^{2+}$ and $Ca^{2+}$ ions are bound close together, being 4.6 Å apart, and are both surrounded by an approximately octahedral coordination shell containing four ligands from the protein and two water molecules. The protein ligands to the $Mn^{2+}$ are the side chains of Glu(8), Asp(10), Asp(19) and His(24), while those to the $Ca^{2+}$ are the side chains of Asp(10), Asn(14), Asp(19) and Tyr(12). Thus it can be seen that two of the protein ligands, Asp(10) and Asp(19) are shared by both metal ions (BECKER et al. 1975).

The saccharide-binding site is a shallow pocket located about 7 Å from the $Ca^{2+}$ site and 12 Å from the $Mn^{2+}$ site (HARDMAN and AINSWORTH 1976), which is surrounded by residues Asn(14), Asp(16), Asp(208), Tyr(12), Tyr(100), Leu(99) and Arg(228). Of interest is that for a reasonable fit in this region, the peptide bond between Ala(207) and Asp(208) must be in the cis configuration. Also, Glu(102), which is the only charged residue on the interior of the molecule, other than the metal ligands, is adjacent to this site.

When the $Mn^{2+}$ and $Ca^{2+}$ ions normally present in Con A are removed, the protein loses its carbohydrate binding capacity. Comparisons between the native and demetallized proteins reveal a number of differences between the structures (REEKE et al. 1978). Most of the changes are in the metal-binding

```
                                  1                 10                  20
a. Concanavalin A                 A D T I V A V E L D T Y P N T D I G D P

                       21              30                  40                  50
a.                     S Y P H I G I D I K S V R S K K T A K W N M Q D G K V G T A

                       51              60                  70                  80
a.                     H I I Y N S V D K R L S A V V S Y P N A D A T S V S Y D V D
b. Favin                                                   ⎧ L T G Y T L S E V V P
c. Lentil lectin                                           ⎪ V T S Y T L N E V V P
d. Pea lectin                                  α-chains   ⎨ V T S Y T L S D V V S
e. Sweet pea lectin                                        ⎪ V T S Y T L N E V V P
f. Vicia cracca Glc lectin                                 ⎩ V T S Y T L S D V V P


                       81              90                  100                 110
a.                     L N D V L P E W V R V G L S A S T G L Y L E T N T I L S W S
b.                     L K D V V P E W V R I G F S A T T G A E Y A T H E V L S W T
c.                     L K D V V P E W V R I G F S A T T G A E F A A Q E V H S W S
d.                     L K D V V P E W V R I G F S A T T G A E Y A A H E V L S W S
e.                     L K D V V P E W V R I G F S........
f.                     L K D V V P E W V R I G F S..........


                       111             120 122                     123         130
a.                     F T S K L K S N S T H Q—————————————————————T D A L H F M F
b.                     F L S E L T G P S N                       ⎧ T D E I T S F S I
c.                     F N S Q L G H T S K S                     ⎪ T E T T S F S I
d.                     F H S E L S G T S S K Q                   ⎪ T E T T S F L I
e.                                                               ⎪ T E T T S F L I
f.                                                               ⎪ T E T T S F L I
g. Vicia cracca Gal NAc lectin                       β-chains  ⎨ T E S T S F S F
h. Soybean lectin                                                ⎪ A E T V S F S W
i. Peanut lectin                                                 ⎪ A E T V S F N F
j. R-Phytohaemagglutinin                                         ⎪ A S E T S F S F
k. L-Phytohaemagglutinin                                         ⎩ S N D I Y F N F


                       131             140                 150                 160
a.                     N E F S K D Q K D L I L Q G D A T T G T N G N L E L T R V S
b.                     P K F R P D Q P N L I F Q G G G Y T - T K E K L T L T K A -
c.                     T K F S P D Q Q N L I F Q G D G Y............
d.                     T K F S P D Q Q N L I F Q G N G Y............
e.                     T K F S P D Q Q N L I F Q G D G Y.......
f.                     T K F S P D Q Q N L I F Q G D G Y.....
g.                     T E F N Q N Q Q N L I L E G D A T.........
h.                     N K F V P K E P D M I L E G D A I.........
i.                     N S F S E G N P A I N F Q G D V T..................
j.                     E R F N E T - - N L I L Q R D A S.....
k.                     E R F N E T - - N L I L Q R D A S............

                       161             170                 180                 190
a.                     S N G S P E G S S V G R A L F Y A P V H I W E S S A T V S A
b.                     - - V - - K - N T V G R A L Y S L P I H I W D SET G N V A D

                       191             200                 210                 220
a.                     F E A T F A F L I K S P D S H P A D G I A F F I S N I D S S
b.                     F Q T T F I F V I D A PNG Y N V A D G F T F F I A P V D T K

                       221             230           237       1               10
a.                     I P S G S T G R L L G L F P D A N       a d t i v a v e l d
b.                     P Q T G G G Y S W L G V F Y N G K D Y D K T A Q T V A V E F D

                       11              20                  30                  40
a.                     t y p n t d i g d p s y p h i g i d i k s v r s k k t a k w
b.                     T F Y N A A - W D P SNGKR H I G I D V N T I K S I S T K S W

                       41              50                  60                69
a.                     n m q d g k v g t a h i i y n s v d k r l s a v v s y p n
b.                     N L Q N G E E A H V A I S F N A T T N V L S V T L L Y P N
```

**Fig. 15.** Sequence homologies between various plant lectins. *a* Concanavalin A, *Canavalia ensiformis* (CUNNINGHAM et al. 1975); *b* Favin, *Vicia faba* (CUNNINGHAM et al. 1979); *c* Lentil, *Lens culinaris* (FORIERS et al. 1978); *d* Pea, *Pisum sativum* (VAN DRIESSCHE et al. 1976, RICHARDSON et al. 1978); *e* Sweet pea, *Lathyrus odoratus* (KOLBERG et al. 1980); *f, g Vicia cracca* (BAUMAN et al. 1979); *h* Soyabean, *Glycine max; i* Peanut, *Arachis hypogaea; j, k* Kidney bean phytohemagglutinins, *Phaseolus vulgaris* (FORIERS et al. 1977)

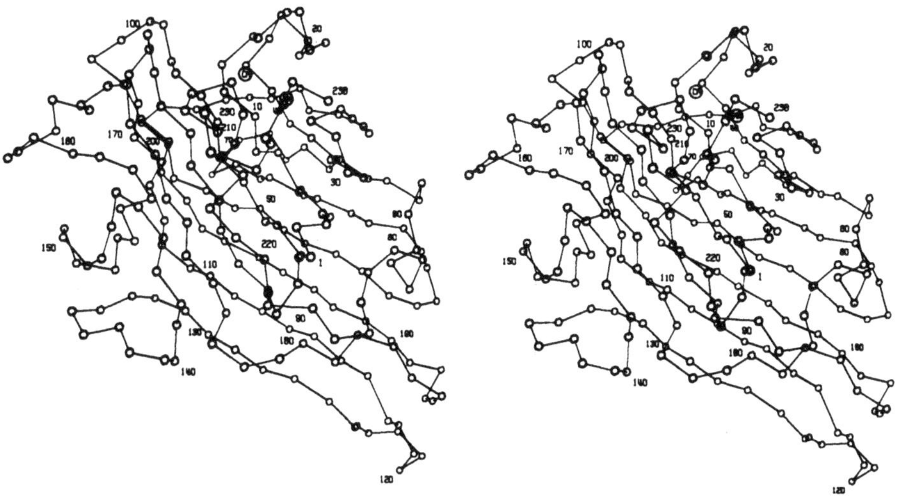

**Fig. 16.** Stereo drawing of the polypeptide chain of Concanavalin A. (EDELMAN et al. 1972)

region, particularly in the loop of polypeptide chain formed by residues 12–22, causing this region to become more accessible to solvent, and the S2 site to become almost totally unrecognizable. Also, REEKE et al. (1978) propose a cis-trans isomerization of the configuration of the peptide bond between Ala(207) and Asp(208) accompanying loss of metals. They postulate that this may be a link between the metal-binding sites and conformational changes in the saccharide-binding site which prevent saccharide binding in the demetallized protein; however, conflicting results have been obtained by SHOHAM et al. (1979).

### 4.1.2 Favin and Related Lectins

Favin is a lectin isolated from fava beans, *Vicia faba,* which resembles Con A in its ability to bind glucose and mannose, but unlike Con A it does not contain any metal-binding sites but does have bound carbohydrate (HEMPERLY et al. 1979). Also, unlike Con A, favin is not a single polypeptide chain, but consists of two chains of molecular weights 5,600 and 20,000, although the total molecular weight is very similar to that of Con A. The amino acid sequences of both chains have been determined (HEMPERLY et al. 1979, CUNNINGHAM et al. 1979) and are shown in Fig. 15. Lectins from a variety of other legumes have been described which are apparently homologous to favin in having a similar two-chain structure and saccharide-binding specificity. Partial structural data is available for several of these lectins and is also shown in Fig. 15. (References are given in the figure legend.) These sequence data show this group of legume lectins to be a homologous group of proteins. Of note, however, is the homology reported by CUNNINGHAM et al. (1979) between this group of two-chain lectins and the single-chain lectin Con A. The small chains are extremely homologous to the 70–122 region of Con A while the larger chains are homologous to the contiguous 123 region of Con A. Comparing the complete sequence of favin with Con A, however, shows that even further homology exists and that the C-terminal region of its larger chain is homologous to the N-terminal region of Con A. This unusual homology, which was noted by CUNNINGHAM et al. (1979), has been termed by these authors a "circular permutation" of sequences. Of further interest is that the naturally occurring break point found occasionally in Con A preparations (WANG et al. 1975) coincides with the break between the two independent favin chains.

Thus the important saccharide-binding site residues which are located in the N-terminal region of Con A, and were thought to be absent in the two-chain legume lectins, may now be expected to be found in the C-terminal region of the larger chains.

Crystals of favin suitable for diffraction studies have been reported (WANG et al. 1974), but the completion of a structural analysis to confirm this hypothesis is still awaited.

### 4.1.3 Wheat Germ Agglutinin

A lectin has been purified from wheat germ which is able to agglutinate a variety of animal cells including both malignant and normally dividing cells. These observed activities are inhibited by N-acetyl-D-glucosamine and its (1→4)-linked oligomers. Under normal physiological conditions the protein is a dimer, but at low pH it dissociates into two identical subunits of molecular weight, 17,000; each protomer binds specific saccharides with equal affinity at two different locations. The molecule is devoid of bound carbohydrate and does not require metal ions for activity. Amino acid analysis shows it to be particularly rich in half-cystine residues; in the intact molecule these are present as 16 disulphide bonds per protomer. (See WRIGHT 1977 for references.)

Although there is no complete amino acid sequence available, the three-dimensional structure of the molecule has been described (WRIGHT 1977). This analysis shows (see Fig. 17) the molecule to consist of a single polypeptide chain of 164 residues. This protomer consists of four distinct domains, each of 41 residues folded in an irregular fashion with very little ordered secondary structure. The folding patterns observed between domains are very similar, but not identical.

Each domain is extensively cross-linked by four disulphide bridges in homologous locations, which account for the great stability of the molecule in solution. This structural homology, and the partial sequence homology derived from the disulphide bond arrangements suggest that the wheat germ agglutinin polypeptide chain has evolved as a result of gene quadruplication followed by divergent changes. While the locations of the disulphides are not in general homologous with those of other plant proteins which are also rich in disulphides (e.g., certain protease inhibitors, Sect. 4.2), the arrangement of the half-cystine residues in the peptide hevein (see DAYHOFF 1972), which is found in the latex from rubber trees suggests a possible homology.

The sugar-binding sites of the agglutinin have also been examined in the crystalline state. The non-covalent complex with N-acetyl-D-neuraminic acid shows two strong binding sites on the agglutinin dimer (WRIGHT 1980a). These

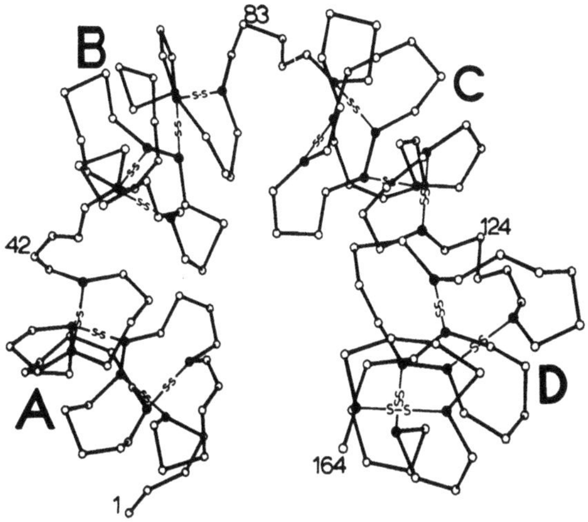


**Fig. 17.** α-Carbon drawing of the polypeptide chain of wheat germ agglutinin. (WRIGHT 1977)

sites are located in crevices at the protomer/protomer interface, with side-chains from B- and C-type domains of opposite protomers contributing to the binding site. In the complex with di-N-acetylglucosamine four binding sites were found (WRIGHT 1980b). In addition to the previous sites, two further sites involving the A- and D-type domains were located, but these were only poorly occupied.

### 4.1.4 Lectins from *Ricinus communis*

It has long been known that extracts of the castor bean seed, *Ricinus communis*, contain in large quantities a highly toxic protein ricin, which also exhibits strong haemagglutinating activity. Recently it was shown that the extracts contain two distinct components, one termed ricin which is toxic and the other an agglutinin, RCA, which is non-toxic. Both these components can be further fractionated to a number of related forms. Both ricin and RCA are glycoproteins consisting of two chains held together by disulphide bonds. It has been shown that only one of the chains in each protein is needed in vitro to cause inactivation of ribosomal activity, and that the other isolated chains each bind 1 mol of D-galactose and are thus responsible for the cell surface-binding activities. In ricin the toxic chain does not show toxic properties in vivo in the separate form, but requires combination with the second chain in order to penetrate the cell cytoplasm (for references see CAWLEY et al. 1978).

CAWLEY et al. (1978) have studied the N-terminal sequences of the individual chains of ricin and RCA. They found that the two pairs of sequences are identical between the two proteins but that the two chains in each protein are different. From gel electrophoretic results, however, they noted that the Ala-chain of ricin was larger than that of RCA and was more basic, presumably as a result of around 30 extra residues, and that it was this difference which was responsible for the toxicity of ricin (CAWLEY et al. 1978).

The complete sequences of both chains of a ricin fraction, ricin D, have recently been determined (YOSHITAKE et al. 1978, FUNATSU et al. 1979). The

```
1                   10                  20                  30                  40
I F P K Q Y P I I N F T T A G A T V Q S Y T N F I R A V R G R L T T G A D V R H
                  |
                  carb

41                  50                  60                  70                  80
E I P V L P N R V G L P I N Q R F I L V E L Q N H A E L S V T L A L S V T N A Y

81                  90                  100                 110                 120
V V G Y R A G N S A Y F F H P D N Q E D A E A I T H L F T D V Q N R Y T F A F G

121                 130                 140                 150                 160
G N Y D R L E Q L A G N L R E N I E L G N G P L E E A I S A L Y Y Y S T G G T Q

161                 170                 180                 190                 200
L P T L A R S F I I C I Q M I S E A A R F Q Y I E G E M R T R I R Y N R R S A P

201                 210                 220                 230                 240
D P S V I T L E N S W G R L S T A I Q E S N Q G A F A S P I Q L Q R D G S K F S

241                 250                 260       265
V Y D V S I L L P I I A M V Y R C A P P P S S Q F
                                ↓
                             Ala-chain

      Ile-chain
1       ↑           10                  20                  30                  40
A D V C M D P E P I V R I V G R N G L C V N V R D G R F N H G N A I Q L W P C K

41                  50                  60                  70                  80
S N T D A N Q L T L K R D N T I R S N G K C L T T Y G Y P S G V Y Y V M I Y D C N

81                  90                  100                 110                 120
T A A T T A D R E I W N N G T I I N P R S S L V L A A T S G N S G T T L T V Q T
                        |
                        carb

121                 130                 140                 150                 160
N I Y A V S Q G P L F T N N T Q P W V T T I V G L Y G L C L Q A N S G Q V V I E
                        |
                        carb

161                 170                 180                 190                 200
D S C S E Y A E Q Q W A L Y A S G N I N P Q Q R R D N C L T S D S N I R E T V V

201                 210                 220                 230                 240
K I L S C G P A S S G E R W M F K N D G T I L N L Y S G L V L D V R A S D P S L

241                 250                 260
K Q I I L Y P L W G H D P N Q L I L P F
```

**Fig. 18.** The amino acid sequence of ricin D (YOSHITAKE et al. 1978, FUNATSU et al. 1979). Carbohydrate is shown as *-carb*. Disulphide bonds link residues 20–39, 62–79, 149–163 and 188–205 within the Ala chain. The interchain bond is shown

Ile-chain which is the toxic component, is a single chain of 265 amino acid residues with carbohydrate bound in one place. It contains two Cys residues, one of which, Cys(257), is involved in an interchain disulphide bond (see Fig. 18). The Ala-chain, which is the galactose-binding component, is a single chain of 260 residues with carbohydrate bound to two places. It contains four intra-chain disulphide bonds and provides Cys(4) to the interchain disulphide bond.

The N-terminal sequence data for two different lectins from the seeds of *Momordica charantia* have been reported (LI 1980). On the basis of this limited sequence data, it is possible that these sequences are homologous with the ricin D the chain, starting from residue 9.

## 4.2 Protease Inhibitors

The occurrence of protease inhibitors in various plant tissues, particularly seeds and tubers, is widespread. However, the physiological function of these inhibitors in plants has been somewhat of a puzzle because no direct evidence for any roles had been established. A few are known to inhibit endogenous proteolytic enzymes, but in most cases inhibition of plant proteases has not been found. Their toxicity arises, however, from the strong inhibition of the animal proteases, particularly those of the digestive system. The inhibitors contain "active sites" for these enzymes which give them their specificities of action. Structural studies on these "active sites" from inhibitors of diverse origin show that these sites appear to have been conserved. This suggests that in part the inhibitory activity may relate to the protein's function and thus survival (see RYAN 1973).

### 4.2.1 Protease Inhibitors from the Leguminoseae

#### *4.2.1.1 Single Site Inhibitors*

Soyabeans contain a variety of protease inhibitors of which two, the Kunitz and Bowman-Birk inhibitors, have been extensively studied. The Kunitz inhibitor consists of a single polypeptide chain (with two disulphide bonds) of molecular weight 21,500 which has inhibitory activity towards trypsin through a single site. Its complete amino acid sequence has been determined (KOIDE and IKENAKA 1973) and consists of 181 residues (Fig. 19). The sequence around the active site was shown to be Arg-Ile-Arg-Phe- and it was shown that the Arg(63)-Ile bond was the site specifically cleaved by trypsin at low pH, while the Arg(65)-Phe bond was unaffected. There is some evidence of repetition of sections of the sequence, although this is not of the simple gene-doubling type (KOIDE and IKENAKA 1973).

Homologous inhibitors have also been isolated from the silktree, *Albizzia julibrissin* (ODANI et al. 1979) and partial sequence data (Fig. 19) showing substantial homology to the soyabean inhibitor has been reported (ODANI et al. 1979, 1980). Whereas soyabean inhibitor is essentially mono-specific to bovine trypsin, the silktree inhibitor A-II inhibits both bovine trypsin and chymotrypsin, while inhibitor B-II inhibits bovine chymotrypsin and porcine elastase (ODANI et al. 1979). The lack of tryptic inhibition by B-II is explained by substitutions to the active site residues 63–64.

The interaction of the Kunitz inhibitor and trypsin has also been studied by X-ray diffraction methods, and SWEET et al. (1974) have reported the crystal structure of the complex of porcine trypsin with this inhibitor at 2.6 Å resolution. These results show that the inhibitor (see Fig. 20) is an approximately spherical

| | | |
|---|---|---|
| | | 1 10 20 30 |
| a. | *Glycine max* | D F V L D N E G N P L E N G G T Y Y I L S D I T A F G G I R |
| b. | *Albizzia julibrissin* AII | K E L L D A D G D I L L N G G ? Y Y I V..... |
| c. | *Albizzia julibrissin* BII | S N L L L D T D G N L L G D G G ? Y Y I L P A.... |
| d. | *Oryza sativa* | A P P P V Y B T Z G H G L S A B G S Y Y V L P A S P G H G G G L |
| | | 31 40 50 60 |
| a. | | A A P T G N E R C P L T V V Q S R N E L D K G I G T I I S P |
| b. | | .......M I W T P |
| c. | | .......R L A S P |
| d. | | T M....... |
| | | 61 70 80 90 |
| a. | | S Y R I R F I A E G H P L S L K F D S F A V I M L C V G I P |
| b. | | S - R I A S L T P A F Y L N ? E F........ |
| c. | | V - L M L T L S P G F P L T I E F........ |
| | | 91 100 110 120 |
| a. | | T E W S V V E D L P E G P A V K I G E N K D A M D G W F R L |
| | | 121 130 140 150 |
| a. | | E R V S D D E F N N Y K L V F C P Q Q A E D D K C G D I G I |
| b. | | .....K D D H C K D L G - |
| c. | | ......N R D C K D L G I |
| | | 151 160 170 180 |
| a. | | S I D D D G H T R R L V V S K N K P L V V Q F Q K L D K E S |
| b. | | S I D D D E N........ |
| c. | | S T D D D N....... |
| | | 181 |
| a. | | L |

**Fig. 19.** The amino acid sequences of "Kunitz-type" trypsin inhibitors. *a* Soyabean, *Glycine max* (KOIDE and IKENAKA 1973); *b, c* Silktree, *Albizzia julibrissin* (ODANI et al. 1979, 1980); *d* Rice, *Oryza sativa* (KATO et al. 1972)

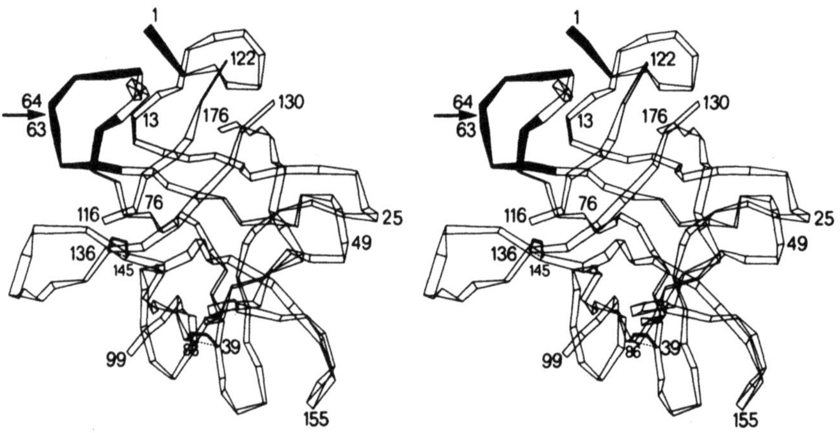


**Fig. 20.** Stereo drawing of the polypeptide chain of soyabean trypsin inhibitor (Kunitz). (SWEET et al. 1974)

molecule which contains no α-helical structure and only some irregular β-pleated sheet structure. The high resolution map of the complex showed that only about 12 of the 181 amino acid residues of the inhibitor make contact with the trypsin molecule. Most of these contacts involve the sequence Ser-Tyr-Arg(63)-Ile-Arg-Phe, which forms a curved loop protruding from the main body of the inhibitor molecule. The side chain of Arg(63) occupied the expected position in the primary specificity pocket of the trypsin.

#### *4.2.1.2 Double Site Inhibitors*

Apart from the Kunitz inhibitor, the other major inhibitor from soyabeans which has been extensively studied is the Bowman-Birk inhibitor. This protein is a single chain of molecular weight about 8,000 which is extensively cross-linked by seven disulphide bonds. It is believed that this leads to a very rigid structure which would explain the unusual stability against acid, proteolytic digestion and heat. The complete sequence has been determined and shown to have 71 residues (Odani and Ikenaka 1972) (Fig. 21) and with a disulphide bond arrangement as follows:

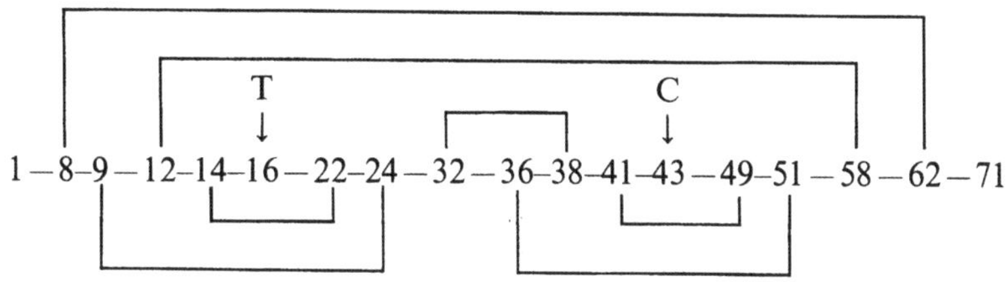


The Bowman-Birk inhibitor differs from the Kunitz inhibitor in showing strong activity against both trypsin and chymotrypsin. Further studies revealed that these activities are located at two completely independent sites. The site for trypsin is Lys-Ser and that for chymotrypsin Leu-Ser (see Odani and Ikenaka 1972). Modification of the chymotryptic site by replacement of the Ser by other amino acids showed that the Ser form was the most active. The Ala, Thr, Val and Leu forms were progressively less active in this order, while a Gly form was essentially inactive (Odani and Ikenaka 1978b).

Inspection of the sequence (Fig. 21) and of the disulphide bond arrangement clearly shows internal homology within the sequence which has been caused by a gene duplication. It is this duplication which has lead to the double-site nature of these inhibitors, and the two active sites are conserved in position between the halves of the molecule.

A wide variety of other homologous double-site inhibitors have been obtained both from soyabeans and from other legumes, and many have had their sequences determined (for references, see legend to Fig. 21). The sequences of these inhibitors (Fig. 21) form a structurally homologous group with the Bowman-Birk inhibitor. Studies on the active sites of these inhibitors show that they are in identical locations in all of them. However, there is variation in the nature of the active site residues and this correlates well with the differing specificites to other enzymes e.g., elastase, which are sometimes found. Hetero-

```
                                            1                 10                  20
a. Glycine max            Bowman-Birk                          D D E S S K P C C D
b.                               C-II       S D H - - S S S - - D D E S S K P C C D
c.                         D-II & E-I       S D Q - - S S S Y D D D E Y S K P C C D
d. Phaseolus lunatus          IV & IV'      S G H H E H S T D Z P S Z S S K P C C B
e. Phaseolus vulgaris         II & II'      (B,H,H,Z,S,S,B,B)Z P S Z S S P P C C B
f. Phaseolus angularis               I      S G H H D E T T D E P S E S S K P C C D
g.                            II & II'      S G H H D E T T D E P S E S S K P C C D
h. Macrotyloma axillare           DE-3        D H H - H S T D E P S E S S K P C C D
i.                                DE-4            H E H S S D E S S S E S S K P C C D
j. Vicia angustifolia                                         G D D V K S A - C C D
k. Cicer arietinum                                                S T T T A - C C D
l. Arachis hypogaea                                             C T B K T Z G R C P

                            21                30                  40                  50
a.                          Q C A C T K S N P P Q C R C S D M R L N S C H S A C K S C I
b.                          L C M C T A S M P P Q C H C A D I R L N S C H S A C D R C A
c.                          L C M C T R S M P P Q C S C E D I R L N S C H S D C K S C M
d.                          H C A C T K S I P P Q C R C T D L R L D S C H S A C K S C I
e.                          I C V C T A S I P P Q C V C T B I R L B S C H S A C K S C M
f.                          Q C S C T K S M P P K C R C S D I R L N S C H S A C K S C A
g.                          Q C - C T K S M P P K C R C S D I R L D S C H S A C K S C A
h.                          E C A C T K S I P P Q C R C T D V R L N S C H S A C S S C V
i.                          L C T C T K S I P P Q C H C N D M R L N S C H S A C K S C I
j.                          T C L C T R S Q P P T................
k.                          S C V C T K S I P P Q C R C N D M..............
l.                          V T Z C - R S B P P Z C R - - - - - - - - - - - - - - - - -

                            51                60                  70                  80
a.                          C A L S Y P A Q C E C V D I T D F C Y E P C K P S - - E D D K E N
b.                          C T R S M P G Q C R C L D T T D F C Y K P C K S S D E D D D
c.                          C T R S Q P G Q C R C L D T N D F C Y K P C K - - - S R D D
d.                          C T L S I P A Q C V C B B I B D F C Y E P C K S S H S D D D N N N
e.                          C T R S M P G K C R C L B T T B Y C Y K S C K S B S G Z B B
f.                          C T Y S I P A K C F C T D I N D F C Y E P C K S S R - D D D W D N
g.                          C T Y S I P A K C F C T D I N D F C Y E P C K S S R - D D D W D N
h.                          C T F S I P A Q C V C V D M K D F C Y A P C K A S H D D
i.                          C A L S E P A Q C F C V D T T D F C Y K S C H N N A E K D
j.                                                ................K E E V I K N
k.
l.                          - - A P P Y F Z C V C V B T F B H C P A S C B S C C T R
```

**Fig. 21.** Sequence homologies between various double-ended protease inhibitors. *a–c* Soyabean, *Glycine max,* Bowman-Birk (ODANI and IKENAKA 1972); *C–II* (ODANI and IKENAKA 1977); *D–II* and *E–I* (ODANI and IKENAKA 1978a); *d* Lima Bean, *Phaseolus lunatus IV, IV′* (STEVENS et al. 1974); *e* Garden bean, *P. vulgaris* (WILSON and LASKOWSKI 1975); *f, g* Adzuki bean, *P. angularis I* (ISHIKAWA et al. 1969); *II, II′* (YOSHIKAWA et al. 1979); *h, i Macrotyloma axillare; DE–3, DE–4* (JOUBERT et al. 1979), *j Vicia angustifolia* (ABE et al. 1979); *k* Chick pea, *Cicer arietinum* (BELEW and EAKER 1976); *l* Peanut, *Arachis hypogaea* (HOCHSTRASSER et al. 1970)

geneity has been observed in some of the sequences, but this possibly reflects mixtures of closely related but distinct components within the plant, as a variety of very similar inhibitors have been obtained from soyabean.

The inhibitor from the peanut, *Arachis hypogaea,* is a single-site inhibitor. Nevertheless, it is closely related to the double-site inhibitors, but lacks any evidence of a gene duplication having occurred. It may thus represent an ancestral form of this group of protease inhibitors.

### 4.2.2 Protease Inhibitors from the Solanaceae

Plants of the Solanaceae have been shown to be a rich source of many inhibitors of proteolytic enzymes (see RYAN 1973). Several of these inhibitors, particularly

```
   1                 10                  20                  30                  40
(K)E F E C N G K L Q W P E L I G V P T K L A K E I I E K Q N S L I S N V H I L L
             K       S                           G                     T     Q   K
             D

   41                50                  60                  70
   N G S P V T M D F R C N R V R L F D D I L G S V V Q I P R V A
               L   Y                   N       Y       D L   V L G
                   L                           N
```

**Fig. 22.** Composite sequence of chymotryptic inhibitor I, variants A–D, from potatoes. (RICHARDSON 1974, RICHARDSON and COSSINS 1974)

those from potatoes, *Solanum tuberosum*, have been studied in detail and structural information has been determined.

Chymotryptic inhibitor I from potatoes is a tetrameric protein composed of a heterogeneous mixture of protomers. Using ion-exchange chromatography, MELVILLE and RYAN (1972) were able to separate the subunits into four main types, A–D. The amino acid sequences of all these four types have been reported (RICHARDSON 1974, RICHARDSON and COSSINS 1974) and shown to be closely related sequences of 71–72 residues, with no bound carbohydrate (Fig. 22). Limited hydrolysis by catalytic amounts of enzymes at low pH has shown that the major reactive site of the inhibitor for chymotrypsin was the Met/Leu(47)-Asp(48) peptide bond (RICHARDSON et al. 1977) and that the protein also contained weak anti-trypsin reactive sites. The reported sequences all showed a high level of heterogeneity. This most probably arises from there being more than four components present, as isoelectric focusing in polyacrylamide gels has allowed as many as ten different components to be resolved (RICHARDSON et al. 1976). Further studies are thus needed to determine which residues in the positions showing heterogeneity belong to each specific component. Despite this, the sequence data which are available show no homology to any of the inhibitors from other sources.

The inhibitor II fraction from potatoes can be separated into two fractions. One of these, IIa, is a potent inhibitor of trypsin while the other fraction, IIb, is a potent inhibitor of chymotrypsin. Despite these specificity differences, both have similar molecular weights, 10,500, both are single-headed in action and both share the same active site bond, Lys-Ser. The sequences of large active site fragments from both these inhibitors have been determined (IWASAKI et al. 1976, 1977) and are shown in Fig. 23. While these fragments do not show any homology to the complete sequence of chymotryptic inhibitor I, they are homologous to each other and to the N-terminal region of another small chymotryptic inhibitor from potatoes (HASS et al. 1976b) as indicated in Fig. 23.

One of the more unusual inhibitors which has been studied from potatoes is the inhibitor to carboxypeptidases A and B. It is a mixture of two polypeptide chains which differ only in their N-terminal sequences, one having an extra Gln residue; both have pyrrolidone carboxylic acid (Pca) as their N-terminal (HASS et al. 1975) (see Fig. 23). The sequence contains three disulphide bonds which link residues 8–24, 12–27 and 18–34 (LEARY et al. 1979). Despite having completely different inhibitory activity, this CPA inhibitor shows some homology to the small chymotrypsin inhibitor (HASS et al. 1976b) (see Fig. 23).

```
                                             1 L        C           10                      20
a. Aubergine chymotryptic                   Pca I  C T N N C A G R K G C S Y F S E D G T F -
   inhibitor
                              1                 10                    20
b. Fragment, potato           S E G S P E N R I C T N N C A G Y K G C N Y N C D T N - - -
   inhibitor IIa
                                              1                10                      20
c. Fragment, potato                      .....I C T N N C A G Y K G C N Y A C P L N D H P
   inhibitor IIb
                                1                  10                    20
d. Potato carboxy-             Pca Q H A D P I C N K P C K T H D D C S G A W F C Q A C W
   peptidase inhibitor
                                               1                10                18
e. Potato chymotrypsin                         P I C T N K C A G Y K G C N Y Y S A........
   inhibitor

                                                    30                      40                      50
a.                            - - - - - - I C K G E S N P E N P K A C P R N C D G R I A Y G I C P L S

                                30                     40                    45
b.                            I A S Y K S V C E G E F D P K - - - - C L R.....

                                              30                     40
c.                            I A - Y K S - C E G E F D P K S K - - C P R........

                               30            39
d.                            N S A R T C G P Y V G
```

**Fig. 23.** Homologies among various low molecular weight inhibitors from the Solanaceae. *a* Aubergine (*Solanum melongea*) chymotryptic inhibitor (RICHARDSON 1979); *b* Active fragment, potato inhibitor IIa (IWASAKI et al. 1976); *c* Active fragment, potato inhibitor IIb (IWASAKI et al. 1977); *d* Potato carboxypeptidase A inhibitor (HASS et al. 1975, LEARY et al. 1979); *e* Potato chymotrysin inhibitor (HASS et al. 1976b)

A variety of chemical modification studies have been used to probe the mode of action of the inhibitor (HASS et al. 1976a). Modification of the side chains of Asp residues 5, 16 and 17, of Lys residues 10 and 13, of His residues 3 and 15 and of Arg(32), and removal of the N-terminal 5 residues all had little effect on the inhibitor activity. However, modification of the α-carboxylate group of Gly(39) led to inactivation and binding to the enzyme protected Tyr(37) from modification. This strongly suggests a tight complex involving the C-terminal of the inhibitor as its active site.

The three-dimensional structure at 2.5 Å resolution of the complex between the inhibitor and carboxypeptidase A has been reported (REES and LIPSCOMB 1980), and in general confirms the conclusions of the chemical modification studies. The entire polypeptide chain can be traced and this clearly shows that the carboxy-terminal of the inhibitor is bound to the active site of the carboxypeptidase. Because of a sharp turn at Pro(36) only residues 37–39 are within the active site. A distinct gap however was observed between Val(38) and Gly(39) and it was suggested that this gap is caused by hydrolysis and subsequent entrapment of Gly(39) in the active site pocket.

Although the inhibitor and wheat germ agglutinin (WRIGHT 1977) share the same pattern of cysteine pairings, the two structures are distinguished by an unusual knot in the inhibitor structure in which the disulphide bond between residues Cys(18) and Cys(34) passes through a loop generated by the other two disulphide bonds (REES and LIPSCOMB 1980).

The complete sequence of the major trypsin inhibitor from aubergine, *Solanum melongena,* has been determined (RICHARDSON 1979). It consists of a single

```
1                 10                    20                    30                    40
S A G T S C V P P(P,S,B,G,C,H)A I L R T G I P G R L P P L Z K T C G I G P R Q V

41                50                    60          65
Z R L Q B L P C P G R R Q L A B M I A Y C P R C R
```

**Fig. 24.** The sequence of a trypsin inhibitor from maize. (HOCHSTRASSER et al. 1970b)

```
                     1                   10                    20                    30
Chain A-1            D E Y K C Y C A D T Y S D C P G F C K K C K A E F G K Y I C

                     31                  41
                     L D L I S P N D C V K

                     1                   11
Chain B-2            (T)A C S E C V C P L Q
```

**Fig. 25.** The sequence of a bromelain inhibitor, Fraction VII, from pineapple stem. (REDDY et al. 1975)

polypeptide chain of 52 residues with the N-terminal blocked as pyrrolidone carboxylic acid (Pca). While this inhibitor shows some homology to the potato inhibitors (Fig. 23) particularly at its N-terminal region, this homology does not extend to the active site which is Arg(38)-Asn(39) in the aubergine inhibitor.

### 4.2.3 Protease Inhibitors from Other Sources

The amino acid sequence of a trypsin inhibitor from maize, *Zea mays,* has been reported (HOCHSTRASSER et al. 1970a). It consists of a single polypeptide chain of 65 residues which forms a trimer in the native state (Fig. 24). Arg(25) is the site of interaction with trypsin, which breaks the chain at this point to give two chains which are linked by disulphide bonds.

A family of inhibitors to the plant enzyme bromelain have been isolated and characterized from pineapple stem, the source of the enzyme itself. All the inhibitors have molecular weights of about 5,600 and contain five disulphide bonds. The primary structure of one of these components, Fraction VII, has been determined (REDDY et al. 1975) and shown to consist of two peptide chains joined by disulphide bonds (Fig. 25). The location of the Cys residues is homologous to those in the double-site legume inhibitors (Fig. 21) which suggests that the bromelain inhibitor may be derived from a single-chain precursor.

An inhibitor from a monocotyledonous plant, rice, *Oryza sativa* has been reported and the N-terminal sequence of this protein (Fig. 19) suggests that it is homologous to the legume single-site inhibitors (KATO et al. 1972).

## 4.3 Other Toxic Proteins

### 4.3.1 Viscotoxin

Viscotoxin is a mixture of small, basic proteins isolated from the European mistletoe, *Viscum album,* which are pharmacologically active for heart muscle.

```
Purothionins

                                     1                 10                20

a. Triticum aestivum   -I            K S C C R S T L G R N C Y N L C R A R G
b. Triticum aestivum   -II           K S C C R T T L G R N C Y N L C R S R G
c. Triticum aestivum                 K S C C K S T L G R N C Y N L C R A R G
d. Hordeum vulgare                   K S C C R S T L G R N C Y N L C R V R G

Viscotoxins

e. Viscum album      A3              K S C C P N T T G R N I Y N A C R L T G
f. Viscum album      B               K S C C P N T T G R N I Y N T C R L G G
g. Viscum album      A2              K S C C P N T T G R N I Y N T C R F G G
h. Viscum album      1-PS            K(S.C.C.P.B.T.T.G)R.B(I.Y.B.T.C)R.F(G.G.

Phoratoxin

i. Phoradendron tomentosum           K S C C P T T T A R N I Y N T C R F G G

                       21             30                  40         46

a.                     A Q K - L C A G V C R C K I S S G L S C P K G F P K
b.                     A Q K - L C S T V C R C K L T S G L S C P K G F P K
c.                     A Q K - L C A N V C R C K L T S G L S C P K D F P K
d.                     A Q K - L C A G V C R C K L T S S G K C P T G F P K

e.                     A P R P T C A K L S G C K I I S G S T C P S Y P D K
f.                     G S R E R C A S L S G C K I I S A S T C P S Y P D K
g.                     G S R E V C A S L S G C K I I S A S T C P S Y P D K
h.                     G.S)R Z(V.C.A)R.I(S.G.C)K.I(I.S.A.S.T.C.P.S.Y.P.B)K

i.                     G S R P V C A K L S G C K I I S G T K C D S G W N H
```

**Fig. 26.** Sequence homologies between various toxins from plant sources. Purothionins (MAK and JONES 1976, JONES and MAK 1977, OHTANI et al. 1977, OZAKI et al. 1980); Viscotoxins (SAMUELSSON et al. 1968, SAMUELSSON and PETTERSSON 1971, OLSON and SAMUELSSON 1972, SAMUELSSON and JAYAWARDENE 1974); Phoratoxin (MELLSTRAND and SAMUELSSON 1974)

The mixture can be separated into three main components, A2, A3 and B by ion-exchange chromatography. The amino acid sequences of these components have been determined (SAMUELSSON et al. 1968, SAMUELSSON and PETTERSSON 1971, OLSON and SAMUELSSON 1972) and shown to be closely related to the sequence of toxin I-PS (from a subspecies of *Viscum album* for which *Pinus silvestris* was the host plant) and to phoratoxin from the Californian mistletoe, *Phoradendron tomentosum* (SAMUELSSON and JAYAWARDENE 1974, MELLSTRAND and SAMUELSSON 1974). These sequences all have 46 residues and form a homologous group of sequences of which 27 of the residues are common to all five sequences (see Fig. 26). There are six Cys residues present as three disulphide bonds in all the sequences.

### 4.3.2 Purothionins

Purothionins are a group of low molecular weight proteins which are extracted from wheat flour as lipoprotein complexes, and are found in hexaploid, tetraploid and diploid species of the genus *Triticum*. The in vivo function of the purothionins is not known, but the proteins are toxic to animals when injected, and to bacteria and yeasts. The protein binds to the cell membrane in yeasts, causing functional changes in the cytoplasmic membranes and consequent inhibition of sugar incorporation (see MAK and JONES 1976 and OHTANI et al. 1977). The purothionins can be separated into three distinct components (JONES and MAK 1977) and the amino acid sequences of these components have been reported (MAK and JONES 1976, JONES and MAK 1977, OHTANI et al. 1977). All sequences consist of a single chain of 45 residues with four disulphide bonds (see Fig. 26), and are very similar, differing by only five to six residues.

A purothionin homologue from barley flour has also been isolated and its amino acid sequence determined (OZAKI et al. 1980). This protein, which shows similar toxic activities as the wheat proteins, also consists of 45 residues including 8 Cys residues linked as disulphide bonds. The sequence shows a high degree of homology with the wheat purothionins (see Fig. 26) and it is assumed that the disulphide bond arrangement 3–39, 4–31, 12–29 and 16–25 is common to all purothionins (OZAKI et al. 1980).

Of particular interest is the homology which exists between the purothionins and the viscotoxins (see Fig. 26). Although the wheat and mistletoe families are phylogenetically distant, the structural similarity of these toxins is clear and may be correlated to a functional similarity, although their in vivo function is as yet uncertain.

## 5 Other Plant Proteins

### 5.1 Seed Storage Proteins

Seed proteins form an important component of animal and human nutrition. The seeds of legumes are of particular interest as the storage globulins, which make up the bulk of the protein of these seeds, contain adequate levels of most essential amino acids, although they have a deficiency in sulphur-containing amino acids. However, the lower quantities of nitrogenous fertilizer required by legumes as a result of the symbiotic nitrogen fixation in the root nodules render legumes particularly important in the demand for protein.

Most of the structural work on storage proteins has been concerned with the determination of amino acid compositions, molecular size and quaternary structure (e.g., see the review by DERBYSHIRE et al. 1976). Nevertheless some complete sequence data is available for storage proteins.

The sequence of the smaller subunit of conglutin from *Lupinus angustifolius* has been determined (ELLEMAN 1977) and found to be a single homogeneous polypeptide chain of 154 amino acid residues (Fig. 27a). The presence of five sulphur-containing amino acids is much greater in proportion than that found in the legumin-like and vicillin-like globulins from lupins and other legumes.

The 11S globulin of cotton seeds consists of three types of subunits. The amino acid sequence of one of these, the carbohydrate-containing subunit C,

```
    1                  10                  20                  30                  40
A.  S T S Y M G S G E I G G A L I T T T H P Y T V L S H S I P E V F T Q V F A N N M

    41                 50                  60                  70                  80
    P K Q A Q V K A V G P F G L C Y D S R K I S G G A P S V D L I L D K N D A V W R

    81                 90                  100                 110                 120
    I S S E N F M V Q A Q D G V S C L G F V D G G V M A R A G I A L G A H H L E E N

    121                130                 140                 150     154
    L B B F D L E R S R V G F N S N S L K S Y G L T C S N L F D L N N P

    1                  10                  20                  30                  40
B.  H N Q W E E H G N N F R G D A E E L V I B S T P R V Q G N Q R L I S F V A B Z R

    41                 50                  60                  70                  80
    V T H K D Q R G Q E S R Q I N G F L E H E N R Z A G V T E A B G L Z Z T F S Z R

    81                 90                  100                 110                 120
    Q F H Q N R K F I Z E N R I P Q A S A R Q N P Q N Q V L Q R Q T F Q S H Q N R Q

    121                130         136
    E G D I V A L G E Q D R S Q Q N

    1                  10                  20                  30                  40
C.  (Z,V,S,S)K K P E G V N T G A G D R H N L K T E W P E L V G K S V E E A K K V I L

    41                 50                  60                  70                  80
    Q D K P E A Q I I V L P V G T I V T M E Y R I D R V R L F V D K L D N I A Q V P R V G
```

**Fig. 27 A–C.** The amino acid sequences of plant seed proteins. **A** Lupin, *Lupinus angustifolius*, conglutin-$\gamma$ (ELLEMAN 1977); **B** Cotton, *Gossypium barbadense*, 11S globulin subunit C (ASATOV et al. 1978); **C** Barley, *Hordeum vulgare* var Hyproly, component SPIIA (SVENDSON et al. 1980)

has been determined (ASATOV et al. 1978). It consists of a single chain of 135 amino acid residues (Fig. 27b) with a single attachment site for carbohydrate at position 21, and lacks any sulphur-containing amino acids.

Two lysine-rich components isolated from a "Hiproly" barley variety have been subject to sequence determination (SVENDSEN et al. 1980). The sequences of both the components, SPIIA and SPIIB were identical (Fig. 27c), except that component SPIIA was longer by 11 residues at the N-terminal. These components are present in minor quantities in various barley lines, but only in "Hiproly" are they present in high quantities. Since they are characterized by a high lysine content, they are in part responsible for the high lysine content in this variety.

The N-terminal sequence data has been presented for the basic subunits of the 11/12S storage proteins of *Vicia faba*, legumin, and *Glycine max*, glycinin (GILROY et al. 1979). DERBYSHIRE et al. (1976) had suggested that the 11/12S protein, legumin, is widely distributed in the Leguminoseae. However, immunological evidence suggested that this protein was absent in members of the tribe Glycineae. Inspection of the limited sequence data, however, shows considerable similarity between the N-terminals of the two proteins despite the high degree of heterogeneity present in both.

```
                            1                 10                  20
Vicia faba (legumin)        G L E E T V C T V K L R L N I G(Q)P(A)R P D L Y N P(Q)A G....
                                      I     A       E     A         A

Glycine max (glycinin-      G V D E N I C T L K L R E N I G Q P(S)R P D L Y N P(Q)A G....
       basic subunits)        I E   T       M R   H H     A R       A
```

The N-terminal residues of the seed globulin of a *Cucurbita* species have also been reported, G-L-E-D-G-T-I ..., and suggest a wider distribution of this basic subunit amongst Dicotyledonae (HARA et al. 1978).

The other seed proteins to be studied are mainly those from wheat, *Triticum aestivum*. The N-terminal sequences of two wheat albumin components have been determined (REDMAN 1976, PETRUCCI et al. 1978).

Albumin (0.19) S ? Z W – M C Y P G Q A F V V P A L P A C R P ...

Albumin (0.28) S G P W S W C D P A T G Y K V S A L T G C R A ...

These components are reported to be strong inhibitors of the mid-gut α-amylase of yellow mealworm larvae, *Tenebrio molitor,* but inactive against other amylases (see REDMAN 1975).

The gliadins of common bread wheat, which make up the major storage protein fraction of grain endosperm, have also been extensively studied, particularly because of their possible involvement in coeliac disease. The gliadins comprise a complex mixture of more than 40 proteins, all sharing similar amino acid compositions and properties. This mixture is normally separated into four fractions, α, β, γ, ω, each containing many components. The question of which component(s) is responsible for the toxic effects in coeliac disease remains controversial (see JOS et al. 1977). However, it has been shown that the α-gliadin fraction was not the unique toxic component and that β- and, to a lesser degree, γ-gliadins also exhibited damaging effects; only the ω-gliadins appear to be completely harmless (JOS et al. 1977). The N-terminal sequences of certain gliadin (KASARDA et al. 1974, PATEY et al. 1975, BIETZ et al. 1977, SHEWRY et al. 1980), and also hordein (SHEWRY et al. 1980a, SCHMIDT and SVENDSEN 1980) components have been reported.

Comparisons between the various gliadin components show a high degree of homology, suggesting only one or two common ancestral genes for all the components. In addition to this sequence data, some limited N-terminal sequences have been reported for unfractionated protein mixtures from various *Triticum, Aegilops* and *Secale* species (AUTRAN et al. 1979).

The structures of certain tryptic peptides from edestin have also been described (DLOUHA et al. 1964), but none shows homology to the other storage protein sequence data.

While several plant storage proteins have been crystallized and had diffraction patterns recorded, in only one case, canavalin, has a study been continued to produce detailed structural information (MCPHERSON 1980). Canavalin forms the classical vicilin fraction of the seeds of the jack bean, *Canavalia ensiformis*. The crystal structure at 3.0 Å resolution clearly shows the subunit structure of the molecule and allows the path of complete polypeptide chain to be tentatively established. It appears that the canavalin subunit is composed almost entirely of $\beta$ structure with virtually no $\alpha$ helix present, which is in agreement with ORD measurements. The canavalin molecule contains one $Zn^{2+}$ ion per hexamer and this was shown to occupy either of two quasi-equivalent positions. Based on the tentative chain assignments, the molecule appears to exhibit several features of structural redundancy and internal symmetry.

### 5.2 Histones

The histones, which are the small basic proteins found in association with DNA, were the first chromosomal proteins to be identified. They are major general structural proteins of chromatin and may act as repressors of template activity. The histones are coded for by repetitous genes in the DNA of all eukaryotic organisms so far studied, although the numbers of copies varies from species to species (see ELGIN and WEINTRAUB 1975). The chromatin of higher eukaryotes contains five main types of histone. Comparisons of the histones from various animals have shown the proteins to have maintained a very stable structure throughout evolution and studies of plant histones H3 and H4 have shown them to be virtually identical to the corresponding animal histones. On the other hand, plants also contain histone fractions which appear to differ substantially from animal histones H1, H2A, H2B. Sequence studies have revealed however that plant histones 2 represent a family of variants of histones H2A and that plant histone 1 represents histone H2B (RODRIGUES et al. 1979).

The first plant histone to have its sequence determined was histone H4 from pea, *Pisum sativum,* seedlings (DELANGE et al. 1969b). It was shown to consist of a single polypeptide chain of 102 residues (Fig. 28) and to be virtually identical to calf thymus histone H4 (DELANGE et al. 1969a), there only being four differences between them. Two of these were genetic differences involving substitutions at positions 60 (Ile in pea, Val in calf) and 77 (Arg in pea, Lys in calf) while the other differences both involved modifications of the proteins after synthesis. Thus residues 8 and 16 were 6% acetylated in total, although it was expected, however, that histone samples obtained at different stages of growth and development might display varying degrees of acetylation and methylation (DELANGE et al. 1969b).

The extreme evolutionary conservation of structure found for histone H4 was also found for histone H3. Determination of the sequence of pea embryo histone (PATTHY et al. 1973) showed it to be a single polypeptide chain of 135 residues (Fig. 28) which differed from calf thymus histone at only four positions. Of interest was position 96, which showed heterogenity in the pea histone, with 60% Ala and 40% Ser being found. This gives definitive evidence for two forms of this histone in this species. As with histone H4, modification of the protein occurs with residues 9 and 27 being methylated; no acetylated lysines were detected however (PATTHY et al. 1973). Subsequently, the N-terminal 46 residues of histone H3 from cycad, *Encephalartos caffer,* pollen were determined (BRANDT et al. 1974) and found to be identical with the pea protein.

A.
1 10 22 30 40
? T (D/E) V ? E T P ? P L.............Y A E M (I/V) S E A (I/V) T S L K E R T G S S
41 50 60 84 90
Z (Y/F) A I A K (F/Y) V E D K H K D K L P ? D F.......(Y/F) K L P ? ? T ? X ? ? ? K.........

B.
1 10 20 30
a. H2A(1) A G R K G G D - R K K A V T R S V K A G L
b. H2A(2) Ac-M D G S K L K K V A A K K F G G P R K K S V T K S I K A G L
c. H2A(3) Ac-M D A S K A K K V A G K K F G G P R K K S V T R S I K A G L

31 40 50 60
a. Q F P V G R I G R Y L K K G R Y A Q R V G S G A P V Y L A A
b. Q F P V G R I..............
c. Q F P V G R I.............

61 69
a. V Y L A A V L E Y.............

C.
Ac...........T Y K I Y I F K V L K Q V H P D I G I S S K A M S I M N S
F I N D I F E K L A G E S A K L A R Y ? ? K P T I T S R E I
Q T S V R L V L G P E A K ? ? V S E G T K A V T K F(T S A) K

D.
1 10 20 30 40
A R T K Q T A R K S T G G K A P R K Q L A T K A A R K S A P A T G G V K K P H R
41 50 60 70 80
F R P G T V A L R E I R K Y Q K S T E L L I R K L P F Q R L V R Z I A Z B F K T
81 90 100 110 120
B L R F Z S S A V S A L Q E A A E A Y L V G L F E D T N L C A I H A K R V T I M
121 130 135
P K D I Q L A R R I R G E R A

E.
1 10 20 30 40
S G R G K G G K G L G K G G A K R H R K V L R D N I Q G I T K P A I R R L A R R
41 50 60 70 80
G G V K R I S G L I Y E E T R G V L K I F L E N V I R D A V T Y T E H A R R K T
81 90 100 102
V T A M D V V Y A L K R Q G R T L Y G F G G

**Fig. 28 A–E.** The amino acid sequences of plant histones. **A** Partial data for maize histone H1 (Hurley and Stout 1980). **B** Partial data for wheat histones H2A (Rodrigues et al. 1979). **C** Partial data for wheat histone H2B (Von Holt et al. 1979). **D** The complete sequence of pea histone H3 (Patty et al. 1973). **E** The complete sequence of pea histone H4 (DeLange et al. 1969b)

The other plant histones to have their sequences studied do not show the same high degree of conservation of structure when compared with other histones. Thus, Rodrigues et al. (1979) have shown that the histone 2 fraction from embryonic tissue of wheat contains a mixture of at least five distinct components. The partial sequence data which has been reported for three of these histones (Rodrigues et al. 1979: Van Holt et al. 1979) shows specific sequence differences between the forms ruling out preparative artefacts as the cause of the different forms (Fig. 28). There exists extensive homology between these plant histones 2 and the sequence of calf thymus histone H2A, and therefore RODRIGUES et al. (1979) have proposed that they consist of a family of H2A variants and should thus be named $H2A_{(1)}$, $H2A_{(2)}$, etc. accordingly. Similarly, partial sequence data on a wheat histone 1 fraction (Fig. 28) has shown that it is homologous to the animal histone H2B proteins (VAN HOLT et al. 1979). However, the amino acid sequence of a maize histone 1 fraction (Fig. 28) shows no homology to these wheat or calf histones, but rather is homologous with rabbit H1 and chicken H5 histones (HURLEY and STOUT 1980).

A prominent feature of all of the structures of histones is that of the skewed distribution of the basic amino acid residues. In histones H2A, H2B, H3 and H4 a predominance of the basic residues occur in clusters at the N-terminal and C-terminal regions, with the intermediate region being dominated by hydrophobic and acidic amino acids (ELGIN and WEINTRAUB 1975). Further, BRANDT et al. (1974) noted that in histone H3 a repetitive unit consisting generally of nine residues, of which the last two are basic residues, can be recognized. It is possible that the varying degrees of methylation and acetylation have prevented crystallization of the proteins for further structural studies on the significance of such distributions.

## 5.3 Leghaemoglobin

The leghaemoglobins are a group of proteins found in effective $N_2$-fixing nodules of legumes formed after infection with *Rhizobium* species. The protein ensures an adequate supply of oxygen to the symbiotic bacterium at a stabilized and very low concentration of free oxygen. The role and properties of leghaemoglobin have been reviewed by APPLEBY (1974).

In the legume root nodule, a very complex relationship exists between the host plant and the symbiont, and at least two proteins, leghaemoglobin and nitrogenase, are produced in large quantity; neither protein, however, is produced when the two organisms are grown independently of each other. DILWORTH (1969) was able to show by crossing the same *Rhizobium* strain to two dissimilar legumes, that it is the host plant which genetically determines the protein for the leghaemoglobin which is produced. The nodule is found to contain several different leghaemoglobins; in soyabean there are four major (a, $c_1$, $c_2$, $c_3$) and four minor (b, $d_1$, $d_2$, $d_3$) components with the ratios of $a/c_{1-3}$ varying with age, whereas a/b and c/d are independent of age (see WHITTAKER et al. 1979).

The amino acid sequences of leghaemoglobins from five different plant sources have been established, including three different variants from soyabean and two variants from lupin (see Fig. 29, which also gives the references). In addition, WHITTAKER et al. (1979) have shown from peptide map studies that the soyabean b variant differs from the a variant only at the N-terminal where the first amino acid was absent and the end amino acid (Ala) was acetylated. They suggested that the d variants were related to the c variants in a similar manner. Because of the difficulties in resolving components, a considerable degree of heterogeneity has been observed in all the sequences, and this makes

```
                           1                  10                            20                            30
a. Soyabean         a      V A F T E K Q D A L V S S S F E A F K A N I P Q Y S V V F Y
b. Soyabean         c      G A F T E K Q D A L V S S S F E A F K A N I P Q Y S V V F Y
c. Soyabean         c2     G A F T D K Q E A L V S S S F E A F K T N I P Q Y S V V F Y
d. Kidney bean      a      G A F T F K Q E A L V N S S W E A F K G N I P Q Y S V V F Y
e. Lupin            I      G V L T D V Q V A L V K S S F E E F N A N I P K N T H R F F
f. Lupin            II     G A L T E S Q A A L V K S S W E E F N A N I P K H T H R F F
g. Broad Bean       I      G - F T E K Q E A L V N S S S Q L F K Q N P S N Y S V L F Y
h. Pea              I      G - F T D K Q E A L V N S S S E - F K Q N L P G Y S V L F Y

                           31                 40                            50                            60
a.                         T S I L E K A P A A K D L F S F L A N P T D - G - V N P K L
b.                         N S I L E K A P A A K D L F S F L A N P T D - G - V N P K L
c.                         T S I L E K A F A V K D L F S F L A N G V N - P - T N P K L
d.                         T S I L E K A P A A K N L F S F L A N G V D - P - T N P K L
e.                         T L V L E I A P G A K D L F S F L K G S S E V P Q N N P D L
f.                         I L V L E I A P A A K D L F S F L K G T S E V P Q N N P E L
g.                         T I I L Q K A P T A K A M F S F L K D S A G V V D S - P K L
h.                         T I I L E K A P A A K G L F S F L K D T A G V E D S - P K L

                           61                 70                            80                            90
a.                         T G H A E K L F A L V R D S A G Q L K A S G T V V A D A - -
b.                         T G H A E K L F A L V R D S A G Q L K A S G T V V A D A - -
c.                         T G H A E K L F G L V R D S A G Q L K A - - T V V A D A - -
d.                         T A H A E S L F G L V R D S A A Q L R A N G A V V A D A - -
e.                         Q A H A G K V F K L T Y E A A I Q L E V N G A V A S D A T L
f.                         Q A H A G K V F K L V Y E A A I Q L E V T G V V A S D A T L
g.                         G A H A E K V F G M V R D S A V Q L R A T G E V V L D G - -
h.                         Q A H A E Q V F G L V R D S A A Q L R T K G E V V L - G N A

                           91                 100                           110                           120
a.                         - A L G S V H A Q K A V T N P E F - V V K E A L L K T I K A
b.                         - A L G S I H A Q K A V T N P E F - V V K E A L L K T I K E
c.                         - A S G S I H A Q K A I T N P E F - V V K E A L L K T I K E
d.                         - A L G S I H S Q K G V S N D Q F L V V K E A L L K T L K Q
e.                         K S L G S V H V S K G V V D A H F P V V K E A I L K T I K E
f.                         K N L G S V H V S K G V A D A H F P V V K E A I L K T I K E
g.                         K D - G S I H I Q K G V L D P H F V V V K E A L L K T I K E
h.                         - T L G A I H V Q K G V T N P H F V V V K E A L L Q T I K K

                           121                130                           140                           150
a.                         A V G D K W S D E L S R A W E V A Y D E L A A A I K A K
b.                         A V G D K W S D E L S S A W E V A Y D E L A A A I K K A F
c.                         A V G D K W S D E L S S A W E V A Y D E L A A A I K K A F
d.                         A V G D K W T D Q L S T A L E L A Y D E L A A A I K K A Y A
e.                         V V G D K W S E E L N T A W T I A Y D E L A I I I K K E M K D A A
f.                         V V G A K W S E E L N S A W T I A Y D E L A I V I K K E M D D A A
g.                         A S G D K W S E E L S A A W E V A Y D G L A T A I K A A
h.                         A S G N N W S E E L N T A W E V A Y D G L A T A I K K A M K T A
```

**Fig. 29.** Sequence homologies between various leghaemoglobins. *a* Soyabean *a, Glycine max* (ELLFOLK and SIEVERS 1971, 1973, 1974); *b* Soyabean *c* (SIEVERS et al. 1978); *c* Soyabean $c_2$ (HURRELL and LEACH 1977); *d* Kidney bean *a, Phaseolus vulgaris* (LEHTOVAARA and ELLFOLK 1975); *e* Lupin I; *f* Lupin II, *Lupinus luteus* (EGOROV et al. 1976, EGOROV et al. 1978); *g* Broad bean I, *Vicia faba* (RICHARDSON et al. 1975); *h* Pea I, *Pisum sativum* (LEHTOVAARA et al. 1980). Extensive heterogeneity has been recorded in several of the sequences; only the most common residue for such positions is shown

comparisons between sequences difficult. For example, SIEVERS et al. (1978) report that soyabean c differs from the $c_2$ variant at only one position, while the sequence of the $c_2$ variant established by HURREL and LEACH (1977) differs substantially, by at least eight positions, from either of these variants. Also, the sequence of broad bean I leghaemoglobin was found to have considerably more heterogeneity than the other sequences (RICHARDSON et al. 1975). It is unclear whether this represents contamination by other distinct components or is due to allelic variation at a single locus. Clearly, further study is required to resolve all these issues.

The general characteristics of leghaemoglobin, including its physicochemical and spectroscopic properties, as well as its size and sequence, all suggest a close similarity to the structures of animal globins. Nevertheless, there are several distinct characteristics of the leghaemoglobins, including its much greater oxygen affinity, its ability to bind long-chain carboxylic acids, and the tendency of the ferric protein to exist as an equilibrium mixture of spin states. (For a review of these topics, see APPLEBY 1974.) VAINSHTEIN et al. (1977) have studied the three-dimensional structure of lupin leghaemoglobin at 2.8 Å resolution. They have shown that the folding pattern is essentially identical to that of the animal globin family, although some distinct small differences exist. For example, the lengths of certain of the helical regions, which account for 82% of the structure, are different to accommodate sequence insertions and deletions. Also the haem pocket is notably more open than that of the animal globins.

This parallel of structure presents many interesting evolutionary and functional questions. If the plant and animal globins share a common ancestor, then as much as 1,500 million years have elapsed since their divergence from this ancestor. However, since the plant globin has been found only in legumes, which have been evolving as a group for probably less than 150 million years, if the globins had such a common ancestor, the plant genes have seemingly been dormant for a long period of time.

### 5.4 Sweet Proteins

A variety of plants yield proteins with unusual taste properties. (See INGLETT and MAY 1968.) Some, for example the "taste-modifying protein" that has been isolated from the miracle fruit, *Synsepalum dulcificum,* modify taste so that sour substances taste sweet. Others, for example monellin from serendipity berries, *Dioscoreophyllum cumminsii,* and thaumatin from *Thaumatococcus daniellii,* are intensely sweet, being over $10^5$ times sweeter than sucrose on a molar basis. Both these sweet proteins have been subject to structural investigations.

The amino acid sequence of monellin has been determined (BOHAK and LI 1976, HUDSON and BIEMANN 1976, FRANK and ZUBER 1976) and has shown the molecule to consist of two non-identical chains of 50 and 42 residues (Fig. 30), and to lack bound carbohydrate. This structure raises the possibility that monellin may be derived from a larger precursor and is thus consistent with the observation (INGLETT and MAY 1969) that proteolytic enzymes liberate a sweet material from a high molecular weight fraction extracted from serendipity berries.

Monellin has also been crystallized, and a structural analysis is in progress (WLODAWER and HODGSON 1975). Until this completed structure is available, chemical methods have been used to investigate the sweet properties of the protein. Thus, [$^3$H]-labelled methylated monellin has been shown to bind in vitro to taste receptor tissue, and sugars appear to compete to some extent

A.

Chain A:

```
1                   10                  20                  30                  40    44
R E I K G Y E Y Q L Y V V Y A S D K L F R A D I S E D Y K T R G R K L L R F N G P V P P P
```

Chain B:

```
1                   10                  20                  30                  40
G E W E I I D I G P F T Q N L G K F A V D E E N K I G Q Y C R L T F N K V I R P

41                  50
C M K K T I Y E N E
```

B.

```
1                   10                  20                  30                  40
A T F E I V N R C S Y T V W A A A S K G D A A L D A G G R Q L N S G E S W T I N

41                  50                  60                  70                  80
V E P G T N G G K I W A R T D C Y F D D S G S G I C K T G D C G G L L R C K R F

81                  90                  100                 110                 120
G R P P T T L A E F S L N Q Y G K D Y I D I S N I K G F N V P M N F S P T T R G

121                 130                 140                 150                 160
C R G V R C A A D I V G Q C P A K L K A P G G G C N D A C T V F Q T S E Y C C T

161                 170                 180                 190                 200
T G K C G P T E Y S R F F K R L C P D A F S Y V L D K P T T V T C P G S S N Y R

201         207
V T F C P T A
```

**Fig. 30 A, B.** The sequences of sweet proteins from plants. **A** Monellin (BOHAK and LI 1976, HUDSON and BIEMANN 1976, FRANK and ZUBER 1976). **B** Thaumatin (IYENGAR et al. 1979)

for the binding sites (CAGAN and MORRIS 1979). The sweetness of the protein requires the undissociated molecule; neither of the individual subunits is sweet by itself, nor did either of them block the sweetness of the intact molecule. The individual subunits do recombine slowly to give intact monellin with an accompanying return of sweet properties (BOHAK and LI 1976). Chemical modification studies have shown that either blocking the single free sulphydryl residue, B chain Cys(41), or modifying the single Met residue, B chain Met(42), with CNBr, both lead to a total loss of sweetness (BOHAK and LI 1976). Since these residues are adjacent in the sequence, it is suggested that they are part of, or close to, the binding site of the molecule.

The other sweet proteins to be extensively studied are those extracted from the fruit of *Thaumatococcus daniellii*. Two intensely sweet substances are found, called thaumatin I and II, both being entirely proteinaceous (VAN DER WEL and LOEVE 1972). The amino acid sequence of one of these components, thaumatin I, has been determined (IYENGAR et al. 1979). It consists of a single polypeptide chain of 207 residues (Fig. 30) containing eight disulphide bonds; by contrast, monellin has no disulphide bonds. However, as with monellin, an intact tertiary structure is vital for the sweet properties as the loss of even a single disulphide bond is sufficient to cause loss of sweetness (IYENGAR et al. 1979). Antibodies to thaumatin have been prepared and shown to cross-react to some extent with monellin (HOUGH and EDWARDSON 1978). However, statistical comparisons of the sequences of monellin and thaumatin I failed to indicate any evolutionary relationship between the proteins (IYENGAR et al. 1979).

## 5.5 Pollen Allergens

Although the pollens of almost all wind-pollinated weeds have been implicated in human pollen allergy, the allergenic compounds extracted from the pollens of grasses (Gramineae) and ragweeds (Ambrosieae) have been those most extensively studied (see MARSH 1975). Of these various allergens studied however, detailed structural data has only been determined for the allergen Ra5 from ragweed, *Ambrosia elatior* (MOLE et al. 1975). While Ra5 is only a minor allergen, it is of importance in genetic studies of allergy because of its simple molecular structure. It consists of a single polypeptide chain of 45 residues and has no detectable carbohydrate (Fig. 31). Determination of the sequence revealed previously undetected heterogeneity at position 2, where both Val and Leu were found in a 2:1 ratio. The sequence showed the presence of eight Cys residues, constituting four disulphide bonds, which may give the molecule a fairly rigid three-dimensional structure. MOLE et al. (1975) suggested that a compact three-dimensional structure containing several aromatic groups may be important to the biological activity. The amino acid composition data which is available for several other ragweed and ryegrass allergens (see MARSH 1975) also shows a high Cys content, although further sequence data is needed to establish whether these different allergens possess common structural features which are responsible for their activity.

```
1                  10                    20                    30                    40
L V P C A W A G N V C G E K R A Y C C S D P G R Y C P W Q V V C Y E S S E I C S K K C G K
  L
```

**Fig. 31.** The sequence of the ragweed pollen allergen Ra5

## 5.6 Other Proteins and Recent Studies

Preliminary crystallographic data has been presented for a wide variety of plant proteins, but apart from those already described, none of these studies has at this time led to any detailed structure.

Since the completion of the original manuscript, which included structures published upto the end of 1979, and a subsequent revision to include some papers published in 1980, several additional relevant papers have been published.

Further studies on plant enzymes include the sequence of a peptide from ribulose bisphosphate carboxylase which identifies the activator carbon dioxide site (LORIMER 1981) and details of the sequence of the proteolipid subunit of the ATP synthase from spinach chloroplasts (SEBALD and WACHTER 1980).

The sequences of additional electron transfer proteins, two ferredoxins from *Dunaliella salina* have been reported (HASE et al. 1980).

New data on toxic proteins have been described, including studies on various lectins (BULL et al. 1980, GEBAUER et al. 1981, VILLAFRANCA and ROBERTUS 1981, WRIGHT 1981) and on an inhibitor (HASS and HERMODSON 1981).

Further studies on seed proteins (DRENSKA et al. 1980, OHMIYA et al. 1980, SHEWRY et al. 1980b, SUN et al 1981), leghaemoglobin (SULLIVAN et al. 1981) and an allergen (KLAPPER et al. 1980) have been reported.

In addition, the tertiary structure of the hydrophobic protein crambin has been determined (HENDRICKSON and TEETER 1981) and the sequences of peptides from wheat germ elongation factor 2 (BROWN and BODLEY 1979) and a phytochrome chromophore (LAGARIAS and RAPOPORT 1980) have been reported.

*Acknowledgement.* I would like to acknowledge the invaluable help of Mr. A. Gordon Brown in preparing the figures for the manuscript.

## References

Abe O, Shimokawa Y, Ohata J, Kuromizu K (1979) Isolation and activities of the trypsin-modified *Vicia angustifolia* proteinase inhibitor lacking carboxyl-terminal hexapeptide. Biochem Biophys Acta 568:71–79

Adman ET (1979) A comparison of the structures of electron transfer proteins. Biochim Biophys Acta 549:107–144

Aitken A (1975) Prokaryote-eukaryote relationships and the amino acid sequence of plastocyanin from *Anabaena variabilis*. Biochem J 149:675–683

Aitken A (1979) Purification and primary structure of cytochrome *c*-552 from the Cyanobacterium, *Synechococcus* PCC 6312. Eur J Biochem 101:297–308

Ambler RP (1976) Standards and accuracy in amino acid sequence determination. In: Markham R, Horne RW (eds) Structure-function relationships of proteins. North-Holland Publishing, Amsterdam, pp 1–14

Ambler RP, Bartsch RG (1975) Amino acid sequence similarity between cytochrome f from a blue-green bacterium and algal chloroplasts. Nature (London) 253:285–288

Appleby CA (1974) Leghemoglobin In: Quispel A (ed) The biology of nitrogen fixation. North-Holland Publishing, Amsterdam, pp 521–554

Asatov SI, Yadgarov EG, Yunusov TS, Yuldashev PK (1978) A study of cotton seed globulins. 17. Primary structure of the carbohydrate-containing subunit C of the 11S globulin. Khim Prir Soedin 4:541–542

Autran J-C, Lew EJL, Nimmo CC, Kasarda DD (1979) N-terminal amino acid sequencing of prolamins from wheat and related species. Nature (London) 282:527–529

Baker EN (1977) Structure of Actinidin: Details of the polypeptide chain conformation and active site from an electron density map at 2·8 Å resolution. J Mol Biol 115:263–277

Baker EN (1980) Structure of Actinidin, after refinement at 1.7 Å resolution. J Mol Biol 141:441–484

Baker EN, Dodson EJ (1980) Crystallographic refinement of the structure of actinidin at 1.7 Å resolution by fast Fourier least-squares methods. Acta Cryst A36:559–572

Baker TS, Suh SW, Eisenberg D (1977) Structure of ribulose-1,5-bisphosphate carboxylase-oxygenase: Form III crystals. Proc Natl Acad Sci USA 74:1037–1041

Baumann C, Rüdiger H, Strosberg AD (1979) A comparison of the two lectins from *Vicia cracca*. FEBS Lett 102:216–218

Becker JW, Reeke GN, Wang JL, Cunningham BA, Edelman GM (1975) The covalent and three-dimensional structure of concanavalin A. J Biol Chem 250:1513–1524
Bedbrook JR, Coen DM, Beaton AR, Bogorad L, Rich A (1979) Location of the single gene for the large subunit of ribulose bisphosphate carboxylase on the maize chloroplast chromosome. J Biol Chem 254:905–910
Belew M, Eaker D (1976) The trypsin and chymotrypsin inhibitors in chick peas (*Cicer arietinum* L.). Eur J Biochem 62:499–508
Bell EA (1978) Toxins in seeds. In: Harbourne JB (ed) Biochemical aspects of plant and animal coevolution. Academic Press, London New York, pp 143–161
Benson AM, Yasunobu KT (1969a) Non-heme iron proteins X. The amino acid sequences of ferredoxins from *Leucaena glauca*. J Biol Chem 244:955–963
Benson AM, Yasunobu KT (1969b) Non-heme iron proteins, XI. Some genetic aspects. Proc Natl Acad Sci USA 63:1269–1273
Bergman C, Gandvik E-K, Nyman PO, Strid L (1977) The amino acid sequence of stellacyanin from the lacquer tree. Biochem Biophys Res Commun 77:1052–1059
Bietz JA, Huebner FR, Sanderson JE, Wall JS (1977) Wheat gliadin homology revealed through N-terminal amino acid sequence analysis. Cereal Chem 54:1070–1083
Bohak Z, Li S-L (1976) The structure of monellin and its relation to the sweetness of the protein. Biochim Biophys Acta 427:153–170
Boulter D (1973) Amino acid sequences of cytochrome *c* and plastocyanins in phylogenetic studies of higher plants. Syst Zool 22:549–553
Boulter D (1974) The use of amino acid sequence data in the classification of higher plants. In: Nobel Foundation Series, vol XXV. Almquist & Wiksell Forlag AB, Stockholm, pp 211–216
Boulter D (1977) Present status of the use of amino acid sequence data in plant phylogenetic studies. In: Matsubara H, Yamanaka T (eds) Evolution of protein molecules. Japan Scient Soc Press, Tokyo, pp 243–250
Boulter D, Ramshaw JAM (1972) Structure-function relationships in plant cytochrome *c*. Phytochemistry 11:553–561
Boulter D, Thompson EW, Ramshaw JAM, Richardson M (1970) Higher plant cytochrome *c*. Nature (London) 228:552–554
Boulter D, Ramshaw JAM, Thompson EW, Richardson M, Brown RH (1972) A phylogeny of higher plants based on the amino acid sequences of cytochrome *c* and its biological implications. Proc R Soc London Ser B 181:441–455
Boulter D, Haslett BG, Peacock D, Ramshaw JAM, Scawen MD (1977) Chemistry, function and evolution of plastocyanin. In: Northcote DH (ed) Int Rev Biochem, Plant Biochem II, vol XIII. Univ Park Press, Baltimore, pp 1–40
Boulter D, Gleaves JT, Haslett BG, Peacock D, Jensen U (1978) The relationships of 8 tribes of the compositae as suggested by plastocyanin amino acid sequence data. Phytochemistry 17:1585–1589
Boulter D, Peacock D, Guise A, Gleaves JT, Estabrook G (1979) Relationships between the partial amino acid sequences of plastocyanin from members of ten families of flowering plants. Phytochemistry 18:603–608
Bowes G, Ogren WL, Hageman RH (1971) Phosphoglycolate production catalysed by ribulose diphosphate carboxylase. Biochem Biophys Res Commun 45:716–722
Braithwaite A (1976) Unit cell dimensions of crystalline horseradish peroxidase. J Mol Biol 106:229–230
Brandt WF, Strickland WN, Morgan M, Von Holt C (1974) Comparison of the N-terminal amino acid sequences of histone F3 from a mammal, a bird, a shark, an echinoderm, a mollusc and a plant. FEBS Lett 40:167–172
Brown BA, Bodley JW (1979) Primary structure at the site in beef and wheat elongation factor 2 of ADP-ribosylation by diphtheria toxin. FEBS Lett 103:253–255
Brown RH, Boulter D (1973a) The amino acid sequence of cytochrome *c* from *Allium porrum* L. (leek). Biochem J 131:247–251
Brown RH, Boulter D (1973b) The amino acid sequence of cytochrome *c* from *Nigella damascena* L. (love-in-a-mist). Biochem J 133:251–254
Brown RH, Boulter D (1974)The amino acid sequences of cytochrome *c* from four plant sources. Biochem J 137:93–100

Brown RH, Boulter D (1975) A re-examination of the amino acid sequence data of cytochromes *c* from *Solanum tuberosum* (potato) and *Lycopersicon esculentum.* (tomato). FEBS Lett 51:66–67
Brown RH, Richardson M, Scogin R, Boulter D (1973) The amino acid sequence of cytochrome *c* from *Spinacea oleracea* L. (spinach). Biochem J 131:253–256
Bull H, Li SS-L, Fowler E, Lin TT-S (1980) Isolation and characterisation of cyanogen bromide fragments of the A and B chains of the antitumor toxin Ricin D. Int J Peptide Protein Res 16:208–218
Cagan RH, Morris RW (1979) Biochemical studies of taste sensation: Binding to taste tissue of $^3$H-labeled monellin, a sweet-tasting protein. Proc Natl Acad Sci USA 76:1692–1696
Carne A, Moore CH (1978) The amino acid sequence of the tryptic peptides from actinidin, a proteolytic enzyme from the fruit of *Actinidia chinensis*. Biochem J 173:73–83
Cashmore AR (1979) Reiteration frequency of the gene coding for the small subunit of ribulose-1,5-bisphosphate carboxylase. Cell 17:383–388
Cashmore AR, Broadhurst MK, Gray RE (1978) Cell free synthesis of leaf protein: Identification of an apparent precursor of the small subunit of ribulose-1,5-bisphosphate carboxylase. Proc Natl Acad Sci USA 75:655–659
Cawley DB, Hedblom ML, Houston LL (1978) Homology between ricin and *Ricinus communis* agglutinin: Amino terminal sequence analysis and protein synthesis inhibition studies. Arch Biochem Biophys 190:744–755
Chan P-H, Wildman SG (1972) Chloroplast DNA codes for the primary structure of the large subunit of Fraction I protein. Biochim Biophys Acta 277:677–680
Chapman GV, Colman PM, Freeman HC, Guss JM, Murata M, Norris VA, Ramshaw JAM, Venkatappa MP (1977) Preliminary crystallographic data for a copper-containing protein, plastocyanin. J Mol Biol 110:187–189
Chirgadze YN, Garber MB, Nikonov SV (1977) Crystallographic study of plastocyanins. J Mol Biol 113:443–447
Clarke J, Shannon LM (1976) The isolation and characterization of the glycopeptides from horseradish peroxidase isoenzyme C. Biochim Biophys Acta 427:428–442
Colman PM, Freeman HC, Guss JM, Murata M, Norris VA, Ramshaw JAM, Venkatappa MP (1978) X-ray crystal structure analysis of plastocyanin at 2.7 Å resolution. Nature (London) 272:319–324
Cunningham BA, Wang JL, Waxdal MJ, Edelman GM (1975) The covalent and three-dimensional structure of concanavalin A. J Biol Chem 250:1503–1512
Cunningham BA, Hemperly JJ, Hopp TP, Edelman GM (1979) Favin versus concanavalin A: Circularly permuted amino acid sequences. Proc Natl Acad Sci USA 76:3218–3222
Dayhoff MO (1972) Atlas of protein sequence and structure, vol V. National Biomedical Research Foundation, Silver Spring
DeLange RJ, Fambrough DM, Smith EL, Bonner J (1969a) Calf and pea histone IV. J Biol Chem 244:319–334
DeLange RJ, Fambrough DM, Smith EL, Bonner J (1969b) Calf and pea histone IV. J Biol Chem 244:5669–5679
Derbyshire E, Wright DJ, Boulter D (1976) Legumin and vicilin, storage proteins of legume seeds. Phytochemistry 15:3–24
Dickerson RE, Timkovich R (1975) Cytochromes *c*. In: Boyer PD (ed) The enzymes, vol XI. Academic Press, London New York, pp 397–547
Dilworth MJ (1969) The plant as the genetic determinant of leghaemoglobin production in the legume root nodule. Biochim Biophys Acta 184:432–441
Dlouhá V, Keil B, Sorm F (1964) Structure of the peptides isolated from the tryptic hydrolysate of the A chain of edestin. Collect Czech Chem Commun 29:1835–1850
Drenska AI, Ganchev KD, Ivanov Ch P (1980) Amino acid sequence of zein fractions in the N-terminal region. Dokl Bolg Akad Nauk 33:67–70
Drenth J, Jansonius JN, Koekoek R, Swen HM, Wolthers BG (1968) Structure of papain. Nature (London) 218:929–932
Drenth J, Kalk KH, Swen HM (1976) Binding of chloromethyl ketone substrate analogues to crystalline papain. Biochemistry 15:3731–3738
Driessche Van E, Foriers A, Strosberg AD, Kanarek L (1976) N-terminal sequences of

the $\alpha$ and $\beta$ subunits of the lectin from the garden pea (*Pisum sativum*). FEBS Lett 71:220–222
Dutton JE, Rogers LJ, Haslett BG, Takruri IAH, Gleaves JT, Boulter D (1980) Comparative studies on the properties of two ferredoxins from *Pisum sativum* L. J Exp Bot 31:379–391
Edelman GM, Cunningham BA, Reeke GN, Becker JW, Waxdal MJ, Wang JL (1972) The covalent and three-dimensional structure of concanavalin A. Proc Natl Acad Sci USA 69:2580–2584
Egorov TA, Feigina MY, Kazakov VK, Shakhparonov MI, Mitaleva SI, Ovchinnikov YA (1976) Full amino acid sequence of leghemoglobin I from yellow lupine nodules. Bioorg Khim 2:125–128
Egorov TA, Kazakov VK, Shakhparonov MI, Feigina MY, Kostetskii PV (1978) Primary structure of leghemoglobin II from yellow lupin (*Lupinus luteus* L.) nodules. Bioorg Khim 4:476–480
Elgin SCR, Weintraub H (1975) Chromosomal proteins and chromatin structure. Annu Rev Biochem 44:725–774
Elleman TC (1977) Amino acid sequence of the smaller subunit of conglutin $\gamma$, a storage globulin of *Lupinus angustifolius*. Aust J Biol Sci 30:33–45
Ellfolk N, Sievers G (1971) The primary structure of soybean leghemoglobin. Acta Chem Scand 25:3532–3548
Ellfolk N, Sievers G (1973) The primary structure of soybean leghemoglobin. Acta Chem Scand 27:3986–3992
Ellfolk N, Sievers G (1974) Correction of the amino acid sequence of soybean leghemoglobin *a*. Acta Chem Scand B28:1245–1246
Fee JA (1975) Copper proteins. Systems containing the "blue" copper center. Struct Bonding (Berlin) 23:1–60
Fitch WM, Margoliash E (1967) Construction of phylogenetic trees. Science 155:279–284
Foriers A, Wuilmart C, Sharon N, Strosberg AD (1977) Extensive sequence homologies among lectins from leguminous plants. Biochem Biophys Res Commun 75:980–986
Foriers A, DeNeve R, Kanarek L, Strosberg AD (1978) Common ancestor for concanavalin A and lentil lectin? Proc Natl Acad Sci USA 75:1336–1339
Frank G, Zuber H (1976) The complete amino acid sequences of both subunits of the sweet protein monellin. Hoppe-Seylers Z Physiol Chem 357:585–592
Freeman HC (1979) Elegance in molecular design: The copper site of a photosynthetic electron-transfer protein. J Proc R Soc NSW 112:45–62
Freeman HC, Norris VA, Ramshaw JAM, Wright PE (1978) High resolution proton magnetic resonance studies of plastocyanin. FEBS Lett 86:131–135
Funatsu G, Kimura M, Funatsu M (1979) Primary structure of Ala chain of ricin D. Agric Biol Chem 43:2221–2224
Fukuyama K, Hase T, Matsumoto S, Tsukihara T, Katsube Y, Tanaka N, Kakudo M, Wada K, Matsubara H (1980) Structure of *S. platensis* [2Fe-2s] ferredoxin and evolution of chloroplast-type ferredoxins. Nature (London) 286:522–524
Gebauer G, Schiltz E, Rüdiger H (1981) The amino-acid sequence of the $\alpha$ subunit of the mitogenic lectin from *Vicia sativa*. Eur J Biochem 113:319–325
Gibbons GC, Strøbaek S, Haslett B, Boulter D (1975) The N-terminal amino acid sequence of the small subunit of ribulose-1,5-diphosphate carboxylase from *Nicotiana tabacum*. Experientia 31:1040–1041
Gilroy J, Wright DJ, Boulter D (1979) Homology of basic subunits of legumin from *Glycine max* and *Vicia faba*. Phytochemistry 18:315–316
Goto K, Murachi T, Takahashi N (1976) Structural studies on stem bromelain isolation, characterization and alignment of the cyanogen bromide fragments. FEBS Lett 62:93–95
Goto K, Takahashi N, Murachi T (1980) Structural studies on stem bromelain. Int J Peptide Protein Res 15:335–341
Gray JC (1978) Purification and properties of monomeric cytochrome f from charlock, *Sinapis arvensis* L. Eur J Biochem 82:133–141
Hankins CN, Shannon LM (1978) The physical and enzymatic properties of a phytohemagglutinin from mung beans. J Biol Chem 253:7791–7797
Hara I, Ohmiya M, Matsubara H(1978) Pumpkin (*Cucurbita* sp.) seed globulin III. Compari-

son of subunit structures among seed globulins of various *Cucurbita* species and characterization of peptide components. Plant Cell Physiol 19:237–243

Hardman KD, Ainsworth CF (1976) Structure of the concanavalin A methyl-α-D-mannopyranoside complex at 6-Å resolution. Biochemistry 15:1120–1128

Hase T, Wada K, Matsubara H (1977a) Horsetail (*Equisetum telmateia*) ferredoxins I and II. Amino acid sequences. J Biochem (Tokyo) 82:267–276

Hase T, Wada K, Matsubara H (1977b) Horsetail *(Equisetum arvense)* ferredoxins I and II. Amino acid sequences and gene duplication. J Biochem (Tokyo) 82:277–286

Hase T, Matsubara H, Ben-Amotz A, Rao KK, Hall DO (1980) Purification and sequence determination of two ferredoxins from *Dunaliella salina*. Phytochemistry 19:2065–2070

Hase T, Wakabayashi S, Wada K, Matsubara H, Jüttner F, Rao KK, Fry I, Hall DO (1978) *Cyanidium caldarium* Ferredoxin: A red algal type? FEBS Lett 96:41–44

Haslett BG, Boulter D (1976) The *N*-terminal amino acid sequence of plastocyanin from *Stellaria media* L. Biochem J 153:33–38

Haslett BG, Yarwood A, Evans IM, Boulter D (1976) Studies on the small subunit of Fraction I protein from *Pisum sativum* L. and *Vicia faba* L. Biochim Biophys Acta 420:122–132

Haslett BG, Gleaves T, Boulter D (1977) *N*-terminal amino acid sequences of plastocyanins from various members of the Compositae. Phytochemistry 16:363–365

Haslett BG, Bailey CJ, Ramshaw JAM, Scawen MD, Boulter D (1978a) The amino acid sequence of plastocyanin from *Rumex obtusifolius*. Phytochemistry 17:615–617

Haslett BG, Evans IM, Boulter D (1978b) The amino acid sequence of plastocyanin from *Solanum crispum* using automatic methods. Phytochemistry 17:735–739

Haslett BG, Boulter D, Ramshaw JAM, Scawen MD (1979) Partial amino acid sequences of plastocyanin from members of ten families of the flowering plants: Data Set. Phytochemistry 18:608

Hass GM, Hermodson MA (1981) Amino acid sequence of a carboxypeptidase inhibitor from tomato fruit. Biochemistry 20:2256–2260

Hass GM, Nau H, Biemann K, Grahn DT, Ericsson LH, Neurath H (1975) The amino acid sequence of a carboxypeptidase inhibitor from potatoes. Biochemistry 14:1334–1342

Hass GM, Ako H, Grahn DT, Neurath H (1976a) Carboxypeptidase inhibitor from potatoes. The effects of chemical modifications on inhibitory activity. Biochemistry 15:93–100

Hass GM, Venkatakrishnan R, Ryan CA (1976b) Homologous inhibitors from potato tubers of serine endopeptidases and metallocarboxypeptidases. Proc Natl Acad Sci USA 73:1941–1944

Hemperly JJ, Hopp TP, Becker JW, Cunningham BA (1979) The chemical characterization of favin, a lectin isolated from *Vicia faba*. J Biol Chem 254:6803–6810

Hendrickson WA, Teeter MM (1981) Structure of the hydrophobic protein crambin determined directly from the anomalous scattering of sulphur. Nature (London) 290:107–113

Highfield PE, Ellis RJ (1978) Synthesis and transport of the small subunit of chloroplast ribulose bisphosphate carboxylase. Nature (London) 271:420–424

Hochstrasser K, Illchmann K, Werle E (1970a) Die Aminosäuresequenz des spezifischen Trypsininhibitors aus Maissamen, Charakterisierung als Polymeres. Hoppe-Seylers Z Physiol Chem 351:721–728

Hochstrasser K, Illchmann K, Werle E, Hössl R, Schwarz S (1970b) Die Aminosäuresequenz des Trypsininhibitors aus Samen von *Arachis hypogaea*. Hoppe-Seylers Z Physiol Chem 351:1503–1512

Holt Von C, Strickland WN, Brandt WF, Strickland MS (1979) More histone structures. FEBS Lett 100:201–218

Hough CAM, Edwardson JA (1978) Antibodies to thaumatin as a model of the sweet taste receptor. Nature (London) 271:381–383

Hudson G, Biemann K (1976) Mass spectrometric sequencing of proteins. The structure of subunit I of monellin. Biochem Biophys Res Commun 71:212–220

Hurley CK, Stout JT (1980) Maize histone Hl: A partial structural characterisation. Biochemistry 19:410–416

Hurrell JGR, Leach SJ (1977) The amino acid sequence of soybean leghaemoglobin $c_2$. FEBS Lett 80:23–26
Husain SS, Lowe G (1970) The amino acid sequence around the active-site cysteine and histidine residues, and the buried cysteine residue in ficin. Biochem J 117:333–340
Inglett GE, May JF (1968) Tropical plants with unusual taste properties. Econ Bot 22:326–331
Inglett GE, May JF (1969) Serendipity berries; source of a new intense sweetner. J Food Sci 34:408–411
Ishihara H, Tokahashi N, Oguri S, Tejima S (1979) Complete structure of the carbohydrate moiety of stem bromelain. J Biol Chem 254:10715–10719
Ishikawa C, Nakamura S, Watanabe K, Takahashi K (1979) The amino acid sequence of adzuki bean proteinase inhibitor I. FEBS Lett 99:97–100
Iwasaki T, Kiyohara T, Yoshikawa M (1976) Amino acid sequence of an active fragment of potato proteinase inhibitor IIa. J Biochem (Tokyo) 79:381–391
Iwasaki T, Wada J, Kiyohara T, Yoshikawa M (1977) Amino acid sequence of an active fragment of potato proteinase inhibitor IIb. J Biochem (Tokyo) 82:991–1004
Iyengar RB, Smits P, Van Der Ouderaa F, Van Der Wel H, Van Brouwershaven J, Ravestein P, Richters G, Van Wassenaar PD (1979) The complete amino-acid sequence of the sweet protein thaumatin I. Eur J Biochem 96:193–204
Johal S, Bourque DP, Smith WW, Suh SW, Eisenberg D (1980) Crystallization and characterization of ribulose 1,5-bisphosphate carboxylase/oxygenase from eight plant species. J Biol Chem 255:8873–8880
Jones BL, Mak AS (1977) Amino acid sequences of the two α-purothionins of hexaploid wheat. Cereal Chem 54:511–523
Jos J, Charbonnier L, Mougenot JF, Mosse J, Rey J (1977) Isolation and characterization of the toxic fraction of wheat gliadin in coeliac disease. In: McNicholl B, McCarthy CF, Fottrell PF (eds) Perspectives in coeliac disease. Univ Park Press, Baltimore, pp 75–89
Joubert FJ, Kruger H, Townshend GS, Botes DP (1979) Purification, some properties and the complete primary structures of two protease inhibitors (DE-3 and DE-4) from *Macrotyloma axillare* seed. Eur J Biochem 97:85–91
Kasarda DD, Da Roza DA, Ohms JI (1974) N-terminal sequence of $\alpha_2$-gliadin. Biochim Biophys Acta 351:290–294
Kato I, Tominaga N, Kihara F (1972) Chemical structure of a rice bran proteinase inhibitor. In: Iwai K (ed) Procedings of the 23rd conference on protein structure, pp 53–56
Kawashima N, Wildman SG (1972) Studies on fraction I protein. Biochim Biophys Acta 262:42–49
Kelly J, Ambler RP (1973) The amino acid sequence of plastocyanin from *Chlorella fusca.* Biochem J 143:681–690
Keresztes-Nagy S, Perini F, Margoliash E (1969) Primary structure of alfalfa ferredoxin. J Biol Chem 244:981–995
Klapper DG, Goodfriend L, Capra JD (1980) Amino acid sequence of ragweed allergen Ra3. Biochemistry 19:5729–5734
Koide T, Ikenaka T (1973) Studies on soybean trypsin inhibitors. Eur J Biochem 32:417–431
Kolberg J, Michaelsen TE, Sletten K (1980) Subunit structure and N-terminal sequences of the *Lathyrus odoratus* lectin. FEBS Lett 117:281–283
Lagarias JC, Rapoport H (1980) Chromopeptides from phytochrome. The structure and linkage of the $P_R$ form of the phytochrome chromophore. J Am Chem Soc 102:4821–4828
Laycock MV (1972) The amino acid sequence of cytochrome *c*-553 from the chrysophycean alga *Monochrysis lutheri.* Can J Biochem 50:1311–1325
Laycock MV (1975) The amino acid sequence of cytochrome f from the brown alga *Alaria esculenta* (L.) Grev. Biochem J 149:271–279
Leary TR, Grahn DT, Neurath H, Hass GM (1979) Structure of potato carboxypeptidase inhibitor: disulfide pairing and exposure of aromatic residues. Biochemistry 18:2252–2256
Lehtovaara P, Ellfolk N (1975) The amino-acid sequence of leghemoglobin component a from *Phaseolus vulgaris* (kidney bean) Eur J Biochem 54:577–584

Lehtovaara P, Lappalainen A, Ellfolk N (1980) The amino acid sequence of pea (*Pisum sativum*) leghemoglobin. Biochim Biophys Acta 623:98–106
Li SS-L (1980) Purification and partial characterization of two lectins from *Momordica charantia*. Experientia 36:524–527
Lindqvist Y, Bränden CI (1980) Structure of glycolate oxidase from spinach at a resolution of 5.5 Å. J Mol Biol 143:201–211
Lorimer GH (1981) Ribulosebisphosphate carboxylase: Amino acid sequence of a peptide bearing the activator carbon dioxide. Biochemistry 20:1236–1240
Lynn KR (1979) A purification and some properties of two proteases from papaya latex. Biochim Biophys Acta 569:193–201
Lynn KR, Yaguchi M (1979) N-terminal homology in three cysteinyl proteases from *Papaya* latex. Biochim Biophys Acta 581:363–364
McIntosh L, Poulsen C, Bogorad L (1980) Chloroplast gene sequence for the large subunit of ribulose bisphosphatecarboxylase of maize. Nature (London) 288:556–560
McPherson A (1980) The three-dimensional structure of canavalin at 3.0 Å resolution by X-ray diffraction analysis. J Biol Chem 255:10472–10480
Mak AS, Jones BL (1976) The amino acid sequence of wheat $\beta$-purothionin. Can J Biochem 54:835–842
Margoliash E, Schejter A (1966) Cytochrome *c*. Adv Protein Chem 21:113–286
Marsh DG (1975) Allergens and the genetics of allergy. In: Sela M (ed) The antigens, vol III. Academic Press, London New York, pp 271–359
Martin PG (1979) Amino acid sequence of the small subunit of ribulose-1,5-bisphosphate carboxylase from spinach. Aust J Plant Physiol 6:401–408
Martinez G, Rochat H (1974) The amino acid sequence of cytochrome *c* from *Solanum tuberosum* (potato). FEBS Lett 47:212–217
Matsubara H, Sasaki RM (1968) Spinach ferredoxin: tryptic, chrymotryptic, and thermolytic peptides, and complete amino acid sequence. J Biol Chem 243:1732–1757
Matsubara H, Sasaki RM, Chain RK (1967) The amino acid sequence of spinach ferredoxin. Proc Natl Acad Sci USA 57:439–445
Mazza G, Welinder KG (1980) Covalent structure of turnip peroxidase 7. Eur J Biochem 108:481–489
Meatyard BT, Boulter D (1974) The amino acid sequence of cytochrome *c* from *Enteromorpha intestinalis*. Phytochemistry 13:2777–2782
Meatyard BT, Scawen MD, Ramshaw JAM, Boulter D (1975) Cytochrome *c*'s from *Rhodymenia palmata* and *Porphyra umbilicalis* and the amino acid sequences of their *N*-terminal regions. Phytochemistry 14:1493–1497
Mellstrand ST, Samuelsson G (1974) Phoratoxin, a toxic protein from the mistletoe *Phoradendron tomentosum* subsp. macrophyllum (Loranthaceae): The amino acid sequence. Acta Pharm Suec II:347–360
Melville JC, Ryan CA (1972) Chymotrypsin inhibitor I from potatoes. J Biol Chem 247:3445–3453
Milne PR, Wells JRE, Ambler RP (1974) The amino acid sequence of plastocyanin from french bean (*Phaseolus vulgaris*). Biochem J 143:691–701
Mitchel REJ, Chaiken IM, Smith EL (1970) The complete amino acid sequence of papain. J Biol Chem 245:3485–3492
Mole LE, Goodfriend L, Lapkoff CB, Kehoe JM, Capra JD (1975) The amino acid sequence of ragweed pollen allergen Ra5. Biochemistry 14:1216–1220
Moreira MA, Hermodson MA, Larkins BA, Nielsen NC (1979) Partial characterisation of the acidic and basic polypeptides of glycinin. J Biol Chem 254:9921–9926
Mori E, Morita Y (1978) Primary structure of rice cytochrome *c*. Agric Biol Chem 42:1079–1080
Mori E, Morita Y (1980) Amino acid sequence of cytochrome *c* from rice. J Biochem (Tokyo) 87:249–266
Morita Y, Ida S (1972) A preliminary crystallographic investigation of rice cytochrome *c*. J Mol Biol 71:807–808
Morita Y, Yagi F, Ida S, Asada K, Takahashi M (1973) A preliminary X-ray crystallographic study of spinach cytochrome *c*. FEBS Lett 31:186–188

Nakano K, Wakabayashi S, Hase T, Matsubara H, Fukui T (1978) Amino acid sequence around the pyridoxal 5'-phosphate binding site in potato phosphorylase. J Biochem (Tokyo) 83:1085–1094
Nakano K, Fukui T, Matsubara H (1980a) Sequence homology between potato and rabbit muscle phosphorylases. J Biochem (Tokyo) 87:919–927
Nakano K, Fukui T, Matsubara H (1980b) Structural basis for the difference of the regulatory properties between potato and rabbit muscle phosphorylases. J Biol Chem 255:9255–9261
Odani S, Ikenaka T (1972) Studies on soybean trypsin inhibitors IV. Complete amino acid sequence and the antiproteinase sites of Bowman-Birk soybean proteinase inhibitor. J Biochem (Tokyo) 71:839–848
Odani S, Ikenaka T (1977) Studies on soybean trypsin inhibitors. J Biochem (Tokyo) 82:1523–1531
Odani S, Ikenaka T (1978a) Studies on soybean trypsin inhibitors. J Biochem (Tokyo) 83:737–745
Odani S, Ikenaka T (1978b) Studies on soybean trypsin inhibitors XIV. Change of the inhibitory activity of Bowman-Birk inhibitor upon replacements of the α-chymotrypsin reactive site serine residue by other amino acids. J Biochem (Tokyo) 84:1–9
Odani S, Odani S, Ono T, Ikenaka T (1979) Proteinase inhibitors from a Mimosoideae legume, *Albizzia julibrissin*. J Biochem (Tokyo) 86:1795–1805
Odani S, Ono T, Ikenaka T (1980) The reactive site amino acid sequences of silktree (*Albizzia julibrissin*) seed proteinase inhibitors. J Biochem (Tokyo) 88:297–301
Ohmiya M, Hara I, Matsubara H (1980) Pumpkin (*Cucurbita* sp.) seed globulin IV. Terminal sequences of the acidic and basic peptide chains and identification of a pyroglutamyl peptide chain. Plant Cell Physiol 21:157–167
Ohtani S, Okada T, Yoshizumi H, Kagamiyama H (1977) Complete primary structures of two subunits of purothionin A, a lethal protein for brewers yeast from wheat flour. J Biochem (Tokyo) 82:753–767
Olson T, Samuelsson G (1972) The amino acid sequence of viscotoxin A2 from the European mistletoe (*Viscum album* L., Loranthaceae). Acta Chem Scand 26:585–595
Ozaki Y, Wada K, Hase T, Matsubara H, Nakanishi T, Yoshizumi H (1980) Amino acid sequence of a purothionin homolog from barley flour. J Biochem (Tokyo) 87:549–555
Patey AL, Evans DJ, Tiplady R, Byfield PGH, Matthews EW (1975) Sequence comparisons of γ-gliadin and coeliac-toxic α-gliadin. Lancet II:718
Patthy L, Smith EL, Johnson J (1973) Histone III. The amino acid sequence of pea embryo histone III. J Biol Chem 248:6834–6840
Petrucci T, Sannia G, Parlamenti R, Silano V (1978) Structural studies of wheat monomeric and dimeric protein inhibitors of α-amylase. Biochem J 173: 229–235
Poulsen C, Strøbaek S, Haslett BG (1976) Studies on the primary structure of the small subunit of ribulose-1,5-diphosphate carboxylase. In: Bücher T (ed) Genetics and biogenesis of chloroplasts and mitochondria. Elsevier, Amsterdam, pp 17–24
Poulsen C, Martin B, Svendsen I (1979) Partial amino acid sequence of the large subunit of ribulosebisphosphate carboxylase from barley. Carlsberg Res Commun 44:191–199
Ramshaw JAM, Boulter D (1975) The amino acid sequence of cytochrome *c* from niger-seed, *Guizotia abyssinica*. Phytochemistry 14:1945–1949
Ramshaw JAM, Thompson EW, Boulter D (1970) The amino acid sequence of *Helianthus annuus* L. (Sunflower) cytochrome *c* deduced from chymotryptic peptides. Biochem J 119:535–539
Ramshaw JAM, Richardson M, Boulter D (1971) The amino acid sequence of the cytochrome *c* of *Ginkgo biloba* L. Eur J Biochem 23:475–483
Ramshaw JAM, Scawen MD, Bailey CJ, Boulter D (1974a) The amino acid sequence of plastocyanin from *Solanum tuberosum* L. (potato). Biochem J 139:583–592
Ramshaw JAM, Scawen MD, Boulter D (1974b) The amino acid sequence of plastocyanin from *Vicia faba* L. (broad bean). Biochem J 141:835–843
Ramshaw JAM, Scawen MD, Jones EA, Brown RH, Boulter D (1976) The amino acid sequence of plastocyanin from *Lactuca sativa* (lettuce). Phytochemistry 15:1199–1202

Rao KK, Matsubara H (1970) The amino acid sequence of taro ferredoxin. Biochem Biophys Res Commun 38:500–506
Reddy MN, Keim PS, Heinrikson RL, Kézdy FJ (1975) Primary structural analysis of sulfhydryl protease inhibitors from pineapple stem. J Biol Chem 250:1741–1750
Redman DG (1975) Structural studies on wheat (*Triticum aestivum*) proteins lacking phenylalanine and histidine residues. Biochem J 149:725–732
Redman DG (1976) N-terminal amino acid sequence of wheat proteins that lack phenylalanine and histidine residues. Biochem J 155:193–195
Reeke GN, Becker JW, Edelman GM (1975) The covalent and three-dimensional structure of concanavalin A. J Biol Chem 250:1525–1547
Reeke GN, Becker JW, Edelman GM (1978) Changes in the three-dimensional structure of concanavalin A upon demetallization. Proc Natl Acad Sci USA 75:2286–2290
Rees DC, Lipscomb WN (1980) Structure of the potato inhibitor complex of carboxypeptidase A at 2.5-Å resolution. Proc Natl Acad Sci USA 77:4633–4637
Richardson C, Behnke WD, Freisheim JH, Blumenthal KM (1978) The complete amino acid sequence of the α-subunit of pea lectin, *Pisum sativum*. Biochim Biophys Acta 537:310–319
Richardson M (1974) Chymotryptic inhibitor I from potatoes: the amino acid sequence of subunit A. Biochem J 137:101–112
Richardson M (1979) The complete amino acid sequence and the trypsin reactive (inhibitor) site of the major proteinase inhibitor from the fruits of aubergine (*Solanum melongena* L.). FEBS Lett 104:322–326
Richardson M, Cossins L (1974) Chymotryptic inhibitor I from potatoes: the amino acid sequences of subunits B, C and D. FEBS Lett 45:11–13 (FEBS Lett 52:161 Corrigendum)
Richardson M, Ramshaw JAM, Boulter D (1971) The amino acid sequence of rape (*Brassica napus* L.) cytochrome *c*. Biochem Biophys Acta 251:331–333
Richardson M, Dilworth MJ, Scawen MD (1975) The amino acid sequence of leghaemoglobin I from root nodules of broad bean (*Vicia faba* L.) FEBS Lett 51:33–37
Richardson M. McMillan RT, Barker RDJ (1976) The protomer isoinhibitors of chymotryptic inhibitor I from potatoes. Biochem Soc Trans 4:1107–1108
Richardson M, Barker RDJ, McMillan RT, Cossins LM (1977) Identification of the reactive (inhibitory) sites of chymotryptic inhibitor I from potatoes. Phytochemistry 16:837–839
Rodrigues JDA, Brandt WF, Von Holt C (1979) Plant histone 2 from wheat germ, a family of histone H2A variants. Partial amino acid sequences. Biochim Biophys Acta 578:196–206
Ryan CA (1973) Proteolytic enzymes and their inhibitors in plants. Annu Rev Plant Physiol 24:173–196
Ryden L, Lundgren J-O (1979) On the evolution of blue proteins. Biochimie 61:781–790
Samuelsson G, Jayawardene AL (1974) Isolation and characterization of viscotoxin 1-PS from *Viscum album* subspecies *austriacum* growing on *Pinus silvestris*. Acta Pharm Suec 11:175–184
Samuelsson G, Pettersson BM (1971) The amino acid sequence of viscotoxin B from the european mistletoe (*Viscum album* L., Loranthaceae). Eur J Biochem 21:86–89
Samuelsson G, Seger L, Olson T (1968) The amino acid sequence of oxidized viscotoxin A3 from the European mistletoe (*Viscum album* L, Loranthaceae). Acta Chem Scand 22:2624–2642
Scawen MD, Boulter D (1974) The amino acid sequence of plastocyanin from *Cucurbita pepo* L. (vegetable marrow). Biochem J 143:257–264
Scawen MD, Ramshaw JAM, Boulter D (1974) The amino-acid sequence of plastocyanin from *Sambucus nigra* L. (elder). Eur J Biochem 44:299–303
Scawen MD, Ramshaw JAM, Boulter D (1975) The amino acid sequence of plastocyanin from spinach (*Spinacia oleracea* L.). Biochem J 147:343–349
Scawen MD, Ramshaw JAM, Brown RH, Boulter D (1978) The amino acid sequence of plastocyanin from *Mercurialis perennis* and *Capsella bursa-pastoris*. Phytochemistry 17:901–905
Schiltz E, Palm D, Klein HW (1980) N-terminal sequences of *Escherichia coli* and potato phosphorylase. FEBS Lett 109:59–62

Schloss JV, Stringer CD, Hartman FC (1978) Identification of essential lysyl and cysteinyl residues in spinach ribulose bisphosphate carboxylase/oxygenase modified by the affinity label N-bromoacetylethanolamine phosphate. J Biol Chem 253:5707–5711

Schmidt GW, Devillers-Thiery A, Desruisseaux H, Blobel G, Chua N-H (1979) $NH_2$-terminal amino acid sequences of precursor and mature forms of the ribulose-1,5-bisphosphate carboxylase small subunit from *Chlamydomonas reinhardtii*. J Cell Biol 83:615–622

Schmitt JM, Svendsen I (1980) Amino acid sequences of hordein polypeptides. Carlsberg Res Commun 45:143–148

Scogin R, Richardson M, Boulter D (1972) The amino acid sequence of cytochrome *c* from tomato (*Lycopersicon esculentum* Mill.). Arch Biochem Biophys 150:489–492

Sebald W, Wachter E (1980) Amino acid sequence of the proteolipid subunit of the ATP synthase from spinach chloroplasts. FEBS Lett 122:307–311

Sebald W, Hoppe J, Wachter E (1979) Amino acid sequence of the ATPase proteolipid from mitochondria, chloroplasts and bacteria (wild type and mutants). In: Quagliariello E, Palmieri F, Papa S, Klingenberg M (eds) Function and molecular aspects of biomembrane transport. Elsevier/North Holland, Amsterdam New York, pp 63–74

Sherry AD, Buck AE, Peterson CA (1978) Sugar binding properties of various metal ion induced conformations in concanavalin A. Biochemistry 17:2169–2173

Shewry PR, Autran JC, Nimmo CC, Lew EJL, Kasarda DD (1980a) N-terminal amino acid sequence homology of storage protein components from barley and a diploid wheat. Nature (London) 286:520–522

Shewry PR, March JF, Miflin BJ (1980b). *N*-terminal amino acid sequence of *C* hordein. Phytochemistry 19:2113–2115

Shin M, Yokoyama Z, Abe A, Fukasawa H (1979) Properties of common wheat ferredoxin, and a comparison with ferredoxins from related species of *Triticum* and *Aegilops*. J Biochem (Tokyo) 85:1075–1081

Shoham M, Kalb AJ, Pecht I (1973) Specificity of metal ion interaction with concanavalin A. Biochemistry 12:1914–1917

Shoham M, Yonath A, Sussman JL, Moult J, Traub W, Kalb AJ (1979) Crystal structure of demetallized concanavalin A: the metal binding region. J Mol Biol 131:137–155

Sievers G, Huhtala ML, Eufolk N (1978) The primary structure of soybean (*Glycine max*) leghemoglobin *c*. Acta Chem Scand B32:380–386

Solomon EI, Hare JW, Dooley DM, Dawson JH, Stephens PJ, Gray HB (1980) Spectroscopic studies of stellacyanin, plastocyanin, and azurin. Electronic structure of the blue copper sites. J Am Chem Soc 102:168–178

Stevens FC, Glazer AN, Smith EL (1967) The amino acid sequence of wheat germ cytochrome *c*. J Biol Chem 242:2764–2779

Stevens FC, Wuerz S, Krahn J (1974) Structure function relations in lima bean protease inhibitor. In: Fritz H, Tschesne H, Greene LJ, Truscheit E (eds) Proteinase inhibitors. Springer, Berlin Heidelberg New York, pp 344–354

Stringer CD, Hartman FC (1978) Sequences of two active site peptides from spinach ribulosebisphosphate carboxylase/oxygenase. Biochem Biophys Res Commun 80:1043–1048

Strøbaek S, Gibbons GC, Haslett B, Boulter D, Wildman SG (1976) On the nature of the polymorphism of the small subunit of ribulose-1,5-diphosphate carboxylase in the amphidiploid *Nicotiana tabacum*. Carlsberg Res Commun 41:335–343

Sugeno K, Matsubara H (1969) The amino acid sequence of *Scenedesmus* ferredoxin. J Biol Chem 244:2979–2989

Sullivan D, Brisson N, Goodchild B, Verma DPS, Thomas DY (1981) Molecular cloning and organisation of two leghaemoglobin genomic sequences of soybean. Nature (London) 289:516–518

Sun SM, Slightom JL, Hall TC (1981) Intervening sequences in a plant gene – comparison of the partial sequence of cDNA and genomic DNA of french bean phaseolin. Nature (London) 289:37–41

Svendsen I, Martin B, Jonassen I (1980) Characteristics of Hiproly barley III. Amino acid sequences of two lysine-rich proteins. Carlsberg Res Commun 45:79–85

Sweet RM, Wright HT, Janin J, Chothia CH, Blow DM (1974) Crystal structure of the complex of porcine trypsin with soybean trypsin inhibitor (Kunitz) at 2.6 Å resolution. Biochemistry 13:4212–4228
Takruri I, Boulter D (1979a) The amino acid sequence of ferredoxin from *Triticum aestivum* (wheat). Biochem J 179:373–378
Takruri I, Boulter D (1979b) The amino acid sequence of ferredoxin from *Sambucus nigra*. Phytochemistry 18:1481–1484
Takruri I, Boulter D (1980) The amino acid sequence of ferredoxin from *Brassica napus* (rape). Biochem J 185:239–243
Takruri I, Haslett BG, Boulter D, Andrew PW, Rogers LJ (1978) The amino acid sequence of ferredoxin from the red alga *Porphyra umbilicalis*. Biochem J 173:459–466
Tan CGL, Stevens FC (1971) Amino acid sequence of lima bean protease inhibitor component IV. 2. Isolation and sequence determination of the chymotryptic peptides and the complete amino acid sequence. Eur J Biochem 18:515–523
Thompson EW, Laycock MV, Ramshaw JAM, Boulter D (1970a) The amino acid sequence of *Phaseolus aureus* L. (mung-bean) cytochrome *c*. Biochem J 117:183–192
Thompson EW, Richardson M, Boulter D (1970b) The amino acid sequence of sesame (*Sesamum indicum* L.) and castor (*Ricinus communis* L.) cytochrome *c*. Biochem J 121:439–446
Thompson EW, Notton BA, Richardson M, Boulter D (1971a) The amino acid sequence of cytochrome *c* from *Abutilon theophrasti* Medic. and *Gossypium barbadense* L. (cotton). Biochem J 124:787–791
Thompson EW, Richardson M, Boulter D (1971b) The amino acid sequence of cytochrome *c* from *Cucurbita maxima* L. (pumpkin). Biochem J 124:779–781
Thompson EW, Richardson M, Boulter D (1971c) The amino acid sequence of cytochrome *c* of *Fagopyrum esculentum* Moench (buckwheat) and *Brassica oleracea* L. (cauliflower). Biochem J 124:783–785
Tsunoda JN, Yasunobu KT (1966) The amino acid sequence around the reactive thiol group of chymopapain B. J Biol Chem 241:4610–4615
Vainshtein BK, Arutyunyan EH, Kuranova IP, Borisov VV, Sosfenov NI, Pavlovskii AG, Grebenko AI, Konareva NV, Nekrasov YV (1977) Three-dimensional structure of lupine leghemoglobin with a resolution of 2.8 Å. Dokl Acad Nauk SSSR 233:238–241
Villafranca JE, Robertus JD (1981) Ricin B chain is a product of gene duplication. J Biol Chem 256:554–556
Wakabayashi S, Hase T, Wada K, Matsubara H, Suzuki K, Takaichi S (1978) Amino acid sequences of two ferredoxins from pokeweed, *Phytolacca americana*. J Biochem (Tokyo) 83:1305–1319
Wakabayashi S, Hase T, Wada K, Matsubara H, Suzuki K (1980) Amino acid sequences of two ferredoxins from *Phytolacca esculenta*. J Biochem (Tokyo) 87:227–236
Wallace DG, Brown RH, Boulter D (1973) The amino acid sequence of *Cannabis sativa* cytochrome *c*. Phytochemistry 12:2617–2622
Wang JL, Becker JW, Reeke GN, Edelman GM (1974) Favin, a crystalline lectin from *Vicia faba*. J Mol Biol 88:259–262
Wang JL, Cunningham BA, Waxdal MJ, Edelman GM (1975) The covalent and three-dimensional structure of concanavalin A. J Biol Chem 250:1490–1502
Wel H Van der, Loeve K (1972) Isolation and characterization of thaumatin I and II, the sweet-tasting proteins from *Thaumatococcus daniellii* Benth. Eur J Biochem 31:221–225
Welinder KG (1976) Covalent structure of the glycoprotein horseradish peroxidase (EC 1.11.1.7). FEBS Lett 72:19–23
Welinder KG (1979) Amino acid sequence studies of horseradish peroxidase. Eur J Biochem 96:483–502
Welinder KG, Mazza G (1977) Amino acid sequence of heme-linked, histidine-containing peptides of five peroxidases from horseradish and turnip. Eur J Biochem 73:353–358
Whittaker RG, Moss BA, Appleby CA (1979) Determination of the blocked N-terminal of soybean leghemoglobin b. Biochim Biophys Res Commun 89:552–558
Wildman SG (1979) Aspects of Fraction I protein evolution. Arch Biochem Biophys 196:598–610

Wilson KA, Laskowski M, Sr (1975) The partial amino acid sequence of trypsin inhibitor II from garden bean, *Phaseolus vulgaris*, with the location of the trypsin and elastase-reactive sites. J Biol Chem 250:4261–4267
Wlodawer A, Hodgson KO (1975) Crystallization and crystal data of monellin. Proc Natl Acad Sci USA 72:398–399
Wood PM (1978) Interchangeable copper and iron proteins in algal photosynthesis. Eur J Biochem 87:9–19
Wright CS (1977) The crystal structure of wheat germ agglutinin at 2.2 Å resolution. J Mol Biol 111:439–457
Wright CS (1980a) Location of the *N*-acetyl-D-neuraminic acid binding site in wheat germ agglutinin. J Biol Chem 139:53–60
Wright CS (1980b) Crystallographic elucidation of the saccharide binding mode in wheat germ agglutinin and its biological significance. J Mol Biol 141:267–291
Wright CS (1981) Histidine determination in wheat germ agglutinin isolectin by X-ray diffraction analysis. J Mol Biol 145:453–461
Yoshikawa M, Kiyohara T, Iwasaki T, Ishii Y, Kimura N (1979) Amino acid sequences of proteinase inhibitors II and II′ from adzuki beans. Agric Biol Chem 43:787–796
Yoshitake S, Funatsu G, Funatsu M (1978) Isolation and sequences of peptic peptides, and the complete sequence of Ile chain of ricin D. Agric Biol Chem 42:1267–1274

# 8 Protein Types and Distribution

M.-N. Miège

## 1 Introduction

Proteins, which are the central molecules of cellular processes, are defined chemically as holo- or conjugated proteins. Requirements involved in their isolation and analysis have led to operational classifications based on their solubility characteristics. However, proteins so defined, i.e., as albumins, globulins, prolamins, glutelins, play very different roles in cells, and therefore must be distinguished from a physiological viewpoint also, i.e., as structural proteins, enzymatic proteins, etc.

In this chapter, the distribution of different types of proteins will be considered. In order to characterize proteins, it is necessary to extract and purify them. However, proteins in vivo go through various post-translational modifications before they assume their final form and these aspects, i.e., changes in time and location, are also considered.

The relationship between structure and function will be emphasized, and representative plant proteins such as reserve proteins, enzymes, lectins, cell wall proteins and pollen proteins will be used to illustrate the principles involved. Lastly, brief mention will be made of nutritional aspects when specific plant proteins are of importance as food or feeds.

## 2 Protein Types with Regard to Their Chemistry, Physiology, Histology and Ontogeny

### 2.1 Chemical Types

#### 2.1.1 Holoproteins

The sequence of amino acids in the polypeptide chains of proteins constitutes the primary structure. Further interactions between side groups give additional secondary structures, such as helices or folded sheets. In addition to the secondary structure, the polypeptide chain may coil into a compact molecule by intramolecular covalent linkages, e.g., disulphide bonds, or low-energy linkages (hydrogen, hydrophobic and ionic bonds). Furthermore, polypeptide chains can associate together under specific conditions and form dimers or oligomers of higher order. Associated subunits in oligomeric molecules are called "protomers", whereas "monomer" is a term used to describe the dissociated state

(YON 1969); the subunit (polypeptide) association in oligomeric molecules constitutes the protein's quaternary structure.

These constructional principles, although initially used to describe animal proteins, apply to plant proteins as well. BOULTER (1965) pointed out the large molecular diversity that this constructional uniformity gives rise to: in any one cell there may be tens of thousands of different protein species.

### 2.1.2 Conjugated Proteins

One major reason accounting for the variety of proteins found in nature is their aptitude, through covalent linkages, to combine with non-protein components, e.g., to form nucleoproteins, lipoproteins and glycoproteins. Glycoproteins are of special interest; for further information on them see SHARON (1974).

### 2.1.3 Protein Solubility

Solubilization of proteins, the initial step in their isolation and purification, depends on their chemical composition, since a substance's solubility depends on its affinity with the solvent molecules. Proteins that are rich in amino acids with aliphatic hydrophobic side-chains, e.g., leucine and alanine, are not soluble in water. If, however, this type of protein is also rich in polar amino acids with hydroxyl or amide groups, it is soluble in organic polar solvents such as alcohol. This is the case with prolamins, the name of which is a contraction of the names of their two major amino acids: proline and glutamine. In order to be soluble in water, proteins must be rich in ionizable amino acids, e.g., arginine, lysine, glutamic acid, tryptophane, and albumins and globulins meet this condition. When this type of protein has a high molecular weight, as, for example, with globulins, solubilization in water requires the presence of salts; these ions spread the water dipoles surrounding protein molecules. Three classes of plant proteins can thus be separated according to their behaviour in aqueous solvents: (1) albumins, soluble in pure water, (2) globulins, soluble in dilute saline solutions, (3) prolamins, soluble in alcohol.

A last category of plant proteins, glutelins, are those which are not so easily solubilized. This requires the breakage of molecular disulphide bonds by reducing agents or of electrostatic linkages by alkalis or acids. Glutelins are generally of high molecular weight (up to one million) and this property makes them viscous in solution.

### 2.1.4 Protein Stability to Heat and Cold

It appears that stability of some enzymes to different temperatures is ensured in part by the presence in them of carbohydrate residues. Thus, glycoproteins are more resistant to heat and other denaturing agents than some other proteins (MARSHALL and RABINOWITZ 1975). It has also been shown, for instance, that glycoenzymes such as glycoamylases have a greater stability in low temperature storage conditions than the same enzymes in which constituent carbohydrate residues have been oxidized (PAZUR et al. 1970). Cryoprotection by constituent polysaccharides may contribute to the resistance of certain plants to the cold, but it has been suggested that the main resistance factor to freezing is the presence of proteins rich in hydrophilic amino acids which allow for vital water retention (ROCHAT and THERRIEN 1975). VOLGER and HEBER (1975) have characterized two such cryoprotective polypeptides in the leaves of some cold-resisting species; they are thermostable, soluble, have a molecular weight of 1,000 to 2,000 and are rich in polar amino acids. These polypeptides were 1,000 times more effective in protecting thylakoid membranes against freezing than low molecular weight protective compounds such as sucrose or glycerol.

On the other hand, some proteins are reversibly insolubilized at low temperatures, a well-established phenomenon in animals. The same type of cryoprecipitation is found in many seeds, however (GHETIE and BUZILA 1962, MIÈGE 1970, TULLY and BEEVERS 1976, DAUSSANT et al. 1969).

## 2.2 Metabolic and Structural Proteins

In animals, the borderline between metabolic and structural proteins is quite clear: fibrous proteins are characteristic of supportive tissues, e.g., keratin and collagen, whereas metabolic proteins are globular and soluble in the cytosol

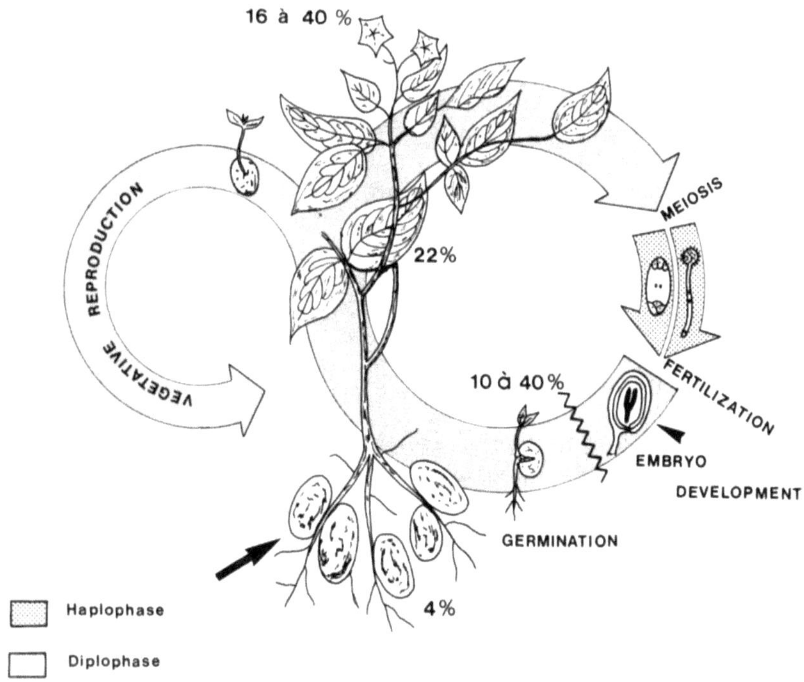


**Fig. 1.** Life cycle of an angiosperm. *Arrows* indicate the occurrence of storage proteins in reserve tissues designed to ensure vegetative or sexual reproduction. *Numbers* indicate average nitrogen amount (% dry weight) in tubers, leaves, pollen and seeds

or in circulatory and secretory fluids. In plants, however, this distinction is not as clear-cut. Some structural proteins, although only slightly soluble, are not fibrous, for example cereal glutelins are non-fibrous proteins which may be considered as structural proteins. Conversely, some structural proteins, e.g., cell wall extensin, appear to have metabolic activity (Ridge and Osborne 1970).

However in plants, as well as in animals, the two fundamental types of protein can be discerned: structural proteins, which are generally high in molecular weight and insoluble, (e.g., glutelin, extensin) and metabolic proteins, which are readily soluble (e.g., enzymes, lectins). Storage proteins constitute a third type of proteins, of much greater prominence in plants than in animals; they are generally of high molecular weight and more or less soluble. Their importance is due to the role that storage plays in both vegetative and sexual multiplication, both relying on the dispersal of dormant organs (bulbs, tubers, seeds) containing reserves (Fig. 1). These reserves initially ensure autonomy of the dispersed organs on their reactivation.

## 3 Storage Proteins of Mature Seeds: Nature, Function and Ultrastructural Localization

The role of storage proteins during seed development has been the subject of much research, some of which has found practical applications in nutrition and agronomy. Seventy percent of the edible proteins produced in the world come from seeds (Spencer and Higgins 1979). One crucial problem humanity will have to face in the coming decades, is that of world nutrition, and its solution depends largely on improving plant cultivation and seed production. The most widely studied seeds, those of cereals and legumes, differ in that cereal seeds have an endosperm (storage tissue), whereas in legumes the embryo usually digests the endosperm before desiccation of the seed and stores its reserves instead in two cotyledonary leaves (Fig. 2).

### 3.1 Storage Proteins in Legume Seeds

What is the exact definition of a reserve protein? Spencer and Higgins (1979) based their definition, at least as far as the Leguminosae are concerned, on their location in the protein bodies of seeds. However, Derbyshire et al. (1976) noted that when a protein has been isolated from a seed, it has rarely been shown to be located in protein bodies in vivo. These authors therefore enlarged the definition of storage proteins and suggested that extracted proteins which represent 5% or more of the total proteins of seeds should be considered initially as potential storage proteins. In members of the Leguminosae, up to 80% of the total proteins in the seed appear to be reserve proteins. The storage role of a seed protein is confirmed by its rapid degradation during seedling establishment.

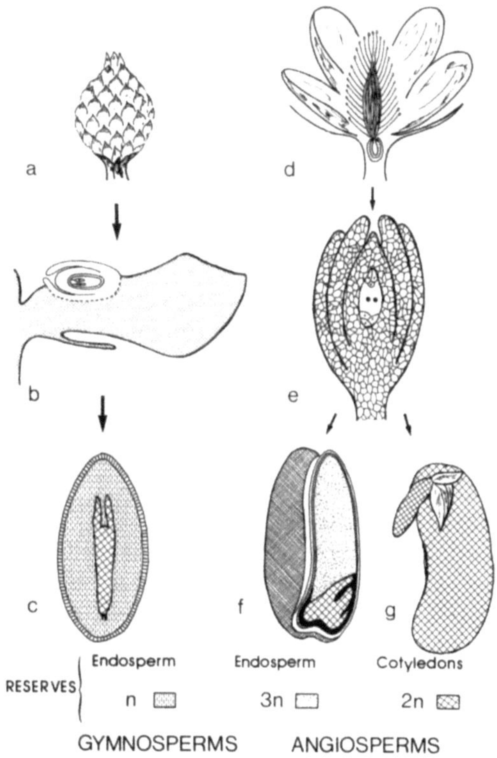


**Fig. 2a–g.** Seed formation in gymnosperms and angiosperms. Each scale of a female cone (**a**) harbours an ovule (**b**) that develops into a seed after fertilisation (**c**). The ovary, central organ of a flower (**d**) encloses the ovule (**e**) that develops after fertilisation into a seed with endosperm (**f**) or without endosperm (**g**)

### 3.1.1 Globulins

Chemical data on reserve proteins of the flowering plant family, Fabaceae, are of great biological interest. These proteins are found only in the seeds and are therefore of particular importance in the study of differential expression of genes (see MILLERD 1975, DERBYSHIRE et al. 1976).

The storage proteins of the Leguminoseae are globulins. They were first referred to as "legumins" but when their heterogeneity was discovered, the term "legumin" was reserved for the main storage globulin of some species (i.e., *Pisum sativum, Vicia faba,* and other members of the Viciae) (CROY et al. 1979).

There are two types of storage globulins in seeds: legumin has a high molecular weight (about 330,000), precipitates from saline extracts at its isoelectric point (pH 4.8) and is not coagulable at high temperatures. The other type of globulin, vicilin, has a molecular weight of about 186,000, an isoelectric point of pH 5.5, is more readily soluble than legumin, is stable in solution at pH 5, but coagulates at a temperature of 95 °C. When analyzed in the ultracentrifuge, legumin and vicilin sediment as apparently unique components with a sedimentation constant of 11–12S for legumin and 7S for vicilin (DERBYSHIRE et al. 1976). However, both legumin and vicilin are heterogeneous (see below).

7 S and 11 S globulins can be partially separated on dextran gels, followed by several different ion-exchange chromatographic steps, which complete their separation. Purification and characterization are monitored by electrophoretic methods including isoelectrofocusing (DERBYSHIRE et al. 1976). Although present separatory methods are adequate for the purification of legume storage proteins, problems due to their association with phytin and other molecules may arise.

### *3.1.1.1 Legumin of Viciae*

Legumin and vicilin have been studied in detail from *Vicia faba, Pisum sativum* and *Glycine max* and the following review will principally examine these data. In addition, in Section 3.1.1.5 the globulins of *Phaseolus* will be considered, as these storage proteins appear to differ from the other vicilin and legumin types described.

Legumin was isolated from extracts of *Vicia faba* at neutral pH by isoelectric precipitation (BAILEY and BOULTER 1970). Dye binding and in vivo radiochemical labelling experiments suggested that legumin, molecular weight 320,000, was made up of three subunits of molecular weight 56,000, 42,000 and 23,000 in a 1:3:6 ratio. In order to obtain a pure legumin fraction, it was necessary to repeat the isoelectric precipitation step and, after each precipitation step, part of the globulins was lost through irreversible insolubilization. WRIGHT and BOULTER (1973 and 1974) were able to overcome this difficulty by means of zonal isoelectric precipitation which consists in absorbing the preparation of crude globulins on Sephadex at the isoelectric point of the legumin (pH 4.7). Vicilin, soluble at this pH, is excluded and legumin is then eluted by raising the pH.

A polymeric constitution was proposed for the molecular form of molecular weight 342,000, consisting of $\alpha$ and $\beta$ subunits of 35,800 and 21,300 respectively:

$$\alpha_6\beta_6 \xrightarrow{\text{sodium dodecyl sulphate}} 6\alpha\beta \xrightarrow[\text{+mercaptoethanol}]{\text{sodium dodecyl sulphate}} 6\alpha+6\beta$$

CROY et al. (1979) established details of the complex structure of *Vicia* legumin after studying it by isoelectrofocusing. The molecule is made up of pairs of acidic subunits (molecular weight about 40,000) and basic subunits (molecular weight about 20,000). KRISHNA et al. (1979) demonstrated a similar structure for the *Pisum* legumin with the occurrence of at least five different pairs of acidic and basic subunits although not necessarily belonging to any one molecule. In comparing the legumin isolated from *Vicia faba* to that of *Pisum sativum*, CROY et al. (1979) showed a similar subunit structure in both legumins with six pairs of subunits held together by disulphide bonds to give molecular weights of 390,000 for *Pisum* legumin and 347,000 for *Vicia* legumin. In addition to the differences in molecular weight, other slight differences were demonstrated: *Pisum* legumin was more acidic at pH values of neutral and above, and had larger acidic subunits than *Vicia* legumin. Nevertheless, the homology between both legumins was confirmed by tryptic peptide maps and by their serological behaviour; the legumin of *Pisum* had all the antigenic determinants of that of *Vicia* and one or more determinant group in addition.

GATEHOUSE et al. (1980) demonstrated further heterogeneity of the *Pisum* legumin. Both acidic and basic subunits were shown to be heterogeneous in charge and molecular weight by two-dimensional analysis employing isoelectric focusing in the first dimension, and sodium dedecyl sulphate/polyacrylamide gel electrophoresis in the second. This heterogeneity could not be attributed to varying amounts of carbohydrate residues, since no carbohydrate could be detected by using both a fluorescent-labelling technique and a sensitive radioactive-labelling technique.

Subunit heterogeneity and the absence of covalently linked carbohydrate were also confirmed by CASEY (1979), who proposed a rapid and powerful means of purifying legumin of *Pisum* seeds by zonal isoelectric precipitation followed by immunoaffinity chromatography on immobilized anti-legumin. Some physicochemical properties of purified *Pisum* legumin were determined, i.e., molecular weight 395,000 calculated from gel filtration data, $S_{20w}12.23$, Stokes radius 71 nm, subunit molecular weights 40,370, 38,800, 37,600, 22,280 and subunit molar ratios (large/small) 1.24 or 0.94 according to the method used.

*Vicia* and *Pisum* are members of the Viciae tribe. The legumin of *Glycine,* a member of the Phaseolae tribe, has also been purified and submitted to physicochemical analysis. It has a similar structure to *Vicia* legumin (CATSIMPOOLAS and WANG 1971). KOSHIYAMA and FUKUSHIMA (1976) showed that *Glycine* legumin is compactly folded in aqueous solution; it has little α-helical structure but has an intramolecular cross β-structure stabilized by hydrophobic and hydrogen bonds. The molecular shape was estimated to be that of an oblate ellipsoid and the globulin exists in a rigid and nearly anhydrous state in solution. WRIGHT and BOULTER (1980) demonstrated that the denaturation parameters of the 11 S proteins (legumins) of *Glycine max* and *Vicia faba* were very similar, whereas those of the 7S proteins of the two taxa were not.

### *3.1.1.2 Vicilin of Viciae*

Vicilin is best characterized from *Vicia faba*. It is a complex multimeric globulin (SCHLESIER et al. 1978a). Ultracentrifugation analysis suggested a sedimentation constant of 7S and a molecular weight of 186,000, i.e., a value smaller than that deduced from the sum of the subunits as revealed in the presence of sodium dodecyl sulphate (219,000) (BAILEY and BOULTER 1972). Furthermore, the molecular weight deduced from the tryptic peptide map and from N-terminal amino acids was only 130,000, indicating that the three or four major polypeptides contain regions of identical amino acid sequences. In view of this chemical evidence, the cistrons which code for the constituent vicilin polypeptides are probably related. Gene duplication with subsequent independent development would explain vicilin's complexity. BAILEY and BOULTER (1972) suggested that multiple genes coding for storage proteins could ensure their fast synthesis during the short period of deposition of reserves and therefore maximize the rate of seed development and represent a selective advantage for the plant.

Such a process of gene duplication and subsequent modifications probably also partly accounts for the variations in globulin structure, between species and even varieties. Because these changes do not affect the seed's viability, it is clear that they meet packing and unpacking ultrastructural requirements as well as supplying nitrogen in a suitable form at germination (DERBYSHIRE et al. 1976).

MANTEUFFEL and SCHOLZ (1975) used two-dimensional electrophoresis in addition to the above purification techniques and showed that vicilin, completely freed from legumin, still exhibited some serological heterogeneity. It contained four subunits with molecular weights of 66,000, 60,000, 56,000 and 36,000. Recent unpublished data of BOULTER et al. indicate that vicilin is a molecule of about 150,000 molecular weight and contains three subunits of about 50,000. The subunits differ slightly in molecular weight and two-dimensional methods showed several different 50,000-type subunits which varied in both charge and molecular weight. Clearly not all the different possible 50,000-type subunits can exist in a single vicilin molecule. Some of the 50,000 subunits are "nicked" and in SDS give rise to lower molecular weight subunits 33,000, 17,000, 13,000, 12,000.

Unlike legumin, vicilin is glycosylated. BAILEY and BOULTER (1972) showed that *Vicia faba* vicilin contains small but significant quantities of neutral sugars (0.5% w/w) but insignificant quantities of hexosamines ($\leqq$0.2% w/w). Due to the presence of covalently linked sugars, vicilins of *Vicia* and *Pisum* are precipitable by concanavalin A, unlike *Vicia* and *Pisum* legumin (CROY et al. 1979), further evidence that legumin does not contain covalently linked sugars of the glucose/mannose/glucosamine type.

BASHA and BEEVERS (1976) reported 0.3% neutral sugar and 0.2% amino sugars in *Pisum sativum* vicilin, but also reported that legumin is glycosylated with 1% neutral sugar and 0.1% amino sugar. CASEY (1979), however, using very sensitive detection methods, could find only negligible amounts of neutral sugars in *Pisum sativum* legumin (<0.1% w/w). It is possible, but unlikely, that Basha and Beevers' results differ due to genetically determined variation in legumin carbohydrate content.

Subunit heterogeneity of vicilin is unlikely to result from in vitro globulin proteolysis by endogeneous proteases since BLAGROVE and GILLESPIE (1978) and others have noted that slow extraction of seed flours gave the same subunit pattern as did immediate extraction with boiling sodium dodecyl sulphate. The proteolysis process, if it occurs, must therefore do so in vivo during seed storage and not in vitro during extraction. In fact, GRANGE (1976, 1980) noted that a start to globulin degradation takes place under certain storage conditions in *Phaseolus vulgaris* seeds.

### *3.1.1.3 Convicilin of Viciae*

Recently a third storage protein, distinct from legumin and vicilin, was isolated from seeds of *Pisum sativum* and analyzed by CROY et al. (1980). Because of its serological cross-reactivity with vicilin, the authors named this new globulin "convicilin" until a systematic nomenclature for seed storage proteins is developped. It is a globulin made of 71,000 molecular weight subunits associated in a tetrameric native molecule of molecular weight 290,000. Convicilin has a distinctive amino acid composition, particularly with respect to sulphur amino acids, having less cysteine and methionine residues than legumin but more than vicilin, which has none. Convicilin gave a serological reaction of non-identity with legumin and is not a glycoprotein. Limited heterogeneity was apparent on isoelectrofocusing, N-terminal analysis and CNBr cleavage. This new storage globulin was located in protein bodies and its accumulation takes place towards the end of the storage protein deposition stage.

### *3.1.1.4 Need of Nomenclature for Globulins*

In the absence of a systematic nomenclature for the seed storage proteins, these are often identified by their sedimentation coefficient. Thus 11S globulins are often present in legume seeds and were found, with very few exceptions in 34 species examined by DANIELSSON (1949). However, on account of the uncertainty of homology between "legumins" of different origins, HALL et al. (1972) preferred not to use the term "legumin", whilst MILLERD (1975) suggested reserving this term for the 11S globulin in members of the Viciae (*Vicia, Cicer, Lens, Lathyrus, Pisum*) and Trifoliae (*Medicago, Trigonella, Ononis, Melilotus*). These objections were based principally on the suggestion that a 11S globulin is absent from *Phaseolus* seeds. However, ERICSON and CHRISPEELS (1973) isolated an 11.3S legumin in *Vigna radiata* (formerly *Phaseolus aureus*) seeds, and DERBYSHIRE and BOULTER (1976) isolated a legumin-like protein from seeds of *Vigna radiata* (*Phaseolus aureus*) and *Phaseolus vulgaris*. This protein was comparable to the legumin found in *Vicia* since it had the same sedimentation coefficient, a similar amino acid composition, the same N-terminal amino acids, a low carbohydrate content and a low solubility at pH 4.7 and ionic strength 0.3. Furthermore, GILROY et al. (1979) have recently shown homology between glycinin, the 11S protein of *Glycine*, a member of the Phaseolae tribe, and legumin of *Vicia faba*. Both proteins have similar molecular weights, similar numbers of subunits, both have di-sulphide bonded acidic and basic subunits, and approximately two thirds of the partial amino acid sequence positions have the same residues.

However, there is only a small amount of 11S globulins in members of Phaseolae (DERBYSHIRE and BOULTER 1976). According to CARASCO et al. (1978), the legumin/vicilin ratio is 4 in seeds of some *Vicia faba* varieties and only 1:9 in *Phaseolus*. 11S globulin is found also in very small amounts in *Vigna unguiculata* seeds. It seems therefore that legumin-like proteins are not essential, during seed development, for the formation of functional protein bodies; in fact GILLESPIE and BLAGROVE (1978) noted the absence of an 11 S globulin in the seeds of *Psophocarpus tetragonolobus* where 6 S and 2 S globulins constitute the protein reserves.

Legumin-like proteins are probably widespread in dicotyledonous plants (see DERBYSHIRE et al. 1976, SCHWENKE et al. 1977, RAO et al. 1978).

### *3.1.1.5 Phaseolus Globulins*

In order to isolate agglutinins from extracts of *Phaseolus vulgaris* seeds, PUSZTAI and WATT (1970) first removed the pH 5 insoluble globulin fraction (legumin-like). The major soluble fraction then corresponded to vicilin and before further purification the fraction contained agglutinating substances. The soluble fraction was purified by precipitation with ammonium sulphate, high voltage electrophoresis, sieving on Sephadex G200 and chromatography on DEAE cellulose to give a purified 7S globulin called glycoprotein II (GP II). It showed only one electrophoresis band at neutral or slight alkaline pH, only one precipitation arc on immunodiffusion, and gave a single band on zonal equilibrium ultracentrifugation. Haemagglutinating activity was eliminated during the purification. At pH values between 2.3 and 3.4, this 7S globulin was constituted mainly of monomers of molecular weight 140,000 which associated in tetramers of molecular weight 560,000 between pH 3.4 and 6.6. Its carbohydrate moiety is formed mainly of D-mannose and D-glucosamine and typically it is poor in sulphur amino acids, especially cyst(e)ine.

A comparable glycoprotein fraction was characterized in the same species by ERICSON and CHRISPEELS (1973). It contains 0.2% of glucosamine and 1% of mannose.

SUN and HALL (1975) extracted globulins from *Phaseolus vulgaris* seeds using an acid solvent (pH 3) in order to prevent the formation of associations between globulin fractions during extraction. A G1 fraction was precipitated from acid extract by dilution, and the remaining soluble fraction was called G2. G1 was first identified as a legumin by MCLEESTER et al. (1973) and then later as the GPII of PUSZTAI and WATT, i.e., purified vicilin, by SUN and HALL (1975).

WRIGHT and BOULTER (1973) applied SUN and HALL's acid extraction method to *Vicia faba* globulins. Two fractions, I and II, were isolated, corresponding to Sun and Hall's G1 and G2. Fraction I was similar to the vicilin fraction by its 7S sedimentation coefficient and subunit pattern obtained in the presence of sodium dodecyl sulphate, but the 11S legumin of *Vicia,* when isolated in non-acid medium, dissociated at pH 3 into 7–8S components. This makes acidic extraction ambiguous and generally not useful, i.e., Gl and vicilin cannot necessarily be equated.

BARKER et al. (1976), using a different variety, reported a *Phaseolus* 7S globulin soluble at pH 4.7, which was comparable to Pusztai and Watt's GP II. It was made of two major subunits of molecular weight 50,000 and 47,000. At pH 6 it associated into an 18S molecular species and at lower temperature precipitated like a cryoprotein.

PUSZTAI and STEWART (1980) have shown GP II of *Phaseolus* to have a molecular weight of 142,000 and to consist of four subunits. Subunits of molecular weight 52,000, 50,000, 47,000 and 43,000 were separated after SDS dissociation and hence the protein population is heterogeneous.

In order to point out the uniqueness of the main globulin of *Phaseolus,* i.e., Pusztai and Watt's GP II, which is strictly neither a legumin nor a vicilin, MURRAY and CRUMP (1979) called it "euphaseolin". However, the problem of nomenclature and of homology of the legume globulins will not be solved until enough amino acid sequence data become available.

### *3.1.1.6 Other Globulins*

GILLESPIE and BLAGROVE (1978) reported the presence of a 2S globulin in *Psophocarpus tetragonolobus* seeds. SCHLESIER et al. (1978b) also found a 2 S globulin in *Vicia narbonensis* which they called "narbonin". It consisted of a single polypeptide chain of molecular weight 33,500 and did not coincide with any of the subunits of legumin or vicilin; furthermore, no serological homology was observed with any of the other seed globulins.

Another globulin of molecular weight about 100,000, studied by SUSHEELAMMA and RAO (1978), showed surface active properties, i.e., a foaming action and film formation on reduction, which is of interest since these play a role in food texture determination.

### 3.1.2 Nutritive Value of Leguminoseae

There is a growing interest in legume seeds as nutritive sources since the seeds have a high protein content (up to 50% of their dry weight). It is for this reason that they have been widely used since prehistoric times in many different cultures, e.g., *Phaseolus* in Brazil and in South America generally, *Dolichos* and *Vigna* in Africa and Asia and *Glycine* in China and S.E. Asia. Their principal deficiency as a protein source is their low content of sulphur amino acids.

Recently, possible new sources of plant proteins have attracted active research interest. MIÈGE et al. (1978), among others, has pointed out the possibilities of using *Psophocarpus tetragonolobus*, whose very large pods enclose big edible seeds which are rich in protein. MIÈGE and MIÈGE (1978) investigated another legume, *Cordeauxia edulis*, which grows in the poorest soils and manages to complete its life-cycle during the brief rainy season in the Ethiopian and Somalian deserts. Thus they are eaten even though their protein content is lower than that of other legume seeds (about 10% of the dry weight) due to a relatively low content of globulins. Correspondingly, the proportion of free amino acids is high and, interestingly, among them is an amino acid rarely found in plants but frequently in animals, namely sarcosin or methylglycine, which is present in considerable amounts (MIÈGE and RAMAN unpublished data).

Deficiency in sulphur amino acids is only one nutritional disadvantage of legume seeds. Protein digestibility, which depends on sensitivity of proteins to digestive enzymes, is another. ROMERO and RYAN (1978) reported cases of almost complete protein resistance, as well as cases of partial resistance, which were enough to diminish greatly the nutritive value of the proteins concerned. The authors noted that of the proteases which were active on bovine serum albumin, only a few were able to hydrolyze the major protein Gl (vicilin type) in *Phaseolus vulgaris*. Although many peptide linkages in this globulin were extremely labile to trypsin action, there were some which were not.

The presence of inhibitors in seeds is yet another possible factor constituting resistance to trypsin action. However, trypsin inhibitors from one source may not inhibit the proteases from another and their precise role in each particular situation must be established. Trypsin inhibitors are usually destroyed on cooking.

Trypsin inhibitors are proteins which are generally rich in cyst(e)ine. It would appear therefore that they are nutritionally favourable due to their sulphur amino acid content, but PHILLIPS and BOULTER (unpublished data) have shown that the cystine of the trypsin inhibitor of *Phaseolus* is largely unused by the rat.

Trypsin inhibitors accumulate in large amounts in some seeds (e.g., *Glycine*). Boulter (1980) suggested that these abundant proteins could play a primary metabolic role in the seed and also be secondarily adapted for storage.

Other protein constituents in seeds, lectins, are often responsible for the lack of nutritive value of a meal; they can even be toxic. However, soaking and cooking seeds remove their toxicity. It is also possible that seed germination improves nutritive value, although Venkataraman et al. (1976), using *Vigna radiata* (*Phaseolus aureus*), *Vigna sinensis* and *Cicer arietinum* seeds, found that seed germination did not have a positive effect on the digestibility coefficient.

Boulter (1980) suggested that improvement in protein quality in legume might most likely be brought about by a changed proportion of the major proteins synthesized rather than by changes in the structure of template mRNA molecules. The nutritional limitation of the proteins of legume seeds is mainly due to the small amounts of sulphur amino acids and amounts of these amino acids vary in different globulin proteins. Thus Boulter noted that if all the vicilin were to be replaced by legumin in *Vicia faba*, the g Met 16 g $N^{-1}$ in the seed would increase from 0.65 to 0.75 and the 1/2 cystine from 1.2 to 1.5 g Cys 16 g $N^{-1}$. Changes in relative proportion of different reserve proteins is also possible in nature, e.g., Boulter (1980) reported that in field beans harvested after an exceptionally dry and warm summer, the 7S:12S globulin ratio was increased.

Much of this new information on seed biochemistry and biology is potentially applicable to programmes for the improvement of protein quality and quantity in crops; legumes are important in this respect.

### 3.1.3 Taxonomic Applications of Protein Structure Comparisons

Classifying plants is complicated by individual variation and made difficult by the fact that species are in a state of flux and the boundaries between them are therefore not clear. The taxonomist continually revises and improves but never finalizes the taxonomic system, always developing a better explanation of the nature and status of taxa.

Ideally, comparative studies should be based on genome differences but a direct comparison of genomes has not been possible although the achievements of genetic engineering and DNA sequencing may change this situation in future. However, genome differences may be compared through the first expression of genome activity, i.e., proteins. With today's powerful methods of analysis, protein characters are increasingly useful to taxonomists, especially in solving difficult problems for which morphological comparisons give no answer. Boulter (1981) discussed the significance of proteins as taxonomic characters by virtue of the information which their structure encodes. However, the genetic code is degenerate and some information is lost when protein structure is investigated, furthermore eukaryote genes are "split", further complicating the relationship between gene and protein. Proteins contain some convergent or parallel amino acid substitutions and suffer to some extent at least from the same limitations as other characters, if used to establish relationships between present-day organisms (Boulter 1981). Due to uncertainty in the significance of any

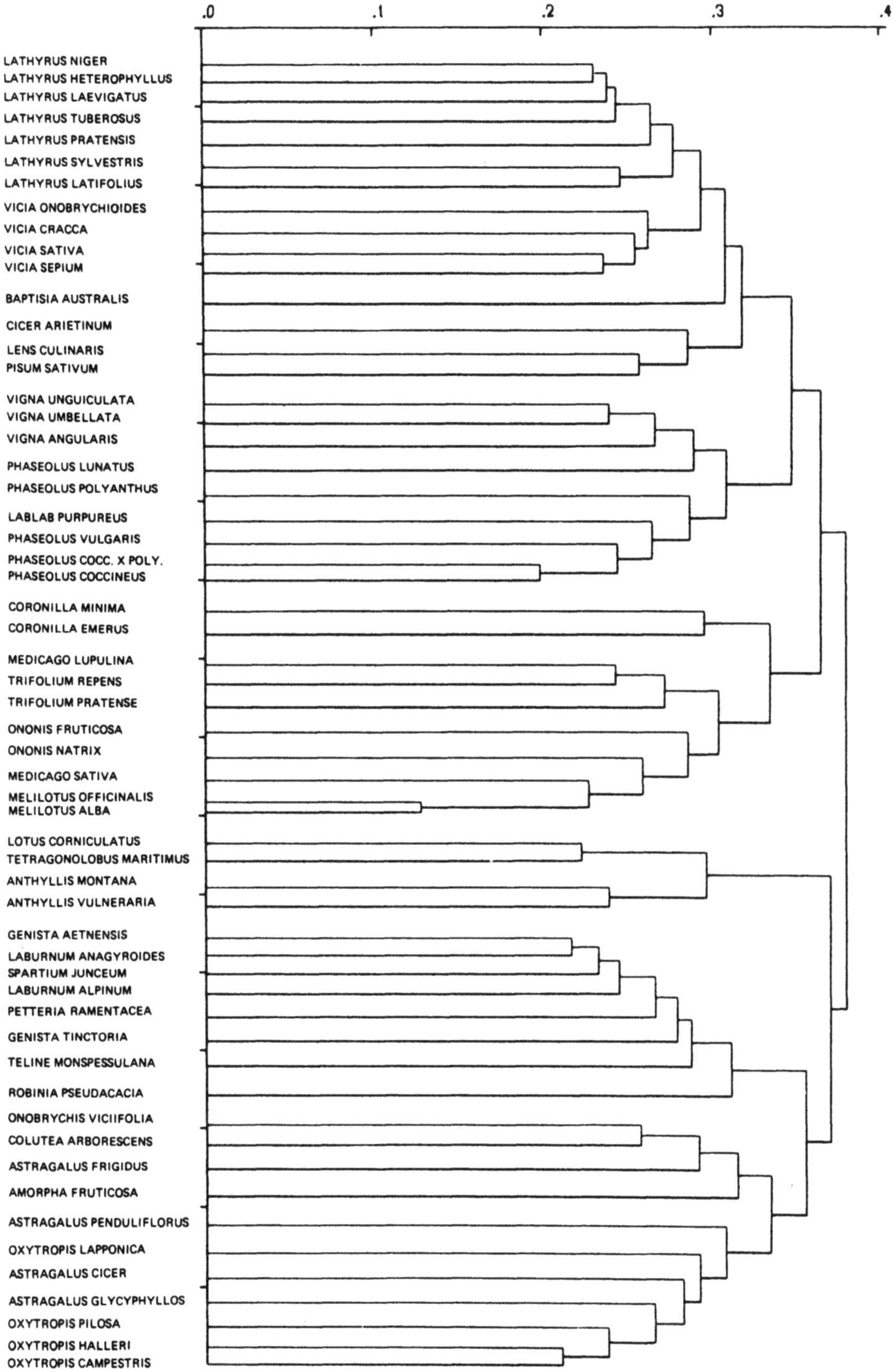


**Fig. 3.** Dendrogram obtained by qualitative protein data from 57 species of Leguminoseae. (Redrawn from MISSET 1977)

taxonomic character, taxonomic importance is often determined a posteriori according to how well it correlates with other characters, and this is true for protein characters also.

Several methods of protein analysis are available to measure taxonomic distances between taxa. BOULTER (1981) ordered them according to their decreasing power of resolution: amino acid sequence data, serological properties, peptide finger prints, iso-focusing patterns, gel electrophoresis patterns, chromatographic and molecular weight data. Amino acid sequences of a particular protein offer the greatest resolution, whereas gel electrophoretic protein patterns, using many proteins, give more information but of poorer resolution. Electrophoretic analysis is the most common way of attempting to resolve taxonomic problems, using protein characters. Its advantage and limitation have been noted by BOULTER (1981). Thus the migration of a protein from two taxa to the same position on a gel does not prove homology and it is far easier to demonstrate that two taxa differ than to prove that they are related.

Amino acid sequence analysis may also prove to be the only means of demonstrating homology between proteins; for example questions as to whether all legumes contain homologous storage proteins and the extent of the differences between homologous proteins in different legumes will not be clear until amino acid sequences of these proteins are known (BOULTER 1981).

MIÈGE J (1975) reported many examples of the usefulness of protein data in classification and also for resolving the geographical origin of species and for revealing the putative parents of hybrids, polyploids, amphiploids. Some examples were reported where palaeoprotein data contributed to a better understanding of evolution (see also BOULTER 1974, 1981, VAUGHAN 1975).

MISSET (1977) analyzed many qualitative and quantitative protein characters from 57 species of the Leguminosae and compared the groups obtained by computer data processing to those known from classical taxonomy. The results obtained with chemical data were more similar to the Engler classification than to that of Hutchinson (Fig. 3). Qualitative characters gave more information than quantitative ones, and the total albumin patterns and albumin isoesterase patterns were the most useful characters of those used.

Another example of the use of protein characters to improve the classification of the Leguminoseae is given by CRISTOFOLINI and CHIAPELLA (1977). They assessed the systematic relationships within the legume tribe Genistae using serological data which were processed by cluster analysis and principal component analysis. A last example of the many which could have been noted is the relationships between *Phaseolus* and *Vigna* genera and between the species of these genera which have been studied by SAHAI and RANA (1977). Electrophoresis of these seed proteins gave data which justified the separation of *Phaseolus* and *Vigna* species into seven clusters.

### 3.2 Storage Proteins of Seeds with Endosperm: Cereals

Protein content in graminaceous caryopses varies from 8% (*Oryza*) to 12% (*Avena*). *Triticum, Hordeum* and *Secale* have average contents ranging from 10 to 11% of the grain dry weight. *Triticum* shows large variations according to type, with protein content of 9 to 14% of the grain's weight. Storage proteins are deposited in the endosperm, of which the outer layers, formed of aleurone cells, are particularly rich in protein and phosphorus reserves. Storage proteins occur also in the embryo, mainly in the scutellum.

The protein reserves are usually primarily prolamins and glutelins which are deposited in the endosperm. Prolamin is usually the major reserve protein, with some exceptions, e.g., in *Avena,* endosperm prolamin constitutes only 13.6% of total proteins (KIM et al. 1979). Globulins are found both in embryo and in endosperm. According to KHAVKIN et al. (1978) the embryo globulins serve as the immediate protein reserves for the growing axial organs of the seedling at the very beginning of germination before the breakdown of endosperm reserve proteins is perceptible. (For further data on chemical characteristics of the cereal proteins, see Chap. 9, this Vol.)

### 3.3 Cellular Localization of Storage Proteins in Seeds

A mature seed is a "dormant" embryo that has a certain amount of reserves and is protected by an envelope. Thus, the essential structural elements of a seed are: seed coat, reserve tissues (endosperm or cotyledons), and the embryo, or, for seeds without endosperm, their non-cotyledonary part, that is the germinative axis. Integuments (seed coats) are not inert protective tissues. During maturation, they are green and hydrated, and play an active part in metabolite transfer. MURRAY (1979) showed that transfer is not a passive process and that integuments transform the metabolites from the phloem that pass through them. The role of integuments seems important during germination also (DAVIES and CHAPMAN 1979).

Storage proteins are found mainly in the reserve tissues and are deposited in organelles called protein bodies.

#### 3.3.1 Protein Bodies in Seeds Without Endosperm (Legumes and Other Classes)

##### *3.3.1.1 Protein Bodies Without Inclusions*

Protein bodies without apparent inclusions have been described in different members of the Leguminoseae (ORY and HENNINGSEN 1969, HARRIS and CHRISPEELS 1975, HARRIS and BOULTER 1976, MIÈGE and MASCHERPA 1976). BRIARTY et al. (1969) described only one protein body type in *Vicia faba,* which had an apparently homogeneous matrix and GRAHAM and GUNNING (1970) localized both legumin and vicilin in *Vicia* protein bodies in situ, using antibodies coupled to fluorescein and ferritin; both globulins occurred in the same protein bodies. However, inclusions are often difficult to visualize as described and discussed by LOTT and BUTTROSE (1978a) (see below).

##### *3.3.1.2 Protein Bodies with Inclusions: Morphological Variations*

Although protein bodies in Leguminoseae seeds and other non-endospermous seeds often do not have inclusions (Fig. 4a), other types as complex as aleurone grains (discussed in Sect. 3.3.2.2) can occur (Fig. 4b). MLODZIANOWSKI (1978) reported homogeneous protein bodies in the adaxial part of *Lupinus luteus*

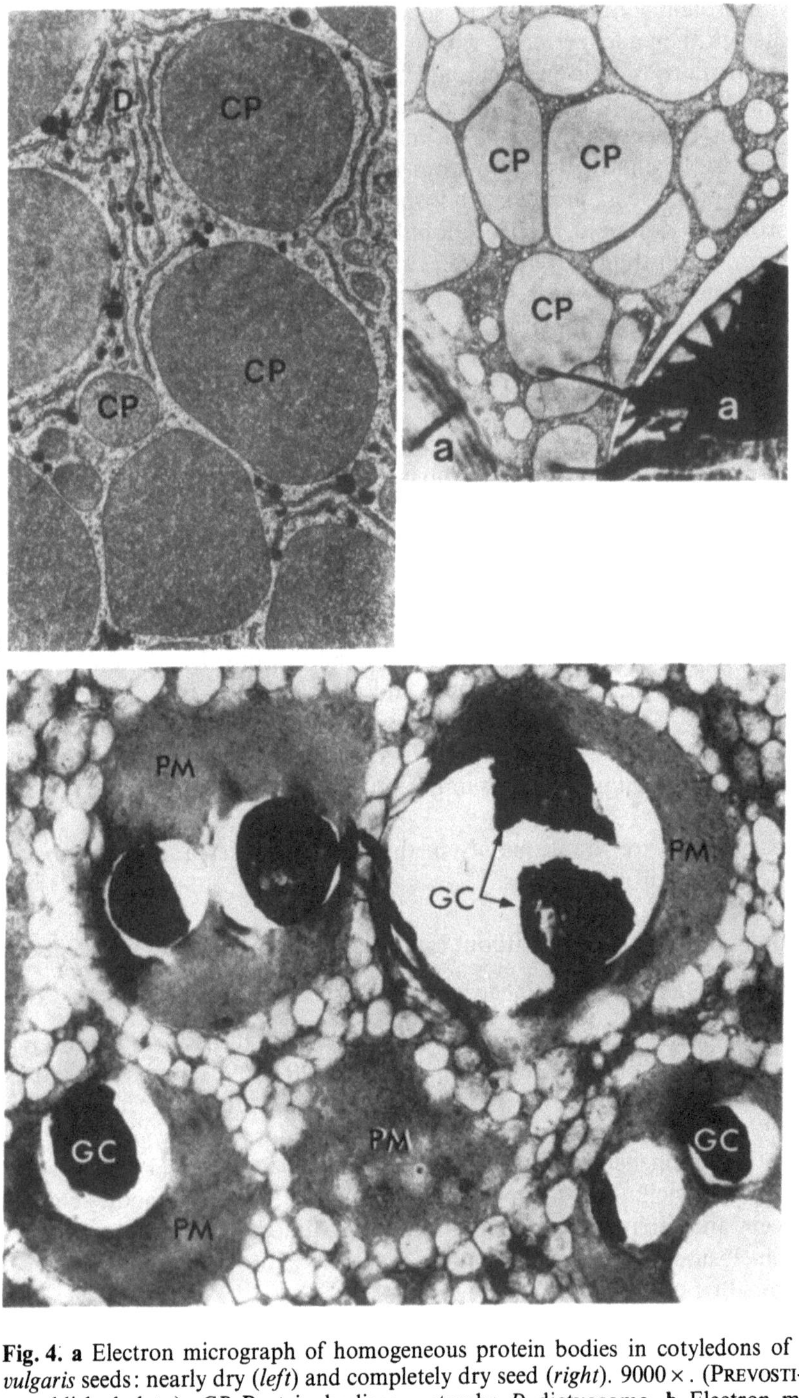


a

b

**Fig. 4. a** Electron micrograph of homogeneous protein bodies in cotyledons of *Phaseolus vulgaris* seeds: nearly dry (*left*) and completely dry seed (*right*). 9000 ×. (PREVOSTI-TOURMEL unpublished data). *CP* Protein bodies; *a* starch; *D* dictyosome. **b** Electron micrograph of protein bodies with globoids in cotyledons of *Cassia* seed. *GC* globoid crystal; *PM* protein matrix 27000 ×. (Courtesy of LOTT and BUTTROSE 1977)

cotyledons, and protein bodies with structure in the abaxial part. LOTT et al. (1971) and LOTT and VOLLMER (1973) have described protein bodies in *Cucurbita maxima* cotyledons, with a protein matrix which encloses protein crystalloids and globoids. The latter had an electron-dense part, presumed phytin-rich, ("crystal globoid"), surrounded by a less electron-dense part ("soft globoid") which was apparently phytin in a non-condensed state. The term "crystal globoid" may give rise to confusion since "crystalloid" and "globoid" are traditionally used to specify distinct inclusions in aleurone grains (see Sect. 3.3.2.2). The use of terms generated from "crystal" should be restricted to structures of a crystalline arrangement as are "crystalloids". "Crystal globoid", the electron-dense part of the globoid, should perhaps be called "hard globoid" in contrast with "soft globoid".

LOTT and BUTTROSE (1977) observed globoids in several Leguminoseae species. In order to detect mineral elements (P, K, Mg, Ca) and so reveal the presence of phytin, they coupled energy dispersive X-ray analysis (EDX) to electron microscopy. They found that globoids are rich in phytin, although phytin may also be present in the protein matrix of some protein bodies. They observed globoids in all species investigated. They were very small in some species, which probably accounts for the fact that authors have failed to report them. Globoids were small and few in number in *Vicia, Glycine, Phaseolus, Pisum, Acacia,* whereas they were large and numerous in *Arachis, Clianthus* and *Cassia*. Thus, size and frequency of globoids does not seem to be a tribal characteristic.

Globoids have been found in many non-leguminous species, e.g., in *Simondsia chinensis, Macadamia integrifolia, Juglans regia, Corylus avellana, Eucalyptus erythrocorys* and in members of Asteraceae and Anacardiaceae (BUTTROSE and LOTT 1978, LOTT and BUTTROSE 1978a and b). The numerous and large globoids in *Eucalyptus* seeds seem to be related to the exceptional richness of the seeds in phosphorus.

#### *3.3.1.3 Variations in Protein Content: Are There Several Types of Protein Bodies?*

GRAHAM and GUNNING (1970) demonstrated the presence of both legumin and vicilin in the same protein bodies in *Vicia faba*. However, KIRK and PYLIOTIS (1976) separated the reserve proteins of *Sinapis alba* seeds (12S and 1.7S) into two fractions (one filterable, the other not) and therefore postulated the existence of more than one protein body type; this seems plausible if we consider there were at least five different tissues which contained protein bodies (epidermal, palisade, spongy parenchyma, vascular and myrosin cells). Protein bodies of cells containing myrosin were not as intensely coloured by protein stains as those of other cells.

### 3.3.2 Protein Bodies in Seeds with Endosperm: Cereal Grains and *Ricinus communis* Seeds

#### *3.3.2.1 Seed Structure*

In endosperm seeds, the embryo and reserve tissues are sometimes protected by a residual nucellus and by the integument. In cereals, the integument and the pericarp fuse, thus ensuring further protection.

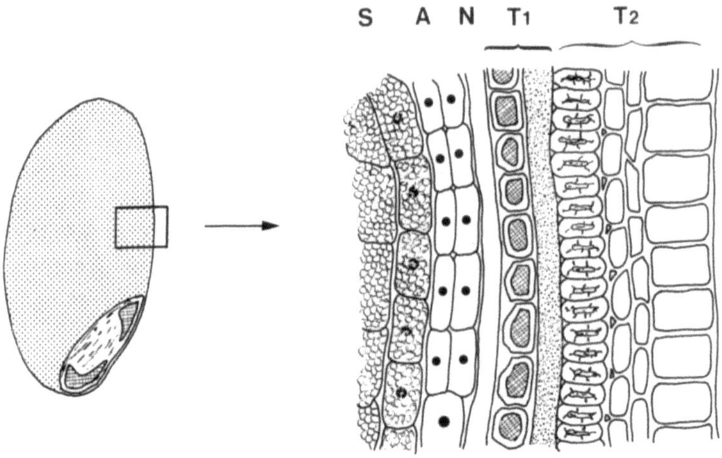


**Fig. 5.** Diagrammatic representation of a cross-section of a seed with endosperm from the inside towards the outside: the large starchy endosperm cells (*S*), the aleurone layer (*A*), nucellus (*N*), the integument formed by the tegmen (*T1*) made of crushed cells, and the testa (*T2*) with sclerous cells sometimes covered, in certain species, by mucilaginous cells. (Adapted from CHAMPAGNAT et al. 1969)

The periphery of endosperm (Fig. 5) consists of a tissue formed of one or several cell layers rich in phosphorus and protein reserves within typical structured particles: the aleurone grains.

BRIARTY et al. (1979), using stereological analysis, have determined the volumes occupied by the different organelles of *Triticum* endosperm cells.

### *3.3.2.2 Aleurone Grains of Aleurone Cells*

At the beginning of the century, GUILLIERMOND (1909), among others, used metachromatic staining with aniline blue dyes to observe aleurone cells. He described numerous granules filling the cells which contained two kinds of inclusions: green crystalloids and red globoids, the red staining being due to phytin which represents 60 to 80% of the globoid dry weight. The term "aleurone grain" was then used to denote any kind of protein body. Today, workers in Eastern European countries continue to use the term "aleurone grain" in this respect. On the other hand, other workers do not even call protein bodies in aleurone cells "aleurone grains" (KOCOŇ et al. 1978). In this chapter, the term "aleurone grain" is used to denote strictly the protein bodies of the aleurone cells, while those of the endosperm are called "protein bodies".

JACOBSEN et al. (1971) described the aleurone grains of *Hordeum vulgare*: each grain was about 4 μm in diameter, was enclosed in a single unit-membrane and contained, in a protein matrix, one to several globoids, up to 3 μm in diameter, and one or rarely two "protein-carbohydrate bodies" which were electron-dense, not membrane-bound and 1–1.5 μm in diameter. The protein-carbohydrate body corresponds to the crystalloid, the green inclusions observed by aniline blue staining in light microscopy (Fig. 6).

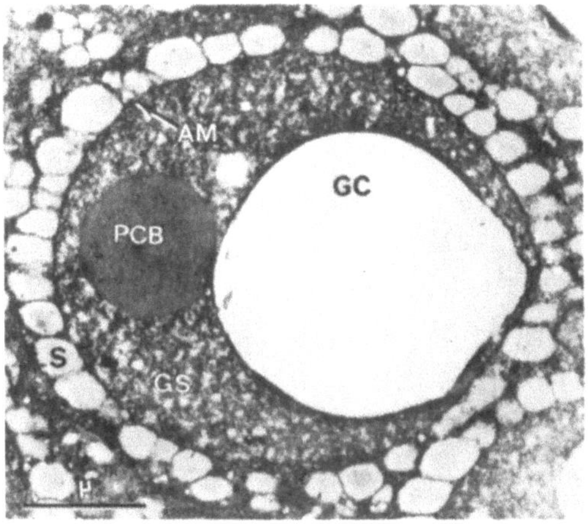


**Fig. 6.** Electron micrograph of aleurone tissue of a *Hordeum vulgare* caryopsis showing a typical aleurone grain containing in a ground substance (*GS*), a protein-carbohydrate body (*PCB*) and a globoid cavity (*GC*) (globoid was shattered during sectioning). Spherosomes (*S*) are packed close to the aleurone membrane (*AM*). (Courtesy of JACOBSEN et al. 1971)

### *3.3.2.3 Protein Bodies in Endosperm*

The main reserve of endosperm is usually prolamin and is localized in protein bodies as was shown in *Zea mays* by WOLF et al. (1967) and BURR and BURR (1979), or in *Hordeum* by SINGH and SASTRY (1978).

Protein body sizes and numbers seem to be related to the prolamin level. Thus BAENZIGER and GLOVER (1977) showed that, in *Zea mays* endosperm, no protein bodies were visible in high lysine mutants, while the mutants with intermediate level of lysine had fewer and generally smaller protein bodies than did normal varieties, and that lysine level was inversely correlated with prolamin content.

The size of protein bodies in endosperm may vary from 0.1 to 10 μm (PERNOLLET 1978). Although some are without inclusions, they can have a structure similar to that of aleurone grains with globoids and a crystalloid in a matrix, as described in *Ricinus communis* by TULLY and BEEVERS (1976) and YOULE and HUANG (1976) (Fig. 7a, b). Tully and Beevers used a non-aqueous isolation medium, then, by passage through aqueous medium, they dissolved the protein matrix and freed the inclusions which were separated on sucrose gradients. Crystalloids sedimented at 1.30 g $ml^{-1}$ and their water-insoluble proteins were extracted by sodium dodecyl sulphate. They were non-glycosylated and the main fraction of molecular weight 65,000 was made up of subunits of molecular weights 32,000 and 15,800 linked by disulphide bridges. As the prolamin of *Ricinus communis* seeds has a molecular weight of 332,000 (DERBYSHIRE et al. 1976), it was suggested that prolamin is a hexamer of the crystalloid's main protein.

In *Ricinus communis* seeds, the albumin fraction represents 40% of the total proteins; YOULE and HUANG (1978) localized this mainly in the water-soluble matrix of endosperm protein bodies and TULLY and BEEVERS (1976) found several glycoprotein lectins, one of these being ricin D, a highly toxic lectin.

Protein bodies without inclusions have been described in developing *Hordeum* caryopses by MIFLIN and SHEWRY (1979) (Fig. 7c, d, e). When globoids are

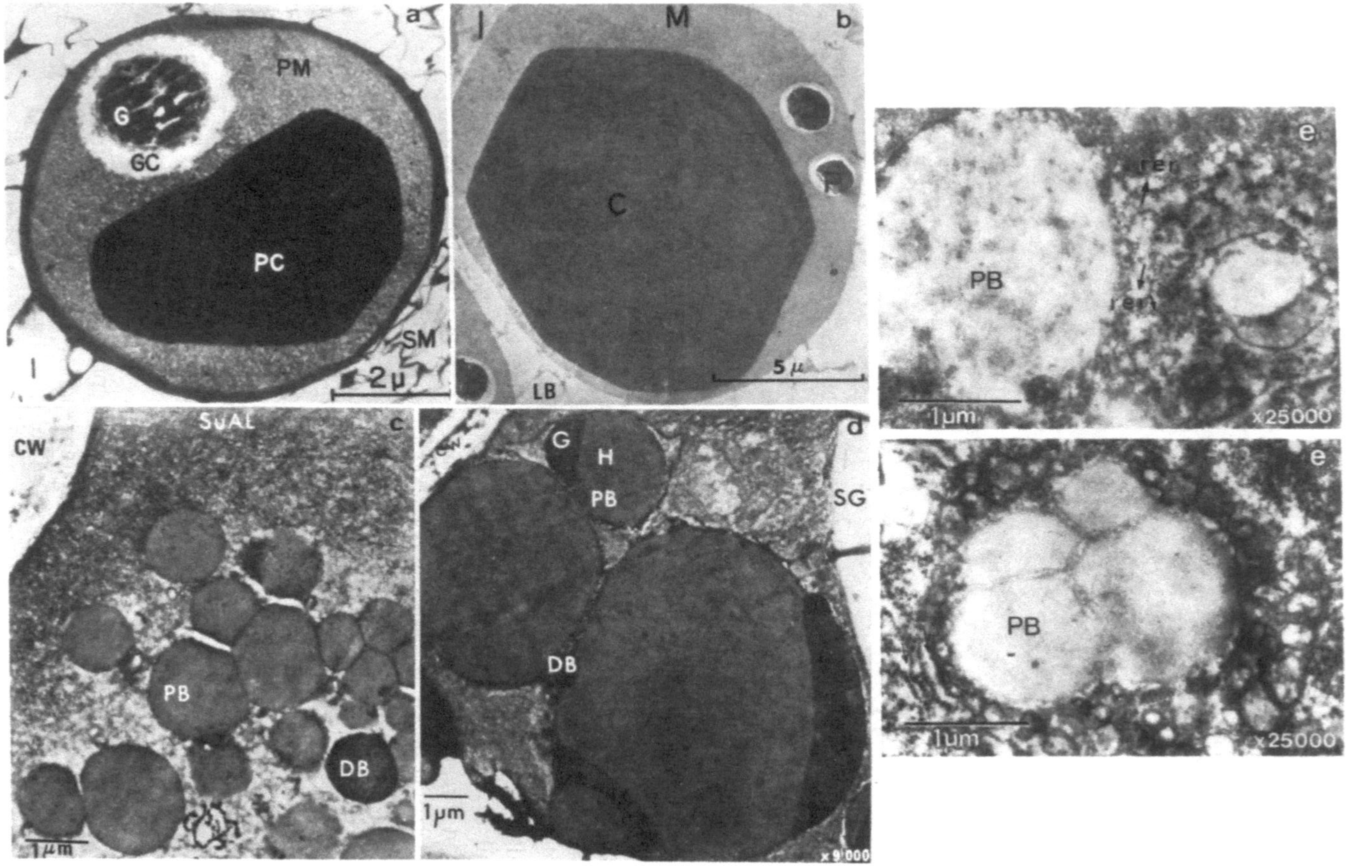


**Fig. 7.** Electron micrographs of protein bodies in the endosperm of *Ricinus communis* dry seeds (**a, b**) and in developing *Hordeum caryopsis* (normal lysine) (**c, d, e**). **a** Globoid, within globoid cavity (*GC*) and the crystalloid (*PC*) are embedded in a proteinaceous matrix (*PM*) surrounded by numerous spherosomes called "spherosome membranes" (*SM*). (Courtesy of TULLY and BEEVERS 1976). **b** Two phytin globoids (*P*) and a large crystalloid (*C*) are embedded into the matrix (*M*). Densely packed lipid bodies (*LB*) are also visible. (Courtesy of YOULE and HUANG 1976). **c** Protein bodies (*PB*) and a darker body (*DB*) appear homogeneous in the subaleurone layer of the endosperm of grain $^3/_4$ mature. Part of the cell wall (*CW*) is visible. **d** In a next cell inwards in the same endosperm, protein bodies are larger with homogeneous (*H*) and granular (*G*) appearance; darker bodies are also present (*DB*); part of a starch grain (*SG*) is visible. **e** In young endosperm: rough endoplasmic reticulum (*rer*) surrounds the central deposite, note the absence of a membrane around it. (Courtesy of MIFLIN and SHEWRY

absent, phytin could be generally located perhaps in the form of a phytin-protein complex (PERNOLLET 1978). The concentric lamellar structures found in deep-lying endosperm protein bodies have been interpreted, but without experimental evidence, as an indication of the presence of alternating phytin–protein deposits (ORY and HENNINGSEN 1969).

Protein bodies with lamellar structure have been described in developing endosperm of *Oryza sativa* by TANAKA et al. (1980). They used an aqueous polymer two-phase system and recovered them at $d = 1.27\ g\ ml^{-1}$. They contained prolamin, whereas globulin and glutelin were in homogeneous protein bodies recovered at $d = 1.29\ g\ ml^{-1}$. BECHTEL and JULIANO (1980) described three kinds of protein bodies both in normal and high protein varieties in *Oryza sativa*, by following sequential protein body formation. Large membrane-bound spherical protein bodies, with a dense centre and a concentric ring, were generated first from the rough endoplasmic reticulum (RER). Then crystallin protein bodies appeared secreted into vacuoles via the Golgi apparatus. Late in development, small spherical homogeneous protein bodies were formed by vesiculation of RER.

Prolamin represents only 5% of the total protein content, whereas glutelin constitutes 80% in *Oryza sativa*. HARRIS and JULIANO (1977) inferred that both glutelin and prolamin are present in protein bodies from the fact that the same ratio of these protein fractions was found in isolated protein bodies as in whole milled proteins. In general, however, glutelins are not localized in protein bodies but constitute an intergranular protein matrix of variable thickness which is practically non-existent in *Oryza sativa*. KOCOŇ et al. (1978) showed that, in eight *Triticum* cultivars, the amount of the matrix is correlated with the protein content of the grain, being almost absent in low protein varieties but well represented in high protein ones (Fig. 8). A similar matrix has been found in *Zea mays* (WOLF et al. 1969), *Hordeum* (TRONIER et al. 1971) and *Sorghum* (SECKINGER and WOLF 1973).

### 3.3.3 The Membranes Surrounding Protein Bodies

Aleurone grains and endosperm protein bodies are surrounded by a single limiting membrane. METTLER and BEEVERS (1979) isolated protein bodies from *Ricinus communis* endosperm in non-aqueous medium, removed the matrix protein by the addition of water and freed the membrane from its adhesion to crystalloids by sonication. Membrane vesicles, sedimented on sucrose gradients in a single zone of $d = 1.21\ g\ ml^{-1}$. Among the constituents of these vesicles were ethanolamine, choline, inositol and proteins, the latter being different from those of the matrix and of the crystalloid with regard to their electrophoretic behaviour.

A limiting membrane has also often been reported surrounding protein bodies in the cotyledons of legumes. PUSZTAI et al. (1979) isolated membranes from protein bodies of cotyledons of *Phaseolus* by centrifugation on sucrose gradients ($d = 1.16\ g\ ml^{-1}$). They were composed of glycolipids and of proteins with polypeptidic subunits of molecular weights 16,000, 20,000, 25,000, 30,000 and 78,000.

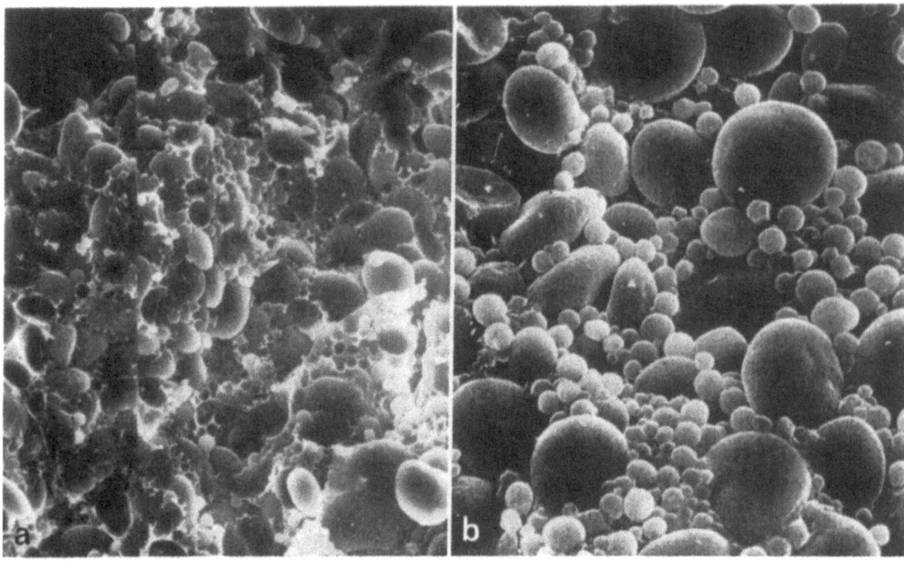


**Fig. 8. a** Cross-section of a *Triticum aestivum* kernel showing the structure of the endosperm in a protein-rich cultivar; intergranular protein matrix is clearly seen. **b** Cross-section of a protein-poor cultivar endosperm, no granular protein matrix is visible, only the thin protein layer coating the starch is seen. (Scanning electron micrograph by courtesy of KOCOŇ et al. 1978)

The polypeptides of molecular weight 25,000 and 78,000 were specific to membranes, but that of 30,000, a constituent of an agglutinating lectin, was also found in the protein body globulin fraction. The above authors attributed a possible role for this lectin in the formation and stabilization of the membrane, since it was synthesized early during seed development, whereas the synthesis of albuminic lectins, located in the matrix, occurred later. PUSZTAI et al. (1978a) attributed an important role to protein body membranes, and particularly to their constitutive proteins, in the exchange of materials with the external surroundings. They noted that some matrix proteins were leached out from isolated protein bodies, following slight environmental modifications in osmotic or ionic composition, and that this occurred without any apparent disruption of the membrane.

The origin of protein body membranes is not clear. Protein bodies may originate from vacuoles and later, at germination, reconstitute other vacuoles. This, if confirmed, would indicate a continuity between tonoplast and protein body membranes. However, in *Ricinus communis,* the membrane density of 1.21 g $ml^{-1}$, as measured by METTLER and BEEVERS (1979), is greater than that of vacuole tonoplast (1.10 g $ml^{-1}$) at the time of germination. If continuity does exist between the vacuole tonoplast at germination and the dry seed protein body membrane, the latter must undergo alterations which imply, for instance, increases in lipid constituents or decreases in protein constituents, or both in order to explain the decrease in tonoplast density.

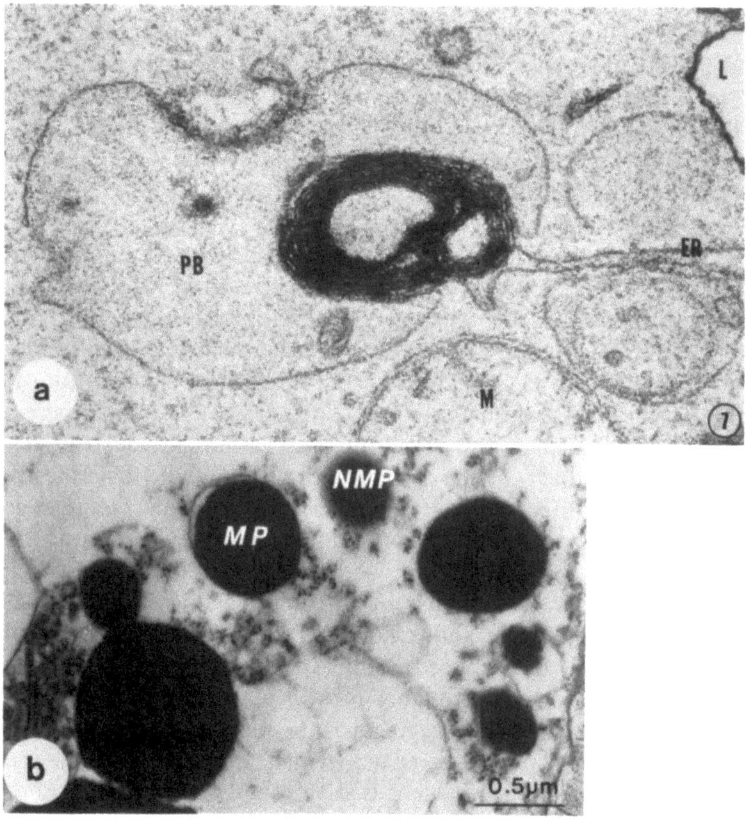


**Fig. 9.** **a** Electron micrograph of a *Zea mays* embryo after 18 h of imbibition of grain in water showing a myelin-like body closely associated with the surface of a protein body. Endoplasmic reticulum (*ER*) is continuous with the myelin-like body (glutaraldehyde-$MnO_4K$ fixation) 60000× (Courtesy of MOLLENHAUER et al. 1978) **b** Micrograph of a cotyledon cell of a 7-week-old embryo of developing *Lupinus albus* seed showing both membrane-bound proteinaceous aggregation (*MP*) and apparently non-membrane-bound proteinaceous aggregation (*NMP*) 25000× (Courtesy of DAVEY and VAN STADEN 1978)

It seems that at the beginning of germination, some protein bodies accumulate membranous materials designed for membrane reconstitution and especially those of the endoplasmic reticulum. MOLLENHAUER et al. (1978) described lamellar formations in protein bodies of *Phaseolus* and *Pisum* cotyledons, and in those of *Zea mays* embryos (Fig. 9a). These myelinic formations appear to result from interactions between lipid bodies and protein bodies. This interpretation is in agreement with the results obtained by KOEHLER and VARNER (1973). They observed a turnover rather than a net synthesis of phospholipids after adding gibberellic acid to isolated *Hordeum* aleurone layers and suggested that GA may cause the mobilization of phospholipids from storage in order to support membrane synthesis.

It is important to note the rapidity with which membranes disappear and reappear in cells of seeds near dryness. Desiccation modifies a seed's physical state, a process that is not yet well understood, but consequent on this modification is an alteration in the membrane system, which is unlikely to be explained away as an artefact of bad fixation due to the seed's dryness. Furthermore, the appearance of lipid vesicles has been noted just before desiccation, when membranes and especially endoplasmic reticulum partially disappears, followed by disappearance of these lipid vesicles after a few hours of imbibition when the membranes again show their typical morphology (NEUMANN and WEBER 1978, PREVOSTI-TOURMEL 1979).

However, there are exceptions to the presence of limiting membranes. For example, DAVEY and VAN STADEN (1978) observed two types of proteinaceous aggregations during cotyledon cell development in *Lupinus albus*. One kind was membrane-bound and closely associated with the rough endoplasmic reticulum, the other type appeared as a proteinaceous aggregation, not membrane-bound, and was present only from 6 to 8 weeks after anthesis (Fig. 9b). The presence of these aggregates and the observation that electron-dense protein-like vesicles are budded off the dictyosomes during protein synthesis suggested that there may be different ways of synthesizing storage proteins in cotyledon cells.

A doubt concerning the continuous existence of limiting membranes also arises with endosperm protein bodies. MIFLIN and SHEWRY (1979) observed protein bodies with no apparent surrounding membrane in the starchy endosperm of a high lysine and a normal lysine cultivar of *Hordeum* at all stages of development. They suggested that storage proteins are formed on the endoplasmic reticulum and probably pass into the lumen where, because of their hydrophobic nature, they aggregate into small particles which are brought together to form larger aggregates which eventually become protein bodies (Fig. 7e). Nevertheless, protein bodies usually appear surrounded by a single limiting membrane in cotyledons as well as in endosperm, the origin of which is still not understood.

### 3.4 Albumins

Albumins are differentiated from globulins by their solubility in "pure" water and by their greater stability in low ionic strength solutions at their isoelectric point. In practice, however, it is difficult to separate the albumins and globulins of seeds. Most enzymes and other metabolic proteins are albumins which do not normally act as storage reserves.

GILLESPIE and BLAGROVE (1978) tried to isolate albumins from *Psophocarpus* seeds in buffered water at pH $<5$ but very few proteins were extracted; the albumins were thought to be present together with some globulins in the fraction which was soluble at pH 5.4.

Potentially, the simplest method for isolating albumins would be to extract seed meals with pure water prior to extraction with saline. However, meals contain ions in sufficient amounts to give an ionic strength which solubilizes considerable amounts of globulins. Non-protein nitrogen, as well as other small molecules, can be eliminated by dialysis of salt extracts, thus causing globulins to precipitate. However, precipitation is progressive and allows interactions which alter the constitution and solubility characteristics of the native proteins. Thus MANEN and MIÈGE (1977) reported associations of globulins and agglutinating albumins by this process. Furthermore the quantity of albumins isolated by dialysis depends on the dialysis process (MIÈGE 1975).

Generally, legume seeds contain 5 to 10% albumins (DERBYSHIRE et al. 1976). They have a higher sulphur amino acid content [cyst(e)ine and especially methionine] than globulins. Albumins isolated from *Pisum* seeds represented 13 to 14% of the total nitrogen and contained components of 78,000, 47,700 and 26,000 molecular weight, made of subunits of 25,000 and 15,500 (GRANT et al. 1976). Protease inhibitors and enzymes are present in the albumin fraction.

Albumins from seeds other than those of legumes have been characterized, for example, the albumin fraction from *Helianthus annuus* seed (Asteraceae) and *Brassica napus* (Brassicaceae) contains low molecular weight (10,000 to 16,000) and very basic proteins (pH > 10) (RAAB and SCHWENKE 1975). Albumins isolated from *Cucurbita maxima* seeds (Cucurbitaceae) were separated into 12 electrophoretic components (at pH 8.5) and into 10 after isoelectrofocusing (pI 4.0 to 9.0) (PICHL 1978).

## 4 Distribution of Enzymes

### 4.1 Definition, Nomenclature

An enzyme is a protein possessing catalytic properties due to its power of specific activation (DIXON and WEBB 1967). Life depends on a complex network of specific enzymes, and enzymology has become a very important subject related to many disciplines. For detailed information the reader is referred to the standard treatises on enzyme biochemistry.

The enormous and ever-increasing number of known enzymes made it imperative to develop a comprehensive scheme of classification and nomenclature. This was produced by the Commission on Enzymes of the International Union of Biochemistry in 1961. The enzymes were classified into groups catalyzing similar processes with subgroups specifying more precisely the catalyzed reaction. In general, each enzyme has a name consisting of the name of the substrate, the name of the reaction and ends with "ase": e.g., alcohol dehydrogenase. In addition to these systematic names, trivial names were agreed upon for practical convenience in some cases. In addition, a system of mathematical symbols was also adopted. Each enzyme number contains four elements, separated by points, the first figure showing to which of the six main classes the particular enzyme belongs, i.e.: 1 oxydoreductase, 2 transferase, 3 hydrolase, 4 lyase, 5 isomerase, 6 ligase (see DIXON and WEBB 1967 for further details).

### 4.2 Cellular Localization of Enzymes

Enzymes are not randomly located in cells, they are either situated in particular intracellular organelles, or are in a soluble form in the cytoplasm. The relationships between the organelle structure and the enzyme systems it encloses is one of the bases of cell physiology. The interested reader may consult other volumes of this series, e.g., for mitochondrial enzymes: Volume 3, for chloroplast enzymes: Volumes 5 and 6, etc.

## 4.3 Multimolecular Forms of Enzymes: Isoenzymes

With the improvement in analytical techniques, especially electrophoretic methods, it has been shown that many enzymes occur in multiple molecular forms in an organism. The term "isozyme" or "isoenzyme" was adopted for such multiple forms (MARKERT 1975), and isoenzymes have been shown to exist in plants as well as in animals.

### 4.3.1 Isoenzymes in Relation to Subcellular Structure

The molecular form and function of an enzyme may differ according to its intracellular distribution. For example TING et al. (1975) have reported that malate dehydrogenase in the cytosol has a role in non-autotrophic $CO_2$ fixation to form malate, in mitochondria in the citric acid cycle, in leaf peroxysomes in photorespiration and in glyoxysomes in the glyoxylate cycle. An example where enzyme polymorphism is related to subcellular localization is that of the isophosphatases of *Pisum sativum* which are different according to whether they are in protein bodies or at the cytoplasm–cell wall interface (BOWEN and BRYANT 1978).

### 4.3.2 Isoenzymes in Relation to Tissue Localization and to Ontogeny

Isoenzyme patterns have often been shown to be specific to organs. For example, POPOV and SERBAN (1977) have investigated the organ specificity of cytochrome oxidase, peroxidase, catalase, acid phosphatase and esterase in 3-day-old seedlings and cotyledons from several populations of *Phaseolus vulgaris*. Peroxidase was shown to be the most organ-specific enzyme, while esterase was the least. Organ-specificity was also demonstrated by MURRAY and COLLIER (1977) for isophosphatase of *Pisum sativum,* whose patterns differed in one or several bands according to whether they were from the cotyledon, axis, seed coat or pod. Different patterns were obtained by PIHAKASKI and PIHAKASKI (1978) with myrosinase from different organs of *Sinapis alba* (cotyledon, hypocotyl, primary root and seed). TAO and KHAN (1975) showed differences in isoperoxidase patterns from different parts of the grain of *Triticum* and observed the appearance, after soaking, of several new isoperoxidases in the embryo.

Enzyme activities are modified during the life cycle of an organism and the modifications in enzyme activities are often accompanied by marked changes in the population of enzymic constituents. Thus, DAUSSANT et al. (1979) have shown the existence of three sets of isoamylase in *Triticum* grains, one during the period of grain development, another one for germination and a third occurring at both stages but in different amounts; the replacement of one set by another involved degradation and de novo syntheses. Many other examples of enzyme changes in relation to ontogeny can be found quoted in other parts of this volume.

### 4.3.3 Isoenzymes in Relation to Environment: Biological Significance

The evolutionary role of enzyme polymorphism may be to provide metabolic flexibility in a changeable environment. Definitive evidence is not yet available, nevertheless some experimental data support the idea that polymorphism among enzyme loci is not selectively neutral, but that the allelic isoenzymes may have physiologically significant differences. Thus SCANDALIOS et al. (1972) and FELDER and SCANDALIOS (1971) have shown that hybrid isoenzymes generated by allelic or non-allelic interactions have improved physicochemical properties when compared to the parental molecules, providing a possible advantage to the individual carrying them.

Very few data are available to answer the question whether or not multimolecular forms of enzymes reflect the most effective enzyme equipment for a given situation. For example, is adaptation to anaerobic conditions reflected in dehydrogenase isoenzyme patterns? In plants, alcohol dehydrogenase (ADH) is a dimeric polymorphic molecule, as shown by SCANDALIOS (1967, 1969) and SCHWARTZ and ENDO (1966). ADH activity was found in the scutellum, embryo and endosperm of *Zea mays* kernel at each stage of development or germination. SCHWARTZ (1971, 1973, 1976) demonstrated the advantage to the seed of the presence of a complete ADH pattern compared to a mutant he obtained which lacked one ADH type. The latter had a significantly lower percentage of germination in anaerobic conditions.

The tolerance to natural anaerobiosis was studied by CRAWFORD and MCMANMON (1968) who compared the ADH activity in roots of tolerant and non-tolerant species to immersion. They observed that tolerant species reacted to immersion either without change or by a decrease in the level of ADH activity, whereas non-tolerant species reacted to such conditions by an increase in level. It seems possible, therefore, that an elevated ADH activity is damaging to the plant under anaerobic conditions.

Another example of the biological significance of isoenzymes is that the resistance mechanism of tobacco leaves against mosaic virus appears to involve a particular catalase isoenzyme (ALEXANDRESCU et al. 1975).

### 4.3.4 Isoenzymes in Relation to Taxonomy

Studies of the electrophoretic variants of enzymes have been applied to taxonomy, for example to investigate the genetic structure of populations, to estimate the degree of genetic similarity between species or to study the distribution of geographical variations in allelic frequencies. (Details can be found in MARKERT 1975, BOULTER 1981, MIÈGE J 1975).

Dehydrogenase, esterase and peroxidase are often used because of the ease with which these enzymes can be detected on electrophoregrams. For example, WALL and WALL (1975) tried to clarify the evolutionary problems of the Mexican *Phaseolus vulgaris–P. coccineus* complex using isoenzyme polymorphism of seed alcohol dehydrogenase and esterase. PAYNE et al. (1973), utilizing malate dehydrogenase of seeds, demonstrated differences among *Coffea* species, varieties and cultivars. THURMAN et al. (1967) found significant differences in formic and glutamic dehydrogenase patterns of 90 Fabaceae species among the 103 examined. SHECHTER and DE WET (1975) used differences in isoenzymes of esterase, malate dehydrogenase and peroxidase to investigate phenetic and phylogenetic relationships in *Sorghum*.

The variations in isoenzyme patterns are often most significant at the specific and infraspecific levels. Thus PAYNE and KOSZYKOWSKI (1978) used quantitative esterase isoenzyme differences to identify *Glycine* cultivars. POPOV et al. (1976) have found variations in acid phosphatase and esterase patterns between *Phaseolus vulgaris* taxa. SINGH and GUPTA (1978) examined 24 genotypes of *Brassica* belonging to *B. campestris* and *B. juncea;* each genotype could be characterized by variations in isoesterase patterns during ontogenesis.

Stem and leaf isoenzymes are also used to reveal genetic differences. FIELDES et al. (1976) found differences in isoperoxidase patterns from the stems of two genotypes of *Linum usitatissimum*. After isolation of the isoenzymes, they detected differences in the Tyr-Try/peptide ratio in the corresponding isoenzymes of the two genotypes. TORRES et al. (1978) used isoglutamate oxaloacetate transaminases, isophosphoglucose isomerases and isophosphoglucomutases as genetic markers to elucidate relationships in *Citrus*. This group includes two subgenera: *Citrus* and *Papeda,* the former including all the edible cultivars. The problem was to distinguish, long before fruiting, nucellar seedlings from those of zygotic origin and it was solved by the use of the above isoenzymes. Populations of the Galapagos tomato, made of a single but polymorphic species: *Lycopersicon cheesmanii*, were analyzed by RICK and FOBES (1975) using variations in allelically determined isoesterases and many other examples could be given of the increasing use of enzyme polymorphism in taxonomic or genetic problems.

## 5 Distribution of Protease Inhibitors in Plants

Many plant tissues contain proteins of low molecular weight (500 to 20,000 approximately) which inhibit proteases. The influence of protease inhibitors on human digestive proteases (trypsin, chymotrypsin, pepsin, etc.) has been widely studied because of the implications of their activity for the nutritive value of the plant containing them. (For details on the structure of trypsin inhibitors and on the mechanisms of inhibition, see ROYER 1975; for a general review on protease inhibitors, see RICHARDSON 1977 and for details of their structure see Chap. 7, this Vol.)

Protease inhibitors are widely distributed in seeds, especially those of cereals and legumes: they represent 5 to 10% of the water-soluble proteins of the grain in *Triticum, Avena, Hordeum* and *Oryza* (MIKOLA and KIRSI 1972). CAMUS and LAPORTE (1974) classified different *Triticum* into six groups according to their content of trypsin inhibitors; varieties of *T. vulgare* contained on average twice as much inhibitory activity as the varieties of *T. durum*.

Some protease inhibitors of legumes are well characterized, e.g., inhibitors of trypsin and chymotrypsin of *Glycine max* (LIENER and KAKADE 1969), of

*Phaseolus lunatus* (PUSZTAI 1967), of *Vigna unguiculata* (VENTURA and XAVIER-FILHO 1966; XAVIER-FILHO and DE SOUZA 1979; ROYER et al. 1974), of *Arachis hypogaea* (BIRK and GERTLER 1971), of *Psophocarpus tetragonolobus* (KORTT 1979), of *Vicia faba* (ORTANDERL and BELITZ 1977).

Besides the seeds, protease inhibitors are also found in leaves (spinach, potato, tomato, tobacco), in roots (radish), in tuber (potato), but there is practically none in fruits (CHEN and MITCHELL 1973).

Certain inhibitors have a well-documented effect on microbial and insect intestinal proteases. This calls attention to their possible defensive role in plants, and RYAN (1973) suggested that inhibitors of chymotrypsin in potato leaves are part of an immune system that protects the plants against insects and microorganisms. In potato tubers, KAISER and BELITZ (1971, 1972, 1973) demonstrated their activity towards serine-dependent proteases of animal and microbial origin, whilst being inactive towards the proteases of the aleurone layer in *Hordeum* grains. The inhibitors of potato tubers have been shown to inhibit growth and proteolytic activity of microorganisms present in infected tubers (SENSER et al. 1974).

Inhibitors are generally not glycoproteins, but some exceptions exist, e.g., a glycoprotein of molecular weight 80,000 in potato tuber inhibits papain, also a trypsin and chymotrypsin inhibitor of molecular weight 9,500 in *Dolichos lablab* seeds contains 20% galactose and 10% hexosamine (FURUSAWA et al. 1974).

The intracellular localization of inhibitors has been examined. SHUMWAY et al. (1976), using immunofluorescence methods and antibodies coupled to ferritin, located the trypsin inhibitors of tomato leaves in intravacuolar protein bodies. CHRISPEELS and BAUMGARTNER (1978) localized trypsin inhibitors both by cytochemistry with fluorescent antibodies and by isopycnic cell fractionation. They found inhibitors associated with the cytoplasm and not with the protein bodies. However, PUSZTAI et al. (1977) reported a rather complex distribution of trypsin-inhibitory activity in cotyledonary cells of *Phaseolus vulgaris;* an appreciable part of activity was associated with the protein body fraction obtained by differential centrifugation.

The genetic control of trypsin-inhibitor synthesis was studied by ORF and HYMOWITZ (1977). They identified in *Glycine* seeds a trypsin inhibitor different from the Künitz inhibitor and demonstrated that the three electrophoretic forms of this protein were controlled by a co-dominant multiple allelic system at a single locus. In potato tubers, isoinhibitors constitute 21 out of the 42 isoelectrofocusing bands and isoinhibitor patterns were characteristic of each of the 12 varieties studied (KAISER et al. 1974).

# 6 Lectins: Nature and Distribution

## 6.1 Occurrence in Plants and Definition

It has been known since 1880 that some seed extracts are able to agglutinate erythrocytes. The proteins responsible were first designated as "agglutinins", but later they were shown to be able to agglutinate lymphocytes and to bind, via receptors, to the surface of other cell types. The term "agglutinin" was later replaced by the term "lectin" when the mechanism of the specific binding of the agglutination with cell receptors was known. The concept of lectin has recently been enlarged and now includes monovalent molecules with one single binding site, and agglutination is no longer considered as a necessary consequence of their action. These considerations led CALLOW (1977) to define them as follows: "lectins are proteins, or more often glycoproteins, that have the ability to interact with different types of animal cells to produce various effects."

The lectin–sugar linkage is specific, so that it is the presence of the specific sugar in a cell component that elicits the response. The specific sugar involved is often inferred from the interference to the binding by some sugars, referred to as "haptens" by analogy with the immunological term (KAUSS 1976). This specificity was first studied with erythrocyte surface sugars, and specific A, A+B, H, M, N reactive lectins were discovered. However, a large number of lectins show no specificity towards erythrocytes and agglutinate them all. Among the well-known lectins, only about 3% are specific to haptens of red blood groups.

SHARON and LIS (1972) have summarized the different properties that lectins may exhibit: sugar binding, polysaccharide and glycoprotein precipitation, specificity towards blood groups, toxicity towards animals, mitogenous action towards lymphocytes and preferential agglutination of malignant cells.

JERMYN and YEOW (1975) developed a still wider concept of lectins since they did not consider effects upon animal cells only, but simply defined lectins as proteins or glycoproteins that interact with sugars by non-covalent linkages.

## 6.2 Structure

Studies of lectins purified from legumes have shown the existence of several kinds of lectin structures. There may be a single subunit, or two or more different subunits which are usually associated by non-covalent forces into a polymeric molecule. More than one type of lectin may be present in a single plant. Thus, in *Vicia cracca,* RÜDIGER (1977) characterized a lectin specific for the A group of erythrocytes which was a tetramer of molecular weight 120,000 formed from protomers of 33,000. Another lectin was also present in *Vicia cracca* seeds which is glucose-specific and without specificity towards erythrocytes, it is made up of 7,000 and 21,000 molecular weight subunits (BAUMANN et al. 1979).

*Phaseolus vulgaris* lectin, or PHA, an extensively studied lectin, was characterized by PUSZTAI and STEWART (1978) as a protomer of molecular weight 119,000 formed from four similar subunits; the protomer is in equilibrium with dimers and oligomers. This is similar to the tetrameric molecular structure proposed by LEAVITT et al. (1977) of molecular weight 115,000 made up of four subunits of molecular weight 33,000.

In *Phaseolus vulgaris* cv. contender, MANEN and MIÈGE (1977) have found five tetrameric isolectins formed of two subunits A and B of molecular weights 35,000 and 34,000 (Fig. 10).

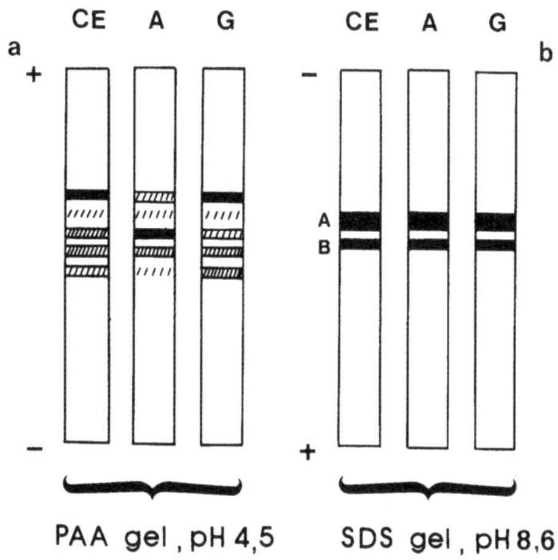


**Fig. 10a, b.** Electrophoretic patterns of lectins from *Phaseolus vulgaris* purified by affinity chromatography from: crude extract (*CE*), albumin fraction (*A*) and globulin fraction (*G*). (The various proportions of the fractions revealed after amido-black staining are shown by the different densities of the *dotted lines*). **a** polyacrylamide (*PAA*) gel 7.5%, pH 4.5, 3 mA/tube, 4 h; **b** samples in sodium dodecyl sulphate (*SDS*) 2.5%, 100° 2 min; run in PAA gel 16% containing SDS 0.1%, pH 8.6, 2 mA/tube 3 h. The polymeric lectins are dissociated into their two subunits A and B

The *Dolichos biflorus* lectin, molecular weight 100,000, dissociates in the presence of sodium dodecyl sulphate into six components of molecular weight 10,000 to 30,000 (RAO et al. 1976). The *Pisum* lectin is also very complex consisting of two subunits of molecular weight 8,000 and 18,000 associated into molecules of molecular weight 68,000 and 134,000 (VAN DRIESSCHE et al. 1976). Other lectins are dimeric, e.g., the lectin of *Lathyrus odoratus* seeds with subunits of molecular weight 37,000 (KOLBERG 1978), and *Vicia sativa* lectin which is a dimer of molecular weight 70,000. There are also examples of low molecular weight tetramers; thus in *Vicia faba,* a tetramer of molecular weight 48,500 is formed of two pairs of subunits of molecular weight 18,800 and 5,500; this tetramer is erythroagglutinating and mitogenic (HORSTMANN et al. 1978). In *Vicia ervilia,* a lectin of molecular weight 60,000 is formed of two pairs of subunits of molecular weight 20,000 and 4,700.

Constituent polypeptide chains are generally characterized by the absence of sulphur amino acids, and by their high content of hydroxylated amino acids.

## 6.3 Glycoprotein Nature, Glycosylation Site, Lectin–Protein Linkages

Lectins are usually glycoproteins, although exceptions exist, the best known of which is Con A. Wheat germ agglutinin (WGA) and the lectins of *Arachis hypogaea* and *Vicia ervilia* are other non-glycoprotein lectins (the latter was isolated by FORNSTEDT and PORATH 1975).

There are usually 2 to 15 sugar residues covalently bound to the protein backbone of lectins. Most commonly these sugars are mannose, glucose, galactose and fucose. Glycosylation of proteins is a common reaction in plants and SHARON and LIS (1979) have shown that the carbohydrate-peptide linkage can be N-glucosidic or O-glucosidic. The N-acetylglucosamine-asparagine linkage is of frequent occurrence in both plant and animal glycoproteins, e.g., in *Glycine* agglutinins. The O-glucosidic linkage occurs in glycoproteins rich in hydroxyproline and this linkage, found in plant cell wall glycoproteins, is also found, for example, in potato tuber lectins.

In glycoprotein lectins, the carbohydrate moiety is, at least in part, usually necessary for lectin activity. Thus Font and Bourrillon (1971) eliminated all lectin activity in the *Robinia pseudoacacia* lectin by oxidation. However, Biroc and Etzler (1978) showed that the integrity of the carbohydrate part of the *Dolichos biflorus* lectin is not absolutely necessary for its activity.

After protein synthesis on polysomes, glycosylation takes place in the lumen of the ER or in the Golgi apparatus where the necessary enzymes are present (see Chap. 4, this Vol.).

Sometimes lectins can be freed from membranes by haptenic sugars: for example, a lectin associated (in vitro) with the mitochondria membranes in *Ricinus communis* endosperm was freed from them by lactose (Bowles et al. 1976, Köhle and Kauss 1979). However, the lectin–membrane association can be stronger, as it is in isolated hypocotyl membranes of *Vigna radiata* (*Phaseolus aureus*), where the lectins can only be freed by the combined effects of a chelating agent and a detergent; in this case, the lectin is probably an integral part of the membrane (Bowles and Kauss 1976).

So far, the true significance of lectin–membrane associations in both animal and plant cells has not been established; Monsigny et al. (1979) interpreted the association of lectins with cytoplasmic Golgi and lysosomal membranes to indicate their possible participation in the transport of glycoconjugates through the cell and also in the internalization of glycoproteins during their passage into membrane-limited structures.

Gleeson and Jermyn (1977) showed, by immunodiffusion methods, that lectin–glycoprotein interactions were inhibited by specific sugars. For instance, Con A reacted with extracts of seeds belonging to 18 species (*Coffea arabica, Lolium perenne* and 16 Leguminoseae species) and that precipitation was prevented by addition of α-D-mannopyranoside, the sugar specific to the Con A linkage. Some of the glycoproteins involved in these reactions were isolated by affinity chromatography and were shown to agglutinate trypsinized human red blood cells. Lectin–lectin interactions also occur. Thus the two *Ricinus communis* agglutinins RCA I and RCA II both contain mannose, and both interact with Con A (Maher and Molday 1977). Rougé and Père (1979) showed the same type of interaction when Con A interacted with lectins from seeds of *Pisum sativum, Lens culinaris, Vicia faba* and *Vicia sativa.*

## 6.4 Different Types of Lectins

### 6.4.1 Agglutinins

*Ricinus communis* seeds contain two proteins: ricin (toxin) and RCA (agglutinin) which cause toxicity and agglutination respectively. RCA is a tetramer of 120,000 molecular weight which reacts with galactose and is formed of two pairs of different A and B chains. Pappenheimer et al. (1974) demonstrated the immunological identity of the RCA B chain with the B chain of the toxic ricin, whereas the RCA A chain lacks some of the antigenic determinants of the ricin A chain. Surolia et al. (1978) showed that the RCA tetramer dissociates on reduction first into subunits of 58,000 and 64,500 molecular weight and then, under the action of sodium dodecyl sulphate, into subunits A and B of 29,000 and 32,000 molecular weight respectively:

$$2A2B \xrightarrow{\text{mercaptoethanol}} AA+BB \xrightarrow{\text{sodium dodecyl sulphate}} A+A+B+B$$

The tetrameric molecule has only two binding sites carried by one single type of subunit, which is unusual although the *Glycine* agglutinin also has only two binding sites (LOTAN et al. 1975). In general, however, agglutinins have one binding site per subunit. Thus PUSZTAI and WATT (1974) isolated tetrameric isolectins from the albumin fraction of *Phaseolus vulgaris;* these isolectins are formed of subunits of molecular weights 30,000 and 35,000 in a ratio 3:1 and each subunit has a binding site. These lectins are erythro- and leucoagglutinating, but have practically no mitogenic activity towards lymphocytes. The *Pisum sativum* agglutinins are composed of two heavy and two light subunits. ENTLICHER and KOCOUREK (1975) showed that the heavy and light subunits A and B of pI 5.9 and 7.0 respectively, hybridize to a C subunit of pI 6.35. A polymeric structure has also been reported for agglutinins from organs other than seeds, thus the agglutinin isolated from the sieve tube sap of *Robinia pseudoacacia* by GIETL et al. (1979) has a molecular weight of 100,000 and consists of four types of subunits.

### 6.4.2 Mitogenic Lectins

A large number of agglutinins are mitogenous towards lymphocytes in vitro, e.g., the agglutinins from *Canavalia ensiformis* (Con A), *Pisum sativum, Lens culinaris, Phaseolus vulgaris*. The latter, of molecular weight 70,000, agglutinates horse, rabbit and human erythrocytes but not calf and sheep erythrocytes, and has mitogenic activity on human peripheral blood lymphocytes (FALASCA et al. 1979).

The mitogenic potential of *Phaseolus vulgaris* lectins varies according to variety. JAFFÉ et al. (1974) separated 20 cultivars of *Phaseolus vulgaris* into four groups according to their mitogenic potential: beans with the most mitogenic lectins were also the most toxic. However, on the whole, *Phaseolus vulgaris* lectins are stronger agglutinins than they are mitogens. LEAVITT et al. (1977) characterized five tetrameric isolectins from *Phaseolus vulgaris* which were more or less agglutinating or mitogenic according to whether the predominant subunit was erythro-agglutinating (E) or mitogenic towards lymphocytes (L). These five isolectins have the following tetrameric structures:

$$L_4 \quad L_3E_1 \quad L_2E_2 \quad L_1E_3 \quad E_4$$

A lectin isolated from *Lathyrus odoratus* by KOLBERG (1978) was strongly mitogenous and also erythro-agglutinating. Mitogens have also been found in *Wistaria floribunda* seeds (Leguminoseae) (BARKER and FARNES 1967) and in *Phytolacca americana* (Phytolaccaceae) (FARNES et al. 1964).

### 6.4.3 Lectins as Enzymes

HANKINS and SHANNON (1978) have found that a highly purified lectin from *Vigna radiata* had α-galactosidase activity. The lectin and enzymatic activities co-purified could not be separated. The lectin and enzyme possessed indistinguishable carbohydrate specificities and exhibited the same properties, thus providing compelling evidence that both lectin and enzymatic activities resided in a single protein species which was a tetrameric glycoprotein of molecular weight 160,000 composed of identical subunits of 45,000.

### 6.4.4 Lectins as Toxins: Ricin and Abrin

Ricin and abrin are lectins which are toxic and distinct from the agglutinins of *Ricinus communis* (Euphorbiaceae) and *Abrus precatorius* (Leguminoseae) respectively. LIN et al. (1970) showed that ricin and abrin ensure therapeutic protection against Ehrlich's ascite tumours in mice and this discovery was to stimulate the application of such toxins in the treatment of human cancer. It seems that these particular toxins show selective activity towards cancerous cells and OLSNES and PIHL (1978) reported that the anticancer effects of these toxins was at least as encouraging as that of the usual cytostatics in medicine. The greater toxic effect on tumour cells seems to be due to a greater penetration of the molecule since they do not carry more binding sites.

Abrin and ricin have molecular weights of 60,000 and 65,000 respectively and are each formed of two polypeptide chains A and B linked by disulphide bridges. Toxic effects depend upon the integrity of toxin's structure and are preserved in hybrid molecules ricin A-abrin B and vice versa. The B-chain (haptomer) anchors the toxin to the cell via linkage with lactose or galactose at a single binding site, and then the A-chain, which carries toxicity (effectomer), penetrates the cell. By binding with the 60S ribosome-subunit, it prevents the association of 60 S–40 S ribosomal subunits and consequently the initiation of protein synthesis (OLSNES and PIHL 1978).

Another plant toxin from the roots of *Adenia digitata* was reported by REFSNES et al. (1977). This binds carbohydrate receptors on the cell surface and its extreme toxicity to animals and man is apparently due to the inhibition of protein synthesis. An inhibitory effect on protein synthesis by other agglutinins was demonstrated by BARBIERI et al. (1979) using in vitro tests on rabbit reticulocyte lysates: agglutinins from *Momordica charantia* and *Crotalaria juncea* and the mitogenic lectin from *Phytolacca americana* were shown to be inhibitory. A similar inhibitory activity was acquired by the agglutinins from *Ricinus communis* and *Vicia cracca* after reduction with 2-mercapto-ethanol.

### 6.4.5 β-Lectins

CLARKE et al. (1975) defined a new class of lectins, β-lectins, which are widespread in plants. They are glycoproteins which interact with the β-glucosyl determinants of a coloured, artificial carbohydrate antigen, Yariv's antigen, in which 4-aminophenyl diazotated glycosides are coupled with phloroglucinol. Their specificity towards β-D-glucopyranosyl differentiates them from the other known lectins. These proteoglucans, of high molecular weight (126,000 for *Brassica napus* lectin) are apparently formed of one single polypeptide chain, since only one N-terminal amino acid has been found. The glycoprotein contains hydroxyproline, and glucosamine, galactose and arabinose residues are the major sugars; the carbohydrate/protein ratio is astonishingly high (8:1 on a molecular basis). β-Lectins are refractory to proteolytic degradation, possibly due to the inaccessibility of the protease to the proteoglucan peptide core (JERMYN and YEOW 1975).

Among 104 families of angiosperms and gymnosperms that were tested, JERMYN and YEOW (1975) found β-lectins in 91, the few cases where these lectins were not detected were probably due to technical difficulties (JERMYN 1975).

## 6.5 Lectin Distribution

### 6.5.1 Detection

Detection of lectins is difficult. Agglutination tests only reveal lectins with specificity corresponding to the binding sites present on the erythrocyte surface and, in any case, require soluble lectins. Immunological detection avoids these difficulties, but can only be made if the lectin antigen is available; if so, radioimmuno-assays can detect very low quantities of lectins (TALBOT and ETZLER 1978).

### 6.5.2 Distribution of Lectins in Different Tissues During Development

Plants contain the maximum amount of lectins during the embryonic phase. During seed development in *Dolichos biflorus*, TALBOT and ETZLER (1978) detected a lectin 27 days after anthesis which rapidly reached the high concentration characteristic of mature seeds; it was concentrated mainly in the cotyledons. GRACIS and ROUGÉ (1977) also noted early lectin appearance in *Vicia sativa* seeds, both in the axis and the cotyledons. Lectins found in the axis rapidly disappear on germination, whereas those of the cotyledons persist for a longer time. These authors detected the lectins by immunoprecipitation and haemagglutination and showed they were found in the seed and young seedling only.

By using haemagglutination, a radioimmuno-assay and an isotope dilution assay, PUEPPKE et al. (1978) found most of the lectin activity in the cotyledons of *Glycine* seeds, but also appreciable levels were detected in the embryo axis and the seed coat. Lectin activity was still present in all of the tissues of the young seedling, but disappeared three weeks after germination. The total lectin activity present in the axis is too small to account for that of the young plant tissues, thus lectins must either be synthesized within each tissue or be transported there from the cotyledons. In *Arachis hypogaea* seedlings between 3 and 9 days old, PUEPPKE (1979) found that more than 90% of the total lectin activity detected in the plant was in the cotyledons, with most of the remainder in hypocotyl, stem and leaves, whereas in plants 21 to 30 days old, the lectin activity was found in all tissues, except in the roots.

SABNIS and HART (1978) have isolated a strongly agglutinating lectin from the phloem exudates of *Cucurbita* where it was present in even larger proportions than in seeds (15 to 20% of total protein). None of the sugars they tested, including those conveyed in the phloem, inhibited the agglutination. The lectin was absent from the seeds, but appeared in the phloem of 5-day-old seedlings and was identical to that in the phloem of adult plants. KAUSS and ZIEGLER (1974) have also reported lectin activity in the sieve tube sap of *Robinia pseudoacacia*. It is possible that this type of lectin has either a transport role or is designed to protect the sugar-rich phloem from bacterial and fungal invasions.

The lectin isolated from the sieve tube sap of *Robinia pseudoacacia* by GIETL et al. (1979) is very similar but not identical to that isolated from the bark of this species: they have different subunits; both molecules have about the same molecular weight (100,000 and 110,000) and the same specificity (N-acetyl-D-galactosamine).

Lectins found in dormant seeds differ in their structure from those that appear during later development phases. Thus ROUGÉ (1977) found agglutinating lectins in *Pisum* seedlings which had no antigenic determinant in common with seed agglutinins; furthermore they did not have the same haptenic specificity and, in this respect, agglutinating lectins found in young pods were also different from those of seeds. PUEPPKE (1979) purified the lectin of *Arachis hypogaea* cotyledons which bind lactose by affinity chromatography and separated this preparation into six isolectins by isoelectrofocusing (pI 5.7 to 6.7). As cotyledons senesced, several of the more basic of these isolectins decreased to undetectable levels, whereas the acidic isolectins remained until 15 days after germination; on the other hand, in 5-day-old roots and hypocotyls, new lactose-binding lectins appeared.

### 6.5.3 Subcellular Distribution

Lectins are mainly located in the protein bodies of the cells of reserve tissues. YOULE and HUANG (1976) identified agglutinin and ricin among the proteins of the protein body matrix in *Ricinus communis* endosperm. TULLY and BEEVERS (1976) located ricin also in the matrix of the endosperm protein bodies. This sequestration of ricin is an important point since, on germination, lectins would remain in the protein bodies which progressively become vacuoles and thus they would not inhibit the cell's ribosomes. BOLLINI and CHRISPEELS (1978), WEBER et al. (1978) and PUSZTAI et al. (1978b) have also located the lectins of some legume seeds (*Phaseolus vulgaris* and *Vicia faba*) in protein bodies. Furthermore, PUSZTAI et al. (1979) found a lectin subunit, associated with the protein body membrane in *Phaseolus vulgaris* seeds. YOSHIDA (1978) isolated two novel lectins from the endoplasmic reticulum fraction of wheat germ, which were different from wheat germ agglutinin (WGA); one had a molecular weight of 4,300 and the other of more than 300,000. Their location suggests a possible role in the binding of ribosomes to endoplasmic reticulum.

CLARKE et al. (1975) localized the Con A and *Phaseolus* agglutinin in cytoplasm sites on sections of *Canavalia ensiformis* and *Phaseolus vulgaris* cotyledons. They also showed that $\beta$-lectins were in the intercellular spaces in the form of clusters of spherical bodies, on cell wall sites, and at the periphery of the cytoplasm associated with the cell membrane. KAUSS and BOWLES (1976) suggested that lectins present in cell walls of *Vigna radiata* take part in the extension process. Thus they could bind polysaccharide molecules non-covalently and more or less reversibly and, under the influence of auxin, an increase in proton concentration in the wall could lower the binding forces between the lectins and their receptor groups. As a consequence, the existing turgor pressure could cause the lectin to jump from one receptor group to the next, thus allowing a controlled movement of the cell wall components relative to each other.

### 6.5.4 Taxonomical Distribution

LEE et al. (1977) examined the agglutination specificity of extracts of 125 taxa from Leguminoseae and Connaraceae and found 75% of species with lectin activity. RÜDIGER (1978) reported the sugar specificity and the type of erythrocyte or other interactive cells for 11 Leguminoseae and 7 non-Leguminoseae species.

Agglutination titres of the sieve tube sap from several tree species belonging to 21 families were reported by GIETL et al. (1979). Most of the species investigated showed agglutinating activities and the most potent was *Robinia pseudoacacia* followed by *Tilia petiolaris*. There were some differences recorded between the sieve tube sap lectins in Fabaceae and those of other plant taxa.

Some variability in lectin content may be characteristic at the specific and infra-specific level. For example, JAFFÉ et al. (1974) distinguished four types of beans among the 21 cultivars of *Phaseolus vulgaris* they screened for haemagglutinating specificity and mitogenic activity. In the 26 cultivars of *Phaseolus vulgaris* studied by KLOZOVÀ and TURKOVÀ (1978), an agglutinating protein was polymorphic and was inherited as a monogenic dominant character. In *Glycine max* seeds, ORF et al. (1978) also found that the presence or absence of the major lectin is controlled by a single dominant gene. Seven varieties of *Arachis hypogaea* and a related species (*A. villosulicarpa*), studied by PUEPPKE (1979), had an isolectin composition that may be characteristic of both the species and the varieties.

FORIERS et al. (1977) compared the primary structure of the lectins from *Pisum, Lens, Glycine, Arachis* and *Phaseolus*. After comparing the first 25 residues of the amino-terminal sequence of subunits, they found extensive homologies ranging from near identity in the case of the $\beta$-chains of *Lens* and *Pisum* lectins to 24% identity between *Glycine* lectin and the L subunit from *Phaseolus* lectin. These extensive homologies between different lectins from a single plant family establish without doubt a common genetic origin for these proteins, and lectins can be grouped in families which have conserved their primary structure though their carbohydrate specificity and some of their biological properties may be different (FORIERS et al. 1977, see also Chap. 7, this Vol.).

### 6.5.5 Possible Roles for the Lectins

On the whole, the role of lectins in plants is not yet fully understood. Because of their large amount in seeds (often 1 to 12% of total proteins), it is sometimes believed that they are storage proteins. In *Vicia sativa* for example, accumulation starts with the synthesis of reserve proteins and they disappear at the same rate as the reserves do (GRACIS and ROUGÉ 1977). However, YOULE and HUANG (1976) noted that, in *Glycine*, lectins are not substantially degraded at germination.

Furthermore, since the binding between lectins and sugars is specific and there is an increasing rate of discovery of new lectins with different properties and localizations, it is now believed that lectins play multiple roles and have a high functional importance in cellular life. This important subject is regularly discussed in many publications (see for example KAUSS 1976, CALLOW 1977, SHARON 1974, SHARON and LIS 1979).

# 7 Tuber Proteins

Tubers are reserve organs born from stems or tuberized roots. Their protein content is generally modest, i.e., 4 to 5% of the dry weight.

## 7.1 Tubers of Leguminoseae

As opposed to conventional tubers, legume tubers can have exceptionally high protein contents; 2 to 20 times greater (Fig. 11a) (FAO 1978). EVANS et al. (1977) reported details of the nitrogen content of four legume tubers; free amino acids accounted for more than half of the total nitrogen (Fig. 11b).

## 7.2 *Solanum tuberosum* Tubers

Potato was first domesticated in America and is now cultivated everywhere in America and Europe. There are numerous potato cultivars and the majority have a protein content of 5% of the dry weight. However, varieties with a lower content (1.2%) as well as varieties with a much higher amount (19.1%) exist (SPLITTSTOESSER 1977).

STEGEMANN et al. (1973) resolved the proteins of potato juice into 40 components by isoelectrofocusing and the patterns were characteristic of varieties. The molecular weight of the protein subunits was uniform (about 20,000) and decreased with the age of the tuber, often due to depolymerization. Some subunits were linked by disulphide bridges and part of the diversity in charge distribution was based on different degrees of amidation. Thus treatment at pH 10 changed the variety-dependent pattern to a picture similar for all varieties.

KAPOOR et al. (1975) separated albumins, globulins, prolamins and glutelins from potato tubers by successive solubilization. The main fraction, obtained by water extraction, was called albumin. All the protein fractions except the prolamin fraction were well balanced in essential amino acids, comparable to the FAO reference protein. Methionine was the limiting amino acid.

ALLEN and NEUBERGER (1973) isolated an unusual lectin from tubers. It had a high cysteine and hydroxyproline content, like wheat germ agglutinin

a

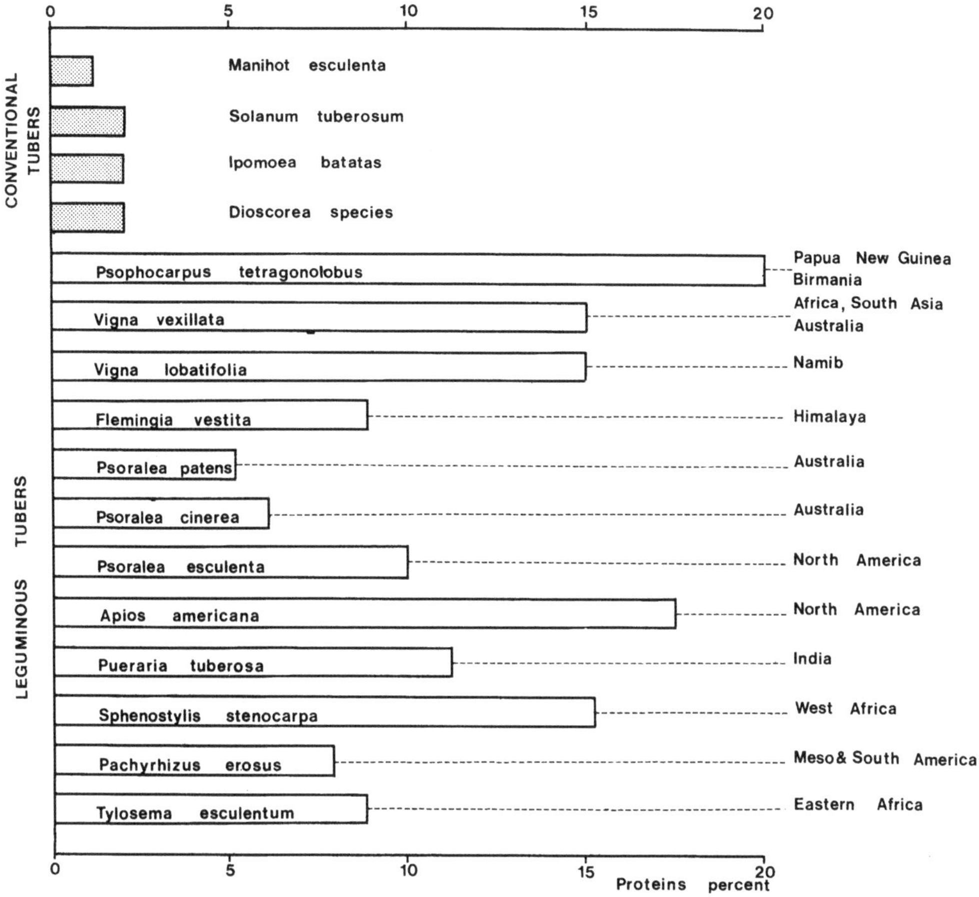


b

| | % Protein in meal (N x 6,25) | % N in meal | % amino-acid N in meal | % N free amino-acid in total N |
|---|---|---|---|---|
| Psophocarpus tetragonolobus (Winged bean) var. "Kano" | 18,1 | 2,90 | 1,66 | 57,2 |
| Psophocarpus tetragonolobus (Winged bean) var. 1 | 15,1 | 2,42 | 1,17 | 48,3 |
| Sphenostylis stenocarpa (African yam bean) | 13,1 | 2,10 | 1,04 | 49,5 |
| Pachyrhisus erosus (Mexican yam bean) | 19,2 | 3,08 | 1,77 | 57,5 |
| Vigna vexillata (Cowpea IV nu 64) | 13,6 | 2,17 | 1,19 | 54,8 |

**Fig. 11.** Protein and nitrogen content of tubers. **a** Protein amount (in % of dry weight) of Leguminoseae tubers, compared to protein amount in conventional tubers. (Redrawn from FAO 1980). **b** Detail of nitrogen content in tubers (dry weight) of four Leguminoseae species, according to EVANS et al. 1977

(WGA); it also had a similar specificity towards N-acetyl-glucosamine, but unlike the wheat protein, the potato lectin was a glycoprotein with 50% sugar content, mainly arabinose.

### 7.3 *Ipomoea batatas* Tubers

Sweet potato is grown in warm countries and is a protein source comparable to that of potato. In most cultivars, tubers have a protein content of 4.5 to 7% of dry weight, with extreme values from 1.7 to 11.8%. Furthermore, the tubers are rich in vitamins. Sweet potato proteins, like potato proteins, show sulphur amino acid and tryptophan limitations, but other essential amino acids are in excess of FAO norms.

### 7.4 *Dioscorea* spp. Tubers

Yam is grown in tropical countries. Its tubers supply proteins in comparable amounts to that of potato and sweet potato (usually 6.6 to 11.2% of dry weight). Some cultivars of *Dioscorea esculenta* have a protein content of 7.9 to 13.4% and the tubers are also rich in vitamin C. Yam has the same sulphur amino acids and tryptophan deficiency as the other tubers described above (SPLITTSTOESSER 1977).

Among *Dioscorea* species, two types must be distinguished: one which is edible (*D. alata, D. esculenta, D. cayenensis,* etc.) and has tubers with few saponosides, the other which has a high amount of saponosides and is used

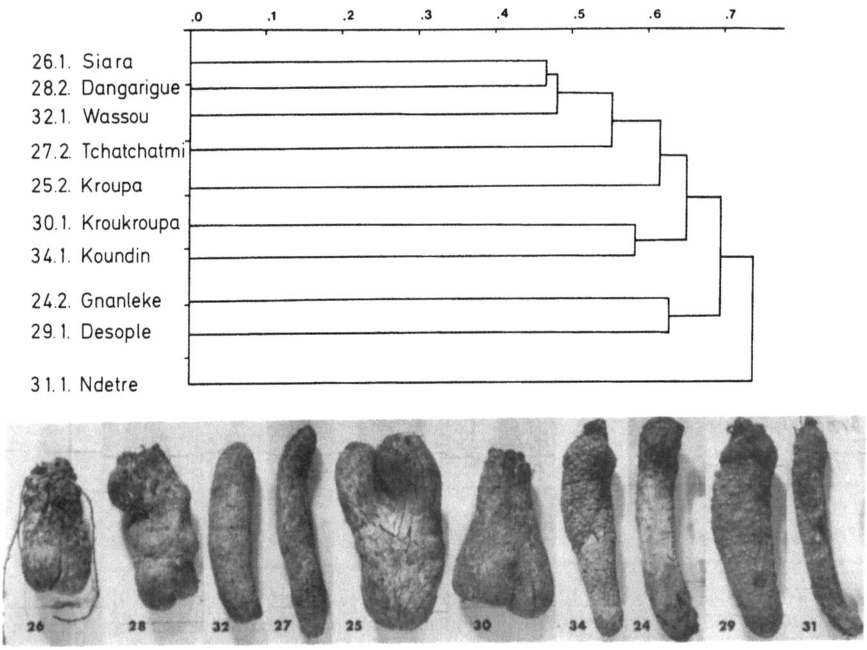


**Fig. 12.** Dendrogram obtained using protein and enzyme characters qualitatively from 10 cultivars of the *Dioscorea cayenensis–D. rotundata* complex. (Redrawn from MIÈGE 1978)

in the pharmaceutical industry for corticoid hemisynthesis (*D. composita, D. deltoidia, D. floribunda*) (SELVARAJ et al. 1975).

MIÈGE (1981) attempted to resolve the taxonomy of the difficult *D. cayenensis–D. rotundata* complex; the protein characters of tubers from ten yam cultivars belonging to the above complex were used as a means of bypassing the problem caused by the absence of seeds in male or sterile cultivars. A dendrogram showing the relationships between the cultivars is given in Fig. 12.

### 7.5 *Manihot esculenta* Tubers

Cassava, also called tapioca, yuca or manioc, is grown in tropical countries where preference is given to the sweet types, poor in cyanogenetic glucosides, rather than bitter types, which harbour some undesirable components. Protein content is low (1.5 to 5.2% of tuber dry weight) and soluble compounds are removed during water treatment designed to free the tuber of its cyanogenetic components. After this treatment, proteins constitute 1.2 to 2.7% of dry weight. However one high protein cultivar has been reported (SPLITTSTOESSER 1977). The poor protein content is not compensated for by a good balance of essential amino acids. Manioc also shows sulphur amino acids and tryptophan limitation, as well as being low in tyrosine and isoleucine (SPLITTSTOESSER 1977, NARTEY and MØLLER 1976, NASSAR and COSTA 1977). Manioc's principal interest correspondingly resides in its starch content, not only as a food, but also for the relatively high yield when starch is transformed into alcohol as an energy source.

### 7.6 *Colocasia esculenta, Xanthosoma sagittifolium, Canna edulis, Maranta arundinacea* Tubers

Taro (*Colocasia esculenta*) is grown in the tropics. Its tubers are poor in proteins: 1 to 4.5% of dry weight, though one cultivar has been found in the mountains of Puerto Rico with 11.7% proteins (MARTIN and SPLITTSTOESSER 1975). Another plant grown in the Pacific, tannier or cocoyam (*Xanthosoma sagittifolium*), is similar to taro but has a higher tuber protein content (5 to 8.9% of dry weight).

Arrow-root comes from tubers of several different species. *Canna edulis* is the arrow-root of Queensland, Australia, with only 1.7% protein in its tubers, whereas *Maranta arundinacea*, grown in India, is richer in proteins (4.6% of dry weight).

Proteins in these tubers have the limitations described above, i.e., for sulphur amino acids, tryptophan, tyrosine, lysine, isoleucine (SPLITTSTOESSER 1977, DANIMIHARDJA and SASTRAPRADJA 1978).

## 8 Proteins in Specialized Structures

### 8.1 Cell Wall Proteins

#### 8.1.1 Extensin

Plant cell walls are composed of cellulose microfibrils enclosed in a non-cellulosic matrix containing one protein. This is rich in an otherwise unusual amino

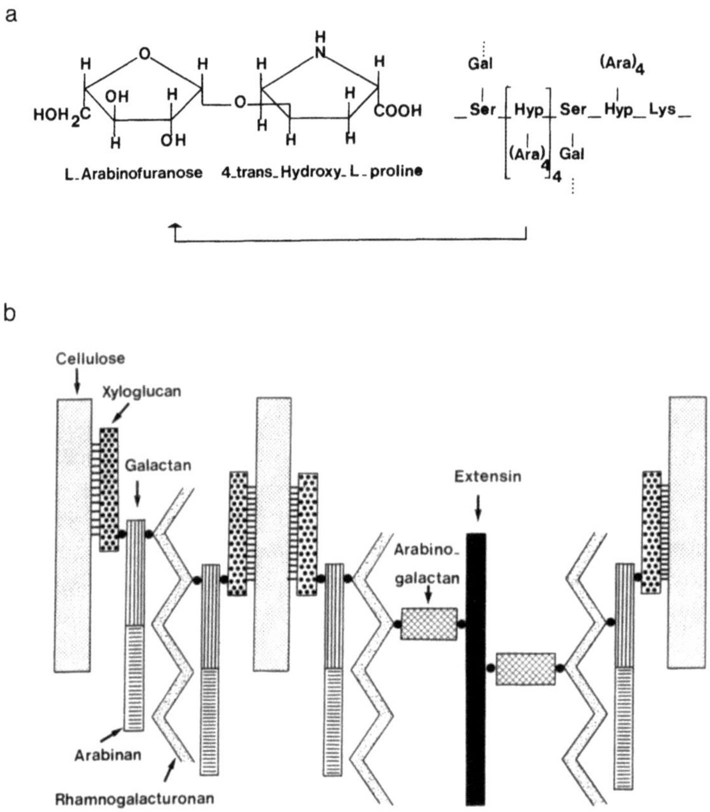


**Fig. 13.** Extensin. **a** Repeating unit of extensin. (Redrawn from BECK and WIECZOREK 1977). **b** Model of attachment of extensin to arabinogalactanes which in turn are linked to cellulose fibres by xyloglucanes. ●: covalent linkage; ≡ : hydrogen bond. (After KEEGSTRA et al. 1973, ALBERSHEIM 1975, adapted from ROBINSON 1977)

acid, hydroxyproline, which constitutes 25% of the amino acid residues and most hydroxyproline residues are glycosylated, via O-glycosidic bonds, to arabinose (LAMPORT 1967). Glycosylation probably occurs in the Golgi apparatus (ROBINSON 1977).

The arabinose hydroxyproline link is acid-labile, but MONRO et al. (1974) were able to solubilize only a part of the cell wall hydroxyproline in *Lupinus* hypocotyl by acid treatment, the resistant part was solubilized by dilute alkali. Accordingly, they suggested that more than one kind of linkage between the hydroxyproline-rich protein and cell wall polysaccharides occurred. The hydroxyproline-tetra-arabinose glycopeptide was shown to be covalently linked to arabinogalactane by the serine OH groups (BECK and WIECZOREK 1977). A scheme for integration of extensin into the complex structure of the cell wall is shown in Fig. 13 according to the model proposed by KEEGSTRA et al. (1973), ALBERSHEIM (1975), ROBINSON (1977). Extensin is an integral part of the cell wall, it is a connecting link in a protein-glucan network that cross-links cellulose fibres through covalent linkages (LAMPORT et al. 1973).

In order to become soluble, cell wall proteins must be detached from this network by breaking covalent bonds. This is done with enzymes, or by drastic chemical means, e.g., alkalis, hydrazine, etc. (SELVENDRAN et al. 1975). ROBINSON (1977) proposed a scheme for destructive fractionation of the cell wall's constitutive material, through progressive solubilization:

Cell walls treated by polygalacturonase
→ Soluble pectic polymers
Residue + 1–4 endoglucanase
→ Soluble hemicelluloses
Residue + pronase → Soluble glycoproteins

Although the function of the cell wall protein is still not completely clear, it is likely that it is involved in controlling extension of the cell wall. MONRO et al. (1974) inferred that extensin was involved in stopping cell elongation since there was much more hydroxyproline glycoprotein in the non-elongating than in the elongating tissues of the *Lupinus* hypocotyl, and SADAVA and CHRISPEELS (1973) observed that proline accumulation necessary for extensin synthesis coincided, in *Pisum* epicotyls, with the cessation of cell elongation.

The action extensin exerts on growth may also influence morphogenesis. BASILE (1969) suggested, for example, that, in liverwort, glycoproteins that are rich in hydroxyproline stop growth and development in highly localized areas at critical moments.

### 8.1.2 Enzymes

In cell wall preparations from *Pisum sativum* epicotyl, RIDGE and OSBORNE (1970) found that peroxidases covalently linked to cell walls were very stable and contained isoenzymes not found in the cytoplasm; these were also different from those linked to cell walls by non-covalent, ionic bonds. The parallel increase, under the influence of ethylene, of the wall hydroxyproline levels and the covalently bound peroxidases led them to suggest that the peroxidases themselves were hydroxyproline-rich proteins. Hydroxyproline-rich proteins with enzyme activity are rare in both plants and animals, and in plants they are probably purely structural components of the cell wall.

## 8.2 Pollen Proteins

The pollen grain is the male gametophyte, protected by two walls, intine and exine, each of quite different origin. The intine belongs to the gametophyte generation and the proteins it contains are synthetized by the haploid genome of the spore. The exine belongs to the sporophyte generation and its proteins come from the tapetum, a diploid parental tissue of the anthers (HESLOP-HARRISON 1975) (Fig. 14).

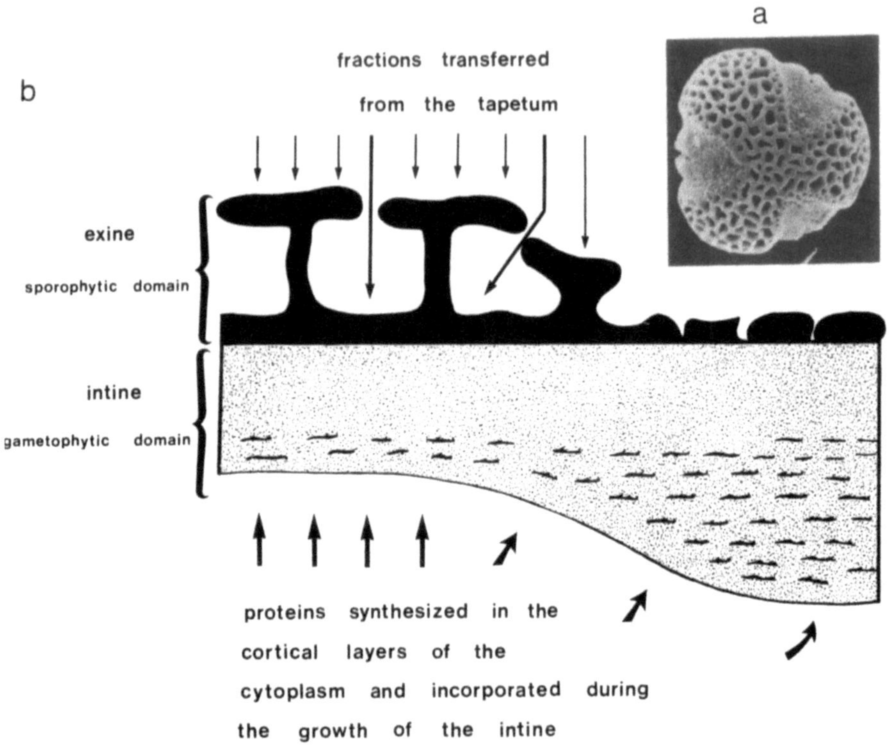


**Fig. 14. a** Exine ornamentation and apertures in a pollen grain from *Cordeauxia edulis*, scanning electron micrograph. (Courtesy of MIÈGE et al. 1978). **b** The two walls of a pollen grain: outwards, exine, of which the proteins come from the sporophyte (tapetum); inwards, intine, of which the proteins are synthetized in the microspore cytoplasm (gametophyte). (After HESLOP-HARRISON 1975)

### 8.2.1 Intine Proteins

Intine proteins are enzymes, mostly concentrated in extracellular lysosomal inclusions, in the intine layers underlying each potential site of pollen tube emergence in pollen grain with apertures (for example Liliaceae, Brassicaceae, Asteraceae, Poaceae). In non-apertural grains, like those of Cornaceae and certain Iridaceae, the entire outer intine contains the extracellular lysosomal inclusions. The enzymes of the lysosomes are sealed by a continuous polysaccharide layer. Under hydration, this layer is disrupted, discharging enzymes and enzyme precursors which degrade the cuticle of the stigmatic papilla after appropriate activation and thus allow the pollen tube to enter (HESLOP-HARRISON 1975).

Acid phosphatase, ribonuclease, non-specific esterase, amylase and protease, the typical enzymes of lysosomes, were characterized in the intine of over 50 angiosperm pollens of all structural types and from 27 families (KNOX and HESLOP-HARRISON 1970). Some of these enzymes are highly polymorphic, thus 15 esterase isoenzymes were recorded in the diffusate of pollen grain of *Oenothera organensis;* 5 of them appeared within 5 min, 8 within 30 min, 9 within 2 h and 13 within 19 h of hydration (MÄKINEN and BREWBAKER 1967).

### 8.2.2 Exine Proteins

Exine, the more or less sculptured outer coating of the pollen grain, was examined cytochemically by HESLOP-HARRISON (1975) who reported a positive PAS-reaction of *Agrostemma githago* pollen exine which is probably attributable to the presence of glycoproteins. The only detectable enzyme activity was that of esterases. The architecture of the exine affects the relative rate of release of proteins from both exine and intine but, in all cases, the exine proteins diffuse out first, usually within seconds of hydration.

Diffusates from both pollen walls of species belonging to Brassicaceae and Asteraceae, when compared immunologically and by electrophoretic techniques, revealed differences that can be generic or even species-specific (HESLOP-HARRISON 1975).

### 8.2.3 Nutritional Value of Pollen

Pollen is a food for bees, and is their only nitrogen source; it is also used by man. All essential amino acids are found in pollen, either free or combined in proteins or glycoproteins. The total nitrogen compounds represent 16 to 40% of dry weight of the pollen (PONS 1972). The proportion of free amino acids relative to total amino acids is approximately 25% and the most abundant amino acid in a free state is proline. Two unusual amino acids, pipecolic acid and taurine, are also present in the unbound form and, among proteins, the albumin fraction is often important; thus in *Zea mays,* it varies from 3.8 to 20.3% of the total proteins (STANLEY and LINSKENS 1974).

This survey has considered those plant proteins which play a predominant role in the plant's life, either continuously through the life cycle (enzymes, inhibitors, lectins, extensin etc.), or at particular times, for example, reserve proteins in seeds and tubers and pollen grain proteins. In the last few years, understanding of plant proteins has increased enormously, especially with regard to those of direct interest for man's needs (i.e., reserves in food, lectins in medicine, etc.). It has not been possible, therefore, to make an exhaustive coverage of the literature. Instead, the main object was to give a synthetic and condensed account of a representative selection of recently acquired knowledge

## References

Albersheim P (1975) The walls of growing plant cells. Sci Am 232(4):80–95

Alexandrescu V, Hagima I, Popov D, Jilăveanu A (1975) Multiple molecular forms of catalase, peroxidase and acid phosphatase in *Lycopersicum* sp. with genetical and induced resistance against tobacco mosaic virus (TMV). Rev Roum Biochim 12:209–212

Allen AK, Neuberger A (1973) The purification and properties of the lectin from potato tubers, a hydroxyproline-containing glycoprotein. Biochem J 135:307–314

Baenziger PS, Glover DV (1977) Protein body size and distribution and protein matrix morphology in various endosperm mutants of *Zea mays* L. Crop Sci 17:415–421

Bailey CJ, Boulter D (1970) The structure of legumin, a storage protein of broad bean (*Vicia faba*) seed. Eur J Biochem 17:460–466
Bailey CJ, Boulter D (1972) The structure of vicilin of *Vicia faba*. Phytochemistry 11:59–64
Barbieri L, Lorenzoni E, Stirpe F (1979) Inhibition of protein synthesis in vitro by a lectin from *Momordica charantia* and by other haemagglutinins. Biochem J 182:633–635
Barker BE, Farnes P (1967) Mitogenic property of *Wistaria floribunda* seeds. Nature (London) 215:659–660
Barker RDJ, Derbyshire E, Yarwood A, Boulter D (1976) Purification and characterization of the major storage proteins of *Phaseolus vulgaris* seeds, and their intracellular and cotyledonary distribution. Phytochemistry 15:751–757
Basha SMM, Beevers L (1976) Glycoprotein metabolism in the cotyledons of *Pisum sativum* during development and germination. Plant Physiol 57:93–97
Basile DV (1969) Inhibition of proline-hydroxyproline metabolism and phenovariation of *Scapania nemorosa*. Bull Torrey Bot Club 96:577–582
Baumann C, Rüdiger H, Strosberg AD (1979) A comparison of the two lectins from *Vicia cracca*. FEBS Lett 102:216–218
Bechtel DB, Juliano BO (1980) Formation of protein bodies in the starchy endosperm of rice (*Oryza sativa* L.): a re-investigation. Ann Bot (London) 45:503–509
Beck E, Wieczorek J (1977) Carbohydrate metabolism. In: Ellenberg H, Esser K, Merxmüller H, Schnepf E, Ziegler H (eds) Progress in botany, vol XXXIX. Springer, Berlin Heidelberg New York, pp 62–82
Birk Y, Gertler A (1971) Chemistry and biology of proteinase inhibitors from soybean and groundnuts. In: Fritz H, Tschesche H (eds) Proceedings of the international research conference on proteinase inhibitors. de Gruyter, Berlin New York, pp 142–148
Biroc SL, Etzler ME (1978) The effect of periodate oxidation and α-mannosidase treatment on *Dolichos biflorus* lectin. Biochim Biophys Acta 544:85–92
Blagrove RJ, Gillespie JM (1978) Variability of the seed globulins of winged bean, *Psophocarpus tetragonolobus* (L.) DC. Aust J Plant Physiol 5:371–375
Bollini R, Chrispeels MJ (1978) Characterization and subcellular localization of vicilin and phytohemagglutinin, the two major reserve proteins of *Phaseolus vulgaris* L. Planta 142:291–298
Boulter D (1965) Protein biosynthesis in higher plants. In: Pridham JB, Swain T (eds) Biosynthetic pathways in higher plants. Academic Press, London New York, pp 101–115
Boulter D (1974) The use of amino acid sequence data in the classification of higher plants. In: Bendz G, Santesson J (eds) Chemistry in botanical classification. Nobel Foundation Symposium, vol XXV. Academic Press, London New York, pp 211–216
Boulter D (1980) Ontogeny and development of biochemical and nutritional attributes in legume seeds. In: Summerfield RJ, Bunting AH (eds) Advances in legume science. Royal Botanic Gardens, Kew, pp 127–134
Boulter D (1981) Proteins of legumes. In: Polhill RM, Raven PH (eds) Advances in legume systematics. Part 2 Royal Botanic Gardens, Kew, 501–512
Bowen ID, Bryant JA (1978) The fine structural localization of p-nitrophenyl phosphatase activity in the storage cells of pea (*Pisum sativum* L.) cotyledons. Protoplasma 97:241–250
Bowles DJ, Kauss H (1976) Characterization, enzymatic and lectin properties of isolated membranes from *Phaseolus aureus*. Biochem Biophys Acta 443:360–374
Bowles DJ, Schnarrenberger C, Kauss H (1976) Lectins as membrane components of mitochondria from *Ricinus communis*. Biochem J 160:375–382
Briarty LG, Coult DA, Boulter D (1969) Protein bodies of developing seeds of *Vicia faba*. J Exp Bot 20:358–372
Briarty LG, Hughes CE, Evers AD (1979) The developing endosperm of wheat. A stereological analysis. Ann Bot (London) 44:641–658
Burr FA, Burr B (1979) Molecular basis of zein protein synthesis in maize endosperm. In: Rubenstein I, Phillips RL, Green CE, Gengenbach BG (eds) The plant seed: development, preservation, and germination. Academic Press, London New York, pp 27–48
Buttrose MS, Lott JNA (1978) Calcium oxalate druse crystals and other inclusions in seed protein bodies: *Eucalyptus* and jojoba. Can J Bot 56:2083–2091

Callow JA (1977) Recognition, resistance and the role of plant lectins in host-parasite interactions. Adv Bot Res 4:1–49
Camus M-C, Laporte J-C (1974) Etude de la teneur et de la répartition des inhibiteurs trypsiques de divers blés durs (*T. durum*) et blés tendres (*T. vulgare*) par une méthode sur film de gélatine. Ann Technol Agric 23:421–431
Carasco JF, Croy R, Derbyshire E, Boulter D (1978) The isolation and characterization of the major poly-peptides of the seed globulin of cowpea (*Vigna unguiculata* L. Walp) and their sequential synthesis in developing seeds. J Exp Bot 29:309–323
Casey R (1979) Immunoaffinity chromatography as a means of purifying legumin from *Pisum* (pea) seeds. Biochem J 177:509–520
Catsimpoolas N, Wang J (1971) Analytical scanning isoelectrofocusing: 5 – Separation of glycinin subunits in urea-dithiothreitol media. Anal Biochem 44:436–444
Champagnat R, Ozenda P, Baillaud L (1969) Biologie végétale. Croissance, morphogenèse, reproduction, Vol III. Masson et Cie, Paris, 510 pp Précis de Sciences Biologiques publiés sous la direction du Professeur PP Grassé
Chen I, Mitchell HL (1973) Trypsin inhibitors in plants. Phytochemistry 12:327–330
Chrispeels MJ, Baumgartner B (1978) Trypsin inhibitors in mung bean cotyledons. Purification, characteristics, subcellular localization, and metabolism. Plant Physiol 61:617–623
Clarke AE, Knox RB, Jermyn MA (1975) Localization of lectins in legume cotyledons. J Cell Sci 19:157–167
Crawford RMM, McManmon M (1968) Inductive responses of alcohol and malic dehydrogenases in relation to flooding tolerance in roots. J Exp Bot 19:435–441
Cristofolini G, Chiapella LF (1977) Serological systematics of the tribe Genisteae (Fabaceae). Taxon 26:43–56
Croy RRD, Derbyshire E, Krishna TG, Boulter D (1979) Legumin of *Pisum sativum* and *Vicia faba*. New Phytol 83:29–35
Croy RRD, Gatehouse JA, Tyler M, Boulter D (1980) The purification and characterization of a third storage protein (convicilin) from the seeds of pea (*Pisum sativum* L.). Biochem J 191:509–516
Danielsson CE (1949) Seed globulins of the Gramineae and Leguminosae. Biochem J 44:387–400
Danimihardja S, Sastrapradja S (1978) Variation of some cultivated and wild talas, *Colocasia esculenta* (L.) Schott in crude protein contents and electrophoretic pattern. Ann Bogor 6:177–186
Daussant J, Neucere NJ, Yatsu LY (1969) Immunochemical studies on *Arachis hypogaea* proteins with particular reference to the reserve proteins. I. Characterization, distribution, and properties of α-arachin and α-conarachin. Plant Physiol 44:471–479
Daussant J, Renard M, Hill RD (1979) Ontogenical evolution of α-amylase in cereal seeds: examples provided by immunochemical studies on wheat and on one *Triticale* cultivar showing shrivelling at maturation. In: Laidman DL, Wyn Jones RG (eds) Recent advances in the biochemistry of cereals. Academic Press, London New York, pp 345–348
Davey JE, van Staden J (1978) Ultrastructural aspects of reserve protein deposition during cotyledonary cell development in *Lupinus albus*. Z Pflanzenphysiol 89:259–271
Davies HV, Chapman JM (1979) The control of food mobilisation in seeds of *Cucumis sativus* L. 1. The influence of embryonic axis and testa on protein and lipid degradation. Planta 146:579–584
Derbyshire E, Boulter D (1976) Isolation of legumin-like protein from *Phaseolus aureus* and *Phaseolus vulgaris*. Phytochemistry 15:411–414
Derbyshire E, Wright DJ, Boulter D (1976) Legumin and vicilin, storage proteins of legume seeds. Phytochemistry 15:3–24
Dixon M, Webb EC (1967) Enzymes, 2nd edn. Longmans, London
Entlicher G, Kocourek J (1975) Studies on phytohemagglutinins. XXIV. Isoelectric point and hybridization of the pea (*Pisum sativum* L.) isophytohemagglutinins. Biochim Biophys Acta 393:165–169
Ericson MC, Chrispeels MJ (1973) Isolation and characterization of glucosamine-containing storage glycoproteins from the cotyledons of *Phaseolus aureus*. Plant Physiol 52:98–104

Evans IM, Boulter D, Eaglesham ARJ, Dart PJ (1977) Protein content and protein quality of tuberous roots of some legumes determined by chemical methods. Qual Plant Foods Hum Nutr 27:275–285
Falasca A, Franceschi C, Rossi CA, Stirpe F (1979) Purification and partial characterization of a mitogenic lectin from *Vicia sativa*. Biochim Biophys Acta 577:71–81
FAO (1980) Cultures conventionnelles et protéines. Tubercules légumineux. Ceres 68: 4–5
Farnes P, Barker BE, Brownhill LE, Fanger H (1964) Mitogenic activity in *Phytolacca americana* (pokeweed) Lancet ii:1100–1101
Felder MR, Scandalios JG (1971) Effects of homozygosity and heterozygosity on certain properties of genetically defined electrophoretic variants of alcohol dehydrogenase isozymes in maize. Mol Gen Genet 111:317–326
Fieldes MA, Bashour N, Deal CL, Tyson H (1976) Isolation of peroxidase isozymes from two flax genotypes by column chromatography. Can J Bot 54:1180–1188
Font J, Bourrillon R (1971) Les constituants glucidiques de la phytoagglutinine de *Robinia pseudoacacia* et leur rôle dans l'activité érythroagglutinante. Biochim Biophys Acta 243:111–116
Foriers A, Wuilmart C, Sharon N, Strosberg AD (1977) Extensive sequence homologies among lectins from leguminous plants. Biochem Biophys Res Commun 75:980–985
Fornstedt N, Porath J (1975) Characterization studies on a new lectin found in seeds of *Vicia ervilia*. FEBS Lett 57:187–191
Furusawa Y, Kurosawa Y, Chuman I (1974) Purification and properties of trypsin inhibitor from Hakuhensu bean (*Dolichos lablab*). Agric Biol Chem 38:1157–1164
Gatehouse JA, Croy RRD, Boulter D (1980) Isoelectric-focusing properties and carbohydrate content of pea (*Pisum sativum*) legumin. Biochem J 185:497–503
Ghetie V, Buzila L (1962) Crioproteinale vegetale. Stud Cercet Biochim 5:65
Gietl C, Kauss H, Ziegler H (1979) Affinity chromatography of a lectin from *Robinia pseudoacacia* L. and demonstration of lectins in sieve-tube sap from other tree species. Planta 144:367–371
Gillespie JM, Blagrove RJ (1978) Isolation and composition of the seed globulins of winged bean, *Psophocarpus tetragonolobus* (L.) DC. Aust J Plant Physiol 5:357–369
Gilroy J, Wright DJ, Boulter D (1979) Homology of basic subunits of legumin from *Glycine max* and *Vicia faba*. Phytochemistry 18:315–316
Gleeson PA, Jermyn MA (1977) Leguminous seed glycoproteins that interact with concanavalin A. Aust J Plant Physiol 4:25–37
Gracis J-P, Rougé P (1977) Evolution des hémagglutinines de *Vicia sativa* L. au cours du cycle de développement de la plante. Bull Soc Bot Fr 124:301–306
Graham TA, Gunning BES (1970) Localization of legumin and vicilin in bean cotyledon cells using fluorescent antibodies. Nature (London) 228:81–82
Grange A (1976) Effets des conditions de conservation sur la structure et l'activité des protéines de graines de *Phaseolus vulgaris* (L.) var. *contender*. Applications taxonomiques. Thèse, Fac Sci Genève, 96 pp
Grange A (1980) Vieillissement des graines de *Phaseolus vulgaris* (L.) var. *contender*. I. Effets sur la germination, la vigueur, la teneur en eau et la variation des formes d'azote. Physiol Vég 18:579–586
Grant DR, Summer AK, Johnson J (1976) An investigation on pea seed albumins. Can Inst Food Sci Technol J 9:84–91
Guilliermond A (1909) Recherches cytologiques sur la germination des graines de quelques graminées et contribution à l'étude des grains d'aleurone. Arch Anat Microsc Morphol Exp 10:141–226
Hall TC, McLeester RC, Bliss FA (1972) Electrophoretic analysis of protein changes during the development of the french bean fruit. Phytochemistry 11:647–649
Hankins CN, Shannon LM (1978) The physical and enzymatic properties of a phytohemagglutinin from mung beans. J Biol Chem 253:7791–7797
Harris N, Boulter D (1976) Protein body formation in cotyledons of developing cowpea (*Vigna unguiculata*) seeds. Ann Bot (London) 40:739–744
Harris N, Chrispeels MJ (1975) Histochemical and biochemical observations on storage

protein metabolism and protein body autolysis in cotyledons of germinating mung beans. Plant Physiol 56:292–299
Harris N, Juliano BO (1977) Ultrastructure of endosperm protein bodies in developing rice grains differing in protein content. Ann Bot (London) 41:1–5
Heslop-Harrison J (1975) The physiology of the pollen grain surface. Proc R Soc London Ser B 190:275–299
Horstmann C, Rudolph A, Schmidt P (1978) Isolation, characterization, and subunit structure of a phytohemagglutinin from seeds of *Vicia faba* L. Biochem Physiol Pflanz 173:311–321
Jacobsen JV, Knox RB, Pyliotis NA (1971) The structure and composition of aleurone grains in the barley aleurone layer. Planta 101:189–209
Jaffé WG, Levy A, González DI (1974) Isolation and partial characterization of bean phytohemagglutinins. Phytochemistry 13:2685–2693
Jermyn MA (1975) Invariance throughout the seed plants of the protein moiety of a lectin glycoprotein. Proc Aust Biochem Soc 8:2
Jermyn MA, Yeow YM (1975) A class of lectins present in the tissues of seed plants. Aust J Plant Physiol 2:501–531
Kaiser K-P, Belitz H-D (1971) Proteinaseninhibitoren in Lebensmitteln. IV. Vorkommen und Isolierung von Trypsin und Chymotrypsininhibitoren in Kartoffeln. Chem Mikrobiol Technol Lebensm 1:1–7
Kaiser K-P, Belitz H-D (1972) Proteinaseninhibitoren. VI. Einige Eigenschaften verschiedener Trypsin- und Chymotrypsininhibitoren der Kartoffel. Chem Mikrobiol Technol Lebensm 1:191–194
Kaiser K-P, Belitz H-D (1973) Specificity of potato isoinhibitors towards various proteolytic enzymes. Z Lebensm Unters Forsch 151:18–22
Kaiser K-P, Bruhn LC, Belitz H-D (1974) Protease inhibitors in potatoes. Protein-, trypsin- and chymotrypsininhibitor patterns by isoelectric focusing in polyacrylamidegel. A rapid method for identification of potato varieties. Z Lebensm Unters Forsch 154:339–347
Kapoor AC, Desborough SL, Li PH (1975) Potato tuber proteins and their nutritional quality. Potato Res 18:469–478
Kauss H (1976) Plant lectins (Phytohemagglutinins). In: Ellenberg H, Esser K, Merxmüller H, Schnepf E, Ziegler H (eds) Progress in botany, vol XXXVIII. Springer, Berlin Heidelberg New York, pp 58–70
Kauss H, Bowles DJ (1976) Some properties of carbohydrate-binding proteins (lectins) solubilized from cell walls of *Phaseolus aureus*. Planta 130:169–174
Kauss H, Ziegler H (1974) Carbohydrate-binding proteins from the sieve-tube sap of *Robinia pseudoacacia* L. Planta 121:197–200
Keegstra K, Talmadge KW, Bauer WD, Albersheim P (1973) The structure of plant cell walls. III A model of the walls of suspension-cultured sycamore cells based on the interconnections of the macromolecular components. Plant Physiol 51:188–196
Khavkin EE, Misharin SI, Markov YY, Peshkova AA (1978) Identification of embryonal antigens of maize: globulins as primary reserve proteins of the embryo. Planta 143:11–20
Kim SI, Pernollet J-C, Mossé J (1979) Evolution des protéines de l'albumen et de l'ultrastructure du caryopse d'*Avena sativa* au cours de la germination. Physiol Veg 17:231–245
Kirk JTO, Pyliotis NA (1976) Cruciferous oilseed proteins: the protein bodies of *Sinapis alba* seed. Aust J Plant Physiol 3:731–746
Klozová E, Turková V (1978) Variability of some seed proteins of the species *Phaseolus vulgaris* and their relationship to phytohaemagglutinating activity. Biol Plant 20:129–134
Knox RB, Heslop-Harrison J (1970) Pollen-wall proteins: localization and enzymic activity. J Cell Sci 6:1–27
Kocoń J, Muszyński S, Sowa W (1978) The ultrastructure of endosperm in wheat (*Triticum aestivum* L.) as revealed by scanning electron microscopy. Bull Acad Pol Sci Ser Sci Biol Cl V 26:5–6
Köhle H, Kauss H (1979) Binding of *Ricinus communis* agglutinin to the mitochondrial inner membrane as an artifact during preparation. Biochem J 184:721–723
Koehler DE, Varner JE (1973) Hormonal control of orthophosphate incorporation into phospholipids of barley aleurone layers. Plant Physiol 52:208–214

Kolberg J (1978) Isolation and partial characterization of a mitogenic lectin from *Lathyrus odoratus* seeds. Acta Pathol Microbiol Scand Sect C 86:99–104

Kortt AA (1979) Isolation and characterization of the trypsin inhibitors from winged bean seed (*Psophocarpus tetragonolobus* (L.) DC). Biochim Biophys Acta 577:371–382

Koshiyama I, Fukushima D (1976) Physico-chemical studies on the 11S globulin in soybean seeds: size and shape determination of the molecule. Int J Pept Protein Res 8:283–289

Krishna TG, Croy RRD, Boulter D (1979) Heterogeneity in subunit composition of the legumin of *Pisum sativum*. Phytochemistry 18:1879–1880

Lamport DTA (1967) Hydroxyproline-O-glycosidic linkage of the plant cell wall glycoprotein extensin. Nature (London) 216:1322–1324

Lamport DTA, Katona L, Roerig S (1973) Galactosylserine in extensin. Biochem J 133:125–131

Leavitt RD, Felsted RL, Bachur NR (1977) Biological and biochemical properties of *Phaseolus vulgaris* isolectins. J Biol Chem 252:2961–2966

Lee DW, Tan GS, Liew FY (1977) A survey of lectins in south-east asian Leguminoseae. Plant Med 31:83–93

Liener IE, Kakade ML (1969) Protease inhibitors. In: Liener IE (ed) Toxic constituents of plant foodstuffs. Academic Press, London New York, pp 7–68

Lin JY, Tserng KY, Chen CC, Lin LT, Tung TC (1970) Abrin and ricin: new anti-tumour substances. Nature (London) 227:292–293

Lotan R, Cacan R, Debray H, Carter WG, Sharon N (1975) On the presence of two types of subunit in soybean agglutinin. FEBS Lett 57:100–103

Lott JNA, Buttrose MS (1977) Globoids in protein bodies of legume seed cotyledons. Aust J Plant Physiol 5:89–111

Lott JNA, Buttrose MS (1978a) Thin sectioning, freeze fracturing, energy dispersive X-ray analysis, and chemical analysis in the study of inclusions in seed protein bodies: almond, Brazil nut, and quandong. Can J Bot 56:2050–2061

Lott JNA, Buttrose MS (1978b) Location of reserves of mineral elements in seed protein bodies: macadamia nut, walnut, and hazel nut. Can J Bot 56:2072–2082

Lott JNA, Vollmer CM (1973) The structure of protein bodies in *Cucurbita maxima* cotyledons. Can J Bot 51:687–688

Lott JNA, Larsen PL, Darley JJ (1971) Protein bodies from the cotyledons of *Cucurbita maxima*. Can J Bot 49:1777–1782

Mäkinen Y, Brewbaker JL (1967) Isoenzyme polymorphism in flowering plants. I. Diffusion of enzymes out of intact pollen grains. Physiol Plant 20:477–482

Maher P, Molday RS (1977) Binding of concanavalin A to *Ricinus communis* agglutinin and its implication in cell-surface labeling studies. FEBS Lett 84:391–394

Manen J-F, Miège M-N (1977) Purification et caractérisation des lectines isolées dans les albumines et les globulines de *Phaseolus vulgaris*. Physiol Veg 15:163–173

Manteuffel R, Scholz G (1975) Studies on seed globulins from legumes. V. Immunoelectrophoretic control of vicilin purification by gel filtration. Biochem Physiol Pflanz 168:277–285

Markert CL (ed) (1975) Isozymes, Vol IV. Genetics and evolution. Academic Press, London New York

Marshall JJ, Rabinowitz ML (1975) Enzyme stabilization by covalent attachment of carbohydrate. Arch Biochem Biophys 167:777–779

Martin FW, Splittstoesser WE (1975) A comparison of total protein and amino acids of tropical roots and tubers. Trop Root Tuber Crops Newslett 8:7–15

Matile P (1975) The lytic compartment of plant cells. Cell biology monographs, Vol I. Springer, Wien New York

McLeester RC, Hall TC, Sun SM, Bliss FA (1973) Comparison of globulin proteins from *Phaseolus vulgaris* with those from *Vicia faba*. Phytochemistry 12:85–93

Mettler IJ, Beevers H (1979) Isolation and characterization of the protein body membrane of castor beans. Plant Physiol 64:506–511

Miège J (1975) Les protéines des graines en taxonomie et phylogénie végétales. In: Miège J (ed) Les protéines des graines. Genèse, nature, fonctions, domaines d'utilisation. Georg, Genève, pp 305–365

Miège J (1981) Etude chimiotaxonomique de dix cultivars de Côte d'Ivoire relevant du complexe *Dioscorea cayenensis-D. rotundata*. Séminaire international sur l'igname, Buea, Cameroun. In: Miège J, Lyonga SN (eds) Yams. Univ Press, Oxford, pp 197–231
Miège J, Miège M-N (1978) *Cordeauxia edulis* – A Caesalpiniaceae of arid zones of East Africa. Caryologic, blastogenic and biochemical features. Potential aspects for nutrition. Econ Bot 32:336–345
Miège J, Crapon de Caprona A, Lacotte D (1978) Caractères séminaux, palynologiques, caryologiques de deux Légumineuses alimentaires: *Cordeauxia edulis* Hemsley et *Psophocarpus tetragonolobus* (L.) DC. Candollea 33:329–347
Miège M-N (1970) Etude des protéines des graines d'une légumineuse: *Lablab niger* Medik. Arch Sci 23:75–149
Miège M-N (1975) Chimie taxonomique: analyse critique de l'utilisation des caractères biochimiques des protéines de graines. Exemple de variabilité d'origine technique apporté par une étude cinétique de la dialyse. Saussurea 6:153–169
Miège M-N, Mascherpa J-M (1976) Isolation and analysis of protein bodies from cotyledons of *Lablab purpureus* and *Phaseolus vulgaris* (Leguminoseae). Physiol Plant 37:229–238
Miflin BJ, Shewry PR (1979) The synthesis of proteins in normal and high lysine barley seeds. In: Laidman DL, Wyn Jones RG (eds) Recent advances in the biochemistry of cereals. Academic Press, London New York, pp 239–273
Mikola J, Kirsi M (1972) Differences between endospermal and embryonal trypsin inhibitors in barley, wheat, and rye. Acta Chem Scand 26:787–795
Millerd A (1975) Biochemistry of legume seed proteins. Annu Rev Plant Physiol 26:53–72
Misset M-T (1977) Contribution à la chimie taxonomique de 57 espèces de Légumineuses. Etudes qualitative et quantitative des protéines de leurs graines. Traitement informatique des données. Saussurea 8:1–18
Mlodzianowski F (1978) The fine structure of protein bodies in lupine cotyledons during the course of seed germination. Z Pflanzenphysiol 86:1–13
Mollenhauer HH, Morré DJ, Jelsema CL (1978) Lamellar bodies as intermediates in endoplasmic reticulum biogenesis in maize (*Zea mays* L.) embryo, bean (*Phaseolus vulgaris* L.) cotyledon, and pea (*Pisum sativum* L.) cotyledon. Bot Gaz 139:1–10
Monro JA, Bailey RW, Penny D (1974) Cell wall hydroxy-proline-polysaccharide associations in *Lupinus* hypocotyls. Phytochemistry 13:375–382
Monsigny M, Kieda C, Roche A-C (1979) Membrane lectins. Biol Cell 36:289–300
Murray DR (1979) Nutritive role of the seedcoats during embryo development in *Pisum sativum* L. Plant Physiol 64:763–769
Murray DR, Collier MD (1977) Acid phosphatase activities in developing seeds of *Pisum sativum* L. Aust J Plant Physiol 4:843–848
Murray DR, Crump JA (1979) Euphaseolin, the predominant reserve globulin of *Phaseolus vulgaris* cotyledons. Z Pflanzenphysiol 94:339–350
Nartey F, Møller BL (1976) Amino acid profiles of cassava seeds (*Manihot esculenta*). Econ Bot 30:419–423
Nassar NMA, Costa CP (1977) Tuber formation and protein content in some wild cassava (Mandioca) species native of central Brazil. Experientia 33:1304–1306
Neumann D, Weber E (1978) Formation of protein bodies in ripening seeds of *Vicia faba* L. Biochem Physiol Pflanz 173:167–180
Olsnes S, Pihl A (1978) Abrin and ricin – two toxic lectins. Trends Biochem Sci 3:7–10
Orf JH, Hymowitz T (1977) Inheritance of a second trypsin inhibitor variant in seed protein of soybeans. Crop Sci 17:811–813
Orf JH, Hymowitz T, Pull SP, Pueppke SG (1978) Inheritance of soybean seed lectin. Crop Sci 18:899–900
Ortanderl H, Belitz H-D (1977) Trypsin-/Chymotrypsin-Inhibitoren in Samen der Puffbohne (*Vicia faba*). Z Lebensm Unters Forsch 163:31–34
Ory RL, Henningsen KW (1969) Enzymes associated with protein bodies isolated from ungerminated barley seeds. Plant Physiol 44:1488–1498
Pappenheimer AM Jr, Olsnes S, Harper AA (1974) Lectins from *Abrus precatorius* and *Ricinus communis*. I – Immunochemical relationships between toxins and agglutinins. J Immunol 113:835–841

Payne RC, Koszykowski TJ (1978) Esterase isoenzyme differences in seed extracts among soybean cultivars. Crop Sci 18:557–559

Payne RC, Oliveira AR, Fairbrothers DE (1973) Disc electrophoretic investigation of *Coffea arabica* and *C. canephora*: general protein and malate dehydrogenase of mature seeds. Biochem Syst 1:59–61

Pazur JH, Knull HR, Simpson DL (1970) Glycoenzymes: a note on the role for the carbohydrate moieties. Biochem Biophys Res Commun 40:110–116

Pernollet J-C (1978) Protein bodies of seeds: ultrastructure, biochemistry, biosynthesis and degradation. Phytochemistry 17:1473–1480

Pichl I (1978) Characterization of albumins isolated from seeds of *Cucurbita maxima* L. Biochem Physiol Pflanz 172:61–66

Pihakaski K, Pihakaski S (1978) Myrosinase in Brassicaceae (Cruciferae). II. Myrosinase activity in different organs of *Sinapis alba* L. J Exp Bot 29:335–345

Pons A (1972) Pollen. In: Encyclopaedia universalis (ed), Vol 13. Paris, pp 247–248

Popov D, Serban M (1977) The significance of molecular heterogeneity of proteins and several enzymes in comparative studies on *Phaseolus vulgaris* taxa. Rev Roum Biochim 14:207–211

Popov D, Serban M, Muresan T, Gutenmacher P (1976) Comparative biochemical studies on populations, inbreds and lines of *Phaseolus vulgaris*. II – Multiple molecular forms of acid phosphatase and esterase. Rev Roum Biochim 13:107–110

Prevosti-Tourmel A-M (1979) Embryogenèse du *Phaseolus vulgaris* L. var. *contender*. Etude macroscopique, microscopique et évolution ultrastructurale de la maturation de la graine. Thèse, Fac Sci Genève, 105 pp, 67 pl

Pueppke SG (1979) Distribution of lectins in the Jumbo Virginia and Spanish varieties of the peanut, *Arachis hypogaea* L. Plant Physiol 64:575–580

Pueppke SG, Bauer WD, Keegstra K, Ferguson AL (1978) Role of lectins in plant-microorganism interactions. Plant Physiol 61:779–784

Pusztai A (1967) Trypsin inhibitors of plant origin, their chemistry and potential role in animal nutrition. Nutr Abstr Rev 37:1–9

Pusztai A, Stewart JC (1978) Isolectins of *Phaseolus vulgaris*. Physicochemical studies. Biochim Biophys Acta 536:38–49

Pusztai A, Stewart JC (1980) Molecular size, subunit structure and microheterogeneity of glycoprotein II from the seeds of kidney bean (*Phaseolus vulgaris* L.) Biochim Biophys Acta 623:418–428

Pusztai A, Watt WB (1970) Glycoprotein II. The isolation and characterization of a major antigenic and non-haemagglutinating glycoprotein from *Phaseolus vulgaris*. Biochim Biophys Acta 207:413–431

Pusztai A, Watt WB (1974) Isolectins of *Phaseolus vulgaris*. A comprehensive study of fractionation. Biochim Biophys Acta 365:57–71

Pusztai A, Croy RRD, Grant G, Watt WB (1977) Compartmentalization in the cotyledonary cells of *Phaseolus vulgaris* L. seeds: a differential sedimentation study. New Phytol 79:61–71

Pusztai A, Croy RRD, Grant G (1978a) Mobilization of the nitrogen reserves of *Phaseolus vulgaris* during the early stages of germination. Abh Akad Wiss Abt Math Naturwiss Tech 1978:133–144

Pusztai A, Stewart JC, Watt WB (1978b) A novel method for the preparation of protein bodies by filtration in high (over 70% w/v) sucrose-containing media. Plant Sci Lett 12:9–15

Pusztai A, Croy RRD, Stewart JS, Watt WB (1979) Protein body membranes of *Phaseolus vulgaris* L. cotyledons: isolation and preliminary characterization of constituent proteins. New Phytol 83:371–378

Raab B, Schwenke KD (1975) Isolierung und Charakterisierung der Albuminhauptfraktionen aus den Samen von Sonnenblumen (*Helianthus annuus* L.) und Raps (*Brassica napus* L.). Nahrung 19:829–833

Rao AG, Urs MK, Rao MSN (1978) Studies on the proteins of mustard seed (*B. juncea*). Can Inst Food Sci Technol J 11:155–161

Rao DN, Hariharan K, Rao DR (1976) Purification and properties of a phytohaemagglutinin from *Dolichos lablab* (Field bean). Lebensm Wiss Technol 9:246–250

Refsnes K, Haylett T, Sandvig K, Olsnes S (1977) Modeccin – A plant toxin inhibiting protein synthesis. Biochem Biophys Res Commun 79:1176–1183

Richardson M (1977) The proteinase inhibitors of plants and micro-organisms. Phytochemistry 16:159–169

Rick CM, Fobes JF (1975) Allozymes of Galapagos tomatoes: polymorphism, geographic distribution, and affinities. Evolution 29:443–457

Ridge I, Osborne DJ (1970) Hydroxyproline and peroxydases in cell walls of *Pisum sativum*: regulation by ethylene. J Exp Bot 21:843–856

Robinson DG (1977) Plant cell wall synthesis. Adv Bot Res 5:89–151

Rochat E, Therrien HP (1975) Etude des protéines des blés résistant, Kharkov, et sensible, Selkirk, au cours de l'endurcissement au froid. 1. Protéines solubles. Can J Bot 53:2411–2416

Romero J, Ryan DS (1978) Susceptibility of the major storage protein of the bean, *Phaseolus vulgaris* L., to in vitro enzymatic hydrolysis. J Agric Food Chem 26:784–788

Rougé P (1977) Sur la présence d'hémagglutinines différentes de celles des graines dans les organes végétatifs du pois (*Pisum sativum* L.). Ann Pharm Fr 35:287–294

Rougé P, Père M (1979) Etude des interactions entre les lectines et les constituants glucidiques (glycoprotéines et glucides solubles) des graines de quelques Légumineuses. Bull Soc Bot Fr 126:387–398

Royer A (1975) Inhibiteurs d'enzymes protéolytiques et protéases des graines. In: Miège J (ed) Les protéines des graines. Genèse, nature, fonctions, domaines d'utilisation. Georg, Genève, pp 159–201

Royer A, Miège M-N, Grange A, Miège J, Mascherpa J-M (1974) Inhibiteurs anti-trypsine et activités protéolytiques des albumines de graine de *Vigna unguiculata*. Planta 119:1–16

Rüdiger H (1977) Purification and properties of blood-group-specific lectins from *Vicia cracca*. Eur J Biochem 72:317–322

Rüdiger H (1978) Lectine, pflanzliche zuckerbindende Proteine. Naturwissenschaften 65:239–244

Ryan CA (1973) Proteolytic enzymes and their inhibitors in plants. Annu Rev Plant Physiol 24:173–196

Sabnis DD, Hart JW (1978) The isolation and some properties of a lectin (haemagglutinin) from *Cucurbita* phloem exudate. Planta 142:97–101

Sadava D, Chrispeels MJ (1973) Hydroxyproline-rich cell wall protein (extensin): role in the cessation of elongation in excised pea epicotyls. Dev Biol 30:49–55

Sahai S, Rana RS (1977) Seed protein homology and elucidation of species relationships in *Phaseolus* and *Vigna* species. New Phytol 79:527–534

Scandalios JG (1967) Genetic control of alcohol dehydrogenase isozymes in maize. Biochem Genet 1:1–8

Scandalios JG (1969) Genetic control of multiple molecular forms of enzymes in plants: a review. Biochem Genet 3:37–79

Scandalios JG, Liu EH, Campeau MA (1972) The effects of intragenic and intergenic complementation on catalase structure and function in maize: a molecular approach to heterosis. Arch Biochem Biophys 153:695–705

Schlesier B, Manteuffel R, Scholz G (1978a) Studies on seed globulins from legumes. VI. Association of vicilin from *Vicia faba* L. Biochem Physiol Pflanz 172:285–290

Schlesier B, Manteuffel R, Rudolph A, Behlke J (1978b) Studies on seed globulins from legumes. VII. Narbonin, a 2S globulin from *Vicia narbonensis* L. Biochem Physiol Pflanz 173:420–428

Schwartz D (1971) Subunit interaction of a temperature-sensitive alcohol dehydrogenase mutant in maize. Genetics 67:515–519

Schwartz D (1973) Comparisons of relative activities of maize $Adh_1$ alleles in heterozygotes – Analyses at the protein (CRM) level. Genetics 74:615–617

Schwartz D (1976) Regulation of expression of Adh genes in maize. Proc Natl Acad Sci USA 73:582–584

Schwatz D, Endo T (1966) Alcohol dehydrogenase polymorphism in maize – Simple and compound loci. Genetics 53:709–715

Schwenke KD, Sologub LP, Braudo EE, Tolstoguzov WB (1977) Über Samenproteine.

9. Mitteilung. Einfluß von Neutralsalzen auf die Löslichkeit des 11-S-Globulins aus Sonnenblumensamen. Nahrung 22:491–503
Seckinger HL, Wolf MJ (1973) Sorghum protein ultrastructure as it relates to composition. Cereal Chem 50:455–465
Selvaraj Y, Chander MS, Bammi RK, Randhawa GS (1975) Comparative study of the chemical constituents of edible and saponin bearing *Dioscorea* tubers. Indian J Hortic 31:78–83
Selvendran RR, Davies AMC, Tidder E (1975) Cell wall glycoproteins and polysaccharides of mature runner beans. Phytochemistry 14:2169–2174
Senser F, Belitz H-D, Kaiser K-P, Santarius K (1974) Suggestion of a protective function of proteinase inhibitors in potatoes: Inhibition of proteolytic activity of microorganisms isolated from spoiled potato tubers. Z Lebensm Unters Forsch 155:100–101
Sharon N (1974) Glycoproteins of higher plants. In: Pridham JB (ed) Plant carbohydrate biochemistry. Academic Press, London New York, pp 235–252
Sharon N, Lis H (1972) Lectins: cell-agglutinating and sugar-specific proteins. Science 177:949–959
Sharon N, Lis H (1979) Comparative biochemistry of plant glycoproteins. Biochem Soc Trans 7:783–799
Shechter Y, De Wet JMJ (1975) Comparative electrophoresis and isoenzyme analysis of seed proteins from cultivated races of sorghum. Am J Bot 62:254–261
Shumway LK, Yang VV, Ryan CA (1976) Evidence for the presence of proteinase inhibitor I in vacuolar protein bodies of plant cells. Planta 129:161–165
Singh U, Sastry LVS (1978) Nature of the protein bodies of barley grains. Indian J Exp Biol 16:943–944
Singh VP, Gupta VK (1978) Electrophoretic variations in *Brassica* with respect to esterase isoenzyme patterns. J Indian Bot Soc 57:146–154
Spencer D, Higgins TJV (1979) Molecular aspects of seed protein biosynthesis. Curr Adv Plant Sci 34:1–15
Splittstoesser WE (1977) Protein quality and quantity of tropical roots and tubers. Hort Science 12:294–298
Stanley RG, Linskens HF (1974) Pollen. Biology, biochemistry, management. Springer, Berlin Heidelberg New York
Stegemann H, Francksen H, Macko V (1973) Potato proteins: genetic and physiological changes, evaluated by one- and two-dimensional PAA-gel techniques. Z Naturforsch 28c:722–732
Sun SM, Hall TC (1975) Solubility characteristics of globulins from *Phaseolus* seeds in regard to their isolation and characterization. J Agric Food Chem 23:184–189
Surolia A, Bachhawat BK, Vithyathil PJ, Podder SK (1978) Unique subunit structure for *Ricinus communis* agglutinin. Indian J Biochem Biophys 15:248–250
Susheelamma NS, Rao MVL (1978) Purification and characterization of the surface active proteins of black gram (*Phaseolus mungo*). Int J Pept Protein Res 12:93–102
Talbot CF, Etzler ME (1978) Development and distribution of *Dolichos biflorus* lectin as measured by radio-immunoassay. Plant Physiol 61:847–850
Tanaka K, Sugimoto T, Ogawa M, Kasai Z (1980) Isolation and characterization of two types of protein bodies in the rice endosperm. Agric Biol Chem 44:1633–1639
Tao K-L, Khan AA (1975) Occurrence of some enzymes in starchy endosperm and hormonal regulation of isoperoxidase in aleurone of wheat. Plant Physiol 56:797–800
Thurman DA, Boulter D, Derbyshire E, Turner BL (1967) Electrophoretic mobilities of formic and glutamic dehydrogenase in the Fabaceae: a systematic survey. New Phytol 66:37–45
Ting IP, Führ I, Curry R, Zschoche WC (1975) Malate dehydrogenase isozymes in plants: preparation, properties and biological significance. In: Markert CL (ed) Isozymes, vol II. Physiological function. Academic Press, London New York, pp 369–384
Torres AM, Soost RK, Diedenhofen U (1978) Leaf isozymes as genetic markers in citrus. Am J Bot 65:869–881
Tronier B, Ory RL, Henningsen KW (1971) Characterization of the fine structure and proteins from barley protein bodies. Phytochemistry 10:1207–1211

Tully RE, Beevers H (1976) Protein bodies of castor bean endosperm. Isolation, fractionation and the characterization of protein components. Plant Physiol 58:710–716
Van Driessche E, Foriers A, Strosberg AD, Kanarek L (1976) N-terminal sequences of the $\alpha$ and $\beta$ subunits of the lectin from the garden pea (*Pisum sativum*). FEBS Lett 71:220–222
Vaughan JG (1975) Proteins and taxonomy. In: Harborne JB, Van Sumere CF (eds) The chemistry and biochemistry of plant proteins. Academic Press, London New York, pp 281–298
Venkataraman LV, Jaya TV, Krishnamurthy KS (1976) Effect of germination on the biological value, digestibility coefficient and net protein utilization of some legume proteins. Nutr Rep Int 13:197–205
Ventura MM, Xavier-Filho J (1966) A trypsin and chymotrypsin inhibitor from black-eyed pea (*Vigna sinensis* L.) I. Purification and partial characterization. Ann Acad Bras Cienc 38:553–566
Volger HG, Heber U (1975) Cryoprotective leaf proteins. Biochim Biophys Acta 412:335–349
Wall JR, Wall SW (1975) Isozyme polymorphisms in the study of evolution in the *Phaseolus vulgaris-P. coccineus* complex of Mexico. In: Markert CL (ed) Isozymes, vol IV. Genetics and evolution. Academic Press, London New York, pp 287–305
Weber E, Manteuffel R, Neumann D (1978) Isolation and characterization of protein bodies of *Vicia faba* seeds. Biochem Physiol Pflanz 172:597–614
Wolf MJ, Khoo U, Seckinger HL (1967) Subcellular structure of endosperm protein in high-lysine and normal corn. Science 157:556–557
Wolf MJ, Khoo U, Seckinger HL (1969) Distribution and subcellular structure of endosperm protein in varieties of ordinary and high-lysine maize. Cereal Chem 46:253–263
Wright DJ, Boulter D (1973) A comparison of acid extracted globulin fractions and vicilin and legumin of *Vicia faba*. Phytochemistry 12:79–84
Wright DJ, Boulter D (1974) Purification and subunit structure of legumin of *Vicia faba* L. (Broad bean). Biochem J 141:413–418
Wright DJ, Boulter D (1980) Differential scanning colorimetric study of meals and constituents of some food grain legumes. J Sci Food Agric 31:1231–1241
Xavier-Filho J, De Souza FDN (1979) Isolation and characterization of a trypsin inhibitor from *Vigna sinensis* seeds. Biol Plant 21:119–126
Yon J (1969) Structure et dynamique conformationnelle des protéines. Hermann, Paris
Yoshida K (1978) Novel lectins in the endoplasmic reticulum of wheat germ and their possible role. Plant Cell Physiol 19:1301–1305
Youle RJ, Huang AHC (1976) Protein bodies from the endosperm of castor bean. Subfractionation, protein components, lectins, and changes during germination. Plant Physiol 58:703–709
Youle RJ, Huang AHC (1978) Evidence that the castor bean allergens are the albumin storage proteins in the protein bodies of castor bean. Plant Physiol 61:1040–1042

# 9 Cereal Storage Proteins: Structure and Role in Agriculture and Food Technology

P.I. PAYNE and A.P. RHODES

## 1 Introduction

Cereals are the fruits of cultivated members of the grass family, the Gramineae. They provide much of the carbohydrate and protein for the majority of people on earth. Wheat, a plant of temperate climates, is widely grown and is the staple food of much of Europe, the nations of European descent and temperate regions of India and China. Rice is grown in the tropics where rain and sunshine are abundant and forms the staple diet of one half of the world's human population. In tropical regions that have a limited rainfall, sorghum and various types of millet replace rice as the staple diet. The remaining major cereal, maize, is grown extensively in tropical and subtropical areas and forms the staple food of the populations in parts of South America, Eastern Europe and East and South Africa.

Cereals are also of prime importance in agriculture for feeding domestic animals. In temperate regions, barley, oats and rye are grown for this purpose. They are grown in conditions where wheat is less profitable; for instance, rye is cultivated on light soils where winters are very cold and oats are grown in cool, wet climates. In tropical and sub-tropical areas, millet and particularly maize are prime sources of animal feeds.

Hence, man relies extensively on cereals as his source of protein, either directly, or indirectly through domestic animals. In this review, the biochemistry of the principal protein complex of the cereal grain, the storage protein, is described. Then follows a progress report of the ways in which the content and the quality of the storage protein is being improved by breeding methods to benefit agriculture. Finally, the importance of the different types of storage protein to the food technologist is discussed. The synthesis of cereal storage protein during grain development is discussed in Chapter 14, this volume.

## 2 Anatomical Structure of the Cereal Grain; Protein Content and Protein Distribution

The grains of different cereals have a common anatomical structure. They consist of a peripheral fruit wall (the pericarp) and a seed within. The seed is composed of a seed wall which is fused to the pericarp, an embryo and endosperm tissue. The outer layer of the endosperm forms the aleurone layer, which is between one and four cells thick, depending upon the cereal. This review deals with the proteins of the dry, mature embryo, the aleurone layer and the inner, starchy endosperm.

The protein content of mature grains of cereals grown under normal conditions with adequate, but not excessive, fertilizer levels varies from between 8 and 17% with some wheats at the higher protein levels and maize and rice at the lower levels. The cereals therefore contain much smaller proportions of proteins than do legume seeds. The protein is distributed unevenly throughout the grain. The embryo and the aleurone layer are the richest in terms of protein concentration for they contain in wheat, for instance, 30 and 20% of protein by weight, respectively (HINTON 1953). The starchy endosperm contains, on average, a much smaller percentage of protein but as it is the largest organ of the grain it contains the bulk of the protein, 75% of it in maize (*Zea mays*) (SHOLLENBERGER and JAEGER 1943) and 70% in wheat (*Triticum*) (HINTON 1952). The protein in the endosperm is unevenly distributed. For instance, HINTON (1953) found that the outer endosperm of wheat contained 13.7% protein, the middle endosperm 8.8% and the inner endosperm 6.2%. In maize, there is additional variation in the distribution of protein in the endosperm. Those cells at the opposite end of the grain to the embryo are lightly pigmented and low in protein. Those near the embryo are highly pigmented, vitreous in texture and high in protein.

## 3 Classification of the Cereal Proteins

The cereal proteins were originally defined by OSBORNE (1907) and his contemporaries on the basis of their solubility properties:
(1) Albumin: soluble in water; (2) Globulin: soluble in salt solutions but insoluble in water; (3) Prolamin: soluble in aqueous alcohols; (4) Glutelin: soluble in dilute acid and alkalis.

These terms for the different groups of proteins are still used today although modern workers use chemicals and extraction procedures which are different from those used at the beginning of the century. This has resulted in changed meanings for these protein groups and unfortunately there are no simple, modern definitions that are acceptable to all. The major problems lie with prolamin and glutelin.

The chief difficulty in defining the prolamin and glutelin proteins by their solubility properties is that there is no distinction between extractability and solubility. Thus, a protein may not be extracted in aqueous alcohol, not because it is insoluble in this solvent, but because the protein is inaccessible, being enclosed in a membrane, complexed with a different macromolecule or being at the centre of a larger-than-average flour particle. Some of these alcohol-soluble but alcohol-inextractable proteins would, in all probability, dissolve in the next solvent to be applied – dilute acid – and so be erroneously described as a glutelin. This carry-over can be partially overcome by sequentially extracting prolamin with several changes of aqueous alcohol. Another problem is that

**Table 1.** Approximate proportions of the protein groups in cereals prepared by differential solubility

| Cereal | Albumin | Globulin | Prolamin | Glutelin (Alkali-soluble) |
|---|---|---|---|---|
| Wheat[a] | 9 | 5 | 40 | 46 |
| Maize[b] | 4 | 2 | 55 | 39 |
| Barley[b] | 13 | 12 | 52 | 23 |
| Oats[c] | 11 | 56 | 9 | 23 |
| Rice[b] | 5 | 10 | 5 | 80 |
| Sorghum[b] | 6 | 10 | 46 | 38 |

[a] Data from ORTH and BUSHUK (1972), [b] data from WHITEHOUSE (1973), [c] data from PETERSON and SMITH (1976)

workers differ in their choice of alcohol, its concentration and in the temperature of extraction, so that fractions containing different populations of proteins will all be described as prolamin. For example, gliadin has been extracted from wheat flour with 70% (v/v) ethanol at 4 °C (CHEN and BUSHUK 1970) with 70% (v/v) ethanol at room temperature (EWART 1975) and 55% (v/v) propan-2-ol in the presence of reductant and with sonication (SHEWRY et al. 1978b).

There are also difficulties in adequately defining glutelin. This protein complex was originally defined as being soluble in both dilute alkali and dilute acid, but subsequent work has shown that alkali is much more efficient than acid. The reason for this is that alkali slowly modifies cystine and lysine residues, the former process causing cleavage of disulphide bonds (NIELSEN et al. 1970) like thiol reducing agents. Solubilization is hence caused by a destrucitve procedure and is unsatisfactory. The inefficiency of acids such as 0.1 M acetic acid for dissolving glutelin has led to the subdivision of this protein complex into acid-soluble glutelin and acid-insoluble glutelin (alternatively described as residue protein) (ORTH and BUSHUK 1972).

For general guidance, Table 1 shows the approximate proportions of the different protein groups extracted from the common cereals.

## 4 The Protein Constituents

Only a brief account of the main properties of each protein group is given, with illustrations from selected studies. More detailed reviews are: wheat (*Triticum aestivum*), KASARDA et al. (1976); maize (*Zea mays*) and sorghum (*Sorghum vulgare*), WALL and PAULIS (1978). Available amino acid sequence data are given in Chapter 7, this volume.

### 4.1 Albumin and Globulin

In all the major cereals but oats (*Avena sativa*), the albumin and globulin proteins of the mature grain (sometimes jointly called the "soluble proteins") are concentrated in the embryo and the aleurone layer. They consist of a complex mixture of proteins, for they will contain most of the enzymes of metabolism from the developing grain which survived grain dehydration (see CHING 1972). Also present are hydrolytic enzymes necessary for germination, their corresponding inhibitors and those proteins which cause the clotting of red blood cells, the phytohaemaglutinins. The aleurone layer and the embryo also contain globulin storage protein deposited in protein bodies. They have similar properties to the storage proteins of legume seeds but are unlike the principal storage proteins of most cereal endosperms.

The storage protein of the rice embryo has been well studied. The chief protein, $\gamma$-globulin, was found to be homogeneous by sedimentation analysis but heterogeneous by chromatography with DEAE Sephadex A-50 (SAWAI and MORITA 1970a). One of its major components, $\gamma_1$, had an $S_{20w}$ of 7.26 and a molecular weight of 200,000 (determined by ultracentrifugation) (SAWAI and MORITA 1970b). It was slightly basic and built up of ten different subunits apparently bound by hydrophobic interactions rather than by disulphide bonds.

The other cereals (barley, rye, oats and maize) also contain proteins of similar size and properties to the $\gamma$-globulin of rice (DANIELSSON 1949) but they have been less well characterized. In wheat, $\gamma$-globulin also forms the principal storage protein of the embryo, although as with rice, a similar protein is also a major component of the aleurone layer (DANIELSSON 1949).

The endosperm of the oat grain is unique amongst the cereals in having globulin as its major reserve protein. PETERSON (1978) showed that the major component had a molecular weight of about 322,000, as determined by sedimentation equilibrium and an $S_{20,w}$ of 12.1. Upon reduction and fractionation by sodium dodecylsulphate polyacrylamide-gel electrophoresis (SDS-PAGE), two major components separated with molecular weights of 21,700 (termed $\alpha$) and 31,700 (termed $\beta$). In addition, a number of minor bands of molecular weight about 56,000 were detected. It was concluded that the globulin molecule is built up of $6\alpha$ and $6\beta$ subunits.

The endosperm globulins of other cereals have been poorly studied because they form only a small proportion of the total protein. DANIELSSON (1949) showed that these proteins from several cereals were much smaller than the 12S globulins of the oat endosperm and the 7S globulins of the rice embryo. He calculated they had an $S_{20,w}$ of 2.5 and a molecular weight of 29,000. It is not clear if these endosperm globulins are metabolic proteins, structural proteins or storage proteins deposited in protein bodies. The amino acid composition is remarkably similar to both the storage protein globulins of the rice embryo and the oat endosperm (Table 2), so a storage function seems the more likely.

The globulin proteins of cereals bear a striking similarity to the storage proteins of legumes. In Table 2, the amino acid composition of the three cereal globulins discussed above is compared with a typical legume storage protein, the 11S globulin of *Vicia faba*, the broad bean. All four compositions are remarkably similar: the acid amides glutamine and asparagine predominate; of the basic amino acids, arginine is abundant and lysine is scarce; the cheapest amino acids to be produced in vivo in energy terms – glycine and alanine

**Table 2.** Amino acid composition of storage globulins

| | $\gamma_1$, Rice[a] | Endosperm, sorghum[b] | Endosperm, oat[c] | 11s, Broad bean[d] |
|---|---|---|---|---|
| Lys | 3.3 | 4.5 | 2.6 | 4.5 |
| His | 2.7 | 2.2 | 2.0 | 2.6 |
| Arg | 8.5 | 8.6 | 6.4 | 7.6 |
| Aspx | 7.2 | 8.2 | 9.9 | 11.3 |
| Glux | 15.6 | 12.9 | 21.4 | 14.9 |
| Gly | 9.2 | 11.8 | 8.2 | 7.4 |
| Ala | 8.2 | 9.3 | 6.4 | 5.5 |
| Val | 6.9 | 6.3 | 5.8 | 5.4 |
| Leu | 6.1 | 6.8 | 7.7 | 7.8 |
| Ile | 4.1 | 3.7 | 4.4 | 4.7 |
| Pro | 4.7 | 5.4 | 5.3 | 8.1 |
| Ser | 7.8 | 7.5 | 6.2 | 6.0 |
| Thr | 2.7 | 5.4 | 3.5 | 4.2 |
| H pro | 3.9 | 0.0 | 0.0 | 0.0 |
| ½ Cys | 1.0 | N | N | 1.5 |
| Met | 0.9 | 1.3 | N | 0.4 |
| Tyr | 2.5 | 2.5 | 3.5 | 3.7 |
| Trp | 0.5 | N | N | 0.8 |
| Phe | 4.3 | 3.4 | 5.6 | 3.8 |

Values stated by the authors were recalculated and expressed as mol % where necessary. Amino acids not determined were tabulated as N. In this and further tables, glutamic acid and glutamine are combined as glux and aspartic acid and asparagine as aspx.

[a] From SAWAI and MORITA (1970b), [b] from JONES and BECKWITH (1970), [c] from PETERSON (1978), [d] from BAILEY and BOULTER (1970)

– are well represented; the polar, neutral serine is common and the most frequent non-polar amino acids are alanine, valine, leucine and proline. As will be discussed in the following sections, the prolamin and glutelin storage proteins of cereals have a characteristically different composition.

There are also other similarities between the globulins of cereals and legumes. Not only is the molecular weight of legumin, the 11S globulin of legumes, similar to that of oat globulin, but in many species it has a similar subunit structure, each molecule consisting of 12 polypeptides, six of one type and six of another (DERBYSHIRE et al. 1976). The $\gamma$ globulin of rice (*Oryza sativa*) has a similar molecular weight and $S_{20,w}$ to vicilin (7S globulin), the other major storage protein of legumes. It, like the 7S globulin of soyabean (*Glycine max*), precipitates out of neutral salt solutions at low temperatures.

In spite of these similarities between the globulin storage proteins of cereals and legumes it would be premature to conclude that they are distantly homologous. Much more information is required on comparative protein structure and upon the constraints of storage-protein evolution.

## 4.2 Prolamin

The prolamin storage proteins of cereals are restricted to the starchy endosperm. As shown in Table 1, they are the principal reserve protein in maize, barley, sorghum and wheat.

**Table 3.** Amino acid composition of cereal prolamin

| | Purified gliadin [a] | Zein [b] | Hordein [c] | Kafirin [d] | Avenin [e] |
|---|---|---|---|---|---|
| Lys | 0.6 | 0.3 | 0.5 | 0.2 | 0.6 |
| His | 1.6 | 0.8 | 0.9 | 0.9 | 0.6 |
| Arg | 2.1 | 1.2 | 1.1 | 1.3 | 2.1 |
| Asx | 3.0 | 5.5 | 1.2 | 6.2 | 1.8 |
| Glx | 32.8 | 23.6 | 33.1 | 23.6 | 33.2 |
| Gly | 3.9 | 2.2 | 3.5 | 1.8 | 2.9 |
| Ala | 3.6 | 12.7 | 2.4 | 15.9 | 8.2 |
| Val | 3.3 | 2.7 | 3.2 | 5.0 | 8.5 |
| Leu | 6.7 | 19.8 | 5.5 | 16.3 | 10.7 |
| Ile | 2.7 | 3.8 | 3.5 | 3.6 | 3.4 |
| Pro | 16.2 | 10.6 | 29.9 | 10.0 | 11.1 |
| Ser | 9.2 | 6.4 | 4.4 | 4.2 | 3.4 |
| Thr | 3.1 | 3.0 | 1.8 | 2.7 | 1.9 |
| ½ Cys | 3.1 | Traces | 1.4 | 0.7 | 4.1 |
| Met | 1.4 | 0.1 | 0.6 | 0.8 | 1.8 |
| Tyr | 2.0 | 3.5 | 1.6 | 3.3 | 1.2 |
| Trp | 0.8 | N | N | N | N |
| Phe | 4.0 | 5.8 | 5.3 | 3.2 | 4.5 |

Compositions stated by the authors were recalculated where necessary and expressed as mol%. Amino acids not determined were tabulated as N.

[a] From PATEY and WALDRON (1976), [b] from GIANAZZA et al. (1977), [c] from LAURIÈRE et al. (1976), [d] from JAMBUNATHAN and MERTZ (1972), [e] from KIM et al. (1978)

Wheat prolamin, usually referred to as gliadin, is generally defined as those proteins which are soluble in 70% (v/v) ethanol at room temperature (KASARDA et al. 1976). When these proteins are fractionated by starch gel electrophoresis at low pH using aluminium lactate buffer, numerous components separate and are grouped into α-, β-, γ-, and ω-gliadins, the latter having the slowest mobilities (WOYCHIK et al. 1961). In addition there are a few rapidly migrating albumins and globulins which are also soluble in aqueous ethanol as well as water. Finally, there are some proteins which remain at the point of application of the sample or streak slightly into the gel and are usually regarded as contaminating glutelin.

The molecular weights of the α-, β- and γ-gliadins (as determined by SDS-PAGE) vary from between 30,000 and 40,000 whereas ω-gliadin is larger, about 70,000 molecular weight (BIETZ and WALL 1972). The mobilities of all these components are essentially unaffected by pretreatment with β-mercaptoethanol, a reagent which cleaves disulphide bonds. This suggests that the gliadins have no subunit structure but consist of single, polypeptide chains and strong support for this comes from (1) gel filtration of gliadin preparations (BIETZ and WALL 1972) and (2) amino acid analysis of individually isolated gliadins (EWART 1975).

The amino acid composition of the gliadin proteins, while possessing certain similarities to the glutelin proteins (see Sect. 4.2) is markedly different to that of the globulin storage proteins. The glutamine content is extremely high, consisting of about one third of all residues (Table 3) and one in every six residues in gliadin is proline. The abundance of this amino acid, plus the significant contribution of leucine and phenylalanine, makes the gliadins more hydrophobic than storage globulin. Another difference is the low levels of arginine in gliadin, and as lysine and the acidic amino acids are also scarce, the gliadin proteins carry almost no ionic charges.

Optical studies on isolated gliadins in solution indicate that they consist of compact, globular structures (WU and DIMLER 1964). Presumably, this tight internal structure is maintained by intra-chain disulphide bonds (from cystine, cysteine being absent), hydrogen bonds (mainly from glutamine) and hydrophobic interactions (mainly from proline).

Recently, two research groups (KASARDA et al. 1974, BIETZ et al. 1977) have determined the N-terminal amino-acid sequences of isolated gliadins. The results obtained so far indicate that the $\alpha$-, $\beta$- and $\gamma$-gliadins fall into two groups that differ in sequence in many regions but have several homologous sequences in common. Within each group, the N-terminal sequences are very similar. The authors conclude that all the $\alpha$, $\beta$, and $\gamma$ gliadins originated from gene duplications and the subsequent mutation of two ancestral genes.

The gliadin fraction as indicated by starch-gel electrophoresis is extremely complex, caused by differences in the numbers of basic amino acids. Thus, WRIGLEY and SHEPHERD (1973) fractionated the gliadins of hexaploid wheat varieties into about 45 components using a two-dimensional electrophoretic system. Gliadin heterogeneity is made use of commercially as a major aid in varietal identification (see Sect. 6.8).

The prolamins of the other cereals, zein of maize, hordein of barley, kafirin of sorghum and avenin of oats, are basically similar to gliadin; they are small proteins (LAURIÈRE et al. 1976), they are heterogeneous with respect to charge (RIGHETTI et al. 1977) and they contain large amounts of glutamine and hydrophobic amino acids. As shown in Table 3, however, there are differences between the amino acid composition of the prolamins. Hordein has a glutamine content similar to that of gliadin but it has about twice as much proline, indicating that it is more hydrophobic. On the other hand avenin, while having a comparable glutamine content, has significantly less proline than either hordein or gliadin. To compensate for this, there are higher levels of the non-polar amino acids leucine and valine. Zein and kafirin have the lower levels of proline found in avenin, about one in every ten residues, but they have also smaller amounts of glutamine, one residue in four as opposed to one in three. The outcome of this for both zein and kafirin is that leucine is the second most common residue after glutamine, the small, non-polar alanine is third and proline is only fourth.

The different amino acid compositions of prolamin from the various cereals causes them to have different overall charges at low pH so they fractionate differently in aluminium lactate-starch gels as shown in Fig. 1. Zein, like gliadin, is often extracted with 70% ethanol at room temperature (PAULIS et al. 1975), but hordein (LAURIÈRE et al. 1976) and avenin (KIM et al. 1978) are extracted with 45% ethanol. Zein (RIGHETTI et al. 1977) and hordein (SHEWRY et al. 1978a) are frequently extracted in aqueous isopropanol and kafirin (JONES and BECKWITH 1970) with aqueous t-butanol.

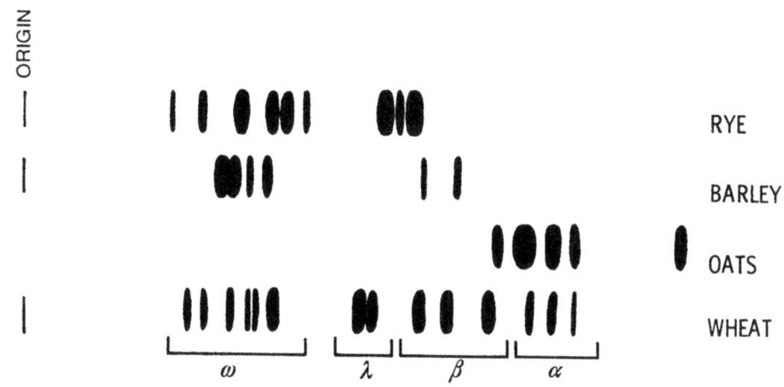


**Fig. 1.** Fractionation of cereal prolamins in aluminium-lactate starch gels; composite diagram drawn from the separations of LAURIÈRE and MOSSÉ (1977), KIM et al. (1978) and AUTRAN et al. (1979). See text for details

Zein migrates essentially as two components by SDS-PAGE, the major of molecular weight 23,000 and the minor, 21,000 (BURR et al. 1978). The zein polypeptides are hence about two-thirds the size of the average gliadin. The mobilities of the kafirin polypeptides are similar to those of zein (PAULIS and WALL 1979), but the major prolamin of rice migrates more rapidly and has an approximate molecular weight of 17,000 (MANDAC and JULIANO 1978). Hordein is more complex, for there are three major components of molecular weight 50,000, 45,000 and 32,000 (SHEWRY et al. 1978a).

BIETZ et al. (1979) determined the N-terminal amino acid sequence of bulk zein. They showed that there was only one, or sometimes two, amino acids present at each of the 33 positions in the polypeptide chain. The various components of zein must therefore have a very similar sequence and the authors concluded that the structural genes had arisen by duplication and mutation of a single ancestral gene, a conclusion reached by others for prolamins of various cereals from less strong evidence. It was further suggested that the major zein of molecular weight 23,000 must have either an internal insertion or a C-terminal extension of about 20 amino acid residues to account for it being larger than the 21,000 molecular weight polypeptide.

In most extraction schemes for maize and barley grain protein, treatment with aqueous alcohol is frequently followed by extraction with the same solvent but with β-mercaptoethanol added. These proteins in maize have been called zein-2 (SODEK and WILSON 1971), glutelin-1 (MOUREAUX and LANDRY 1968), or alcohol-soluble reduced glutelin (PAULIS et al. 1975). We prefer either of the last two terminologies, especially the latter, and accordingly these proteins will be described in the next section (4.3).

## 4.3 Glutelin

The glutelin proteins are the most difficult of all cereal proteins to characterize and define. There are two reasons for this. The first is that glutelin is so difficult to solubilize: the second is that any residual albumin, globulin and prolamin remaining after previous, sequential extraction with water, salt solutions, and aqueous alcohol will almost certainly dissolve in acid and contaminate the glutelin fraction, thus complicating any attempts at its characterization. In spite of this, the distinguishing feature of glutelin in acid solution has been shown to be its large size. It cannot penetrate the pores of aluminium lactate-starch gels (maize: BOUNDY et al. 1967, wheat: ELTON and EWART 1966, barley: LAURIÈRE et al. 1976) and it is excluded from all types of Sephadex column (wheat: MEREDITH and WREN 1966, oats: KIM et al. 1978, rice: SAWAI and MORITA 1968). When, however, a reducing agent is added to the medium such as sodium sulphite (NIELSEN et al. 1962) or the now more widely used $\beta$-mercaptoethanol, there is a marked reduction in the molecular weight of glutelin. Therefore, glutelin must consist of subunits that are principally held together by the disulphide bonds of cystine. EWART (1972) calculated that for the glutelin of wheat (glutenin) there are on average two disulphide bonds between each subunit, suggesting that subunits must link up to form long, linear structures with little or no branching. However, WALL (1979) has suggested that the highly insoluble proportion of glutenin (the residue protein) is extensively cross-linked with disulphide bonds. The way in which the subunits are connected together may differ in various cereals, so accounting, in part, for the differences in their biophysical properties.

Glutelin only occurs as the major storage protein in rice where it constitutes 80% of the total protein (Table 1). Unfortunately it has not been studied extensively. While conforming to the general properties of cereal glutelin, it is particularly insoluble and hence difficult to characterize. JULIANO and BOULTER (1976) reduced and alkylated glutelin and showed by SDS-PAGE that it consisted of three major subunits of molecular weight 38,000, 25,000 and 16,000 in the ratio of 16:3:1. The major, 38,000 component was unique to the glutelin fraction. Rice glutelin has a different amino acid composition from cereal prolamin (Table 3) and wheat glutenin (Table 4). It has much lower levels of glutamine (Table 4), it is somewhat lower in proline, but is higher in the basic amino acids arginine and asparagine and the hydrophobic amino acids phenylalanine, histidine, leucine and valine.

By far the most-studied glutelin is that from wheat, known as glutenin. With the possible exception of rye glutelin, it possesses the unique property of elasticity so important for bread making (see Sect. 6.2). The subunit composition of glutenin varies according to the way in which it is prepared. For instance, glutenin extracted by differential solubility fractionates into over 15 bands by SDS-PAGE after reduction corresponding to subunits which range in molecular weight from 11,600 to 133,000 (BIETZ et al. 1975, ORTH and BUSHUK 1973). In contrast, glutenin prepared by column chromatography in a highly dissociating medium contained many fewer subunits (PAYNE and CORFIELD 1979), the two major polypeptides of molecular weight 64,000 and 71,000 and all small polypeptides of molecular weight less than 30,000 being absent (Fig. 2). Many of the "missing" polypeptides have identical mobilities in gels to some of the major albumin, globulin and prolamin proteins, so it is likely that they are contaminants to the glutelin fraction. The remaining subunits can be divided into three groups on the basis of molecular weight (Fig. 2b); group A subunits

**Table 4.** Amino acid compositions of glutelin and glutelin fractions

| | Wheat | | | Rice glutelin[c] |
|---|---|---|---|---|
| | Glutenin[a] | EtoH-soluble glutenin[b] | EtoH-insoluble glutenin[b] | |
| Lys | 1.5 | 2.7* | 2.8* | 4.8 |
| His | 1.8 | 1.3 | 1.3 | 3.2 |
| Arg | 2.4 | 2.2 | 2.7 | 6.2 |
| Asx | 2.7 | 1.4 | 2.0 | 9.6 |
| Glx | 35.7 | 39.4 | 32.7 | 15.3 |
| Gly | 7.3 | 3.3 | 12.9 | 7.6 |
| Ala | 3.4 | 2.4 | 4.1 | 6.7 |
| Val | 4.2 | 4.3 | 3.5 | 6.3 |
| Leu | 7.0 | 7.5 | 6.1 | 9.5 |
| Ile | 3.3 | 3.7 | 2.3 | 4.1 |
| Pro | 12.9 | 15.0 | 11.0 | 8.9 |
| Ser | 6.7 | 7.2 | 7.0 | 7.2 |
| Thr | 2.9 | 2.4 | 2.6 | 3.4 |
| $\frac{1}{2}$ Cys | N | ** | ** | ** |
| Met | 1.4 | 1.3 | 0.8 | 0.9 |
| Tyr | 2.4 | 1.5 | 4.2 | 2.0 |
| Trp | N | N | N | trace |
| Phe | 4.4 | 4.5 | 2.5 | 4.3 |

Values stated by the authors have been recalculated where necessary as mol%.

* Values include aminoethyl-cysteine
** Values included with lysine
Amino acids not determined are tabulated as N

[a] From ORTH and BUSHUK (1973), [b] from BIETZ and WALL (1973), [c] from TAKEDA et al. (1970)

have molecular weights between 97,000 and 136,000, group B subunits between 41,000 and 48,000 and group C subunits between 31,000 and 35,000. BIETZ and WALL (1973) alkylated the subunits of glutenin after reduction and showed that while the B and C subunits dissolved in 70% ethanol, the A subunits did not. The A subunits appear to be restricted to the glutenin and residue fractions of the wheat grain but the B and C subunits are also constituents of the high molecular weight protein found in the gliadin fraction (see Fig. 2b, d). The C subunits have molecular weights similar to classical gliadin except they are much less heterogeneous (Fig. 2d, e). The amino acid composition of the A subunits is rather distinctive (Table 4), whereas the compositions of the B+C subunits and classical gliadin are broadly similar. The A subunits are much richer in glycine and tyrosine and poorer in glutamine, proline and phenylalanine.

The other cereal glutelins also consist of subunits which can be divided after reduction and alkylation into alcohol-soluble and alcohol-insoluble fractions. In maize, the alcohol-soluble subunits separated by SDS-PAGE into four major bands of molecular weights 23,000, 21,000, 13,500 and 9,600 (GIANAZ-

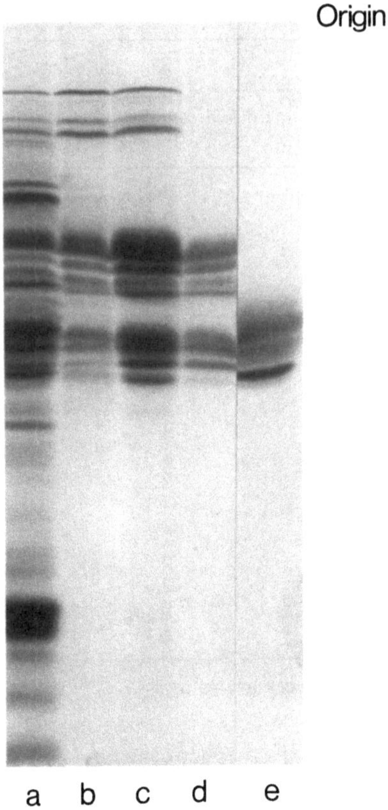


**Fig. 2.** Fractionation of gliadin and reduced glutenin by SDS-PAGE (CORFIELD and PAYNE unpublished results): *a* total protein of flour of the wheat variety Mardler; *b, c* glutenin subunits; *d* high molecular weight gliadin (alternatively called low molecular weight glutenin): *e* classical gliadin

ZA et al. 1976). The two major subunits had identical mobilities to the zein polypeptides but they were present in different proportions. In contrast, the alcohol-insoluble subunits are larger and more heterogeneous in size than the zein polypeptides. The molecular weights of the major subunits are 97,000, 60,000, 55,000 and 34,000 (PAULIS et al. 1975).

LAURIÈRE et al. (1976) fractionated the reduced and alkylated subunits of acid-soluble glutelin from barley by aluminium-lactate starch gels. Five major bands separated and they had identical mobilities to those alcohol-soluble proteins which elute at the void volume of Sephadex columns. This indicates a close similarity between glutelin and high molecular weight prolamin, as has been shown for wheat. When reduced and alkylated hordein was fractionated in this system, only minor components had the same mobilities as the acid-soluble glutelin subunits. The major hordein proteins separated into four bands of slower electrophoretic mobility. When, however, the subunits of acid-soluble glutelin and the polypeptides of hordein were compared by SDS-PAGE (SHEWRY et al. 1978d) the differences between them were much less obvious.

From the preceding descriptions, it is clear that classical prolamin and acid-soluble glutelin, although different from each other, have similar biochemical properties. It is thus possible that they are synthesized from genes which have diverged from a common ancestor. During evolution the three-dimensional structure of prolamin has favoured internal disulphide bonds whereas subunits of glutenin have become linked by inter-molecular disulphide bonds. The similar biochemical properties of prolamin and glutelin after reduction and alkylation have led MIFLIN and SHEWRY (1979) to recommend extracting these proteins together by a mixture of aqueous alcohol and thiol reducing agent at high temperature and describing them collectively as prolamin. However, the original terms are likely to remain, especially glutenin and gliadin of wheat, for the two have different biophysical properties – elasticity and extensibility respectively – and they have contrasting roles in the bread-making and biscuit-making industries (see Sects. 6.2 and 6.3).

# 5 Agricultural Aspects

As described in the introduction, the protein component of cereal grains is important in both human and animal nutrition although it only forms a small proportion of grain dry weight. The percentage of protein in a cereal is arguably sufficient to satisfy the protein requirements of man when adequate calories are obtained (ALTSCHUL 1965, SUKHATME 1975). It is, however, inadequate for supporting the growth of intensively reared farm animals. Furthermore the amino acid composition of cereal protein is not ideally suited for the diet of humans or animals. The essential amino acid most deficient in all cereals is lysine, though in maize tryptophan is co-limiting. The second is usually threonine.

Attempts at improving both the content and the quality of cereal proteins have been beset with problems. Protein concentrations in cereal grains can be raised by agronomic practices such as the late application of nitrogenous fertilizers but, in nearly all cases, this causes a lowering in the lysine content of the protein, thus reducing its quality (e.g., ZOSCHKE 1970). Increases in protein content achieved by breeding often produce corresponding decreases in the yield of the crop (RHODES and JENKINS 1978) and this is unacceptable. Examples of the ways in which plant breeders, biochemists and physiologists are currently collaborating to overcome these difficulties to improve protein content and quality are described below.

## 5.1 Breeding High-Protein Wheats

VOGEL et al. (1973) examined over 20,000 entries of The World Wheat Collection maintained by the United States Department of Agriculture (USDA), for protein content and lysine percentage. The protein content varied from 7% to 25% in the 12,613 common wheats, and of this 17% variation, 5% was due to variation in genetic components. One of the high-protein varieties was Atlas 66. Its genes for high protein have been transferred by breeding programmes into agronomically acceptable, high-quality North American Wheats. One such variety is

Lancota, released in 1975. In field experiments it had a similar yield to the commonly grown variety Centurk, but contained 2% more protein. The extra protein was in the starchy endosperm and so was present in white fluor milled from the grains.

Another high-protein variety Nap Hal has also been crossed with Atlas 66. Some of the progeny were significantly higher in protein content than either parent, suggesting that different genes were involved and their effects were additive (JOHNSON et al. 1973). As yet, however, the progeny tested have lower yields than the control variety Centurk. The high-protein trait was claimed to be caused by elevated nitrate reductase activity, increased nitrogen absorption by the roots and a more efficient translocation of nitrogen from the maternal plant to the grain (JOHNSON et al. 1973).

The lysine content of wheat and other cereals is strongly affected by protein content. Wheats of between 7% and 15% protein show a negative relationship with lysine content which drops from 4% to 3% (lysine as a percentage of protein). Above the 16% level of protein, the lysine percentage remains constant. JOHNSON et al. (1975) calculated that the genetic component of total lysine varied in the World Wheat Collection by no more than 0.5%. This is only one third of the amount required to bring lysine into reasonable balance with the other essential amino acids in wheat proteins (JOHNSON and MATTERN 1978). In assessing lysine content, it is essential to take into account the much wider variation in protein content. JOHNSON and MATTERN (1978) used several of the high-lysine varieties in breeding programmes. The two most promising lines, Nap Hal (also high in protein) and CI 13,449, had elevated lysine levels which were principally located in the aleurone layer and the starchy endosperm respectively. Crosses between the two have produced progeny which are higher in lysine than either parent. In some lines, this is combined with the high protein content of Nap Hal and high grain yield.

### 5.2 High-Lysine Barley

Considerable effort over the past decade has been directed towards the production of a high-lysine, high-yielding barley (*Hordeum vulgare*) suitable for feeding to monogastric animals. MUNCK et al. (1969) screened the entire USDA world collection of barleys and found Hiproly, a variety high in both lysine and protein. The increased lysine in Hiproly is due mainly to increased levels of a number of water-soluble polypeptides (TALLBERG 1973, RHODES and GILL 1980). The hordein fraction in this variety is only slightly reduced in amount. Unfortunately, the yield of Hiproly is only about one third that of commercial varieties. Extensive breeding programmes have been undertaken using Hiproly and high-lysine lines are now yielding 85%–95% of the standard. The remaining shortfall in yields was due to smaller seed size, problems with sterility and poor performance in adverse weather conditions (PERSSON 1975).

Mutants of barley varieties, induced by chemical methods or atomic irradiation, have been obtained which are high in lysine; the Risø mutants from Denmark (KØIE and DOLL 1979), Notch-1 and Notch-2 from India (BALARAVI et al. 1976) and from Italy, Lys 95 and Lys 449 (Di Fonzo and Stanca 1977). In these mutants, in contrast to Hiproly, high lysine content is caused by a reduced accumulation of hordein. Unfortunately, starch accumulation is also reduced, producing shrivelled grains and low crop yields. Several of these mutants have been used in breeding programmes in attempts to produce high-yielding, high-lysine barleys. Results with the mutant highest in lysine, Risø 1508, have been disappointing so far because it has not been possible to separate the high-lysine character from shrunken seed and thus low grain yield (Rhodes 1978).

Promising results have been obtained from a cross between Risø mutant 7 and Hiproly × Mona 5 (Hagberg et al. 1979). Double recessives for high lysine content were selected from this cross which have 35% more lysine than normal varieties and well-filled grains. There is some optimism that this material will be of practical importance in breeding an adapted high-lysine barley.

### 5.3 High-Lysine Maize

A number of mutants have been produced in maize with improved lysine content in the grain. The most widely studied of these has been Opaque-2 (Mertz et al. 1964). This mutant contains twice as much lysine as normal maize and more tryptophan. The improved amino acid composition of this mutant is due to a 50% reduction in the lysine-deficient zein and an increase in the albumin, globulin and glutelin fractions. The Opaque-2 allele reduces the rate of accumulation of zein but its effect is unequally distributed among the various zein components. Isoelectric focussing has shown that the zeins which have alkaline isoelectric points are preferentially repressed (Soave et al. 1976). As well as being lower in yield, the endosperm texture is floury rather than vitreous, giving it poor milling quality.

Recently, Opaque-2 has been crossed with high-yielding maize genotypes possessing a hard-milling endosperm and considerable progress has been made in combining high lysine with good milling quality. Several hard endosperm Opaque-2 progeny developed at the International Maize and Wheat Improvement Centre, Mexico (CIMMYT) (Vasal et al. 1979) have good yield potential, satisfactory milling characteristics, but lysine levels slightly lower than in Opaque-2. Thus it is possible that high-quality protein in maize may be in significant commercial production in the near future.

### 5.4 Protein Improvement in Rice

Rice grain has high levels of lysine (3.5% to 4% of protein) compared with the other cereals. This is due to the low prolamin content in normal rice so there is little prospect of improving the lysine content further by the methods used for the other cereals. However, it may be possible to improve protein content which is lower than in most cereals.

An extensive breeding programme has been established at the International Rice Research Institute in The Philippines and part of it is designed for the improvement of protein content. The most advanced high protein line (IR 2153-338-3) had a mean protein content of 8.9% compared with the control at 7.8% with only slightly reduced yield (COFFMAN and JULIANO 1979). Currently, the search for a high-protein rice is less important than breeding for disease and pest resistance and tolerance to environmental stresses.

### 5.5 High-Lysine Sorghum

In 1973, SINGH and AXTELL (1973) reported the discovery of two Ethiopian lines of sorghum (*Sorghum vulgare*), IS 11167 and IS 11758, that were high in both lysine and protein content. Line IS 11167 had 15.7% protein of which 3.33% was lysine, and IS 11758 had 17.2% protein and 3.13% lysine in protein. This compares to 12.7% protein for normal sorghums with 2.05% as lysine. Another high-lysine sorghum, P721, has been obtained by chemical mutation (MOHAN and AXTELL 1975).

Both of the Ethiopian high-lysine genotypes have higher levels of salt-soluble proteins and reduced levels of prolamin and acid-soluble glutelin. Unfortunately, these sorghums have shrunken, floury kernels analogous to Opaque-2 maize and Risø 1508 barley, but recently SINGH (1976) has obtained high-lysine kernels which are of a more vitreous nature. Although the protein content of these modified kernels was, on average, lower than the normal and high-lysine parents, the lysine values were comparable to the high-lysine parents. If high-lysine sorghums with acceptable yield and grain quality can be produced it could have a significant impact on the world hunger problem, as they are grown in the marginal grain-producing areas of the world where the population is most in need of extra protein.

### 5.6 Triticale

Triticale is an intergeneric hybrid between wheat and rye; hybridization between hexaploid wheat and diploid rye gave octaploid triticales and hybridization between tetraploid wheat and diploid rye gave hexaploid triticales. The latter hybrids are higher-yielding and thus more important. Triticales inherit some of the bread-making properties of wheat and some of the disease resistance of rye. However, there are problems with the crop, such as poor fertility, grain shrivelling and susceptibility to yellow rust and ergot. The protein content of triticale is higher than that in wheat and of superior nutritional quality, for it has a higher content of lysine. The solubility distribution and amino acid composition of the endosperm protein of triticale is generally intermediate between those of its wheat and rye parents and there is as yet no biochemical evidence that rye and wheat genomes interact to produce new proteins.

Recent work with three French triticales showed that bread-making quality was generally intermediate between rye (*Secale cereale*) (Petkus) and wheat (Zenith), but up to 20% triticale flour could be added to wheat flour without deleterious effect, giving bread with improved flavour and better keeping quality

(SAURER and DORMAN 1979). Triticales with yields equal to or better than those of high-yielding bread and durum wheats have been produced (CIMMYT Report 1977), but the yields tend to be less stable from year to year than those of the parents.

# 6 Technological Aspects

All cereal grains contain a large proportion of starch. To make them digestible and acceptable for human consumption they must be cooked. A common form of cooking is by boiling; for instance, rice after shelling and polishing, and oats, maize and millet after milling to form porridge. For wheat, however, a series of sophisticated manufactured products have been developed, such as bread, macaroni and chapatis, all requiring different amounts and types of endosperm protein. These products will be discussed in turn. A more detailed account is given by KENT (1975).

## 6.1 Milling Quality of Wheats

Wheats are described as either "hard" or "soft" according to the way in which the endosperm breaks down during milling; some proteins, as yet unidentified, play an important part in this process. In hard wheats there is a strong adhesion between starch and protein in the endosperm. Fragmentation of the endosperm consequently occurs along the lines of the cell boundaries and a coarse, gritty, free-flowing flour is produced. In soft wheats, where adhesion between starch and protein is low, the endosperm fractures in a random way and a very fine, slightly sticky flour is produced. Mechanical damage to starch granules is much greater in hard wheats and this increases the flour's capacity to absorb and retain water.

## 6.2 Bread

Bread is made from flour (mainly wheat but sometimes rye), yeast, salt and water. There are three stages in its manufacture: (1) mixing and development of the dough; (2) aeration of the dough and (3) oven-baking of the dough.

During dough mixing, the glutenin proteins are rendered more soluble (TSEN 1969) by a breakdown of disulphide bonds. The mechanism of breakdown is not established but it could occur by SH/SS interchange reactions (BLOKSMA 1975) or by the involvement of free radicals generated during mixing (GRAVELAND et al. 1980). During the subsequent resting period, disulphide bonds reform and, with the addition of hydrogen bonds and hydrophobic interactions, a continuous gluten matrix forms throughout the dough. The molecular architecture of dough is not understood, even in general terms. One possibility, based on scanning electron microscopy of isolated components, is that glutenin and starch bind tightly together to form a rigid, latticed structure. Gliadin, which forms thin sheets like zein (see Sect. 6.7) blocks the holes in this structure and enables the dough to retain the gas released by yeast during the fermentation process and so to rise. A dough which is suitable for conversion into bread must be resilient and elastic, properties conferred by glutenin, and extensible, a property conferred by gliadin. During the baking of the dough, the gas must be retained until starch and protein has coagulated.

Varieties of wheat differ in their bread-making qualities. With respect to protein, this is caused by differences in protein content and differences in the types of glutenins and gliadins present. Many good-quality wheats are currently grown in North America, whereas the bulk of those grown in Western Europe are too low in protein content and have an unsuitable type of glutenin which is not elastic enough. This results in poor gas retention either during fermentation, when the dough rises poorly, or during baking, when the loaf collapses. Currently, a big effort is being made to determine what makes a glutenin either suitable or unsuitable for bread making (HUEBNER and WALL 1976, PAYNE et al. 1979) with the object of helping the breeder develop new varieties of improved bread-making quality.

Both hard-milling and soft-milling wheats can be used for conversion into bread. However, hard-milling varieties are generally preferred because the flour produced absorbs and retains more water. Thus, bread baked from hard wheats contains less flour per unit volume (an important commercial consideration) and stales more slowly. The proteins which confer hardness of milling are different from those which confer dough strength and loaf volume and the genes determining these characters are located on different chromosomes (KONZAK 1977, LAW et al. 1978).

### 6.3 Biscuits (Cookies)

Biscuits are best made from a low-protein, soft-milling wheat whose glutenin has low elasticity. The most important criterion of a biscuit dough is extensibility. Thus, a bread-making wheat is quite unsuitable for biscuit making and vice versa. Soft-milling wheats are preferred for biscuit making because their lower water content means (1) less energy is needed to dry the biscuits during cooking and (2) shallower water gradients during cooking, reducing the chances of cracking. Cracking is also more likely to occur with hard wheats because of the granular texture of its flour.

### 6.4 Pasta Products (Macaroni, Spaghetti, Vermicelli and Noodles)

Pasta products can only be made from wheats which are hard-milling. The most suitable are durum wheats, very high in protein but usually grown in low-yielding environments. However, hard hexaploid wheats from North America, most suitable for bread making, can also be used.

Extreme hardness is required because the grain is milled to coarse particles rather than fine flour. The product, semolina, can absorb large quantities of water and is the starting material for all pasta products.

### 6.5 Breakfast Cereals from Wheat

The wheat grain may be processed in different ways to produce different breakfast products (KENT 1975). At least two of them are dependent on the protein moiety of the grain. Shredded wheat is, like biscuits, made from soft wheat. The grain is boiled to gelatinize the starch and then cooled. It is passed through shredding rollers; one roller is smooth and the other has circular grooves. Long parallel threads are formed.

Puffed wheat can, in contrast, only be produced from hard wheat. The grain is fed into a pressure chamber and then heated by injection of steam until the pressure increases to 14.0 kg $cm^{-2}$ (1.38 $MN/m^2$). The pressure is suddenly released and expansion of water vapour causes the grains to increase in volume to several times their original size. For satisfactory puffing, the material at the moment before expansion requires cohesion to prevent shattering and elasticity from glutenin to permit controlled expansion. If soft wheats were used for this process they would shatter.

### 6.6 Chapatis

Chapatis, a kind of unleavened bread, are the major form in which wheat is consumed in India, Pakistan, and parts of Tibet and China. They are made from a dough which is worked, rested and baked. Water absorption of the flour should be high, so a hard wheat is required. A strong gluten is not required, unlike leavened bread, but it should be stronger than gluten which is optimal for biscuit making.

### 6.7 Used of Zein Films

When ethanol solutions of zein are evaporated, transparent films of protein are formed. This property of zein makes it of use, after chemical modification, in the pharmaceutical and food industries (WALL and PAULIS 1978). Medical pills coated with zein cause the contents to be released slowly into the stomach after swallowing. Confectionery coated with zein possesses attractive glaze and is impervious to moisture and so does not become sticky. Zein has additionally been used as a bonding agent for cork bottle tops and on photographic film. It may also have some potential in the textile fibre industry.

### 6.8 Identification of Cereal Varieties

The gliadin proteins of the wheat grain are, as previously discussed, very heterogeneous. Furthermore, different varieties of wheat have different combinations

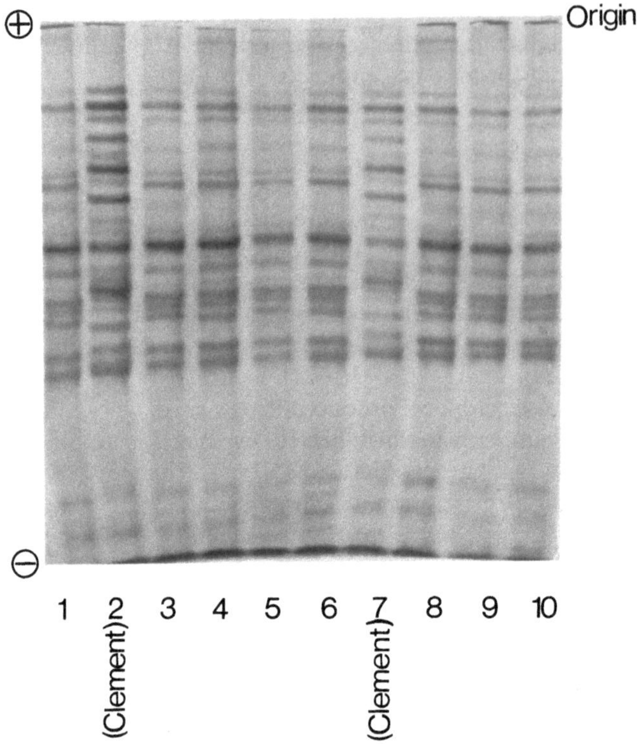


**Fig. 3.** Fractionation on an aluminium-lactate starch gel of the gliadin proteins from 10 individual grains selected at random from a stock variety of Copain. From the fractionation pattern, 8 grains were identified as Copain (slots *1, 3–6, 8–10*) but two were actually Clement (slots *2* and *7*). (Unpublished results of Dr. J.R.S. ELLIS)

of gliadins. This finding has recently been exploited and methods have been developed to identify varieties by the gliadin proteins they contain. Protein extracts from wheat flours are fractionated by gel electrophoresis at low pH to form a distinct one-dimensional fingerprint pattern. This fingerprinting procedure is becoming a valuable asset to the milling and bread-making concerns of Europe and it is by far the most reliable procedure for identifying the varieties which comprise a delivery of grain. The need for such identification has arisen in recent years because of the increasing proportion of European-grown wheat included in flour that is used for making bread. Unfortunately, only a few of the varieties grown in Europe are suitable for bread making and some widely grown varieties, such as "Clement" and "Maris Huntsman", are positively undesirable.

In the most acceptable form of the gliadin fingerprinting test in Western Europe (AUTRAN and BOURDET 1975, ELLIS and BEMINSTER 1977) about 50 grains are taken at random from a load of wheat. The proteins are extracted from each grain and fractionated. The results of such an experiment are obtained in less than 48 h and they will give (1) the principal varieties present and (2) the level of contamination of harmful varieties. For instance, if 10 grains of the 50 are shown to be of the variety "Clement", the true proportion of this variety in the original sample would be (at the 95% confidence level) between 11% and 34%. If no grains of "Clement" are detected, then the maximum contribution of this variety would be 7%. Figure 3 gives an example of this type of analysis. In a bulk sample of grain of "Copain", 2 of the 10 grains analyzed were actually shown to be "Clement".

A big advantage of the method is that the fingerprinting pattern is independent of the location where the variety was grown, the year of harvest or the level of protein in the grain (ELLIS and BEMINSTER 1977). There are a few varieties which cannot be distinguished by gliadin fingerprinting. Several of these have been shown to be distinguishable by fractionation of the proteins soluble in 55% isopropanol-β-mercaptoethanol by SDS-polyacrylamide-gel electrophoresis (SHEWRY et al. 1978b).

Although varietal identification for purposes of bread making is virtually restricted to wheat, the identification of all cereals is of agricultural importance. The seedsman is required by law in Europe to supply seed of the correct variety and needs to be able to check the purity of his basic stocks. A limited degree of identification can be obtained from the morphology of the grain, but biochemical characters are needed for complete identification. There is likely to be a development of protein identification techniques in the coming years for cereals other than wheat. Indeed, SHEWRY et al. (1978c) have already developed methods for the identification of single or half grains of barley by SDS-PAGE of the hordeins.

*Acknowledgement.* We are most grateful to Dr. J.R.S. Ellis (Rank Hovis MacDougall Research, High Wycombe, U.K.) for supplying Fig. 3.

## References

Altschul AM (1965) Proteins: their chemistry and politics. Chapman and Hall, London

Autran JC, Bourdet A (1975) L'identification des variétés du blé; établissement d'un tableau général de determination fondé sur le diagramme électrophorétique des gliadins du grain. Ann Amelior Plant 25:277–301

Autran JC, Lew E J-L, Nimmo CC, Kasarda DD (1979) N-terminal amino acid sequencing of prolamins from wheat and related species. Nature (London) 282:527–529

Bailey CJ, Boulter D (1979) The structure of legumin, a storage protein of broad bean (*Vicia faba*) seed. Eur J Biochem 17:460–466

Balaravi SP, Bansal HC, Eggum BO, Bhaskaran S (1976) Characterisation of induced high protein and high lysine mutants in barley. J Sci Food Agric 27:545–552

Bietz JA, Wall JS (1972) Wheat gluten subunits: molecular weights determined by sodium dodecyl sulphate-polyacrylamide gel electrophoresis. Cereal Chem 49:416–430

Bietz JA, Wall JS (1973) Isolation and characterisation of gliadin-like subunits from glutenin. Cereal Chem 50:537–547

Bietz JA, Shepherd KW, Wall JS (1975) Single-kernel analysis of glutenin: use in wheat genetics and breeding. Cereal Chem 52:513–532

Bietz JA, Huebner FR, Sanderson JE, Wall JS (1977) Wheat gliadin homology revealed through N-terminal amino acid sequence analysis. Cereal Chem 54:1070–1083

Bietz JA, Paulis JW, Wall JS (1979) Zein subunit homology revealed through amino-terminal sequence analysis. Cereal Chem 56:327–332

Bloksma AH (1975) Thiol and disulphide groups in dough rheology. Cereal Chem 52:170r–183r
Boundy JA, Woychik JH, Dimler RJ, Wall JS (1967) Protein composition of dent, waxy and high-amylase corns. Cereal Chem 44:160–169
Burr B, Burr FA, Rubenstein I, Simon MN (1978) Purification and translation of zein messenger RNA from maize endosperm protein bodies. Proc Natl Acad Sci USA 75:696–700
Chen CH, Bushuk W (1970) The nature of proteins in triticale and its parental species 1. Solubility characteristics and amino acid composition of endosperm proteins. Can J Plant Sci 50:9–14
Ching TM (1972) Metabolism of germinating seeds. In: Kozlowski TT (ed) Seed biology, Vol 2, Academic Press, New York, pp 103–218
CIMMYT report on wheat improvement 1977. CIMMYT, Mexico
Coffman WR, Juliano BO (1979) Seed protein improvement in rice. In: Seed improvement in cereals and grain legumes Vol 2. Int AEA, Vienna, pp 261–276
Danielsson CE (1949) Seed globulins of the Gramineae and Leguminosae. Biochem J 44:387–400
Derbyshire E, Wright DJ, Boulter D (1976) Legumin and vicilin, storage proteins of legume seeds. Phytochemistry 15:3–24
Di Fonzo N, Stanca AM (1977) EMS derived barley mutants with increased lysine content. Genet Agrar 31:401–409
Ellis JRS, Beminster CH (1977) Identification of UK wheat varieties by starch gel electrophoresis of gliadin proteins. J Natl Inst Agric Bot 14:221–231
Elton GAH, Ewart JAD (1966) Glutenins and gliadins: electrophoretic studies. J Sci Food Agric 17:34–38
Ewart JAD (1972) A modified hypothesis for the structure and rheology of glutelins. J Sci Food Agric 23:687–699
Ewart JAD (1975) Isolation of a Cappelle-Desprez gliadin. J Sci Food Agric 26:1021–1025
Gianazza E, Righetti PG, Pioli F, Galante E, Soave C (1976) Size and charge heterogeneity of zein in normal and opaque-2 maize endosperms. Maydica 21:1–17
Gianazza E, Viglienghi V, Righetti PG, Salamini F, Soave C (1977) Amino acid composition of zein molecular components. Phytochemistry 16:315–317
Graveland A, Bosveld P, Lichtendonk WJ, Moonen JHE (1980) Superoxide involvement in the reduction of disulphide bonds of wheat gel proteins. Biochem Biophys Res Commun 93:1189–1195
Hagberg H, Persson G, Ekman R, Karlsson K-E, Tallberg AM, Stoy V, Bertholdsson N-O, Mounla H, Johansson H (1979) The Svalöv protein quality breeding programme. In: Seed protein improvement in cereals and grain legumes Vol 11, Int AEA, Vienna, pp 303–313
Hinton JJC (1952) The structure of cereal grains. In: Bate-Smith EC, Morris TN (eds) Food science: A symposium on quality and preservation of foods. Cambridge Univ Press
Hinton JJC (1953) The distribution of protein in the maize kernel in comparison with that in wheat. Cereal Chem 30:441–445
Huebner FR, Wall JS (1976) Fractionation and quantitative differences of glutenin from wheat varieties varying in baking quality. Cereal Chem 53:258–269
Jambunathan R, Mertz ET (1972) Amino acid compositions of whole kernel and endosperm fractions in sorghum. In: Axtell JD, Oswalt DL (eds) Research progress report on inheritance and improvement of protein quality in Sorghum bicolor. Purdue Univ Indiana, pp 43–56
Johnson VA, Mattern PJ (1978) Improvement of wheat protein quality and quantity by breeding. In: Friedman M (ed) Nutritional improvement of food and feed proteins. Plenum Press, New York, pp 301–316
Johnson VA, Mattern PJ, Schmidt JW, Stroike JE (1973) Genetic advances in wheat quality and composition. In: Sears ER, Sears LMS (eds) Proceedings of the 4th international wheat genetics symp, Columbia, Missouri, pp 547–556
Johnson VA, Mattern PJ, Vogel KP (1975) Cultural, genetic and other factors affecting

quality of wheat In: Spicer A (ed) Bread. Applied Science Publishers, Barking, England, pp 127–140
Jones RW, Beckwith AC (1970) Proximate composition and proteins of three grain sorghum hybrids and their dry-mill fractions. J Agr Food Chem 18:33–36
Juliano BO, Boulter D (1976) Extraction and composition of rice endosperm glutelin. Phytochemistry 15:1601–1606
Kasarda DD, Da Roza DA, Ohms JI (1974) N-terminal sequence of $\alpha_2$-gliadin. Biochim Biophys Acta 351:290–294
Kasarda DD, Bernardin JE, Nimmo CC (1976) Wheat Proteins. In: Pomeranz Y (ed) Advances in cereal science and technology, Vol 1. Am Assoc Cereal Chem, St Paul, Minnesota, pp 158–236
Kent NL (1975) Technology of cereals, 2nd edn. Pergamon Press, Oxford New York Toronto
Kim SI, Charbonnier L, Mossé J (1978) Heterogeneity of avenin, the oat prolamin. Fractionation, molecular weight and amino acid composition. Biochim Biophys Acta 537:22–30
Køie B, Doll H (1979) Protein and carbohydrate components in the Risø high lysine barley mutants. In: Seed Protein improvement in cereals and grain legumes Vol. 1. Int AEA, Vienna, pp 205–214
Konzak CF (1977) Genetic control of the content, amino acid composition, and processing properties of proteins in wheat. Adv Genet 19:407–581
Laurière M, Mossé J (1977) Selective extraction and purification of barley alcohol soluble proteins and their analysis on starch gel electrophoresis. In: Miflin BJ, Shewry PR (eds) Techniques for the separation of barley and maize proteins. Commission of the European Communities, Luxembourg, pp 49–60
Laurière M, Charbonnier L, Mossé J (1976) Nature et fractionnement des protéines de l'orge extraites par l'ethanol, l'isopropanol et le n-propanol a des titres différents. Biochimie 58:1235–1245
Law CN, Young CF, Brown JWS, Snape JW, Worland AJ (1978) The study of grain protein control in wheat using whole chromosome substitution lines. In: Seed protein improvement by nuclear techniques. Int AEA, Vienna, pp 483–502
Mandac BE, Juliano BO (1978) Properties of prolamin in mature and developing rice grain. Phytochemistry 17:611–614
Meredith OB, Wren JJ (1966) Determination of molecular-weight distribution in wheat-flour proteins by extraction and gel filtration in a dissociating medium. Cereal Chem 43:169–186
Mertz ET, Bates LS, Nelson OE (1964) Mutant gene that changes protein composition and increases lysine content of maize endosperm. Science 145:279–280
Miflin BJ, Shewry PR (1979) The biology and biochemistry of cereal seed prolamins. In: Seed protein improvement in cereals and grain legumes, Vol 1. Int AEA, Vienna, pp 137–158
Moureaux T, Landry J (1968) Extraction sélective des proteins du grain de mais en particulier de la fraction "glutelines". CR Acad Sci 266:2302–2305
Mohan DD, Axtell JD (1975) Dimethyl sulfate-induced high-lysine mutants in sorghum. Proc 9th Bienn Grain Sorghum Res Util Conf, Lubbock, Tex
Munck L, Karlsson K-E, Hagberg A (1969) Bättre näringsvärde hos spannmalsprotein. J Swed Seed Ass 79:196–205
Nielsen HC, Babcock GE, Senti FR (1962) Molecular weight studies on glutenin before and after disulphide-bond splitting. Arch Biochem Biophys 96:252–258
Nielsen HC, Paulis JW, James C, Wall JS (1970) Extraction and structure studies on corn glutelin proteins. Cereal Chem 47:501–512
Orth RA, Bushuk W (1972) A comparative study of the proteins of wheats of diverse baking qualities. Cereal Chem 49:268–275
Orth RA, Bushuk W (1973) Studies of glutenin. III Identification of subunits coded by the D-genome and their relation to breadmaking quality. Cereal Chem 50:680–687
Osborne TB (1907) The proteins of the wheat kernel. Carnegie Inst Washington Publ 84, Judd and Detweiler, Washington, USA

Patey AL, Waldron NM (1976) Gliadin proteins from Maris Widgeon wheat. J Sci Food Agric 27:838–842
Payne PI, Corfield KG (1979) Subunit composition of wheat glutenin proteins, isolated by gel filtration in a dissociating medium. Planta 145:83–88
Payne PI, Corfield KG, Blackman JA (1979) Identification of a high-molecular-weight subunit of glutenin whose presence correlates with breadmaking quality in wheats of related pedigree. Theor Appl Genet 55:153–159
Paulis JW, Wall JS (1979) Distribution and electrophoretic properties of alcohol-soluble proteins in normal and high-lysine sorghums. Cereal Chem 56:20–23
Paulis JW, Bietz JA, Wall JS (1975) Corn protein subunits: molecular weights determined by sodium dodecyl sulphate-polyacrylamide gel electrophoresis. Agric Food Chem 23:197–201
Peterson DM (1978) Subunit structure and composition of oat seed globulin. Plant Physiol 62:506–509
Peterson DM, Smith D (1976) Changes in nitrogen and carbohydrate fractions in developing oat groats. Crop Sci 16:67–71
Persson G (1975) The barley protein project at Svälof. In: Breeding for seed protein improvement using nuclear techniques. Int AEA, Vienna, pp 91–97
Rhodes AP (1978) Amino acid and carbohydrate accumulation in the grains of the high lysine barley mutant Risø 1508 and its parent Bomi. In: Miflin BJ, Zoschke M (eds) Carbohydrate and protein synthesis. Commission of the European Communities, Luxembourg, pp 123–126
Rhodes AP, Gill AA (1980) Fractionation and amino acid analysis of the salt soluble protein fractions of normal and high lysine barleys. J Sci Food Agric 31:467–473
Rhodes AP, Jenkins G (1978) Improving the protein quality of cereals, grain legumes and oilseeds by breeding. In: Norton G (ed) Plant Proteins. Butterworths, London, pp 207–226
Righetti PG, Gianazza E, Viotti A, Soave C (1977) Heterogeneity of storage proteins in maize. Planta 136:115–123
Saurer W, Dorman A (1979) Einige Untersuchungen über die Backeigenschaften von Triticale. Mitt Schweiz Landwirtsch 27:69–75
Sawai H, Morita Y (1968) Studies on rice glutelin I Isolation and purification of glutelin from rice endosperm. Agric Biol Chem 32:76–80
Sawai H, Morita Y (1970a) Studies on $\gamma$-globulin of rice embryo II. Separation of three components of $\gamma$-globulin by ion exchange chromatography. Agric Biol Chem 34:53–60
Sawai H, Morita Y (1970b) Studies on $\gamma$-globulin of rice embryo III Molecular dimension and chemical composition of $\gamma_1$-globulin. Agric Biol Chem 34:61–67
Shewry PR, Ellis JRS, Pratt HM, Miflin BJ (1978a) A comparison of methods for the extraction and separation of hordein fractions from 29 barley varieties. J Sci Food Agric 29:433–441
Shewry PR, Faulks AJ, Pratt HM, Miflin BJ (1978b) The varietal identification of single seeds of wheat by sodium dodecyl-sulphate polyacrylamide gel electrophoresis of gliadin. J Sci Food Agric 29:847–849
Shewry PR, Pratt HM, Miflin BJ (1978c) Varietal identification of single seeds of barley by analysis of hordein polypeptides. J Sci Food Agric 29:587–596
Shewry PR, Hill JM, Pratt HM, Leggatt MM, Miflin BJ (1978d) An evaluation of techniques for the extraction of hordein and glutelin from barley seed and a comparison of the protein composition of Bomi and Risø 1508. J Exp Bot 29:677–691
Shollenberger JH, Jaeger CM (1943) Corn – its products and uses. Northern Regional Research Laboratory, Peoria, Ill, USA, Bulletin
Singh R, Axtell JD (1973) High lysine mutant gene (hl) that improves protein quality and biological value of grain sorghum. Crop Sci 13:535–539
Singh SP (1976) Modified vitreous endosperm recombinants from crosses of normal and high-lysine sorghum. Crop Sci 16:296–297
Soave C, Righetti PG, Lorenzoni C, Gentinetta E, Salamini F (1976) Expressivity of the *opaque*-2 gene at the level of zein molecular components. Maydica 21:61–75
Sodek L, Wilson CM (1971) Amino acid composition of proteins isolated from normal,

*opaque*-2 and *floury*-2 corn endosperms by a modified Osborne procedure. J Agric Food Chem 19:1144–1150

Sukhatme PV (1975) Human protein needs and the relative role of energy and protein in meeting them. In: Steele F, Bourne A (eds) The man/food equation. Academic Press, London New York, pp 53–75

Takeda M, Namba Y, Nunokawa Y (1970) Heterogeneity of rice glutelin. Agr Biol Chem 34:473–475

Tallberg AM (1973) Ultrastructure and protein composition in high lysine barley mutants. Hereditas 75:195–200

Tsen CC (1969)Effects of oxidising and reducing agents on changes of flour proteins during dough mixing. Cereal Chem 46:435–442

Vasal SK, Villegas E, Bauer R (1979) Present status of breeding quality protein maize. In: Seed protein improvement in cereals and grain legumes. Vol 11. Int AEA, Vienna, pp 127–150

Vogel KP, Johnson VA, Mattern PJ (1973) Results of systematic analyses for protein and lysine composition of common wheats (*Triticum aestivum* L.) in the USDA world collection. Nebr Res Bull 258:27 pp

Wall JS (1979) The role of wheat proteins in determining baking quality. In: Laidman DL, Wyn Jones RG (eds) Recent advances in the biochemistry of cereals. Academic Press, London New York, pp 275–311

Wall JS, Paulis JR (1978) Corn and sorghum grain proteins. In: Pomeranz Y (ed) Advances in cereal science technology, Vol II. Am Assoc Cereal Chem, St Paul, Minnesota, pp 135–219

Whitehouse RNH (1973) The potential of cereal grain crops for protein production. In: Jones JGW (ed) The biological efficiency of protein production. Cambridge Univ Press, UK, pp 83–99

Woychik JH, Boundy JA, Dimler RJ (1961) Starch gel electrophoresis of wheat gluten proteins using concentrated urea. Arch Biochem Biophys 94:477–482

Wrigley CW, Shepherd KW (1973) Electrofocusing of grain proteins from wheat genotypes. Ann NY Acad Sci 209:154–162

Wu YV, Dimler RJ (1964) Conformational studies of wheat gluten, glutenin, and gliadin in urea solutions at various pH's. Arch Biochem Biophys 107:435–440

Zoschke M (1970) Effect of additional nitrogen at later growth stages on protein content and quality in barley. In: Improving plant protein by nuclear techniques. Int AEA, Vienna, pp 345–356

# 10 Biochemistry and Physiology of Leaf Proteins

R.C. HUFFAKER

## 1 Introduction

Leaf proteins function not only as catalysts but also as major storage sinks of nitrogen. Some proteins such as ribulose bisphosphate carboxylase (RuBP-Case) act as both a catalyst and a storage protein. The turnover characteristic of individual proteins depends a great deal on their intracellular location and their accessibility to leaf proteases. Proteins may be sequestered within an organelle such as the chloroplast and thereby be protected from cytoplasmic proteases. As a result, the synthesis phase and degradative phase of some proteins may also be sequentially separated.

Cytoplasmic proteins may be under constant turnover, depending on their susceptibility to proteolysis. In vivo degradation rate in mammalian cells seems to be related to molecular weight and charge: proteins with a higher molecular weight had shorter half-lives and those with low isoelectric points degraded faster than those with neutral or basic points (DICE and GOLDBERG 1975).

Information is just beginning to appear on susceptibility of plant proteins to proteolytic degradation as a function of the above parameters. ACTON and GUPTA (1979) showed that the half-life and isoelectric points for five different plant proteins were weakly correlated. A highly significant correlation was shown between the logarithm of the subunit size and the half-time ($t_{1/2}$) of degradation.

Workers are beginning to study the in vivo enzymatic activity of several leaf proteins. The relationship between in vivo activity and enzyme concentration suggests that in vivo regulation may often be more important than the concentration of the protein catalyst. In addition to allosteric effectors, the sources of energy that drive certain enzymatic reactions are important regulators. The activity of some cytoplasmic enzymes may be driven by glycolysis and by systems that shuttle ATP and reducing power from the chloroplast or mitochondria to the cytoplasm. For example, several biochemical pathways are integrated in the regulation of nitrate reductase (NR) activity. Since photosynthetic $CO_2$ fixation supports both the synthesis and activity of nitrate reductase, the turnover of RuBPCase can strongly affect nitrate reductase (NR).

The leaf storage proteins are extremely important in the maturation, reproduction, and final seed yields of plants. Nitrogen is a main factor that limits photosynthetic capacity and seed yield as the plant matures. Final seed yields often depend on the proteolysis of stored leaf N and its translocation to the seed (DALLING et al. 1975, HAGEMAN and LAMBERT 1981). Proteolysis of storage protein can be induced by environmental factors such as limitations in N supply, water, light, and temperature.

This presentation deals with only those proteins for which information is available about their turnover, turnover in relation to in vivo enzymatic activity, and where activity and turnover of several proteins are interconnected. Attempts are made to relate those characteristics to genetic control, and, where possible, to final crop yields.

## 2 Ribulose Bisphosphate Carboxylase (RuBPCase)

RuBPCase occupies a key role in the metabolism of plant leaves: it functions as both a catalyst and a storage protein. It catalyzes photosynthetic $CO_2$ fixation (carboxylase activity) and furnishes the first product for photorespiration (oxygenase activity). Information is rapidly accumulating regarding the relationship between RuBPCase and other biochemical processes, such as nitrate assimilation and its effect on biomass production. Because of its high concentration in the leaf, it serves as a major storage protein. It constitutes 40% to 80% of the total soluble leaf protein of soyabean (*Glycine max*), cereals, or alfalfa (*Medicago sativa*), or 10% to 20% in corn (*Zea mays*), depending on the species and environmental conditions such as N supply and light. It shows many of the characteristics expected of a storage protein and it is assembled and stored in the chloroplast. RuBPCase can be degraded whenever the plant needs the stored nitrogen, sulfur, or carbon because of changing environmental conditions, aging, or newly developing sink demands (e.g., seed formation) (HUFFAKER and MILLER 1978, WITTENBACH 1979, FRIEDRICH and HUFFAKER 1980).

The ready response of RuBPCase to environmental conditions and leaf aging allows for many different approaches to studying its catalytic, storage, and mobilization characteristics. Large-scale RuBPCase synthesis can be initiated by transferring dark-grown plants into light (KLEINKOPF et al. 1970); conversely, RuBPCase can be lost either by placing plants in darkness (PETERSON et al. 1973) or during leaf aging (WITTENBACH 1979, FRIEDRICH and HUFFAKER 1980).

*Biosynthesis of RuBPCase.* The biosynthesis of RuBPCase is complicated, because RuBPCase is made up of two different subunits – a large (molecular weight ca. 48,000 to 55,000) and a small (molecular weight ca. 14,000) and each is synthesized in a different cellular location. The large subunit is synthesized on 70S polyribosomes in the chloroplast (HARTLEY et al. 1975, ALSCHER et al. 1976) and the small subunit precursor on 80S polyribosomes in the cytoplasm (CHUA and SCHMIDT 1978, GOODING et al. 1973, HIGHFIELD and ELLIS 1978). The small subunit precursor is then transported into the chloroplast, where it and the large subunit are combined into the native enzyme (HIGHFIELD and ELLIS 1978, CHUA and SCHMIDT 1978). A detailed account of the synthesis of RuBPCase is contained in Chapter 16, Volume 14B this series.

*Factors Influencing Synthesis.* RuBPCase is synthesized in plants in darkness; the amount synthesized depends on stored energy reserves (HUFFAKER et al. 1966). When barley plants grown in the dark are put into light, the concentration

**Table 1.** Effect of N nutrition on RuBPCase, total soluble protein, and the proportion of total soluble protein as RuBPCase in barley leaves. Plants received the prescribed amount of $NO_3^-$-N in a complete nutrient solution. Total soluble protein and RuBPCase protein (by reaction with specific antibody [FRIEDRICH and HUFFAKER 1980]) were determined from primary leaves three days after collar formation (12 days after planting). (FRIEDRICH and HUFFAKER unpublished information)

| N Level (μmol $NO_3^-$-N/ plant/day) | RuBPCase (mg $dm^{-2}$) | Soluble protein minus RuBPCase (mg $dm^{-2}$) | Total soluble protein (mg $dm^{-2}$) | Proportion RuBPCase (%) |
|---|---|---|---|---|
| 2.5 | 13.4 (±0.8) | 12.8 | 26.2 (±0.7) | 51.5 (±2.2) |
| 25.0 | 26.2 (±0.7) | 12.6 | 38.8 (±1.3) | 67.5 (±2.3) |

Numbers in parentheses are standard errors of the mean.

of RuBPCase increases (KLEINKOPF et al. 1970). The increase is probably related to the onset of photosynthetic capacity. The final concentration of RuBPCase remains remarkably steady until the leaf goes through senescence, during which RuBPCase disappears quite rapidly (PETERSON et al. 1973, FRIEDRICH and HUFFAKER 1980). RuBPCase concentration depends on environmental conditions imposed on the plant, such as light, N supply, and leaf age (Table 1, BJÖRKMAN 1968, FRIEDRICH and HUFFAKER 1980). BJÖRKMAN showed that plants grown in shade had much less in vitro RuBPCase activity than did those grown under higher light. Similar results were obtained using sugar beet leaves (N. TERRY personal communication). FRIEDRICH and HUFFAKER (1980) showed that under a photon flux of 400 μE $m^{-2}s^{-1}$, RuBPCase made up 50% of the total soluble protein; at 550 μE $m^{-2}s^{-1}$, RuBPCase made up 81% of the total. Under low light conditions, neither low nor high N concentration in the nutrient solution affected the lower RuBPCase activity present. Under high light, high N supply increased in vitro RuBPCase activity almost threefold (MEDINA 1971). These results again confirm that RuBPCase, as expected of a storage protein, reaches a steady concentration that depends on the nutrients and energy supplies available.

*RuBPCase Turnover.* The characteristics of RuBPCase turnover in barley leaves were determined during three different phases: during its synthesis to a steady concentration, after it achieved a steady concentration, and during induced and natural leaf senescence.

*Turnover During Synthesis.* Synthesis of RuBPCase was initiated by placing leaves detached from dark-grown plants into light with their bases placed in [$^{14}$C]-amino acids. After several hours the labeled amino acids were replaced with cold amino acids. Leaves were harvested during greening, homogenized, and the RuBPCase concentration then determined by reaction with a specific antibody. The [$^{14}$C]-content of the RuBPCase remained quite constant during the time that its synthesis continued in the presence of the cold amino acids, showing the absence of degradation during this period of synthesis (Table 2).

Nothing is yet known concerning turnover of RuBPCase subunits before assembly into native RuBPCase. HIGHFIELD and ELLIS (1978) and CHUA and

**Table 2.** Lack of turnover of RuBPCase during greening (SMITH et al. 1974). Barley leaves were labeled with [$^{14}$C]-amino acids. The leaves were then placed in unlabeled amino acid solution in light. After homogenization of leaves [$^{14}$C]-RuBPCase was assayed by reaction with a specific antibody from rabbits

| Hours in light | cpm in RuBPCase/5 leaves |
|---|---|
| 0 | 675 |
| 2 | 627 |
| 4 | 715 |
| 6 | 707 |
| 8 | 685 |
| 10 | 768 |
| 12 | 654 |

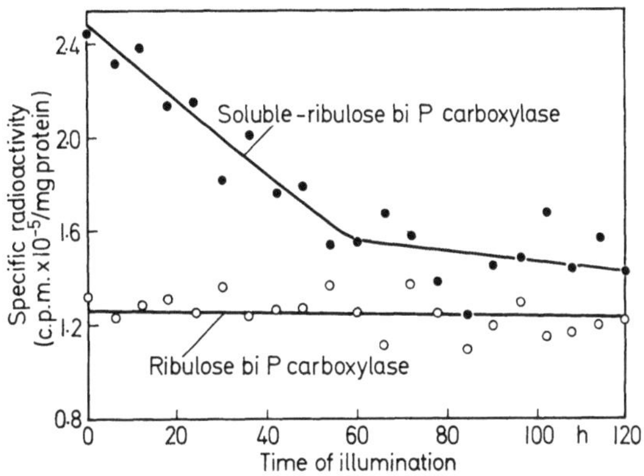


**Fig. 1.** Specific radioactivity of RuBPCase from intact barley leaves (PETERSON et al. 1973). Barley plants were radiolabeled with [$^{14}$C]$O_2$ for 6 h during rapid synthesis of RuBPCase. The plants were then placed in light in unlabeled $CO_2$ and RuBPCase was followed by reaction with specific antibody. Non-RuBPCase protein was determined by subtracting the concentrations of RuBPCase protein from that of total protein

SCHMIDT (1978) showed that the small subunit was synthesized first as a precursor – molecular weight about 3,500 larger than the authentic small subunit. The precursor is apparently processed by a specific proteinase, either as it crosses the chloroplast membrane or after it is inside. Since the small subunit is apparently synthesized in the cytoplasm, is it turned over? Does the precursor fragment affect the degradation of the small subunit? Since the large subunit is synthesized in the chloroplast, its turnover may be slow as long as the chloroplast remains intact.

*Turnover During Steady Concentration.* In an experiment reported by PETERSON et al. (1973), leaves from green [$^{14}$C]-labeled barley plants were assayed for RuBPCase over several days (Fig. 1). RuBPCase concentration and specific radioactivity remained quite constant during the time course, indicating that little turnover occurred. In contrast, cytoplasmic protein had high turnover rates, although its concentration also remained quite constant. Evidence for the lack of turnover depended on detecting changes in specific radioactivity of a highly labeled RuBPCase. This method may not detect a low rate of turnover.

Other studies reveal little if any turnover of RuBPCase in barley leaves after it reached a steady concentration. ALSCHER et al. (1976), showed that messenger activity for the large subunit of RuBPCase was not associated with barley polyribosomes in an in vitro heterologous protein-synthesizing system after RuBPCase attained a steady concentration. The in vitro heterologous system synthesized the large subunit when polyribosomes were isolated from barley leaves during a period of rapid synthesis of RuBPCase (ALSCHER et al. 1976). RNA synthesis decreased and incorporation of labeled amino acids into RuBPCase greatly decreased after RuBPCase reached a steady concentration (PATTERSON and SMILLIE 1971). BRADY and SCOTT (1977) reported that little incorporation of labeled amino acids into RuBPCase occurred after the steady level of RuBPCase was achieved in wheat leaves. ZUCKER (1971) observed that little incorporation of amino acids into Fraction I protein (see below) occurred in green *Xanthium* leaf discs, but large amounts were incorporated into phenylalanine ammonia lyase. Fully expanded *Perilla* leaves incorporated very little radioactivity into Fraction I protein when treated with $[^{14}C]O_2$ (KANNANGARA and WOOLHOUSE 1968). Fraction I protein may be considered to be crude RuBPCase (KAWASHIMA and WILDMAN 1970).

RuBPCase in corn, however, did not exhibit such stability. SIMPSON (1978) reported that corn leaf RuBPCase had a 3-day half-life. RuBPCase constituted only about 10% of the total soluble protein in the corn leaves and may not be as important a leaf storage protein as it is in the species described above.

*Turnover During Senescence.* Although little turnover of RuBPCase is detected during its initial synthesis or well after it achieves a steady concentration, the opposite occurs in many plants during leaf senescence. Early investigators showed that Fraction I protein was lost during leaf senescence of tobacco (*Nicotiana tabacum*) (DORNER et al. 1957, KAWASHIMA et al. 1967, KAWASHIMA and MITAKE 1969, KAWASHIMA and WILDMAN 1970) and *Perilla* (KANNANGARA and WOOLHOUSE 1968). Later the loss of both RuBPCase protein and RuBPCase in vitro activity in leaves that were induced to senescence by changes in the environment was followed (Fig. 2). Many studies show that leaf senescence may be initiated, increased or decreased in rate by changing light, temperature, water, N level, or other factors (THOMAS and STODDART 1980). Senescence has been induced by placing either detached leaves or whole barley plants in darkness. RuBPCase protein was lost similarly from detached barley (*Hordeum vulgare*) leaves (PETERSON et al. 1973) and whole wheat (*Triticum esculentum*) plants (WITTENBACH 1978) when these were placed in darkness. During an ensuing 3-day dark period, barley leaves lost 60% of their total soluble protein (PETERSON et al. 1973). In a similar study, wheat leaves lost 60% of their total soluble protein over a 6-day dark period (WITTENBACH 1978). RuBPCase accounted for 90% of the soluble protein lost in barley leaves and 80% of that lost in wheat leaves.

*Reversibility of RuBPCase Loss.* When barley plants that had been in darkness for 3 days and had lost 90% of their RuBPCase protein were placed back into light, they totally recovered their soluble protein content (Fig. 2). RuBPCase accounted for 90% of the increased soluble protein in light. WITTENBACH (1977) showed that the protein loss during a dark treatment of wheat seedlings was reversible if the plants were placed back into light in 2 days. After 2 days protein loss was irreversible.

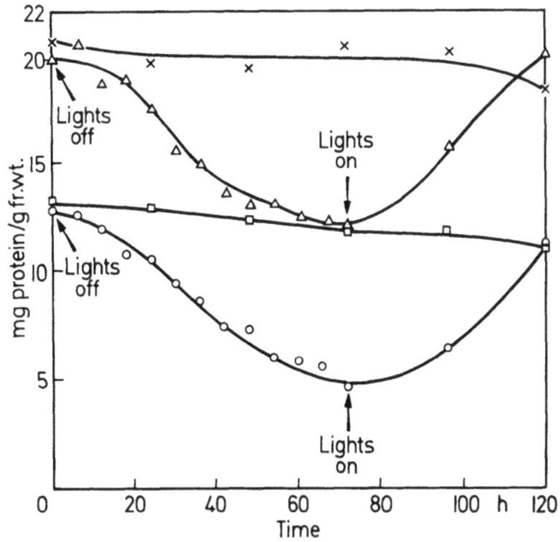


**Fig. 2.** The effect of light and dark on barley leaf RuBPCase and total soluble protein. Barley seedlings grown in light for 6 days were placed under light and dark for various periods (PETERSON et al. 1973). △, × total soluble protein; □, ○ RuBPCase protein; ×, □ continuous light; △, ○ 72 h dark then 48 h light. RuBPCase was determined by reaction with specific antibody

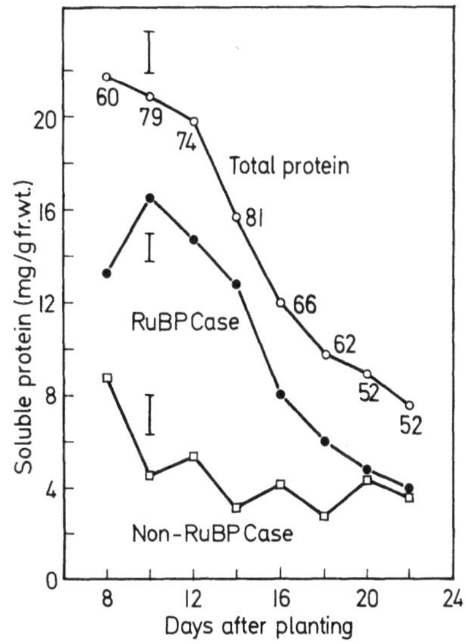


**Fig. 3.** Total protein, RuBPCase protein, and non-RuBPCase protein in primary barley leaves (FRIEDRICH and HUFFAKER 1980). *Numbers* represent the proportion made up by RuBPCase. Barley plants were grown in a growth chamber using a 15-h day at 24 °C, a 9-h night at 12.5 °C, and a constant relative humidity of 85%. RuBPCase was determined by reaction with rabbit anti-RuBPCase. Standard deviation indicated by bar line

The reversibility of protein loss poses interesting questions. The plants were placed in darkness when RuBPCase was at a steady concentration and synthesis was either stopped or very slow. That indicates that the messages for synthesis of the large and, possibly, the small subunits may either not be present or be present in low concentrations. To recover, the plant must apparently be induced to resynthesize the necessary mRNA. Because chlorophyll also is lost when the plants are put into darkness, must new chloroplasts or chlorophyll be synthesized as well?

*RuBPCase Loss During Leaf Aging.* Primary barley leaves grown under optimum conditions in a growth chamber showed their greatest concentration of RuBPCase at about 10 days of age (Fig. 3). Thereafter, RuBPCase steadily decreased until, at 22 days of age, the leaves had lost 76%. WITTENBACH (1979) reported changes in RuBPCase in the flag and the second leaf of field-grown winter wheat beginning one week before anthesis and lasting until harvest. RuBPCase constituted about 45% and 58% of the total soluble protein in the flag and the second leaf, respectively, until senescence, when RuBPCase protein was again more rapidly lost.

*Soluble Protein Other Than RuBPCase.* Non-RuBPCase protein remained at a steady concentration while detached primary barley leaves senesced in darkness (PETERSON and HUFFAKER 1975). These leaves incorporated large amounts of stable amino acids into the non-RuBPCase protein, which indicates an appreciable turnover (Fig. 1). Likewise, the concentration of non-RuBPCase protein remained quite constant in intact primary leaves of barley as they aged (Fig. 3). In both cases this happened at a time when RuBPCase was rapidly lost.

*Synthesis of RuBPCase During Degradation.* Synthesis during degradation was detected by placing detached barley leaves base down into tritiated amino acids in darkness (PETERSEN and HUFFAKER 1975). Leaves were periodically homogenized and the RuBPCase protein isolated by reaction with specific antibody. A linear but low incorporation of amino acids into RuBPCase occurred while it was being rapidly degraded. In contrast, soluble protein other than RuBPCase incorporated much higher quantities of amino acids while the concentration remained constant.

*RuBPCase Loss and Other Senescence Symptoms.* Loss of RuBPCase from senescing leaves often occurs simultaneously with a loss of chlorophyll, a decrease in rate of $CO_2$ exchange, and increased stomatal closure (PETERSON and HUFFAKER 1975, WITTENBACH 1979, FRIEDRICH and HUFFAKER 1980). THIMANN and SATLER (1979) proposed recently that stomatal closure is responsible for leaf senescence. THOMAS and STODDART (1975) and HALL et al. (1978) recently separated the loss of RuBPCase from chlorophyll during leaf senescence. THOMAS showed that a mutant of *Festuca pratensis* Huds., that loses very little of its chlorophyll during senescence, still loses RuBPCase. The wild type lost RuBPCase and chlorophyll simultaneously. Mutant chloroplasts had lost the stroma matrix, but the thylakoid membranes remained loose and unstacked. HALL et al. (1978) reported that chlorophyll disappeared more rapidly from the flag leaf of wheat than did RuBPCase. On the other hand, WITTENBACH (1979) showed nearly equal rates of disappearance of each in the flag leaf of field-grown wheat as it senesced.

Physiological parameters measured during the development and senescence of primary leaves on normally growing barley plants fell into three classes, based on rate of change (FRIEDRICH and HUFFAKER 1980). Transpiration and stomatal resistance (Class I) changed faster than any other parameter. Included in Class II were decreases in RuBPCase protein, true photosynthesis, and mesophyll resistance to $CO_2$ uptake. Class III comprised chlorophyll loss and increases in proteolytic activity.

*RuBPCase Loss and Proteolytic Activity.* In vitro endopeptidase activity greatly increases in detached barley leaves placed in darkness (PETERSON and HUFFAKER 1975, HUFFAKER and MILLER 1978), whereas it remains quite constant in leaves on intact plants senescing in darkness (Fig. 4). Protein loss was almost as great in the intact as in the detached leaves. Endopeptidase activity is already sufficiently high in non-senescing leaves to account for RuBPCase degradation during senescence (PETERSON and HUFFAKER 1975, HUFFAKER and MILLER 1978, FRIEDRICH and HUFFAKER 1980). BEEVERS (1968) showed that caseolytic activity was not correlated with loss of protein in *Nasturtium* leaf discs. Development of high levels of in vitro endopeptidase activity was not required in detached corn leaves (SOONG et al. 1977) or in detached oat leaves (VAN LOON and HAVERKORT 1977) for in vivo protein to degrade. It appears that neither new proteinases nor increased amounts of proteinases are required to account for either the loss of RuBPCase or turnover of cytoplasmic protein during leaf senescence. FRIEDRICH and HUFFAKER (1980) showed that primary leaves from barley seedlings growing under normal conditions developed increased proteolytic activity before other senescence symptoms were detected and before RuBPCase began to disappear. In fact, RuBPCase was at its greatest concentration in the leaves at a time when proteolytic activity was nearly maximum.

*Artificial Proteolytic Substrates.* Much work has been done using artificial substrates to identify different proteases, aminopeptidases, carboxypeptidases, and endopeptidases, and their changes during germination and leaf development (FELLER et al. 1977, MIKOLA and KOLEHMAINEN 1972, PETERSON and HUFFAKER 1975). Recent success in purifying leaf proteases indicates that the use of artificial substrates may not indicate proteolytic activity. L-leucyl-β-naphthylamide (LNA) has been used as a substrate for leucine aminopeptidase (LAP), and α-N-benzoyl-dl-arginine-β-naphthylamide (BANA) has been used as a substrate indicative of endopeptidase activity. Partial purification of barley leaf proteases using electrophoresis separated the enzymes that showed activity against the artificial substrates from the proteinases that showed activity against true protein substrate (Miller and Huffaker unpublished data). The protein bands that exhibited activity against protein substrates had no activity against LNA or BANA. Furthermore, the band that showed activity against BANA had no activity against true protein substrates. The bands showing activity against LNA showed no activity against gelatin, but perhaps the methods prevented detection of low activities (Table 3).

Activity of cell-free extracts against N-carbobenzoxy-L-tyrosine-*P*-nitrophenyl ester (CTN) did not increase during senescence of detached leaves, as did activity against azocasein (PETERSON and HUFFAKER 1975) or purified RuBPCase (THOMAS and HUFFAKER 1980). In the presence of cycloheximide, the leaves continuously lost activity against CTN, which has been considered to be a substrate for carboxypeptides. Activity against α-naphthylacetate constantly decreased during senescence; no effect of either cycloheximide or chloramphenicol was detected.

*RuBPCase Loss in Intact Versus Detached Leaves.* Many studies on senescence have employed detached leaves, or leaf discs, but there are some important

**Table 3.** Separation of activity against artificial substrates from activity against a protein (MILLER and HUFFAKER unpublished information). Electrophoresis was done according to ORNSTEIN (1964) and DAVIS (1964). The pH of stacking gel was 8.9; development was at pH 9.5

| Activity assayed | | Relative mobility |
|---|---|---|
| LAP | Band 1 | 0.36 |
| | Band 2 | 0.39 |
| BANAase | | 0.52 |
| Gelatinase | | 0.47 |

*BANA and LNA assays:* After electrophoresis of crude extracts, the polyacrylamide gels were extruded from the glass tubes into 15-ml test tubes containing 0.5 M Tris-maleate or K-phosphate buffer, pH 5.4. The test tubes were kept on ice for 1 h. The gels were transferred to tubes containing 40 μg/ml BANA, 50 μg/ml Fast Black K salt, and 0.2 M Tris-maleate or K-phosphate buffer, pH 5.4. BANA and LNA were dissolved in warm dimethylsulphoxide. Final concentration of DMSO in the reaction mixture was 0.1% (MELVILLE and SCANDALIOS 1972). After incubation at 40 °C for 15 min, the reaction was stopped by transferring the gels to test tubes containing a destaining solution of 10% isopropanol and 10% acetic acid. LAP activity was visualized by substituting LNA for BANA and incubating at room temperature. Gels were scanned at 540 nm. See text for abbreviations.

Proteolytic bands were detected by slicing gels longitudinally, placing them on a thin layer of agar-gelatin, and incubating them at 40 °C for 2 h. The disc gels were removed and the zymograms were developed by pouring a solution of acidified $HgCl_2$ over the thin layer. Cleared areas are a result of proteolytic digestion. (HUFFAKER and MILLER 1978).

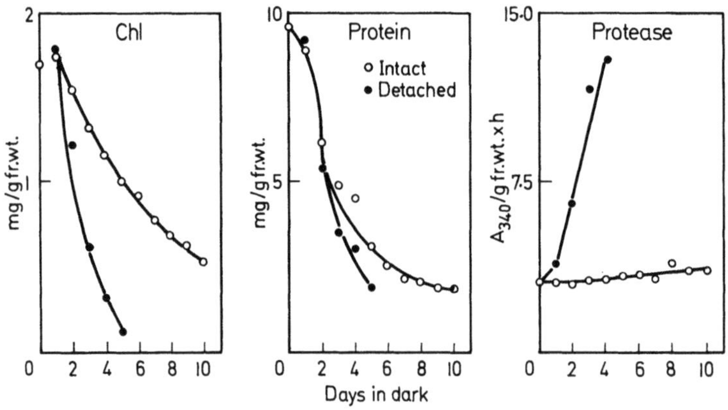


**Fig. 4.** Time course for chlorophyll and protein concentrations and proteolytic activity in leaves from intact and detached leaves of barley. Plants were grown in a chamber at 28 °C and 60% relative humidity, at 540 μ Einsteins $s^{-1}$ $m^{-2}$ to 7 days of age, and then were placed in darkness. Leaves were harvested and ground in 0.1 M K phosphate buffer, pH 6, and 1 mM dithiothreitol (DTT), the homogenate was filtered through cheesecloth, and the filtrate was centrifuged at 30,000 *g* for 20 min. Aliquots of the supernatant were assayed for total protein by the Lowry method and for proteolytic activity against azocasein. Chlorophyll (Chl) was extracted from the pellet with 80% acetone and determined according to Arnon (HUFFAKER and MILLER 1978)

differences and similarities between them and intact leaves. In general, the symptomology of RuBPCase and chlorophyll losses are similar but faster in the detached leaves (Fig. 4). They greatly diverge in development of new proteolytic activity. Whereas the intact leaf maintains its normal complement of endopeptidases, the detached leaf develops new proteinases (HUFFAKER and MILLER 1978). The additional proteolytic activities were identified by separation during isoelectric focusing on acrylamide gels. Because cycloheximide inhibited the appearance of the new proteolytic bands, they may be caused by de novo synthesis.

*Intracellular Location of RuBPCase Degradation.* The site of degradation of RuBPCase is unresolved. Evidence exists that the degradation may be regulated by factors in the cytoplasm. Conversely, recent studies indicate that a proteolytic activity may be associated with RuBPCase. Evidence of the former is that treating detached barley leaves with cycloheximide almost totally prevented the onset of typical symptoms of senescence during a 3-day period (PETERSON and HUFFAKER 1975). RuBPCase protein and chlorophyll were maintained. Proteolytic activity did not increase; rather, a loss in proteolytic activity resulted, which indicated turnover of the proteolytic system. Chloramphenicol had little effect on the onset of any of the symptoms of senescence. Cycloheximide inhibited the synthesis of new proteolysis in detached barley leaves, which was identified by isoelectric focusing (HUFFAKER and MILLER 1978). In addition, THOMAS (1974) and MARTIN and THIMANN (1972) showed that cycloheximide – not chloramphenicol – inhibited the loss of chlorophyll in *Lolium* and oat (*Avena sativa*) leaves.

Before the onset of senescence, the chloroplast seems quite protected from hydrolytic activity. Sequestered in the chloroplast, RuBPCase may be protected from degradation as long as the chloroplast membrane maintains integrity. Hypothetically, if the chloroplast membrane degenerates during senescence, then RuBPCase may leak out or proteinases may leak in. If so, degradation of both RuBPCase and the chloroplast may be closely linked. On that basis, the synthesis, release or activation of a specific hydrolase that is inhibited by cycloheximide may be initiated in the cytoplasm. CHOE and THIMANN (1975) showed that chloroplasts isolated from oat leaves lost chlorophyll and protein more slowly than did leaves. PEOPLES (1980), on the other hand, reported proteolytic activity associated with barley leaf chloroplasts, and HAMPP and DE FILIPPIS (1980) identified two plastid proteinase activities in oat leaves.

Conversely, we found that RuBPCase purified from whole barley leaves or from cytoplasm-free chloroplasts consistently had a detectable level of endogenous endopeptidase activity (THOMAS and HUFFAKER 1980). The rate of self-hydrolysis of RuBPCase isolated from chloroplasts was 0.36% per h. That rate would account for the complete hydrolysis of RuBPCase in the primary barley leaf in about 12 days, a time very similar to the rate of in vivo loss of RuBPCase (THOMAS and HUFFAKER 1980, FRIEDRICH and HUFFAKER 1980). This raises the possibility that an endopeptidase capable of degrading RuBPCase is associated with RuBPCase itself, or is associated with the chloroplast such that it attaches to RuBPCase during isolation and purification. Although RuBP-

Case-associated proteolysis was least active at pH 8.0, it could increase as decreasing photosynthesis causes the pH in the chloroplast to drop (THOMAS and STODDART 1980).

The manner of association of RuBPCase with the endopeptidase is unclear. The association is not broken by the several steps in the procedure to purify RuBPCase (THOMAS and HUFFAKER 1980, Miller and Huffaker unpublished results). BSA was included in the procedure to isolate chloroplasts from protoplasts to help trap proteases. If it is not merely a contamination, then the protease activity associated with RuBPCase becomes very important.

*Degradation Products.* Products of endopeptidic hydrolysis were determined by incubating [$^{14}$C]-labeled RuBPCase protein alone and with a cell-free extract from barley leaves (THOMAS and HUFFAKER 1980). The products were detected by SDS gel electrophoresis and quantitative fluorography. The large subunit (55 kD) was degraded into fragments that had molecular weights of 50,000, 42,200, 37,300, 33,100, 22,400 and 18,800. The large subunit was cleaved at several points, which produced a mixture of polypeptides. The small subunits were stable much longer than were the large. Little was learned about hydrolysis of the small subunit since its peak became broader and more diffuse with time. However, the degradation rates were quite different for the large and small subunits.

*RuBPCase and Photosynthetic $CO_2$ Fixation.* When RuBPCase was first isolated, its in vitro activity was deemed too low and its $K_m$ too high to account for photosynthetic $CO_2$ fixation. Even so, amounts of in vitro RuBPCase activity and photosynthetic $CO_2$ fixation were correlated (BJÖRKMAN 1968). Since then the $K_m$ of RuBPCase has been determined to be much lower (10 to 20 μM) (JENSEN and BAHR 1977), because its true substrate, $CO_2$, (COOPER et al. 1969) was taken into account. In fact, the concentration of active sites of RuBPCase in the chloroplast is 2 to 3 mM (G. LORIMER personal communication). Later LAING et al. (1974) showed that RuBPCase activity could account for photosynthetic $CO_2$ fixation.

Under some conditions RuBPCase may not limit the rate of photosynthesis. The concentration of RuBPCase and its catalytic function were followed during leaf senescence, when rapid changes were occurring, in order to further define the relation between them (FRIEDRICH and HUFFAKER 1980). The decreases in RuBPCase concentration and rate of true photosynthesis were highly correlated in senescing primary leaves of intact barley plants growing under normal conditions (cf. Figs. 3 and 5). The rate of true photosynthesis per mg of RuBPCase increased during the early stages of leaf senescence (Table 4). This indicates that RuBPCase may have been in excess originally. JENSEN and BAHR (1977) point out that both RuBPCase concentration and its in vivo regulation may influence photosynthetic $CO_2$ fixation.

*RuBPCase and Biomass Production.* Photosynthetic activity during seed filling is closely related to grain yields (FISCHER and KOHN 1977, LUPTON et al. 1974). Photosynthetic $CO_2$ fixation furnishes the reduced carbon compounds and supplies energy for reducing $N_2$ in the root nodules of legumes. Photosynthetic

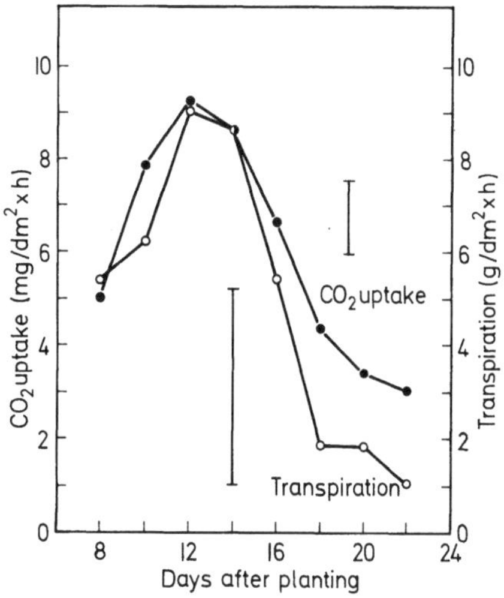


**Fig. 5.** True photosynthesis or $CO_2$ uptake in senescing barley primary leaves (FRIEDRICH and HUFFAKER 1980). Bars represent $LSD_{-0.05}$ for each parameter. The uptake of $CO_2$ was measured in a growth chamber with a dual isotope diffusion parameter as described by TING and HANSCOM (1977)

**Table 4.** Photosynthesis per mg RuBPCase (P/RuBPCase), the ratio of photosynthesis to transpiration (P/T), and the concentration of $CO_2$[a] in the intercellular air spaces [$CO_2^{IAS}$] in senescing barley leaves. (FRIEDRICH and HUFFAKER 1980)

| Days after planting | P/RuBPCase ($\mu$mol $mg^{-1} min^{-1}$) | P/T (mg $g^{-1}$) | [$CO_2^{IAS}$] ($\mu l\, l^{-1}$) |
|---|---|---|---|
| 8 | 88 | 1.08 | 250 |
| 10 | 110 | 1.46 | 244 |
| 12 | 144 | 1.20 | 248 |
| 14 | 156 | 1.30 | 245 |
| 16 | 193 | 1.66 | 237 |
| 18 | 166 | 2.61 | 222 |
| 20 | 168 | 2.02 | 232 |
| 22 | 178 | 3.12 | 210 |
| LSD, 0.05 | 34 | 1.07 | 20 |

[a] External $CO_2$ concentration [$CO_2^{Ext}$] was 268 $\mu l\, l^{-1}$. [$CO_2^{IAS}$] = [$CO_2^{Ext}$] minus (photosynthesis) · (stomatal_resistance)

electron flow in leaves can directly furnish electrons to reduce nitrite (NEYRA and HAGEMAN 1974, MIFLIN 1974) and ammonium (WALLSGROVE et al. 1979). A portion of the energy for reduction of nitrate can be supplied via shuttling of early products of photosynthesis between the chloroplast and cytoplasm of leaf cells (KLEPPER et al. 1971, STOCKING and LARSON 1969, HEBER 1974). In fact, the steady-state rate of nitrate reduction in barley plants in light is twice that in darkness (ASLAM et al. 1979).

When the production of reduced N cannot meet the requirement of the seed, leaf protein is metabolized to meet the N needs. This can occur with a non-legume if soil N becomes limiting. In a legume (described below) it can also occur when photosynthate does not support the sink requirements of both seeds and nodules. RuBPCase is among the first major proteins lost as the leaf moves into senescence. As RuBPCase is lost, photosynthetic $CO_2$ fixation concurrently decreases (FRIEDRICH and HUFFAKER 1980, WITTENBACH 1979). SINCLAIR and DEWIT (1976) called this a "self-destruct" phenomenon in soyabean. WITTENBACH et al. (1980) recently reported that photosynthetic $CO_2$ fixation began to decrease in soyabean leaves somewhat before RuBPCase protein did; both then declined together.

Because of its high protein concentration, soyabeans require a high rate of N supply to the seed. The seed requirement for N is apparently so great that the nodules cannot supply enough; N must therefore be supplied from vegetative parts (SINCLAIR and DEWIT 1976). The resulting loss in photosynthetic $CO_2$ fixation deprives the root nodules of energy, causing their degradation. This limits the period of seed development and filling and, hence, decreases the yield potential.

HARDY et al. (1971) showed that enhanced photosynthate supply extended the exponential phase of $N_2$ fixation, which resulted in increased soyabean yields. When seeds were removed from the soyabean plant, RuBPCase and total protein remained at higher levels in the leaves (MONDAL et al. 1978). A field study at the University of California at Davis (UCD) showed that RuBPCase protein in soyabean leaves were correlated with nitrogenase activity and $H_2$ production over a season (HUFFAKER and MILLER 1978, in cooperation with D.A. Phillips). SINCLAIR and DEWIT (1976) proposed that the exponential phase of $N_2$ fixation might be extended by partitioning more photosynthate to support $N_2$ fixation.

As a test of this hypothesis, five variants that had green leaves and brown mature pods were selected from several thousand plants in a soyabean nursery at UCD. Seeds from one of these produced 14 plants in a growth chamber, five of which showed delayed leaf senescence (ABU-SHAKRA et al. 1978). As expected, these plants had higher concentrations of chlorophyll, RuBPCase protein, and nitrogenase activity than did the senescing control plants. Plants having the delayed senescence characteristic did not show decreased seed yields. Delayed leaf senescence seems to be inheritable. These plants may be better able to proportionally partition photosynthate to maintain the energy needs of both the seed and the nodules.

HELSEL and FREY (1978) reported that an oat variety showing delayed leaf senescence had higher yields. Maintaining production of reduced N may protect RuBPCase from degradation. Much genetic variability also exists in that characteristic. FRIEDRICH and SCHRADER (1979; FRIEDRICH et al. 1979) showed that their corn variety senesced even in the presence of external N and that the degradation of leaf protein still furnished almost all of the reduced N for the seeds.

*Inheritance of RuBPCase.* Isolation of the component polypeptides of the large and small subunits ofRuBPCase has allowed sophisticated studies on the inheritance of RuBPCase. The procedure involves dissociating purified RuBPCase into the large and small subunits by 8 M urea and subsequently isoelectric-focusing the S-carboxymethylated polypeptides (KUNG et al. 1974). It was shown for both tobacco and wheat that inheritance of the large subunit is controlled by the maternal parent species; hence, chloroplast DNA codes for the large subunit (KUNG et al. 1974). The small subunit is inherited from both parent species, neither being dominant, which shows that nuclear DNA codes for the small subunit. Genetic studies agree well with others on the intracellular locations for synthesis of the two subunits and their subsequent assembly in the chloroplast. Such studies have shown probable origins of present species of both tobacco and wheat (CHEN et al. 1975).

Attempts to dissociate the enzyme into the large and small subunits and then assemble them again have not been very successful (NISHIMURA and AKAZAWA 1974). Genetic studies have allowed the creation of different combinations of variable subunits in order to compare the effects of function and structure (KUNG and RHODES 1978).

## 3 Proteinase Inhibitors

Endogenous leaf proteins inhibiting the digestive proteases of insects and microorganisms seem highly suited to be protective agents. Two protease inhibitors inhibiting animal, insect, and microorganism proteases are induced in large quantities in tomato leaves that have been attacked by chewing insects, physically wounded, detached, or treated with the wound hormone PIIF (protease inhibitor-inducing factor). The inhibitors strongly reduce the activities of proteases similar to trypsin and chymotrypsin but do not inhibit plant proteases.

*Synthesis.* The increase in level of protease inhibitor I was due to protein synthesis, since [$^{14}$C]-lysine was preferentially incorporated into it during the period of increase (RYAN 1968). Inhibition in the synthesis in potato (*Solanum tuberosum*) leaves by cycloheximide but not chloramphenicol implicated cytoplasmic ribosomes (RYAN and HUISMAN 1970).

*Intracellular Location.* SHUMWAY et al. (1976) showed that protease inhibitor I is present in tomato (*Lysopersicon esculentum*) leaves in membraneless protein bodies, i.e., probably in vacuoles. Later WALKER-SIMMONS and RYAN (1977a) isolated vauoles from tomato leaves and showed that they contained both protease inhibitors I and II. The authors suggest that it seems advantageous to sequester the protease inhibitors in the vacuole, where they are not subject to degradation (GUSTAFSON and RYAN 1976) but can readily dissolve when the vacuole is broken.

*Induction by Wound Hormone (PIIF).* The synthesis of the protease inhibitors is induced by a wound hormone, PIIF (protein inhibitor-inducing factor) (RYAN 1974), released from wounded leaf tissue. It is translocated to other tissues, where it induces the protease inhibitors. The release of PIIF is both light- and temperature-dependent (WALKER-SIMMONS and RYAN 1977b). Darkness or temperatures below 22 °C greatly decrease the accumulation of the protease inhibitors. The lower temperature affects the accumulation process much more than it affects the transport of PIIF.

PIIF or PIIF-like substances are present in a wide variety of plants. They were found in 37 of 39 species selected from among 20 families of the four major divisions of plants (MCFARLAND and RYAN 1974). WALKER-SIMMONS and RYAN (1977b) applied tomato PIIF to detached leaves from 23 species and found 10 of them produced trypsin inhibitors. They propose that the inhibitors may be classified as allelochemics (naturally occurring chemicals affecting the growth or behavior of another species) (WHITTAKER and FEENY 1971), and their presence in plants determined by environmental pressures during evolution.

# 4 Nitrate Reductase (NR)

Nitrate reductase (NR) is the first enzyme catalyst in the assimilation of nitrate to ammonium (Fig. 6). Since the level of NR activity is affected profoundly by nitrate concentration and different environmental conditions, its respective induction or loss can be followed by providing or removing nitrate, and by placing the plant in darkness or under less-than-favorable environmental conditions. Unfortunately, most of the reported studies concerning turnover have followed NR activity instead of NR protein because of the difficulty of maintaining an active enzyme during purification. Purification methods that appear very promising have been developed recently; they include affinity chromatography (SOLOMONSON 1975). Future research should elucidate the regulation of the NR protein concentration in plants.

Work is also progressing to determine regulation of the in vivo activity of NR as well as nitrite reductase (NiR) and enzymes of $NH_4^+$ assimilation. Methods have been developed to determine the actual in vivo activity of NR and the in vivo steady-state concentrations of $NO_3^-$ and $NO_2^-$ (CHANTAROTWONG et al. 1976, ASLAM et al. 1979). The amount of nitrate reduction occurring in leaves and roots, its regulation, and how this affects the energy relationships of plants is also being elucidated.

*Synthesis of NR.* Nitrate reductase can be induced in leaves in darkness (BEEVERS et al. 1965, ASLAM et al. 1973); the amount induced is dependent on the reserve carbohydrate level. The addition of glucose can greatly increase induction of NR (ASLAM et al. 1973). Induction of NR is greatest in light for several reasons. Leaf polyribosomes, required for NR induction, are increased in light (TRAVIS et al. 1970). Polysomes from plants grown in light are more active in driving in vitro protein synthesis than are those from plants grown in darkness (TRAVIS et al. 1970). Light can also increase the flux of nitrate through the cytoplasm by affecting both root absorption and translocation of nitrate to the leaves (HALLMARK and HUFFAKER 1978), and by facilitating the release of nitrate from the storage to the metabolic pool (cytoplasm) (ASLAM et al. 1976). The induction of NR is largely regulated by the flux of nitrate in the metabolic pool (ASLAM et al. 1976, SHANER and BOYER 1976). JOLLY and TOLBERT (1978) have proposed that light may inactivate an inhibitor of NR from soyabean leaves.

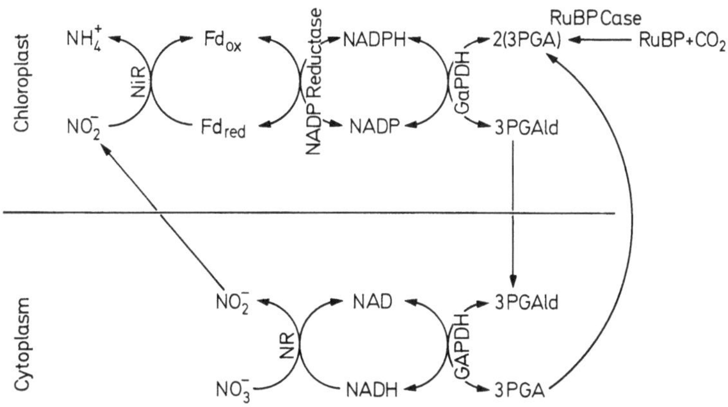


**Fig. 6.** Nitrate assimilation in green leaves in light

*Degradation of NR.* The loss of NR from plant leaves can be followed by changing the environmental conditions, such as darkness (TRAVIS et al. 1969), removal of external nitrate, heat (MATTAS and PAULI 1965, ONWUEME et al. 1971), and water stress (HUFFAKER et al. 1970, MORILLA et al. 1973). Low temperature also greatly decreased the loss of NR. Inhibitors of protein synthesis were applied to induced barley leaves in the hope that the rate of synthesis would be reduced and degradation would continue. Instead of an increased rate of loss of NR activity in the presence of cycloheximide, the loss of NR activity was greatly decreased (Fig. 7). The half-life for NR was 12 and 35 h, respectively, for control and cycloheximide-treated plants. (D-)Chloramphenicol had no effect on the loss of NR. Perhaps the inhibitor decreased the synthesis of both NR and proteinase(s), degrading it. Inactivation of phenylalanine ammonialyase (PAL) also can be prevented by cycloheximide (HYODO and YANG 1971). If the proteolytic system is being turned over as well, then the loss of NR may decrease. Alternatively, cycloheximide may have prevented the synthesis of an activator.

The above results, along with the demonstrated possibilities of inhibitors and activators, confounds the interpretation of NR turnover studies, especially since enzyme activity but not enzyme protein was followed. Experiments of ZIELKE and FILNER (1971) who used suspended tobacco cells overcame many of the problems listed above. They used triple-labeling procedures to show simultaneous synthesis and degradation of NR. Pre-existing protein was labeled with [$^{14}$C]-arginine and $^{15}NO_3$ to increase the buoyant density. Next, the cells were transferred to [$^3$H]-arginine and [$^{14}$N]$O_3$. The [$^3$H]-label showed the effect of the pre-existing pools of [$^{15}$N] amino acids on the density of the newly synthesized protein. The labels were followed during initial synthesis of NR, during its steady-state concentration, and during net degradation. During each phase, isopicnic equilibrium centrifugation was carried out using the isolated soluble protein. At steady-state concentration of NR, its buoyant density shifted from [$^{15}$N]-NR toward [$^{14}$N]-NR. Pre-existing protein labeled with $^{14}$C and new protein labeled with $^3$H turned over. The results showed NR turnover during each of the three phases at a rate ($t_{1/2}$) estimated at 4.3 h.

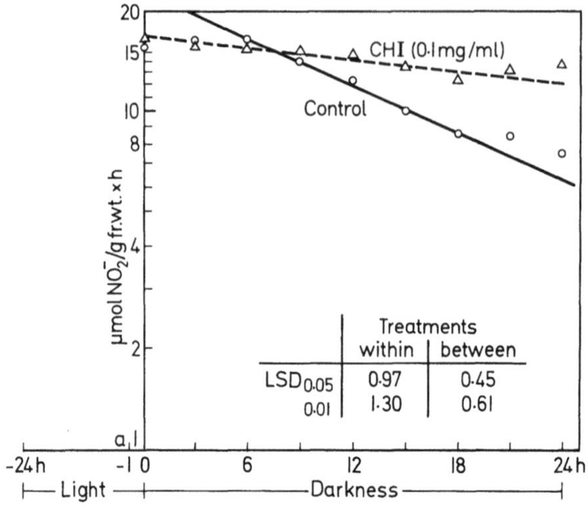


**Fig. 7.** Effect of cycloheximide (CHI) on the loss of nitrate reductase activity from intact barley leaves in darkness (TRAVIS et al. 1969). After 24 h of light, cycloheximide was applied three successive times at 20-min intervals. Applications began at *a* (−1 h) on graph. Plants were then placed in darkness in order to follow the loss of nitrate reductase activity. k (control) = 0.056 (h darkness)$^{-1}$; $t_{1/2}$ (control) = 12.4 h darkness. k (cycloheximide) = 0.020 (h darkness)$^{-1}$; $t_{1/2}$ (cycloheximide) = 34.7 h darkness

Other reported $t_{1/2}$ values for NR turnover are 2.6 h for radish (*Raphanus sativus*) (ACTON and GUPTA 1979) and about 4 h (SCHRADER et al. 1968) and 6 h (ASLAM and OAKS 1976) for young corn leaves.

*Relationship of In Vitro and In Vivo Activities of NR.* Several studies have shown that NR is in excess in leaves (CHANTAROTWONG et al. 1976, HALLMARK and HUFFAKER 1978). In barley leaves assimilating nitrate from a 1 mM solution, extractable NR activity was about twofold greater than in vivo reduction of nitrate. HALLMARK and HUFFAKER (1978) showed that in vivo reduction of nitrate may increase, even if the internal concentration of leaf NR (estimated by tissue-slice technique) was decreasing. This was done by equilibrating Sudan grass (*Sorghum vulgare sudanensis*) seedlings for 24 h at increasing temperatures before tissue-slice assays (Fig. 8) and in vivo assays of actual $NO_3^-$ uptake and reduction (Fig. 9) were begun. As the temperature increased, tissue-slice NR decreased (Fig. 8). Conversely, the in vivo rate of nitrate reduction increased as the temperature increased (Fig. 9). This indicates that NR was in sufficient excess even at the higher temperatures to allow for an increased rate of nitrate reduction with increasing temperature. Environment (in this case, temperature) had exactly opposite effects on the in vivo activity of NR and on its inactivation. Previous studies showed that extractable NR activity decreased as the temperature increased (ONWUEME et al. 1971, MATTAS and PAULI 1965, SCHRADER et al. 1968) and during water stress (HUFFAKER et al. 1970, SHANER and BOYER 1976). Decreased NR activity could possibly be due to increased rate of degradation, decreased rate of synthesis, inactivation by endogenous inhibitors (JOLLY and

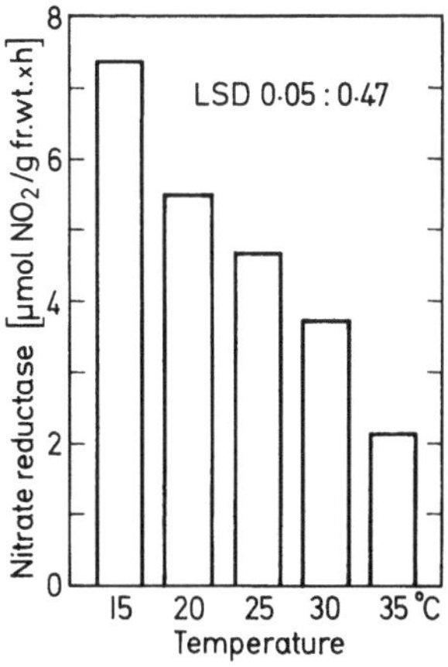


**Fig. 8.** Effect of equilibration for 24 h at different temperatures on NR activity in tissue slices from Sudan grass leaves (HALLMARK and HUFFAKER 1978)

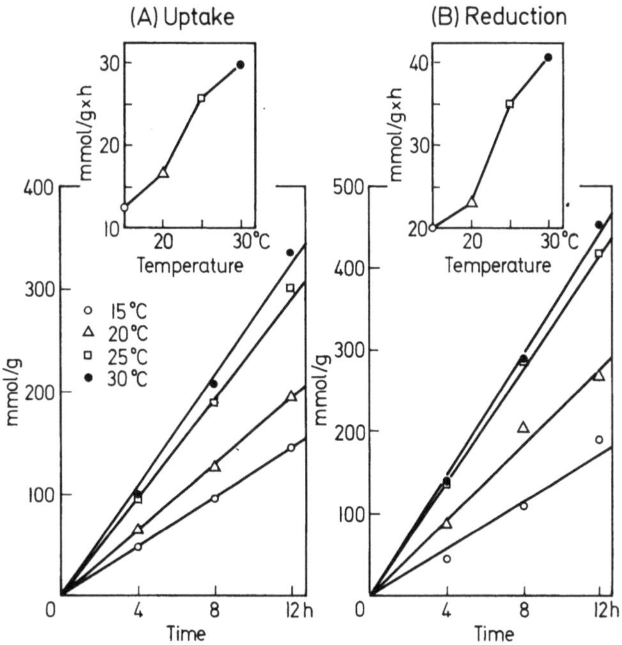


**Fig. 9.** Nitrate uptake (**A**) and reduction (**B**) by Sudan grass seedlings as a function of temperature after equilibration at 0.4 mM $KNO_3$ at the indicated temperatures for 24 h (HALLMARK and HUFFAKER 1978). Nitrate uptake was determined as that disappearing from the external solution and reduction determined by subtracting the internal amount of nitrate from that taken up (CHANTAROTWONG et al. 1976)

TOLBERT 1978) or a change in relative turnover rates of protein activators or inhibitors.

*Regulation by Endogenous Inhibitors.* JOLLY and TOLBERT (1978) isolated and purified a protein inhibitor of NR activity from soyabean leaves in darkness. The inhibitor was inactivated by light but was reactivated during 24 h of darkness. The leaf concentration of the inhibitor was the same in either light or darkness. NR activity in suspended soyabean leaf cells ceased in the presence of the inhibitor. A similar result was reported for increased activity of NR in *Chlorella* in the light (TISCHNER and HÜTTERMANN 1978). Increased NR activity was ascribed to a light-mediated activation rather than to de novo protein synthesis.

The presence of inhibitors and activators may vary in different plant species. The soyabean NR from dark-treated leaves was not active until an inhibitor was removed (JOLLY and TOLBERT 1978). On the other hand, NR from either roots or leaves of barley plants grown in darkness is active in vitro (ASLAM et al. 1973, CHANTAROTWONG et al. 1976, TRAVIS et al. 1970). Results with barley leaves show that the in vivo rate of nitrate reduction in light or dark was regulated by carbohydrate supply (ASLAM et al. 1979). Although future investigations may show the presence of endogenous inhibitors in barley leaves, the energy status of the leaf seems more important as a regulator of in vivo nitrate reduction.

*In Vivo Regulation of Nitrate Reduction.* Nitrate is the main form of N available to plants in an aerobic environment. Plants are capable of reducing nitrate to the level of amino acids in both roots and leaves. Because the assimilation of a molecule of nitrate to the level of glutamate, requires 10 electrons, it is very costly to the plant. Like $CO_2$, N is required in mass quantities; hence, a great deal of energy is expended in nitrate assimilation. The ability of the plant to assimilate nitrate in leaves gives the plant a great advantage, because nitrate assimilation is linked both directly and indirectly to photosynthetic electron flow. It is estimated that cereal plants assimilate about 20 to 30% of the nitrate in their roots and 70 to 80% in their leaves (ASLAM and HUFFAKER unpublished results).

*Intracellular Location of Enzymes.* The compartmentation of the enzymes involved in nitrate assimilation greatly affects the regulation of the processes involved. Their location affects feedback inhibition and the sources of energy available to drive the reactions.

Strong evidence favors a cytoplasmic location for NR (SCHRADER et al. 1967, RITENAUR et al. 1967, TRAVIS et al. 1969). Its synthesis is inhibited by cycloheximide, and inhibitor of cytoplasmic 80S ribosomes, but not by chloramphenicol, an inhibitor of 70S ribosomes (SCHRADER et al. 1967, TRAVIS et al. 1969). Nitrite reductase (NiR) is in the chloroplast (RITENAUR et al. 1967, SCHRADER et al. 1967, WALLSGROVE et al. 1979). A loose association between NR and an organelle, such as chloroplasts, has not been ruled out. Ammonium assimilation enzymes are in both the chloroplast and the cytoplasm (WALLSGROVE et al. 1979).

*Effects of Light and Dark.* Although many conflicting reports exist, recent work shows clearly that nitrate, nitrite, and ammonium can be assimilated in both darkness and light. Carbohydrate furnishes the necessary reducing power in darkness, presumably through glycolysis, at the level of NAD-specific glyceraldehyde-3-phosphate dehydrogenase (BEEVERS and HAGEMAN 1969). A malate-oxalacetate shuttle system between the mitochondria and cytoplasm could also be driven by NAD-specific malate dehydrogenase (JESCHKE 1976, SCHEIBE and BECK 1979). Both reactions would furnish NADH, which is apparently the preferred reductant of nitrate reductase (BEEVERS and HAGEMAN 1969). Using a method that traced the flux of nitrate through barley leaves in darkness, ASLAM et al. (1979) showed that in vivo NR activity decreased as carbohydrate concentration decreased.

In light, the rate of nitrate reduction by barley was two times greater than in darkness; increased nitrate reduction in light required concomitant fixation of $CO_2$. Apparently, recently fixed products of photosynthetic $CO_2$ fixation shuttle between the chloroplast and cytoplasm (HEBER 1974, STOCKING and LARSON 1969) and may furnish additional reducing power to support the greater rate of nitrate reduction (KLEPPER et al. 1971). In this case, NR activity could be closely linked to both RuBPCase activity and noncyclic photosynthetic electron flow through an appropriate triosephosphate shuttling system. A similar mechanism involving transporting oxalacetate and malate between the chloroplast and cytoplasm [oxidation and reduction being catalyzed by NADP-malate-dehydrogenase (MDH) and NAD-MDH] could also transport photosynthetically derived reducing equivalents to the cytoplasm (ANDERSON and HOUSE 1979). The chloroplastic NADP-malate dehydrogenase is rapidly activated in vivo by light (ANDERSON and AVRON 1976); a portion of both the activator and the enzyme are bound to the chloroplast lamellae (SCHEIBE and BECK 1979).

*Metabolic and Storage Pools.* The nitrate pools within plant leaves profoundly affect both the in vivo activity of NR and its steady-state concentration. Evidence for different pools of nitrate was found in tobacco cells, barley aleurone layers, and corn leaf sections (HEIMER and FILNER 1971, FERRARI et al. 1973). After external nitrate was removed, the plant systems ceased nitrate reduction after about 1 h, even though most of the nitrate taken up was still present. That NR was still active was shown by prompt nitrate reduction upon the addition of external nitrate. ASLAM et al. (1976) showed the presence of different nitrate pools in barley leaves. The induction of barley leaf NR was directly proportional to the size of the metabolic pool. The metabolic pool appears to be cytoplasmic and regulates both NR induction (ASLAM et al. 1976, CHANTAROTWONG et al. 1976, FERRARI et al. 1973, HALLMARK and HUFFAKER 1978, SHANER and BOYER 1976) and its in vivo activity (CHANTAROTWONG et al. 1976, ASLAM et al. 1976, HALLMARK and HUFFAKER 1978). The size of the metabolic pool is regulated both by uptake from the roots and by the release of nitrate from the storage pool (vacuole?).

Light greatly facilitated the movement of nitrate from the storage pool into the metabolic pool of barley leaves, whereas glucose had no effect (ASLAM et al. 1976). The presence of the two pools helps explain an earlier observation (BURSTRÖM 1943) that light was required for barley leaves to reduce nitrate in vivo. Detached leaves preloaded with nitrate reduced nitrate only in light. Light was most likely needed to move nitrate from the storage to the metabolic pool, where it could be reduced. Detached barley leaves placed in a nitrate solution in darkness readily reduced nitrate, since a flux of nitrate moves through the metabolic pool (ASLAM et al. 1979).

*Product Inhibition of NR Synthesis and Activity.* Product inhibition of NR synthesis or in vivo activity in leaves does not appear to be important. As shown in Fig. 6, the products of nitrate assimilation ($NO_2^-$ and $NH_4^+$) are probably assimilated in the chloroplast, although ammonium can also be assimilated in the cytoplasm (WALLSGROVE et al. 1979). Furthermore, nitrite assimilation is very efficient; the steady-state concentration of nitrite is very low (ASLAM et al. 1979) and although found in greater concentrations than nitrite, ammonium steady-state concentrations are also quite low (GOYAL and HUFFAKER unpublished results). Hence, compartmentation and efficient assimilation of nitrite and ammonium largely decreases feedback inhibition of NR in leaves.

*NR Mutants of Barley Leaves.* The ability to select for mutants in specific metabolic pathways has greatly facilitated studies of enzyme regulation in bacteria. Relatively few plant mutations have been defined biochemically because of genetic complexity, the long regeneration times involved, and the laborious techniques required. WARNER et al. (1977) made an important contribution in isolating NR-deficient mutants of barley. The mutants were obtained by treating "Steptoe" barley seeds with sodium azide. The treated seeds were grown to maturity in the field and the $M_2$ seeds were bulk-harvested. The $M_2$ seeds were then planted; at 7 days of age a portion of the leaf material was assayed for NR activity. The plants identified as NR mutants (low in vivo NR activity) were grown to maturity and the seed was harvested. Two of these mutants also had lower levels of NR-associated cytochrome c reductase activity (WRAY and FILNER 1970). These mutants were then used to determine the effect of greatly reduced NR activity on grain yield and protein concentration (described below).

*Amount of Leaf NR and Grain Yields.* If a biochemical marker could be found for the rate-limiting step in converting nitrate to reduced products, the selection of superior grain varieties would be greatly facilitated. Since NR is the first enzyme in the nitrate assimilation pathway and overall may be the rate-limiting step ($NO_3^-$ accumulates to much higher levels than either $NO_2^-$ or $NH_4^+$), its activity has been compared across wheat cultivars.

CROY and HAGEMAN (1970) studied the relationship of NR activity to high protein production using 32 wheat varieties. Of the 32 varieties 16 had high NR activity and 13 of the 32 produced high protein. Of the 16 high NR varieties, 9 were also high protein-producing. DECKARD et al. (1973) also found good correlations between amount of NR activity in corn leaves and grain yield and protein. KLEPPER (1975) reported a frequency distribution of NR activity in a population of Centurk (low-protein) and Atlas 66 (high-protein) wheat varieties. The mean NR activity of the Atlas 66 population was significantly greater than that of Centurk.

Other investigators have found difficulties in relying on a single assay system to indicate the end result of complex biochemical and physiological processes

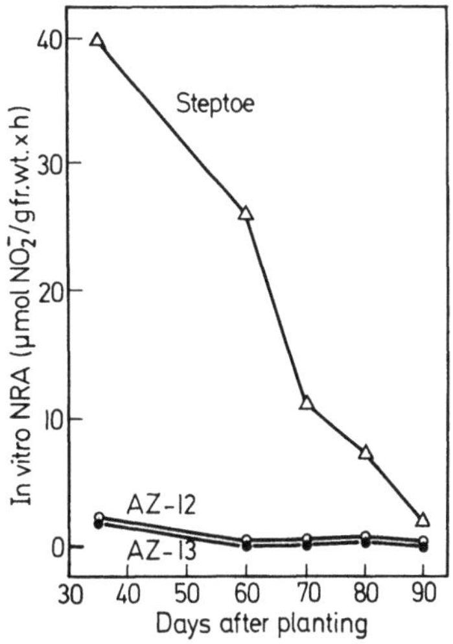


**Fig. 10.** In vitro NR activity from field-grown barley leaves (OH et al. 1980)

($NO_3^-$ reduction in the leaf to grain yield and protein). OH et al. (1980) compared NR activity against grain yield and protein using a control and two nitrate-reductase-deficient barley mutants grown in the field (WARNER et al. 1977). Although mutant in vitro NR activity was less than 10% of that in the control (Fig. 10), no significant difference occurred among total vegetative dry weight, total reduced N, and percent grain protein (Table 5). Both mutants yielded about 20% less grain than did the control. RAO et al. (1977) compared yields and protein concentrations of two wheat varieties that differed widely in amount of NR activity. Both varieties, Anza (lower NR) and UC 44 (higher NR), had similar grain yields and grain protein concentration. The varieties differed in ability to mobilize the reduced N in the straw and translocate it to the grain. The translocation efficiency of the two varieties also varied with the concentration of applied N.

Another complication is that recent studies have shown that NR activity in vivo may proceed at levels well below its maximum capability at physiological levels of nitrate (CHANTAROTWONG et al. 1976, HALLMARK and HUFFAKER 1978). The regulation imposed on the in vivo activity of NR by carbohydrate level, photosynthetic $CO_2$ fixation (ASLAM et al. 1979) and uptake (CHANTAROTWONG et al. 1976, ASLAM et al. 1979) may be more important than the amount of enzyme available.

*Inheritance of NR Activity.* The inheritance of leaf NR activity in corn and wheat has been studied. In corn, the $F_1$ hybrid of B14 × Oh43 had a heterotic level of NR (WARNER et al. 1969). Inbreds B14 and Oh43 differed at two loci controlling NR activity. Oh43 had higher rates of both in vivo synthesis and degradation than B14. Apparently both synthesis and degradation regulated the level of NR in these varieties. The $F_1$ hybrid was similar to B14 in its rate of enzyme synthesis and similar to Oh43 in its in vivo rate of degradation; the final result was a heterotic level of NR activity.

**Table 5.** Grain yield and protein content of Steptoe and two NR-deficient barley mutants: Az 12 and Az 13 (OH et al. 1980)

| Genotype | Grain protein (%) | Grain yield (kg $ha^{-1}$) |
|---|---|---|
| Az 12 | 9.8a[a] | 5168a |
| Az 13 | 9.5a | 5080a |
| Steptoe | 9.0a | 6429b |

[a] means with the same letter within a column do not differ at P=0.05 (Duncan's Multiple-Range Test)

The wheat variety UC 44-111 had twice the in vitro NR activity of Anza (RAO et al. 1977, GALLAGHER et al. 1980). On the basis of in vitro NR activity in the parents and crosses through to the $F_6$ generation, it was concluded that the two genotypes differed, by alleles at a single locus ($N_{ra}/n_{ra}$) for NR activity. The allele determining the higher activity in UC 44-111 was dominant. It was recognized that many factors may regulate the amount of NR activity, such as rates of synthesis and degradation, $NO_3^-$ supply and NR induction, energy supply, and the $K_m$ of NR for nitrate. Although several genes may regulate the amount of NR activity, the results indicated that the genotypes differed by one major gene.

## 5 Nitrite Reductase (NiR)

Although extracted NiR activity seems more stable than NR activity, little is known about its turnover characteristics. Much conflicting information is appearing about the regulation of NiR activity.

*Regulation of Activity.* Many workers studying nitrite reduction in leaves have concluded that nitrite reductase is active only in light (SAWNEY et al. 1978a, b, CANVIN and ATKINS 1974). Others report that nitrite can be reduced in darkness but only under aerobic conditions (JONES and SHEARD 1978). Using two different methods, ASLAM et al. (1979) showed that nitrite was assimilated in darkness. In one method, nitrite reduction was followed by determining a balance sheet for the processes involved in its assimilation. The difference between nitrite uptake and the internal leaf concentration of nitrite at a given time determined the amount of nitrite reduced. The second method traced the flux of [$^{15}$N]-nitrite into reduced products (Table 6). Both methods were in close agreement. Over 99% of the nitrite taken up was accounted for in reduced amino N. Hence, leaf nitrite reductase is extremely active and efficient in either darkness or light. In darkness the translocation of glycolytic intermediates such as glyceraldehyde-3-phosphate or of dicarboxylic acids such as malate into the chloroplast could furnish electrons to reduce NADP. NADP reductase

**Table 6.** Assimilation of $[^{15}N]O_2$-N into reduced nitrogen in carbohydrate-rich and carbohydrate-deficient excised leaves in light and dark

| Treatment | Uptake | Reduction | Incorporation into reduced-N |
|---|---|---|---|
| | μmol g dry wt$^{-1}$ 24 h$^{-1}$ | | |
| a) Carbohydrate-rich | | | |
| Light | 155 | 155 | 164 |
| Dark | 125 | 125 | 129 |
| b) Carbohydrate-deficient | | | |
| Light | 171 | 171 | 182 |
| Dark | 124 | 105 | 110 |

Seedlings were grown in nitrogen-free Hoagland's solution under continuous light as described previously. After 6 days, one set of seedlings was placed in darkness to deplete carbohydrates, and the other set was left in light. Thirty-six hours later, 10 leaves 10 cm long were excised from the tip, placed base down in small glass vials containing 10 ml of 5 mM $Na[^{15}N]O_2$ (95% $^{15}N$), and incubated in light or dark for 24 h. Nitrite uptake was measured as that disappearing from solution. Nitrite reduction was determined by subtracting the total amount accumulated by the leaves from the total uptake, as described before. Incorporation of $[^{15}N]O_2$-N into reduced-nitrogen compounds was calculated from total reduced nitrogen and atom % excess of $^{15}N$ in reduced nitrogen according to the following formula: $[^{15}N]O_2$-N incorporation (μmol g dry wt$^{-1}$) = [total reduced nitrogen [μg g dry wt$^{-1}$)] [$^1/_{100}$ of atm % excess $^{15}N$] + [molecular wt of N]. Total nitrogen was determined after Kjeldahl digestion in sulfuric acid. The digestion mixture was steam-distilled in the presence of concentrated NaOH, and the distillate was collected in boric acid containing methyl red indicator. The distillate was then titrated against standard 0.05 M $H_2SO_4$, and nitrogen content was calculated on a dry weight basis. After titration, the distillate was dried at 60°–65 °C and the isotopic concentration of $^{15}N$ was determined by mass spectrometry.

could then allow for the transfer of electrons to ferridoxin, which in turn could drive NiR activity. Such reactions could serve as an important detoxification function as well. We recently showed that carbohydrate-depleted barley leaves that had become incapable of reducing nitrate still reduced 85% of the nitrite they had taken up (Aslam and Huffaker unpublished results). The leaves appear capable of utilizing electrons from a host of degradation products to ultimately assimilate nitrite.

## 6 Phenylalanine Ammonia-Lyase (PAL)

The regulation of PAL in plant systems has received much attention because of its interesting and complicated controls and its possible relation to the plant's defense mechanisms. PAL is the first enzyme in the synthetic pathway of production of phenols, lignin, flavonoids, and phytoalexins. PAL catalyzes the conversion of phenylalanine to cinnamic acid.

*Synthesis.* PAL synthesis can be induced in some plant tissues by phytochrome (ZUCKER 1972, WELLMAN and SCHOPFER 1975), photosynthetic reactions (ZUCKER 1972), ethylene (HYODO and YANG 1971), naphthylacetic acid and kinetin (DUDLEY and NORTHCOTE 1979), and wounding (FRIEND et al. 1971). DNA intercalating compounds induced PAL and pisatin (an isoflavanoid) in pea endocarp (HADWIGER and SCHWOCHAU 1971). Inhibition of PAL induction by actinomycin D and 6-methylpurine indicated that newly synthesized RNA was required for the induction. This indicates a sophisticated genetic control for induction of specific phytoalexins in legumes to protect against disease (HADWIGER and SCHWOCHAU 1967). Increased PAL activity and lignification of wounded areas may also be a form of defense (FRIEND et al. 1971).

On the other hand, induction of PAL synthesis has been linked to ethylene-induced russet spotting in lettuce (*Lactuca sativa*) leaves, which is caused by the subsequent oxidation of phenolics to quinones (HYODO et al. 1978).

*Inactivation.* The inactivation of PAL in gherkin seedlings has been ascribed to the synthesis of protein inactivators (ENGELSMA and VAN BRUGGEN 1971) and to degradation in *Xanthium* leaf discs in darkness (ZUCKER 1971). ZUCKER detected very little turnover of PAL in light. WELLMAN and SCHOPFER (1975) obtained excellent evidence of PAL turnover in suspended parsley cell cultures by incorporating $^{15}N$ into the protein and subjecting it to isopycnic centrifugation. In darkness (non-inducing conditions), $^{15}N$ was incorporated into PAL; a band broadening occurred with time. This indicates simultaneous synthesis and degradation. Under inducing conditions (red light), $^{15}N$ was again incorporated into PAL but band broadening did not occur.

*In Vitro Synthesis.* SCHRODER et al. (1977) accomplished in vitro synthesis of PAL using polyribosomal RNA from parsley cell cultures in a rabbit reticulocyte lysate. They found polysomal mRNA increased in the parsley cells along with an increase in PAL activity. Conversely, DUDLEY and NORTHCOTE (1979), who used an in vitro PAL synthesizing system from *Phaseolus vulgaris* cells, estimated the concentration of PAL mRNA in *Phaseolus vulgaris* cells by extracting the polyribosomes during the inductive phase of PAL synthesis. The polyribosomes were then used to support in vitro protein synthesis. Their results indicated that induction of PAL synthesis did not require new synthesis of mRNA.

## References

Abu-Shakra SS, Phillips DA, Huffaker RC (1978) Nitrogen fixation and delayed leaf senescence in soybeans. Science 199:973–975

Acton GJ, Gupta S (1979) A relationship between protein degradation rates *in vivo*, isoelectric points, and molecular weights obtained by using density labeling. Biochem J 184:367–377

Alscher R, Smith MA, Peterson LW, Huffaker RC, Criddle RS (1976) *In vitro* synthesis of the large subunit of ribulose diphosphate carboxylase on 70S ribosomes. Arch Biochem Biophys 174:216–225

Anderson LE, Avron M (1976) Light modulation of enzyme activity in chloroplasts. Plant Physiol 57:209–213

Anderson JW, House CM (1979) Polarographic study of dicarboxylic acid-dependent export of reducing equivalents from illuminated chloroplasts. Plant Physiol 64:1064–1069
Aslam M, Huffaker RC, Rains DW, Rao KP (1979) Influence of light and ambient carbon dioxide concentration on nitrate assimilation by intact barley seedlings. Plant Physiol 63:1205–1209
Aslam M, Huffaker RC, Travis RL (1973) The interaction of respiration and photosynthesis in induction of nitrate reductase activity. Plant Physiol 52:137–141
Aslam M, Oaks A, Huffaker RC (1976) Effect of light and glucose on the induction of nitrate reductase and on the distribution of nitrate in etiolated barley leaves. Plant Physiol 58:588–591
Beevers L (1968) Growth regulator control of senescence in leaf discs of nasturtium (*Tropaeolum majus*). In: Wightman F, Setterfield G (eds) Biochemistry and physiology of plant growth substances. Runge Press, Ottawa, Canada, pp 1417–1435
Beevers L, Hageman RH (1969) Nitrate reduction in higher plants. Ann Rev Plant Physiol 20:495–522
Beevers L, Schrader LE, Flesher D, Hageman RH (1965) The role of light and nitrate in the induction of nitrate reductase in radish cotyledons and maize seedlings. Plant Physiol 40:691–698
Björkmann O (1968) Carboxydismutase activity in shade-adapted species of higher plants. Carnegie Inst Year Book 67:487–488
Brady CJ, Scott NS (1977) Chloroplast polyribosomes and synthesis of Fraction I protein in the developing wheat leaf. Aust J Plant Physiol 4:327–335
Burström H (1943) Photosynthesis and assimilation of nitrate by wheat leaves. Lantbrukshoegsk Ann 11:1–50
Canvin DT, Atkins CA (1974) Nitrate, nitrite and ammonia assimilation by leaves: effect of light, carbon dioxide and oxygen. Planta 116:207–224
Chantarotwong W, Huffaker RC, Miller BL, Granstedt RC (1976) *In vivo* nitrate reduction in relation to nitrate uptake, nitrate content, and *in vitro* nitrate reductase activity in intact barley seedlings. Plant Physiol 57:519–522
Chen K, Gray JC, Wildman SG (1975) Fraction I protein and the origin of polyploid wheats. Science 190:1304–1306
Choe HJ, Thimann KV (1975) The metabolism of oat leaves during senescence. II. The senescence of isolated chloroplasts. Plant Physiol 55:828–834
Chua N-H, Schmidt GW (1979) *In vitro* synthesis, transport, and assembly of ribulose 1,5-bisphosphate carboxylase subunits. In: Siegelman NW, Hind G (eds) Photosynthetic carbon assimilation. Plenum Press, New York, pp 325–347
Cooper TG, Filmer D, Wishnick M, Lane MD (1969) The active species of "$CO_2$" utilized by ribulose diphosphate carboxylase. J Biol Chem 244:1081–1083
Croy LI, Hageman RH (1970) Relationship of nitrate reductase activity to grain protein production in wheat. Crop Sci 10:280–285
Dalling MJ, Halloran JM, Wilson HJ (1975) The relation between nitrate reductase activity and grain nitrogen productivity in wheat. Aus J Agric Res 26:1–10
Davis BJ (1964) Disc electrophoresis. Ann NY Acad Sci 121:404–427
Deckard EL, Lambert RJ, Hageman RH (1973) Nitrate reductase activity in corn leaves as related to yields of grain and grain protein. Crop Sci 13:343–350
Dice JF, Goldberg AL (1975) A statistical analysis of the relationship between degradative rates and molecular weights of proteins. Arch Biochem Biophys 170:213–219
Dorner RW, Kahn A, Wildman SG (1957) The proteins of green leaves. VII. Synthesis and decay of the cytoplasmic proteins during the life of the tobacco leaf. J Biol Chem 229:945–952
Dudley K, Northcote DH (1979) Regulation of induction of phenylalanine ammonia-lyase in suspension cultures of *Phaseolus vulgaris*. Planta 146:433–440
Engelsma G, van Bruggen JMH (1971) Ethylene production and enzyme induction in excised plant tissues. Plant Physiol 48:94–96
Feller UK, Soong T-ST, Hageman RH (1977) Leaf proteolytic activities and senescence during grain development of field grown corn (*Zea mays* L.). Plant Physiol 59:290–294

Ferrari TE, Yoder OC, Filner P (1973) Anaerobic nitrite production by plant cells and tissues: evidence for two nitrate pools. Plant Physiol 51:423–431
Fischer RA, Kohn GD (1977) The relationship of grain yield to vegetative growth and post-flowering leaf area in the wheat crop under conditions of limited soil moisture. Aust J Agric Res 17:281–295
Friedrich JW, Huffaker RC (1980) Photosynthesis, leaf resistances, and ribulose-1,5-bisphosphate carboxylase degradation in senescing barley leaves. Plant Physiol 65:1103–1107
Friedrich JW, Schrader LE (1979) N deprivation in maize during grain-filling. II. Remobilization of $^{15}N$ and $^{35}S$ and the relationship between N and S accumulation. Agron J 71:466–472
Friedrich JW, Schrader LE, Nordheim EV (1979) N deprivation in maize during grain-filling. I. Accumulation of dry matter, nitrate-N, and sulfate-S. Agron J 71:461–465
Friend J, Reynolds SB, Aveyard MA (1971) Phenolic metabolism in potato tuber tissue after infection with *Phytophthors infestans*. Biochem J 124:29P
Gallagher LW, Soliman KM, Qualset CO, Huffaker RC, Rains DW (1980) Major gene control of nitrate reductase activity in common wheat. Crop Sci (In press)
Gooding LR, Roy H, Jagendorf AT (1973) Immunological identification of nascent subunits in wheat ribulose diphosphate carboxylase on ribosomes of both chloroplast and cytoplasmic origin. Arch Biochem Biophys 159:324–335
Gustafson G, Ryan CA (1976) Specificity of protein turnover in tomato leaves. J Biol Chem 251:7004–7010
Hadwiger LA, Schwochau ME (1967) Heat resistance responses – an induction hypothesis. Phytopathology 59:223–227
Hadwiger LA, Schwochau ME (1971) Specificity of deoxyribonucleic acid intercolating compounds in the control of phenylalanine ammonia lyase and pisatin levels. Plant Physiol 47:346–351
Hageman RH, Lambert RJ (1981) Recurrent divergent and mass selection in maize with physiological and biochemical traits: preliminary and projected application. In: Lyons JM, Valentine RC, Phillips DA, Rains DW, Huffaker RC (eds) Enhancing biological production of ammonia from atmospheric N and soil nitrate. Plenum Press, New York, in press
Hall NP, Keys AJ, Merrett MJ (1978) Ribulose-1,5-diphosphate carboxylase protein during flag leaf senescence. J Exp Bot 29:31–37
Hallmark WB, Huffaker RC (1978) The influence of ambient nitrate, temperature, and light on nitrate assimilation in Sudangrass seedlings. Physiol Plant 44:147–152
Hampp R, De Filippis LF (1980) Plastid protease activity and prolamellar body transformation during greening. Plant Physiol 65:663–668
Hardy RWF, Burns RC, Herbert RR, Holsten RD, Jackson EK (1971) In: Lie TA, Muldner EG (eds) Biological nitrogen fixation in natural and agricultural habitats. Nijhoff, The Hague, Netherlands, 561 pp
Hartley MR, Wheeler A, Ellis RJ (1975) Protein synthesis in chloroplasts. V. Translation of messenger RNA for the large subunit of Fraction I protein in a heterologous cell-free system. J Mol Biol 91:67–77
Heber U (1974) Metabolite exchange between chloroplasts and cytoplasm. Annu Rev Plant Physiol 25:393–421
Heimer YH, Filner P (1971) Regulation of the nitrate assimilation pathway in cultured tobacco cells. III. The nitrate uptake system. Biochim Biophys Acta 230:362–372
Helsel DB, Frey KJ (1978) Grain yield variations in oats associated with differences in leaf area duration among oat lines. Crop Sci 18:765–769
Highfield PE, Ellis RJ (1978) Synthesis and transport of the small subunit of chloroplast ribulose bisphosphate carboxylase. Nature (London) 271:420–424
Huffaker RC, Miller BL (1978) Reutilization of ribulose bisphosphate carboxylase. In: Siegelman HW, Hind E (eds) Photosynthetic carbon assimilation. Plenum Press, New York, pp 307–324
Huffaker RC, Obendorf RL, Keller CN, Kleinkopf GE (1966) Effects of light intensity on photosynthetic carboxylative phase enzymes and chlorophyll synthesis in greening leaves of *Hordeum vulgare* L. Plant Physiol 41:913–918

Huffaker RC, Radin T, Kleinkopf GE, Cox EL (1970) Effects of mild water stress on enzymes of nitrate assimilation and of the carboxylative phase of photosynthesis in barley. Crop Sci 10:471–474

Hyodo H, Yang SF (1971) Ethylene-enhanced synthesis of phenylalanine ammonia-lyase in pea seedlings. Plant Physiol 47:765–770

Hyodo H, Kuroda H, Yang SF (1978) Induction of phenylalanine ammonia lyase and increase in phenolics in lettuce leaves in relation to the development of russet spotting caused by ethylene. Plant Physiol 62:31–35

Jensen RG, Bahr JT (1977) Ribulose 1,5-bisphosphate carboxylase-oxygenase. Annu Rev Plant Physiol 28:379–400

Jeschke WD (1976) Ionic relations of leaf cells. In: Lüttge U, Pitman MG (eds) Transport in Plants. II. Part B. Tissues and Organs. Springer, Berlin Heidelberg New York, pp 161–194

Jolly SO, Tolbert NE (1978) NADH-nitrate reductase inhibitor from soybean leaves. Plant Physiol 62:197–203

Jones RW, Sheard RW (1978) Accumulation and stability of nitrite in intact aerial leaves. Plant Sci Lett 11:285

Kannangara CG, Woolhouse HW (1968) Changes in the enzyme activity of soluble protein fractions in the course of foliar senescence in *Perilla frutescens* (L.) Britt. New Phytol 67:533–542

Kawashima N, Mitake T (1969) Studies on protein metabolism in higher plants. VI. Changes in ribulose diphosphate carboxylase activity and Fraction I protein content in tobacco leaves with age. Agric Biol Chem 33:539–543

Kawashima N, Wildman SG (1970) Fraction I protein. Annu Rev Plant Physiol 21:325–358

Kawashima N, Imai A, Tamaki E (1967) Studies on protein metabolism in higher plant leaves. III. Changes in the soluble protein components with leaf growth. Plant Cell Physiol 8:447–458

Kleinkopf GE, Huffaker RC, Matheson A (1970) Light induced *de novo* synthesis of ribulose 1,5-diphosphate carboxylase in greening leaves of barley. Plant Physiol 46:416–418

Klepper LA (1975) Nitrate assimilation enzymes and seed protein in wheat. Proc Second Int Winter Wheat Conf, Zagreb, Yugoslavia, pp 334–340

Klepper L, Flesher D, Hageman RH (1971) Generation of reduced nicotinamide adenine dinucleotide for nitrate reduction in green leaves. Plant Physiol 48:580–590

Kung SD, Rhodes PR (1978) Interaction of chloroplast and nuclear genomes in regulating RuBP carboxylase activity. In: Siegelman HW, Hind G (eds) Photosynthetic carbon assimilation. Plenum Press, New York, pp 307–324

Kung SD, Sakano K, Wildman SG (1974) Multiple peptide composition of the large and small subunits of *Nicotiana tabaccum*. Fraction I. Protein ascertained by fingerprinting and electrofocusing. Biochim Biophys Acta 365:138–147

Laing WA, Ogren WL, Hageman RH (1974) Regulation of soybean net photosynthetic $CO_2$ fixation by the interaction of $CO_2$, $O_2$ and ribulose 1,5-diphosphate carboxylase. Plant Physiol 54:678–685

Lupton FGH, Oliver RH, Ruckenbauer P (1974) An analysis of the factors determining yield in crosses between semidwarf and taller wheat varieties. J Agric Sci 82:481–496

Martin C, Thimann KV (1972) The role of protein synthesis in the senescence of leaves. I. The formation of protein. Plant Physiol 49:64–71

Mattas RE, Pauli AW (1965) Trends in nitrate reduction and nitrogen fractions in young corn (*Zea mays* L.) plants during heat and moisture stress. Crop Sci 5:181–184

McFarland D, Ryan CA (1974) Proteinase inhibitor-inducing factor in plant leaves. Plant Physiol 54:706–708

Medina E (1971) Effect of nitrogen supply and light intensity during growth on the photosynthetic capacity and carboxydismutase activity of leaves of *Atriplex patula* spp. *hastata*. Carnegie Inst Year Book 70:551–559

Melville JC, Scandalios JG (1972) Maize endopeptidase: genetic control, chemical characterization, relationship to an endogenous trypsin inhibitor. Biochem Gen 7:15–31

Miflin BJ (1974) Nitrite reduction in leaves: studies on isolated chloroplasts. Planta 116:187–196

Mikola J, Kolehmainen L (1972) Localization and activity of various peptidases in germinating barley. Planta 104:167–177
Mondal MH, Brun WA, Brenner ML (1978) Effects of sink removal on photosynthesis and senescence in leaves of soybean (*Glycine max* L.) plants. Plant Physiol 61:394–397
Morilla CA, Boyer JS, Hageman RH (1973) Nitrate reductase activity and polyribosomal content of corn (*Zea mays* L.) having low leaf water potentials. Plant Physiol 51:817–824
Neyra CA, Hageman RH (1974) Dependence of nitrite reduction on electron transport in chloroplasts. Plant Physiol 54:480–483
Nishimura M, Akazawa T (1974) Studies on spinach leaf ribulose-bisphosphate carboxylase. Carboxylase and oxygenase reaction examined by immunochemical methods. Biochemistry 13:2277–2281
Oh TY, Warner RL, Kleinhofs A (1980) Effect of nitrate reductase-deficiency upon growth, yield, and protein in barley. Crop Sci 20:487–490
Onwueme IC, Laude HM, Huffaker RC (1971) Nitrate reductase activity in relation to heat stress in barley seedlings. Crop Sci 11:195–200
Ornstein L (1964) Disc electrophoresis. I. Background and theory. Ann New York Acad Sci 121:321–349
Patterson BD, Smillie RM (1971) Developmental changes in ribosomal ribonucleic acid and Fraction I protein in wheat leaves. Plant Physiol 47:196–198
Peoples MB (1980) Investigations of the degradative processes involved in the turnover of ribulose-1,5-bisphosphate carboxylase in wheat leaves. PhD thesis, Univ Melbourne
Peterson LW, Huffaker RC (1975) Loss of ribulose 1,5-diphosphate carboxylase and increase in proteolytic activity during senescence of detached primary barley leaves. Plant Physiol 55:1009–1015
Peterson LW, Kleinkopf GE, Huffaker RC (1973) Evidence for lack of turnover of ribulose 1,5-diphosphate carboxylase in barley leaves. Plant Physiol 51:1042–1045
Rao KP, Rains DW, Qualset CO, Huffaker RC (1977) Nitrogen nutrition and grain protein in two spring wheat genotypes differing in nitrate reductase activity. Crop Sci 17:283–286
Ritenaur GL, Joy KW, Bunning J, Hageman RH (1967) Intracellular localization of nitrate reductase, nitrite reductase and glutamic dehydrogenase in green leaf tissue. Plant Physiol 42:233–237
Ryan CA (1968) Synthesis of chymotrypsin inhibitor I protein in potato leaflets induced by detachment. Plant Physiol 43:1859–1865
Ryan CA (1974) Assay and biochemical properties of the proteinase inhibitor-inducing factor, a wound hormone. Plant Physiol 54:328–332
Ryan CA, Huisman W (1970) The regulation of synthesis and storage of chymotrypsin inhibitor I in leaves of potato and tomato plants. Plant Physiol 45:484–489
Sawney SK, Naik MS, Nicholas DJD (1978a) Regulation of NADH supply for nitrate reduction in green plants via photosynthesis and mitochondrial respiration. Biochem Biophys Res Commun 81:1209–1216
Sawney SK, Naik MS, Nicholas DJD (1978b) Regulation of nitrate reduction by light, ATP and mitochondrial respiration in wheat leaves. Nature (London) 272:647–648
Scheibe R, Beck E (1979) On the mechanism of activation by light of the NADP-dependent malate dehydrogenases in spinach chloroplasts. Plant Physiol 64:744–748
Schrader LE, Beevers L, Hageman RH (1967) Differential effects of chloramphenicol on the induction of nitrate and nitrite reductase in green leaf tissue. Biochem Biophys Res Commun 26:14–17
Schrader LE, Ritenour GL, Eilrich GL, Hageman RH (1968) Some characteristics of nitrate reductase from higher plants. Plant Physiol 43:930–940
Schroder J, Betz B, Halbrock K (1977) Messenger RNA controlled increase of phenylalanine ammonia lyase activity in parsley. Light-independent induction by dilution of cell suspension cultures into water. Plant Physiol 60:440–445
Shaner DL, Boyer JS (1976) Nitrate reductase activity in maize (*Zea mays* L.) leaves. II. Regulation by nitrate flux. Plant Physiol 58:499–504
Shumway LK, Yang V, Ryan CA (1976) Evidence for presence of proteinase inhibitor I in vacuolar protein bodies of plant cells. Plant 129:161–165
Simpson E (1978) Biochemical and genetic studies of the synthesis and degradation of

RuBP carboxylase. In: Siegelman HW, Hind G (eds) Photosynthetic carbon assimilation. Plenum Press, New York, pp 113–125

Sinclair TR, deWit CT (1976) Analysis of the carbon and nitrogen limitations to soybean yield. Agron J 68:319–324

Smith MA, Criddle RS, Peterson LW, Huffaker RC (1974) Synthesis and assembly of ribulose bisphosphate carboxylase enzyme during greening of barley plants. Arch Biochem Biophys 165:494–504

Solomonson LP (1975) Purification of NADH-nitrate reductase by affinity chromatography. Plant Physiol 56:853–855

Soong T-ST, Feller UK, Hageman RH (1977) Changes in activities of proteolytic enzymes during senescence of detached corn (*Zea mays* L.) leaves as function of physiological age. Plant Physiol Suppl 59:112

Stocking CR, Larson S (1969) A chloroplast cytoplasmic shuttle and the reduction of extraplastic NAD. Biochem Biophys Res Commun 38:278–282

Thimann KV, Satler SO (1979) Relation between senescence and stomatal opening: senescence in darkness. Proc Natl Acad Sci USA 76:2770–2773

Thomas H (1974) Regulation of alanine aminotransferase in leaves of *Lolium temulentum* during senescence. Z Pflanzenphysiol 74:208–218

Thomas H, Huffaker RC (1980) Hydrolysis of radioactively labeled ribulose-1,5-bisphosphate carboxylase by an endopeptidase from the primary leaf of barley seedlings. Plant Sci Lett 20:251–262

Thomas H, Stoddart JL (1975) Separation of chlorophyll degradation from other senescence processes in leaves of a mutant genotype of meadow fescue (*Festuca pratensis* L.). Plant Physiol 56:438–441

Thomas H, Stoddart JL (1980) Leaf senescence. Annu Rev Plant Physiol 31:83–111

Ting IP, Hanscom Z (1977) Induction of acid metabolism in *Portulacaria afrea*. Plant Physiol 59:511–514

Tischner R, Hüttermann A (1978) Light-mediated activation of nitrate reductase in synchronous *Chlorella*. Plant Physiol 62:284–286

Travis RL, Jordan WR, Huffaker RC (1969) Evidence for an inactivating system of nitrate reductase in *Hordeum vulgare* L. during darkness that requires protein synthesis. Plant Physiol 44:1150–1156

Travis RL, Huffaker RC, Key JL (1970) Light-induced development of polyribosomes and the induction of nitrate reductase in corn leaves. Plant Physiol 46:800–805

van Loon LC, Haverkort AJ (1977) No increase in protease activity during senescence of oat leaves. Plant Physiol Suppl 59:113

Walker-Simmons M, Ryan CA (1977a) Immunological identification of proteinase inhibitors I and II in isolated tomato leaf vacuoles. Plant Physiol 60:61–63

Walker-Simmons M, Ryan CA (1977b) Wound-induced accumulation of trypsin inhibitor activities in plant leaves. Plant Physiol 59:437–439

Wallsgrove RM, Lea PJ, Miflin BJ (1979) Distribution of the enzymes of nitrogen assimulation within the pea leaf cell. Plant Physiol 63:232–236

Warner RL, Hageman RH, Dudley JW, Lambert RJ (1969) Inheritance of nitrate reductase activity in *Zea mays* L. Proc Natl Acad Sci USA 62:785–792

Warner RL, Lin CJ, Kleinhofs A (1977) Nitrate reductase-deficient mutants in barley. Nature (London) 269:406–407

Wellman E, Schopfer P (1975) Phytochrome-mediated *de novo* synthesis of phenylalanine ammonia-lyase in cell suspension cultures of parsley. Plant Physiol 55:822–827

Whittaker RH, Feeny PP (1971) Allelochemics: chemical interactions between species. Science 171:757–770

Wittenbach VA (1977) Induced senescence of intact wheat seedlings and its reversibility. Plant Physiol 59:1039–1042

Wittenbach VA (1978) Breakdown of ribulose bisphosphate carboxylase and change in proteolytic activity during dark-induced senescence of wheat seedlings. Plant Physiol 62:604–608

Wittenbach VA (1979) Ribulose bisphosphate carboxylase and proteolytic activity in wheat leaves from anthesis through senescence. Plant Physiol 64:884–887

Wittenbach VA, Ackerson RC, Giaquinta RT, Hebert RR (1980) Changes in photosynthesis, ribulose bisphosphate carboxylase, proteolytic activity, and ultrastructure of soybean leaves during senescence. Crop Sci 20:225–231
Wray JL, Filner P (1970) Structural and functional relationship of enzyme activities induced by nitrate in barley. Biochem J 119:715–725
Zielke HR, Filner P (1971) Synthesis and turnover of nitrate reductase induced by nitrate in cultured tobacco cells. J Biol Chem 246:1772–1779
Zucker M (1971) Induction of phenylalanine ammonia-lyase in *Xanthium* leaf discs. Increased inactivation in darkness. Plant Physiol 47:442–444
Zucker M (1972) Light and enzymes. Annu Rev Plant Physiol 23:133–156

# 11 Microtubule Proteins and P-Proteins

D.D. SABNIS and J.W. HART

## 1 General Introduction

There are no known functional or structural relationships between tubulins and P-proteins, even though both groups of proteins share the general characteristic of appearing as linear macromolecular assemblies. After actin, tubulin is possibly the second most commonly occurring protein of the non-specialized eukaryote cell; P-protein on the other hand is the major intracellular polymeric component of a highly specialized plant cell, the sieve tube. Each system offers a model situation, regulated temporally and spatially, for studying the assembly of a relatively few distinct components into highly ordered cellular structures.

In this article, the two systems are treated separately. Because two distinct areas of investigation are being described, considerable selectivity has been used in literature citation, particularly in the case of the large number of publications dealing with tubulin; much use is made of recent reviews, and articles are quoted, not so much to assign priority as to allow access to the pertinent literature. The protein components of another group of fibrillar structures, microfilaments, have been dealt with in Volume 7 of this series.

## 2 Tubulin and Associated Proteins

### 2.1 Introduction

Tubulin is recognized as the major protein component of ciliary and cytoplasmic microtubules. Accounts of microtubule morphology and behaviour are given in a recent monograph (DUSTIN 1978), in several books (SLEIGH 1974, INOUÉ and STEPHENS 1975, ROST and GIFFORD 1977), in the proceedings of recent symposia (SOIFER 1975, BORGERS and DEBRABANDER 1975, GOLDMAN et al. 1976) and in reviews (SNYDER and MCINTOSH 1976, STEPHENS and EDDS 1976, MOHRI 1976, KIRSCHNER 1978). A few articles pertain particularly to the structure and functioning of microtubules in plant cells (PICKETT-HEAPS 1974, HEPLER 1976, HEPLER and PALEVITZ 1974, HART and SABNIS 1976c, 1977, FILNER and YADAV 1979, see also ROST and GIFFORD 1977).

Microtubules (MT's) are ubiquitous cellular organelles consisting of regular helical assemblies of tubulin subunits and various other proteins. In certain systems, these non-tubulin components can be seen as appendages projecting from the MT and even linking MT's to each other or to other cellular structures.

MT's seem to be involved in several types of movement, including extracellular flagellar or ciliary motion and intracellular movements as in mitosis or vesicle transport. Both in these processes, and in their involvement in the generation and maintenance of cell shape, MT's may act in a cytoskeletal role rather than as a source of motive force. Throughout cellular development, different populations of MT's are involved in a variety of processes, their appearance and assembly being strictly regulated in time and space. The concept of the MT as a labile structure must be qualified by several lines of evidence which indicate that there are different MT classes: stable (e.g., flagellar MT's) or labile (e.g., cytoplasmic MT's of animal cells) depending upon their ability to respond reversibly to certain depolymerizing conditions or drugs; furthermore, MT's can occur individually, as in the cytoplasm, in loosely organized associations as in the mitotic spindle, or in highly organized complexes as in the axoneme.

The protein tubulin is of ancient origin. Studies of immunological reactivity, peptide mapping and tests of co-polymerization of tubulin from different sources have indicated that it is a highly conserved moiety. However, recent evidence suggests that there is indeed a degree of heterogeneity among tubulins. Furthermore, the accessory proteins, involved in the assembly and functioning of the MT, vary among MT populations.

## 2.2 Tubulin

### 2.2.1 Methodology

Early investigations were hampered by the lack of a method for recognition of a soluble MT component even when such relatively rich sources as flagella or the mitotic apparatus were examined. These difficulties were overcome to some extent by the finding that tubulin was the direct target for the antimitotic drug, colchicine (ADELMAN et al. 1968) and colchicine binding (WILSON et al. 1974) was accepted as a routine assay for tubulin. Tubulin has also been isolated by affinity chromatography in which colchicine is chemically coupled to an inert support material (SCHMITT and LITTAUER 1974, SANDOVAL and CUATRECASAS 1976a, b). However, colchicine-binding activity in vitro has a rapid decay rate; furthermore, its applicability to plant material may be limited by the great stability of plant MT's and a low binding affinity for colchicine (HART and SABNIS 1976a, b).

The development of techniques for the reversible polymerization in vitro of MT's (WEISENBERG 1972) gave rise to new methods of isolating tubulin, based on repeated cycles of polymerization/depolymerization, coupled to centrifugation procedures to separate assembled MT's and non-specific aggregates. A variety of factors influence MT assembly (Sect. 2.4) and this technique is practised in the presence (SHELANSKI et al. 1973) or absence (BORISY et al. 1974) of specific stabilizing agents. Such procedures are limited to preparations sufficiently rich in tubulin to allow polymerization (CLAYTON et al. 1979). Radiolabelling coupled to polymerization has been used to isolate fungal tubulin (SHEIR-

NEISS et al. 1976, WATER and KLEINSMITH 1976) and putative higher plant tubulin (SLABAS et al. 1980); radiolabelled tubulin was isolated by co-polymerization with carrier MT's from brain.

Since tubulin is a highly conserved protein, it is only weakly antigenic. However, immunofluorescent techniques have been powerful tools in visualizing and locating MT proteins in situ (LLOYD et al. 1979).

Use of a particular purification procedure such as polymerization is not in itself enough to guarantee the purity or homogeneity of a tubulin preparation. A routinely employed check involves electrophoresis on urea-sodium dodecyl sulphate polyacrylamide gels (BRYAN 1974). A particular difficulty in dealing with a multicomponent protein system is that of determining whether certain proteins are present because of a definite functional association, or merely as contaminants. For MT protein preparations which show polymerizability, certain standards have been proposed (BULINSKI and BORISY 1979): the system should not require any heterologous polymerization factors (Sect. 2.4); the polymerization should be reversible and should be carried through several cycles; tubulin and associated proteins should maintain a quantitative relationship during such cycling.

### 2.2.2 Distribution

MT's can vary markedly in number both among organisms and with the stage of cellular development (ADELMAN et al. 1968), with the typical interphase plant cell exhibitng a very sparse population (HART and SABNIS 1977). However, the MT's actually observed in a particular cell may represent only 25% of the total extractable tubulin (PIPELEERS et al. 1977, ANDERSON 1979). MT's have traditionally been envisaged (particularly in animal cells) as existing in a dynamic equilibrium with a soluble pool of tubulin subunits. There are many examples which suggest that upon depolymerization of MT's the subunits return to a pool from which new MT structures may be assembled (BLOODGOOD 1974). In addition to occurring in a soluble pool or as MT's, tubulin has been identified in membranes of brain cells (BHATTACHARYYA and WOLFF 1976).

Tubulins isolated from a variety of sources are very similar in their properties, including amino acid composition and peptide finger-prints (STEPHENS and EDDS 1976). Additionally tubulins from such disparate sources as *Chlamydomonas* flagella and chick brain have been shown to co-assemble in vitro (BINDER et al. 1975). Membrane-tubulin isolated from brain copolymerizes with cytoplasmic tubulin (BHATTACHARYYA and WOLFF 1976).

### 2.2.3 Physical and Chemical Properties

#### *2.2.3.1 The Subunit*

Analysis of tubulins by sodium dodecylsulphate polyacrylamide gel electrophoresis (SDS-PAGE) shows two closely spaced but distinct protein bands corresponding to molecular weight (MW) $\sim$55,000 and referred to as the $\alpha$- and $\beta$-polypeptide chains of tubulin. The greater mobility of the $\beta$-chain is thought

**Table 1.** Amino acid composition of samples of sea-urchin tubulin. (LUDUENA and WOODWARD 1973)

| Amino acid | mol/55,000 MW | | Amino acid | mol/55,000 MW | |
|---|---|---|---|---|---|
| | α | β | | α | β |
| Lys | 24.3 | 21.1 | Gly | 40.7 | 32.4 |
| His | 14.2 | 11.3 | Ala | 40.5 | 34.4 |
| Arg | 26.1 | 23.4 | Val | 33.1 | 33.1 |
| Cys[a] | 9.8 | 6.9 | Met | 9.8 | 16.6 |
| Asp | 48.4 | 54.5 | Ile | 24.6 | 18.9 |
| Thr | 29.4 | 31.5 | Leu | 35.5 | 36.4 |
| Ser | 20.9 | 24.2 | Tyr | 17.6 | 16.8 |
| Glu | 69.0 | 69.2 | Phe | 21.1 | 23.6 |
| Pro | 25.3 | 27.7 | Trp | ND | ND |

[a] Cysteine determined as carboxymethylcysteine. ND not determined.
Figures represent the average of two analyses.

to be due to charge differences on the molecule. The distinction between the α- and β-chains has been confirmed by comparisons of amino acid composition, peptide maps and immunological reactivities (KIRSCHNER 1978, DUSTIN 1978).

The 6S tubulin subunit involved in MT assembly is a dimeric molecule with a MW of 110,000 to 120,000. Evidence that this subunit is an αβ-heterodimer comes from studies of trypsin-produced fragments (KIRSCHNER 1978) and from findings that the heterodimer is the most common form to appear in the presence of a cross-linking reagent (LUDUENA et al. 1977). The 8 nm dimer is responsible for the repeat periods of 4 nm and 8 nm seen in diffraction analysis of the MT (AMOS et al. 1976). Its shape has been described as approximating a prolate ellipsoid with an axial ratio of 5 to 7 (BRYAN 1974). A model derived from diffraction studies of MT structure depicts the dimeric subunit as elongated and bi-lobed (ERICKSON 1976). The tertiary structure of the dimer has been analyzed through circular dichroism and exhibits 48% random coil, 22% α-helix and 30% β-sheet (VENTILLA et al. 1972). The number of exposed sulphydryl residues varies (4 to 7 per 55,000 molecular weight sub-unit), and such changes may have a role in subunit assembly (MELLON and REBHUN 1976).

Tubulin is an acidic protein (Table 1, from LUDUENA and WOODWARD 1973) but has been only partially sequenced at the N-terminal (LUDUENA and WOODWARD 1975) and C-terminal regions (LU and ELZINGA 1978). Tubulins from such disparate sources as chick brain and sea urchin sperm flagella (Table 2, from LUDUENA and WOODWARD 1975) show, in the first 25 N-terminal amino acids, no differences in the α-tubulins and only small differences in the β-tubulins, suggesting a high degree of conservation of the molecule during evolution. However, detailed study of tubulins from sea urchin show that α- and β-chains to be distinct (STEPHENS 1978): α-chains typically contain 12–13 His, 8–10 Met, 8–10 Cys, 39–41 Ala; β-chains contain 10–11 His, 15–16 Met, 7–8 Cys, 32–36 Ala.

**Table 2.** Amino acid sequences of $NH_2$-terminal regions of α- and β-tubulins from chick brain and outer doublet microtubules of sea-urchin sperm. (LUDUENA and WOODWARD 1975)

| | |
|---|---|
| | 1 2 3 4 5 6 7 8 9 10 |
| Chick brain α-tubulin | Met-Arg-Glx-Ser?Ile -Ser?Ile -His-Val-Thr- |
| Sea-urchin α-tubulin | Met-Arg-Glu-Ser?Ile -Ser?Ile -His-Val-Thr- |
| Chick brain β-tubulin | Met-Arg-Glu-Ile -Val-His-Ile -Gln-Ala-Thr- |
| Sea-urchin β-tubulin | Met-Arg-Glu-Ile -Val-His-Met-Glx-Ala-Thr- |
| | 11 12 13 14 15 16 17 18 19 20 |
| Chick brain α-tubulin | Gln-Ala-Thr-Val-Gln-Ile-Thr-Asx-Ala-Ser? |
| Sea-urchin α-tubulin | Glx-Ala-Thr-Val-Glx-Ile-Thr-Asx-Ala-Ser? |
| Chick brain β-tubulin | Gln-Ser-Thr-Asx-Gln-Ile-Thr-Ala - ? -Phe- |
| Sea-urchin β-tubulin | Glx-Ser-Thr-Asx-Glx-Ile-Thr-Ala - ? -Phe- |
| | 21 22 23 24 25 |
| Chick brain α-tubulin | ? -Glx-Leu-Tyr-Ser? |
| Sea-urchin α-tubulin | ? -Glx-Leu-Tyr-Ala? |
| Chick brain β-tubulin | Trp?Glx-Val-Ile -Ser? |
| Sea-urchin β-tubulin | ? - ? -Val-Ile -Ser? |

### *2.2.3.2 Microheterogeneity*

Despite the general homogeneity shown in the properties of tubulins from different sources, consideration of such features as variation in MT protofilament number (DUSTIN 1978), the restricted attachment of associated proteins to certain tubulin subunits within the MT (KIM et al. 1979), and the differing stability of various classes of MT, suggests diversity at some level in the tubulin molecule. There is a certain amount of direct evidence for heterogeneity among tubulins.

Electrofocussing of *Chlamydomonas* flagellar outer doublet tubulin resulted in 4 to 5 protein bands (WITMAN et al. 1972). Electrophoretic differences in the α-chains of cytoplasmic and flagellar tubulin from different organisms (OLMSTED et al. 1971) or even the same organism (BIBRING et al. 1976) have been observed. The α-chain of brain tubulin has been resolved into two peaks, α, and $\alpha_2$, on a hydroxyapatite column (LU and ELZINGA 1977). An organ-specific, distinct β-chain has been reported in *Drosophila* testis (KEMPHUES et al. 1979). Multiple forms of electrophoretically distinct α- and β-tubulins from rat brain have been distinguished (MAROTTA et al. 1978, FORGUE and DAHL 1979). Immunologically different pools of tubulin, correlated with different stages of cellular development, have been observed in the amoeba, *Naegleria* (FULTON and SIMPSON 1976). Detailed analysis of amino acid composition and peptide maps of tubulins from a variety of MT types in sea urchin (STEPHENS 1978) indicated that despite an overall similarity between various types of α- and β-chain, there were significant differences between homologous chains; in general, it would seem that the tubulin from stable (flagellar) MT's had a higher basic amino acid content than that from labile (cytoplasmic) MT. It was concluded that cytoplasmic, ciliary and flagellar MT's were constructed from subunits drawn from different soluble pools. The tubulins may be analogous to the immunoglobulins, with constant and variable regions, the latter lending specificity to location or functioning (STEPHENS 1978).

### 2.2.3.3 Interaction with Ligands

*Nucleotides.* Early studies established that two molecules of guanosine nucleotides (GTP or GDP) were bound per tubulin dimer. The roles of these nucleotides during MT assembly and functioning are still controversial (see KIRSCHNER 1978). In summary, one site (the N-site) binds GTP alone in a non-exchangeable fashion. The other site (the E-site) exchangeably binds GTP or GDP with a dissociation constant of $10^{-6}$–$10^{-7}$ $1 mol^{-1}$. Thus, the tubulin dimer can exist in the forms N-GTP/E-GTP or N-GTP/E-GDP. In this latter state, the subunit is inactive and will not assemble. Nucleotides like GTP or ATP promote assembly, by E-GDP being either exchanged for GTP or being phosphorylated by an associated nucleoside diphosphate kinase. Early reports indicated that E-GTP hydrolysis was necessary for assembly, since non-hydrolyzable analogues of GTP inhibited assembly. However, this is thought to have been due to the continued tight binding of GPD at the E site (KIRSCHNER 1978) and subsequent experiments have successfully accomplished assembly in the presence of the analogues. GTP at the E-site *is*, nevertheless, hydrolyzed during assembly, although whether this GTPase activity is (DAVID-PFEUTY et al. 1979) or is not (KIRSCHNER 1978) necessary for the process is still controversial. Thus, the E-site is involved in GTP/GDP transformations. The N-site is remarkably stable; early reports of its involvement in transphosphorylation seem incorrect (ZEEBURG and CAPLOW 1978) and this site is considered to play some structural role (KIRSCHNER 1978).

*Ions.* The initial success of tubulin polymerization in vitro (WEISENBERG 1972) was correlated with the absence of calcium from the medium. Subsequent work (ROSENFELD et al. 1976) showed that calcium inhibition of MT assembly was potentiated by magnesium, e.g., at 5 mM $Mg^{2+}$ inhibition can be obtained by micromolar levels of calcium. There are many reports (summarized in DUSTIN 1978, KIRSCHNER 1978) that calcium can regulate both MT appearance and cellular processes involving MT's. Tubulin preparations have also been shown to bind both calcium and magnesium directly. Additionally, calcium may exert effects on MT's by acting indirectly through other protein systems, e.g., a Ca-dependent protease specific for proteins assisting MT assembly has been reported (SANDOVAL and WEBER 1978). Thus, in vitro and in vivo studies are in agreement that, directly or indirectly, calcium is a prime regulator of MT behaviour.

*Drugs.* Naturally occurring drugs and synthetic chemicals have been extremely important tools in investigating MT structure and function and in many cases have been shown to interact directly with tubulin. In this section, only a brief overview of the situation will be attempted. The most well-studied group of chemicals in this respect are the anti-mitotic drugs including colchicine, vinblastine, podophyllotoxin and griseofulvin (WILSON et al. 1974). The tubulin dimer hydrophobically binds one molecule of colchicine (or colcemid). The reversible reaction involves a configurational change in the subunit (GARLAND 1978) and is thought to occur in the vicinity of cysteine residues (SCHMITT and KRAM 1978) which may explain the correlation between thiol oxidation and in vitro decay of colchicine-binding activity (GILLESPIE 1975). The affinity of animal tubulin for colchicine ($2 \times 10^6\ \mathrm{l\,mol^{-1}}$) is not reflected in plant preparations, necessitating the use of the inactive lumicolchicine analogue in control experiments (HART and SABNIS 1976b). Competition experiments have indicated that podophyllotoxin binds to the same site as colchicine (WILSON et al. 1974). Colchicine-binding activity is stabilized by the alkaloid vinblastine (BRODIE et al. 1979). Vinblastine itself binds to distinct sites: in vivo it induces paracrystals of tubulin to form; in vitro it precipitates tubulin aggregates, by what is thought to be action at a non-specific site (WILSON et al. 1974). Griseofulvin also stabilizes colchicine-binding activity (WILSON et al. 1974) and exerts its anti-mitotic effect by binding to a specific tubulin site (WEHLAND et al. 1977). The effects of herbicides such as trifluvalin, oryzolin, carbamates and amiprophosmethyl, on MT-processes are reviewed by FILNER and YADAV (1979): the anti-MT action in many cases is correlated with an ability to bind to tubulin. However, in such poorly characterized situations, an ability to bind to tubulin need not necessarily indicate the mechanism of cellular action; for example, the respiratory poison, rotenone, has been shown also to bind to tubulin and to inhibit MT-assembly (MARSHALL and HIMES 1978).

#### *2.2.3.4 Other Constituents of Tubulin*

*Polysaccharide.* Conflicting evidence concerning the presence of carbohydrate material in tubulin preparations has resulted from difficulties associated with purification. An early report claimed a significant proportion of oligosaccharide and amino sugars was associated with each tubulin dimer (MARGOLIS et al. 1972). It has been suggested that the "clear zone" consistently seen in electron micrographs of MT's may be a mucopolysaccharide shell (DUSTIN 1978). An improved method of purification resulted in a tubulin preparation with only 1.2 molecules of neutral sugar per dimer (EIPPER 1972). However, the situation may be complicated by the source or type of tubulin. For example, synaptosomal tubulin but not axoplasmic (FEIT and SHELANKSI 1975), and ciliary membrane tubulin but not axonemal (STEPHENS 1977), have been claimed to be glycoproteins. This has led to the suggestion (STEPHENS 1978) that "membrane (associated?) tubulin" is characterized by the presence of a polysaccharide component.

*Lipid.* The evidence for the presence of lipid associated with tubulin is less substantial and indirect (DUSTIN 1978, MOHRI 1976). Tubulin preparations have been reported to contain a mixture of phospholipids and lecithin (DALEO et al. 1974, EIPPER 1975). In addition, diglyceride kinase activity has been identified in tubulin preparations. Treatment of tubulin with phospholipase A removes the ability of the subunits to assemble. MT proteins have been reported to interact quantitatively with phospholipid vesicles (CARON and BERLIN 1979).

## 2.3 Proteins Associated with Tubulin

### 2.3.1 Introduction

In many cells, MT's are decorated with fine filaments or side-arms which may or may not connect them to other MT, membranes or organelles (McINTOSH 1974). In cilia and flagella, side-arms extending from the outer doublets or central pair are complex (WARNER 1976, SATIR and OJAKIAN 1979): isozymes of the ATPase dynein have been associated with some of these axonemal structures (see below). Similar lateral bridges have been observed between mitotic spindle MT's (HEPLER et al. 1970) and may be related to the movement of chromosomes (McINTOSH et al. 1969, BAJER and MOLÈ-BAJER 1975). Other lateral extensions may be important in the MT-associated movements of various particles and organelles (MURPHY 1975). Cross-linked MT's have been observed in higher plant cells (BURGESS (1970) and bridges may extend between MT's and plasma membrane (ROBARDS and KIDWAI 1972). In general, these side-arms may be important in the assembly, stabilization or functioning of MT's, and their structure and activity may vary depending on MT location and cell type.

Purified tubulin is often accompanied on gel electrophoresis columns by several proteins of higher molecular weight (55,000 to 350,000) (SANDOVAL and CUATRECASAS 1976a, SLOBODA et al. 1976a). Some of these MT-associated proteins (MAP's) have been reported to stimulate MT assembly (SLOBODA et al. 1976b), to stabilize assembled tubulin (SLOBODA and ROSENBAUM 1979), or to possess enzyme activity (see below). Consequently much attention has recently been devoted to identifying and characterizing MAP's, ascribing specific MAP's to characteristic side-arms and cross-links, and to uncovering the interaction between MAP's and tubulin in MT assembly and function.

Although many cellular proteins bind to MT's, only a few proteins, termed HMW (for high molecular weight 250,000 to 350,000) and LMW (for low molecular weight 30,000 to 35,000), exhibit a high affinity for tubulin (BERKOWITZ et al. 1977). This raises the problem of distinguishing protein contaminants from proteins genuinely associated with MT's in vivo. BERKOWITZ et al. (1977) have characterized the co-purifying proteins of brain MT's and have suggested a useful differentiation. They divide the proteins into two groups designated Q-MAP's (HMW and LMW) or NQ-MAP's, according to whether the proteins remain in constant quantitative ratio to tubulin during several purification cycles or become removed during purification. BULINSKY and BORISY (1979) add two further criteria for the identification of MAP's; namely, stimulation of MT assembly when added to pure tubulin and co-sedimentation with tubulin in reconstitution assays.

### 2.3.2 Proteins Involved in Assembly

MT's assembled in vitro from homogenates of chick, calf, rat or porcine brain contain two high molecular weight MAP's, HMW-MAP$_1$ (molecular weight 350,000) and HMW-MAP$_2$ (molecular weight 300,000), in addition to the $\alpha$- and $\beta$-subunits of tubulin (SLOBODA et al. 1976a). These HMW-MAP's are integral components of MT's assembled in vitro since they have been shown to copurify with the tubulin in a stoichiometric ratio through several cycles of assembly in vitro (BORISY et al. 1974, SLOBODA et al. 1975, BERKOWITZ et al. 1977). Tubulin alone will assemble into smooth-walled MT's at relatively high protein concentrations (DENTLER et al. 1975). In the presence of the HMW-MAP fraction, the tubulin assembles at concentrations less than 1 mg/ml, and the MAP's increase both the rate of assembly and the total mass of MT's produced (SLOBODA et al. 1976a, b). The MT's assembled in the presence of MAP's had a filamentous coating similar to that observed on MT's assembled from unfractionated preparations. $MAP_1$ and $MAP_2$ comprise between 6% and 20% by weight of the total MT protein preparation. Indirect immunofluorescence using antibodies raised specifically against HMW-MAP's from rat brain showed that these proteins are located in the mitotic spindle of mouse 3T3 fibroblasts, and thus are not restricted to brain MT's (SHERLINE and SCHIAVONE 1978).

However, HMW-$MAP_1$, and -$MAP_2$ are not present in HeLa cells (BULINSKI and BORISY 1979). Instead, copurified Q-MAP's of molecular weight 120,000 and 210,000 were present and stimulated MT assembly. MT's from rat glial cells contained low levels of accessory proteins (WICHE et al. 1979). Hence the importance of HMW-MAP's to non-neural cells is questionable.

Tau, $\tau$, has been described as a MT-associated protein which co-purifies with brain tubulin through several cycles of assembly–disassembly, (WEINGARTEN et al. 1975, CLEVELAND et al. 1977a, b). The protein, purified to near homogeneity, was separated by electrophoresis into four bands corresponding to MW's 55,000 to 70,000. The tau proteins strongly promote MT assembly under polymerizing conditions. A single 68,000 molecular weight heat-stable protein, designated "tubulin assembly protein" (TAP) has been purified from tau fractions and shown to stimulate MT assembly and to have a high affinity for tubulin (LOCKWOOD 1978). Recently, strong evidence has been presented to indicate that this fraction (TAP) is a component of neurofilaments which contaminate some tubulin preparations (BERKOWITZ et al. 1977, RUNGE et al. 1979).

Stimulation of tubulin assembly as a MAP criterion must be applied with caution, since several basic proteins, such as RNase, have been shown to stimulate assembly of tubulin at quite low concentrations (ERICKSON 1976).

### 2.3.3 Enzymes Associated with Tubulin

The curved arms extending from the outer doublets of cilia and flagella are high molecular weight proteins (MW 350,000 to 800,000) with ATPase activity. Since the enzyme is believed to generate the motive forces for ciliary beating it has been termed dynein (GIBBONS and ROWE 1965, for further details see Stephens 1974, Gibbons et al. 1976, DUSTIN 1978). Extracted dynein recombines with MT's in the presence of $Mg^{2+}$ or $Ca^{2+}$. GIBBONS et al. (1976) have electrophoretically fractionated four dynein bands from sea urchin flagella (labelled A–D). The A-band, with properties similar to dynein-1, has a molecular weight approximately 800,000 and may be comprised of two subunits. The D-band (dynein-2) has a molecular weight of 650,000. BACCETTI et al. (1979), using axonemes naturally lacking either the inner or outer arms on the ciliary doublets, were able to ascribe the dynein A band to the outer arm and the dynein B band to the inner arm; dynein-D was ascribed to the $\gamma$-links.

Phosphorylation of tubulin occurs in vitro and in vivo (EIPPER 1972, 1974, PIPERNO and LUCK 1974). The site of phosphorylation is a serine residue: tubulin phosphorylated in vivo contains an average of five phosphoserine residues/monomer (IKEDA and Steiner 1979). GTP ogr ATP can serve as effective phosphate donors. Specific phosphorylation of the $\beta$-subunit of tubulin has been reported (EIPPER 1972, 1974), of $\alpha$-tubulin in the outer doublets of *Chlamydomonas* flagella (PIPERNO and LUCK 1974), or both tubulins (RAPPAPORT et al. 1976). SLOBODA et al. (1975) showed that $MAP_2$ was preferentially phosphorylated. Similarly, other co-polymerizing minor constituents of MT preparations may serve as kinase substrates (SANDOVAL and CUATRECASAS 1976b). Cyclic-AMP dependent and independent protein kinase activity has been frequently reported in tubulin preparations (GOODMAN et al. 1970, SOIFER et al. 1972). It has been claimed that c-AMP-dependent kinase activity is an intrinsic property of tubulin itself (GOODMAN et al. 1970, IKEDA and STEINER 1979): other workers have separated the activity from tubulin (EIPPER 1974, SANDOVAL and CUATRECASAS 1976b).

In addition to dynein, tubulin preparations from *Chlamydomonas* flagella have been reported to contain a variety of enzymes, including a low MW $Ca^{2+}$-specific ATPase (WATANABE and FLAVIN 1973), GTPase activity (JACOBS et al. 1974), adenylate kinase and GDP kinase activity (WATANABE and FLAVIN 1973, KOBAYASHI 1975). The association with tubulin of a phospholipase C-type enzyme (QUINN 1973) and of a diglyceride kinase (DALEO et al. 1974) have given rise to the suggestion that associated lipids are essential for some of the functional properties of tubulin. Post-translational modification of tubulin (Sect. 2.4.3.1) through tyrosylation of its carboxyl terminus involves tubulin-tyrosine ligase activity (NATH and FLAVIN 1979, THOMPSON et al. 1979).

In conclusion, it is becoming clear that MAP's vary with the source, location and function of MT's.

## 2.4 Tubulin Assembly

### 2.4.1 Introduction

The regulated assembly and disassembly of tubulin is probably its most important biological feature and is fundamental to its role in the cell. Within the cell, assembly is regulated both in space, e.g., MT involvement in the orientation of the mitotic spindle and prediction of the plane of cell division (HEPLER 1976, FILNER and YADAV 1979), and in time, e.g., the controlled appearance of flagellar or cytoplasmic MT's correlated with different phases of cell development.

Originally, it seemed as though only tubulin from certain types of cytoplasmic (labile) MT's could be polymerized in vitro, but recent studies have demonstrated that this also is a property of preparations from flagella (KURIYAMA 1976, BINDER and ROSENBAUM 1978). Early attempts to polymerize tubulin in vitro by varying the ion and thiol content of the isolation medium were only partially successful in giving stable, non-MT aggregates (see KIRSCHNER 1978). The importance of low calcium levels was demonstrated in the first, reversible assembly system (WEISENBERG 1972) in which chelating agents seemed the crucial factor. A variety of chemicals are now known to induce or stabilize MT formation. Some presumably engender hydrophobic interaction, e.g., polyethylene glycol (HERZOG and WEBER 1978), glycerol, sucrose and dimethyl sulfoxide (HIMES et al. 1976). From the action of others, an electrostatic effect can be inferred, e.g., polycations like basic proteins or DEAE-dextran (ERICKSON 1976, ERICKSON and SCOTT 1977). Besides MT's, the list of structures formed includes such "polymorphs" as rings and spirals, sheets, ribbons, crystals and macrotubules (DUSTIN 1978).

### 2.4.2 In Vitro Studies

#### *2.4.2.1 Factors Influencing Assembly*

The assembly of tubulin subunits into structures resembling MT's occurs optimally at pH 6.9 and 37 °C. However, the influences of specific conditions on the process are difficult to treat in isolation because of the probable variety of pathways taken by the process and the high degree of interaction between factors. For example, the necessary threshold concentration of tubulin subunits for assembly is 1 mg $ml^{-1}$, but this can be much lower in the presence of accessory proteins (SLOBODA and ROSENBAUM 1979). The process is highly temperature-dependent, with low temperatures leading to depolymerization. Studies on the influence of temperature on the endothermic process (see DUSTIN 1978) and the subsequent derivations of enthalpy changes may be confounded by the extent to which the process is MAP's-assisted. The inhibitory influence of calcium on the assembly process (WEISENBERG 1972) is influenced by the presence of magnesium (ROSENFELD et al. 1976), which itself is required at low concentrations but is inhibitory at high concentrations. Nucleoside triphosphates are required for assembly, GTP and ATP having the greatest effect on the process: GTP is tightly bound at the N-site; the presence of GDP at the E-site results in a subunit inactive in assembly (Sect. 2.2.3.3.).

The assembly proteins of the MAP group have a profound influence on the process. Their presence is not absolutely necessary for tubulin polymerization, although the assembly of so-called "smooth" MT's requires the presence either of a stabilizing agent like polyethylene glycol (HERZOG and WEBER 1978) or a high concentration of tubulin (SLOBODA and ROSENBAUM 1979). Thus, MAP's may affect the overall MT assembly reaction by stabilizing the assembled tubulin (SLOBODA and ROSENBAUM 1979). Besides influencing the rate and extent of the process, HMW-MAP's are themselves incorporated into the assembled MT: such assembled "hairy" MT's in vitro exhibit a striking axial periodicity of 32 nm in the location of $MAP_2$ appendages (KIM et al. 1979). Examination of the sequence of association of non-tubulin proteins during the assembly process suggests that $MAP_2$ may function preferentially in the early stages of assembly (STEARNS and BROWN 1979; Fig. 1).

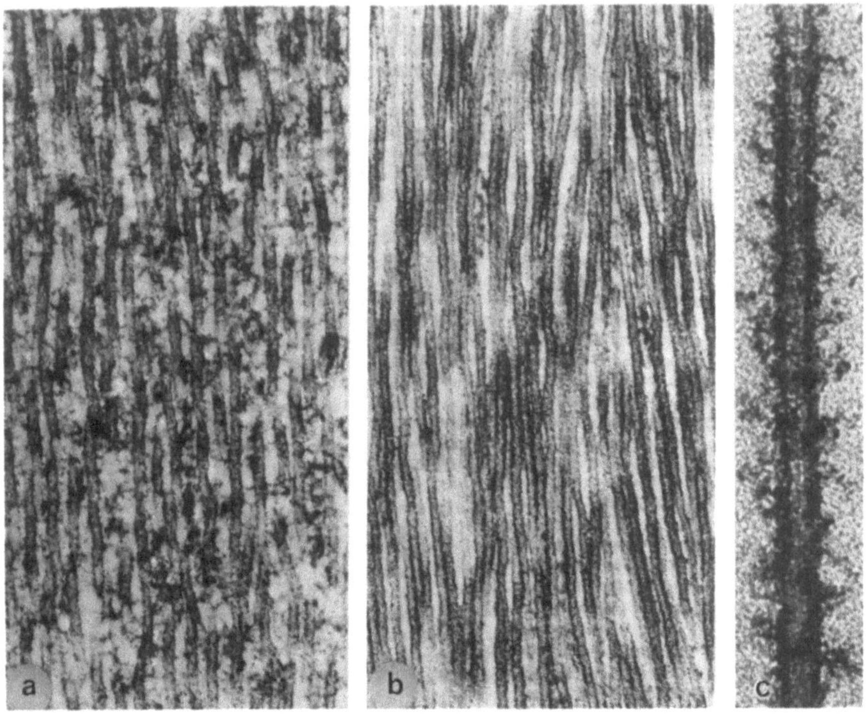


**Fig. 1.** Microtubules assembled in the presence (**a**) and absence (**b**) of high molecular weight MAP's. Filamentous coating on microtubules was only observed in the presence of MAP's. (**c**) Microtubules assembled in the presence of $MAP_2$ show axial periodicity of filamentous projections (**a** and **b** from SLOBODA et al. 1976a; **c** from KIM et al. 1979)

### *2.4.2.2 Pathways of Assembly*

In such in vitro polymerization systems, a variety of oligomeric structures have been observed under the electron microscope and sedimented in the ultracentrifuge (KIRSCHNER 1978, VALLEE and BORISY 1978). The details of size and structure seem to vary with the particular operating procedure of a laboratory, although there is agreement that such oligomers represent intermediates in MT assembly under certain conditions; removal of such structures can prevent assembly. In Borisy's laboratory, where tubulin is prepared in the absence of glycerol as a stabilizing agent, a range of structures have been observed (VALLEE and BORISY 1978): an oligomer with a sedimentation coefficient of 18S whose structure is unclear; a 20S component which is equated with a single ring of dimers with HMW-MAP projections; and, after tryptic digestion of the HMW fraction, a 39S oligomer. These rings are thought to be composed of 13 tubulin subunits. In Kirschner's laboratory, the predominant oligomer observed was a 36S double structure with an outer (48 nm) and inner (36 nm) ring. The rings are composed of from 20 to 25 subunits (KIRSCHNER 1978).

The role of such structures in assembly is debatable. Borisy considers the double rings to be actual segments of MT helix, to which tubulin dimers can accrete. Kirschner suggests that rings represent coiled protofilaments of subunits banded longitudinally; assembly would take place through these structures uncoiling to associate laterally, with tubule-closure occurring when 13 protofilaments had associated. Micrographs both of ribbons of 5 to 13 protofilaments and of MT's with frayed ends have been used to support this view (KIRSCHNER 1978). While it is generally agreed that HMW-proteins are involved in the formation of rings, these latter structures themselves may not be obligatory intermediates in assembly of MT's (VALLEE and BORISY 1978, STEARNS and BROWN 1979).

### 2.4.3 In Vivo Assembly

Some of the in vitro studies may not be relevant to MT assembly in the cell, for a variety of reasons. As indicated earlier, the significance of MAP's and rings is unclear; further, it is becoming evident that different types of MAP components are involved in different types of MT's (Sect. 2.3.2). However, difficulties of an even more fundamental nature involve possible post-translational modifications of tubulin and the role of specific cellular sites in MT initiation.

*Post-Translational Modifications.* A certain amount of the microheterogeneity of tubulin is thought to arise from various types of post-translational change (STEPHENS 1978, MAROTTA et al. 1978). Lack of success in detecting methylhistidine or methyllysine has led to the conclusion that methylation is probably not involved in tubulin modification. Phosphate incorporation seems a general property of MT protein (IKEDA and STEINER 1979) leading to phosphorylserine residues on one or both monomer chains (Sect. 2.3.3). Although phosphorylation could function as a regulatory mechanism, there is as yet no demonstration of a direct correlation between such a modification and tubulin function. Tyrosylation through the reversible addition of tyrosine to the carboxy-terminus, glutamate, of the α-chain can be demonstrated on certain populations of tubulin and is thought to be of functional significance (NATH and FLAVIN 1979). Glycosylation is thought to be another modification, functionally specific in terms of its possible characterization of membrane-associated tubulin (STEPHENS 1978). Thus while these changes are known to have no affect on assembly in vitro, post-translational modification is very likely to be a method whereby the cell constructs different subunit pools with different specificities for future location or function of assembled MT's.

*Cellular Initiation Sites.* Studies of tubulin assembly in vitro have clearly shown that assembly starts from some type of nucleating centre, whether this be a tubulin oligomer $\pm$ MAP's or a MT itself. In the cell, MT's are often seen to be linked to, or even grow from, specific cellular structures or regions (DUSTIN 1978). Structures such as centrioles, basal bodies, kinetochores and certain cell membranes are known to act as such microtubule-organizing centres (MTOC) both in vivo and in vitro (see KIRSCHNER 1978, GOULD and BORISY 1978, TELZER and ROSENBAUM 1979). In higher plant cells, there is evidence that cortical MT's are initiated in specific regions (GUNNING et al. 1978). The origins and components of MTOC's remain obscure and it is not clear to what extent they act as accumulations of tubulin subunits, oligomers or associated proteins. Again, the question of polarity of growth may be a further complexity; in vitro, elongation of MT's is predominantly from one end, although assembly at both ends can occur at higher tubulin concentrations or after longer incubation periods (SNELL et al. 1974). In the flagellum, growth of the outer doublets seems to occur at the distal end, while the central pair MT's grow by proximal addition of dimers (DENTLER and ROSENBAUM 1977).

Thus, growth of MT's in the cell is regulated not only temporally but spatially. Factors such as the presence of nucleotides and the ionic environment are no doubt of importance, but the differential location of different classes of MT throughout the cell may well involve the interaction of populations of modified subunits with specific types of initiation centre.

## 2.5 Plant Tubulin: Special Considerations

Studies of MT in higher plant cells have been especially difficult because of both the relative stability of plant MT's and their general paucity in the plant cell (HART and SABNIS 1976a, 1977, DUSTIN 1978). To be effective on plant mitosis or other MT-dependent processes, drugs such as colchicine are often required in concentrations 1000-fold in excess of those effective on animal cells (DEYSSON 1968, DUSTIN 1978, LLOYD et al. 1980). Physical agents such as low temperature or high pressure also have a much less dramatic effect on plant MT (HARDHAM and GUNNING 1978). The basis of this stability is unknown, but may lie in the tubulin moiety itself, in the properties of plant MAP's, or in the cross-linking of MT's with one another and with membranes.

The number of plant sources from which tubulins have been isolated and identified is few. Axonemal tubulins from *Chlamydomonas* co-polymerize with, and display immunochemical and electrophoretic similarities to, brain tubulins (SNELL et al. 1974, Rosenbaum et al. 1975). Co-polymerization with brain tubulin has also been used to isolate tubulins from yeast (WATER and KLEINSMITH 1976, BAUM et al. 1978, CLAYTON et al. 1979), *Aspergillus* (SHEIR-NEISS et al. 1976) and higher plants (SLABAS et al. 1980). Similarly, isolated spindle pole bodies of *Saccharomyces* nucleate assembly of porcine brain tubulin (HYAMS and BORISY 1978). The evidence suggests that tubulins are highly conserved proteins and that conformational similarities permit the co-assembly of tubulins from widely disparate sources. This view is supported by the recent immunofluorescent visualization of MT's in higher plant protoplasts and cells (LLOYD et al. 1979, 1980) which indicated antigenic homology between higher plant and porcine brain tubulins.

Nevertheless, differences in the tubulin molecule may exist. BAUM et al. (1978) report lower molecular weights of 45,000 and 46,000 for yeast tubulins and CLAYTON et al. (1979) could find no electrophoretic band corresponding to β-tubulin. Microheterogeneity was demonstrated to be associated with both α- and β-subunits in *Chlamydomonas* preparations of tubulin (WITMAN et al. 1972, PIPERNO and LUCK 1976). With regard to the characteristic colchicine-binding activity of tubulin, examinations of plant preparations have met with limited success. Several reports have indicated that colchicine- or colcemid-binding to proteins from plant sources is low or non-existent, and that the activity is unstable (HABER et al. 1972, HART and SABNIS 1973, 1976b, 1976c, RUBIN and COUSINS 1976, FLANAGAN and WARR 1977, CAPPUCINELLI and HAMES 1978, BAUM et al. 1978). However, in a water-mould, colchicine-binding activity was associated with a trichloracetic acid-stable component of molecular weight 30,000 (OLSON 1973), supporting the view that more than one colchicine-binding protein may be present in plants (HOTTA and SHEPARD 1973) and that this property may not be useful for identifying plant tubulin.

In addition to possible differences in the tubulin molecule itself, specialized MAP's differing from those of brain preparations are associated with *Chlamydomonas* tubulins (PIPERNO and LUCK 1976, BLOODGOOD and ROSENBAUM 1976). Furthermore, the resistance of plant MT's to colchicine has been attributed to stabilization by MAP's rather than to intrinsic differences in plant tubulin (LLOYD et al. 1979, 1980). Cross-bridging of MT's to one another and to the plasma membrane has been frequently reported (BURGESS 1970, HEPLER et al. 1970, ROBARDS and KIDWAI 1972, HARDHAM and Gunning 1978) and may be responsible for this increased stability.

A cytoskeletal role for MT's has been repeatedly postulated (see DUSTIN 1978). The interaction of actin-like microfilaments underlying the plasma membrane with skeletal MT's has also been envisaged for animal cells (NICOLSON 1976). Certainly, the distribution of plant cytoplasmic MT's compatible with the suggestions that MT's shape the plasma membrane against turgor pressure (SCHNEPF 1974, SCHNEPF et al. 1978) or otherwise orient the assembly of the plant cell wall (HEATH 1974). Possessed of a rigid wall, the plant cell does not undergo the pleiomorphic transformations of many animal cells and may, therefore, have less need for shifting equilibria between tubulin polymers and subunit pools and, hence, for a high degree of lability in its microtubule population.

### 2.6 Conclusions

Considerable progress has been made in the study of the structural components of MT's. The biochemistry of the tubulin molecule and its relation to the sub-structure of the MT are relatively well understood. The general principles governing tubulin assembly, and the roles of associated proteins and nucleotides in the process, are becoming clearer. However, the cellular mechanisms controlling the strictly regulated assembly of MT's are far from clear. Similarly, study of the interactions of MT components with other cellular organelles is only just beginning.

Consideration of the specificity of role and location of different types of MT within the cell suggest that the question of heterogeneity among tubulins and their associated proteins may assume greater significance. It is already apparent that tubulins can differ to a certain degree both in primary structure and in modification to the molecule after its synthesis. The evidence that different

MAP's are associated with different types of microtubule will no doubt be greatly extended.

Knowledge about MT's and tubulin has been gained almost wholly from studies of animals and lower organisms. MT's of the higher plant cell exhibit characteristic differences, particularly regarding their cytoplasmic stability. Such stability may reflect a changed emphasis in the role of MT's in the interphase plant cell.

## 3 P-Proteins

### 3.1 Introduction

The term P-protein was introduced (ESAU and CRONSHAW 1967) to describe discrete and ultrastructurally characteristic aggregations of proteinaceous material found in the sieve elements of phloem. The initial emphasis upon ultrastructural characterization of the material is reflected in several recent reviews (PARTHASARATHY 1975, CRONSHAW 1975, EVERT 1977). Only in the last decade have any biochemical investigations into P-protein been carried out (ESCHRICH and HEYSER 1975).

P-protein is located in a unique environment. The mature, enucleate sieve element is possibly the most specialized, highly differentiated, living plant cell. Its much-reduced cytoplasm results from a selective, controlled form of autolysis during differentiation and is characterized not only by the structures classed as "P-protein", but also by a variety of plastid inclusions and various forms of stacked, convoluted and cisternoid smooth endoplasmic reticulum (CRONSHAW 1975, PARTHASARATHY 1975). The cellular sap is equally unusual in its alkalinity (pH 7–8), high solute potential and viscosity (equivalent to 10% sucrose) and its high content of potassium and phosphate ions (ZIEGLER 1975).

The concept of P-protein as a single entity is no longer tenable, a situation which gives rise to difficulties of definition or description. Ultrastructural studies have recorded wide polymorphism in the P-protein structures of different species. Biochemical investigations have highlighted the variety of different major proteins present in sieve elements, proteins which have not yet been characterized by any common features of size, charge or biochemical activity. Most present hypotheses of P-protein function are not compatible with this heterogeneity in composition.

### 3.2 Distribution

P-proteins seem to be a regular feature of the sieve elements of all dicotyledonous species. The occurrence of such structures in monocotyledenous plants is more sporadic. P-protein has been reported in genera such as *Dioscorea, Elodea, Tradescantia, Secale* and *Avena* but seems to be absent in *Zea, Hordeum* and

*Triticum* (see CRONSHAW 1975). Furthermore, in certain monocotyledenous plants (ESAU and Gill 1973) including palms (PARTHASARATHY 1974) it may be present in some sieve elements but absent in others in the same organ. The sieve cells of gymnosperms frequently contain membranous aggregates (LAMOUREUX 1975) but such features are generally considered not to represent P-protein (CRONSHAW 1975). Similarly, with regard to the translocatory systems of lower vascular plants, filamentous structures have been observed in *Selaginella* and *Isoetes*, but their relationship to P-proteins is unclear (BEHNKE 1975).

Although P-protein structures have been observed in companion cells and in phloem parenchyma (ESAU 1969) the sieve element is the main cell type carrying P-protein. The protein content of sap exuding from severed phloem of members of the Curcurbitaceae (Sect. 3.4.1) can reach levels of 10–30 $mg\,ml^{-1}$ of exudate. This exudate has been shown to derive specifically from the sieve element and its protein content to consist mainly of P-protein; other sieve element structures such as membranes and endoplasmic reticulum do not seem to be extruded (ESCHRICH et al. 1971). The distribution of P-protein within the sieve element, and particularly its presence around the pores of the sieve plate, is a subject of considerable controversy. Most investigators agree upon the need for some form of anchorage for structures within the translocation stream (PARTHASARATHY 1975, CRONSHAW 1975), although the presence of P-protein in exudate suggests such anchorage must be of a labile, non-covalent nature. However, whether the system of fibrils and tubules which often make up the P-protein complement is dispersed throughout the lumen of the sieve element (CRONSHAW and ANDERSON 1971), or is distributed parietally (EVERT et al. 1973) is still debatable and may be impossible to resolve by available techniques (JOHNSON 1978).

## 3.3 Ultrastructural Studies

### 3.3.1 Morphology

The ultrastructural polymorphism of P-proteins encompasses structures that are granular, tubular and fibrillar, or even crystalloid. They may be linear or branched and can form tight, coherent bodies or appear loose and frayed (Fig. 2).

*In Vivo*. Tubular and fibrillar components have been observed in the sieve elements of a large number of species (see CRONSHAW 1975). The tubular forms seem less common than fibrils and range in size from 10 nm to 24 nm in diameter, varying both among species (see ZEE 1969) and also within a single plant (PARTHASARATHY and MÜHLETHALER 1969). In *Nicotiana*, tubules of 24 nm diameter have been reported (CRONSHAW et al. 1973); in *Ulmus*, the tubules have a diameter of 18 nm (EVERT and DESHPANDE 1969) and in *Cucurbita*, tubules of both 18 nm and 24 nm have been observed (CRONSHAW and ESAU 1968a, b). The 18–24 nm P-protein tubules in the sieve elements of *Coleus* have been reported to decrease progressively to 10 nm diameter during maturation (STEER and NEWCOMB 1969). Tubules of 16 nm diameter have been reported in some

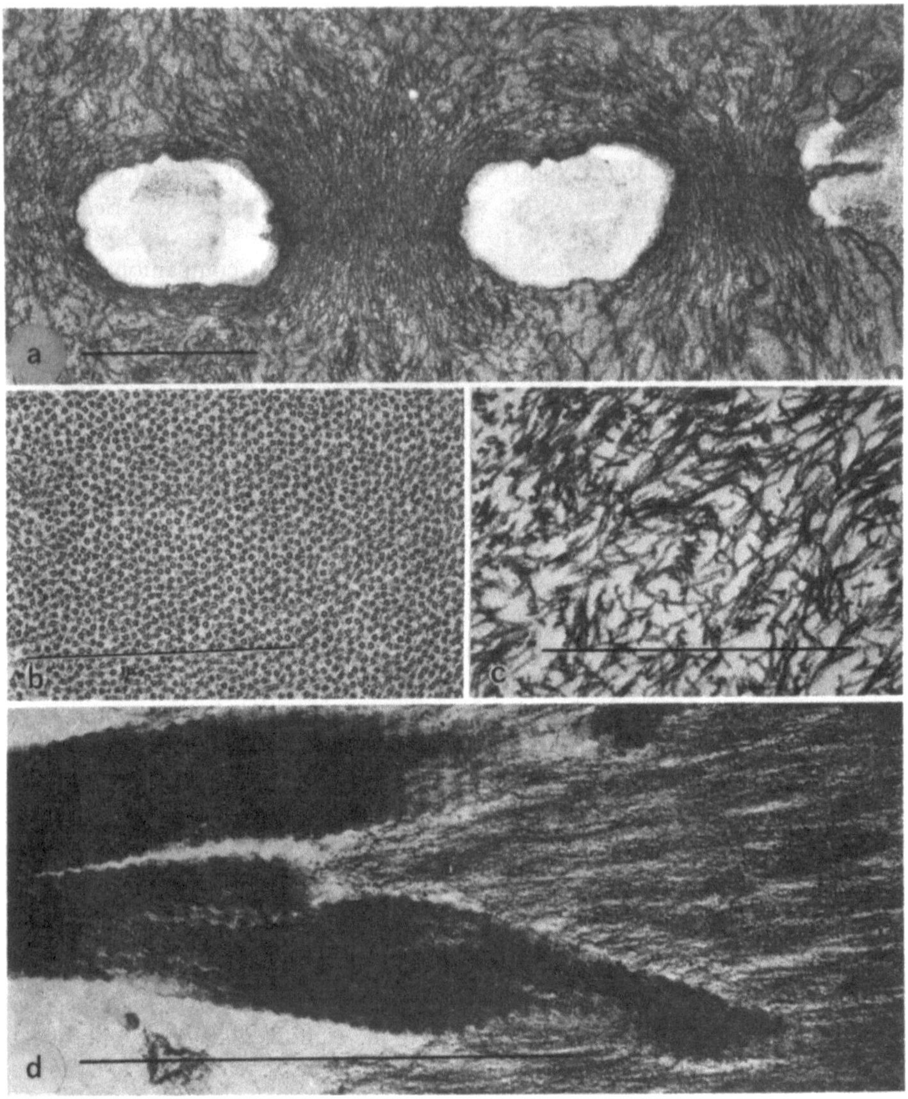


**Fig. 2a–d.** Morphological diversity of P-protein filaments. **a** Fine filaments in sieve pores of *Nelumbo nucifera*. (ESAU 1975). **b** Tubular P-protein in young sieve elements of *Nicotiana tabacum*. (CRONSHAW 1975). **c** Striated filaments in mature sieve elements of *Gossypium hirsutum*. (ESAU 1978b). **d** Partly dissociated, crystalline P-protein body in mature sieve element of *Phaseolus vulgaris* (ESAU 1978a)

legumes (PALEVITZ and NEWCOMB 1971). Certain types of tubule display spoke-like appendages and seem to be composed of six sub-fibrils (PARTHASARATHY and MÜHLETHALER 1969, CRONSHAW et al. 1973).

The more common fibrillar forms of P-protein similarly show variation in dimensions. In *Musa*, filaments of 16–20 nm were observed in immature cells and 8–12 nm in fully differentiated sieve elements (BEHNKE 1969). The sieve elements of *Ricinus* carry only striated filaments of 15 nm and larger fibrils of 17.5 nm (CRONSHAW 1975). Striated fibrils were 15 nm wide in *Nicotiana,* but only 6 nm in *Cucurbita* (CRONSHAW et al. 1973). In many legumes, these polymeric forms often seem to be aggregated into bundles that form crystalloid structures (ESAU 1978a).

*In Vitro.* Phloem exudate has only been obtained from a limited number of species. Various forms of P-protein structure have been observed under the EM in negatively stained preparations of such exudates. Filamentous structures seem to be the most frequently reported form, although the dimensions vary, not only among species but also according to investigator:

| Exudate source | Structure | Author |
|---|---|---|
| *Cucurbita* | 23 nm tubules | CRONSHAW et al. (1973) |
| | 8 nm filaments | CRONSHAW et al. (1973) |
| | 4 nm and 9 nm filaments | KOLLMAN et al. (1970) |
| *Nicotiana* | 4 nm and 9 nm filaments | KOLLMAN et al. (1970) |
| | 30 nm "filament" | CRONSHAW et al. (1973) |
| *Ricinus* | 20 nm and 2–8 nm filaments | WILLIAMSON (1972) |
| | 20 nm and 14 nm filaments | STONE and CRONSHAW (1972) |

The closest correspondence of structures in exudate with forms observed in vivo in sections of the same plant, seems to be the 20 nm and 14 nm filaments seen in *Ricinus* exudate (STONE and CRONSHAW 1973).

### 3.3.2 Ontogeny

A feature of P-protein, important to its ultrastructural definition, is its presence in immature sieve elements before there are any signs of cytoplasmic degradation. It arises in the cytoplasm in the form of P-protein bodies which consists of ellipsoidal polymorphic aggregates of P-protein not bounded by a membrane. The P-protein may be tubular, as in *Nicotiana* (CRONSHAW and ESAU 1967) and *Coleus* (STEER and NEWCOMB 1969), granular as in Cucurbita (CRONSHAW and ESAU 1968b), fibrillar as in *Ricinus* (CRONSHAW 1975) or display a regular, para-crystalline form as in certain legumes (WERGIN and NEWCOMB 1970, ESAU 1978a). The numbers and types of P-protein bodies within a cell may also vary (PARTHASARATHY 1975).

Although the bodies are often seen to be closely associated with endoplasmic reticulum (EVERT and DESHPANDE 1969), details of the synthesis of P-protein and origins of the bodies are lacking. There may also be different pathways of assembly and development in different species, correlating with the different forms of P-protein which have been observed. P-protein is first observed in *Phaseolus* as "fibrillar precursory material" (ESAU 1978a), and in immature elements of *Gossypium* as "nascent P-protein" (ESAU 1978b); it consists of amorphous masses associated with anastomosing fibrils which are thought to coalesce to form the indistinct tubules of the P-protein body. The origins and growth of the crystalline bodies seen in *Phaseolus* sieve elements, and their relation to the tubular P-protein is not clear (ESAU 1978a). Development of the P-protein body continues with the addition of more tubules and the appearance of spoke-like projections from the tubules in *Curcurbita* (CRONSHAW 1975). The number of P-protein bodies in the cell also increases. Later in sieve element differentiation, the nucleus disappears and the tonoplast breaks down. Simultaneously with, or slightly after, these events (PARTHASARATHY 1975, CRONSHAW 1975), the P-protein bodies of *Coleus* and *Nicotiana* disperse to release tubules and striated fibrils. In *Cucurbita* and *Ricinus*, this stage can be preceded by a fusion of the granular or fibrillar P-protein bodies. In certain legumes, the bodies may only partially disperse, leading to the retention of para-crystalline structures in the mature sieve element (PALEVITZ and NEWCOMB 1971, ESAU 1978a). Further transformations of P-protein have been suggested to occur in the loosening and stretching of tubular structures to form fibrils (CRONSHAW 1975, ESAU 1978b).

### 3.3.3 Histochemistry

There have been several histochemical investigations of the sieve element at both the light and electron microscope levels (see EVERT 1977, ESCHRICH and HEYSER 1975). It has been demonstrated that the filamentous material in the sieve tubes of *Nymphoides* contains no polysaccharide components (FREUNDLICH 1974). Various workers have affirmed the proteinaceous nature of the fibrillar structures with histochemical techniques (ESAU 1969, ILKER and CURRIER 1975, OBERHÄUSER and KOLLMAN 1977). A characteristic feature of P-proteins from cucurbit exudate is their high content of thiol groups (Sect. 3.4.2). In this respect, although gymnosperms are considered on ultrastructural grounds not to contain P-protein, it may be significant that the material in the sieve areas and intracellular strands of conifer sieve cells (and only in mature sieve cells) showed a marked response to tests for sulphydryl groups (SAUTER 1972).

A list of enzymes occurring in the sieve element has been compiled by ESCHRICH and HEYSER (1975). These authors also point out the difficulties, arising from "surge effects", in assigning a location to enzymes in this tissue, and conclude that most of the enzymes probably originate in the companion cells. Investigations of enzyme activities associated with actual P-protein structures have produced conflicting results. Particular attention has been paid to acid phosphatase and ATPase activities. Acid phosphatase is present in exudate and has been shown to be associated with P-protein in vivo (CATESSON 1973), specifically on dispersed filaments (BENTWOOD and CRONSHAW 1976). Other investigators have failed to detect such activity (FIGIER 1968). ATPase activity has been reported in association with dispersed, mature P-protein structures in *Cucurbita* and *Nicotiana* (GILDER and CRONSHAW 1973, 1974). However, no ATPase activity is biochemically detectable in *Cucurbita* exudate, even though P-proteins are present (ESCHRICH et al. 1971, WEBER et al. 1974). No ATPase activity could be located on P-protein structures of *Tetragonia* (YAPA and SPAN-

NER 1974) or *Acer* (CATESSON 1973) and, further, ATPase activity was completely lacking throughout the sieve elements of *Pisum* (BENTWOOD and CRONSHAW 1978).

## 3.4 Biochemical Studies

### 3.4.1 Methodology

Biochemical studies on the structural proteins in sieve tubes have been largely limited to analyses of exudate from cucurbit species. Although many tree species also exude from incisions in the bark (see ZIEGLER 1975), it is uncertain whether P-proteins are extruded with the sap.

Attempts to extract P-proteins from isolated phloem strands of *Heracleum mantegazzianum* have met with little success. SLOAN (1977) extracted and fractionated a very complex mixture of phloem proteins using SDS-PAGE. However, there were no significant differences between similar extracts of phloem and xylem tissues, suggesting that P-proteins are not readily solubilized. The basicity and lectin-like activity of many P-proteins (Sect. 3.4.2 and 3.4.3) make it likely that they can readily bind to cell debris (wall components) during extraction. Thus, when exudate from *Cucurbita maxima* was homogenized with stem tissue, the exudate P-proteins could not be recovered by conventional high salt/pH extraction methods (HART and SABNIS unpublished data).

Exudate welling from the severed phloem of cucurbits may be collected in micropipets (BEYENBACH et al. 1974) or, alternatively, the excised stems may be immersed in buffer solutions containing thiol reagents (SABNIS and HART 1976). The greatest volumes are collected from incisions in the rind of curcurbit fruits, giving volumes of several millilitres (WEBER et al. 1974). Similar diagonal incisions in the stems of *Ricinus communis* also yield considerable volumes of phloem sap (HALL and BAKER 1972). Electron micrographs have shown large numbers of filaments in exudate from species of *Cucurbita, Cucumis, Ricinus* and *Nicotiana* (see Sect. 3.3.1 for references). ESCHRICH et al. (1971) have demonstrated in *Cucurbita* that exudate flows only from mature sieve tubes and therefore truly represents the sieve tube content. Contamination by proteins from damaged cells of the cut stem or fruit has been shown to be negligible (MCEUEN 1979).

Exudate from some species (*C. maxima, C. pepo, Cucumis sativus*) gels rapidly on exposure to the atmosphere (WALKER 1972, BEYENBACH et al. 1974); with other plants, there may be little or no gelling of the phloem sap. Gelling and precipitation can be prevented by collecting exudate into buffer containing thiol reagents. Routine fractionation and purification procedures include molecular sieve and ion exchange chromatography in columns equilibrated with buffers containing thiol reagents (BEYENBACH et al. 1974, WEBER et al. 1974, KLEINIG et al. 1975). Resolution of protein subunits has been accomplished using SDS-PAGE (WEBER and KLEINIG 1971, NUSKE and ESCHRICH 1976, SABNIS and HART 1976, 1979). Isoelectric points have been determined employing column or gel isoelectric focussing (HEYSER et al. 1974, SABNIS and HART 1976).

Antibodies have been successfully raised against total exudate proteins of *C. maxima* (WEBER et al. 1974) and Ouchterlony's immunodiffusion assay used to assess antigenic similarities between purified P-protein fractions.

Lectin activity has been assayed with a standard haemagglutination procedure (KAUSS and ZIEGLER 1974, SABNIS and HART 1978, GIETL et al. 1979), and individual lectins isolated either by ion exchange chromatography (SABNIS and HART 1978) or by affinity chromatography using N-acetylgalactosamine or chitin oligosaccharides linked to Sepharose (GIETL et al. 1979, ALLEN 1979).

## 3.4.2 Physical and Chemical Properties

### *3.4.2.1 The Subunits*

P-protein subunits range in MW from 10,000 to 130,000. Within the three families, eight genera and ten species so far examined using SDS-PAGE (Fig. 3), the subunits are clustered approximately into seven groups according to molecular weight. Characteristic features of this grouping include the large number of low MW polypeptides (10,000–17,000) together with the common occurrence of subunits of unusually high MW (80,000–130,000). Exudate from each plant species contains three to five major subunits, each representing 10%–60% of total exudate protein. Additionally, several minor subunits are present (BEYENBACH et al. 1974, SABNIS and HART 1976, 1979 and unpublished data, SLOAN et al. 1976, WEBER and KLEINIG 1971). The subunits could not be resolved into smaller fragments by treatment with higher concentrations of thiol reagents, or with 8M urea or 6M guanidine hydrochloride (BEYENBACH et al. 1974, SABNIS and HART 1976).

Exudate from *Cucurbita maxima* contains two major subunits which, between them, comprise 80% of total protein. These subunits have been isolated by ammonium sulphate fractionation, DEAE cellulose chromatography and gel filtration (BEYENBACH et al. 1974) and the amino acid composition determined. The results indicated that both subunits were similar in composition, with relatively high proportions of lysine. Major differences were found in the amounts of glycine, alanine and half-cystine. It was calculated that the smaller subunit contained 6 half-cystine residues and the larger subunit 14 half-cystine residues per molecule.

In agreement with the high lysine content, both major subunits from *C. maxima* are strongly basic (BEYENBACH et al. 1974, SABNIS and HART 1976). Isoelectric points in excess of 9.5 were indicated by electrophoresis on cellulose acetate strips (BEYENBACH et al. 1974), while column isoelectric focussing of the exudate proteins provided three prominent protein peaks of pI 9.8, 9.4 and 9.2 (NUSKE and ESCHRICH 1976). Ion exchange chromatography on columns of CM-Sepharose followed by SDS-PAGE of eluted fractions has indicated that several major proteins in exudate from *C. sativus, C. melo, C. pepo* and *Momordica charantia* are highly basic (Fig. 3). In *Lagenaria vulgaris* (bottle gourd) preliminary results indicate that at least three minor proteins are basic. The very basic nature of the subunits makes it probable that the P-protein polymers carry a net positive charge even at the relatively high pH 8.0 of the sieve tube sap.

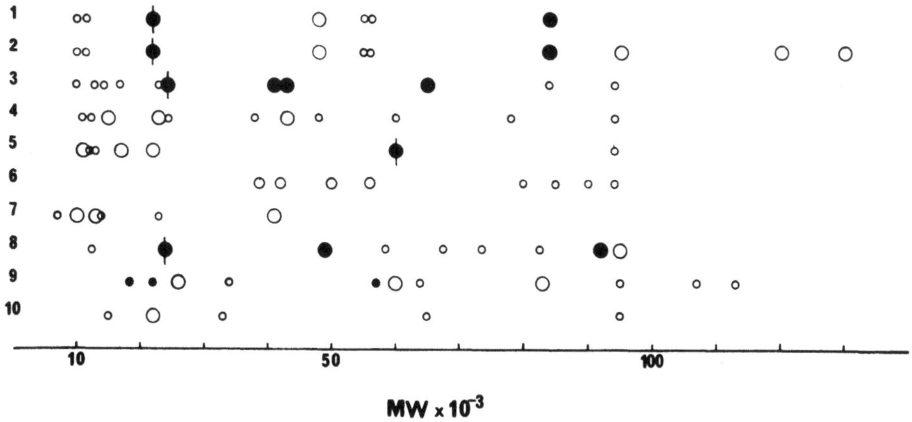


**Fig. 3.** Distribution according to molecular weight of P-protein subunits in phloem exudate from various species as resolved by SDS-PAGE. Species: *1 Cucurbita maxima; 2 C. pepo; 3 Cucumis sativus; 4 C. melo; 5 Bryonia dioica; 6 Acer pseudoplatanus; 7 Luffa acutangula; 8 Momordica charantia; 9 Lagenaria vulgaris; 10 Ricinus communis. Large symbols* represent major subunits (>10% of total exudate protein). *Closed circles* signify proteins known to be basic. *Vertical bars* indicate subunits with known lectin activity

Exudate from *C. maxima* or *C. sativus,* when collected into buffer and allowed to gel, forms a dense network of fibrils (CRONSHAW et al. 1973, CRONSHAW 1975). KLEINIG et al. (1975) have reported that of the two major basic proteins in exudate from *C. maxima,* the larger subunit formed filaments 6–7 nm diameter upon removal of SH-protecting agents by dialysis or by aeration of the preparations. Filament formation was partially reversible upon re-addition of mercaptoethanol. The smaller subunit formed amorphous aggregates under conditions of low thiol reagent concentrations. This work, however, did not provide any comparative data to indicate whether the filaments formed in vitro bore any close resemblance to filaments seen in vivo. The relationships and subunit compositions of the various structures from sieve tubes have yet to be established.

### *3.4.2.2 Subunit Heterogeneity*

A consideration of Fig. 3 reveals that, despite the grouping of protein subunits and the common occurrence of basic proteins, the detailed pattern of subunit size and charge is very different for each species and genus. A comparative analysis of exudate from 17 cultivars, 5 species and 3 genera of the Cucurbitaceae, and from *Acer pseudoplatanus* and *Ricinus communis* has been described (SABNIS and HART 1976, 1979, SLOAN et al. 1976). Both SDS-PAGE and gel isoelectric focussing demonstrated that every genus has a completely distinctive protein pattern or fingerprint. With only one or two exceptions, there are no major subunits in common among the genera examined (Fig. 3). Where some pairs of proteins appear to possess similar molecular weight, they bear an opposite net change at neutral pH (cf. the major subunits of molecular weight 41,000 in *Luffa* and *Cucumis,* Fig. 3). There may be some minor proteins in common.

Comparison of species within a genus indicates a much greater degree of similarity (SABNIS and HART 1979). The major proteins of *Cucurbita maxima* are present in *C. pepo,* but the latter species possesses several additional high molecular weight subunits in relatively high concentrations. Similarly, *Cucumis sativus* shares much of the protein complement of *C. melo,* although some differences are apparent. Several cultivars of each of these four species were also examined and the subunits of P-protein found to be identical, save for those of *C. pepo* which showed some variation in high molecular weight subunits.

The protein pattern for each cultivar, species and genus is remarkably consistent. Even the relative concentrations of each protein subunit remains constant irrespective of whether exudate is obtained from hypocotyl, stem or fruit (WEBER et al. 1974, SABNIS and HART 1978). The protein finger-prints of *Cucurbita* and *Cucumis* are fully established in seedling exudate within 6 days of sowing (NUSKE and ESCHRICH 1976, SABNIS and HART 1979). Conditions which alter the normal pattern of seedling growth, such as germination and growth in total darkness, had no effect on the protein complement of exudate. There is no change following the transition from the vegetative to the flowering phase. Thus, the characteristic pattern of phloem exudate proteins is established early in the growth of the seedling and is uninfluenced by widely disparate developmental or environmental changes, suggesting that it may represent some fundamental feature of sieve tube development and function. The species-dependent heterogeneity of protein subunits must, therefore, possess some intrinsic importance.

Past interpretations of electron micrographs (Sect. 3.3.1) have promulgated the view that "P-protein" represents a distinctive intracellular component of the sieve element, composed of one or more characteristic and similar protein subunits. It has been suggested that conformational changes in P-protein, extending to the assembly of helical or striated fibrils, tubules or paracrystalline and granular bodies whose elementary dimensions vary widely, reflect the responses of similar subunits to differences in environmental conditions within the cell (PALEVITZ and NEWCOMB 1971, CRONSHAW et al. 1973). The protein heterogeneity indicated by biochemical studies now strongly indicates that at least interspecific differences in ultrastructure reflect major variations in subunit composition. This heterogeneity also extends to the phloem haemagglutinins (lectins) so far investigated (Sect. 3.4.3).

### *3.4.2.3 Interactions with Ions and Nucleotides*

*Calcium.* Early reports agree on the low concentrations of calcium in phloem exudate of several plants (mainly trees) including *Cucurbita maxima* (see ZIEGLER 1975). Several lines of evidence, including tracer studies with $^{45}Ca^{2+}$, also indicate the immobility of calcium in the sieve tube. Furthermore, KING and ZEEVART (1974) reported that cut petioles of *Chenopodium rubrum* and *Perilla crispa* showed enhanced exudation from the phloem in the presence of EDTA or EGTA. In view of the regulatory role played by calcium in ordering changes in conformation and activity of several enzymes and structural proteins (reviewed by KRETSINGER 1976) it is possible that calcium may play a similar role in regulating assembly or functioning of P-protein (MCEUEN 1979).

An early report showed the irreversible precipitation of *Cucurbita* exudate proteins by calcium at 20 mM (KLEINIG et al. 1971 a). However, this precipitation at such high concentrations of calcium appears to be a non-specific, ionic effect since similar results can be obtained with a variety of known proteins (WILSON et al. 1970, SLOAN 1977). Furthermore, the enhancement of exudation by chelating agents was not found in other plants such as *Populus deltoides* (DICKSON 1977). Nevertheless, the possibility of calcium interaction with components of the sieve element has been revived by the recent demonstration of high molecular weight calcium-binding protein(s) in exudate from several cucurbit species (MCEUEN 1979). These proteins are present in low concentrations and are distinct from the major protein subunits of the exudate.

*Nucleotides.* Several reports, covering more than twenty herbaceous and woody species (GARDNER and PEEL 1972, HALL and Baker 1972, KLUGE and ZIEGLER 1964, KLUGE et al. 1970, ZIEGLER and KLUGE 1962), have indicated unusually high concentrations of ATP in the sieve tube sap, ranging from 35 to 1,335 μg/ml exudate (ZIEGLER 1975). Evidence from studies of $[^{32}P]O_4^{3-}$ incorporation into ATP and other compounds (BIELESKI 1969, BECKER et al. 1971) suggests that ATP in the sieve tube is in a state of high turnover while the concentration is held constant (ZIEGLER 1975). However, no binding of nucleotides to exudate proteins could be demonstrated (KLEINIG et al. 1971 b), nor could ATPase activity be detected in exudate (KOLLMAN et al. 1970, ESCHRICH et al. 1971). Therefore, whether P-protein itself interacts in any way with nucleotides is uncertain.

### 3.4.3 Lectin Activity

Evidence for the occurrence of lectin activity in sieve tube sap (*Robinia pseudoacacia*) was first reported by KAUSS and ZIEGLER (1974). Since lectins (phytohaemagglutinins) have an affinity for carbohydrate groups and may be highly specific in their interactions (CALLOW 1976, LIENER 1976, KAUSS 1976) the functions and properties of lectins in a sugar-conducting channel are of obvious interest. KAUSS and ZIEGLER showed that centrifuged sieve tube sap agglutinated rabbit erythrocytes at a titre of 1/256. The most potent saccharide to inhibit agglutination was N-acetyl-D-galactosamine. SABNIS and HART (1978) described high haemagglutinating activity in exudate from three cucurbit species, *Cucumis sativus, C. melo* and *Cucurbita maxima*. The lectin from *C. maxima* was isolated by cation exchange chromatography and identified by SDS-PAGE as one of the major exudate subunits with a molecular weight of 22,000. The phloem lectin could not be isolated from *Cucurbita* seeds. However, it constituted 15%–25% of total exudate protein. It showed haemagglutinating activity at concentrations as low as 0.1 μg/ml. Simple sugars, including those transported in the phloem of this species, did not interact with agglutination. ALLEN (1979) showed that a lectin from *C. pepo* exudate strongly interacted with chitin oligosaccharides, glycopeptides from soyabean (*Glycine max*) agglutinin, and fetuin. Thus, although N-acetylglucosamine itself was only a weak ligand, there is evidence that the lectin interacts with internal N-acetylglucosamine residues in the hapten molecule. ALLEN purified the *C. pepo* lectin by affinity chromatography on chitin oligosaccharides covalently attached to Sepharose. In agreement with the data of SABNIS and HART (1978, 1979), the lectin was shown to consist of a single polypeptide chain of molecular weight ~20,000 and may thus be identical to the lectin from *C. maxima*. Lectins from *Cucumis sativus* and *C. melo* were also shown to strongly interact with NN′-diacetylchitobiose. The lectin from *C. sativus* has been isolated (Fig. 3, SABNIS and HART 1979 and unpublished data) and has a molecular weight of 24,500. Like the *Cucurbita* lectins, it is basic, as is the lectin from *Momordica charantia*. A lectin from sieve tube sap of *Bryonia dioica* is also basic, but has a molecular weight of ~60,000 (SABNIS and HART 1979). Exudate from every species so far analyzed possesses only one protein subunit with lectin activity. The ubiquity of phloem lectins has been recently demonstrated by GIETL et al. (1979), who examined sieve tube sap from 63 tree species belonging to 21 families. With the exception of members of the Aceraceae and Oleaceae, samples from all the species agglutinated trypsinized rabbit erythrocytes. Preparations from species of *Acer* showed pronounced haemolytic activity which may have masked haemagglutination. In contrast to the inhibition of *Robinia* exudate lectin by N-acetylgalactosamine, lectins from the other tree species were strongly inhibited by chitin oligosaccharides and by glycoproteins containing N-acetylglucosamine, resembling in this respect the cucubit lectins. Thus, in spite of heterogeneity in molecular weight and charge, the majority of sieve tube lectins examined so far appear to have an affinity for the same sugar residues. The possible significance of lectin activity in the sieve tube is considered in Section 3.5.

### 3.5 Role of P-Proteins: Hypotheses

The wide distribution of P-proteins among the dicotyledonous and many monocotyledonous plants demands some explanation of their function that is consistent with their distribution among species, location within the sieve element, ultrastructure and biochemical properties. Investigation of their ultrastructure and arrangement have given rise to several conflicting hypotheses which assign either active or passive roles to P-protein in the functioning of the sieve element. Recent biochemical investigations make many of these ideas seem unreasonable (SABNIS AND HART 1979).

*Active Roles.* Superficial ultrastructural similarities between P-protein filaments or tubules and F-actin or microtubules prompted early suggestions that P-protein may actively participate in translocation as some form of contractile protein (FENSOM 1975). However, it has become clear that there is complete dissimilarity in molecular weight, charge amino acid composition, ATPase activity and response to drugs between P-proteins and known contractile proteins (WILLIAMSON 1972, BEYENBACH et al. 1974, PALEVITZ and HEPLER 1975, SABNIS and HART 1974, 1979). An active role for P-protein in translocation via an involvement in surface flow phenomena (LEE 1972, RICHMOND and WARDLAW 1976) has been criticized on biophysical grounds (AIKMAN and WILDON 1978); furthermore, the multiplicity of substances transported in the phloem, and therefore the numerous types of interacting site required, render such a process unlikely to be a major driving force in translocation (ZIEGLER 1975). The postulates of electro-osmosis (SPANNER 1978) require that the machinery, P-protein filaments, bears a significant negative charge. However, at least in cucurbit species, it has been established that the major P-proteins are strongly basic; furthermore, the relative amounts of acidic and basic proteins vary in exudate from different genera and several monocotyledonous plants are totally lacking in P-protein.

*Passive Roles.* Broad similarities in size, charge and antigenicity between P-proteins of *C. maxima* and a ribosomal protein fraction led to the suggestion that P-proteins are formed during cytoplasmic degradation accompanying sieve element maturation (WEBER et al. 1974). However, P-protein bodies appear early in sieve element ontogeny, before any visible signs of nuclear or cytoplasmic degeneration; they also appear in such undegenerated cells as phloem parenchyma or companion cells. Additionally, the large molecular weight of certain subunits and the pronounced differences in the P-protein complements of different species argue against the possibility that they represent modified products of any particular protein. Similarly, the range of subunit sizes and the differences between species render it unlikely that P-proteins represent storage proteins. Cucurbit seed globulins from various species are electrophoretically very similar to each other, but differ from P-proteins (HARA et al. 1978). The early establishment of the P-protein complement in the seedling, followed by its stability through flowering to senescence in a wide range of environments also suggest a storage role is unlikely. With the translocation stream continuously flowing through the sieve tube system, there is obviously a need for some form of wound-sealing mechanism, a role proposed many times for P-proteins (PARTHASARATHY 1975). However, the benefits accruing from the assembly of large amounts of protein complexes, permanently anchored in the path of the translocation stream, seem doubtful when an efficient and reversible callose-synthesizing system already exists for this purpose (ESCHRICH 1975).

*Roles in Recognition Processes.* A major feature of P-proteins at present seems to be their heterogeneity, with different protein complements being characteristic of different species. Since in a highly specialized tissue like phloem it is unlikely that the different P-proteins fulfill different roles, it may be that the heterogeneity is an essential feature of the role itself and that P-proteins are involved in some form of recognition process. Such a role may also be related to the seemingly constant feature of lectin activity in phloem sap. It is possible to envisage various types of situation involving recognition; these include proposals that the phloem lectins may be involved in some form of host–parasite interaction (KAUSS and ZIEGLER 1974, ALLEN 1979) or that the lectin may be involved in localizing or anchoring the P-protein assemblies within the translocation stream (SABNIS and HART 1978). Neither of these proposals resolves the problem of P-protein heterogeneity. A role for P-proteins in some form of recognition between plants would give reason to the existence of different P-protein complements and perhaps also to their location in the translocation system. However, until more is known about the significance in nature of interactions between the vegetative parts of plants, and indeed about whether P-proteins are located exclusively in the phloem, such concepts are entirely speculative.

Thus, the presence of P-proteins in the sieve tubes of most angiosperms, in constant and specific patterns, established early in the life of the seedling and comprising subunits whose large size, assembly and maintenance must represent a considerable expenditure of energy, still provides a considerable enigma.

## 4 General Conclusions

There are no obvious biological or functional connections between tubulins and P-proteins, although their superficially similar ultrastructure has promoted misleading comparisons. Consideration of this common morphological form prompts two concluding remarks. First, both microtubules and P-protein fibrils were discovered and characterized as structures seen under the electron microscope, but lack a clearly identifiable cellular function. In the case of MT and tubulins, their ubiquitous distribution in eukaryotes and their association with a large variety of cellular processes has permitted a general (though not universal) agreement that they are involved in a cytoskeletal role. On the other hand, P-proteins have a much more restricted distribution in a highly specialized cell type, and have offered fewer clues to their function.

The second point concerns the significance of tubular or fibrillar structures. It was pointed out some time ago (CRANE 1950) that self-assembly of asymmetric but identical subunits must lead to a helical conformation, and, further (PAULING 1953) that, depending on the resolution of observation, such a structure would appear as some form of beaded or striated tubule or fibril. The occurrence of such structures in biological systems is well documented (KUSHNER 1969), but the numbers of different types of such structural polymers in any cell or tissue appears to be limited. However, the situation in the sieve element where different proteins assemble into fibrillar structures characteristic for a species is quite extraordinary. The significance of this phenomenon is not evident, but must lie outside simple thermodynamic considerations.

## References


Adelman MR, Borisy GG, Shelanski ML, Weisenberg RC, Taylor EW (1968) Cytoplasmic filaments and tubules. Fed Proc 27:1186–1193

Aikman DP, Wildon DC (1978) Phloem transport: The surface flow hypothesis. J Exp Bot 29:387–393

Allen AK (1979) A lectin from the exudate of the fruit of the vegetable marrow (*Cucurbita pepo*) that has a specificity for β-1,4-linked N-acetylglucosamine oligosaccharides. Biochem J 183:133–137

Amos LA, Linck RW, Klug A (1976) Molecular structure of flagellar microtubules. In: Goldman R, Pollard T, Rosenbaum JL (eds) Cell motility. Book C. Cold Spring Habror Lab, pp 847–867
Anderson PJ (1979) The structure and amount of tubulin in cells and tissues. J Biol Chem 254:2168–2171
Baccetti B, Burrini AG, Dalai R, Pallini V (1979) The dynein electrophoretic bands in axonemes naturally lacking the inner or the outer arm. J Cell Biol 80:334–340
Bajer A, Molè-Bajer J (1975) Lateral movements in the spindle and the mechanism of mitosis. In: Inoué S, Stephens RE (eds) Molecules and cell movement. Raven Press, New York, pp 77–96
Baum P, Thorner J, Honig L (1978) Identification of tubulin from the yeast *Saccharomyces cerevisiae*. Proc Natl Acad Sci USA 75:4962–4966
Becker D, Kluge M, Ziegler H (1971) Der Einbau von $^{32}PO_4{}^{2-}$ in organische Verbindungen durch Siebröhrensaft. Planta 99:154–162
Behnke H-D (1969) Über den Feinbau und die Ausbreitung der Siebröhren-Plasmafilamente und über Bau und Differenzierung der Siebporen bei einigen Monocotylen und bei *Nuphar*. Protoplasma 68:377–402
Behnke H-D (1975) Phloem tissue and sieve elements in algae, mosses and ferns. In: Aronoff S, Dainty J, Gorham PR, Srivastava LM, Swanson CA (eds) Phloem transport. NATO Adv Study Inst Ser, Plenum Press, New York, London, pp 187–210
Bentwood BJ, Cronshaw J (1976) Biochemistry and cytochemical localisation of acid phosphatase in the phloem of *Nicotiana tabacum*. Planta 130:97–104
Bentwood BJ, Cronshaw J (1978) Cytochemical localisation of adenosine triphosphatase in the phloem of *Pisum sativum* and its relation to the function of transfer cells. Planta 140:111–120
Berkowitz SA, Katagiri J, Binder HK, Williams RC Jnr (1977) Separation and characterization of microtubule proteins from calf brain. Biochemistry 16:5610–5617
Beyenbach J, Weber C, Kleinig H (1974) Sieve tube proteins from *Cucurbita maxima*. Planta 119:113–124
Bhattacharyya B, Wolff J (1976) Polymerisation of membrane tubulin. Nature (London) 264:576–577
Bibring T, Baxandall J, Denslow S, Walker B (1976) Heterogeneity of the α subunit of tubulin within a single organism. J Cell Biol 69:301–312
Bieleski RL (1969) Phosphorus compounds in translocating phloem. Plant Physiol 44:497–502
Binder LI, Rosenbaum JL (1978) In vitro assembly of flagellar outer doublet tubulin. J Cell Biol 79:500–515
Binder LI, Dentler WK, Rosenbaum JL (1975) Assembly of chick brain tubulin onto flagellar microtubules from *Chlamydomonas* and sea urchin sperm. Proc Natl Acad Sci USA 72:1122–1126
Bloodgood RA (1974) Resorption of organelles containing microtubules. Cytobios 9:142–161
Bloodgood RA, Rosenbaum JL (1976) Initiation of brain tubulin aasembly by a higher molecular weight flagellar protein factor. J Cell Biol 71:322–331
Borgers M, DeBrabander M (eds) (1975) Microtubules and microtubule inhibitors. North-Holland, Amsterdam Oxford
Borisy GG, Olmsted JB, Marcum JM, Allen C (1974) Microtubule assembly in vitro. Fed Proc 33:167–174
Brodie AE, Potter J, Reed DJ (1979) Effects of vinblastine, oncodazole, porcarbazine, chlorambucil and bleomycin *in vivo* on colchicine binding activity of tubulin. Life Sci 24:1547–1554
Bryan J (1974) Biochemical properties of microtubules. Fed Proc 33:152–157
Bulinski JC, Borisy GG (1970) Self-assembly of microtubules in extracts of cultured HeLa cells and the identification of HeLa microtubule-associated proteins. Proc Natl Acad Sci USA 76:293–297
Burgess J (1970) The occurrence of cross-linked microtubules in a higher plant cell. Planta 92:25–28

Callow JA (1976) Plant lectins. In: Smith H (ed) Commentaries in plant science. Pergamon Press, Oxford, pp 221–233
Cappucinelli P, Hames BD (1978) Characterization of colchicine-binding activity in *Dictyostelium discoideum*. Biochem J 169:499–504
Caron JM, Berlin RD (1979) Interaction of microtubule proteins with phospholipid vesicles. J Cell Biol 81:665–672
Catesson AM (1973) Observations cytochimique sur les tubes criblés de quelques angiospermes. J Microsc 16:95–104
Clayton L, Pogson CI, Gull K (1979) Microtubule proteins in the yeast *Saccharomyces cerevisiae*. FEBS Lett 106:67–70
Cleveland DW, Hwo S-Y, Kirschner MW (1977a) Purification of tau, a microtubule-associated protein that induces assembly of microtubules from purified tubulin. J Mol Biol 116:207–225
Cleveland DW, Hwo S-Y, Kirschner MW (1977b) Physical and chemical properties of purified tau factor and the role of tau in microtubule assembly. J Mol Biol 116:227–247
Crane HR (1950) Principles and problems of biological growth. Sci Mon 70:376–389
Cronshaw J (1975) P-proteins. In: Aronoff S, Dainty J, Gorham PR, Srivastava LM, Swanson CA (eds) Phloem transport. NATO Adv Study Inst Ser, Plenum Press, New York London, pp 79–147
Cronshaw J, Anderson R (1971) Phloem differentiation in tobacco pith culture. J Ultrastruct Res 34:244–259
Cronshaw J, Esau K (1967) Tubullar and fibrillar components of mature and differentiating sieve elements. J Cell Biol 34:801–816
Cronshaw J, Esau K (1968a) P-protein in the phloem of *Cucurbita*. I. The development of P-protein bodies. J Cell Biol 38:25–39
Cronshaw J, Esau K (1968b) P-protein in the phloem of *Cucurbita*. II. The P-protein of mature sieve elements. J Cell Biol 38:292–303
Cronshaw J, Gilder J, Stone D (1973) Fine structural studies of P-proteins in *Cucurbita, Cucumis* and *Nicotiana*. J Ultrastruct Res 45:192–205
Daleo GR, Piras MM, Piras R (1974) The presence of phospholipids and diglyceride kinase activity in microtubules from different tissues. Biochem Biophys Res Commun 61:1043–1050
David-Pfeuty T, Simon C, Pantaloni D (1979) Effect of antimitotic drugs on tubulin GTPase activity and self-assembly. J Biol Chem 254:11696–11702
Dentler WL, Rosenbaum JL (1977) Flagellar elongation and shortening in *Chlamydomonas*. III. Structures attached to the tips of flagellar microtubules and their relationship to the directionality of flagellar microtubule assembly. J Cell Biol 74:747–759
Dentler WL, Granett S, Rosenbaum JL (1975) Ultrastrucutral localization of the high molecular proteins associated with *in vitro*-assembled brain microtubules. J Cell Biol 65:237–241
Deysson G (1968) Antimitotic substances. Int Rev Cytol 24:99–148
Dickson RE (1977) EDTA – promoted exudation of $^{14}$C-labelled compounds from detached cottonwood and bean leaves as related to translocation. Can J Bot 7:277–284
Dustin P (1978) Microtubules. Springer, Berlin Heidelberg New York
Eipper BA (1972) Rat brain microtubule protein: purification and determination of covalently bound phosphate and carbohydrate. Proc Natl Acad Sci USA 69:2283–2287
Eipper BA (1974) Properties of rat brain tubulin. J Biol Chem 249:1407–1416
Eipper BA (1975) Purification of rat brain tubulin. Ann NY Acad Sci 253:239–246
Erickson HP (1976) Facilitation of microtubule assembly by polycations. In: Goldman R, Pollard T, Rosenbaum J (eds) Cell motility. Cold Spring Harbor Lab, pp 1069–1080
Erickson HP, Scott B (1977) Microtubule assembly in DEAE dextran: effect of charge density and MW of the polycation. Biophys J 17:274a
Esau K (1969) The phloem. In: Zimmermann W, Ozenda P, Wulff HD (eds) Handbuch der Pflanzenanatomie Bd V, Teil 2. Borntraeger, Berlin
Esau K (1975) The ploem of *Nelumbo nucifera* Gaertn. Ann Bot (London) 39:901–913
Esau K (1978a) Developmental features of the primary phloem in *Phaseolus vulgaris* L. Ann Bot (London) 42:1–13

Esau K (1978b) The protein inclusions in sieve elements of cotton (*Gossypium hirsutum* L.). J Ultrastruct Res 63:224–235
Esau K, Cronshaw J (1967) Tubular components in cells of healthy and tobacco mosaic virus-infected *Nicotiana*. Virology 33:26–35
Esau K, Gill RH (1973) Correlation in differentiation of protophloem sieve elements of *Allium cepa* root. J Ultrastruct Res 44:310–328
Eschrich W (1975) Sealing systems in phloem. In: Zimmermann MH, Milburn JA (eds) Transport in plants. I. Phloem transport. Encyclopedia of plant physiology, new series, Vol 1. Springer, Berlin Heidelberg New York, pp 39–56
Eschrich W, Heyser W (1975) Biochemistry of phloem constituents. In: Zimmerman MH, Milburn JA (eds) Transport in plants. I. Phloem transport. Encyclopedia of plant physiology, new series Vol 1. Springer, Berlin Heidelberg New York, pp 101–136
Eschrich W, Evert RF, Heyser W (1971) Proteins of the sieve-exudate of *Cucurbita maxima*. Planta 100:208–221
Evert RF (1977) Phloem structure and histochemistry. Ann Rev Physiol 28:199–222
Evert RF, Deshpande BP (1969) Electron microscope investigation of sieve element ontogeny and structure in *Ulmus americana*. Protoplasma 68:403–432
Evert RF, Eschrich W, Eichorn SE (1973) P-protein distribution in mature sieve elements of *Cucurbita maxima*. Planta 109:193–210
Feit H, Shelanski ML (1975) Is tubulin a glycoprotein? Biochem Biophys Res Commun 66:920–927
Fensom DS (1975) Other possible mechanisms. In: Zimmermann MH, Milburn JA (eds) Transport in plants. I Phloem transport. Encyclopedia of plant physiology new series Vol 1. Springer, Berlin Heidelberg New York, pp 354–366
Figier J (1968) Localisation infrastructurale de la phosphomonoestérase acide dans la stipule de *Vicia faba* L. au niveau du nectaire. Planta 83:60–79
Filner P, Yadav NS (1979) Role of microtubules in intracellular movements. In: Haupt W, Feinleib ME (eds) Physiology of movements. Encyclopedia of plant physiology, new series Vol 7. Springer, Berlin Heidelberg New York, pp 95–113
Flanagan D, Warr JR (1977) Colchicine binding of a high-speed supernatant of *Chlamydomonas reinhardi*. FEBS Lett 80:14–18
Forgue ST, Dahl JL (1979) Rat brain tubulin: subunit heterogeneity and phosphorylation. J Neurochem 32:1015–1025
Freundlich A (1974) No polysaccharide demonstrated in filamentous structures in sieve elements by Thiery's periodic acid-thiocarbohyrazide-silver proteinate method for electron microscopy. Planta 118:85–87
Fulton C, Simpson PA (1976) Selective synthesis and utilisation of flagellar tubulin. The multi-tubulin hypothesis. In: Goldman R, Pollard T, Rosenbaum JL (eds) Cell motility. Book C. Cold Spring Harbor Lab, pp 987–1005
Gardner DC, Peel AJ (1972) Some observations on the role of ATP in sieve tube translocation. Planta 107:217–226
Garland DL (1978) Kinetics and mechanisms of colchicine binding to tubulin: evidence for ligand-induced conformational change. Biochemistry 17:4266–4271
Gibbons IR, Rowe AJ (1965) Dynein: a protein with adenosine triphosphatase activity from cilia. Science 149:424–426
Gibbons IR, Fronk E, Gibbons BH, Ogawa K (1976) Multiple forms of dynein in sea urchin sperm flagella. In: Goldman R, Pollard T, Rosenbaum J (eds) Cell motility. Cold Spring Harbor Lab, pp 915–932
Gietl C, Krauss H, Ziegler H (1979) Affinity chromatography of a lectin from *Robinia pseudoacacia* L. and demonstration of lectins in sieve tube sap from other tree species. Planta 144:367–372
Gilder J, Cronshaw J (1973a) Adenosine triphosphatase in the phloem of *Cucurbita*. Planta 110:189–204
Gilder J, Cronshaw J (1973b) The distribution of adenosine triphosphatase activity in differentiating and mature phloem cells of *Nicotiana tabacum* and its relationship to phloem transport. J Ultrastruct Res 44:388–404

Gilder J, Cronshaw J (1974) A biochemical and cytochemical study of adenosine triphosphatase activity in the phloem of *Nicotiana tabacum*. J Cell Biol 60:221–235
Gillespie E (1975) The mechanism of tubulin breakdown in vitro. FEBS Lett 58:119–121
Goldman R, Pollard T, Rosenbaum J (eds) (1976) Cell motility. Book C Cold Spring Harbor Lab, pp 839–1373
Goodman DBP, Rasmussen H, Dibella F, Guthrow CE Jnr (1970) Cyclic adenosine 3′-5′-monophosphate-stimulated phosphorylation of isolated neurotubule subunits. Proc Natl Acad Sci USA 67:652–659
Gould RR, Borisy GG (1978) Quantitative initiation of microtubule assembly by chromosomes of CHO cells. Exp Cell Res 113:369–374
Gunning BES, Hardham AR, Hughes JE (1978) Evidence for initiation of microtubules in discrete regions of the cell cortex in *Azolla* root-tip cells, and an hypothesis on the development of cortical arrays of microtubules. Planta 143:161–179
Haber JE, Peloquin JG, Halvorson HO, Borisy GG (1972) Colcemid inhibition of cell growth and the characterization of a colcemid binding activity in *Saccharomyces cerevisiae*. J Cell Biol 58:355–367
Hall SM, Baker A (1972) The chemical composition of *Ricinus* phloem exudate. Planta 106:131–140
Hara I, Ohmiya M, Matsubara H (1978) Pumpkin (*Cucurbita* sp.) seed globulin. III. Comparison of subunit structures among seed globulins of various *Cucurbita* species and characterization of peptide components. Plant Cell Physiol 19:237–243
Hardham AR, Gunning BES (1978) Structure of cortical microtubule arrays in plant cells. J Cell Biol 77:14–34
Hart JW, Sabins DD (1973) Colchicine binding protein from phloem and xylem of a higher plant. Planta 109:147–152
Hart JW, Sabnis DD (1976a) Colchicine and plant microtubules: a critical evaluation. Curr Adv Plant Sci 26:1095–1104
Hart JW, Sabnis DD (1976b) Binding of colchicine and lumicolchicine to components in plant extracts. Phytochemistry 15:1897–1901
Hart JW, Sabnis DD (1976c) Colchicine binding activity in extracts of higher plants. J Exp Bot 27:1353–1360
Hart JW, Sabnis DD (1977) Microtubules. In: Smith H (ed) The molecular biology of plant cells. Blackwell, Oxford, pp 160–181
Heath IP (1974) A unified hypothesis for the role of membrane bound enzyme complexes and microtubules in plant cell wall synthesis. J Theor Biol 48:445–449
Hepler PK (1976) Plant microtubules. In: Bonner J, Varner JE (eds) Plant biochemistry. Academic Press, New York London, pp 147–187
Hepler PK, McIntosh JR, Cleveland S (1970) Intermicrotubule bridges in mitotic spindle apparatus. J Cell Biol 45:438–444
Hepler PK, Palevitz BA (1974) Microtubules and microfilaments. Ann Rev Plant Physiol 25:309–362
Herzog W, Weber K (1978) Microtubule formation by pure tubulin in vitro. The influence of dextran and poly(ethylene glycol). Eur J Biochem 91:249–254
Heyser W, Eschrich W, Huttermann A, Evert FR, Burchardt R, Fritz E, Heyser R (1974) Phosphodiesterase in sieve-tube exudate of *Cucurbita maxima*. Z Pflanzenphysiol 71:413–423
Himes RH, Burton PR, Kersey RN, Pierson GB (1976) Brain tubulin polymerization in the absence of "microtubule-associated proteins". Proc Natl Acad Sci USA 73:4397–4399
Hotta Y, Shepard J (1973) Biochemical aspects of colchicine action on meiotic cells. Mol Gen Genet 122:243–260
Hyams JS, Borisy GG (1978) Nucleation of microtubules in vitro by isolated spindle pole bodies of the yeast *Saccharomyces cerevisiae*. J Cell Biol 78:401–414
Ikeda Y, Steiner M (1979) Phosphorylation and protein kinase activity of platelet tubulin. J Biol Chem 254:66–74
Ilker R, Currier HP (1975) Histochemical studies of an inclusion body and P-protein in phloem of *Xylosma congestum*. Protoplasma 85:127–132

Inoué S, Stephens RE (eds) (1975) Molecules and cell movement. Vol 30. Soc Gen Physiol Series, Raven Press, New York
Jacobs M, Smith H, Taylor EW (1974) Tubulin: nucleotide binding and enzymic activity. J Biol 89:455–468
Johnson RPC (1978) The microscopy of P-protein filaments in freeze-etched sieve pores. Brownian motion limits resolution of their positions. Planta 143:191–205
Kauss H (1976) Plant lectins (phytohaemagglutinins). Fortschr Bot 36:58–70
Kauss H, Ziegler H (1974) Carbohydrate-binding proteins from the sieve-tube sap of *Robinia pseudoacacia* L. Planta 121:197–200
Kemphues KJ, Raff RA, Kaufman TC, Raff EC (1979) Mutation in a structural gene for a β-tubulin specific to testis in *Drosophila melanogaster*. Proc Natl Acad Sci USA 76:3991–3995
Kim H, Binder LI, Rosenbaum JL (1979) The periodic association of $MAP_2$ with brain microtubules in vitro. J Cell Biol 80:266–276
King RW, Zeevart JAD (1974) Enhancement of phloem exudation from cut petioles by chelating agents. Plant Physiol 53:96–103
Kirschner MW (1978) Microtubule assembly and nucleation. Int Rev Cytol 54:1–71
Kleinig H, Dörr I, Kollman R (1971a) Vinblastine-induced precipitation of phloem proteins *in vitro*. Protoplasma 73:293–302
Kleinig H, Dörr I, Weber C, Kollman R (1971b) Filamentous proteins from plant sieve tubes. Nature New Biol 229:152–153
Kleinig H, Thones J, Dörr I, Kollman R (1975) Filament formation *in vitro* of a sieve tube protein from *Cucurbita maxima* and *Cucurbita pepo*. Planta 127:163–170
Kluge M, Ziegler H (1964) Der ATP-Gehalt der Siebröhrensäfte von Laubbäumen. Planta 61:167–177
Kluge M, Becker D, Ziegler H (1970) Untersuchungen über ATP und andere organische Phosphorverbindungen im Siebröhrensaft von *Yucca flaccida* und *Salix triandra*. Planta 91:68–79
Kobayashi T (1975) Dephosphorylation of tubulin-bound guanosine triphosphate during microtubule assembly. J Biochem 77:1193–1198
Kollman R, Dörr I, Kleinig H (1970) Protein filaments – structural components of the phloem exudate. I. Observations with *Cucurbita* and *Nicotiana*. Plant 95:86–94
Kretsinger RH (1976) Calcium-binding proteins. Ann Rev Biochem 45:239–265
Kuriyama R (1976) In vitro polymerisation of flagellar and ciliary outer fibre tubulin into microtubules. J Biochem 80:153–166
Kushner DJ (1969) Self assembly of biological strucutes. Bacteriol Rev 33:302–345
Lamoureux CH (1975) Phloem tissue in angiosperms and gymnosperms. In: Aronoff S, Dainty J, Gorham PR, Srivastava LM, Swanson CA (eds) Phloem transport. NATO Adv Study Inst Ser, Pleum Press, New York London, pp 1–20
Lee DR (1972) The possible significance of filaments in sieve elements. Nature (London) 235:266
Liener IE (1976) Phytohemagglutinins (Phytolectins). Ann Rev Plant Physiol 27:291–319
Lloyd CW, Slabas AR, Powell AJ, MacDonald G, Badley RA (1979) Cytoplasmic microtubules of higher plant cells visualised with antitubulin antibodies. Nature (London) 279:239–241
Lloyd CW, Slabas AR, Powell AJ, Lowe SB (1980) Microtubules, protoplasts and plant cell shape. An immunofluorescent study. Planta 147:500–506
Lockwood AH (1978) Tubulin assembly protein: immunochemical and immunofluorescent studies on its function and distribution in microtubules and cultured cells. Cell 13:613–628
Lu RC, Elzinga M (1977) Chromatographic resolution of the subunits of calf brain tubulin. Anal Biochem 77:243–250
Lu RC, Elzinga M (1978) The primary structure of tubulin. Sequences of the carboxyl terminus and seven other cyanogen bromide peptides from the α-chain. Biochem Biophys Acta 537:320–328
Luduena RF, Woodward DO (1973) Isolation and partial characterisation of α- and β-

tubulin from outer doublets of sea urchin sperm and microtubules of chick embryo brain. Proc Natl Acad Sci USA 70:3594–3598
Luduena RF, Woodward DO (1975) α- and β-tubulin: separation and partial sequence analysis. Ann NY Acad Sci 253:272–283
Luduena RF, Shooter EM, Wilson L (1977) Structure of the tubulin dimer. J Biol Chem 252:7006–7014
McEuen AR (1979) Studies on calcium and a calcium-binding protein in the sieve tube exudate of *Cucurbita maxima* and related species. Ph. D. Thesis Aberdeen Univ
McIntosh JR (1974) Bridges between microtubules. J Cell Biol 61:166–187
McIntosh JR, Hepler PK, Wie DG van (1969) Model for mitosis. Nature (London) 224:659–663
Margolis RK, Margolis RU, Shelanski ML (1972) The carbohydrate composition of brain microtubule protein. Biochem Biophys Res Commun 47:432–437
Marotta CA, Harris JL, Gilbert JM (1978) Characterisation of multiple forms of brain tubulin subunits. J Neurochem 30:1431–1440
Marshall LE, Himes RH (1978) Rotenone inhibition of tubulin self assembly. Biochim Biophys Acta 543:590–594
Mellon MG, Rebhun LI (1976) Sulfhydryls and *in vitro* polymerisation of tubulin. J Cell Biol 70:226–238
Mohri H (1976) The function of tubulin in motile systems. Biochim Biophys Acta 456:85–127
Murphy DB (1975) The mechanism of microtubule-dependent movement of pigment granules in teleost chromatophores. Ann NY Acad Sci 253:692–701
Nath J, Flavin M (1979) Tubulin tyrosylation *in vivo* and changes accompanying differentiation of cultured neuroblastoma-glioma hybrid cells. J Biol Chem 254:11505–11510
Nicolson GL (1976) Transmembrane control of the receptors on normal and tumor cells. I. Cytoplasmic influence over cell surface components. Biochim Biophys Acta 457:57–108
Nuske J, Eschrich W (1976) Synthesis of P-protein in mature phloem of *Cucurbita maxima*. Planta 132:109–118
Oberhäuser R, Kollman R (1977) Cytochemische Charakterisierung des sogenannten „Freien Nucleolus" als Proteinkörper in den Siebelementen von *Passiflora coerulea*. Z Pflanzenphysiol 84:61–75
Olmsted JB, Witman GB, Carlson K, Rosenbaum JL (1971) Comparison of the microtubule proteins of neuroblastoma cells, brain and *Chlamydomonas* flagella. Proc Natl Acad Sci USA 68:2273–2277
Olson LW (1973) A low molecular weight colchicine-binding protein from the aquatic phycomycete *Allomyces neo-moniliformis*. Arch Mikrobiol 91:281–286
Palevitz BA, Hepler PK (1975) Is P-protein actin-like? Not yet. Planta 125:261–271
Palevitz BA, Newcomb EH (1971) The ultrastructure and development of tubular and crystalline P-protein in the sieve elements of certain papilionaceous legumes. Protoplasma 72:399–426
Parthasarathy MV (1974) Ultrastructure of phloem in palms. I. Immature sieve elements and parenchymatic elements. Protoplasma 79:59–91
Parthasarathy MV (1975) Sieve element structure. In: Zimmermann MH, Milburn JA (eds) Transport in plants. I. Phloem transport. Encyclopedia of plant physiology, new series Vol. 1. Springer, Berlin Heidelberg New York, pp 3–38
Parthasarathy MV, Mühlethaler K (1969) Ultrastructure of protein tubules in differentiating sieve elements. Cytobiologie 1:17–36
Pauling L (1953) Protein interactions. Aggregation of globular proteins. Disc Faraday Soc 13:170–176
Pickett-Heaps JD (1974) Plant microtubules. In: Robards AW (ed) Dynamic aspects of plant ultrastructure. McGraw-Hill, New York, London, pp 219–225
Pipeleers DG, Pipeleers-Marichal MA, Sherline P, Kipnis DM (1977) A sensitive method for measuring polymerised and depolymerised forms of tubulin in tissues. J Cell Biol 74:341–350
Piperno G, Luck D (1974) Isolation of phosphorylated tubulin from *Chlamydomonas* flagella. J Cell Biol 63:271a

Piperno G, Luck DJ (1976) Phosphorylation of axonemal proteins in *Chlamydomonas reinhardtii*. J Biol Chem 251:2161–2167
Quinn PJ (1973) The association between phosphatidylinositol phosphodiesterase activity and a specific subunit of microtubular protein in rat brain. Biochem J 133:273–281
Rappaport L, Leterrier JF, Virion A, Nunez J (1976) Phosphorylation of microtubule-associated proteins. Eur J Biochem 62:539–550
Richmond P, Wardlaw IF (1976) On the translocation of sugar: van der Waals' forces and surface flow. Aust J Plant Physiol 3:545–549
Robards AW, Kidwai P (1972) Microtubules and microfibrils in xylem fibres during secondary cell wall formation. Cytobiologie 6:1–21
Rosenbaum JL, Binder LI, Granett S, Dentler WL, Snell W, Sloboda R, Haimo L (1975) Directionality and rate of assembly of chick brain tubulin onto pieces of neurotubules, flagellar axonomes and basal bodies. Ann NY Acad Sci 253:147–177
Rosenfeld AC, Zackroff RV, Weisenberg RC (1976) Magnesium stimulation of calcium binding to tubulin and calcium induced depolymerisation of microtubules. FEBS Lett 65:144–147
Rost TL, Gifford EM (eds) (1977) Mechanisms and control of cell division. Dowden, Hutchinson and Ross, Stroudsburg, PA
Rubin RW, Cousins EH (1976) Isolation of a tubulin-like protein from *Phaseolus*. Phytochemistry 15:1837–1839
Runge MS, Detrich III HW, Williams Jr RC (1979)Identification of the major 68000 – dalton protein of microtubule preparations as a 10 nm filament protein and its effects on microtubule assembly in vitro. Biochemistry 18:1689–1698
Sabnis DD, Hart JW (1974) Studies on the possible occurrence of actomyosin-like proteins in phloem. Planta 118:271–281
Sabnis DD, Hart JW (1976) A comparative analysis of phloem exudate proteins from *Cucumis melo*, *Cucumis sativus* and *Cucurbita maxima* by polyacrylamide gel electrophoresis and isoelectric focusing. Planta 130:211–218
Sabnis DD, Hart JW (1978) The isolation and some properties of a lectin (haemagglutinin) from *Cucurbita* phloem exudate. Planta 142:97–101
Sabnis DD, Hart JW (1979) Heterogeneity in phloem protein complements from different species. Consequences to hypothesis concerned with P-protein function. Planta 145:459–466
Sandoval IV, Cuatrecasas P (1976a) Proteins associated with tubulin. Biochem Biophys Res Commun 68:169–177
Sandoval IV, Cuatrecasas P (1976b) Protein kinase associated with tubulin: affinity chromatography and properties. Biochemistry 15:3424–3432
Sandoval IV, Weber K (1978) Calcium induced inactivation of microtubule formation in brain extracts. Presence of a calcium-dependent protease acting on polymerisation-stimulating microtubule-associated proteins. Eur J Niochem 92:463–470
Satir P, Ojakian GK (1979) Plant Cilia. In: Haupt W, Feinleib ME (eds) Physiology of Movements. Encyclopedia of plant physiology, new series, vol 7. Springer, Berlin Heidelberg New York
Sauter JJ (1972) Cytochemical demonstration of sulfhydryl disulfide-containing proteins in sieve elements of conifers. Naturwissenschaften 10:470
Schmitt H, Kram R (1978) Binding of antimitotic drugs around cysteine residues of tubulin. Exp Cell Res 115:408–411
Schmitt H, Littauer YZ (1974) Tubulin. In: Jakoby WB, Wilchek M (eds) Methods in enzymol affinity techniques. Enzyme purification Vol 34 part B Academic Press, New York London, pp 623–627
Schnepf E (1974) Microtubules and cell wall formation. Port Acta Biol Ser A 14:451–461
Schnepf E, Stein U, Deichgräber G (1978) Structure, function and development of the peristome of the moss, *Rhacopilum tomentosum*, with special reference to the problem of microfibril orientation by microtubules. Protoplasma 97:221–240
Sheir-Neiss G, Nardi RV, Gealt MA, Morris NR (1976) Tubulin-like protein from *Aspergillus nidulans*. Biochem Biophys Res Commun 69:285–290
Shelanski ML, Gaskin F, Cantor CR (1973) Microtubule assembly in the absence of added nucleotides. Proc Natl Acad Sci USA 70:765–768

Sherline P, Schiavone K (1978) High molecular weight MAPs are part of the mitotic spindle. J Cell Biol 77:pp R9–R12
Slabas AR, MacDonald G, Lloyd CW (1980) Selective purification of plant proteins which co-polymerise with mammalian microtubules. FEBS Lett 110:77–79
Sleigh MA (ed) (1974) Cilia and flagella. Academic Press, New York London
Sloan RT (1977) A biochemical study of P-protein. Ph. D. Thesis, Aberdeen Univ
Sloan RT, Sabnis DD, Hart JW (1976) The heterogeneity of phloem exudate proteins from different plants: A comparative survey of ten plants using polyacrylamide gel electrophoresis. Planta 132:97–102
Sloboda RD, Rosenbaum JL (1979) Decoration and stabilization of intact, smooth-walled microtubules with microtubule-associated proteins. Biochemistry 18:48–55
Sloboda RD, Rudolph SA, Rosenbaum JL (1975) Cyclic AMP-dependent endogenous phosphorylation of a microtubule-associated protein. Proc Natl Acad Sci USA 72:177–181
Sloboda RD, Dentler WL, Bloodgood RA, Teizer BR, Granett S, Rosenbaum JL (1976a) Microtubule-associated proteins (MAPS) and the assembly of microtubules in vitro. In: Goldman R, Pollard T, Rosenbaum J (eds) Cell motility. Cold Spring Harbor Lab, pp 1171–1212
Sloboda RD, Dentler WL, Rosenbaum JL (1976b) Microtubule-associated proteins and the stimulation of tubulin assembly in vitro. Biochemistry 15:4497–4505
Snell WJ, Dentler WL, Haimo LT, Binder LI, Rosenbaum JL (1974) Assembly of chick brain tubulin onto isolated basal bodies of *Chlamydomonas reinhardi*. Science 185:357–360
Snyder JA, McIntosh JR (1976) Biochemistry and physiology of microtubules. Ann Rev Biochem 45:699–720
Soifer D (ed) (1975) The biology of cytoplasmic microtubules. Ann NY Acad Sci 253:pp 1–848
Soifer D, Laszlo AH, Scotto JM (1972) Enzymatic activity in tubulin preparations I. Intrinsic protein kinase activity in lyophilized preparations of tubulin from porcine brain. Biochim Biophys Acta 271:182–192
Spanner DC (1978) Sieve plate pores, open or occluded? A critical review. Plant Cell Environ 1:7–20
Stearns MV, Brown DL (1979) Purification of a microtubule-associated protein based on its preferential association with tubulin during microtubule initiation. FEBS Lett 101:15–20
Steer MW, Newcomb EH (1969) Development and dispersal of P-protein in the phloem of *Coleus blumei* Benth. J Cell Sci 4:155–169
Stephens RE (1974) Enzymatic and structural proteins of the axoneme. In: Sleigh MA (ed) Cilia and Flagella. Academic Press, New York London, pp 39–78
Stephens RE (1977) Major membrane protein differences in cilia and flagella: evidence for a membrane associated tubulin. Biochemistry 16:2047–2058
Stephens RE (1978) Primary structural differences among tubulin subunits from flagella, cilia and cytoplasm. Biochemistry 17:2882–2991
Stephens RE, Edds KT (1976) Microtubules: structure, chemistry and function. Physiol Rev 56:709–777
Stone DL, Cronshaw J (1973) Fine structure of P-protein filaments from *Ricinus communis*. Planta 113:193–206
Telzer BR, Rosenbaum JL (1979) Cell cycle dependent in vitro assembly of microtubules onto the pericentriolar material of HeLa cells. J Cell Biol 81:484–489
Thompson WC, Deanin GG, Gordon MW (1979) Intact microtubules are required for rapid turnover of carboxyl-termminal tyrosine of alpha-tubulin in cell cultures. Proc Natl Acad Sci USA 76:1318–1322
Walker TS (1972) The purification and some properties of a protein causing gelling in phloem sieve tube exudate from *Cucurbita pepo*. Biochim Biophys Acta 257:433–444
Warner FD (1976) Ciliary intermicrotubule bridges. J Cell Sci 20:101–114
Watanabe T, Flavin M (1973) Two types of adenosine triphosphatase from flagella of *Chlamydomonas reinhardii*. Biochem Biophys Res Commun 52:195–201

Water RD, Kleinsmith LJ (1976) α- and β-tubulin in yeast. Biochem Biophys Res Commun 70:704–708
Weber C, Kleinig H (1971) Molecular weights of *Cucurbita* sieve tube proteins. Planta 99:179–182
Weber C, Frank WW, Kartenbeck J (1974) Structure and biochemistry of phloem proteins isolated from *Cucurbita maxima*. Exp Cell Res 87:79–106
Wehland J, Herzog W, Weber K (1977) Interaction of griseofulvin with microtubules, microtubule protein and tubulin. J Molec Biol 111:329–342
Weingarten MD, Lockwood AH, Hwo S, Kirschner MW (1975) A protein factor essential for microtubule assembly. Proc Natl Acad Sci USA 72:1858–1862
Weisenberg RC (1972) Microtubule formation in vitro in solutions containing low calcium concentrations. Science 177:1104–1105
Wergin WP, Newcomb EH (1970) Formation and dispersal of crystalline P-protein in sieve elements of soybean (*Glycine max* L.). Protoplasma 71:365–388
Wiche G, Honig LS, Cole RD (1979) Microtubule protein preparations from $C_6$ glial cells and their spontaneous polymer formation. J Cell Biol 80:553–563
Williamson RE (1972) An investigation of the contractile protein hypothesis of phloem translocation. Planta 106:149–157
Wilson L, Bryan J, Ruby A, Mazia D (1970) Precipitation of proteins by vinblastine and calcium ions. Proc Natl Acad Sci USA 66:807–814
Wilson L, Bamburg JR, Mizel SB, Gisham LM, Creswell KM (1974) Interaction of drugs with microtubule proteins. Fed Proc 33:158–166
Witman GB, Carlson K, Berliner J, Rosenbaum JL (1972) *Chlamydomonas* flagella I Isolation and electrophoretic analysis of microtubules matrix, membranes, mastigonemes. J Cell Biol 54:507–539
Vallee RB, Borisy GG (1978) The non-tubulin component of microtubule protein oligomers. Effect on self-association and hydrodynamic properties. J Biol Chem 253:2834–2845
Ventilla M, Cantor CR, Shelanski M (1972) A circular dichroism study of microtubule protein. Biochemistry 11:1554–1561
Yapa PAJ, Spanner DC (1974) Localization of adenosine triphosphatase activity in mature sieve elements of *Tetragonia*. Planta 117:321–328
Zee S-Y (1969) Fine structure of differentiating sieve elements of *Vicia faba*. Aust J Bot 17:441–456
Zeeberg B, Caplow M (1978) Reactions of tubulin-associated guanine nucleotides. J Biol Chem 253:1984–1990
Ziegler H (1975) Nature of transported substances. In: Zimmermann MH, Milburn JA (eds) Transport in plants I. Phloem transport. Encyclopedia of plant physiology, new series Vol 1. Springer, Berlin Heidelberg New York, pp 59–100
Ziegler H, Kluge M (1962) Die Nucleinsäuren und ihre Bausteine im Siebröhrensaft on *Robinia pseudoacacia*. Planta 58:144–153

# 12 Plant Peptides

C.F. HIGGINS and J.W. PAYNE

## 1 Introduction

Despite many early reports of peptide-like compounds in plant tissues (for reviews of these early studies see, BRICAS and FROMAGEOT 1953, SYNGE 1959, 1968, WALEY 1966), this class of compounds has received little attention in recent years. Indeed, the most recent review of the subject was published over 10 years ago (SYNGE 1968). The information pertaining to plant peptides is widely scattered in the literature and an exhaustive coverage will not be attempted here. Rather, a general survey will be made to illustrate the types of peptide found in plant tissues and to allow consideration of their possible functions. In addition, some speculations will be offered on possible roles which peptides might serve in plant cells, in the hope of stimulating greater consideration of this relatively neglected group of compounds.

For present purposes the arbitrary division between peptides and polypeptides will be drawn at a molecular weight of about 1,500. Plant peptides may be conveniently considered in two cateogies. (1) Peptides which have a unique structure or function. Such specific peptides commonly contain unusual amino acids or peptide linkages, not found in proteins or peptides derived from them. However, no account will be taken here of compounds, e.g., the "peptide alkaloids", which, although they may contain peptide linkages, otherwise bear little or no resemblance to protein-derived peptides. (2) Peptides produced as intermediates in the synthesis and degradation of proteins. These would be expected to consist of a mixture of small peptides (i.e., a peptide pool) with an overall composition similar to that of the parent proteins.

In this section, standard three-letter abbreviations for amino acids and amino acid residues will be used (INTERNATIONAL UNION of BIOCHEMISTRY 1978). In the case of D-amino acids and D-amino acid residues, the appropriate prefix will be used, but for the L-isomers the prefix will usually be omitted.

## 2 Peptides with a Specific Structure or Function

Peptides have been shown to serve many important functions in both animal cells and microorganisms (MATTHEWS and PAYNE 1975, PAYNE 1980). In contrast, little is known about the corresponding roles which peptides might serve in plant tissues. Nevertheless, with the identification of ever-increasing numbers of such compounds there seems no reason to suppose that peptides will prove any less important in plants than in other organisms.

## 2.1 Glutathione

Glutathione (γ-Glu-Cys-Gly) can occur in a reduced or in an oxidized (disulphide) form. The two are readily interconverted by the enzyme glutathione reductase or by numerous redox reagents.

Both glutathione and glutathione reductase seem to be ubiquitous. They have been identified in monocotyledonous (CONN and VENNESLAND 1951, TKACHUK 1970), and dicotyledonous plants (MAPSON and ISHERWOOD 1963, RENNENBERG 1976) and in algae (TSANG and SCHIFF 1978) and occur in mitochondria (YOUNG and CONN 1956), chloroplasts (FOYER and HALLIWELL 1976, SCHAEDLE and BASSHAM 1977) and the cytoplasm (WIRTH and LATZKO 1978). In addition, the enzymes for glutathione synthesis have been isolated from plant tissues (WEBSTER 1953).

Despite considerable investigation and speculation, the physiological role, or roles, of glutathione still remain obscure. Many functions associated with the maintenance of optimal redox conditions within the cell have been proposed (see MEISTER 1975, MEISTER and TATE 1976, for general reviews), although these have generally been based upon observations of the activity of glutathione in animal and microbial cells. However, a number of functions specific to plant cells have also been proposed.

When considering the functions of glutathione in plant tissues it is interesting to note that a homologue, γ-Glu-Cys-β-Ala (homoglutathione), has been isolated from *Phaseolus* (CARNEGIE 1963). It is present at 20 times of level of glutathione and shows similar patterns of activity. It may therefore serve the same functions as glutathione in certain plant tissues.

*Cytoplasmic Functions.* Glutathione may be secreted in considerable quantities by cultured plant cells, yet its presence in the extracellular medium is often inhibitory to growth (RENNENBERG 1976). It has been suggested this may arise from interference with cytokinin-induced mitosis and protein synthesis in cultured cells, although no mechanism has been proposed (RENNENBERG 1978).

The formation of disulphide bonds within and between proteins is often considered responsible for certain frost damage in plant cells (LEVITT 1972). Glutathione may, therefore, impart cold-tolerance by maintaining sulphydryl groups in their reduced state; such a role has been proposed, based on observations on the varying levels of glutathione in spruce needles throughout the year (ESTERBAUER and GRILL 1978). Glutathione might also impart salt tolerance in halophytes by a similar mechanism (SHEVYAKOVA and LOSHADKINA 1965).

Nutritional roles for glutathione have also been proposed: for example, as a cofactor in the reduction of inorganic sulphate (TSANG and SCHIFF 1978), in the production of the numerous γ-glutamyl derivatives found in plant cells (Sect. 2.2), and in the storage (RENNENBERG and BERGMANN 1979) and transport (RENNENBERG et al. 1979) of reduced sulphur. In addition, it has been demonstrated (RENNENBERG et al. 1980) that 5-oxo-prolinase participates in the degradation of glutathione in tobacco cells, as it does in animal and microbial cells (MEISTER and TATE 1976).

*Role in the Chloroplast.* Glutathione may be responsible for maintaining reducing conditions within the chloroplast and, thus, the stability or activation of enzymes containing sulphydryl groups (SCHAEDLE and BASSHAM 1977, HALLIWELL and FOYER 1978, BUCHANAN et al. 1979). A number of functions for glutathione, specific to the chloroplast, have also been proposed. In particular, it has been suggested that the ratio of the reduced and oxidized forms of glutathione may be involved in regulating those Calvin cycle enzymes whose activity varies in the light and dark (WOLOSIUK and BUCHANAN 1977), although this view has been disputed on the grounds that the reduction/oxidation of glutathione is too slow to account for the sudden changes in Calvin cycle activity which are observed (HALLIWELL and FOYER 1978, WIRTH and LATZKO 1978). In addition, glutathione may also serve a protective role in the removal of hydrogen peroxide produced during photosynthesis. This could be achieved directly, mediated by glutathione peroxidase (FLOHÉ and MENZEL 1971), or via the intermediate oxidation and reduction of ascorbic acid (FOYER and HALLIWELL 1976, HALLIWELL and FOYER 1978).

*Membrane Transport.* Glutathione is a key compound in the γ-glutamyl cycle which it has been suggested may function by a group-translocation mechanism in the transport of amino acids and/or peptides in mammalian cells (MEISTER and TATE 1976, PRUSINER et al. 1976, GRIFFITH and MEISTER 1979) and in yeast (MOOZ 1979, OSUJI 1979, PENNICKX et al. 1980). It is interesting to speculate that glutathione may be involved in a similar function in plant cells. Indeed, a number of enzymes of the cycle have been isolated from higher plant tissues (MESSER and OTTESEN 1965, GOORE and THOMPSON 1967, MAZELIS and CREVELING 1978, see also MEISTER 1980), and the complete cycle seems to operate in marine phytoplankton, which are known to be extremely active in amino acid transport (KURELEC et al. 1977).

## 2.2 γ-Glutamyl Peptides

An increasing number of γ-glutamyl peptides continue to be isolated from plant tissues, yet such compounds generally seem to be absent from animal and microbial cells. The majority of these compounds are dipeptides although a few tripeptides have also been identified (see for example, FOWDEN 1964, MEISTER 1965). In addition to the glutamate residue they can contain either protein amino acids or non-protein amino acids such as willardiine (KRISTENSEN and LARSEN 1974) and lathyrene (HATANAKA and KANEKO 1978).

γ-Glutamyl peptides frequently occur in considerable quantities, especially in seeds and storage organs. For example, 34% of the non-protein amino nitrogen of kidney bean (*Phaseolus vulgaris*) seeds is present as γ-glutamyl-S-methyl-L-cysteine (ZACHARIUS et al. 1959). This feature, together with evidence for their degradation on germination, makes it almost certain that these peptides play an important role in the storage of nitrogen and/or sulphur. However, this is not always so, for many of the known γ-glutamyl peptides are only ever present in small quantities and could not adequately fulfil such a function.

Certain γ-glutamyl peptides contain toxic moieties which might serve a defensive function. For example, the toxin of *Lathyrus* seeds is a γ-glutamyl peptide (Schilling and Strong 1955), and hypoglycin, the toxic non-protein amino acid of *Blighia* fruits, is also maintained in the cell as a γ-glutamyl peptide (von Holt et al. 1956). Similarly, the lachrymatory and odoriferous products of onion and garlic may be present as γ-glutamyl peptides, particularly in dormant tissues (Whitaker 1976).

Although it seems likely that glutathione, together with the enzyme γ-glutamyl-transpeptidase, participate in the synthesis of at least some of these γ-glutamyl peptides, no such role has yet been demonstrated in vivo. Indeed, it has not even been shown that the synthesis of the majority of these peptides is enzymic. Thus, although it seems certain that some γ-glutamyl peptides do serve specific functions, it is possible that many are non-functional, arising as a result of low specificity in certain enzymic reactions, or even by non-enzymic mechanisms.

### 2.3 Algal Peptides

Many specific peptides have been isolated from seaweeds. Eisenine, a tripeptide from *Eisenia bicyclis,* has the structure <Glu-Gln-Ala (Kaneko et al. 1957) although it is possible that the pyroglutamate residue was formed from glutamine during extraction. A similar peptide, fastigiatin, <Glu-Gln-Gln, has been isolated from *Pelvetia* (Dekker et al. 1949) and a tripeptide containing alanine, aspartic acid and glutamic acid obtained from *Undaria pinnatifida* (Lee et al. 1962). Further reports of peptides, isolated from brown (Channing and Young 1953, Morita 1955, Takagi et al. 1973), green (Miyazawa et al. 1976) and red seaweeds (Young and Smith 1958, Miyazawa and Ito 1974) have also appeared. These peptides generally contain arginine and/or asparagine and glutamine, suggesting a role in nitrogen storage, although no definite function has yet emerged.

Peptides have also been isolated from freshwater algae. Pro-Val-diketopiperazine and Pro-Leu-diketopiperazine are released by species of *Scenedesmus* (Luedemann et al. 1961). Many green algal peptides are rich in arginine. Arg-Gln has been identified in *Cladophora* (Makisumi 1959), at least seven small peptides containing arginine, glutamic acid and aspartic acid have been isolated from *Chlorella* (Kanazawa 1964, Kanazawa et al. 1965) and several more from *Euglena* (Schantz et al. 1975). These peptides may replace arginine-rich histones known to be absent from certain green algae.

Blue-green algae have long been known to release small peptides, although the chain lengths have never been accurately determined (Fogg 1952, Jones and Stewart 1969). Often a very high proportion of the total assimilated nitrogen is excreted in this form (Stewart 1980). The structures of these compounds are not known, although they may be tripeptides (WALSBY 1974). Blue-green algal blooms may also release toxins, at least some of which are peptides; the toxic peptide of *Microcystis* has a molecular weight of 1,200 to 2,600 (BISHOP et al. 1959).

## 2.4 Peptides in the Phloem and Xylem

Both phloem and xylem exudates contain high levels of free amino acids (PATE 1976a, HIGGINS and PAYNE 1980). Early reports of peptides in the phloem have been explained mainly on the basis of the high levels of protein now known to be present in phloem exudates. However, the non-protein-derived peptides, alanylaminobutyric acid and glycylketoglutaric acid, have been identified in the phloem (PATE 1976a) and evidence has also been presented indicating a role for glutathione in the transport of reduced sulphur in this tissue (RENNENBERG et al. 1979).

The xylem contains much lower levels of nitrogen than the phloem, yet peptides have been reported in the xylem exudates of several species (WOLFFGANG and MOTHES 1953, POLLARD and SPROSTON 1954, BOLLARD 1957, FEJER and KONYA 1958, KHACHIDZE 1975) and MIETTINEN (1959) proposed that alanine absorbed by pea roots may be incorporated into a specific peptide-like compound, transported in the xylem and subsequently degraded. However, the techniques employed in these early studies may have identified protein material rather than small peptides. More specifically, PATE (1965) and RENNENBERG et al. (1979) have reported that glutathione, present in the xylem, may be involved in sulphur transport.

Thus, despite extensive investigations into the composition of phloem and xylem exudates, there is little evidence of an important role for peptides. However, most studies of phloem and xylem exudates have been concerned with specific compounds. In many cases a considerable proportion of the nitrogen may actually have been present as a general peptide pool, with no single peptide being present in detectable quantities. Situations in which long-distance transport of such a pool might be important can be envisaged, particularly during the bulk movement of protein-degradation products (e.g., leaf senescence, seed germination). In addition, the transport of amino acids conjugated as peptides could offer protection from metabolic degradation by enzymes known to be present in the phloem. Thus, a significant role for peptides in phloem and xylem transport, at least in certain situations, still remains a possibility.

## 2.5 Symbiosis

Within the plant kingdom, many examples of symbiotic associations are known in which nitrogen is transferred from one species to another. With the demonstration that peptide transport plays an important role in the movement of nitrogen across plant cell membranes (Sect. 4), a role for peptides in symbiotic associations might also be considered. One particular advantage of this (especially in peptides with unusual linkages) would be the protection of transported nitrogen from transformation until it reaches its metabolic sink. It seems inherently unlikely that a mixture of amino acids and small peptides, such as might arise from proteolysis, would be the form in which nitrogen is transferred. The incorporation of fixed nitrogen into a limited number of chemical species, specifically for transport between symbionts, would be more efficient, and most available evidence, although admittedly rather limited, indicates this to be the case.

The form in which nitrogen is transported between symbionts has only been studied in a limited number of species in which either inorganic nitrogen or amino acids, particularly alanine, seem to be important (SMITH 1974, PATE 1976b). However, specific peptides might well serve a similar role in some of the many other diverse symbiotic associations; a certain amount of evidence is available to support this view. Peptide-like material has been detected in the root nodules of clover (BUTLER and BATHURST 1958) and *Alnus* (LEAF et al. 1958) although its nature was not investigated in detail. In three species of the lichen *Peltigera,* the blue-green algal symbiont *Nostoc* releases peptides in vitro and in vivo which may be utilized by the fungal partner (MILLBANK 1974, 1976). These peptides are two to ten residues long and similar to those released by the free-living alga. However, in liverwort (*L. hepatica*) thalli, symbiotic *Nostoc* cells apparently releases only a very small proportion of its fixed nitrogen as peptide, the majority being as free $NH_4^+$ (STEWART and ROGERS 1977). The cause of this difference, and the possible role of peptides in nitrogen transfer between symbionts deserves more attention (see STEWART 1980).

## 2.6 Peptide Hormones

Although a number of peptides affect plant growth at very low concentrations (Sect. 2.7), no peptide has yet been identified in healthy plant tissue which regulates growth under normal conditions. This is despite the wide variety of regulatory and hormonal functions served by peptides in animal tissues (BUTT 1975), and some considerable effort invested in the isolation and identification of plant hormones. Speculations that the proposed flowering hormone, florigen, may be a peptide (COLLINS et al. 1963) have not, as yet, been substantiated, and most reports of peptide hormones have turned out to be peptide-bound derivatives of known plant growth regulators.

The conjugation of IAA with peptides in vivo has been reported (JERCHEL and STAAB-MÜLLER 1954, WINTER and STREET 1963), although it is not clear how large these peptides are, or whether covalent bonds might be involved. These complexes may represent the "protein-bound" derivatives isolated by a number of other groups (e.g., WINTER and THIMANN 1966). However, exogenously supplied IAA can be rapidly conjugated in a peptide linkage with aspartic acid or glutamic acid, especially by legumes (ANDREAE and GOOD 1955, THURMANN and STREET 1962). These derivatives, which also occur naturally (KLÄMBT 1960), are hormonally inactive and are believed to participate in the regulation of auxin activity.

Methionine is generally considered to be the biological precursor of ethylene. However, the production of ethylene by pea extracts is much more rapid from methionyl peptides than from free methionine, and has a more physiological pH optimum, leading to the suggestion that one or more peptides may be involved in ethylene synthesis (KU and LEOPOLD 1970).

Although no gibberellin-amino acid or gibberellin-peptide derivatives have been positively identified in plant tissues, their existence has been postulated on theoretical grounds (SEMBDNER 1974), and a gibberellic acid-peptide derivative has been tentatively identified (NADEAU and RAPPAPORT 1974). Formation of this inactive derivative is enhanced by abscisic acid, consistent with the inhibitory effects exerted by abscisic acid on gibberellin activity. Certain synthetic amino acid and peptide derivatives of gibberellic acid show very little gibberellin activity and are not hydrolyzed by plant tissues (SEMBDNER et al. 1976), leading to the proposal that such derivatives may serve to inactivate, and thus regulate, the levels of the hormone in plant tissues.

## 2.7 Peptides and Plant Pathology

Peptides are frequently associated with plant diseases, exerting toxic or hormone-like effects on plant growth. In many cases the peptide appears to be produced by an infecting agent, though it is not always clear whether the host or pathogen is responsible.

Several bacterial phytotoxins have been identified as peptides. Tabtoxin, produced by *Pseudomonas tabaci* during "wildfire" infection of tobacco plants, is a dipeptide of tabtoxinine, a non-protein amino acid, and threonine (TAYLOR et al. 1972). Syringomycin, the phytotoxin of *Pseudomonas synringae,* a pathogen of stone-fruit trees, is also a peptide, containing nine amino acid residues, although its precise structure is unknown (BACKMAN and DEVAY 1971). Phaseolotoxin, from *Pseudomonas phaseolicola,* which causes halo blight of beans (*Phaseolus*), is a tripeptide with the structure $\delta$-N-phosphosulphamyl-ornithyl-alanyl-homoarginine (MITCHELL 1976). CHANG and LIN (1977) have reported the presence of a peptide specifically in plant tissues infected with *Agrobacterium* (crown-gall tumours). It seems likely that the linear nature of these peptides, allowing them to be recognized by general peptide transport systems, plays an important role in the ability of the otherwise impermeant toxic moieties to enter into the host cell, according to the "smugglin" principle (see PAYNE 1980). Indeed, evidence in favour of this view has been presented for phaseolotoxin (STASKAWICZ and PANOPOULOS 1980).

Many phytopathogenic fungi also produce peptide toxins. However, unlike the bacterial toxins, the majority of those which have been characterized seem to be cyclic. The pathogenic fungus, *Cylindrocladium* produces a cyclic tetrapeptide toxin (HIROTA et al. 1973) and the toxins of *Altenaria,* which cause apple leaf blotch are host-specific cyclic depsipeptides (UENO et al. 1977). Similarly, the host-specific toxins of *Helminthosporium victoriae* (PRINGLE and BRAUN 1958), *H. carbonum* (PRINGLE 1971) and *Periconia circinata* (PRINGLE and SCHAFFER 1966), which infect various cereals, all appear to be small peptides, probably cyclic. The cyclic peptide, malformin (cyclo-D-Cys-D-Cys-L-Val-D-Leu-L-Ile), produced by *Aspergillus*, causes curvature of bean and corn (*Zea mays*) roots at very low concentrations (BODANSZKY and STAHL 1974), and lycomarismin, a peptide produced by *Fusarium*, induces tomato (*Lycopersicum esculentum*) leaf curl and wilt (ROBERT et al. 1962). Small peptide-like compounds have also been reported present during plant viral infections (REINDEL and BIENENFELD 1956) and in nematode-induced gall tumours (CHANG and LIN 1977). Thus, although the mechanism of action of these toxins is generally unknown, it seems clear that, even at very low concentrations, peptides can exert a considerable effect on plant growth.

### 2.8 Miscellaneous Peptides

A number of other peptides have been reported in plant tissues, although little is known of their functions. Evolidine, a cyclic heptapeptide (cyclo-Ser-Phe-Leu-Pro-Val-Asn-Leu) has been identified in *Evodia xanthoxyloides* (LAW et al. 1958), the tetrapeptide methyl ester Pro-Leu-Phe-Val-OMe, is found in linseed (*Linum usitatissimum*) oil (KAUFMANN and TOBSCHIRBEL 1959) and D-Ala-Gly has been isolated from rice (*Oryzae sativus*) leaves (YAMAUCHI et al. 1979). Peptides may also affect the taste and aroma of various foodstuffs, for example, the sweet palm (*Yucca flaccuda*) (SHIMOKOMAKI et al. 1975), cocoa beans (MOHR et al. 1976), and soyabean (*Glycine max*) extracts (OKA and NAGATA 1974).

Peptide-nucleotide complexes have often been isolated from plant tissues (see WALEY 1966, SYNGE 1968). Little is known of their function, although they seem to promote plant growth and have been implicated in protein synthesis and cell division (GORYUNOVA et al. 1974, 1977, VOLODIN 1975, GERASIMENKO and GORYUNOVA 1978). The peptide moieties of these compounds have not been characterized in detail, but in at least some cases they may be rather large (PUSHEVA and KHOREVA 1977). At least some of these compounds seem likely to be tRNA-peptide complexes removed from the ribosome during extraction.

## 3 Peptides as Metabolic Intermediates: The Peptide Pool

### 3.1 Introduction

The concept of a peptide pool is not new (WALEY 1966). However, although it has been widely assumed that peptides will be produced as intermediates in protein metabolism, the possibility that these intermediates might accumulate to form a significant pool has rarely been considered. There seem to be a number of reasons for this neglect. First, although it is accepted that peptides will be produced as intermediates during protein synthesis and degradation, it is usually considered that they will be metabolized so rapidly that a significant pool will never accumulate. Second, in the absence of any obvious function, an incentive for investigation is lacking; however, the recent demonstration of a peptide transport system in plant cells (Sect. 4) provides one possible function. Third, there are methodological problems. In a peptide pool any one defined peptide will be present in vanishingly small amounts, and thus, the total complement of peptides in the pool must be detected, often a difficult proposition in the presence of large quantities of protein and free amino acids.

All plant cells are likely to exhibit protein turnover (HUFFAKER and PETERSON 1974) and might therefore be expected to contain a peptide pool, however small. There is a certain amount of evidence to support this view, although few reports of peptide pools can be regarded as unambiguous. This is generally a reflection of the rather uncritical experimental approaches which have been employed in the past. Even those relatively few cases in which peptides have been more rigorously identified do not always give an adequate picture of the nature of the peptide pool; the peptides have rarely been characterized, and the methods of extraction and purification seem unlikely to result in a complete or representative sample of the total pool.

### 3.2 Tissues of Rapid Protein Synthesis

Early suggestions that peptides might be directly incorporated into proteins led to a number of studies of "peptides" in growing tissues. For example, a large "peptide" pool was reported to increase 20-fold during the elongation of developing root tips (MORGAN and REITH 1954). Similarly, "peptides" were reported in ripening pea (*Pisum sativum*) cotyledons (RAAKE 1951) and cotton seeds (KING and LEFFLER 1979) during periods of rapid protein synthesis; the pools decreased as protein synthesis reached completion. "Peptide" fractions have also been reported in growing pea seedlings (LAWRENCE et al. 1959, BEEVERS and GUERNSEY 1966, PRIKHOD'KO et al. 1975), soyabeans (DUKE et al. 1978) and barley embryos (FOLKES and YEMM 1958).

None of the above reports can now be regarded as conclusive evidence for a peptide pool, although this is certainly a possible explanation for many of the data. More critically, small peptides have been identified in the growing leaves of barley (*Hordeum vulgare*) (HENDRY and STOBART 1977), although in this case no attempt was made to characterize the pool.

### 3.3 Mature Tissues

Tissues in which protein synthesis and degradation are minimal have been least thoroughly studied. Most reports of peptides are rather vague. Thus, ten peptides of 3–12 residues were reported in cotton seed (YULDASHEV et al. 1970) and about ten peptides of less than ten residues in celery (*Apium graveolens*) root (CURI et al. 1973). Peptide-like compounds also leach from seeds, possibly to stimulate colonization by microorganisms (VANČURA and HANZLIKOVA 1972). The stems of many species may also contain a wide range of peptide-like compounds (CHANG et al. 1975) and it has been suggested that salt-tolerant plants may contain an unusually large peptide pool (PRIKHOD'KO 1979, PRIKHOD'KO et al. 1979). Specific peptides, Ala-Gly isolated from rice (TSUMURA et al. 1977), and both Ala-Ala and Gly-Ala from *Leptadenia* (DHAWAN and SINGH 1976), may represent part of a peptide pool or may serve more specific functions.

SYNGE (1951) reported high levels of non-protein, "bound" amino acids in ryegrass leaves. Most of these compounds eventually proved to be N-acyl amino acids rather than peptides (SYNGE and WOOD 1958), but a residual pool of small peptides remained, and although probably rather unrepresentative of the total peptide pool, this remains one of the few convincing reports of a peptide pool to date (CARNEGIE 1961). A similar pool in bean leaves (BAGDASARIAN et al. 1964) probably represents rather larger polypeptides.

### 3.4 Tissues Showing Rapid Protein Hydrolysis

Germinating seeds have received most consideration, although there has been one report of peptides in senescing barley leaves (HENDRY and STOBART 1977).

During germination of *Vicia* a large pool of soluble, non-protein nitrogen was found in the cotyledons (BOULTER and BARBER 1963). Although this was not positively identified as comprising peptide material, the pool did contain at least two defined peptides, one a dipeptide of glycine and alanine, the other containing alanine, aspartic acid and cysteine, and it seems likely that much of the rest of the pool also consisted of peptides. Similar fractions, possibly containing peptides, have been reported in the cotyledons of germinating peas (LAWRENCE et al. 1959, BEEVERS and GUERNSEY 1966), cucumber (*Cucumis sativus*) (BECKER et al. 1978), and castor beans (*Ricinus communis*) (STEWART and BEEVERS 1967), although no such pool was detected in germinating soyabean cotyledons (DUKE et al. 1978).

Similar evidence exists for the presence of a peptide pool in the endosperm of germinating cereal grains. In barley (FOLKES and YEMM 1958), and maize (INGLE et al. 1964, OAKS and BEEVERS 1964, MOREAUX 1979) a pool of soluble nitrogen (probably consisting largely of peptides) increases to a maximum and then decreases as germination proceeds. In addition, "peptides", as well as amino acids, leach from isolated maize endosperm (OAKS and BEEVERS 1964). A hexapeptide has been isolated and partially characterized from wheat (BIEBER and CLAGETT 1956) and a pool of peptides, averaging six to eight residues, has been reported in wheat flour (GRANT and WANG 1972).

The least ambiguous evidence for the presence of a significant peptide pool in plant tissues comes from recent studies of germinating barley grains (HIGGINS 1979, HIGGINS and PAYNE 1981). Peptides were separated from tissue extracts, which also contained amino acids and proteins, by use of ion-exchange and gel-exclusion chromatography, and were positively identified using a variety of complementary analytical techniques. A pool of small peptides of two to six amino acid residues was identified in the endosperm and shown to have a similar (though not identical) overall composition to the seed storage proteins. The size of this pool was comparable with the free amino acid pool of the endosperm and changed markedly during germination (Fig. 1). It seems likely

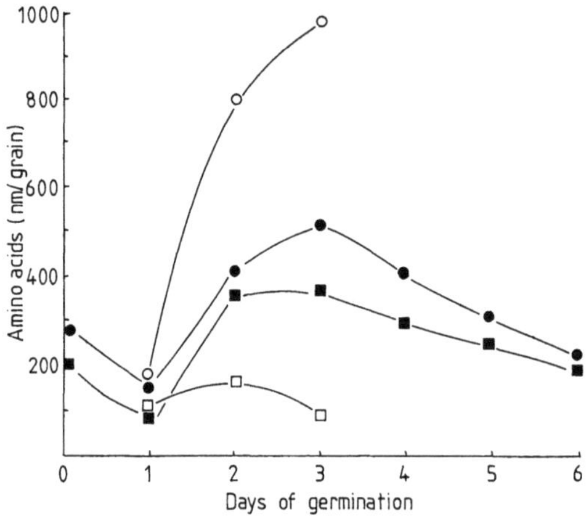


**Fig. 1.** Changes in the peptide and free amino acid pools of barley endosperm and embryo during germination. Embryo amino acids (○) and peptides (□); endosperm amino acids (●) and peptides (■). Peptide concentrations are expressed as nmol peptide-bound amino acids per grain

that this pool of peptides plays an important role in the transfer of nitrogen from the endosperm to the embryo during germination (Sect. 4). In addition, a smaller pool of peptides was found in the growing embryo during the first few days of germination; this pool showed differences in composition from that of the endosperm.

Thus, at least in germinating barley grains, there is now good evidence for a peptide pool in plant tissues. In many other systems there is also evidence that peptides may be relatively abundant. The peptides present in the barley endosperm seem to be important in nitrogen transport (Sect. 4). If in other tissues peptide pools also prove to be of a similar order of magnitude to the free amino acid pools it would be pertinent to consider other possible functions. Does the peptide pool have any specific metabolic or regulatory function? Is the peptide pool important in regulating the supply of amino acids for protein synthesis or other metabolic processes? Is the pool undergoing rapid turnover or are the peptides sequestered away from metabolic activity? Thus, if peptidases were compartmentalized away from peptides the latter could serve to protect amino acids from catabolism. Clearly, there is considerable scope here for further investigation.

## 4 Peptide Transport

In both mammalian cells and microorganisms the absorption of peptides can be of considerable nutritional importance, and may confer a number of inherent advantages over the uptake of free amino acids (MATTHEWS and PAYNE 1980). Peptide transport systems in the mammalian gut (MATTHEWS 1975) and in many microbial species (PAYNE 1980) have been identified and characterized in considerable detail. However, only recently has a similar role for peptides in the nutrition of plant cells been identified.

### 4.1 Algae

Intact peptide uptake has never been demonstrated amongst the algae. Peptides can stimulate the growth of certain algal species (JANKEVICIUS et al. 1972), and *Chlorella* is able to utilize several dipeptides as a nitrogen source (BOLLARD 1966), but the form in which the peptides are absorbed has not been investigated; uptake might be subsequent to extracellular hydrolysis. Despite the fact that NORTH (1975) has proposed, although by no means conclusively demonstrated, that many marine phytoplankton are unable to utilize peptides present in sea water, it seems probable, considering the wide variety of algal species and their habitats, that many will be found to possess peptide transport systems.

### 4.2 Higher Plants

It has been known for many years that peptides can affect the growth of plant tissues cultured in vitro (see HIGGINS and PAYNE 1980 for early references). Moreover, it has been shown that several different defined peptides can serve as sole source of nitrogen for plant growth in duckweed (*Lemna*) fronds (BOLLARD 1966), *Atropa* callus (SALONEN and SIMOLA 1977), and the sundew (*Drosera*) (SIMOLA 1978). However, in none of these cases was any attempt made to determine whether the peptides were hydrolyzed before or after absorption.

The first, though rather fortuitous, demonstration of intact peptide transport in plants came from studies on the absorption of nutrients from pitchers of the carnivorous plant, *Sarracenia* (PLUMMER and KETHLEY 1964). Three peptides, DL-Ala-DL-Asp, DL-Ala-DL-Met, and DL-Ala-DL-Leu were not hydrolyzed by the pitcher fluid but were taken up intact into the leaves, where they remained uncleaved for as long as 4–5 days. No attempt was made to characterize the mechanism by which these peptides were absorbed.

Only recently has an active peptide transport system been described and characterized in a plant tissue. During the germination of barley grains large amounts of nitrogen, initially present as storage proteins, must be transferred from the endosperm to the growing embryo, across an absorptive organ, the scutellum (Fig. 2). Although it has generally been assumed that proteolysis reaches completion in the endosperm, and that only amino acids are absorbed by the scutellum, the scutellum is also able to absorb peptides (HIGGINS and

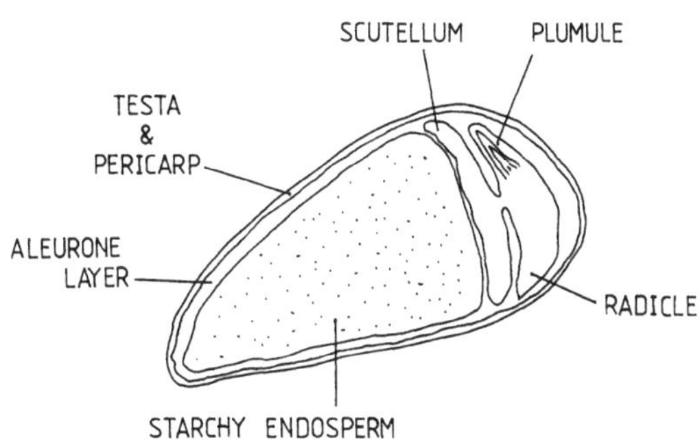


**Fig. 2.** Diagram illustrating the structure of a dormant barley grain

PAYNE 1977a, b, SOPANEN et al. 1977). Peptide uptake is mediated by an active transport system, able to accumulate peptides intact against a concentration gradient (HIGGINS and PAYNE 1977b). Uptake shows a pH optimum of about pH 4, similar to the pH of barley endosperm. Once absorbed, peptides are rapidly hydrolyzed to amino acids by the endogenous embryonic peptidases (HIGGINS and PAYNE 1978a, b, SOPANEN et al. 1978). A single system seems to exist, capable of transporting both di- and small oligopeptides (HIGGINS and PAYNE 1978b) and with little specificity for the amino acid side chains (HIGGINS and PAYNE 1978a, and unpublished results). The aspects of peptide structure which confer affinity for the transport system have been investigated in some detail and show many similarities to the requirements for peptide transport in bacterial, fungal and mammalian cells (HIGGINS and PAYNE 1978c, HIGGINS, 1979, MATTHEWS and PAYNE 1980). Peptide transport activity is low in ungerminated grains, increasing to a maximum at about three days of germination (SOPANEN 1979, C.F. HIGGINS and J.W. PAYNE, unpublished results) and is rather more rapid than the uptake of free amino acids (HIGGINS and PAYNE 1978a). In addition, during germination, a pool of peptides is present in the endosperm at a concentration similar to the measured $K_m$ for the transport system, and thus suitable for the efficient operation of the peptide transport system (HIGGINS and PAYNE 1981). Furthermore, the rate of peptide transport at these peptide concentrations is sufficient to account for the rate at which nitrogen is transferred to the embryo to support growth. Thus, it seems that peptides are at least as important as amino acids in the transfer of nitrogen from the endosperm to the growing embryo of barley grains. Preliminary evidence also indicates the existence of similar active peptide transport systems in the scutella of wheat (*Triticum*), oats (*Avena*), maize and sorghum (*Sorghum*) (C.F. HIGGINS and J.W. PAYNE, unpublished results).

In view of the widespread importance of peptides in the nutrition of animals and microorganisms (MATTHEWS and PAYNE 1980), and the recent demonstration of a peptide transport system in a plant tissue, it is pertinent to consider other possible roles for peptide transport within the plant kingdom. Unlike other organisms in which peptide transport has been characterized, higher plants are generally autotrophic. Thus, the utilization of exogenous sources of organic nitrogen will normally be unimportant. However, certain situations may be envisaged where this could be potentially useful.

Thus, although nitrogen is usually taken up by plant roots in inorganic form, reports of amino acid transport by plant roots have appeared (WRIGHT 1962, WATSON and FOWDEN 1975, SOLDAL and NISSEN 1978); in such cases peptides might also be involved, especially in the roots and leaves of aquatic plants growing in highly eutrophic conditions. Indeed, it has been reported that the duckweed is able to utilize peptides as a nitrogen source (BOLLARD 1966).

Peptide transport may be important in certain parasitic interactions in which nitrogen from the host plant is absorbed and utilized by the parasite. Such a mechanism would be most appropriate in species which secrete proteases and absorb the hydrolytic products. Carnivorous plants might also be expected to absorb nitrogen from their prey in the form of peptides. Considerable indirect evidence indicates this to be so (discussed in HIGGINS and PAYNE 1980), and in one case intact peptide uptake has actually been demonstrated (PLUMMER and KETHLEY 1964).

Transport of peptides within the plant may also be important. Long-distance movement in the phloem and xylem has already been considered (Sect. 2.4), yet considerable transfer of nutrients must also occur at the cell-to-cell level, via the apoplast or symplast. Cells from a wide variety of plant tissues are able to absorb amino acids, often, if not always, via specific transport systems (see HIGGINS and PAYNE 1980 for a general review). In many situations, particularly during periods of rapid proteolysis (e.g., leaf senescence), it may be more efficient for peptides to be transported. Indeed, preliminary evidence that bean leaf cells possess a peptide transport system has been obtained, using the tripeptide phytotoxin, phaseolotoxin (B.J. STASKAWICZ and N.J. PANOPOULOS personal communication).

Intracellular peptide transport across organellar membranes might also be of importance. There is now good evidence that many proteolytic enzymes are compartmentalized within the vacuolar system of the cell (MATILE 1978). Thus, peptides may have to cross the vacuolar membrane prior to hydrolysis and in the scutellum of germinating barley grains preliminary evidence has been obtained indicating that this is indeed so (HIGGINS 1979). Similarly, during the germination of both monocotyledonous and dicotyledonous seeds, protein hydrolysis products must pass out of the protein bodies; again, it would be energetically more efficient if peptides rather than free amino acids were the chemical species transported.

Thus, many situations may be envisaged in which peptide transport may be important in plant cells; certainly, in situations in which amino acid transport is known, or believed to be important, such a possibility should always be considered.

## 5 Concluding Remarks

From the foregoing discussion it can be seen that, while there are many reports of peptides in plant cells, few definitive studies have been undertaken. Peptides have been implicated in many plant processes, yet rarely has a role been clearly established. In many other situations the possibility that peptides might be of importance has not even been considered. While this may be justified in many cases, it is hoped that this attempt to bring together much of the literature concerning plant peptides will lead to a renewed consideration of this diverse group of compounds and to a reappraisal of some of the many interesting questions which remain unanswered.

## References

Andreae WA, Good NE (1955) The formation of indole-acetylaspartic acid in pea seedings. Plant Physiol 30:380–382

Backman PA, deVay JE (1971) Studies on the mode of action and biogenesis of the phytotoxin syringomycin. Physiol Plant Pathology 1:215–233
Bagdasarian M, Matheson NA, Synge RLM, Youngson MA (1964) New procedures for isolating polypeptides and proteins from plant tissues. Biochem J 91:91–105
Becker WM, Leaver CJ, Weir EM, Riezman H (1978) Regulation of glyoxysomal enzymes during germination of cucumber I. Plant Physiol 62:542–549
Beevers L, Guernsey FS (1966) Changes in some nitrogenous components during the germination of pea seeds. Plant Physiol 41:1455–1458
Bieber L, Clagett CO (1956) Determination of the amino acid sequence in a durum wheat peptide. Proc ND Acad Sci 10:31–35
Bishop CT, Anet EFLJ, Gorham PR (1959) Isolation and identification of the fast-death factor in *Microcystis aeruginosa* NRC-1. Can J Biochem Physiol 37:453–471
Bodanszky M, Stahl G (1974) The structure and synthesis of malformin. Proc Natl Acad Sci USA 71:2791–2794
Bollard EG (1957) Nitrogenous compounds in tracheal sap of woody members of the family Rosaceae. Aust J Biol Sci 10:288–291
Bollard EG (1966) A comparative study of the ability of organic nitrogenous compounds to serve as sole sources of nitrogen for the growth of plants. Plant Soil 25:153–166
Boulter D, Barber JT (1963) Amino acid metabolism in germinating seeds of *Vicia faba* L. in relation to their biology. New Phytol 62:301–316
Bricas E, Fromageot C (1953) Naturally occurring peptides. Adv Protein Chem 8:1–125
Buchanan BB, Cranford NA, Wolosiuk RA (1979) Activation of plant acid phosphatases by oxidized glutathione and dehydroascorbate. Plant Sci Lett 14:245–249
Butler GW, Bathurst NO (1958) Free and bound amino acids in legume root nodules: bound $\gamma$-aminobutyric acid in the genus *Trifolium*. Aust J Biol Sci 11:529–537
Butt WR (1975) Hormone chemistry, Vol. 1, 2nd edn. Ellis Horwood, Chichester
Carnegie PR (1961) Bound amino acids of ryegrass: the isolation of amphoteric peptide-like substances of low molecular weight. Biochem J 78:697–707
Carnegie PR (1963) Isolation of a homologue of glutathione and other acidic peptides from seedlings of *Phaseolus aureus*. Biochem J 89:459–471
Chang CC, Lin BY (1977) Accumulation of a simple peptide and some Pauly-positive compounds in crown-gall tumours induced by *Agrobacterium tumefaciens* strains IIBV7 and 181. Bot Bull Acad Sin 18:82–87
Chang CC, Lin BY, Hu S-P (1975) Observations on some simple peptides and two unknown Pauly-positive components in normal plants, crown gall tumors, and tomato nematode gall. Bot Bull Acad Sin 16:101–114
Channing DM, Young GT (1953) Amino acids and peptides. X. The nitrogenous constituents of some marine algae. J Chem Soc:2481–2491
Collins WT, Salisbury FB, Ross CW (1963) Growth regulators and flowering III. Antimetabolites. Planta 60:131–144
Conn EE, Vennesland B (1951) Glutathione reductase of wheat germ. J Biol Chem 192:17–28
Curi J, Grega B, Jendzelovsky J, Adam J (1973) Presence of ninhydrin positive substances in the ethanolic extract of the celery root. Pol'nohospodarstvo 19:328–336
Dekker CA, Stone D, Fruton JS (1949) A peptide from a marine alga. J Biol Chem 181:719–729
Dhawan AK, Singh H (1976) Free pools of amino acids and sugars in *Leptadinia pyrotechnica* F. Curr Sci 45:198
Duke SH, Schrader LE, Miller MG, Niece RL (1978) Low temperature effects on soybean (*Glycine max* L. Merr. cv. Wells). Free amino acid pools during germination. Plant Physiol 62:642–647
Esterbauer H, Grill D (1978) Seasonal variations of glutathione and glutathione reductase in needles of *Picea abies*. Plant Physiol 61:119–121
Fejer D, Konya E (1958) Occurrence of two additional peptides in the fine sap of corn. Naturwissenschaften 45:387–388
Flohé L, Menzel H (1971) The influence of glutathione upon light-induced high-amplitude

swelling and liquid peroxide formation of spinach chloroplasts. Plant Cell Physiol 12:325–333

Fogg GE (1952) The production of extracellular nitrogenous substances by a blue-green alga. Proc R Soc London Ser B 139:372–397

Folkes BF, Yemm EW (1958) The respiration of barley plants X. Respiration and the metabolism of amino-acids and proteins in germinating grain. New Phytol 57:106–131

Fowden L (1964) The chemistry and metabolism of recently isolated amino acids. Annu Rev Biochem 33:173–204

Foyer CH, Halliwell B (1976) The presence of glutathione and glutathione reductase in chloroplasts: a proposed role in ascorbic acid metabolism. Planta 133:21–25

Gerasimenko LM, Goryunova SV (1978) Effect of sulfur-containing nucleotide-peptides on morphology and nucleic acid content of *Anabaena cylindrica*. Microbiology 47:709–712

Goore MY, Thompson JF (1967) γ-Glutamyl transpeptidase from kidney bean fruit. Biochim Biophys Acta 132:15–26

Goryunova SV, Gerasimenko LM, Pusheva MA (1974) Comparative study of the action of a polynucleotide-peptide complex and colchicine on cells of a synchronous culture of *Hydrodictyon reticulatum* Lagerh. Microbiology 43:224–227

Goryunova SV, Gerasimenko LM, Khoreva SL (1977) Effect of a sulfur-containing nucleotide-peptide complex on the growth and development of different algal species. Mikrobiologiya 46:1082–1086

Grant DR, Wang CC (1972) Dialysable components resulting from proteolytic activity in extracts of wheat flour. Cereal Chem 49:201–207

Griffith OW, Meister A (1979) Glutathione: Interorgan-translocation, turnover, and metabolism. Proc Natl Acad Sci USA 76:5606–5610

Halliwell B, Foyer CH (1978) Properties and physiological function of glutathione reductase purified from spinach leaves by affinity chromatography. Planta 139:9–17

Hatanaka S-L, Kaneko S (1978) γ-Glutamyl-L-lathyrene from *Lathyrus japonicus*. Phytochemistry 17:2027

Hendry GAF, Stobart AK (1977) Metabolism of protein, peptides, and amino acids in ageing etiolated barley leaves. Phytochemistry 16:1339–1346

Higgins CF (1979) Peptide transport by embryos of germinating barley (*Hordeum vulgare*). PhD Thesis, Univ Durham

Higgins CF, Payne JW (1977a) Peptide transport by germinating barley embryos. Planta 134:205–206

Higgins CF, Payne JW (1977b) Characterization of active peptide transport by germinating barley embryos: effects of pH and metabolic inhibitors. Planta 136:71–76

Higgins CF, Payne JW (1978a) Peptide transport by germinating barley embryos: uptake of physiological di- and oligopeptides. Planta 138:211–215

Higgins CF, Payne JW (1978b) Peptide transport by germinating barley embryos: evidence for a single common carrier for di- and oligopeptides. Planta 138:217–221

Higgins CF, Payne JW (1978c) Stereospecificity of peptide transport by germinating barley embryos. Planta 142:299–305

Higgins CF, Payne JW (1980) The uptake and utilization of amino acids and peptides by higher plants. In: Payne JW (ed) Microorganisms and nitrogen sources. Wiley, Chichester New York, pp 609–639

Higgins CF, Payne JW (1981) The peptide pools of germinating barley grains: relation to hydrolysis and transport of storage proteins. Plant Physiol submitted for publication

Hirota A, Suzuki A, Tamura S (1973) Characterization of four amino acid constituting Cyl-2, a metabolite from *Cylindrocladium scoparium*. Agric Biol Chem (Tokyo) 37:1185–1189

Huffaker RC, Peterson LW (1974) Protein turnover in plants and possible means of its regulation. Annu Rev Plant Physiol 25:363–392

Ingle J, Beevers L, Hageman RH (1964) Metabolic changes associated with the germination of corn. Plant Physiol 39:735–740

International Union of Biochemistry: Biochemical Nomenclature and Related Documents (1978), Spottiswood Ballantyne, London, pp 64–84

Jankevicius K, Budriene S, Baranauskiene A, Lubianskiene V, Jankaviciute G, Kiselyte T, Biveinis J (1972) Free amino acids in freshwater plankton and its medium. Liet TSR Mokslu Akad Darb Ser C:3–17
Jerchel D, Staab-Müller R (1954) Analytical characteristics and growth activity of homologues and peptides of indole-acetic acid. Z Naturforsch 9b:411–415
Jones K, Stewart WDP (1969) Nitrogen turnover in marine and brackish habitats III. The production of extracellular nitrogen by *Calothrix scopulorum*. Mar Biol Assoc UK 49:475–488
Kanazawa T (1964) Changes of amino acid composition of *Chlorella* during their life cycle. Plant Cell Physiol 5:333–354
Kanazawa T, Kanazawa K, Morimura Y (1965) New arginine-containing peptides isolated from *Chlorella* cells. Plant Cell Physiol 6:631–643
Kaneko T, Shiba T, Watarai S, Imai S, Shimada T, Ueno K (1957) Synthesis of eisenine. Chem Ind (London):986–987
Kaufmann HP, Tobschirbel A (1959) An oligopeptide from linseed. Chem Ber 92:2805–2809
Khachidze OT (1975) Peptides in the vegetative organs and berries of grape plants and their formation path. In: Oparin AI (ed) Vopr Biokhim Vinograda Vina, Tr Uses Konf 2nd. Moscow, USSR, pp 118–122
King EE, Leffler HR (1979) Nature and patterns of proteins during cotton seed development. Plant Physiol 63:260–263
Klämbt HD (1960) Indole-3-acetylaspartic acid, a naturally occurring indole derivative. Naturwissenschaften 47:398
Kristensen I, Larsen PO (1974) γ-Glutamyl willardiine and γ-glutamylphenylalanylwillardiine from seeds of *Fagus silvatica*. Phytochemistry 13:2799–2802
Ku HS, Leopold AC (1970) Ethylene formation from peptides of methionine. Biochem Biophys Res Commun 41:1155–1160
Kurelec B, Rijavec M, Britvic S, Müller WEG, Zahn RK (1977) Phytoplankton: presence of γ-glutamyl cycle enzymes. Comp Biochem Physiol 56B:415–419
Law HD, Millar IT, Springall HD, Birch AJ (1958) The structure of evolidine. Proc Chem Soc London 198
Lawrence JM, Day KM, Stephenson JE (1959) Nitrogen metabolism in pea seedlings. Plant Physiol 34:668–674
Leaf G, Gardner IC, Bond G (1958) Observations on the composition and metabolism of the nitrogen-fixing root nodules of *Alnus*. J Exp Bot 9:320–331
Lee TY, Kwon TW, Lee CY (1962) The presence of a new peptide in the brown alga, *Undaria pinnatifida*. J Korean Chem Soc 6:84–87
Levitt J (1972) Responses of plants to environmental stress. Academic Press, New York
Luedemann G, Charney W, Woyciesjes A, Petterson E, Peckham WD, Gentles MJ, Marshall H, Herzog HL (1961) Microbiological transformation of steroids. IX. J Org Chem 26:4128–4130
Makisumi S (1959) Occurrence of arginylglutamine in green alga, *Cladophora* species. J Biochem 46:63–71
Mapson LW, Isherwood FA (1963) Glutathione reductase from germinated peas. Biochem J 86:173–191
Matile P (1978) Biochemistry and function of vacuoles. Annu Rev Plant Physiol 29:193–213
Matthews DM (1975) Intestinal absorption of peptides. Physiol Rev 55:537–608
Matthews DM, Payne JW (1975) Occurrence and biological activities of peptides. In: Matthews DM, Payne JW (eds) Peptide transport in protein nutrition. North-Holland/Elsevier, Amsterdam Oxford New York, pp 392–464
Matthews DM, Payne JW (1980) Transmembrane transport of small peptides. Curr Top Membr Transp 14:331–425
Mazelis M, Creveling RK (1978) 5-Oxoprolinase (L-pyroglutamate hydrolase) in higher plants. Plant Physiol 62:798–801
Meister A (1965) Biochemistry of the amino acids, 2nd edn. Academic Press, New York London
Meister A (1975) Biochemistry of glutathione. In: Greenberg DM (ed) Metabolic pathways, 3rd edn., Vol 7. Academic Press, New York London, pp 101–188

Meister A (1980) Possible relation of the $\gamma$-glutamyl cycle to amino acid and peptide transport in microorganisms. In: Payne JW (ed) Microorganisms and nitrogen sources. Wiley, Chichester, New York, pp 493–509

Meister A, Tate SS (1976) Glutathione and related $\gamma$-glutamyl compounds: biosynthesis and utilization. Annu Rev Biochem 45:559–604

Messer M, Ottesen M (1965) Isolation and properties of glutamine cyclotransferase of dried papaya latex. Carlsberg Res Commun 35:1–24

Miettinen JK (1959) Assimilation of amino acids in higher plants. Symp Soc Exp Biol 13:210–229

Millbank JW (1974) Nitrogen metabolism in lichens V. The forms of nitrogen released by the blue-green phycobiont in *Peltigera* spp. New Phytol 73:1171–1181

Millbank JW (1976) Aspects of nitrogen metabolism in lichens. In: Brown DH, Hawksworth DL, Bailey RH (eds) Lichenology: progress and problems. Academic Press, London New York

Mitchell RE (1976) Isolation and structure of a chlorosis-inducing toxin of *Pseudomonas phaseolicola*. Phytochemistry 15:1941–1947

Miyazawa K, Ito K (1974) Isolation of a new peptide L-citrullinyl-L-arginine from a red alga *Gratoloupia turuturu*. Bull Jpn Soc Sci Fisher 40:815–818

Miyazawa K, Ito K, Matsumoto F (1976) Amino acids and peptides in seven species of green marine algae. Hiroshima Daigaku Sui-Chikusangakubu Kiyo 15:161–169

Mohr W, Landschreiber E, Severin T (1976) Specificity of cocoa aroma. Fette Seifen Anstrichm 78:88–95

Mooz ED (1979) Association of glutathione synthetase deficiency and diminished amino acid transport in yeast. Biochem Biophys Res Commun 90:1221–1226

Morgan C, Reith WS (1954) The compositions and quantitative relations of protein and related fractions in developing root cells. J Exp Bot 5:119–135

Morita M (1955) Sapid components of *Laminariae* (brown seaweeds) III. Nippon Kagaku Zasshi 76:692–694

Moureaux T (1979) Protein breakdown and protease properties of germinating maize endosperm. Phytochemistry 18:1113–1117

Nadeau R, Rappaport I (1974) An amphotoeric conjugate of tritiated gibberellin-$A_1$ from barley aleurone layers. Plant Physiol 54:809–812

North BB (1975) Primary amines in California coastal waters: utilization by phytoplankton. Limnol Oceanogr 20:20–27

Oaks A, Beevers H (1964) The requirement for organic nitrogen in *Zea mays* embryos. Plant Physiol 39:37–43

Oka S, Nagata K (1974) Isolation and characterization of acidic peptides in soy sauce. Agric Biol Chem (Tokyo) 38:1185–1194

Osuji GO (1979) Glutathione turnover and amino acid uptake in yeast. FEBS Lett 105:283–285

Pate JS (1965) Roots as organs of assimilation of sulphate. Science 149:547–548

Pate JS (1976a) Nutrients and metabolites of fluids recovered from xylem and phloem: significance in relation to long-distance transport in plants. In: Wardlaw IF, Passioura JB (eds) Transport and transfer processes in plants. Academic Press, New York London, pp 253–281

Pate JS (1976b) Transport in symbiotic systems fixing nitrogen. In: Lüttge U, Pitman MG (eds) Encyclopedia of plant physiology, New Series, Vol 2B. Springer, Berlin Heidelberg New York, pp 278–303

Payne JW (ed) (1980) Microorganisms and nitrogen sources. Wiley, Chichester New York

Penninckx M, Jaspers C, Wiame JM (1980) Glutathione metabolism in relation to the amino-acid permeation system of *Saccharomyces cerevisiae*. Eur J Biochem 104:119–123

Plummer GL, Kethley JB (1964) Foliar absorption of amino acids, peptides and other nutrients by the pitcher plant, *Sarracenia flava*. Bot Gaz (Chicago) 125:245–260

Pollard JK, Sproston T (1954) Nitrogenous constituents of sap exuded from the sapwood of *Acer saccharum*. Plant Physiol 29:360–364

Prikhod'ko LS (1979) Effect of medium salting with NaCl on peptide composition of cotton plant roots. Fiziol Biok 11:373–378

Prikhod'ko LS, Klyshev LK, Amirzhanova GM (1975) Effect of salinization of the medium on synthesis and qualitative composition of peptides in pea plants. Sov Plant Physiol 22:470–475
Prikhod'ko LS, Frantsev AP, Klyshev LK (1979) Study of the peptide pool in *Salicornia*. Sov Plant Physiol 26:63–70
Pringle RB (1971) Amino acid composition of the host-specific toxin of *Helminthosporium carbonum*. Plant Physiol 48:756–759
Pringle RB, Braun AC (1958) Constitution of the toxin of *Helminthosporium victoriae*. Nature (London) 181:1205–1206
Pringle RB, Schaffer RP (1966) Amino acid composition of a crystalline host-specific toxin. Phytopathology 56:1149–1151
Prusiner S, Doak CW, Kirk G (1976) Novel mechanism for group-translocation: substrate-product reutilization by $\gamma$-glutamyl transpeptidase in peptide and amino acid transport. J Cell Physiol 89:853–864
Pusheva MA, Khoreva SL (1977) Amino acid composition of polynucleotide-peptide complexes isolated from algae. Microbiology 46:49–52
Raake ID (1951) Protein synthesis in ripening pea seeds. Biochem J 66:101–110
Reindel F, Bienenfeld W (1956) Differences in the qualitative composition of the proteins and peptides from the leaf juice of healthy and leafroll infected potato plants. Hoppe-Seyler's Z Physiol Chem 303:262–271
Rennenberg H (1976) Glutathione in conditioned media of tobacco suspension cultures. Phytochemistry 15:1433–1434
Rennenberg H (1978) Influence of glutathione on duration of growth of cytokinin dependent soybean callus tissue. Z Pflanzenphysiol 88:273–277
Rennenberg H, Bergmann L (1979) Influences of ammonia and sulfate on the production of glutathione in suspension cultures of *Nicotiana tabacum*. Z Pflanzenphysiol 92:133–142
Rennenberg H, Schmitz K, Bergmann L (1979) Long-distance transport of sulfur in *Nicotiana tabacum*. Planta 147:57–62
Rennenberg H, Steinkamp R, Polle A (1980) Evidence for the participation of a 5-Oxoprolinase in degradation of glutathione in *Nicotiana tabacum*. Z Naturforsch 35c:708–712
Robert M, Barbier M, Lederer E, Roux L, Biemann K, Vetter W (1962) Two new natural phytotoxins: aspergillomarasmine A and B, and their relation to lycomarismine and its derivatives. Bull Soc Chim Fr:187–188
Salonen M-L, Simola LK (1977) Dipeptides and amino acids as nitrogen sources for the callus of *Atropa belladonna*. Physiol Plant 41:55–58
Schaedle M, Bassham JA (1977) Chloroplast glutathione reductase. Plant Physiol 59:1011–1012
Schantz R, Schantz ML, Duranton H (1975) Changes in amino acid and peptide composition of *Euglena gracilis* cells during chloroplast development. Plant Sci Lett 5:313–324
Schilling ED, Strong FM (1955) Isolation, structure and synthesis of a Lathyrus factor from *L. odoratus*. J Am Chem Soc 77:2843–2845
Sembdner G (1974) Conjugates of plant hormones. In: Schreiber K, Schütte H, Sembdner G (eds) Biochemistry and chemistry of plant growth regulators. Inst Plant Biochem, Halle, pp 283–302
Sembdner G, Borgmann E, Schneider G, Liebisch HW, Miersch O, Adam G, Lischewski M, Schreiber K (1976) Biological activity of some conjugated gibberellins. Planta 132:249–257
Shevyakova NI, Loshadkina AP (1965) Variation of the sulfhydryl group content in plants under conditions of salinization. Sov Plant Physiol 12:280–286
Shimokomaki M, Abdala C, Franca JF, Draetta IS, Figueiredo IB, Angelucci E (1975) Comparative studies between hearts of sweet palm (*Euterpe edulis* and *E. oleracea*) and the bitter species (*Syagrus oleracea*). Colet Inst Tecnol Aliment 6:69–80
Simola LK (1978) Dipeptides as nitrogen sources for *Drosera rotundifolia* in aseptic culture. Physiol Plant 44:315–318
Smith DC (1974) Transport from symbiotic algae and symbiotic chloroplasts to host cells. Symp Soc Exp Biol 28:485–520

Soldal T, Nissen P (1978) Multiphasic uptake of amino acids by barley roots. Physiol Plant 43:181–188

Sopanen T (1979) Development of peptide transport activity in barley scutellum during germination. Plant Physiol 64:570–574

Sopanen T, Burston D, Matthews DM (1977) Uptake of small peptides by the scutellum of germinating barley. FEBS Lett 79:4–7

Sopanen T, Burston D, Taylor E, Matthews DM (1978) Uptake of glycylglycine by the scutellum of germinating barley grain. Plant Physiol 61:630–633

Staskawicz BJ, Panopoulos NJ (1980) Phaseolotoxin transport in *Escherichia coli* and *Salmonella typhimurium* via the oligopeptide permease. J Bacteriol 142:474–479

Stewart CR, Beevers H (1967) Gluconeogenesis from amino acids in germinating castor bean endosperm and its role in transport to the embryo. Plant Physiol 42:1587–1595

Stewart WDP (1980) Transport and utilization of nitrogen sources by algae. In: Payne JW (ed) Microorganisms and nitrogen sources. John Wiley Chichester, New York, pp 577–607

Stewart WDP, Rogers GA (1977) The cyanophyte-hepatic symbiosis. New Phytol 78:459–471

Synge RLM (1951) Non-protein nitrogenous constituents of rye grass: ionophoretic fractionation and isolation of a 'bound amino acid' fraction. Biochem J 49:642–650

Synge RLM (1959) Non-protein chemically bound forms of amino acids in plants. Symp Soc Exp Biol 13:345–352

Synge RLM (1968) Occurrence in plants of amino acid residues chemically bound otherwise than in proteins. Annu Rev Plant Physiol 19:113–136

Synge RLM, Wood JC (1958) Bound amino acids in protein-free extracts of Italian ryegrass. Biochem J 70:321–329

Takagi M, Iida A, Murayama H, Soma S (1973) Isolation and some chemical properties of a new peptide, analipine, from a brown alga, *Analipis japonicus*. Bull Jpn Soc Sci Fisher 39:961–967

Taylor PA, Schnoes HK, Durbin RD (1972) Characterization of chlorosis-inducing toxins from a plant pathogenic *Pseudomonas* sp. Biochim Biophys Acta 286:107–117

Thurmann DA, Street HE (1962) Metabolism of some indole auxins in excised tomato roots. J Exp Bot 13:369–377

Tkachuk R (1970) L-crysteinylglycine: its occurrence and identification. Can J Biochem 48:1029–1036

Tsang ML-S, Schiff JA (1978) Studies of sulfate utilization by algae. 18. Plant Sci Lett 11:177–183

Tsumura A, Komamura M, Kobayashi H (1977) Existence of a free peptide, alanylglycine, in wild and cultivated rice plants. Nippon Dojo-Hiryogaku Zasshi 48:101–102

Ueno T, Nakashima T, Uemoto M, Fukami H, Less N, Izumiya N (1977) Mass spectrometry of *Alternaria mali* toxins and related cyclodepsipeptides. Biomed Mass Spectrometry 4:134–142

Vančura V, Hanzlikova A (1972) Root exudates of plants. IV. Plant Soil 36:271–282

Volodin BB (1975) Participation of sulfur in processes of multiplication of certain blue-green algae. Sov Plant Physiol 22:255–258

von Holt C, Leppla W, Kronar B, von Holt L (1956) The chemical characteristics of the hypoglycines. Naturwissenschaften 43:279

Waley SG (1966) Naturally occurring peptides. Adv Protein Chem 21:1–112

Walsby AE (1974) The extracellular products of *Anabaena cylindrica* Lemm. II. Br Phycol J 9:383–391

Watson R, Fowden L (1975) The uptake of phenylalanine and tyrosine by seedling root tips. Phytochemistry 14:1181–1186

Webster GC (1953) Peptide-bond synthesis in higher plants. I. The synthesis of glutathione. Arch Biochem Biophys 47:241–250

Whitaker JR (1976) Development of flavor, odor and pungency in onion and garlic. Adv Food Res 22:73–133

Winter A, Street HE (1963) A new natural auxin isolated from 'staled' root culture medium. Nature (London) 198:1283–1288

Winter A, Thimann KV (1966) Bound indoleacetic acid in *Avena* coleoptiles. Plant Physiol 41:335–342
Wirth E, Latzko E (1978) Partial purification and properties of spinach leaf glutathione reductase. Z Pflanzenphysiol 89:69–75
Wolffgang H, Mothes K (1953) Papierchromatographische Untersuchungen in pflanzlichen Blutungssäften. Naturwissenschaften 40:606
Wolosiuk RA, Buchanan BB (1977) Thioredoxin and glutathione regulate photosynthesis in chloroplasts. Nature (London) 266:565–567
Wright DE (1962) Amino acid uptake by plant roots. Arch Biochem Biophys 97:174–180
Yamauchi M, Ohashi J, Ohira K (1979) Occurrence of D-alanylglycine in rice leaf blades. Plant Cell Phys 20:671–673
Young EG, Smith DG (1958) Amino acids, peptides and proteins of Irish Moss, *Chondrus crispus*. J Biol Chem 233:406–410
Young LCT, Conn EE (1956) The reduction and oxidation of glutathione by plant mitochondria. Plant Physiol 31:205–211
Yuldashev AK, Tuichev AV, Ibragimov AP (1970) Isolation and separation of cottonseed peptides on ion-exchange resins. Vop Med Khim Biokhim Gorm Deistviya Foziol Aktiv Veshchestv Radiats 111–113
Zacharius RM, Morris CJ, Thompson JF (1959) The isolation and characterization of γ-L-glutamyl-S-methyl-L-cysteine from kidney beans (*Phaseolus vulgaris*). Arch Biochem Biophys 80:199–209

# 13 Immunology

R. MANTEUFFEL

## 1 Introduction

The immune system of higher vertebrates possesses the fascinating ability to distinguish between "self" and "non-self" and can produce, in response to a challenge, antibodies which will react specifically with the antigen that induced their formation.

A heterogeneous antibody population is produced in response to an antigen which although specific to the antigen consists of subpopulations of immunoglobulins which interact with different regions (antigenic determinants) on the surface of the antigen. These subpopulations exist in different amounts and with different affinities for each antigenic grouping.

The majority of antibodies (IgG) have two antibody combining sites of identical specificity at each end of a Y-shaped molecule and therefore they are able to cross-link antigenic material into complexes. The reaction of an antigen with its corresponding antibody is reversible and can be viewed as a special class of protein-protein interactions. An antibody will react only with the antigen to which it is directed or with certain structurally closely related substances (cross-reaction).

Large protein molecules can have several antigenic determinants on their surface.

SELA (1969) distinguished between sequential determinants due to an amino acid sequence in a random coil form and conformational determinants due to the steric conformation of the antigenic macromolecule. Antibodies to native proteins appear to be directed mostly against conformational determinants, whilst those produced in response to denaturated proteins are predominantly directed towards sequential determinants (CHUA and BLOMBERG 1979).

Extensive discussion of both the general structure of the immunoglobulins and the detailed structure of their combining sites can be found in a recent review (CAPRA and EDMUNDSON 1977). Discussions on the general principles of the molecular location of antigenic determinants in protein antigens can be found in REICHLIN (1977), ARNON and SELA (1969) and RÜDE (1971).

In all immunological analyses the main requirements that any immune serum must satisfy are a high titre and a high specificity against the antigen to be detected.

Before using an antiserum, it is essential to characterize its specificity by highly sensitive techniques, because ill-defined antisera often give confusing results from which erroneous conclusions may be drawn. No all-purpose method of immunization has been described and, in practice, nearly every laboratory uses somewhat different procedures, depending on the antigen, the recipient animal species and the aim of immunization. It is not possible to predict which substances will be antigenic in which animals and this question is answered empirically. Two types of antisera can be distinguished: those which react with several proteins (multispecific antisera) and those which are specific for one protein only (monospecific antisera).

One way, besides immunization with pure antigens, in which monospecific antibodies are obtained, is to isolate precipitating antigen–antibody complexes and use these as immunizing material (KRØLL and ANDERSEN 1976).

It is generally accepted that multiple injections over prolonged periods give immune sera which have more pronounced cross-reactivity than those obtained after shorter periods of immunization.

Another problem associated with the production of multispecific antisera is the different immunogenicities of proteins in a mixture; some proteins may be better immunogens, independent of their amount, than others, thus complicating the subsequent use of the antiserum for the detection of all the components in a protein mixture.

Useful schedules of immunization using several plant protein sources are reviewed in HARBOE and INGILD (1973), DAUSSANT (1975) and CATSIMPOOLAS (1978).

No ideal method can be recommended for the extraction of the plant proteins used as antigenic materials and it is necessary to establish the optimum conditions for each protein, or even for the same protein if extracted at different developmental stages of the plant.

A variety of agents are often included to protect the protein from inactivation or digestion (e.g., dithiothreitol, 2-mercaptoethanol, ethylenediaminetetra-acetic acid, glycerol, stabilizing ions, protease inhibitors etc.). It is also known that several low molecular weight substances such as fats, organic acids, tannins, phenolic compounds, various pigments etc. are released on grinding the plant material and may then interact with proteins, resulting in protein insolubility or modification (SABIR et al. 1974, LOLAS and MARKAKIS 1975, SYNGE 1975, HEJGAARD and SØRENSEN 1975, LOOMIS and BATTAILE 1966). Addition of polyvinylpyrrolidone or polyethyleneglycol, reducing agents, $CaCl_2$, inhibitors of polyphenol oxidases or increasing the pH of the extraction buffer are often used effectively to prevent interaction between proteins and undesirable substances (for more details see DAUSSANT 1975, CATSIMPOOLAS 1978, GRABAR 1975).

As a working principle it should always be assumed that an antigen which is pure by biochemical criteria may not necessarily be immunologically pure and this should be tested for by one or more of the following methods: (1) immunoprecipitation in solution; (2) immunodiffusion in gels; (3) immunoelectrophoresis; (4) immunoadsorption; (5) haemagglutination and inhibition of haemagglutination; (6) complement fixation; (7) radioimmunoassay; (8) enzyme immunoassay and (9) immunohistology.

The object of this section is to summarize, using selected examples, the immunological data on plant proteins, including their detection, identification, structural and functional characterization, biosynthesis and degradation, cellular and subcellular location and evolution. In order to restrict the size of this survey, the methodological principles of immunological techniques and their application in protein research will not be considered here, since numerous detailed manuals are available which should be consulted for information on these topics (OUCHTERLONY 1968, AXELSEN 1975, AXELSEN et al. 1973, RUOSLATHI 1976, WEIR 1967, WILLIAM and CHASE 1967a, b, 1971, 1976, 1977, KABAT and MAYER 1961, FETEANU 1978, NAIRN 1976).

## 2 Immunology of Soluble Enzymes

Enzyme amounts are frequently measured in activity units rather than directly as proteins. Activity measurements are useful, but values may vary due to the presence of activators or inhibitors, or by pH, ionic, or dilution changes, rather than by differences in enzyme quantity. It is therefore desirable to measure protein as well as activity and this is possible by the combination of immunological techniques with the determination of enzymic activity (LANZEROTTI and GULLINO 1972).

Enzymes possess, beside the biochemically detectable specificity for their substrates, antigenicity which provides an additional criterion for identification.

The interaction between enzymes and their respective antibodies generally leads to a reduction in enzyme activity. The enzyme may be completely inhibited, partially inhibited or in exceptional cases enzyme activity is stimulated (ARNON 1973).

Measurements of enzymes in solution have been carried out by immunotitration of enzyme activity with increasing volumes of antiserum. Usually the volume of antiserum required to inhibit enzyme activity completely is that required to quantitatively precipitate the enzyme (MAYER and WALKER 1978).

Enzymes which retain their activity in the enzyme–anti-enzyme complex can also be specifically characterized by immunodiffusion or immunoelectrophoretic methods using chromogenic substrates and the same methods which have been developed for the analytical chemistry of enzymes and for histochemical purposes (URIEL 1963).

Immunochemical studies have been used to study enzyme activity, the nature of antigenic determinants on enzymes, the relationship between enzymes and apoenzymes, multiple forms of enzymes, allostery and enzyme evolution. Furthermore, sensitive immunological techniques are being used to investigate the regulating mechanisms of biosynthesis and degradation of enzymes (for review see GRABAR 1975, DAUSSANT 1975, CATSIMPOOLAS 1978).

*Malate Dehydrogenase (MDH).* In seedlings of fat-storing plants and in green leaves there are several unique isoenzymes of MDH with distinct intracellular distribution and biological function (KHAVKIN et al. 1972). No immunological differences were detected in the sensitivity to antibodies between these MDH types derived from glyoxysomes and peroxisomes (HUANG et al. 1974, ZSCHOCHE and TING 1973). A complete lack of cross-reactivity was observed, however, between microbody isoenzymes and mitochondrial or cytoplasmic isoenzymes using antibodies to glyoxysomal MDH (WAINWRIGHT and TING 1976) and between microbody or cytoplasmic isoenzymes and mitochondrial isoenzyme (WALK and HOCK 1976) using antibodies to mitochondrial MDH. The finding of complete immunological identity between the microbody isoenzymes at all developmental stages of *Cucumis sativus* seedlings suggest that the isoenzyme remains unaltered, despite the transition from glyoxysomal to peroxisomal function that occurs during greening of the cotyledons (WAINWRIGHT and TING 1976).

By means of radial immunodiffusion, the activity of the glyoxysomal isoenzyme, which could not be detected in dry seeds of *Citrullus vulgaris,* was found to increase during early germination and to decline thereafter. From the biochemical and immunological evidence it was concluded that the marked increase of glyoxysomal activity in germinating cotyledons was due to de novo synthesis of the isoenzyme (WALK and HOCK 1977).

The changes of mitochondrial MDH biosynthesis during the germination of watermelon seedlings was assayed by means of quantitative radial immunodiffusion. The biochemical and immunological results are evidence for the de novo synthesis of mitochondrial MDH and its relatively slow turnover in germinating seeds (WALK and HOCK 1976).

*Glyceraldehyde-3-Phosphate Dehydrogenases (GPD).* CERFF and CHAMBERS (1979) compared the isoenzymes of chloroplast NADP-linked GPD from *Sinapis alba* and *Hordeum vulgare* on the basis of cross-reactivity. The isoenzymes showed immunochemical identity with no spurring or crossing-over. These results are in agreement with data obtained by subunit composition, tryptic peptide mapping and amino acid analyses and suggest that both isoenzymes share one common subunit of identical structure and that structural differences between the two subunits are restricted to a short terminal sequence. The absence of cross-reactivity between cytoplasmic NAD-specific GPD and chloroplast NADP-linked GPD confirm an earlier report by McGOWAN and GIBBS (1974) that the NAD-specific enzyme does not cross-react with antisera raised against the NADP-linked enzyme from peas (*Pisum sativum*). The comparison of tryptic peptide maps and the serological behaviour of the two enzymes demonstrates that the enzymes are coded by separate genes and thus refutes the "incorporation" hypothesis (CERFF 1974).

*Glycolate Oxidase.* CODD and SCHMID (1972) characterized the plant glycolate oxidizing enzyme serologically by using pure anti-tobacco enzyme. The antibody was also tested against enzyme preparations from *Euglena gracilis,* taken as an example of algal glycolate dehydrogenase and against *Chlorella vulgaris,* which has glycolate oxidase activity (CODD and SCHMID 1972). Whereas the algal enzymes were identical and showed partial identity with the tobacco enzyme by immunodiffusion, the antiserum only inhibited the glycolate oxidase of the tobacco and *Chlorella* extracts but did not affect the dehydrogenase activity of *Euglena.*

*Peroxidase.* Although only one molecular form of peroxidase was detected serologically in seeds of different *Triticum* species during germination (DAUSSANT and ABBOTT 1969), some modification of this enzyme occurred during the early stages of *Zea mays* seedling development (ALEXANDRESCU and CĂLIN 1969). Several distinct molecular species of peroxidase from one species have been also distinguished in leaf extracts of *Triticum aestivum, Secale cereale, Zea mays* and *Avena sativa,* all of which were antigenically different from one another with the exception of the rye enzyme, which was identical with that from wheat using the anti-wheat leaf serum. However, some isoenzymes of these plant genera were able to cross-react (ALEXANDRESCU and HAGIMA 1973). Antigenic heterogeneity was also detected for peroxidase extracted from crown gall tumour of *Datura stramonium* (DAUSSANT et al. 1971). Immunochemical characterization of the peroxidase from green and etiolated pea plants revealed antigenic similarity between pea and horseradish peroxidases (BAKARDIEVA and DEMIREVSKA-KEPOVA 1976). Peroxidase has been identified as a product released by peanut cells in suspension culture by using immunological assays (VAN HUYSTEE 1976). KAHLEM (1976) demonstrated the occurrence of specific isoperoxidases characteristic of microsporogenesis and tapetum differentiation histoimmunologically and not at other sites of peroxidase activity of male flowers. He proposed therefore that these isoenzymes constitute an early and specific marker of male organogenesis in higher plants.

*Nitrate Reductase.* NADH nitrate reductase from higher plants is very unstable both in vivo and in vitro, but the enzyme which has lost its enzymic activity retains its immunochemical properties. By using the methods described by GRAF et al. (1975), it may therefore be possible to quantify the inactive protein, and this technique may have application especially where several different causes for the loss of catalytic activity are suspected.

*Lipoxygenase.* TROP et al. (1974) have studied the extent of the antigenic cross-reaction among the lipoxygenases from soyabean (Leguminosae) with those of the potato and egg-plant (Solanaceae). The antibodies strongly inhibited the ability of the enzymes to catalyze fatty acid oxidation and all three enzymes had common antigenic determinants, but they were not immunologically identical. Although the catalytic activity was abolished by heat treatment or proteolytic digestion, some of the antigenic determinants were stable under the same conditions, since heat-inactivated enzymes, or peptides, produced by proteolytic digestion, partially reactivated the enzyme which had been previously inhibited by antibodies. From these results it is likely that the active site is not identical with any of the antigenic determinants, but it is certainly close to one or several of them. Thus the reaction with antibody causes steric hindrance to the active site.

*Proteolytic Enzymes.* The proteases, ficin, bromelain, chymopapain and papain, comprise a group of plant thiol enzymes of similar molecular weight. Ficin, like bromelain and papain (PRAGER et al. 1962), will attach spontaneously to erythrocytes which can be used in the passive haemagglutination technique.

Anti-ficin does not abolish the enzymatic activity of ficin (BARRETT and WHITEAKER 1977) and immunological cross-reactions indicate an appreciable degree of surface homology between fruit and stem bromelain (SASAKI et al. 1973, IIDA et al. 1973). Papain and chymopapain not only have similar enzymatic activities but also exhibit immunological cross-reactivity (EDER and ARNON 1973). Fractionation of the total antibody population of sera produced against papain and chymopapain separately into fractions of common and noncommon antibodies may be achieved by cross-adsorption of each of these two antisera with an immunosorbent prepared from the heterologous enzyme (ARNON and SHAPIRA 1967). The two common antibody fractions are about equally capable of inhibiting the enzymatic activity of either papain or chymopapain. In contrast, the two noncommon antibody fractions, although they react with their respective homologous antigens, do not cross-react with the heterologous enzyme. The presence of common antigenic determinants among the four kinds of thiol proteases was also demonstrated by isolating a particular antibody population which could combine with all four enzyme immunoadsorbents. The anti-stem bromelain antibody cross-reacting with fruit bromelain was found to inhibit efficiently the catalytic activities not only of stem and fruit bromelain, but also of papain and ficin. It was concluded that this similarity involved the active site and its proximate regions, but the extent of these common areas varies in different enzymes according to their phylogenetic relationship (KATO and SASAKI 1974). Immunological data corroborate the homology of the amino acid sequences of the active site peptides from the papain, ficin and stem bromelain (HUSAIN and LOWE 1970). Papain and ficin, which are derived from plants in the same class but in different orders, cross-reacted, but stem and fruit bromelain showed greater cross-reactivity (SASAKI et al. 1973). There was also a close immunochemical relationship between pinguinain, another thiol protease of plant origin, and fruit bromelain, but not apparently between pinguinain and papain. It was speculated that no antigenic binding sites are located at the active site of the enzyme because modifications at this site caused by sulphydryl group specific reagents did not alter its immunochemical properties and combination of antibodies with the enzyme did not abolish its enzymic activity on low molecular weight substrates. It can be inferred that the lysine residues in pinguinain play a significant role in its immunochemical reactivity because marked changes are introduced by modification of the ε-amino residues (TORO-GOYCO and RODRIGUEZ-COSTAS 1976).

The vicilin peptidohydrolase of germinating mung beans (*Phaseolus mungo*) has been localized intracellularly by an indirect immunofluorescence technique (BAUMGARTNER et al. 1978). The results are consistent with the interpretation that the enzyme is synthesized de novo in the cytoplasm of storage parenchyma cells and subsequently transported to the protein bodies. The vicilin peptidohydrolase was not present in leaves or seeds during the early stages of germination (CHRISPEELS et al. 1976, BAUMGARTNER and CHRISPEELS 1977). Studies of the serological cross-reactivity between the vicilin peptidohydrolases from species of the genera *Vigna* and *Phaseolus* (CHRISPEELS and BAUMGARTNER 1978) support the reassignment of *Phaseolus aureus* and *Phaseolus mungo* to the genus *Vigna*, confirming the results of KLOZ (1971) obtained using a serological comparison of albumins and globulins from the genera *Vigna* and *Phaseolus*.

An endopeptidase which hydrolyzed polypeptides at hydrophobic amino acid residues was isolated and serologically characterized from yellowing and green leaves of *Nicotiana tabacum* (HOCHKEPPEL 1973). Treatment of the lamellar system of chloroplasts with the endopeptidase caused hydrolysis of the proteins of the thylakoid membrane followed by an acceleration of chlorophyll degradation by photooxidation.

Chromogenic substrates have been used for immunochemical identification of proteolytic enzymes in germinating barley (TRONIER et al. 1974). It was suggested that the increased peptidase activity during barley germination was not due to increased synthesis of the enzyme already present, but more likely due to de novo synthesis of a new isoenzyme. In crossed immunoelectrophoresis, four immunochemically distinct aminopeptidases were observed in barley and malt proteins by HEJGAARD and BØG-HANSEN (1974).

*Amylases. α-Amylase* activity undergoes drastic changes during the life of seeds. It is high at the early stage of seed development, drops rapidly during seed maturation and finally increases again during seed germination (DAUSSANT and RENARD 1972). The changes in α-amylase isoenzyme pattern have been studied by rocket immunoelectrophoresis (HEJGAARD and BØG-HANSEN 1974), crossed immunoelectrophoresis (BØG-HANSEN and DAUSSANT 1974), and immunoabsorption techniques (DAUSSANT and SKAKOUN 1974, DAUSSANT and CARFANTAN 1975).

There were antigenic differences between the α-amylases of developing seeds and those which bear the bulk of amylolytic activity at different stages of germination, and more than 98% of enzymic activity detected in germinating seeds was associated with the new types of α-amylase isoenzymes not found in developing seeds (DAUSSANT et al. 1974). α-Amylases from germinating seeds were antigenically heterogenous, showing a reaction of semi-identity between the two antigens with one isoenzyme appearing before the other during germination. The changed amounts of the two α-amylase isoenzymes during germination were the result of de novo synthesis and degradation (DAUSSANT et al. 1974, BØG-HANSEN and DAUSSANT 1974). The complete lack of reactivity between the immune serum directed against α-amylases from germinating seeds and those extracted from ungerminated seeds indicated that there was no preformed precursor of the germination isoenzymes among the proteins from mature seeds (DAUSSANT and CORVAZIER 1970). Amylolytic activities in cereal seeds, coleoptiles and in the first leaves is also due to enzymes other than those of the germinating seed endosperm (ALEXANDRESCU and MIHĂILESCU 1973). In response to gibberellic acid treatment, as well as during germination, there are two immunologically distinct α-amylase isoenzymes made by and released from aleurone cells into the medium (JACOBSEN and KNOX 1973). Specific localization of α-amylase in aleurone cells has been reported using immunohistochemical techniques (JACOBSEN and KNOX 1974). α-Amlyase could not be detected by immunofluorescence in sections of aleurone from non-inbibed grains, but after inbibition or in response to gibberellic acid treatment, α-amylase occurred in "patches" throughout the cytoplasm. Protein bodies did not contain α-amylase (TRONIER and ORY 1970) and JONES and CHEN (1976) used immunohistological techniques to show that α-amylase synthesis is associated with the rough endoplasmic reticulum in aleurone cells. OKITA et al. (1979) studied the α-amylase products synthesized by in vitro translation systems using immunoprecipitation of translation products followed by gel electrophoresis and proteolysis. Both a precursor and the α-amylase itself were formed using polysomes from wheat aleurone as mRNA source, while only the precursor was synthesized in a mRNA-dependent translation system. Antigenic relationships have been detected between α-amylases of barley, wheat, rye, oats, maize and rice (DAUSSANT and GRABAR 1966, ALEXANDRESCU et al. 1975a, b, ALEXANDRESCU and MIHĂILESCU 1973).

Analysis of seven different cultivars of *Phaseolus vulgaris* suggested that although the relative amounts may vary, the α-amylase inhibitor proteins are immunologically identical. It was also concluded that α-amylase inhibitors are synthesized in the final stage of seed maturation. Leaves of all plants are devoid of these inhibitors (PICK and WÖBER 1978).

*β-Amylase* displays several types of polymorphism (NUMMI et al. 1965). The free β-amylase consists of components of differing molecular size and is extractable with water or saline solution, whereas the bound β-amylase can only be liberated enzymically or by reducing agents. Since the several components of free β-amylase and those of the bound form have the same antigenic properties (DAUSSANT et al. 1966, DAUSSANT 1975), it was suggested that the enzyme exists in several different states of aggregation, either with or without other constituents. HEJGAARD (1976) observed the association of β-amylase with an immunochemically distinct protein and concluded that the association was via disulphide bonds, but that homopolymer aggregates may also exist. The fact that β-amylase of larger molecular size disappears during germination, while the amount of a smaller enzymic unit increases, has led to the proposal of a "splitting-enzyme" mechanism in germinating seeds, which releases the bound enzyme without destroying its activity (NUMMI et al. 1970). Although the antigenicity of β-amylase in germinating seeds becomes slightly modified, the greater part of the antigenicity of the initial β-amylase is maintained (TRONIER and ORY 1970, DAUSSANT 1975, 1978).

By combining the technique of labelled amino acid incorporation into proteins and immunoelectrophoresis it was demonstrated that the modification of β-amylase was not due to de novo synthesis of enzyme molecules but to the activation of a pre-existing enzyme precursor (DAUSSANT and CORVAZIER 1970). In contrast to barley, where immunochemical localization of the bound β-amylase was shown in the protein bodies (TRONIER and ORY 1970), it has recently been demonstrated in rice, using indirect immunofluorescence, that the β-amylase is uniformly associated with the periphery of starch granules in the starchy endosperm cells (OKAMOTO and AKAZAWA 1979).

Comparative studies have been made of β-amylase of grasses by means of immunochemical analysis (NUMMI et al. 1970, ALEXANDRESCU et al. 1975b). A technique combining immunoabsorption, electrophoresis and enzymatic staining reactions in the same gel medium has been developed in order to identify β-amylase in the presence of α-amylase by DAUSSANT and CARFANTAN (1975) and DAUSSANT and MACGREGOR (1979).

*Cellulase.* Two immunologically different forms of cellulase are induced as a result of auxin treatment of dark-grown pea epicotyls (BYRNE et al. 1975), and an immunocytochemical method has been developed using ferritin-conjugated antibodies to determine their intracellular localization (BAL et al. 1976). The results of both cytochemical and immunocytochemical observations suggested that auxin-induced cellolytic activities were confined to the inner surface of the cell wall and to endoplasmatic reticulum vesicles. These observations support the general contention that cellulase has a functional role in cell wall metabolism in growing cells.

*β-Glucosidase.* The β-glucosidase hydrolyzing coniferin in spruce (*Picea abies*) seedlings was localized at the inner layer of the secondary cell wall of all hypocotyl cells by using immunofluorescence; activity appeared greater in the epidermal layer and in the vascular bundles, and these are the tissues in which lignification first appears during seedling development. β-Glucosidase activity was absent in spruce seeds (MARCINOWSKI et al. 1979).

*α-Mannosidase*. The large degree of immunological identity between α-mannosidase I and II from *Phaseolus vulgaris* has made it possible to purify both enzyme forms by affinity chromatography using an immunoadsorbent made from anti-α-mannosidase I antiserum. Once established, this is a fast and highly specific method (PAUS 1976).

*Ribonuclease*. In order to check the cellular compartmention of RNase BAUMGARTNER and MATILE (1976) attempted an immunocytochemical localization of RNase using a specific antiserum to the purified enzyme. It is suggested that the breakdown of RNA was not initially determined by catabolic enzyme activity, but that tonoplast appeared to be responsible for the rate of degradation by controlling the transport of cytoplasmic material into the vacuole. Thus in their experiments a diffuse, occasionally granular or reticular, RNase-specific fluorescence appeared over the central vacuole.

*Esterase*. Affinity electrophoresis and crossed immunoaffinoelectrophoresis using Concanavalin A have shown that the esterase and acid phosphatase from barley malt are glycoprotein enzymes containing glucose or mannose; there were indications of micro-heterogeneity within the carbohydrate moiety of the barley esterase (BØG-HANSEN et al. 1974).

KHAVKIN et al. (1972) were able to identify esterase in the immunopherograms of corn root extracts by using the formazan technique.

*Urease*. *Canavalia* anti-urease antiserum can precipitate soyabean (*Glycine max*) urease activity (KIRK and SUMNER 1932). Soyabean and jack bean urease are serologically related but also contain unique antigenic determinants revealed by affinity chromatography of antibodies to soyabean seed urease on Sepharose 4 B containing covalently linked jack bean urease (POLACCO and HAVIR 1979).

It has been demonstrated that soyabean tissue culture produces an urease indistinguishable from the seed enzyme on the basis of electrophoretic pattern, gel filtration and immunoinhibition. However, it appeared that another ureolytic enzyme may also be produced by cultured cells. Serohistological studies showed that urease activity was associated with sites both inside and outside the storage parenchyma cells (MURRAY and KNOX 1977). Inside the cells the fluorescent-labelled antibodies were attached to small spherical granules clustered around the periphery of the cells. Outside the cells, fluorescence was seen within cell walls, but mainly in spherical granules within the intercellular spaces. At late stages of seed germination, most of the detectable urease activity was extracellular.

*Ribulose-1,5-Bisphosphate Carboxylase (RuBP Carboxylase)*. Due to the antigenic similarity of the enzymes of a wide variety of plants (DORNER et al. 1958,

KAWASHIMA et al. 1968), immobilized antibody columns provide a rapid and efficient method for the isolation of both holoenzyme and subunits (GRAY and WILDMAN 1976, GATENBY 1978, FEIERABEND and WILDNER 1978). Thus, antibodies elicited against the holoenzyme also reacted with both the isolated large and small subunits, as did the subunit-specific antibodies with the holoenzyme. There is no immunological correspondence between the large and the small subunit (GOODING et al. 1973, NISHIMURA and AKAZAWA 1974a, GRAY and KEKWICK 1974, BOWIEN and MAYER 1978, SANO et al. 1979). The reconstitution of large and small subunits to form the holoprotein was demonstrated by its immunological response to the antisera raised against large and small subunits (NISHIMURA and AKAZAWA 1974b). Electron microscopic examination of antibody–carboxylase complexes confirmed that the large subunits are located in the central two layers of the four-layered enzyme molecule, whereas the two outer layers consist of small subunits (BOWIEN and MAYER 1978). The observation that enzyme activity is inhibited by antiserum directed against sites on the large subunit but not by antiserum directed against sites on the small subunit clearly show that the large subunit carries the catalytic site of the enzyme (NISHIMURA and AKAZAWA 1974a, GRAY and KEKWICK 1974, TAKABE and AKAZAWA 1973, SUGIYAMA et al. 1970). The potent inhibitory effect of anti-large subunit serum on the oxygenase reaction means that the large subunit also has the catalytic site for the oxygenase reaction (JOHAL and BOURQUE 1979, NISHIMURA and AKAZAWA 1974a).

Immunological comparisons of the RuBP carboxylase level in several $C_3$ and $C_4$ species showed that the $C_3$ species had a three- to sixfold higher concentration of the protein than the $C_4$ species (KU et al. 1979). Antibodies raised to RuBP carboxylase have been used to determine the intracellular location of the enzyme in $C_3$ and $C_4$ species by immunofluorescence (HATTERSLEY et al. 1976). In $C_3$ plants, specific fluorescence is associated with chloroplasts and leaf chlorenchymatous cells, whilst in species with $C_4$ leaf anatomy the enzyme is located almost exclusively in bundle-sheath cell chloroplasts. Similar results were obtained by cell fractionation followed by quantitative immunoassay (KU et al. 1979, MATSUMOTO et al. 1977). Leaves of $C_3/C_4$ hybrid individuals exhibit a $C_3$ antibody labelling response (HATTERSLEY et al. 1977).

SPREY also showed immunologically that the lamellae-bound crystals of spinach (*Spinacia oleracea*) chloroplasts contain crystalline RuBP carboxylase, presumably in an inactive state (SPREY 1976, SPREY and LAMBERT 1977).

Immunological data are available for RuBP carboxylase from many diverse plants and these could be used to study the phylogenetic relationships between species (MCFADDEN and TABITA 1974, KAWASHIMA et al. 1971, MURPHY 1978, GRAY 1977). Antisera raised against the small subunits from different species do not cross-react, but those against the large subunit do, thus demonstrating the evolutionary conservation of the antigenic determinant sites of the latter (GRAY and KEKWICK 1974, GATENBY 1978). Since the large subunits of RuBP carboxylase are inherited from the maternal parent, whereas the small subunits are inherited from both parents, the serological analysis of this single protein allows the identification not only of the parent species, but also of the direction of the cross giving rise to an allopolyploid, provided that the parent species are known and their RuBP carboxylase can be distinguished serologically (GRAY 1978). HIRAL (1977) demonstrated that the RuBP carboxylase from interspecific $F_1$ hybrid plants could be highly heterogeneous in charge due to random assembly of different kinds of small subunit polypeptides during formation of RuBP carboxylase macromolecules.

*Malate Synthase.* The data of LORD and BOWDEN (1978) suggest that during castor bean germination, a rapid de novo synthesis of malate synthase takes place exclusively on bound ribosomes in the endosperm tissue and the changes in intracellular distribution of the enzyme are consistent with a route of intracellular transport of this protein via the endoplasmic reticulum before its ultimate sequestration in its functional cellular compartment, the glyoxysome. When released nascent polypeptide chains from both, free and bound polysome preparations were precipitated with anti-glyoxysomal malate synthase serum, the radioactive malate synthase was exclusively associated with bound polysomes. Convincing evidence for the initial segregation of the enzyme by the endoplasmic reticulum was obtained, when the [$^{35}$S]-methionine-labelling kinetic behaviour of malate synthase was followed by specifically precipitating this protein from crude extracts of microsomes and glyoxysomes using anti-glyoxysomal malate synthase serum. The microsomal malate synthase was labelled before the glyoxysomal enzyme; during a cold chase period labelled malate synthase was rapidly lost from the microsomal fraction and was quantitatively recovered in the glyoxysomal fraction.

KÖLLER et al. (1979) demonstrated that glyoxysomal proteins were already present in small amounts in dry seeds of cucumber. The enzymatic and immunological properties of malate synthase were not distinguishable from those of enzymes assigned to the glyoxysomes of fully developed cotyledons.

*Citrate Synthetase.* Citrate synthetase is present in mitochondria and glyoxysomes of seedlings of fat-storing seeds and both enzymes have the same immunological and biochemical behaviour (HUANG et al. 1974). Citrate synthetase was also immunologically detected in dry seeds of cucumbers (KÖLLER et al. 1979).

*RNA-Polymerase.* Antibodies directed against hexaploid wheat RNA polymerase II react with the enzymes purified from dicotyledonous plants as well as those of monocotyledonous species, but they do not cross-react with yeast RNA polymerase II or *Escherichia coli* RNA polymerase (JENDRISAK and GUILFOYLE 1978). These studies indicate a highly conserved but very complex molecular structure.

*Isocitrate Lyase (ICL).* The purity of enzyme preparation from cucumber seedlings was checked by antiserum raised in rabbits against a purified enzyme. The glyoxysomal enzyme was immunologically identical with the ICL from crude tissue extracts. Identity was also observed between ICL from glyoxysomes of etiolated cotyledons and the enzyme from illuminated cotyledons. ICL could be detected in dry seeds and was immunologically identical to the enzyme from fat-degrading cotyledons (KÖLLER et al. 1979). Although the enzyme showed no selective binding to lectins with N-acetylgalactosamine or N-acetylglucosamine specificity, it was shown that ICL was a glycoprotein by Schiff staining and by incorporation of [$^{3}$H]-glucosamine in vivo into a protein which could be precipitated with antibodies to isocitrate lyase (FREVERT and KINDL 1978).

# 3 Immunology of Biologically Active Proteins Other Than Enzymes

## 3.1 Lectins (Phytohaemagglutinins)

Lectins are specific sugar-binding proteins, some of which agglutinate not only red blood cells but many other types (SHARON 1977, LIS and SHARON 1973, AGRELL 1966, ZECH 1966, see Chap. 7, this Vol. for information on amino acid sequence data and Chap. 8, this Vol. for information on their distribution). When lectins with multiple combining sites bind to carbohydrate molecules on cell surfaces, they connect large numbers of cells, causing them to agglutinate. The binding of lectins to sugars is often very specific, but quite weak and reversible.

The reaction between a lectin and its specific polysaccharide, glycoprotein or glycolipid is analogous in most respects to that of an antibody–antigen system.

Like the reaction of an antibody with an antigen, if a precipitate is formed by the reaction between lectin and the corresponding receptor it will usually dissolve when one of the reactants is present in excess (SO and GOLDSTEIN 1972). Lectins, like antibodies, can be inhibited competitively by specific sugars which act as haptens. The most important difference between them is that antibodies are produced specifically in response to a foreign substance and their specificity is broad, encompassing not only sugars. On the other hand, the naturally occurring lectins require no preceding immunogenic stimulation; they are present in vivo as constituent proteins and have a specificity directed only towards single carbohydrate units which are usually, but not invariably, terminal (BIRD et al. 1971). All antibodies are structurally related to one another, whereas lectins are structurally different (ALLEN et al. 1973, MATSUMOTO and OSAWA 1970, SELA et al. 1977).

Recently there have been numerous reports of the possibility of combining lectins and serohistological reactions (ROTH and BINDER 1978, MAHER and MOLDAY 1977, FRANZ et al. 1979).

A survey of about 3,000 plant species revealed that over 1,000 of them displayed haemagglutinating activity (LIENER 1976, SHARON 1977). The distribution of lectins in plant is covered in Chapter 8, this Volume, but studies on some lectins will be reported here as examples of the types of investigation undertaken using serological methods.

In most cases, several similarly active compounds, called isolectins, can be separated. The isolectins obtained from a single plant source have identical or very similar specificities and are immunologically similar, although lectins of different specificities which do not cross-react may occur in the same plant (e.g., BAUMANN et al. 1979). Immunological and structural studies of lectins isolated from Leguminosae (see TOMS and WESTERN 1971 for review) suggest the existence of two main groups of proteins with respect to their carbohydrate-binding specificities (BAUMANN et al. 1979, HANKINS et al. 1979). The overwhelming majority of the legume lectins belong to the galactose-specific group and there is extensive immunological homology among them. Only peanut and soyabean agglutinins fail to show cross-reactivity. Initial immunological studies of HOWARD et al. (HOWARD et al. 1979) indicated that many of the major plant lectins are immunologically related and, moreover, that there may be a unique portion of these proteins that is evolutionarily highly conserved. The data reported on the immunological relationships between phytohaemagglutinins from distant taxonomical species are too scarce, however, to warrant definite conclusions.

The first lectin to be purified was Con A from seeds of *Canavalia ensiformis* (see Chap. 7, this Vol. for structural details). Large quantities (23%–28% of the protein) are deposited in the seeds (HAGUE 1975) and antibodies to it can be raised in rabbits (NORTHOFT et al. 1978). Antibodies against Con A from seeds of *Canavalia ensiformis* do not discriminate among the lectins of other *Canavalia* species (HAGUE 1975), but do not cross-react with *Phaseolus vulgaris* agglutinins (GALBRAITH and GOLDSTEIN 1972). Some of the glycoprotein lectins (e.g., *Phaseolus vulgaris, Phaseolus lunatus* and *Glycine max* lectins) have been reported to interact with Con A by its carbohydrate-binding ability (JAFFÉ 1978).

The soyabean seed lectin is one of the few lectins in which the number of binding sites is not equal to the number of subunits (LIENER 1976) and the structure and properties of soyabean lectin have been extensively characterized (LOTAN et al. 1975). Four immunological homologous isolectins have been separated from soyabean (CATSIMPOOLAS and MEYER 1969). The intracellular localization of soyabean lectin has been determined in thin sections of different parts of the seed using anti-soyabean lectin antisera labelled with gold granules; it is uniformly distributed in the protein bodies of the cotyledon cells (HORISBERGER and VONLANTHEN 1980).

Antiserum against soyabean lectin was used for the quantitative estimation of the latter by LIENER and ROSE (1953). The soyabean lectin was found to be released from the seeds during inbibition (FOUNTAIN et al. 1977).

The peanut agglutinin has been called "T antigen" (UHLENBRUCK et al. 1969) because its activity resembles that of the naturally occurring "T antibody", which detects a heterophile antigen on neuraminidase-treated erythrocytes. The structure, specificity and combining site of peanut lectin have been characterized (PEREIRA et al. 1976). More recently it was reported that peanut lectin is not a single molecular species, but rather a mixture of up to six (NEWMAN 1977) or seven (PUEPPKE 1979) isolectins, all of which are erythroagglutinins. Most of the lectin activity in young seedlings was found in the cotyledons and levels in roots were very low, and detectable for relatively few days after germination (as in other legumes). Antibodies prepared against seed isolectins reacted with hypocotyl lectin preparations (PUEPPKE 1979).

Among monocotyledonous plants, wheat germ agglutinins have been extensively characterized (NAGATA and BURGER 1974) although uncertainty exists as to the exact number of binding sites on the molecule (LEVINE et al. 1972). Recently novel lectins were isolated from the endoplasmic reticulum of wheat germ. These lectins were serologically different from Con A although their binding specificity was similar (YOSHIDA 1978). They are also different to the usual wheat germ agglutinin, which specifically binds to the N-acetyl-glucosamine residues of carbohydrate moieties (NAGATA and BURGER 1974). Similar lectins are present in the endoplasmic reticulum of tobacco leaves and yeast (YOSHIDA 1978).

A rice bran lectin has also been purified to homogeneity and dissociated into two nonidentical subunits which did not show agglutinating activity. The lectin was mitogenic against mouse and human lymphocytes (TSUDA 1979). The biological function of lectins in nature, whether in plants or in other organisms, is still a mystery and different lectin types are still being described.

Speculation on their physiological role in plants includes the following: that they protect plants against bacterial or fungal attack (GIETL and ZIEGLER 1979, SHARON 1977); play a role in the recognition of symbiotic bacteria by the host (LOTAN et al. 1975, HAMBLIN and KENT 1973, BOHLOOL and SCHMIDT 1974, BHUVANESWARI et al. 1977, KATO et al. 1979); serve as storage proteins (BOLLINI and CHRISPEELS 1978, HAGUE 1975); play a role in transport and storage of sugars (BOWLES and KAUSS 1975); attach glycoproteins to enzymes in organized multi-enzyme systems (LIENER 1976); provide membrane recognition sites for the regulation of the dynamics of internal membrane compartmentation and are involved in intracellular communication (BOWLES 1979, BOWLES et al. 1976, BOWLES and KAUSS 1975); play some role in the binding of ribosomes to the endoplasmic reticulum (YOSHIDA 1978); play a key role in the development and differentiation of embryonic plant cells either by controlling cell division (NAGL 1972, SHARON and LIS 1972) or by mediation of cell-extension growth (RAYLE 1973, HOWARD et al. 1972).

Whatever their role in nature there is no doubt that lectins will continue to serve (1) as important and versatile tools for structural studies of macromolecules containing carbohydrate linkages (SHARON 1977); (2) as specific reagents for the isolation of carbohydrate containing macromolecules (SHARON and LIS 1972); and (3) as histochemical tools for cytological studies on the in situ organization of the cell surface (FRANZ et al. 1979, ROTH and BINDER 1978) and chromosomes (KURTH et al. 1979).

### 3.2 Protease Inhibitors

Inhibitors of proteolytic enzymes are widely distributed among plant species (RYAN 1973) and tend to accumulate in storage structures (CHIEN and MITCHELL 1973) such as tubers (RYAN and SANTARIUS 1976), fruits (KANAMORI et al. 1976), and seeds (RICHARDSON 1977). The inhibitors strongly inhibit animal and microorganism proteases but do not usually inhibit plant proteases. Plant protease inhibitors are a complex class of several structurally related or distinct molecular species which can usually be found within the same tissue.

*Kunitz Trypsin Inhibitor* (see also Chap. 7, this Vol.). Immunochemical methods for the detection and quantitation of the Kunitz trypsin inhibitor were developed by CATSIMPOOLAS and LEUTHNER (1969). The Kunitz trypsin inhibitor occurs in two electrophoretically distinct but immunochemically identical forms, and its content does not change dramatically during seed germination (FREED and

RYAN 1978a). Since a rapid decrease in Kunitz trypsin inhibitor content does not take place during germination it seems improbable that this inhibitor functions as a storage protein in the seed (FREED and RYAN 1978b).

*Bowman-Birk Trypsin Inhibitor and Related Inhibitors.* In soyabean there is a group of five protease inhibitors, four closely immunologically related to one other and having a more distant relationship to the fifth, the well-known Bowman-Birk inhibitor. Although the four isoinhibitors are closely related immunologically, they vary quantitatively in inhibitory power and molecular weight (HWANG et al. 1977). Neither the *Phaseolus lunatus* trypsin inhibitor nor the Kunitz soyabean trypsin inhibitor competes with the five protease inhibitors in binding reactions to their corresponding antibodies. Radioimmunoassay showed only the cotyledons to contain a significant amount of protease inhibitor, and during inbibition and early germination the protease inhibitors were released and as germination proceeded there was an accompanying disappearance of the protease inhibitors from the seedlings (HWANG et al. 1978).

*Potato Chymotrypsin Inhibitor I and II* (see also Chap. 7, this Vol.). Various protease inhibitors have been investigated in potatoes (*Solanum tuberosum*). Contrary to the results obtained using immunodiffusion (MELVILLE and RYAN 1972), microcomplement fixation assays showed significant differences among the four protomer types of chymotrypsin inhibitor I and significant differences were found among inhibitor I tetramers that were reconstituted individually from the four subunit types (GURUSIDDAIAH et al. 1972). It was suggested that the differences between tuber and leaf inhibitor I reflect a different ratio of protomers in the molecules and that variations in protomer composition may be tissue related (GURUSIDDAIAH et al. 1972). The presence of protease inhibitor I has been followed immunologically in the tissues of the potato plant during growth and development (RYAN et al. 1968). Radial immunodiffusion studies have also led to the discovery that the level of protease inhibitor I in *Lycopersicum* and *Solanum* leaves increased when plants were wounded either mechanically or by insects (GREEN and RYAN 1972). Protein bodies induced in tomato leaf cells by wounding were shown to contain the inhibitor by using immunocytochemistry (SHUNWAY et al. 1976). The immunological relationships of protease inhibitor I of different plants, established by microcomplement fixation assays (GURUSIDDAIAH et al. 1972) are in general agreement with those using the double diffusion method (TUCKER 1969), but the results do not allow a clear definition of taxonomic relationships between the members of the Solanaceae family. Inhibitor I from tomato is immunological more similar to potato than to the other genera of the Solanaceae and *Atropa, Physalis* and *Nicotiana* have indices of dissimilarity for inhibitor I which all indicate similarity to potato. *Datura* and *Petunia* possess inhibitor I proteins which have a much weaker reaction with the potato inhibitor I antiserum than all of the other genera and therefore are presumed to have less antigenic determinants than those of the potato inhibitor.

Although the protease inhibitor II types from potato varieties differ in their physicochemical and inhibitor properties, they exhibit close immunological relationships. Based on this evidence therefore it has been suggested that all of the five inhibitors should be designated inhibitor II (RYAN and SANTARIUS 1976). Protease inhibitor II also accumulated in large quantities in tomato leaves which had been mechanically wounded, attacked by chewing insects, or detached and supplied with the proteinase inhibitor-inducing factor (RYAN 1974). The inhibitors I and II were found almost entirely by means of immunological techniques to be in isolated vacuoles of tomato leaves induced to accumulate inhibitors (WALKER-SIMMONS and RYAN 1977).

Several possible functions for these inhibitors have been suggested, including roles as storage proteins, regulators of endogenous proteases and defence directed against insect, microbial, or delocalized endogenous proteases. However, the physiological function of protease inhibitors is still poorly understood (RYAN 1978).

### 3.3 Phytochrome

Phytochrome, the chromoprotein which serves as the photoreceptor for a wide range of photomorphogenic responses in plants, exists in a physiologically inactive, red-absorbing form (Pr) and in a physiologically active, far red-absorbing form (Pfr) which are antigenically identical (PRATT 1973, RICE and BRIGGS 1973a). Immunoprecipitates still showed photoreactivity, although the Pfr peak at 725 nm was suppressed and the Pr peak at 667 nm had been shifted to 660 nm (RICE and BRIGGS 1973b). Antigens present in extracts of light-exposed tissue were found to be immunologically identical to those obtained from similar extracts of etiolated tissue (PRATT et al. 1974) and the loss of spectral activity of phytochrome roughly paralleled the loss of antigenically active phytochrome in the process referred to as the "destruction reaction" (PRATT et al. 1974, PRATT and COLEMAN 1971, 1974, HUNT and PRATT 1979a). Thus, essentially complete degradation of the phytochrome molecule appears to take place during destruction. Immunocytochemical assays of destruction showed that different regions of seedlings have different destruction kinetics (COLEMAN and PRATT 1974a). It is now well established that native phytochrome is unusually susceptible to endogenous proteolysis (CUNDIFF and PRATT 1973). HUNT and PRATT (1979b) have developed a phytochrome immunoaffinity purification procedure that yields undegraded phytochrome of high purity which is indistinguishable from conventionally prepared phytochrome and so made possible many physicochemical studies that previously were not so because of the difficulty in obtaining sufficient homogeneous phytochrome by conventional methods. A phytochrome radioimmunoassay has also been developed which is not affected by the presence of other pigments and will detect non-chromophore-containing phytochrome and phytochrome fragments with a sensitivity of about 2 ng of phytochrome (HUNT and PRATT 1979a). The protein moiety of phytochrome was not present in substantial quantities in dry grains of grasses but was synthesized de novo during seedling germination. The complete absence of any immunologically active material suggests that the protein does not appear to be present in the embryo in a preformed state requiring only attachment of the chromophore for normal spectral activity (COLEMAN and PRATT 1974a). Preliminary results have been reported using an indirect antibody-labelling method employing peroxidase as the ultimate label to localize phytochrome in grass seedlings at both the light- and an electron microscope level (COLEMAN and PRATT 1974b, PRATT and COLEMAN 1971, PRATT and COLEMAN 1974). The pattern of phytochrome distribution in eliolated shoots varies widely, although, the distribution of phytochrome is highly specific with respect both to organs and to cell types within an organ for any given plant species (COLEMAN and PRATT 1974b, PRATT and COLEMAN 1974). Near the tip of the shoot, for example, phytochrome is found only in the parenchyma cells of the coleoptile, while farther back from the tip it is found only in the epidermal cells and vascular strands of the same organ (PRATT and COLEMAN 1971).

Electron microscope localization studies indicate its association with the nuclear membrane and with the interior of mitochondria and amyloplasts, as well as being present throughout the cytoplasm (COLEMAN and PRATT 1974b). The differences observed between phytochrome preparations from different plant sources suggested a correlation between phylogenetic distance and immunological dissimilarity (RICE and BRIGGS 1973b, PRATT 1973). Using double diffusion and microcomplement fixation assays, oat cultivars were immunologically indistinguishable, whereas other grasses, although identical in the double diffusion test, showed decreased activity in the microcomplement fixation assay. Pea phytochrome which had partial identity in the double diffusion test showed no activity against antiserum directed to oat phytochrome in the complement fixation assay.

### 3.4 Leghaemoglobin (Lg)

Current understanding of the role of Lg in symbiotic nitrogen fixation suggests that its intracellular location must be such that it can serve as an efficient oxygen carrier and maintain an optimal partial pressure of oxygen within the nodule in order that ATP synthesis and nitrogen fixation can occur simultaneously. VERMA and BAL (1976) using ferritin-labelled antibodies against soyabean Lg found that it is localized in the host cell cytoplasm and does not appear to penetrate the membrane enclosing the bacterioids (VERMA et al. 1979). The immunoreactivity of the released nascent polypeptide chains from free or membrane-bound polysomes and immunoprecipitation of the in vitro translation products using messenger templates from the same sources revealed that Lg is synthesized preferentially on the free polysomes in the host cell cytoplasm (VERMA and BAL 1976). Since the mobility of authentic Lg corresponded exactly to that of the immunoprecipitated translation product, it may be concluded that apo-Lg is not formed as a larger precursor molecule (KONIECZNY and LEGOCKI 1978). The presence of Lg could be detected by rocket immunoelectrophoresis 3 days after infection with *Rhizobium* and was followed by the appearance of nitrogenase activity 7 days later.

In the case of Lg, the antibodies can still recognize the determinant regions after tryptic digestion, 75% of the antigenicity remaining. Therefore peptide synthesis and radioimmunoadsorption on the same support media could be used to delineate the antigenically reactive regions of Lg. The soyabean Lg molecule has at least five different antigenic determinants (HURRELL et al. 1978). The extent of immunological cross-reactivity among different plant Lg's was best demonstrated using the Farr radioimmunoassay procedure.

In summary, amino acid substitutions in the Lg family are conformationally, but not immunochemically, conservative (HURRELL et al. 1976). An extensive comparison of immunological cross-reactivities between Lg from members of the Fabaceae (Leguminosae) has allowed the construction of a phylogenetic tree for this family (HURRELL et al. 1977).

Recently LEGOCKI and VERMA (1979) have reported the presence of a protein, Nodulin-35, in soyabean root nodules which was formed as a result of infection with *Rhizobium,* irrespective of whether or not the nodules were effective in nitrogen fixation. Nodulin-35 was synthesized by the host plant cells as shown by immunoprecipitation of the translation products of the nodule polysomes, but although it is a nodule-specific protein, its appearance was not related to nitrogen fixation.

### 3.5 Proteins Related to Incompatibility Reactions

Immunological and biochemical techniques have demonstrated that there are specific proteins in pollen and styles which are involved in the incompatibility reaction (NELSON and BURR 1973). MĀKINEN and LEWIS (1962) found that antisera against pollen containing allele $S_6$ also reacts with pollen with a self-compatible mutant allele, $S_6$ 1, and observed that the antigens diffused from the pollen. KNOX and HESLOP-HARRISON (1971), KNOX et al. (1972) and KNOX (1973) have shown that the pollen-grain walls of all flowering plants so far tested contain non-enzymic proteins and it has been suggested that some of these may be concerned with the control of incompatibility reactions, acting as recognition substances. In contrast, however, NASRALLAH and WALLACE (1967a, b) injected rabbits with stigma homogenates of *Brassica oleracea* plants of known S-allele constitution. Some of the resulting antisera contained antibody reactivity which appeared to be specific for the S-allele of the homozygous plant and they called the antigen the S-protein. The stigma of heterozygous plants contained the expected two antigens of the S-alleles present. The S-proteins were not detectable in the pollen or in any other plant parts, and only appeared in the young flower buds just before anthesis (NASRALLAH et al. 1969, 1972). However, before the technique can be used as an alternative method to the routine S-allele diagnostic tests employed at present, some method of increasing the immunogenicity of the S-proteins will have to be found (SEDGLEY 1974).

### 3.6 Proteins of Unknown Nature Associated with Development in Plants

It is generally thought that development in plants involves an orderly sequence of gene activation and repression, giving rise to change in the pattern of protein synthesized. Immunological comparisons of different organs have shown that there are protein antigens specific for one organ and others common to several organs or even to the whole plant (for review see DAUSSANT 1975). Immunochemical studies show that changes in antigen spectra accompany the growth and differentiation of plant cells in vitro (BOUTENKO and VOLODARKSY 1968). WRIGHT (1960) demonstrated qualitative and quantitative changes in the soluble antigenic proteins during the differentiation of wheat coleoptiles. Immunological techniques show that leaves and inflorescences of plants contain specific antigenic proteins (KAHLEM 1973). A preliminary study of the protein complement of the shoot apical bud showed that its antigenic protein pattern is altered during the transition to flowering (PIERARD et al. 1977). A specific leaf antigen has been detected in the female plant of *Mercurialis annua* and the action of cytokinins on its synthesis has been studied (DURAND-RIVIÈRES 1969, DURAND and DURAND-RIVIÈRES 1969). It has been shown that seedlings contain half as many antigens again as are found in the embryonic axis and that leaves contain no antigenic proteins identical with those from seeds and seedlings, but only some partially identical ones (ALEXANDRESCU 1974). In ripening and dormant seeds the decline in respiration was accompanied by the disappearance of several mitochondrial antigens. The reactivation of mitochondria during development was related to significant alterations in the patterns of mitochondrial antigens; qualitative changes were observed during inbibition while quantitative changes mostly occurred during subsequent germination (IVANOV and KHAVKIN 1976). KHAVKIN et al. (1977) have described a characteristic group of embryonic antigens and followed the pattern of their sequential accumulation in the developing maize caryopsis. They comprised at least two classes of proteins with apparently different physiological functions. Two predominant proteins, because of their transient nature, seemed to act as specific reserve proteins of embryonal tissues; while the others were constitutive and were also maintained in callus culture (KHAVKIN et al. 1978a, 1978b). Several antigenic proteins were found in the primary root of maize seedlings and in suspension cultures of cells that were absent from epidermal and cortical cells; they were characteristic of a well-defined tissue complex, the stele (KHAVKIN et al. 1980).

In most cases, the identity of soluble proteins associated with developmental processes in plants is only characterized by their antigenicity, but probably the majority of them possess enzymic activity.

The allergens deposited in the pollen grains of many plants have been indentified as proteins and characterized extensively (MARSH 1975, BELIN 1972, AUGUSTIN 1959, WEEK and LØWENSTEIN 1975, HUBSCHER and EISEN 1972, LEE et al. 1979, LEE and DICKINSON 1979, DAUSSANT et al. 1976). Presumably the allergenicity of the pollen antigens is no more than an unfortunate accident, but their significance in the biology of plants is obscure.

## 4 Immunology of Storage Proteins (see also Chaps. 7, 8, 9, 14, this Vol.)

The seed storage proteins are organ-specific, i.e. they are immunologically detectable only in seeds and principally in their storage tissue and absent from the leaves and roots of the older plant (MILLERD 1975, DERBYSHIRE et al. 1976, DAUSSANT 1978, EWART 1978). Other storage proteins are specific to the storage tissues in which they are found.

### 4.1 Storage Proteins of Legume Seeds (see also Chaps. 7, 8, this Vol.)

A number of biochemical and immunochemical criteria suggest that the storage proteins of many legumes are closely related (MILLERD 1975, DERBYSHIRE et al. 1976). DUDMAN and MILLERD (1975) and KLOZ (1971) demonstrated that proteins, immunologically related to the storage proteins, legumin and vicilin, to be widely distributed in the tribes Fabeae and Trifolieae and also to be present in members of the tribe Ononidea, Podalyrieae and Loteae. Legumin-like and vicilin-like proteins were not detected immunologically in the tribes Sophoreae, Dalbergieae, Genisteae and Phaseoleae. In contrast to KLOZ (1971), the results of DUDMAN and MILLERD (1975) indicate that it is legumin rather than vicilin which is the most consistent storage protein in legumes. The protein bodies of various *Vicia* and *Pisum* varieties contain some proteins which are antigenically different to legumin and vicilin. Crossed immunoelectrophoresis suggests that broad bean or pea cotyledons may contain as many as seven different globulin components (OLSEN 1978, MANTEUFFEL and SCHOLZ 1975, MILLERD et al. 1978, GULDAGER 1978). In all cases no interaction between any of these precipitated proteins was observed, e.g., the boundaries of the peaks crossed clearly, without fusion, deviation or spur formation. However, further work is required to clarify the situation before these additional compounds can be identified as storage proteins.

The effect of different extraction and fractionation procedures, heat treatment, pH shift etc. on storage proteins can be monitored also by immunological techniques (OLSEN 1978, GULDAGER 1978, CATSIMPOOLAS et al. 1967, MANTEUFFEL and SCHOLZ 1975).

Immunoaffinity chromatography on immobilized antibodies has been shown to be a generally applicable method for the purification of undegraded pure proteins from a range of plants with immunologically related storage proteins (CASEY 1979).

Legumin and vicilin have no determinants in common and are completely different proteins (SCHOLZ et al. 1974, MILLERD et al. 1971). The results of CROY et al. (1979), in contrast to the findings of DUDMAN and MILLERD (1975), indicate that pea legumin has all the determinants of broad bean legumin and one or more in addition. GULDAGER (1978) suggested that legumin of the pea is composed of two slightly different proteins which have a very high degree of common antigenic structure, since the legumin peak shows assymmetry with several shoulders on crossed immunoelectrophoresis. This phenomenon could not be demonstrated with *Vicia* legumin (MANTEUFFEL and SCHOLZ 1975). Antibodies produced against the holoprotein react with both the acid and basic subunits of broad bean legumin; there was no immunological correspondence between the two types of subunits. Using immunological techniques it was possible to demonstrate the reconstitution of isolated acid and basic subunits to the holoprotein (SCHLESIER and MANTEUFFEL unpublished). DOMONEY et al. (1980) have developed a highly sensitive enzyme-linked immunosorbent assay for the quantitative detection of legumin. These workers found that legumin was more stable to heat treatment and acid precipitation (OLSEN 1978) than vicilin. Comparisons of the legumins of the other members of the Vicieae, i.e., *Lathyrus, Lens* and *Cicer,* are less complete, but it is clear that they have proteins with similar immunological properties and compositions (CROY et al. 1979).

WEBER et al. (1981) demonstrated a partial identity between the vicilins of broad bean and peas in contrast to the results of DUDMAN and MILLERD (1975), who claimed them to be immunologically identical. The vicilin peak obtained by crossed immunoelectrophoresis by MANTEUFFEL and SCHOLZ (1975) was more or less assymetric with several sub-precipitin lines indicating possible micro-heterogeneity. A possible explanation for peak asymmetry may be association of vicilin (SCHLESIER et al. 1978). During immunoelectrophoresis associated vicilin had an increased anodic mobility, thereby perhaps explaining the asymmetry of the precipitin arc as a consequence of overlapping vicilin types of different degrees of association. Vicilin may be separated into two different components by affinity chromatography on Con A-Sepharose, indicating further heterogeneity with respect to glycosylation of this protein (MILLERD et al. 1979, WEBER et al. 1981, DAVEY and DUDMAN 1979). In contrast to THOMSON et al. (MILLERD et al. 1978, THOMSON et al. 1978, 1979), we suggest that only one precipitin peak is formed involving a series of heterogenous but related forms of vicilin. All other precipitin peaks of the immunopherogram formed by the storage proteins of broad beans or peas are distinct entities without common antigenic determinants.

Protein degradation in germinating seeds of broad beans has been investigated by means of immunological methods (LICHTENFELD et al. 1979). In agreement with the results of other investigators (CATSIMPOOLAS et al. 1968a, b, DAUSSANT et al. 1969b), changes in the relative mobility of storage proteins on germination were demonstrated. However, changes in the amido-nitrogen content of the proteins were not responsible for the shift in mobility. Legumin was degraded earlier and faster than vicilin in agreement with the findings of BASHA and BEEVERS (1975). Degradation of vicilin by proteolytic plant enzymes as well as by papain or trypsin was monitored immunologically using vicilin-specific antisera. Vicilin gave rise to two fragments, only one of which gave a serological response, and was shown to persist for several days in cotyledons on germination (BUZILĂ 1975).

The storage proteins of soyabean seeds have been subjected to very extensive immunological studies (for review see CATSIMPOOLAS 1978) and shown to consist of at least four immunological distinct components, i.e., glycinin, $\gamma$-conglycinin, $\beta$-conglycinin, and $\alpha$-conglycinin (CATSIMPOOLAS and EKENSTAM 1969). Glycinin, $\gamma$-conglycinin, and $\beta$-conglycinin but not $\alpha$-conglycinin are located in the protein bodies of cotyledon cells (CATSIMPOOLAS et al. 1968a, KOSHIYAMA 1972). The most stable form of glycinin, the major component of soyabean storage proteins, is the dimer (CATSIMPOOLAS 1969). Glycinin retains its immunological activity after heating, although complement fixation assays indicated some reduction in antigenic sites in heated samples which was more pronounced with higher temperatures (CATSIMPOOLAS et al. 1971). KOSHIYAMA (1972) has developed a simple isolation procedure for a 7S soyabean globulin immunologically identified as $\gamma$-conglycinin. This globulin seems to be one of the major antigenic components in the soyabean globulin fraction, judging by its antigenic strength after immunization of the whole globulin fraction, although it was found to constitute only 3% of the total globulin protein content (OCHIAI-YANAGI et al. 1978). A second 7S component of soyabean proteins, $\beta$-conglycinin, forms a number of polymers which have been shown to be immunologically identical by loop formation of their precipitin arcs (CATSIMPOOLAS and EKENSTAM 1969). Its isolated subunits exhibit antigen–antibody reaction with antisera against the native $\beta$-conglycinin (THANH and SHIBASAKI 1977).

It was concluded that $\alpha$-conglycinin, the 2S globulin of soyabean, was not identical with the Kunitz trypsin inhibitor although it exhibited trypsin inhibitor activity and had a molecular weight similar to the inhibitor (CATSIMPOOLAS et al. 1968b). During germination of soyabean seed, glycinin increased in electrophoretic mobility, followed by an apparent decrease in the amount of protein; it was degraded more slowly than the 7S storage proteins (CATSIMPOOLAS et al. 1968a, b).

DAUSSANT et al. (1969b) and NEUCERE (1978a, b) studied $\alpha$-arachin and $\alpha$-conarachin, the two major proteins of *Arachis* seeds and found a twofold higher concentration of $\alpha$-arachin in the cotyledons than in the embryo axes (NEUCERE and ORY 1970).

While $\alpha$-arachin is located in the protein bodies, $\alpha$-conarachin appears to be dispersed in the cytoplasm of the storage cells (JACKS et al. 1972). SINGH and DIECKERT (1973) reported a scheme for the isolation of substantial amounts of relatively pure arachin and used immunological methods for recognizing impurities in the isolates. The elongated double precipitin arcs for $\alpha$-arachin indicated the presence of polymeric forms each with the same antigenicity (NEUCERE and CHERRY 1975, DAUSSANT et al. 1969a). Conformational changes in $\alpha$-arachin were definitely induced by heating although the effect on antigenicity was much smaller than those found with other peanut proteins (NEUCERE 1974, NEUCERE et al. 1969). Identification of $\alpha$-conarachin showed two immunologically distinct proteins called $\alpha_1$- and $\alpha_2$-conarachin (NEUCERE and CHERRY 1975, DAUSSANT et al. 1968a) with the former predominating. Semiquantitation of $\alpha$-arachin during early stages of embryogenesis indicated a progressive synthesis during the course of seed development (NEUCERE and YATSU 1975). On the other hand, an extensive anodic shift in the electrophoretic mobility of $\alpha$-arachin and a slight anodic shift in the migration of $\alpha_1$-conarachin was observed during early germination (DAUSSANT et al. 1969b). $\alpha$-Arachin appeared to be one of the first detectable proteins to undergo catabolism during the very early phases of germination. A comparative immunological survey of the seed proteins in 36 species of the genus *Arachis* showed evidence of protein synthesis in some wild species comparable to that in cultivated *Arachis hypogaea,* although both qualitative and quantitative differences were observed. All wild species of peanut contained much less $\alpha$-arachin than the cultivated ones (NEUCERE and CHERRY 1975).

Kloz et al. (KLOZ 1971, KLOZ AND KLOZOVÂ 1974, KLOZOVÂ et al. 1976) investigated the taxonomic distribution and variability of the main storage proteins in the genus *Phaseolus* and found that the so-called American endemics formed a relatively close group which was distinct from the other species of the genus. No immunological correspondence was detectable between the 7S component of the tribe Phaseoleae and the vicilin of the tribe Vicieae (KLOZ 1971).

The seed proteins of the tribe Lupineae were also considerably different from those of the tribe Vicieae (KLOZ 1971). According to NOWACKI and PRUS-GŁOWACKI (1971) different species of the genus *Lupinus* formed a serological closely related group, in which the following members were most distinct: *Lupinus luteus, Lupinus albus, Lupinus angustifolius* and *Lupinus perennis*. American members, however, did not differ from one another to such a degree as did the Mediterranean species. The intercrossing behaviour of species of the genus *Lupinus* affords a very similar picture to that based on immunoelectrophoretic separations (NOWACKI and PRUS-GŁOWACKI 1971, PRUS-GŁOWACKI 1975).

### 4.2 Storage Proteins of Cereal Grains (see also Chap. 9, this Vol.)

Since antigen–antibody reaction only takes place in aqueous medium, the main storage proteins of cereals, soluble in alcohol or acid solutions, cannot be quantitatively assayed by the usual serological techniques. Furthermore, the drastic procedures employed in their extractions may alter the antigenic properties of the components.

The experiments of GRABAR and his colleagues (GRABAR et al. 1965, BENHAMOU-GLYNN et al. 1965) indicated that the gliadins of wheat gave a single immunochemical response suggesting a close structural relationship within different members of the group. It would appear, however, that the $\alpha$-, $\beta$- and $\gamma$-gliadins contain immunologically distinct proteins (ELTON and EWART 1963, NIMMO and O'SULLIVAN 1967, BOOTH and EWART 1970). The $\gamma$-gliadin formed three strong precipitin lines, a finding in agreement with that of HUEBNER et al. (1967). The $\beta$-gliadin gave one strong line and two weaker ones, the latter fused with those of $\gamma$-gliadin, and $\alpha$-gliadin gave two well-defined lines which ran into the main $\beta$-line. In addition, BECKWITH and HEINER (1966), using antibodies present in the sera of coeliac patients to study gluten proteins, found that one of the antigenic determinants, present in both $\alpha$- and $\beta$-gliadin, was missing in $\gamma$-gliadin. A sample of $\omega$-gliadin gave no serological reaction (BOOTH and EWART 1970). These results indicate that gliadin fractions are incompletely separated but contain immunologically distinct proteins.

The water solubility of gliadins can be increased by reacting them with N-carboxy-D,L-alanine anhydride (BENHAMOU-GLYNN et al. 1965). The fact that these gliadin derivatives were able to react with antibodies directed against the native protein implied that lysine residues were not involved in the antigenic determinants of gliadin. Methylation, however, reduced the antigenicity of gliadin to a low level (BECKWITH and HEINER 1966), indicating that aspartic and glutamic acid and/or their amides are required unmodified for serological activity. This hypothesis was confirmed by MAURER et al. (1964) who demonstrated with the aid of passive cutaneous anaphylaxis that gliadin cross-reacts with an antiserum with dominant specificity towards glutamic acid residues. Whereas reducing and blocking the disulphide bonds of gliadin abolished its ability to react with an antiserum directed against native gliadin (ESCRIBANO and GRABAR 1966), antigenicity was restored almost completely if reduced gliadins were allowed to reoxidize in dilute solution. Heating of the protein solution to near 100 °C had no noticeable effect on the immunological properties of gliadins (ELTON and EWART 1963, SZABOLCS et al. 1978).

ELTON and EWART (1963) and EWART (1966) used anti-gliadin with the double diffusion technique to examine the serological relationships of cereal prolamins. Wheat flours gave a strong reaction and at least five antigens were common to the eight varieties tested. Rye flour showed a closely similar antigenic reaction to that of wheat, but barley appeared to contain only a few wheat antigens, two of these being in trace quantities. Oat and maize did not react to any significant extent although a slight reaction was observed in earlier work with anti-gluten (ELTON and EWART 1963). Individual gliadins within a given variety are genetically related, although their tertiary structures as judged by immunological tests show characteristic differences (BOOTH and EWART 1970).

WELLS and OSBORNE (1911) concluded that wheat gliadin and glutelin have a reacting group in common, but gliadin has extra antigens since animals sensitized to glutelin could be saturated with gliadin to become unreactive to glutelin, whereas the reverse process did not happen. GRABAR and his colleagues (GRABAR et al. 1965, ESCRIBANO 1966) concluded as a result of their extensive immunoelectrophoretic studies on gliadin and glutelin that it was impossible to distinguish between these proteins. In additional, BENHAMOU-GLYNN et al. (1965) also suggested that there is at least one common antigenic determinant among the gliadins and glutelins, and that possibly both proteins are closely related in their structure.

The possibility that entrapped gliadin is responsible for the serological reaction of the glutelin fraction is difficult to dismiss since biochemical methods have shown that both protein fractions are not easy to separate completely from one another (WOYCHIK et al. 1961, EWART 1966). In addition, the bulk of the glutelins are insoluble and, therefore not amenable to serological techniques. As discussed in Chapter 9, this Volume the prolamin and glutelin storage proteins of cereals have a characteristically different amino acid composition.

Serological studies of water-soluble proteins from wheat have shown that a small part of the globulins are water-soluble and can contaminate the albumin fraction. The soluble proteins of equivalent mobility from different wheat varieties been shown to have some structural features in common (NIMMO and O'SULLIVAN 1967). HALL (1959) used immunoelectrophoresis to study the relationships between the patterns of soluble proteins of wheat, rye and *Triticale*. The latter had all the proteins of its parents except that one of the rye proteins appeared to be missing. The soluble proteins of an intergeneric hybrid, *Loliofestuca*, also had all the proteins of the parent species, although the species with the higher chromosome number had a greater influence quantitatively on the protein composition (PRUS-GŁOWACKI et al. 1971). The strengths of the serological affinities of the soluble proteins of species belonging to different genera of the family Triticeae were extensively studied by SMITH (1972) and ANIOŁ (1976a, b) are their values were confirmed by the inter-crossing behaviour of the same species (NOWACKI et al. 1972). The serological technique showed significant structural resemblances among the soluble proteins of the probable three ancestors of wheat, namely *Triticum monococcum*, *Aegilops speltoides*, and *Aegilops squarrosa*, which contributed to the A, B, and D genomes, respectively (ANIOL 1974a, b).

The storage proteins of barley also probably exist in vivo as extremely large aggregates. Antisera directed against six electrophoretically separated components of hordein have been used to study the antigenic character of some hordein components and of the total prolamins of several cereals (KLING 1975). Components of the hordein fraction of each species showed reactions of identity with one exception, confirming similar results obtained by "fingerprinting" (BOOTH and EWART 1969). Gliadin of rye and wheat showed cross-reactions while avenin, the prolamin of oats, did not. Structural similarities also existed between the hordein components of the prolamins of *Festuca pratensis* and *Lolium multiflorum*, while the prolamins of *Poa pratensis* and *Phleum pratense* seemed to have few or no features in common with the hordein of barley. A great number of the barley antigens belonged to both the albumin and globulin fractions according to the classical definition of OSBORNE (1924); some were found in both fractions and also some constituents of the hordein fraction appear to be partially soluble in water (NUMMI 1963, DJURTOFT and HILL 1965, HILL and DJURTOFT 1964). It has been possible by using immunoelectrophoresis to distinguish a total of 21 components in the fraction containing the "salt-soluble" proteins of barley (HILL and DJURTOFT 1964). ARU and MIKK (1965) concluded that soluble proteins of different barley varieties were immunologically similar.

It was suggested by ANIOŁ and NOWACKI (1973) that information on the serological relationships between proteins within the genus *Hordeum* was likely to be helpful in the selection of partners in heterotic breeding. The antigenic analysis of maize inbred lines and their hybrids has also been used to study heterosis and to investigate the possibilities for its prognosis (DIMITROV et al. 1974).

The serological relationship between proteins is also useful in practical applications related to the control of the origin of proteins in food products; thus antibodies reacting with barley proteins and not with wheat proteins were used to check the source of flour (LUIZZI and ANGELETTI 1969). Furthermore, by using immune serum specific for maize and rice proteins, it was possible to indicate whether maize or rice was employed during the preparation of beer samples (DORNHAUSER 1967, 1972, SCHUSTER and DORNHAUSER 1967a, b, c).

$\gamma$-Globulins in rice seeds have been investigated immunochemically (HORIKOSHI and MORITA 1975). $\gamma_1$-Globulin was found to be immunochemically distinct from $\gamma_3$-globulin while $\gamma_2$-globulin was identical with $\gamma_3$-globulin. By means of the fluorescent antibody technique it was shown that both $\gamma$-globulins were localized in the scutellum and aleurone cells although their concentration was inversely related in these tissues. During seed development, $\gamma_1$- and $\gamma_3$-globulins were synthesized at about the same rate; the rate of disappearance of the latter was greater than that of the former during germination.

### 4.3 Storage Proteins of Other Crops

The members of this artificial group are taxonomically unrelated and differ not only in the structure of their storage organs, but also in the content and composition of their storage proteins. Immunological studies have been undertaken in order to establish the association-dissociation behaviour of the 12S rapeseed (*Brassica*) globulin (GILL and TUNG 1978). It was suggested that this protein self-associates to form dimers that are immunologically identical to the monomer. The subunit of the 12S protein with the lowest molecular weight must have been located on the surface of the dimer since this glycopeptide was immunoresponsive whereas the other subunits showed no serological reactions. YOULE and HUANG (1978b) showed that the albumins of castor bean are the well characterized CB-1A allergens. Similarly, the 2S albumins of cotton seeds have also been shown to be identical to the cotton allergens by immuno-cross-reactivity (YOULE and HUANG 1979). YOULE and HUANG (1978a, b) suggested, in view of the abundance, ubiquitous occurrence, amino acid composition, sub-organelle compartmentation and developmental properties of albumins in various seeds, that albumins could be important seed storage proteins. Approximately 70%–80% of tuber protein of potato is storage protein and, as in seeds, this protein fraction is heterogeneous and highly amidated. The tuber proteins from immature potato cultivars were cross-reacted with other cultivars and the immunoresponses were identical among all cultivars. Furthermore, the interrelationships of other proteins in mature tubers have been checked by immuno-techniques (STEGEMANN et al. 1973, STEGEMANN 1977, 1979). It is known that much of the protein of fleshy fruits can be expressed in their sap. By testing immunosera with concentrated orange and lemon juice, substantial differences between the proteins of these two citrus fruits were demonstrated. Practically, the immunological technique can be used satisfactorily for the analysis of commercial products since pasteurization does not alter the protein pattern of the juice (CANTAGALLI et al. 1972).

The utility of the serological characterization of seed proteins in the study of taxonomic relationships, hybridization, and polyploidy of plants other than legumes and cereals has been demonstrated by several investigators (JOHNSON et al. 1967, PICKERING and FAIRBROTHERS 1970, HILLEBRAND and FAIRBROTHERS 1970a, b, CHERRY et al. 1971, FAIRBROTHERS 1977, ZIEGENFUS and CLARKSON 1971, HOUTS and Hillebrand 1976). In most cases, however, these investigations have used a mixture of salt-soluble seed proteins which were not characterized. It should, however, be noted that where evidence of phylogenetic relationships is available from morphological, cytogenetic and biochemical studies, serological analyses usually support the conclusions drawn from these other approaches.

## 5 Immunological Properties of Integral Proteins of Sub-Cellular Structures

The immunochemical analysis of integral proteins of sub-cellular structures pose a number of methodological problems. Membrane proteins have been found to be immunogenic both in situ and after solubilization. Antibodies may also be found after immunization with proteins solubilized by detergents, which cross-react with the protein in its native conformation (for review see BJERRUM 1977). Antibodies directed against sub-cellular structures may be used as specific precipitating or agglutinating reagents, which, in the case of enzymic antigens can also act as specific inhibitors.

Since hydrophilic antibodies do not penetrate membranes, the reaction of a specific antibody with a protein as manifested by agglutination, adsorption or inhibition localizes the reacting component at the surface of the sub-cellular structure. Several methods are available for the solubilization of bound proteins, but not all of them are suitable in cases where subsequent immunochemical analysis is to be performed (BJERRUM 1977).

The solubilizer should liberate the proteins from sub-cellular structures and keep them in solution, but at the same time it should not destroy their antigenic structure. Furthermore, the solubilizing agent, if present during the immunochemical analysis, must not denature the antibody, interfere with the antigen–antibody reaction or otherwise alter the immunological reaction. Solubilization with non-ionic detergents seems at present to be the method of choice, because these do not affect the antigen–antibody reaction to any appreciable extent. Unfortunately, however, although detergents dissolve many, they may not dissolve all sub-cellular structures. In contrast, ionic detergents, which are the more commonly used solubilizers of sub-cellular structures, bind both hydrophilic and hydrophobic areas of the proteins, resulting in an unfolding of the polypeptide chain with possible loss of function and antigenicity. Furthermore, subsequent immunological analysis cannot be performed in the presence of free dodecyl sulphate because it dissociates the antigen–antibody complex and can give rise to artefactual precipitin lines by non-specific precipitation of serum lipo-proteins.

The review of BJERRUM (1977) should be consulted for further information on this topic, although the survey considers only membrane proteins of animal cells. It is obvious that antisera to integral proteins of sub-cellular structures should be carefully characterized before use in structural and functional investigations of any system.

In this respect, the crossed immunoelectrophoresis technique originally designed by CONVERSE and PAPERMASTER (1975) and modified by CHUA and BLOMBERG (1979) is particularly useful.

The most thoroughly investigated and best-characterized sub-cellular structures of plants are the thylakoids of chloroplast. Early immunochemical investigations of thylakoid membranes were concerned primarily with peripheral membrane proteins which were easily extracted in soluble form and were therefore available for antibody production and subsequent immunological studies. On the other hand, structural membrane proteins can be solubilized only by the addition of detergents and the methodological problems mentioned above therefore arise in their analysis.

Extensive discussions of thylakoid structure and the effects of antiserum on membrane-bound electron transport enzymes are dealt with in the reviews of MÜHLETHALER (1977) and BERZBORN and LOCKAU (1977).

Another plant system which has been investigated immunologically are the glyoxysomes of germinating *Ricinus* seedlings (HOCK 1974, GONZALEZ and BEEVERS 1976, BOWDEN and LORD 1977). Double gel diffusion as well as crossed immunoelectrophoresis using antibodies against purified glyoxysomal membranes from watermelon cotyledons and glyoxysomes or crude cotyledon extracts gave a single fused immunoprecipitation band. The cross-reaction of mitochondria and endoplasmic reticulum with anti-glyoxysomal antibodies reflects the synthesis and destruction of membrane material belonging to several organelles (HOCK 1974), supporting the hypothesis of KAGAWA et al. (1973) that components of glyoxysomal and mitochondrial membranes originate in a "light membrane" fraction derived from the endoplasmic reticulum. The serological comparison of endoplasmic reticulum and glyoxysomal proteins of castor bean endosperm (BOWDEN and LORD 1977) established that antigenic determinants common to both endosperm endoplasmic reticulum and glyoxysomal membranes reside in all the major polypeptide components which characterize these membranes. In addition, the results confirmed that common antigenic determinants were present in the soluble glyoxysomal matrix and the soluble components obtained from the microsomal fraction. BOWDEN-BONNETT (1979) described the isolation of mRNA from germinating castor bean endosperm which was actively translated in a cell-free protein-synthesizing system. Immunoprecipitation in the presence of antibodies raised against the total glyoxysomal matrix proteins revealed that these proteins accounted for 15% to 20% of the total translational products. By using the same method it was revealed that protein synthesis during germination was largely, perhaps completely, dependent on newly transcribed mRNA (ROBERTS and LORD 1979). Together these data support a model for microbody biogenesis in which glyoxysomal proteins are synthesized on ribosomes attached to the endoplasmic reticulum, vectorially discharged across the endoplasmic reticulum membrane, collecting in dilating cisternae which ultimately vesiculate to form isolated, membrane-bound organelles.

Most of the proteins of chromatin do not have an assayable function which allows one to follow their fate during the various developmental stages of a tissue or during the life cycle of a cell. Serological reactions can conceivably be used therefore to follow conformational changes or structural rearrangements occurring in chromatin or chromosomes (for review see BUSTIN 1976). The microcomplement fixation test has been used to identify rye histone fractions and to estimate the extent of structural differences between the corresponding histone fractions of rye and calf (ROLAND and PALLOTTA 1978). There is far less similarity found between rye and calf H 1 than between rye and calf H2As and H2Bs. The near structural identity of the H 3 histones was confirmed by their small immunological distance which indicated an amino acid sequence difference of less than 2% between the rye and calf proteins. However, it should be pointed out that the calculated immunological distances were only approximations; the sources of errors were discussed by ROLAND and PALLOTTA (1978). Until now there have been no studies of plant chromosomal structures using immunological techniques.

It was suggested by HARTMANN et al. (1973) that antibodies could also be employed specifically to label proteins on the surface of protoplasts. Antibodies raised against protoplasts caused agglutination of the majority of protoplasts, while control sera caused no reaction (HARTMANN et al. 1973, STROBEL and HESS 1974). The degree of antibody cross-reactivity between protoplasts isolated from different plant species seems to provide a means of assessing the degree of structural similarity between sites on protoplast membranes. The immune serum, which presumably contained complement, lyzed the protoplasts unless it was heat-treated prior to use (HARTMANN et al. 1973). In contrast, the findings of LARKIN (1977) suggest that the agglutination observed is not produced by the binding of specific antibodies to protoplast membrane antigens. Instead, the binding molecules on the protoplast membrane appeared to be $\beta$-lectins, as defined by JERMYN and YEOW (1975), and the protoplast agglutination would have resulted from their binding to non-specific serum molecules bearing accessible $\beta$-D-glycosyl residues. The results presented, however, do not necessarily mean that no immunological reaction occurred between specific antibodies and antigenic determinants on the protoplast membrane, but that their effect may have been masked by the $\beta$-lectin activity.

Using lectins instead of antibodies as surface markers offers a new methodological tool for studying the antigenic composition of plant membranes. The previous finding that Con A agglutinates protoplasts of higher plants (GLIMELIUS et al. 1974, BURGESS and LINSTEAD 1976) can be regarded as indirect evidence for the existence of Con A-binding sites on the plasmalemma of plant cells. Furthermore, the specific binding of Con A as visualized in the electron microscope by haemocyanin (WILLIAMSON et al. 1976) or peroxidase-diaminobenzidine staining (GLIMELIUS et al. 1978b) demonstrated that glycoproteins and/or other sugar-containing substances with glucose and/or mannose residues are present on the outer surface of the plasmalemma. Clumping and agglutination was only obtained at room temperature, whereas incubation in the cold or with prefixed protoplasts yielded more homogeneous binding of Con A and no agglutination. Therefore GLIMELIUS et al. (1978a, b) suggested that the plant plasma membrane, like that of a variety of animal cells, complies with the fluid mosaic model proposed by SINGER and NICOLSON (1972). However, it was pointed out by SHEPARD and MOORE (1978) that the agglutination may be artefactual in the sense that the membrane saccharides to which the Con A binds are not necessarily endogenous, but may have become attached as a result of the extraction procedure. Clearly, the binding of cellulase-released sugars to membranes is an important limitation to the use of enzyme-macerated plant tissue for studies on endogenous membrane glycoproteins.

## 6 Concluding Remarks

The individuality of proteins, arising from their specific amino acid sequences and their conformational structures, is the reason that immunochemistry is such a specific and sensitive discriminative tool in protein research. In some immunological techniques the sensitivity is in the picogram range. Very little is known, however, about the structure, composition and number of the antigenic determinants on native proteins. Furthermore, it must be emphasized that the results obtained with immunological techniques are totally dependent on the specificity and quality of the antibodies. Therefore the antiserum must be carefully checked by sensitive methods before use. In this respect the use of isolated immunoglobulins of antisera purified by immunoadsorbents is an advantage; several proteins present in unfractionated antiserum may cause artefacts by interactions with the antigen. The choice of the immunological technique will depend on the purpose of the investigation, the nature of the antigen and the specificity of the antiserum. The benefit of the combination of physicochemical separation methods and immunological analysis followed by specific characterization reactions allows a protein to be characterized according to its physicochemical properties, immunochemical specificity and chemical or biological reactivities. Immunological techniques may be used for (1) the identification and quantitative determination of a protein in a mixture containing other protein constituents; (2) the analysis of the purity of proteins during their isolation; (3) the characterization of modifications to protein structures; (4) the study of the structural relationships between proteins; (5) the recognition of the appearance or disappearance of proteins during plant development and during tissue morphogenesis or as a consequence of environmental conditions or of pathological changes; (6) the localization of proteins in tissues, cells or in sub-cellular particles at the ultrastructural level.

The aim of this chapter has been to summarize the immunological data on plant proteins and to illustrate the wide range of applications of immunological methods to problems of fundamental and applied research.

## References


Agrell IPS (1966) The mitogenic action of phosphate and phytohemagglutinin on free-living amebae. Exp Cell Res 43:691–694

Alexandrescu V (1974) Water-soluble proteins from maize in the first growing stages. Electrophoretic and immunoelectrophoretic study. Rev Roum Biochim 11:77–85

Alexandrescu V, Călin I (1969) Electrophoretic and immunoelectrophoretic study of maize isoperoxidasis. Rev Roum Biochim 6:171–178

Alexandrescu V, Hagima I (1973) Peroxidase in leaves of some cereals immunochemical study. Rev Roum Biochim 10:15–21

Alexandrescu V, Mihăilsecu F (1973) Immunochemical investigations on germinated seeds endosperm α-amylase of some cereals. Rev Roum Biochim 10:89–94

Alexandrescu V, Mihăilescu F, Păun L (1975a) Amylases in the endosperm of wheat, rye and triticale germinated seeds. I Electrophoretic and immunoelectrophoretic investigations. Rev Roum Biochim 12:3–6

Alexandrescu V, Mihăilescu F, Păun L (1975b) Amylases in the endosperms of wheat, rye and triticale germinated seeds. II Immunological and immunochemical investigations. Rev Roum Biochim 12:61–66

Allen LW, Neuberger A, Sharon N (1973) The purification, composition and specificity of wheat-germ agglutinin. Biochem J 131:155–162

Anioł A (1974a) Is *Triticum macha* ssp. *tubalicum* var. *sublets-chumicum* an ancient *Triticale*? Z Pflanzenzuecht 72:226–232

Anioł A (1974b) A serological investigation of wheat evolution. Z Pflanzenzuecht 73:194–203

Anioł A (1976a) Serological studies within the tribe *Triticeae*. VIII. Serological affinity between genera. Genet Pol 17:523–529

Anioł A (1976b) Serological studies in the tribe *Triticeae*. VII. Serological affinity within the genus *Elymus*. Genet Pol 17:343–351

Anioł A, Nowacki E (1973) Serological relationships within *Hordeum* genus. Genet Pol 14:255–267

Arnon R (1973) Immunochemistry of enzymes. In: Sela M (ed) The antigen, Vol I. Academic Press, London New York, pp 87–159

Arnon R, Sela M (1969) Antibodies to an unique region in lysozyme provoked by a synthetic antigen conjugate. Proc Natl Acad Sci USA 62:163–170
Arnon R, Shapira E (1967) Antibodies to papain. A selective fractionation according to inhibitory capacity. Biochemistry 6:3942–3950
Aru LH, Mikk HT (1965) O serologicheskoi identifikatsii rastilet'nykhkelkev. Fiziol Rast 12:182–184
Augustin R (1959) Grass pollen allergens. II Antigen-antibody precipitation patterns in gel; their interpretation as a serological problem and in relation to skin reactivity. Immunology 2:148–169
Axelsen NH (ed) (1975) Quantitative immunoelectrophoresis. Scand Immunol Suppl 2, Universitetsforlaget, Oslo Bergen Tromsø
Axelsen NH, Krøll J, Weeke B (eds) (1973) A manual of quantitative immunoelectrophoresis. Scand J Immunol Suppl 1, Universitetsforlaget, Oslo Bergen Tromsø
Bakardieva NT, Dimirevska-Kepova K (1976) Immunochemical characteristic of peroxidase from green and etiolated pea plants, enriched by calcium and copper ions. Fiziol Rast II 4:28–37
Bal AK, Verma DPS, Byrne H, Maclachlan GA (1976) Subcellular localization of celluloses in auxin-treated pea. J Biol Chem 69:97–105
Barrett JT, Whiteaker RS (1977) Serological studies with the ficin-antificin system. In: Colombo JP, Frei J, Greengard O, Knox WE (eds) Enzyme, vol XXII. S Karger, Basel, pp 266–269
Basha SMM, Beevers L (1975) The development of proteolytic activity and protein degradation during germination of *Pisum sativum* L. Planta 124:77–87
Baumann C, Rüdiger H, Strossberg AD (1979) A comparison of the two lectins from *Vicia cracca*. FEBS Lett 102:216–218
Baumgartner B, Chrispeels MJ (1977) Purification and characterization of vicilin peptidohydrolase, the major endopeptidase in the cotyledons of mung bean seedlings. Eur J Biochem 77:223–233
Baumgartner B, Matile Ph (1976) Immunocytochemical localization of acid ribonuclease in morning glory flower tissue. Biochem Physiol Pflanz 170:279–285
Baumgartner B, Tokuyasu KT, Chrispeels MJ (1978) Localization of vicilin peptidohydrolase in the cotyledons of mung bean seedlings by immunofluorescence microscopy. J Cell Biol 79:10–19
Beckwith AC, Heiner DC (1966) An immunological study of wheat gluten proteins and derivatives. Arch Biochem Biophys 117:239–247
Belin L (1972) Separation and characterization of birch pollen antigens with special reference to the allergenic components. Int Arch Allergy Appl Immunol 42:329–342
Benhamou-Glynn N, Escribano M-J, Grabar P (1965) Study of gluten proteins by means of immunochemical methods. Bull Soc Chim Biol 47:141–156
Berzborn RJ, Lockau W (1977) Antibodies. In: Trebst A, Avron M (eds) Encyclopedia of plant physiology, vol V. Photosynthesis I. Springer, Berlin Heidelberg New York, pp 283–296
Bhuvaneswari TV, Pueppke SG, Bauer WD (1977) Role of lectins in plant-microorganism interactions. I. Binding of soybean lectin to rhizobia. Plant Physiol 60:486–491
Bird GWG, Uhlenbruck G, Pardoe GJ (1971) Serochemical studies of the specificity of some plant and animal agglutinins. Bibl Haematol 38:58–64
Bjerrum OJ (1977) Immunological investigations of membrane proteins. A methodological survey with emphasis placed on immunoprecipitation in gels. Biochim Biophys Acta 472:135–195
Bøg-Hansen TC, Daussant J (1974) Immunochemical quantitation of isoenzymes. α-amylase isoenzymes in barley malt. Anal Biochem 61:522–527
Bøg-Hansen TC, Brogren C-H, McMurrough IC (1974) Identification of enzymes as glycoproteins containing glucose or mannose. J Inst Brew 80:443–446
Bohlool BB, Schmidt EL (1974) Lectins: a possible basis for specificity in the *Rhizobium* – legume root nodule symbiosis. Science 185:269–271
Bollini R, Chrispeels MJ (1978) Characterization and subcellular localization of vicilin and phytohemagglutinin, the two major reserve proteins of *Phaseolus vulgaris* L. Planta 142:291–298

Booth MR, Ewart JAD (1969) Studies on four components of wheat gliadins. Biochim Biophys Acta 181:226–233
Booth MR, Ewart JAD (1970) Relationship between wheat proteins. J Sci Food Agric 21:187–192
Boutenko RG, Volodarsky AD (1968) Analyse immunochimique de la différenciation cellulaire dans les tissus de culture de tabac. Physiol Veg 6:299–309
Bowden L, Lord JM (1977) Serological and developmental relationships between endoplasmic reticulum and glyoxysomal proteins of castor bean endosperm. Planta 134:267–272
Bowden-Bonnett L (1979) Isolation and cell-free translation of total messenger RNA from germinating castor bean endosperm. Plant Physiol 63:769–773
Bowien B, Mayer F (1978) Further studies on the quaternary structure of D-Ribulose-1,5-biphosphate carboxylase from *Alcaligenes eutrophus*. Eur J Biochem 88:97–197
Bowles DJ (1979) Lectins as membrane components: Implications of lectin-receptor interaction. FEBS Lett 102:1–3
Bowles DJ, Kauss H (1975) Carbohydrate-binding proteins from cellular membranes of plant tissue. Plant Sci Lett 4:411–418
Bowles DJ, Schnarrenberger C, Kauss H (1976) Lectins as membrane components of mitochondria from *Ricinus communis*. Biochem J 160:375–382
Burgess J, Linstead PJ (1976) Ultrastructural studies of the binding of concanavalin A to the plasmalemma of higher plant protoplasts. Planta 130:73–79
Bustin M (1976) Chromatin structure and specificity revealed by immunological techniques. FEBS Lett 70:1–10
Buzilă L (1975) Hydrolysis with proteolytic enzymes of vicilin from pea seeds. Rev Roum Biochim 12:7–10
Byrne H, Christou NV, Verma DPS, Maclachlan GA (1975) Purification and characterization of two cellulases from auxin-treated pea epicotyls. J Biol Chem 250:1012–1018
Cantagalli P, Forconi V, Gagnoni G, Pieri J (1972) Immunochemical behaviour of the proteins of the orange. J Sci Food Agric 23:905–910
Capra DJ, Edmundson AB (1977) The antibody combining site. Sci Am 236:50–59
Casey R (1979) Immunoaffinity chromatography as a means of purifying legumin from *Pisum* (pea) seeds. Biochem J 177:509–520
Catsimpoolas N (1969) Isolation of glycinin subunits by isoelectric focusing in urea-mercaptoethanol. FEBS Lett 4:259–261
Catsimpoolas N (1978) Immunological properties of soybean proteins. In: Catsimpoolas N (ed) Immunological aspects of foods. Avi Publishing Company Inc, Westport Connecticut, pp 37–59
Catsimpoolas N, Ekenstam C (1969) Isolation of alpha, beta, and gamma conglycinins. Arch Biochem Biophys 127:338–345
Catsimpoolas N, Leuthner E (1969) Immunochemical methods for detection and quantitation of Kunitz soybean trypsin inhibitor. Anal Biochem 31:437–447
Catsimpoolas N, Meyer EW (1969) Isolation of soybean hemagglutinin and demonstration of multiple forms. Arch Biochem Biophys 132:279–284
Catsimpoolas N, Rogers DA, Cirde SJ, Meyer EW (1967) Purification and structural studies of the 11 S component of soybean proteins. Cereal Chem 44:631–637
Catsimpoolas N, Campbell TG, Meyer EW (1968a) Immunochemical study of changes in reserve proteins of germinating soybean seeds. Plant Physiol 43:799–805
Catsimpoolas N, Ekenstam C, Rogers DA, Meyer EW (1968b) Protein subunits in dormant and germinating seeds. Biochim Biophys Acta 168:122–131
Catsimpoolas N, Kenney JA, Meyer EW (1971) The effect of thermal denaturation on the antigenicity of glycinin. Biochim Biophys Acta 229:451–458
Cerff R (1974) Inhibitor-dependent, reciprocal changes in the activities of glyceraldehyde-3-phosphate dehydrogenases in *Sinapis alba* cotyledons. Z Pflanzenphysiol 73:109–118
Cerff R, Chambers SE (1979) Subunit structure of higher plant glyceraldehyde-3-phosphate dehydrogenases. J Biol Chem 254:6094–6098
Chien J, Mitchell HL (1973) Trypsin inhibitors in plants. Phytochemistry 12:327–330
Cherry JP, Katterman FRM, Endrizzi JE (1971) A comparative study of seed proteins of allopolyploids *Gossypium* by gel electrophoresis. Can J Genet Cytol 13:155–158

Chrispeels MJ, Baumgartner B (1978) Serological evidence confirming the assignment of *Phaseolus aureus* and *P. mungo* to the genus *Vigna*. Phytochemistry 17:125–126
Chrispeels MJ, Baumgartner B, Harris N (1976) Regulation of reserve protein metabolism in the cotyledons of mung bean seedlings. Proc Natl Acad Sci USA 73:3168–3172
Chua N-H, Blomberg F (1979) Immunochemical studies of thylakoid membrane polypeptides from spinach and *Chlamydomonas reinhardtii*. J Biol Chem 254:215–223
Codd GA, Schmid GH (1972) Serological characterization of the glycolate oxidizing enzymes from *Tobacco*, *Euglena gracilis*, and a yellow mutant of *Chlorella vulgaris*. Plant Physiol 50:769–773
Coleman RA, Pratt LH (1974a) Phytochrome: Immunocytochemical assay of synthesis and destruction. Planta 119:221–231
Coleman RA, Pratt LH (1974b) Subcellular localization of the redabsorbing form of phytochrome by immunocytochemistry. Planta 121:119–131
Converse CA, Papermaster DS (1975) Membrane protein analysis by two-dimensional immunoelectrophoresis.Science 189:469–472
Croy RRD, Derbyshire E, Krishna TG, Boulter D (1979) Legumin of *Pisum sativum* and *Vicia faba*. New Phytol 83:29–35
Cundiff SC, Pratt LH (1973) Immunological determination of the relationship between large and small sizes of phytochrome. Plant Physiol 51:210–213
Daussant J (1975) Immunochemical investigations of plant proteins. In: Harborne JB, van Sumere CF (eds) The chemistry and biochemistry of plant proteins. Academic Press, London New York, pp 31–69
Daussant J (1978) Immunochemistry of barley seed proteins. In: Catsimpoolas N (ed) Immunological aspects of foods. Avi Publishing Company, Westport Connecticut, pp 60–86
Daussant J, Abbott DC (1969) Immunochemical study of changes in the soluble proteins of wheat during germination. J Sci Food Agric 20:631–637
Daussant J, Carfantan N (1975) Electro-immunoabsorption in gel, application to enzyme studies (α- and β-amylases from barley). J Immunol Methods 8:373–382
Daussant J, Corvazier P (1970) Biosynthesis and modifications of α- and β-amylases in germinating wheat seeds. FEBS Lett 7:191–194
Daussant J, Grabar P (1966) Comparaison immunologique des α-amylases extraites de céréales. Ann Inst Pasteur Paris Suppl 110:79–83
Daussant J, MacGregor AW (1979) Combined immunoabsorption and isoelectric focusing of barley and malt amylases in polyacrylamide gel. Anal Biochem 93:261–266
Daussant J, Renard M (1972) Immunochemical comparison of α-amylase in developing and germinating wheat seeds. FEBS Lett 22:301–304
Daussant J, Skakoun A (1974) Combination of absorption technique and α-amylase activity determination in the same gel medium. J Immunol Methods 4:127–133
Daussant J, Grabar P, Nummi M (1966) β-amylase. II. Identification des différentes β-amylases de l'orge et du malt. Proc 10th Eur Brew Conv, Stockholm 1965. Elsevier, Amsterdam, pp 52–69
Daussant J, Neucere NJ, Yatsu LY (1969a) Immunochemical studies on *Arachis hypogaea* proteins with particular reference to the reserve proteins. I. Characterization, distribution, and properties of α-arachin and β-conarachin. Plant Physiol 44:471–479
Daussant J, Neucere NJ, Conkerton EJ (1969b) Immunochemical studies on *Arachis hypogaea* proteins with particular reference to the reserve proteins. II. Protein modification during germination. Plant Physiol 44:480–484
Daussant J, Roussoux J, Manigault P (1971) Caractérisations immunochimiques de deux auxine oxydases extraites de tumeurs végétales. FEBS Lett 14:245–250
Daussant J, Skakoun A, Niku-Paavola ML (1974) Immunochemical study on barley α-amylases. J Inst Brew 80:55–58
Daussant J, Ory RL, Layton LL (1976) Characterization of proteins and allergens in germinating castor seeds by immunochemical techniques. J Agric Food Chem 24:103–107
Davey RA, Dudman WF (1979) The carbohydrate of storage glycoproteins from seeds of *Pisum sativum*: Characterization and distribution on component polypeptides. Aust J Plant Physiol 6:435–447

Derbyshire E, Wright DJ, Boulter D (1976) Legumin and vicilin, storage proteins of legume seeds. Phytochemistry 15:3–24
Dimitrov P, Nashkova O, Petkova S, Nashkov D, Marinkov D (1974) Immunochemical prognosis of heterosis in *Zea mays*. Theor Appl Genet 45:91–95
Djurtoft R, Hill RJ (1965) Immunoelectrophoretic studies of proteins in barley, malt, beer and beer haze preparations. Proc 10th Eur Brew Conv, Stockholm, 1965. Elsevier, Amsterdam, pp 137–146
Domoney C, Davies DR, Casey R (1980) The initiation of legumin synthesis in immature embryos of *Pisum sativum* L. grown in vivo and in vitro. Planta 149:454–460
Dorner RW, Kahn A, Wildman S (1958) Proteins of green leaves. VIII The distribution of fraction I protein in the plant kingdom as detected by precipitin and ultracentrifugal analyses. Biochim Biophys Acta 29:240–245
Dornhauser S (1967) Immunologische Untersuchungen über die Veränderung der salzlöslichen Eiweiß-Fraktionen von der Gerste bis zum Bier unter Variation des Mälzungsverfahrens sowie Untersuchungen über Rohfruchtbiere. Proc Eur Brew Conv II, Elsevier, Amsterdam, pp 323–325
Dornhauser S (1972) Nachweisversuche an enzymatisch stabilisierten Bieren mit immunologischen Methoden. Brauwissenschaft 25:189–192
Dudman WF, Millerd A (1975) Immunochemical behaviour of legumin and vicilin from *Vicia faba*: a survey of related proteins in the *Leguminosae* subfamily *Faboideae*. Biochem Syst Ecol 3:25–33
Durand B, Durand-Rivières R (1969) Cytokinines et régulation de la synthése d'une protéine antigénique spécifique du sexe femelle chez une plante dioique *Mercurialis annua* L. CR Acad Sci Ser D 269:1639–1641
Durand-Rivières RCR (1969) Mise en évidence d'une protéine antigénique spécifique dans les mérestémes et less feuilles femelles de *Mercurialis annua* L. CR Acad Sci Ser D 268:2046–2048
Eder J, Arnon R (1973) Structural and functional comparison of antibodies to common and specific determinants of papain and chymopapain. Immunochemistry 10:535–543
Elton GAH, Ewart JAD (1963) Immunological comparison of cereal proteins. J Sci Food Agric 14:750–758
Escribano M-J (1966) Application of immunochemical methods to the study of insoluble wheat proteins. Getreide Mehl Brot 12:134–136
Escribano M-J, Grabar P (1966) Immunochemical study of the insoluble proteins of wheat flour after fission of disulphide bonds. Ann Inst Pasteur Paris Suppl III. 110:84–88
Ewart JAD (1966) Cereal proteins: immunological studies. J Sci Food Agric 17:279–284
Ewart JAD (1978) Immunochemistry of wheat proteins. In: Catsimpoolas N (ed) Immunological aspects of foods. Avi Publishing Company Inc, Westport Connecticut, pp 87–116
Fairbrothers DE (1977) Perspectives in plant serotaxonomy. Ann Mo Bot Gard 64:147–160
Feierabend J, Wildner G (1978) Formation of the small subunit in the absence of the large subunit of ribulose 1,5-biphosphate carboxylase in 70S ribosome-deficient rye leaves. Arch Biochem Biophys 186:283–291
Feteanu A (ed) (1978) Labelled antibodies in biology and medicine. Abacus Press and McGraw-Hill Int Book Co
Fountain DW, Foard DE, Replogle WD, Yang WK (1977) Lectin release by soybean seeds. Science 197:1185–1187
Franz H, Bergmann P, Ziska P (1979) Combination of immunological and lectin reactions in affinity histochemistry: Proposition of the term affinitin. Histochemistry 59:335–342
Freed RC, Ryan DS (1978a) Note on modification of the Kunitz soybean trypsin inhibitor during seed germination. Cereal Chem 55:534–538
Freed RC, Ryan DS (1978b) Changes in Kunitz trypsin inhibitor during germination of soybean – An immunoelectrophoresis assay system. J Food Sci 43:1316–1319
Frevert J, Kindl H (1978) Plant microbody proteins. Purification and glycoprotein nature of glyoxysomal isocitrate lyase from *Cucumber* cotyledons. Eur J Biochem 92:35–43
Galbraith W, Goldstein IJ (1970) Phytohemagglutinins: A new class of metalloproteins. Isolation, purification, and some properties of the lectin from *Phaseolus vulgaris*. FEBS Lett 9:197–201

Galbraith W, Goldstein IJ (1972) Phytohemagglutinin of the lima bean (*Phaseolus lunatus*). Isolation, characterization, and interaction with type A blood substance. Biochemistry 11:3976–3984

Gatenby AA (1978) A comparison of the polypeptide isoelectric points and antigenic determinant sites of the large subunit of fraction I protein from *Lycopersicon esculentum, Nicotiana tabacum* and *Petunia hybrida*. Biochim Biophys Acta 534:169–172

Gietl CH, Ziegler H (1979) Lectins in the excretion of intact roots. Naturwissenschaften 66:161–164

Gill TA, Tung MA (1978) Electrophoretic and immunochemical properties of the 12 S rapeseed protein. Cereal Chem 55:809–817

Glimelius K, Wallin A, Eriksson T (1974) Agglutinating effects of Concanavalin A on isolated protoplasts of *Daucus carota*. Physiol Plant 31:225–230

Glimelius K, Wallin A, Eriksson T (1978a) Ultrastructural visualization of sites binding Concanavalin A on the cell membrane of *Daucus carota*. Protoplasma 97:291–300

Glimelius K, Wallin A, Eriksson T (1978b) Ultrastructural visualization of sites binding Concanavalin A on the cell membrane of *Daucus carota*. Protoplasma 97:291–300

Gonzalez E, Beevers H (1976) Role of the endoplasmic reticulum in glyoxysome formation in castor bean endosperm. Plant Physiol 57:406–409

Gooding LR, Roy H, Jagendorf AF (1973) Immunological identification of nascent subunits of wheat ribulose diphosphate carboxylase on ribosomes of both chloroplast and cytoplasmic origin. Arch Biochem Biophys 159:324–335

Grabar P (1975) Immunological methods in tissue analysis. J Immunol Methods 7:305–326

Grabar P, Escribano M-J, Benhamou N, Daussant J (1965) Immunochemical study of wheat, barley, and malt proteins. J Agric Food Chem 13:392–398

Graf L, Notton BA, Hewitt EJ (1975) Serological estimation of spinach nitrate reductase. Phytochemistry 14:1241–1243

Gray JC (1977) Serological relationship of fraction I proteins from species in the genus *Nicotiana*. Plant Syst Evol 128:53–69

Gray JC (1978) Serological reactions of fraction I proteins from interspecific hybrids on the genus *Nicotiana*. Plant Syst Evol 129:177–183

Gray JC, Kekwick RGO (1974) The synthesis of the small subunit of ribulose 1,5-biphosphate carboxylase in the french bean *Phaseolus vulgaris*. Eur J Biochem 44:491–500

Gray JC, Wildman SG (1976) A specific immunoabsorbent for the isolation of fraction I protein. Plant Sci Lett 6:91–96

Green TR, Ryan CA (1972) Wound-induced proteinase inhibitor in plant leaves. A possible defense mechanism against insects. Science 175:776–777

Guldager P (1978) Immunoelectrophoretic analysis of seed proteins from *Pisum sativum* L. Theor Appl Genet 53:241–250

Gurusiddaiah S, Kuo T, Ryan CA (1972) Immunological comparisons of chymotrypsin inhibitor I among several genera of the *Solanaceae*. Plant Physiol 50:627–631

Hall O (1959) Immuno-electrophoretic analyses of allopolyploid ryewheat and its parental species. Hereditas 45:495–504

Hamblin J, Kent SP (1973) Possible role of phytohemagglutinin in *Phaseolus vulgaris* L. Nature (London) 245:28–30

Hankins CN, Kindinger JI, Shannon LM (1979) Legume lectins. I. Immunological cross-reactions between the enzymic lectin from mung beans and other well characterized legume lectins. Plant Physiol 64:104–107

Hague DR (1975) Studies of storage proteins of higher plants. I. Concanavalin A from three species of the genus *Canavalia*. Plant Physiol 55:636–642

Harboe N, Ingild A (1973) Immunization, isolation of immunoglobulins, estimation of antibody titre. In: Axelsen NH, Krøll J, Weeke B (eds) A manual of quantitative immunoelectrophoresis. Scand J Immunol Suppl 1 Universitetsforlaget, Oslo Bergen Tromsø, pp 161–164

Hartmann JY, Kao KN, Gamborg OL, Miller RA (1973) Immunological methods for the agglutination of protoplasts from cell suspension cultures of different genera. Planta 112:45–56

Hattersley PW, Watson L, Osmond CB (1976) Metabolic transport of leaves of $C_4$ plants:

specification and speculation. In: Transport and transfer processes in plants. Academic Press, London New York, pp 191–201
Hattersley PW, Watson L, Osmond CB (1977) In situ immunofluorescent labelling of ribulose-1,5-biphosphate carboxylase in $C_3$ and $C_4$ plant leaves. Aust J Plant Physiol 4:523–539
Hejgaard J (1976) Free and protein-bound $\beta$-amylases of barley grain. Characterization by two-dimensional immunoelectrophoresis. Physiol Plant 38:293–299
Hejgaard J, Bøg-Hansen TC (1974) Quantitative immunoelectrophoresis of barley and malt proteins. J Inst Brew 80:436–442
Hejgaard J, Sørensen SB (1975) Characterization of a protein-rich beer fraction by two-dimensional immunoelectrophoresis techniques. Compt Rend Trav Lab Carlsberg 40:187–203
Hill RJ, Djurtoft R (1964) Some immunoelectrophoretic studies on barley proteins. J Inst Brew 70:416–424
Hillebrand GP, Fairbrothers DE (1970a) Phytoserological systematic survey of the *Caprifoliaceae*. Brittonia 22:125–133
Hillebrand GP, Fairbrothers DE (1970b) Serological investigation of the systematic position of the *Caprifoliaceae*. I. Correspondence with selected *Rubiaceae* and *Cornaceae*. Am J Bot 57:810–815
Hiral A (1977) Random assembly of different kinds of small subunit polypeptides during formation of fraction I protein macromolecules. Proc Natl Acad Sci USA 74:3443–3445
Hochkeppel H-K (1973) Isolierung einer Endopeptidase aus alternden Tabakblättern und ihre Beziehung zum Vergilben. Z Pflanzenphysiol 69:329–343
Hock B (1974) Antikörper gegen Glyoxysomenmembranen. Planta 115:271–280
Horikoshi M, Morita Y (1975) Localization of $\gamma$-globulin in rice seed and changes in $\gamma$-globulin content during seed development and germination. Agric Biol Chem 39:2309–2314
Horisberger M, Vonlanthen M (1980) Ultrastructural localization of soybean agglutinin on this sections of *Glycine max* (soybean) *var. Altona* by the gold method. Histochemistry 65:181–186
Houts KP, Hillebrand GR (1976) An electrophoretic and serological investigations of seed proteins in *Galeopsis tetrahit* L. (*Labiatae*) and its putative parental species. Am J Bot 63:156–165
Howard CN, Kindinger JI, Shannon LM (1979) Conservation of antigenic determinants among different seed lectins. Arch Biochem Biophys 192:457–465
Howard IK, Sage HJ, Horton CB (1972) Studies on the appearance and location of hemagglutinins from common lentil during the life cycle of the plant. Arch Biochem Biophys 149:323–326
Huang AHC, Bowman PhD, Beevers H (1974) Immunological and biochemical studies on isozymes of malate dehydrogenase and citrate synthetase in castor bean glyoxysomes. Plant Physiol 54:364–368
Hubscher T, Eisen AH (1972) Localization of ragweed antigens in the intact ragweed pollen grain. Int Arch Allergy Appl Immunol 42:466–473
Huebner FR, Rothfus JA, Wall JS (1967) Isolation and chemical comparison of different $\gamma$-gliadins from hard red winter wheat flour. Cereal Chem 44:221–226
Hunt RE, Pratt LH (1979a) Phytochrome radioimmunoassay. Plant Physiol 64:327–331
Hunt RE, Pratt LH (1979b) Phytochrome immunoaffinity purification. Plant Physiol 64:332–336
Hurrell JGR, Nicola NA, Broughton WJ, Dilworth MJ, Minasian E, Leach SJ (1976) Comparative structural and immunochemical properties of leghemoglobins. Eur J Biochem 66:389–399
Hurrell JGR, Thulborn KR, Broughton WJ, Dilworth MJ, Leach SJ (1977) Leghemoglobins: Immunochemistry and phylogenetic relationships. FEBS Lett 84:244–246
Hurrell JGR, Smith JA, Leach SJ (1978) The detection of five antigenically reactive regions in the soybean leghemoglobin molecule. Immunochemistry 15:297–302
Husain SS, Lowe GC (1970) The amino acid sequence around the active-site cysteine and histidine residues of stem bromelain. Biochem J 117:341–346

Huystee van RB (1976) Immunological studies on proteins released by a peanut (*Arachis hypogaea* L.) suspension culture. Bot Gaz 137:325–329
Hwang DL, Lin K-T, Yang W-K, Ford DE (1977) Purification, partial characterization, and immunological relationships of multiple low molecular weight protease inhibitors of soybean. Biochim Biophys Acta 495:369–382
Hwang DL, Yang W-K, Ford DE (1978) Rapid release of protease inhibitors from soybeans. Immunochemical quantitation and parallels with lectins. Plant Physiol 61:30–34
Iida S, Sasaki M, Ota S (1973) Immunological cross-reaction between thiol proteases of plant origin: stem and fruit bromelains. J Biochem 73:377–386
Ivanov VN, Khavkin EE (1976) Protein patterns of developing mitochondria at the onset of germination in maize (*Zea mays* L.). FEBS Lett 65:383–385
Jacks TJ, Neucere NJ, Yatsu LY (1972) Characterization of proteins from subcellular fractions of peanuts. J Am Peanut Res E Duc Assoc 4:195–205
Jacobsen JV, Knox RB (1973) Cytochemical localization and antigenicity of α-amylase in barley aleurone tissue. Planta 112:213–224
Jacobsen JV, Knox RB (1974) The proteins released by isolated barley aleurone layers before and after gibberellic-acid treatment. Planta 115:193–206
Jaffé WG (1978) Immunology of plant agglutinins. In: Catsimpoolas N (ed) Immunological aspects of foods. Avi Publishing Company Inc, Westport Connecticut, pp 170–180
Jaffé WG, Lery A, Gonzalez DI (1974) Isolation and partial characterization of bean phytohemagglutinins. Phytochemistry 13:2685–2693
Jendrisak J, Guilfoyle TJ (1978) Eukaryotic RNA polymerases: comparative subunit structures, immunological properties, and α-amanitin sensitivities of the class II enzymes from higher plants. Biochemistry 17:1322–1327
Jermyn MA, Yeow YM (1975) A class of lectins present in the tissue of seed plants. Aust J Plant Physiol 2:501–531
Johal S, Bourque DP (1979) Crystalline ribulose 1,5-biphosphate carboxylase-oxygenase from spinach. Science 204:75–77
Johnson BL, Barnhart D, Hall O (1967) Analysis of genome and species relationships in the polyploid wheats by protein electrophoresis. Am J Bot 54:1089–1098
Jones RL, Chen R (1976) Immunohistochemical localization of α-amylase in barley aleurone cells. J Cell Sci 20:183–198
Kabat EA, Mayer MM (eds) (1967) Experimental immunochemistry. Thomas Springfield, Illinois
Kagawa T, Lord JM, Beevers H (1973) The origin and turnover of organelle membranes in castor bean endosperm. Plant Physiol 51:61–65
Kahlem G (1973) Proteins and development in a dioecious plant: *Mercurialis annua* L. Z Pflanzenphysiol 69:377–380
Kahlem G (1976) Isolation and localization by histoimmunology of isoperoxidases specific for male flowers of the dioecious species (*Mercurialis annua* L.). Dev Biol 50:58–67
Kanamori M, Ibuki F, Tashiro M, Yamada M, Mioyshi M (1976) Purification and partial characterization of a proteinase inhibitor isolated from egg plant exocarp. Biochim Biophys Acta 439:398–405
Kato G, Maruyama Y, Nakamura M (1979) Role of lectins and lipo-polysaccharides in the recognition of specific legume – *Rhizobium* symbiosis. Agric Biol Chem 43:1085–1092
Kato T, Sasaki M (1974) Biological significance and localization of antigenic determinant common to thiol proteases of plant origin. J Biochem 76:1021–1030
Kawashima N, Imai A, Tamaki E (1968) Immunological comparison of fraction I proteins from various plants. Agric Biol Chem 32:535–536
Kawashima N, Kwok S-Y, Wildman SG (1971) Studies on fraction-I protein. III. Comparison of the primary structure of the large and small subunits obtained from five species of *Nicotiana*. Biochim Biophys Acta 236:578–586
Khavkin EE, Kohl J-G, Misharin SI, Iwanow WN (1972) Enzymatische Identifikation der Antigene der wachsenden Wurzelzellen von *Zea mays* L. Biochem Physiol Pflanz 163:308–315
Khavkin EE, Misharin SJ, Ivanov VN (1977) Embryonal antigens in maize caryopses: The temporal order of antigen accumulation during embryogenesis. Planta 135:225–231

Khavkin EE, Misharin SI, Markov YY, Peshkova AA (1978a) Identification of embryonal antigens of maize: Globulins as primary reserve proteins of the embryo. Planta 143:11–20
Khavkin EE, Misharin SJ, Monastyreva LE, Polikarpochkina RT, Sokhorzhevskaia TB (1978b) Specific proteins maintained in maize callus cultures. Z Pflanzenphysiol 86:273–277
Khavkin EE, Markov EY, Misharin SJ (1980) Evidence for proteins specific for vascular elements in intact and cultured tissues and cells of maize. Planta 148:116–123
Kirk J, Sumner JB (1932) Immunological identity of soy and jack bean urease. Proc Soc Exp Biol Med 29:712–713
Kling H (1975) Immunochemische Untersuchungen an Prolaminen. Z Pflanzenphysiol 76:155–162
Kloz J (1971) Serology of the *Leguminosae*. In: Harborne JB, Boutler D, Turner BL (eds) Chemotaxonomy of the *Leguminosae*. Academic Press, London New York, pp 309–366
Kloz J, Klozová E (1974) The protein euphaseolin in *Phaseolinae* – a chemotaxonomical study. Biol Plant 16:290–300
Klozová E, Kloz J, Winfield PJ (1976) A typical composition of seed proteins in cultivars of *Phaseolus vulgaris* L. Biol Plant 18:200–205
Knox RB (1973) Pollen wall proteins: Pollen-stigma interactions in ragweed and cosmos (*Compositae*). J Cell Sci 12:421–443
Knox RB, Heslop-Harrison J (1971) Pollen-wall proteins: the fate of intine-held antigens on the stigma in compatible and incompatible pollinations of *Phalaris tuberosa* L. J Cell Sci 9:239–251
Knox RB, Willing RR, Ashford AE (1972) Pollen-wall proteins; role as recognition substances in interspecific incompatibility in poplars. Nature (London) 237:381–383
Köller W, Frevert J, Kind H (1979) Albumins, glyoxysomal enzymes and globulins in dry seeds of *Cucumis sativus*: Qualitative and quantitative analysis. Hoppe-Seyler's Z Physiol Chem 360:167–176
Konieczny A, Legocki AB (1978) Isolation and in vitro translation of leghaemoglobin mRNA from yellow lupin root nodules. Acta Biochem Pol 25:379–390
Koshiyama I (1972) A never method for isolation of the 7 S globulin in soybean seeds. Agric Biol Chem 36:2255–2257
Krøll J, Andersen MM (1976) Specific antisera produced by immunization with precipitin lines. J Immunol Methods 13:125–130
Ku MSB, Schmitt MR, Edwards GE (1979) Quantitative determination of RuBP carboxylase – oxygenase protein in leaves of several $C_3$ and $C_4$ plants. J Exp Bot 30:89–98
Kurth PD, Bustin M, Moudrianakis EN (1979) Concanavalin A binds to puffs in polytene chromosomes. Nature (London) 279:448–450
Lanzerotti PM, Gullino PM (1972) Immunochemical quantitation of enzymes using multispecific antisera. Anal Biochem 50:344–353
Larkin PJ (1977) Plant protoplast agglutination and membrane-bound $\beta$-lectins. J Cell Sci 26:31–46
Lee YS, Dickinson DB (1979) Characterization of pollen antigens from *Ambrosia* L. (*Compositae*) and related taxa by immunoelectrophoresis and radial immunodiffusion. Am J Bot 66:245–252
Lee YS, Dickinsin DB, Schlager D, Velu JG (1979) Antigen E content of pollen from individual plants of short ragweed (*Ambrosia artemisiifolia*). J Allergy Clin Immunol 63:336–339
Legocki RP, Verma DPS (1979) A nodule-specific plant protein (Nodulin-35) from soybean. Science 205:190–193
Levine D, Kaplan MJ, Greenaway PJ (1972) The purification and characterization of wheat-germ agglutinin. Biochem J 129:847–856
Lichtenfeld C, Manteuffel R, Müntz K, Neumann D, Scholz G, Weber E (1979) Protein degradation and proteolytic activities in germinating field beans (*Vicia faba, var. minor*). Biochem Physiol Pflanz 174:255–274
Liener IE (1976) Phytohemagglutinins (Phytolectins). Annu Rev Plant Physiol 27:291–319

Liener IE, Rose JE (1953) Soyin, a toxic protein from the soybean. III. Immunochemical properties. Proc Soc Exp Biol Med 83:539–545

Lis H, Sharon N (1973) The biochemistry of plant lectins (Phytohemagglutinins). Annu Rev Biochem 42:541–574

Lolas GM, Markakis P (1975) Phytic acid and other phosphorus compounds of bean (*Phaseolus vulgaris* L.). J Agric Food Chem 23:13–15

Loomis WD, Battaile J (1966) Plant phenolic compounds and the isolation of plant enzymes. Phytochemistry 5:423–438

Lord MJ, Bowden L (1978) Evidence that glyoxysomal malate synthase is segregated by the endoplasmic reticulum. Plant Physiol 61:266–270

Lotan R, Cacan R, Cacan M, Debray H, Carter WG, Sharon N (1975) On the presence of two types of subunit in soybean agglutinin. FEBS Lett 57:100–103

Luizzi A, Angeletti PU (1969) Application of immunodiffusion in detecting the presence of barley in wheat flour. J Sci Food Agric 20:207–209

Maher P, Molday RS (1977) Binding of concanavalin A to *Ricinus communis* agglutinin and its implication in cell-surface labeling studies. FEBS Lett 84:391–394

Mākinen YLA, Lewis D (1962) Immunological analysis of incompatibility (S) proteins and of cross-reacting material in a self-compatible mutant of *Oenothera organensis*. Genet Res 3:352–363

Manteuffel R, Scholz G (1975) Studies on seed globulins from legumes. V. Immunoelectrophoretic control of vicilin purification by gel filtration. Biochem Physiol Pflanz 168:277–285

Marcinowski S, Falk H, Hammer DK, Hoyer B, Grisebach H (1979) Appearance and localization of a β-glucosidase hydrolyzing coniferin in spruce (*Picea abies*) seedlings. Planta 144:161–165

Marsh DG (1975) Allergens and the genetics of allergy. In: Sela M (ed) The antigens, vol III. Academic Press, London New York, pp 271–295

Matsumoto J, Osawa T (1970) Purification and characterization of a cytisus-type anti-M(O) phytohemagglutinin from *Ulex europeus* seeds. Arch Biochem Biophys 140:484–491

Matsumoto K, Nishimura M, Akazawa T (1977) Ribulose-1,5-biphosphate carboxylase in the bundle sheath cells of maize leaf. Plant Cell Physiol 18:1281–1290

Maurer PH, Gerulat BF, Pinchuk P (1964) Antigenicity of polypeptides (poly-α-amino acids). J Biol Chem 239:922–929

Mayer RJ, Walker JH (1978) Techniques in enzyme and protein immunochemistry. In: Kornberg HL, Metcalfe JC, Northcote DH, Pogson CJ, Tipton KF (eds) Techniques in life science. Techniques in protein and enzyme biochemistry, vol B1/II. Elsevier/North-Holland, Biomedical Press, pp 1–32

McFadden BA, Tabita FR (1974) D-ribulose-1,5-diphosphate carboxylase and the evolution of autotrophy. Bio-Systems 6:93–112

McGowan RE, Gibbs M (1974) Comparative enzymology of the glyceraldehyde 3-phosphate dehydrogenase from *Pisum sativum*. Plant Physiol 54:312–319

Melville JC, Ryan CA (1972) Chymotrypsin inhibitor I from potatoes. J Biol Chem 247:3445–3453

Millerd A (1975) Biochemistry of legume seed proteins. Annu Rev Plant Physiol 26:53–72

Millerd A, Simon M, Stern H (1971) Legumin synthesis in developing cotyledons of *Vicia faba* L. Plant Physiol 48:419–425

Millerd A, Thomson JA, Schroeder HE (1978) Cotyledonary storage proteins in *Pisum sativum*. III. Patterns of accumulation during development. Aust J Plant Physiol 5:519–534

Millerd A, Thomson JA, Randall PJ (1979) Heterogeneity of sulphur content in the storage proteins of pea cotyledons. Planta 146:463–466

Mühlethaler K (1977) Introduction to structure and function of the photosynthesis apparates. In: Trebst M, Avron M (eds) Photosynthetic electron transport and photophosphorylation, vol V. Springer, Berlin Heidelberg New York, pp 503–521

Murphy TM (1978) Immunochemical comparisons of ribulose-biphosphate carboxylase using anti-sera to tobacco and spinach enzymes. Phytochemistry 17:439–443

Murray DR, Knox RB (1977) Immunofluorescent localization of urea in the cotyledons of jack bean, *Canavalia ensiformis*. J Cell Sci 26:9–18
Nagata Y, Burger MM (1974) Wheat germ agglutinin, molecular characteristics and specificity for sugar binding. J Biol Chem 249:3116–3121
Nagl W (1972) Phytohemagglutinin: Transitory enhancement of growth in *Phaseolus* and *Allium*. Planta 106:269–272
Nairn PC (ed) (1976) Fluorescent protein tracing, 4th edn. Edu Livingstone, Edinburgh London
Nasrallah ME, Wallace DH (1967a) Immunochemical detection of antigens in self-incompatibility genotypes of cabbage. Nature (London) 213:700–701
Nasrallah ME, Wallace DH (1967b) Immunogenetics of self-incompatibility in *Brassica oleracea* L. Heredity 22:519–527
Nasrallah ME, Barber JT, Wallace DH (1969) Self-incompatibility proteins in plants: detection genetics and possible mode of action. Heredity 24:23–27
Nasrallah ME, Wallace DH, Savo RM (1972) Genotype, protein, phenotype relationships in self-incompatibility of *Brassica*. Genet Res 20:151–160
Nelson OE, Burr B (1973) Biochemical genetics of higher plants. Annu Rev Plant Physiol 24:493–518
Neucere NJ (1969) Isolation of α-arachin, the major peanut globulin. Anal Biochem 27:15–24
Neucere NJ (1974) Antigenic and electrophoretic changes of α-arachin after heating in vitro. J Agric Food Chem 22:146–148
Neucere NJ (1978a) Aminopeptidase activity associated with $\alpha_1$-conarachin (peanut protein). Phytochemistry 17:546–548
Neucere NJ (1978b) Immunochemistry of peanut proteins. In: Catsimpoolas N (ed) Immunological aspects of foods. Avi Publishing Company Inc, Westport Connecticut, pp 117–151
Neucere NJ, Cherry JO (1975) An immunochemical survey of proteins in species of *Arachis*. Peanut Sci 2:66–72
Neucere NJ, Ory RL (1970) Physicochemical studies on the proteins of the peanut cotyledon and embryonic axis. Plant Physiol 45:616–619
Neucere NJ, Yatsu LY (1975) Genesis of storage protein synthesis in the developing peanut seed. Peanut Sci 2:38–41
Neucere NJ, Ory RL, Carney WB (1969) Effect of roasting on the stability of peanut proteins. J Agric Food Chem 17:25–28
Newman RA (1977) Heterogeneity among the anti-$T_F$ lectins derived from *Arachis hypogaea*. Hoppe-Seyler's Z Physiol Chem 358:1517–1520
Nimmo CC, O'Sullivan MT (1967) Immunochemical comparisons of antigenic proteins of durum and hard red spring wheats. Cereal Chem 44:584–591
Nishimura M, Akazawa T (1974a) Studies on spinach leaf ribulosebiphosphate carboxylase. Carboxylase and oxygenase reaction examined by immunochemical methods. Biochemistry 13:2277–2281
Nishimura M, Akazawa T (1974b) Reconstitution of spinach ribulose-1,5-diphosphate carboxylase from separated subunits. Biochem Biophys Res Commun 59:584–590
Northoft H, Jungfer H, Resch K (1978) The effect of anticon A on the binding of con A to lymphocytes. Exp Cell Res 115:151–158
Nowacki E, Prus-Głowacki W (1971) Differentiation of protein fractions in species and varieties of the genus *Lupinus* with the use of serological methods. Genet Pol 12:245–260
Nowacki E, Anioł A, Bieber D (1972) An attempted cross of *Zea mays* and *Coix lacryma jobi* and the serological relationships of these species. Bull Acad Pol Sci XX:695–698
Nummi M (1963) Fractionation of barley globulins on dextran gel columns. Acta Chem Scand 17:527–529
Nummi M, Vilhunen R, Enari T-M (1965) β-amylase: I. β-amylases of different molecular size in barley and malt. Proc 10th Congr Eur Brew Conv, Stockholm 1965, Elsevier, Amsterdam, pp 52–61
Nummi M, Daussant J, Niku-Paalova ML, Kalsta H, Enari T-M (1970) Comparative

immunological and chromatographic study of some plant β-amylases. J Sci Food Agric 21:258–260
Ochiai-Yanagi S, Fukazawa C, Harada K (1978) Formation of storage protein components during soybean seed development. Agric Biol Chem 42:697–702
Okamoto K, Akazawa T (1979) Enzymic mechanism of starch breakdown in germinating rice seeds. Plant Physiol 64:337–340
Okita TW, Decaleya R, Rappaport L (1979) Synthesis of a possibile precursor of α-amylase in wheat aleurone cells. Plant Physiol 63:195–200
Olsen HS (1978) Faba bean protein for human consumption. In: Adler-Nissen J (ed) Biochemical aspects of new protein food, vol 44 A3. Pergamon Press, Oxford New York, pp 31–42
Osborne TB (ed) (1924) The vegetable proteins, 2nd edn. London
Ouchterlony O (ed) (1968) Handbook of immunodiffusion and immunoelectrophoresis. Publ Ann Arbor Sci Publ Inc
Paus E (1976) Immunoadsorbent affinity purification of the two enzyme forms of α-mannosidase from *Phaseolus vulgaris*. FEBS Lett 72:39–42
Pereira MEA, Kabat EA, Lotan R, Sharon N (1976) Immunochemical studies on the specificity of the peanut (*Arachis hypogaea*) agglutinin. Carbohydr Res 51:107–118
Pick K-H, Wöber G (1978) Proteinaceous α-amylase inhibitor from beans (*Phaseolus vulgaris*). Immunological characterization. Hoppe-Seyler's Z Physiol Chem 359:1379–1384
Pickering JL, Fairbrothers DE (1970) A serological comparison of *Umbelliferae* subfamilies. Am J Bot 57:988–992
Pierard D, Jacmard A, Bernier G (1977) Changes in the protein composition of the shoot apical bud of *Sinapis alba* in transition of flowering. Physiol Plant 41:254–258
Polacco JC, Havir EA (1979) Comparisons of soybean urease isolated from seed and tissue culture. J Biol Chem 254:1707–1715
Prager MD, Fetcher MA, Efron K (1962) Mechanism of the immunohematologic effect of papain and related enzymes. J Immunol 89:834–840
Pratt LH (1973) Comparative immunochemistry of phytochrome. Plant Physiol 51:203–209
Pratt LH, Coleman RA (1971) Immunocytochemical localization of phytochrome. Proc Natl Acad Sci USA 86:2431–2435
Pratt LH, Coleman RA (1974) Phytochrome distribution in etiolated grass seedlings as assayed by an indirect antibody-labelling method. Am J Bot 61:195–202
Pratt LH, Kidd GH, Coleman RA (1974) An immunochemical characterization of the phytochrome destruction reaction. Biochim Biophys Acta 365:93–107
Prus-Głowacki W (1975) Changes of protein fractions in the ontogenesis of four Lupin species studied by immunological methods. I. Differences in the seed protein fractions of the studied *Lupin* species and varieties. Genet Pol 16:37–46
Prus-Głowacki W, Sulinowski S, Nowacki E (1971) Immunoelectrophoretic studies of *Lolium-Festuca* alloploid and its parental species. Biochem Physiol Pflanz 162:417–426
Pueppke SG (1979) Distribution of lectins in the jumbo virginia and spanish varieties of the peanut, *Arachis hypogaea* L. Plant Physiol 64:575–580
Rayle DL (1973) Auxin-induced hydrogen ion secretion in *Avena* coleoptiles and its implications. Planta 114:63–73
Reichlin M (1977) Use of antibody in the study of protein structure. In: Needleman SB (ed) Molecular biology, biochemistry and biophysics. Advanced methods in protein sequence determination, vol XXV. Springer, Berlin Heidelberg New York, pp 55–185
Rice HV, Briggs WR (1973a) Partial characterization of oat and rye phytochrome. Plant Physiol 51:927–938
Rice HV, Briggs WR (1973b) Immunochemistry of phytochrome. Plant Physiol 51:939–945
Richardson M (1977) The proteinase inhibitors of plants and micro-organisms. Phytochemistry 16:159–169
Roberts LM, Lord JM (1979) Developmental changes in the activity of messenger RNA isolated from germinating castor bean endosperm. Plant Physiol 64:630–634
Roland B, Pallotta D (1978) An immunological comparison of rye and calf histones. Can J Biochem 56:1021–1027

Roth J, Binder M (1978) Colloidal gold, ferritin and peroxidase as markers for electron microscopic double labeling lectin techniques. J Histochem Cytochem 26:163–169
Rüde E (1971) Antigens and immunogenicity. FEBS Lett 17:6–10
Ruoslathi E (ed) (1976) Immunoadsorbents in protein purification. Universitetsforlaget, Oslo
Ryan CA (1973) Proteolytic enzymes and their inhibitors in plants. Annu Rev Plant Physiol 24:173–196
Ryan CA (1974) Assay and properties of the proteinase inhibitor inducing factor, a wound hormone. Plant Physiol 54:328–332
Ryan CA (1978) Immunology of plant proteinase inhibitors. In: Catsimpoolas N (ed) Immunological aspects of foods. Avi Publishing Company Inc, Westport Connecticut, pp 182–198
Ryan CA, Santarius K (1976) Immunological similarities of proteinase inhibitors from potatoes. Plant Physiol 58:683–685
Ryan CA, Huisman OC, Van Denburgh RW (1968) Transitory aspects of a single protein in tissues of *Solanum tuberosum* and its coincidence with the establishment of new growth. Plant Physiol 43:589–596
Sabir MA, Sosulki FW, Finlayson AJ (1974) Chlorogenic acid. Protein interactions in sunflower. Agric Food Chem 22:575–578
Sano H, Spaeth E, Burton WG (1979) Messenger RNA of the large subunit of ribulose-1,5-biphosphate carboxylase from *Chlamydomonas reinhardi*. Eur J Biochem 93:173–180
Sasaki M, Kato T, Iida S (1973) Antigenic determinant common to four kinds of thiol proteases of plant origin. J Biochem (Tokyo) 74:635–637
Schlesier B, Manteuffel R, Scholz G (1978) Studies on seed globulins from legumes. VI. Association of vicilin from *Vivia faba* L. Biochem Physiol Pflanz 172:285–290
Scholz G, Richter J, Manteuffel R (1974) Studies on seed globulins from legumes. I. Separation and purification of legumin and vicilin from *Vicia faba* L. by zone precipitation. Biochem Physiol Pflanz 166:163–172
Schuster K, Dornhauser SC (1967a) Auftrennung und Spezifizierung der bei den technologischen Vorgängen der Bierbereitung auftretenden salzlöslichen Proteine von Gerste und Rohfrucht durch immunologische und physico-chemische Methoden. Brauwissenschaft 20:135–144
Schuster K, Dornhauser S (1967b) Auftrennung und Spezifizierung der bei den technologischen Vorgängen der Bierbereitung auftretenden salzlöslichen Proteine von Gerste und Rohfrucht durch immunologische und physico-chemische Methoden. Brauwissenschaft 20:209–214
Schuster K, Dornhauser S (1967c) Auftrennung und Spezifizierung der bei den technologischen Vorgängen der Bierbereitung auftretenden salzlöslichen Proteine von Gerste und Rohfrucht durch immunologische und physico-chemische Methoden. Brauwissenschaft 20:234–247
Sedgley M (1974) Assessment of serological techniques for S-allele identification in *Brassica oleracea*. Euphytica 23:543–551
Sela B-A, Lis H, Sharon N, Sachs L (1977) Isolectins from wax bean with differential agglutination of normal and transformed mammalian cells. Biochem Biophys Acta 310:273–277
Sela M (1969) Antigenicity: Some molecular aspects. Science 166:1365–1374
Sharon N (1977) Lectins. Am Sci 236:108–119
Sharon L, Lis H (1972) Cell-agglutinating and sugar-specific proteins. Science 177:949–959
Shepard DV, Moore KG (1978) Concanavalin A – mediated agglutination of plant plastids. Planta 138:35–39
Shunway LK, Yang VV, Ryan CA (1976) Evidence for the presence of proteinase inhibitor I in vacuolar protein bodies of plant cells. Planta 129:161–165
Singh J, Dieckert JW (1973) Isolation and partial characterization of arachin – P 6. Prep Biochem 3:53–72
Singer SJ, Nicolson GL (1972) The fluid mosaic model of the structure of the cell membranes. Science 175:720–731
Smith PM (1972) Serological and species relationships in annual bromes (*Bromus* L. sect *Bromus*). Ann Bot 36:1–30

So LL, Goldstein IJ (1972) Protein-carbohydrate interaction. IV. Application of the quantitative precipitin method to polysaccharide-concanavalin A interaction. J Biol Chem 242:1617–1622

Sprey B (1976) Intrathylakoidal occurrence of ribulose 1,5-diphosphate carboxylase in spinach chloroplasts. Z Pflanzenphysiol 78:85–89

Sprey B, Lambert C (1977) Lamellae-bound inclusions in isolated spinach chloroplasts. II. Identification and composition. Z Pflanzenphysiol 83:227–247

Stegemann H (1977) Plant proteins evaluated by two-dimensional methods. In: Graesslin D, Radola BJ (eds) Electrofocusing and isotachophoresis. Walter de Gruyter & Co, Berlin New York, pp 385–394

Stegemann H (1979) Indicator proteins in potato and maize for use in taxonomy and physiology. Gel-electrophoretic patterns. In: Müntz K (ed) Seed proteins of dicotyledonous plants. Proc Symp, Gatersleben 1977. Academie Verlag, Berlin, pp 217–224

Stegemann H, Frankensen H, Macko V (1973) Potato proteins: genetic and physiological changes evaluted by one- and two-dimensional PAA-gel-techniques. Z Naturforsch 28c:722–733

Strobel GA, HessWM (1974) Evidence for the presence of the toxin-binding protein on the plasma membrane of sugarcane cells. Proc Natl Acad Sci USA 71:1413–1417

Sugiyama T, Matsumoto C, Akazawa T (1970) Structure and function of chloroplast proteins. XI. Dissociation of spinach leaf ribulose-1,5-diphosphate carboxylase by urea. J Biochem (Tokyo) 68:821–831

Synge RLM (1975) Polyphenole in Pflanzen. Naturwiss Rundsch 28:204–208

Szabolcs M, Csorba S, Hauk M (1978) Eigenschaften und Antigenität der aus Brot isolierten Gluteneiweiße. Acta Paediatr Acad Sci Hung 19:125–135

Takabe T, Akazawa T (1973) Catalytic role of subunit A in ribulose diphosphate carboxylase from *Chromatium* strain D. Arch Biochem Biophys 157:303–308

Thanh VH, Shibasaki K (1977) β-conglycinin from soybean proteins. Isolation and immunological and physico-chemical properties of the monomeric form. Biochim Biophys Acta 490:370–384

Thomson JA, Schroeder HE, Dudman WF (1978) Cotyledonary storage proteins in *Pisum sativum*. I. Molecular heterogeneity. Aust J Plant Physiol 5:263–279

Thomson JA, Millerd A, Schroeder HE (1979) Genotype-dependent patterns of accumulation of seed storage proteins in *Pisum*. In: International atomic energy agency. Seed protein improvement in cereals and grain legumes, vol I. Vienna, pp 231–240

Toms GL, Western A (1971) Phytohemagglutinins. In: Harborne JB, Boulter D, Turner BL (eds) Chemotaxonomy of the legumes. Academic Press, London New York, pp 367–462

Toro-Goyco E, Rodriguez-Costas J (1976) Immunochemical studies on pinguinain, a sulfhydryl plant protease. Arch Biochem Biophys 175:359–366

Tronier B, Ory RL (1970) Association of bound β-amylase with protein bodies in barley. J Inst Brew 47:464–471

Tronier B, Ory RL, Djurtoft RJ (1974) Immunochemical identification of neutral peptide hydrolases in dormant and germinating barley grains. Int J Peptide Protein Res 6:13–19

Trop M, Grossman S, Veg Z (1974) The antigenicity of lipoxygenase from various plant sources. Ann Bot 38:783–794

Tsuda M (1979) Purification and characterization of a lectin from rice bran. J Biochem 86:1451–1461

Tucker WG (1969) Serotaxonomy of the *Solanaceae*: a preliminary survey. Ann Bot 33:1–23

Uhlenbruck G, Pardoe GI, Bird GWG (1969) On the specificity of lectins with a broad agglutination spectrum. 2. The nature of the T-antigen and the specific receptors for *Arachis hypogaea* lectin. Z Immunitaetsforsch Allerg Klin Immunol 138:423–433

Uriel J (1963) Characterization of enzymes in specific immune-precipitates. Ann New York Acad Sci 103:956–979

Verma DPS, Bal AK (1976) Intracellular site of synthesis and localization of leghemoglobin in root nodules. Proc Natl Acad Sci USA 73:3843–3847

Verma DPS, Ball S, Guérin C, Wanamaker L (1979) Leghemoglobin biosynthesis in soybean root nodules. Characterization of the nascent and released peptides and the relative rate of synthesis of the major leghemoglobins. Biochemistry 18:476–483

Wainwright JM, Ting JP (1976) Microbody malate dehydrogenase isoenzyme in cotyledons of *Cucumis sativus* L. during development. Plant Physiol 58:447–452
Walk R-A, Hock B (1976) Mitochondrial malate dehydrogenase of watermelon cotyledons: Time course and mode of enzyme activity changes during germination. Planta 129:27–32
Walk R-A, Hock B (1977) Glyoxysomal malate dehydrogenase of watermelon cotyledons: De novo synthesis on cytoplasmic ribosomes. Planta 134:277–285
Walk R-A, Hock B (1978) Cell-free synthesis of glyoxysomal malate dehydrogenase. Biochem Biophys Res Commun 81:636–643
Walker-Simmons M, Ryan CA (1977) Immunological identification of proteinase inhibitors I and II in isolated tomato leaf vacuoles. Plant Physiol 60:61–63
Weber E, Manteuffel R, Jakubek MF, Neumann D (1981) Comparative studies on protein bodies and storage proteins of *Pisum sativum* L. and *Vica faba* L. Biochem Physiol Pflanz 176:342–356
Weeke B, Løwenstein H (1975) Quantitative immunoelectrophoresis used in analysis of allergen extracts and diagnosis of allergy. Int Arch Allergy Appl Immunol 49:74–78
Weir DM (ed) (1967) Handbook of experimental immunology. Blackwell, Oxford
Wells G, Osborne TB (1911) The biological reactions of the vegetable proteins. I. Anaphylaxis. J Infect Dis 8:66–124
William CA, Chase MW (eds) (1967a)Methods of immunology, vol I. Preparation of antigens and antibodies. Academic Press, London New York
William CA, Chase MW (eds) (1967b) Methods of immunology, vol II. Physical and chemical methods. Academic Press, London New York
William CA, Chase MW (eds) (1973) Methods of immunology, vol III. Reaction of antibodies with soluble antigens. Academic Press, London New York
William CA, Chase MW (eds) (1976) Methods of immunology, vol IV. Agglutination, complement, neutralization and inhibition. Academic Press, London New York
William CA, Chase MW (eds) (1977) Methods of immunology, vol V. Antigen-antibody reactions in vivo. Academic Press, London New York
Williamson FA, Fowke LC, Constable FC, Gamborg OL (1976) Labelling of Concanavalin A sites on the plasma membrane of soybean protoplasts. Protoplasma 89:305–316
Woychik JH, Boundy JA, Dimler RJ (1961) Starch gel electrophoresis of wheat gluten proteins with concentrated urea. Arch Biochem Biophys 94:477–482
Wright STC (1960) Occurrence of an organ specific antigen associated with the microsome fraction of plant cells and its possible significance in the process of cellular differentiation. Nature (London) 185:82–85
Youle RJ, Huang AHC (1978a) Albumin storage proteins in the protein bodies of castor bean. Plant Physiol 61:13–16
Youle RJ, Huang AHC (1978b) Evidence that the castor bean allergens are the albumin storage proteins in the protein bodies of castor bean. Plant Physiol 61:1040–1042
Youle RC, Huang AHC (1979) Albumin storage protein and allergens in cotton seeds. J Agric Food Chem 27:500–503
Yoshida K (1978) Novel lectins in the endoplasmic reticulum of wheat germ and their possible role. Plant Cell Physiol 19:1301–1305
Zech L (1966) The effect of phytohemagglutinin on growth of some protozoa. Exp Cell Res 44:312–320
Ziegenfus TT, Clarkson RB (1971) A comparison of the soluble seed proteins of certain *Acer* species. Can J Bot 49:1951–1957
Zschoche WC, Ting IP (1973) Purification and properties of microbody malate dehydrogenase from *Spinacia oleracea* leaf tissue. Arch Biochem Biophys 159:767–776

# II. Nucleic Acids and Proteins in Relation to Specific Plant Physiological Processes

# 14 Seed Development

K. MÜNTZ

## 1 Introduction

In general, seeds are the product of a sexual reproduction process. In most cases the offspring do not develop continuously after seed formation but ontogenesis is retarded by a resting period due to the action of exogenous and endogenous factors. During the last period of seed formation, the maturation period, seeds acquire a structural and physiological state which combines a high degree of resistance to environmental damage with very low levels of metabolic activity. This is the basis of survival during a period when offspring are separated from the mother plant but have not yet rooted at another place. In general, seed development covers the ontogenic period from zygote formation until seed separation from the mother plant.

Ontogenesis of the young plant proceeds heterotrophically from zygote formation until an autotrophically living seedling has developed. The mother plant provides nourishment until the seeds have entered the resting state and have separated from the placenta. Subsequently, the seeds depend on storage compounds that have been accumulated in specialized tissue during seed formation, and which are also derived originally from the mother plant. Specialization of storage tissues, biosynthesis and accumulation of reserve compounds represent main events during seed development. In addition, prerequisites for the reactivation of reserves during germination and seedling development are also formed alongside the storage compounds.

Since at the onset of the resting state metabolic activity is drastically reduced by dessication, embryo and storage tissues are furnished with factors needed to reestablish physiological activity when germination starts. Formation of the embryo proper and of those factors necessary for reactivation of the metabolism represents another main event in seed development.

In view of the extensive literature on the subject, only selected papers published subsequently to 1957 will be reviewed in the present chapter. For earlier references Vol. VIII of the first edition of the Encyclopedia should be consulted at the review of MCKEE (1958).

# 2 Changes in Structure and Composition of Seeds During Development

## 2.1 Embryogenesis and Seed Growth

In general, seed development starts with the double fertilization characterizing the sexual propagation of spermatophytes, which leads to the formation of the triploid endosperm nucleus and the diploid embryonic nucleus. Whereas the seed coat develops from the integuments after fertilization, the endosperm nucleus gives rise to the endosperm and the egg cell forms the zygote from which embryogenesis proceeds.

Embryogenesis can be considered in three stages (DURE 1975, MÜNTZ 1978a, b). First, divisions of the zygote result in the formation of the suspensor and basal cells, as well as in the development of the proembryo, e.g., in *Pisum sativum* (COOPER 1938a, b), *Phaseolus vulgaris* L. (BROWN 1917), Cruciferae (NORTON et al. 1978) or cereals (WARDLAW 1955). The first externally visible event of proembryo differentiation is cotyledon formation, which in dicotyledons normally completes the proembryo stage and leads to a heart-shaped embryo, whereas in cereals the scutellum then develops from the single cotyledon. In dicotyledons that store their reserves in the cotyledons and digest the endosperm during seed maturation, e.g., crop legumes, crucifers, cotton (*Gossypium hirsutum*), pumpkin (*Cucurbita maxima*), stage 1 is now completed. Seeds that store their reserve in a persisting endosperm, like cereals and castor bean (*Ricinus communis*), exhibit high mitotic activity in this tissue up to the end of stage 1. Afterwards, nuclear and cell divisions stop in the endosperm and continue only in the embryo (EVERS 1970). This first embryonic stage is characterized by embryonic cells with high mitotic activity in the future storage tissue, whereas the size of the embryo increases only slightly (Fig. 1). At the end of embryonic stage 1, mitotic activity of the storage tissue ceases and cell divisions continue only in the embryo proper (LOEWENBERG 1955, BRIARTY et al. 1969, MILLERD and WHITFELD 1973, SMITH 1973, MANTEUFFEL et al. 1976).

The second stage of embryogenesis is characterized by rapid cell elongation growth. Whereas during the first stage the seed coat occupied the major part of the seed, and endosperm plus embryo only represented a small percentage of the total size and fresh weight of the seed, now the storage tissues become the predominant component. Fresh weight increase and size growth are linarly correlated in many seeds during stage 2 of embryogenesis. At the same time cell specialization occurs with the formation of starch grains, protein bodies, and lipid vacuoles, which are the sites of reserve deposition in the storage tissue. The embryo grows only slowly during this period and differentiates into radicule, plumule, and embryonic axis. Embryonic growth is due to both cell division and elongation and most of the cells remain meristematic.

Finally, during the third stage of embryogenesis, seed dessiccation takes place and storage tissue, as well as the embryo, enters the resting state. Cells of the storage tissues are densely filled with storage organelles. The water content declines to values lower than 15%.

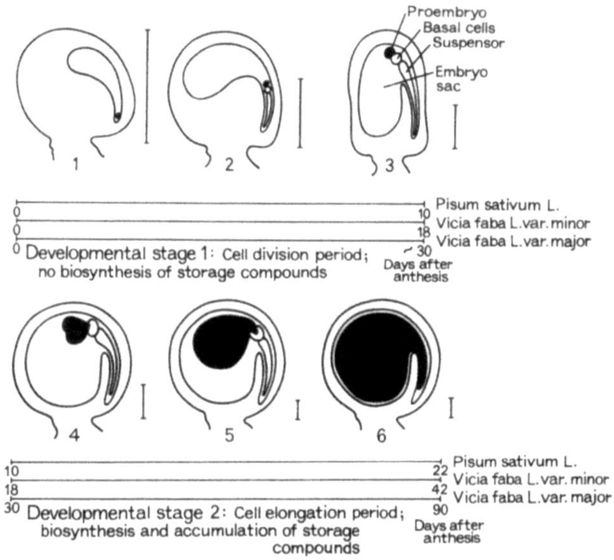


**Fig. 1.** Comparative timetable of legume seed development (drawings of developing pea seeds from MARINOS 1970); the *bars* indicate a natural seed length of 1 mm. Timing of developmental stages according to MILLERD et al. (1975) for *Pisum,* peas, MANTEUFFEL et al. (1976) for *Vicia faba* var. *minor* field beans, and BRIARTY et al. (1969) for *Vicia faba* var. *major* broad beans

Seed growth characters such as fresh weight, dry matter or cotyledon length follow a characteristic sigmoid growth curve (Figs. 2, 3). The slow growth at the beginning of seed development corresponds to stage 1 of embryogenesis. The subsequent logarithmic increase of the growth rate reflects the change from stage 1 to stage 2 followed by the approximately linear shape of the growth curve during stage 2. At the end of this period the growth rate declines and the dry weight remains approximately constant, which is characteristic of the beginning of stage 3 of embryogenesis (MCKEE et al. 1955a, CARR and SKENE 1961, JENNINGS and MORTON 1963a, INGLE et al. 1965, BOULTER and DAVIS 1968, MILLERD and WHITFELD 1973, MANTEUFFEL et al. 1976, BERGFELD et al. 1980). Frequently, a final decline in dry matter of the seeds has been recorded. The shape of the seed growth curve is mainly determined by the development of the storage tissue.

## 2.2 General Compositional Changes During Seed Development

### 2.2.1 DNA

Diploid embryonic cells and triploid endosperm cells contain species-specific amounts of DNA which in general do not change during stage 1 (Figs. 2, 3). Consequently, the increase of DNA per seed parallels the increase of cell number in embryo and endosperm. The DNA concentration remains more

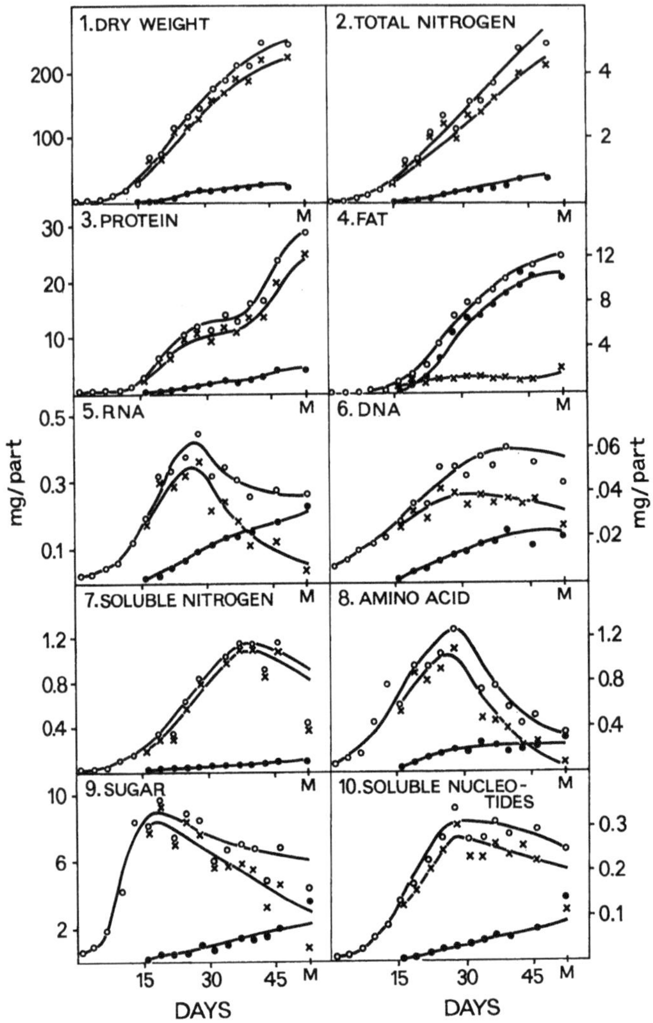


**Fig. 2.** Compositional changes in the whole kernel (○—○), endosperm (×—×), and embryo (●—●) of developing *Zea mays* grains (INGLE et al. 1965)

or less constant and meristematic cells generally exhibit constant volume ratios of cell nucleus to the cytoplasm. In many seeds, after cell divisions have ceased in the storage tissues, e.g., endosperm or cotyledons, the DNA content increases only a little due to cell division in the embryo proper up to the end of embryogenesis. Crop legumes, like peanuts (*Arachis hypogaea*), field beans, broad beans (*Vicia faba*) or peas (*Pisum sativum*), however, may exhibit continued DNA accumulation after the end of mitotic activity in the cotyledons.

## 2.2.2 RNA

The amount of RNA per cell also remains more or less constant as far as stage 1 is concerned (Figs. 2, 3) but it increases during the stage of rapid

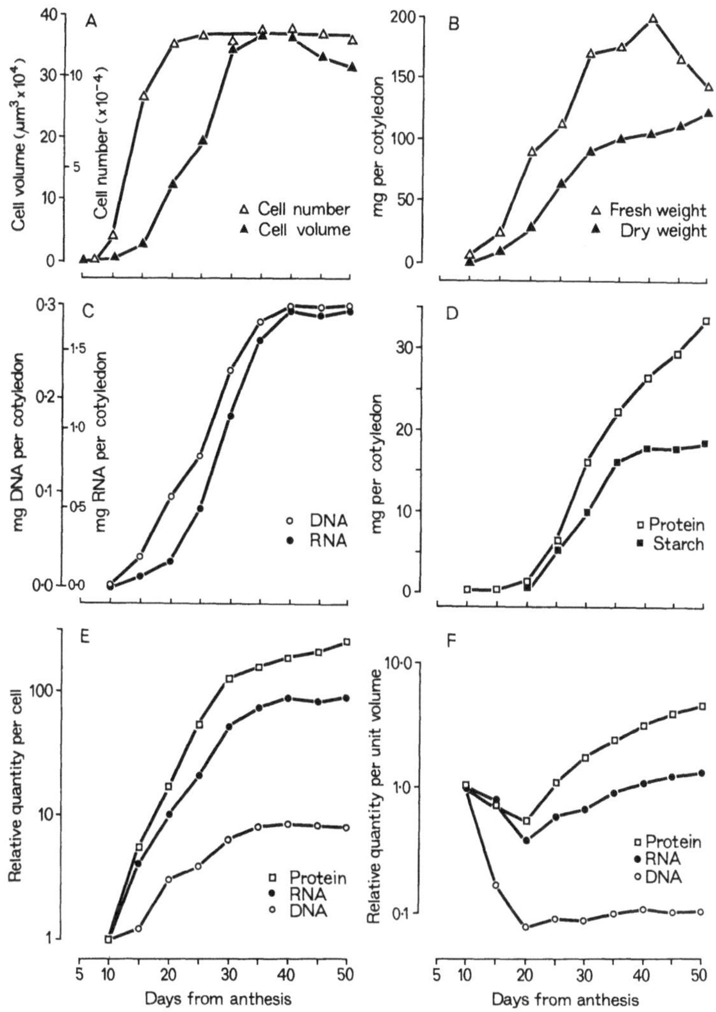


**Fig. 3A–F.** Changes of: **A** cell number and cell volume; **B** fresh weight and dry weight; **C** DNA and RNA; **D** protein and starch; **E** DNA, RNA, and protein per cell, and **F** per unit cell volume, in developing cotyledons of *Pisum arvense* L. (SMITH 1973)

cell elongation growth and accumulation of storage compounds (JENNINGS and MORTON 1963b, INGLE et al. 1965, MILLERD 1973, MILLERD and WHITFELD 1973, SCHARPÉ and VAN PARIJS 1973, SMITH 1973, WALBOT 1971, 1973, MANTEUFFEL et al. 1976, PÜCHEL et al. 1979). At first therefore the increase in RNA per seed is based on the increase in cell number during stage 1. Later on in cotyledons with continued DNA synthesis the RNA content also increases during stage 2, but it comes to a constant level much earlier than the end of this stage. In cereals, where the DNA content per seed remains constant from the end of stage 1, only a small increase of RNA has been registered in the seeds during stage 2. The RNA concentration per unit dry or fresh matter decreases during stage 2 since other components, particularly the storage compounds, accumulate at high rates, to become the predominant components of the seeds by the end of this period. In many dicotyledons the RNA content per seed remains constant during stage 3, but in cereals (JENNINGS and MORTON 1963b wheat (*Triticum aestivum*); INGLE et al. 1965 corn (*Zea mays*)) a decrease in the RNA content of the endosperm has been recorded, whereas the RNA quantity remains constant in the embryo.

### 2.2.3 Total Nitrogen

Kjeldahl analyses show a slow increase of total and insoluble nitrogen on a per-organ basis to be characteristic of stage 1. Nitrogen accumulation exhibits a sharp rise when stage 2 starts and in most seeds nitrogen accumulation proceeds linearly up to the end of this phase. During stage 3 the amount of total insoluble nitrogen remains constant on a per-seed basis or decreases slightly. Insoluble nitrogen has been taken to reflect the situation with proteins, though this fraction also includes nucleic acids, but only as a very low percentage.

### 2.2.4 Soluble Nitrogen

The quantity of soluble nitrogen remains approximately constant during grain development of wheat on a per-seed basis (JENNINGS and MORTON 1963a, c). The same has been found with legumes, e.g., *Vicia faba* L. var. *major* (BOULTER and DAVIS 1968), although INGLE et al. (1965) report an increase in soluble nitrogen of developing maize kernels in the endosperm as well as in the embryo, but the increase is much lower in the embryo than in the endosperm. Compositional changes in various nitrogen compounds of different maize kernel constituents are presented in Fig. 2.

The ratio of non-protein or soluble nitrogen to protein or insoluble nitrogen is frequently high during stage 1 before storage protein synthesis starts, but it declines during stage 2 parallel to the increasing storage protein accumulation (JENNINGS and MORTON 1963a, c, BOULTER and DAVIS 1968). In wheat non-protein nitrogen amounts for 45% of the total nitrogen at day 8 (stage 1) and comes down to only 15% at day 19 (beginning of stage 2), finally reaching 3% at maturity (JENNINGS and MORTON 1963a). Since the amount of protein per grain, per seed and per storage tissue increases during stage 2, but the quantity of soluble or non-protein nitrogen remains more or less constant or even decreases at this time, the changes in the ratio of soluble to unsoluble nitrogen are mainly based on the increase in protein.

Not only proteins are accumulated during stage 2, but also large amounts of starch or lipids, depending on the final composition of the seed investigated. These compositional changes induce sharp decreases in the concentration of soluble nitrogen compounds in the dry and fresh matter, of the corresponding tissue, grain or seed (MCKEE et al. 1955a, b, RAACKE 1957a, b, JENNINGS and MORTON 1963a, c, BOULTER and DAVIS 1968, DONOVAN 1979). The final concentration of soluble nitrogen amounts to less than 10% of the total nitrogen. In cases where a transient rise of the soluble nitrogen concentration has been observed in the seeds or storage tissue authors have attributed this to the transfer of non-protein nitrogen to the site of later storage protein synthesis (BOULTER and DAVIS 1968).

### 2.2.5 Protein Nitrogen

Protein nitrogen content initially rises slowly on a per-seed basis. This increase is based on the growing number of embryonic cells up to the pro-embryo stage in seeds which store their reserves in the cotyledon mesophyll, whereas in seeds which accumulate reserves in the endosperm it is based on the formation of the endosperm cells. At the time of change from stage 1 to stage 2 the rate of protein nitrogen accumulation per seed and per storage organ rises until it becomes constant during the main period of storage compound formation (Figs. 2, 3). The accumulation rate decreases and falls to zero when stage 2 finishes and the seeds enter stage 3. From this time until full maturity is reached the protein nitrogen content per seed remains constant or even decreases a little (LOEWENBERG 1955, MCKEE et al. 1955a, RAACKE 1957a, BILS and HOWELL 1963, JENNINGS and MORTON 1963a, INGLE et al. 1965, PROKOFIEV et al. 1967, BOULTER and DAVIS 1968, FLINN and PATE 1968, FINLAYSON and CHRIST 1971, NORTON and HARRIS 1975, MANTEUFFEL et al. 1976, Donovan 1979). Different patterns of change in the protein nitrogen concentration have been observed with different seeds depending on their composition. In some oil seeds like rape (*Brassica napus*), protein concentration is highest during the initial part of stage 2, since during stage 2 lipids are preferentially accumulated and the relative amount of protein declines continuously (NORTON and HARRIS 1975).

In grain legumes that exhibit a high final protein concentration in the mature seeds, protein nitrogen increases not only per seed but also on a dry weight basis (MCKEE et al. 1955a, RAACKE 1957a, BILS and HOWELL 1963, BOULTER and DAVIS 1968) or else the protein nitrogen concentration remains at least constant throughout the time of seed development as with *Phaseolus vulgaris* L. (WOLLGIEHN 1960). Initially during the period of reserve accumulation in wheat grains the protein concentration is somewhat diminished, but afterwards remains fairly constant throughout the whole grain-filling period (JENNINGS and MORTON 1963a), whereas barley (*Hordeum vulgare*) grains show a rise in the percentage of protein of the dry matter during grain development (IVANKO 1971).

## 2.3 Changes in the Cell Structure

The cell structures of the wheat embryo and of the embryo axis of peas have been compared during embryogenesis (SETTERFIELD et al. 1959). No significant differences were recorded in the subcellular structures of the embryos of these representatives of the monocotyledons and dicotyledons, respectively. The cells were embryonic or poorly specialized. Sometimes oil bodies and starch grains, as well as protein bodies, were found, indicating some storage material accumulation in the embryo. In protein bodies of the embryo of *Zea mays* L. the same pattern of zein polypeptides was observed as in protein bodies of the endosperm (TSAI 1979).

In general, three striking features are always observed in storage cells:

a) The development of large amounts of RER membranes, which disappear when the seeds enter stage 3;
b) a transient existence of large vacuoles, which become visible at the change from stage 1 to stage 2 but disappear during stage 2 of seed development.
c) the appearance of specialized storage organelles for carbohydrate, lipid, and protein deposition, respectively, the starch grains, lipid vacuoles, and protein bodies.

### 2.3.1 Endoplasmic Reticulum and Dictyosomes

The proliferation of the RER in the cells of developing wheat endosperm was first observed by BUTTROSE (1963a, b), who postulated the existence of a continuous reticular membrane system composed of the ER membranes in connection with the nuclear envelope, an observation which was later repeated by MARES et al. (1976). Highly developed RER systems (Fig. 4a) are also present in the endosperm cells of maize (KHOO and WOLF 1970) and barley (WETTSTEIN 1975) during the time of rapid storage compound synthesis. Analogous formation of RER has been demonstrated in the legume cotyledon cells of *Pisum sativum* L. (BAIN and MERCER 1966), *Phaseolus vulgaris* L. (ÖPIK 1968), *Vicia faba* L. (BRIARTY et al. 1969, NEUMANN and WEBER 1978), *Arachis hypogaea* L. (DIECKERT and DIECKERT 1972), and *Lupinus* (DAVEY and VAN STADEN 1978), as well as in other dicotyledonous species i.e., *Gossypium hirsutum* L. (YATSU 1965, ENGLEMAN 1966, DIECKERT and DIECKERT 1972), *Sinapis alba* L. (REST and VAUGHAN 1972, BERGFELD et al. 1980, SCHOPFER et al. 1980), *Brassica sp.* (NORTON et al. 1978), *Capsella bursa-pastoris* L. (DIECKERT and DIECKERT 1972) and *Ricinus communis* L. (SOBOLEV et al. 1968). The generalization seems to be justified that in monocotyledonous, as well as in dicotyledonous plants, the ER proliferation and especially the formation of RER occur concomitantly with the formation and deposition of storage compounds in the seed storage tissue. After seed maturation, cells of the storage tissue appear to be almost free of polysomes and RER in all cases. The high protein synthesis activity of the storage period can be specifically related to the ER and polysome proliferation.

There are always many dictyosomes (Fig. 4a) segregating electron-dense protein-containing vesicles present in the cytoplasm of storage cells during the period of highly developed RER and active storage compound formation (BUTTROSE 1963a, BAIN and Mercer 1966, BRIARTY et al. 1969, KHOO and WOLF 1970, DIECKERT and DIECKERT 1972, DAVEY and VAN STADEN 1978, NEUMANN and WEBER 1978, CRAIG et al. 1979a, BERGFELD et al. 1980). HARRIS (1978, 1979), using semi-thin sections of *Vicia faba* cotyledons, could distinguish two different types of ER, the cisternal ER (CER) with attached polysomes and the smooth tubular ER (TER), with the narrower tubules interconnected to the dictyosomes. Areas of the nuclear envelope adjacent to CER exhibit a reduced number of nuclear envelope pores, whereas TER lay adjacent to sites with a series of nuclear pores (HARRIS 1978).

Structural features of the cells of storage tissue have supported the idea that RER, as well as the dictyosomes, could participate in storage protein formation. HARRIS (1979) suggested that the CER could be the site of globulin synthesis in developing broad bean cotyledons, whereas the newly synthesized polypeptides could be glycosylated in the dictyosomes after passing along TER.

### 2.3.2 Vacuoles

The formation of vacuoles has frequently been observed in cells of developing storage tissues from different plants (BAIN and MERCER 1966, ENGLEMAN 1966, ÖPIK 1968, BRIARTY et al. 1969, KHOO and WOLF 1970, DAVEY and VAN STADEN 1978, CRAIG et al. 1979a, 1980a). In *Sinapis alba* L. a large central vacuole is formed from smaller ones during cotyledon development. Later this central vacuole subdivides, thus generating again small vacuolar organelles (REST and VAUGHAN 1972). Subdivision of large vacuoles has also been observed in *Phaseolus* cotyledon cells (ÖPIK 1968). At the beginning of storage protein deposition, lumps of proteinaceous materials have been found to appear attached to the tonoplast of these vacuoles (BAIN and MERCER 1966, ÖPIK 1968, BRIARTY et al. 1969, DIECKERT and DIECKERT 1972, NEUMANN and WEBER 1978). In *Vicia faba* L., var. *minor* (NEUMANN and WEBER 1978), *Gossypium hirsutum* L. and *Capsella bursa-pastoris* L. (DIECKERT and DIECKERT 1972) these lumps (Fig. 4b) have been shown to be proteins by pronase digestion in thin sections. In some plants the vacuoles cannot be traced through the whole period of storage protein formation in electron micrographs, but seem to disappear more or less at the time when the protein bodies are formed (BAIN and MERCER 1966, BRIARTY et al. 1969, NEUMANN and WEBER 1978, CRAIG et al. 1979a, 1980a). These findings led to the suggestion that protein bodies could be generated by the transformation of vacuoles (see Chap. 8, this Vol.). In *Sinapis* this type of protein body formation is most probable (BERGFELD et al. 1980), whereas in other cases, especially in legumes, this origin for protein bodies is less evident.

### 2.3.3 Protein Bodies

In mature seeds storage proteins are restricted to the protein bodies (see Fig. 4c, d from CRAIG et al. 1979b). Protein bodies develop parallel to the storage protein accumulation during stage 2. Three main functions have been attributed to these protein storaging organelles: (a) The deposition inside special membrane-bound organelles protects the respective proteins against enzymatic degradation during seed formation; (b) inside the protein bodies special environmental conditions may be maintained that are needed for packaging and storing of protein reserves. These two functions presume that during germination the respective conditions have to be reversed for storage protein reactivation. (c) Protein bodies represent a part of the lytic compartment of the storage tissue cells (MATILE 1968, 1975). First experimental evidence was published supporting the suggestion of Matile (WILDEN et al. 1980) with *Vigna radiata* L. Since

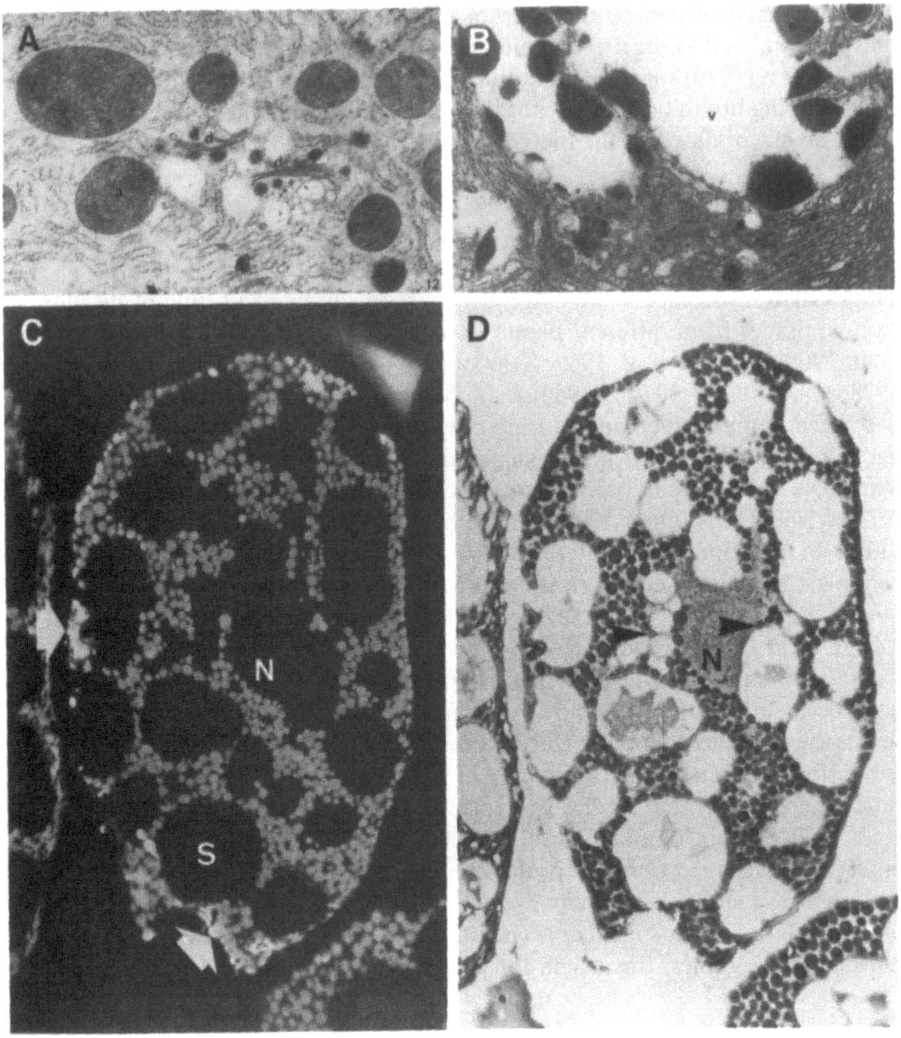


**Fig. 4A–D.** Structural characteristics of protein accumulating cells from the cotyledons of legume seeds. **A** dictyosomes, which segregate electron-dense vesicles (*d*), protein bodies (*a*), and abundant rER in cotyledon mesophyll cells from developing field beans (*Vicia faba* L. var. *minor*) 36 days after flowering; ×32,400; **B** vacuoles (*v*) containing protein lumps in cotyledon mesophyll cells from developing field beans 31 days after flowering; ×13,400 (**A** and **B** from NEUMANN and WEBER 1978); **C** and **D** light micrographs of cotyledon mesophyll cells from developing peas after fluorescent antibody-labelling (**C**) or staining (**D**) of the globulin containing protein bodies (*arrows*); *N* nucleus; *S* starch grains (CRAIG et al. 1979b)

problems of protein bodies were recently reviewed by PERNOLLET (1978) and WEBER and NEUMANN (1980) and a detailed discussion by MIÈGE was included in this volume (Chap. 8), the use of these papers is recommended for further information.

### 2.3.4 Stereological Analysis of Cell Structures During Storage Tissue Development

#### *2.3.4.1 Legumes*

Briarty (1973) introduced stereological methods for investigations of the quantitative changes in the subcellular structures during storage protein deposition in cotyledons of *Phaseolus vulgaris* L. Similar attempts have been made with *Pisum* cotyledons (Craig et al. 1979a) and the endosperm of *Triticum aestivum* L. (Briarty et al. 1979). In *Phaseolus* cotyledons protein deposition is initiated at the 40-mg fresh weight stage. Previous to the dramatic increase in the amount of RER, the surface-to-volume ratio of the ER rises sharply from the 20- to the 40-mg stage. This has been interpreted as ER flattening before RER amplification starts. The surface of the total vacuolar compartment remains constant throughout the period of cotyledon size increase from 20 to 120 mg fresh weight, although the surface-to-volume ratio of the vacuoles increases, since a few large vacuoles, which are present in the cell at the beginning, develop into numerous small sized vacuoles at the end of this period; there is a decrease in the cell volume percentage occupied by vacuoles from 57% at the beginning to 33% at the end. Sometimes structures were observed in electron micrographs which were interpreted as membrane continuities between ER and tonoplast. No similar findings were reported by Craig et al. (1979a) using pea cotyledons, although in general a similar pattern of subcellular structure development of RER and vacuoles was recorded. At day 8 of pea seed development, when storage protein formation starts, vacuoles occupy 75% of the total cell volume in storage parenchyma, with a total volume of 41,000 $\mu m^3$ and a mean vacuole diameter of 39 μm. The entire vacuole surface amounts to 5,500 $\mu m^2$. By day 20 the cells contain approximately 175,000 protein bodies per cell with a total volume of 91,500 $\mu m^3$, a total surface of 550,000 $\mu m^2$, and a mean protein body diameter of 1 μm. The total protein body volume of a parenchyma cell corresponds to 20% of the total cell volume. The doubling of the cell diameter that occurs from day 10 to 15 after flowering is accompanied by a decrease in the vacuole diameter of approximately 40 μm to approximately 2–3 μm. Protein bodies have been observed to fill completely between day 16 and 20 of pea cotyledon development.

#### *2.3.4.2 Cereals*

In wheat endosperm cell division stops at day 16–20 after anthesis and afterwards mainly cell enlargement follows (Briarty et al. 1979). Parallel with cell elongation growth, vacuolar protein deposits become visible in the wheat endosperm cells. During the same period the volume of RER per cell increases by the factor of approximately 5, accompanied by a corresponding rise in the RER surface per cell. This change in the amount of RER is preceded by a dramatic increase of the surface-to-volume ratio of the RER comparable to that observed with *Phaseolus* cotyledons (Briarty 1973), indicating a RER vesicle flattening. The total volume of the vacuole per cell remains fairly constant throughout the whole time of endosperm development, but the surface-to-volume ratio approximately doubles between day 6 and 22 after anthesis, indicating that larger vacuoles become replaced by smaller ones. Whether these are generated from the previous large vacuoles or are newly generated smaller vacuoles, while the larger ones degenerate remains unresolved. Although histogenesis of cereal endosperm differs widely from that of legume cotyledon mesophyll (e.g., cf. EVERS 1970 for wheat, with COOPER, 1938a, b, and REEVE, 1948 for peas) stereological analysis of cell development reveals very similar events.

# 3 Nucleic Acids

## 3.1 DNA Metabolism

In 1960 WOLLGIEHN published a paper on the relation between nucleic acids and protein metabolism in developing seeds of *Pisum sativum* L. and *Phaseolus vulgaris* L. The results were not related to developmental stages 1 and 2. However, the results can be related to these and WOLLGIEHN showed that DNA phosphorus continued to accumulate during stage 2 of *Pisum sativum* seed development, whereas in *Phaseolus vulgaris* the cell elongation period seems to be characterized by a constant amount of DNA phosphorus per seed. Similar investigations with *Vicia faba* L. var. *major* (WHEELER and BOULTER 1967) also revealed a continued increase in the DNA content per seed at a period when the dry weight of the seeds was increasing linearly, but the increased DNA was interpreted in terms of cell division. ÖPIK (1965), however, supposed that DNA multiplication occurred within the cell nucleus in *Phaseolus vulgaris* L. cotyledons since she observed an increase in nuclear size in the parenchyma cells. Seeds which store their reserves in the endosperm cells always contain at least triploid nuclei. However, in maize endosperm cells, DNA contents up to 24 C have been found (SWIFT 1950).

Calculations of the amount of DNA per cell based on estimations of cell numbers and DNA quantity per seed led to the conclusion that in developing cotyledons of *Pisum sativum* L. (SMITH 1973, SCHARPÉ and VAN PARIJS 1973, MILLERD and SPENCER 1974, SCHARPÉ and VAN PARIJS 1974, DAVIES and BREWSTER 1975, DAVIES 1976), of *Vicia faba* L. var. *major* (MILLERD 1973, MILLERD and WHITFELD 1973), *Vicia faba* L. var. *minor* (MANTEUFFEL et al. 1976) and *Arachis hypogaea* L. (ALDANA et al. 1972) DNA accumulation takes place in the nuclei of the storage tissue of the cotyledons after mitoses have ceased. In the absence of cell division radioactive thymidine is incorporated only into nuclear DNA at this time in the cotyledons of *Vicia faba* L. var. *major,* of *Trigonella foenum graecum* L. and *Pisum sativum* L. (MILLERD 1973). Cytophotometry, as well as chemical analysis of DNA quantities per nucleus, always results in whole-number multiples of the basic 2 C DNA amount per cell, suggesting that endoreduplication of the whole genome has occurred (SCHARPÉ and VAN PARIJS 1973, SMITH 1973, MILLERD and WHITFELD 1973, MILLERD and SPENCER 1974, MANTEUFFEL et al. 1976, DAVIES 1977). Reassociation kinetics of DNA from diploid cotyledon cells of *Vicia faba* L. var. *major* before cell division stops was compared with the kinetics of DNA from cells of stage 2 and did not reveal any difference, indicating, within the limits of this method, that no selective gene amplification had taken place, but rather duplication of the entire genome (MILLERD and WHITFELD 1973). These results are supported by comparisons of the buoyant density of nuclear DNA from cells of stage 1 and 2, both of which gave bands of density $1{,}696 \pm 0.001\ \mathrm{g\,cm^{-3}}$ and no satellite bands. Very high endoreduplication frequencies have been found in the suspensor cells of *Phaseolus vulgaris* L. (NAGL 1974), and *Phaseolus coccineus* L. (WALBOT et al. 1972, BRADY 1973). These cells arise from the zygote, as do the cells of the embryo including the cotyledons. The *Phaseolus* suspensor contains polytene chromosomes. Both polyploidy and polyteny occur at this stage in the cotyledon cells of *Pisum sativum* L. (MARKS and DAVIES 1979).

SCHARPÉ and VAN PARIJS (1973) and BROEKAERT and VAN PARIJS (1978) suggested that endoreduplication allows an increase of transcriptionally active templates for the proteins necessitated by the elevated rates of storage protein synthesis in *Pisum sativum* L. This interpretation was mainly based on the finding that increased RNA synthesis starts some days later than DNA accumulation, and protein synthesis rate rises some days after the rate of RNA synthesis had increased. MILLERD and SPENCER (1974), however, did not observe any difference in the starting time of the increased rates of DNA and RNA synthesis in developing cotyledons of *Pisum sativum* L. Taking the amount of the two nucleic acids in diploid cells of stage 1 as a basis, the quantities of DNA and RNA rose by the same factor until maturation. Identical results have been obtained with developing cotyledons of *Vicia faba* L. var. *major* (MILLERD 1973, MILLERD and WHITFELD 1973), and *Vicia faba* L. var. *minor* (MANTEUFFEL et al. 1976). Whereas the first doubling of nuclear DNA needed 3 days and was paralleled by a corresponding rise in the rate of RNA accumulation in cotyledons of *Vicia faba* L. var. *minor*, a second doubling in the amount of genomic DNA over a 10-day period did not lead to another increase in the rate of RNA synthesis (MANTEUFFEL et al. 1976). The accelerating synthesis effect of template multiplication would therefore be restricted to the first cycle of endoreplication, provided that it really does form the basis of increased RNA accumulation.

The DNA per unit cell volume is constant throughout the period of stage 2 with concomitant accumulation of storage compounds (SMITH 1973 and DAVIES 1976 with *Pisum sativum* L., DAVIES 1977 with *Vicia* species). The same ratio has been found in meristematic cells at the beginning of the cell division stage, a transient decrease being observed only towards the end of the cell division period when the rate of cell growth in relation to the cell division frequency becomes somewhat greater than at the beginning in developing *Pisum sativum* L. cotyledons (SMITH 1973). As a consequence of this constancy, cell size and DNA content per cell show a strong positive correlation (DAVIES 1976, 1977). Since seed size is positively correlated with cell number and also with cell volume, DNA content and size of the seeds are also positively correlated in peas and *Vicia faba* L. beans. Rates of DNA synthesis may be different depending on the final seed size (DAVIES 1976). In most of the *Pisum sativum* varieties investigated (CULLIS and DAVIES 1975) no differences were found in the percentage of rDNA per total DNA whether root tip cells with the basic 2 C amount of DNA, total seedlings or the highly endoreduplicated cotyledon cells were compared. However, one variety with an extremely low 2 C content per meristematic cell was found to show selective rDNA amplification in the endoreplicated cotyledon tissue, and in this way the mature seeds had the same rDNA percentage of the total DNA as was found in the cotyledons of the other varieties. This finding was discussed in favour of the assumption that a minimum quantity of rDNA must be present in the storage cells of legume cotyledons for the high translation activity needed during storage protein deposition. However, more comparative work is needed to verify this suggestion.

During seed development the embryo axis of legumes as well as the embryo of cereal caryopses contain cells with DNA at the 2 C or 4 C level. DNA polymerase has been extracted and fractionated from wheat germ of ungerminated caryopsis (CASTROVIEJO et al. 1979). Three different DNA polymerase fractions were separated chromatographically named A, B, and C, respectively. Fraction B exhibits characters similar to the mitochondrial DNA polymerase isolated from the same source, but fractions A and C were probably of nuclear origin.

### 3.2 Protein Genes and Their Inheritance in Seeds

Some storage protein genes have been localized by classical and molecular genetic techniques in wheat, maize, and barley genome. In *Zea mays* L. kernels, zein genes have been demonstrated on chromosomes 4, 7, and 10 (SOAVE et al. 1978, 1979) and genetic heterogeneity of the zein polypeptides (RIGHETTI et al. 1977, VIOTTI et al. 1978b, MELCHER and FRAIJ 1979, HANDA et al. 1979) is at least in part due to different structural genes. Some of the genes are inherited in linkage groups, e.g., genes Zp 1, Zp 2, Zp 3, and Zp 16, all coding for different polypeptides of the 19,000 zein, and numbered according to their banding pattern in isoelectric focussing gels, are localized in a linkage group on chromosome 7, where the opaque-2 gene is also situated (SOAVE et al. 1978, 1979). The zein genes exhibit monogenic intermediary inheritance (SOAVE et al. 1979). Experiments in which cDNA produced by reverse transcription of poly(A)$^+$-RNA corresponding to zein mRNA had been hybridized to excess of mRNA revealed at least 15 different zein mRNA's, each of which seems to exist in tenfold reiteration (VIOTTI et al. 1979). The opaque-2 and floury-2 mutants of maize, which are poor in prolamin, show deficiencies in the regulatory mechanism of prolamin formation but not in the zein genes themselves (JONES et al. 1977a, b, JONES 1978), and influence the zein of Mr 19,000 and zein of Mr 23,000 quantitatively to different degrees (SOAVE et al. 1979).

Hordein consists of at least three classes of polypeptide chains, named A, B, and C (KØIE et al. 1976), of which B plus C represent approximately 98% of the total hordein content. The corresponding genes of B and C are localized on chromosome 5 of the barley genome at separate loci (ORAM et al. 1975, SHEWRY et al. 1978, THOMSON and DOLL 1979). The polypeptides corresponding to these loci are codominantly inherited (ORAM et al. 1975). Different regulatory mutants have been found also with barley, e.g., Hiproly and Risø 1508 (MUNCK et al. 1969, DOLL et al. 1974), which have greatly decreased prolamin contents in the endosperm. The genes are localized on two different sites of chromosome 7 (MIFLIN and SHEWRY 1979, THOMSON and DOLL 1979).

The situation is much more complicated with the hexaploid wheats. The main gliadin groups are designated alpha, beta, gamma and omega (KASARDA et al. 1976, MECHAM et al. 1978). Up to 46 different gliadin polypeptides have been distinguished by two-dimensional gel electrophoresis (WRIGLEY and SHEPHERD 1973). The genes corresponding to 16 of the major and 14 of the minor polypeptides have been localized on chromosomes of the homologous groups 1 and 6 (SHEPHERD 1968, WRIGLEY and SHEPHERD 1973, THOMSON and DOLL 1979).

In comparison to cereals, no similar detailed results on the inheritance of storage protein genes of legumes are available. Monogenic determination and intermediary inheritance seem to govern the storage globulins of *Pisum sativum* L. (GOTTSCHALK and MÜLLER 1970, THOMSON and SCHROEDER 1978, THOMSON and DOLL 1979), and *Phaseolus vulgaris* L. (ROMERO et al. 1975).

Differential activation of maternal and paternal loci has been reported during seed development of *Pisum sativum* L. based on comparative analyses of globulin patterns (DAVIES 1973), but these results were not confirmed by other authors working with *Pisum sativum* L. (THOMSON and SCHROEDER 1978), *Phaseolus vulgaris* L. (ROMERO et al. 1975, HALL et al. 1977), and *Vicia faba* L., var. *minor* (MÜNTZ and SILHENGST 1979).

## 3.3 RNA Metabolism

### 3.3.1 RNA Polymerase Activity

RNA polymerases have preferentially been investigated from germinating seeds and developing seedlings (e.g., LIN et al. 1975, JENDRISAK 1980; for review see DUDA 1976). Comparatively high RNA polymerase activities have been measured in extracts from mature dry wheat germ (MAZUS and BUCHOWICZ 1972). Cell nuclei of dry wheat germs contain all three classes of DNA-dependent RNA polymerases, and after purification RNA polymerase II exhibits the typical subunit pattern for this eukaryotic enzyme on SDS gels (HAHN and SERVOS 1980). Biosynthesis of all three classes of RNA has been demonstrated in developing seeds of different plants species (MILLERD and WHITFELD 1973, PÜCHEL et al. 1979).

MILLERD and SPENCER (1974) isolated cell nuclei from developing *Pisum sativum* L. cotyledons at different stages before and after cell division had ceased (day 9 after anthesis) and measured their endogenous RNA-synthesizing activity; there was a continuous decline in the amount of RNA synthesized per μg of DNA from day 9 onwards. Addition of exogenous RNA polymerase from *Escherichia coli* to chromatin preparations from different developmental stages of the pea cotyledons led to constant values in the incorporation of radioactive precursors of RNA synthesis per unit of DNA quantity. These results were interpreted to indicate that template availability did not limit the transcription activity but that RNA polymerase activity is the constraint. The decrease of endogenous RNA synthesis activity per unit of DNA was a result of the rise in DNA amount per cell nucleus by endoreplication. Furthermore, a decrease of RNA synthesis per cell and per cotyledon was observed during stage 2. The RNA synthesis activity of the isolated cell nuclei was sensitive to actinomycin D and ethidium bromide, but α-amantinin only inhibited RNA synthesis by about 10%, i.e., activities mainly reflected that of RNA polymerase I to form rRNA. CULLIS (1976, 1978) extended these investigations to different pea cultivars and environmental conditions, measuring the chromatin-bound RNA polymerase activity which was taken to reflect the activity of RNA polymerase I. Only one out of the three cultivars he compared gave results similar to those obtained by MILLERD and SPENCER (1974), when seed development took place at 30 °C. With the two other cultivars, RNA polymerase activity per cell and per cotyledon increased after stage 2 had started and the DNA quantity per nucleus increased (Fig. 5). Later, during stage 2 a decreasing RNA polymerase activity was found. In accordance with MILLERD's and SPENCER's findings the RNA polymerase activity per unit of DNA showed a sharp decrease from the beginning of the DNA endoreplication period, but the increase in DNA was more pronounced than the decrease in the relative RNA polymerase activity on a DNA basis. This explained the rise of RNA polymerase activity per cell up to the stage when the cotyledons had already reached two thirds of their final fresh weight. The RNA polymerase activity was closely correlated with the rate of increase of RNA content per cell and the ratio of RNA per cell to fresh weight per cell was approximately constant in all three varieties throughout the development.

The transcription activity is influenced by the ambient temperature during seed development. Seeds of *Pisum sativum* growing at 30 °C develop much more rapidly than at 15 °C, but the latter reach a final fresh weight per cotyledon that is 50% higher. Nevertheless, cell numbers per cotyledon, DNA and RNA per unit fresh weight are the same. Whereas in the 30 °C seeds the RNA polymerase activity per μg DNA had its maximum at the time when the cell elongation period starts concomitantly with DNA endoreduplication and declines sharply afterwards, RNA polymerase activity in the 15 °C seeds increases up to the beginning of the cell elongation stage but subsequently remains at this elevated level for some time before it declines slowly (Fig. 6). In both types of seeds the maximum RNA polymerase activity could not be increased by the addition of homologous RNA polymerase I. In 30 °C seeds the RNA polymerase activity per unit DNA decreased during stage 2 and this decrease could be compensated for by the addition of exogenous RNA polymerase I. With 15 °C seeds this was not possible as long as the endogenous RNA polymerase level remained at its maximum value. These results may be interpreted to mean that RNA polymerase is the limiting factor in 30 °C seeds, whereas in 15 °C seeds template availability determined the transcription activity. An alternative explanation is that template availability was the same in both types of seeds, but differences in the specific activity of the bound RNA polymerase, developed during the preceding phase of cotyledon development, occur.

**Fig. 5A–J.** Changes of RNA polymerase activity in cotyledons of developing pea seeds (*Pisum sativum* L.) from three different cultivars (CULLIS 1976). Cell number (**A, B** and **C**); RNA polymerase activity per unit DNA (**D, E** and **F**); and RNA polymerase activity per cell (**G, H** and **J**) of the cultivars JI 813 (**A, D** and **G**), JI 430 (**B, E** and **H**), and JI 181 (**C, F** and **J**)

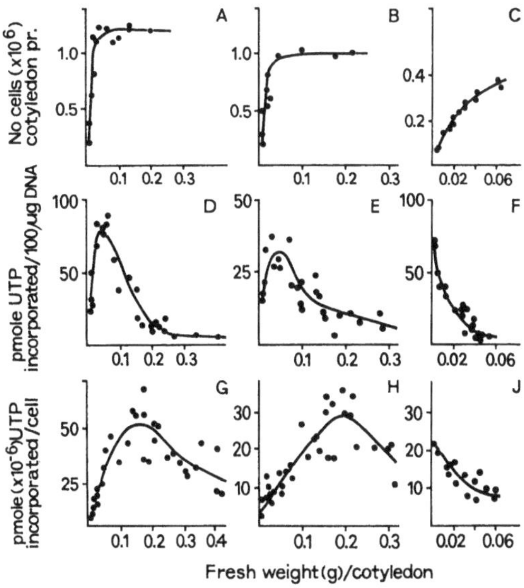


**Fig. 6A, B.** Influence of cultivation temperature on the RNA polymerase activity per unit DNA in cotyledons of developing peas (*Pisum sativum* L.); *closed symbols* endogenous; *open symbols* endogenous plus exogenous RNA polymerase activity (CULLIS 1978). **A** cotyledons at 30 °C; **B** cotyledons at 15 °C

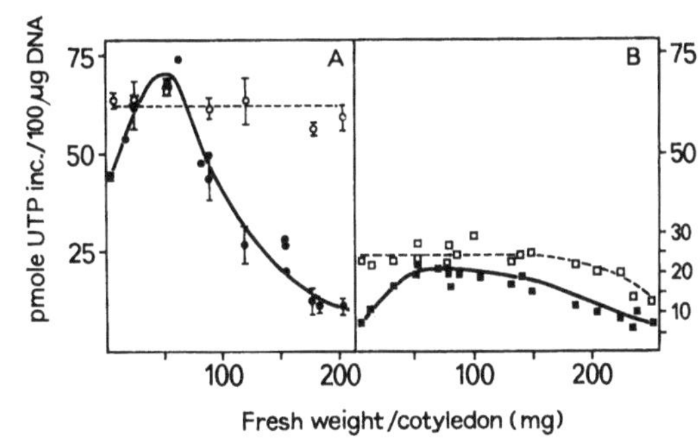


### 3.3.2 Metabolism of Different Types of RNA in Developing Seeds

Maximum RNA content of storage tissues on a per-cell as well as on a per-organ basis is reached at the beginning of or during stage 2, when maximum storage reserves are formed. Subsequently, the RNA content exhibits different patterns according to the species investigated. In many legumes the amount of RNA per cotyledon and per cell remains unchanged or only little decreased after reaching its maximum value (GALITZ and HOWELL 1965, VECHER and MATOSHKO 1965, WHEELER and BOULTER 1967, MILLERD 1973, MILLERD and WHITFELD 1973, POULSON and BEEVERS 1973, WALBOT 1971, 1973, MANTEUFFEL et al. 1976, PÜCHEL et al. 1979). Peanuts seem to represent an exception, since after coming to their maximum, the RNA content per cotyledon decreases and only at the end of seed development does the RNA content rise again to values approximately similar to the maximum. In the axis continuous increase of the RNA quantity has been observed (ALDANA et al. 1972). Cereals generally show a decrease in RNA content per endosperm after the maximum RNA content has been reached at the beginning of stage 2 (JENNINGS and MORTON 1963b, INGLE et al.

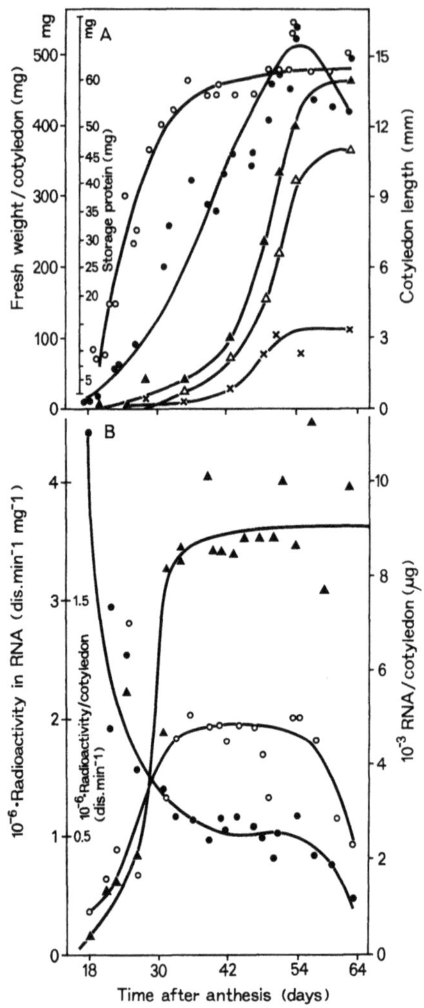


**Fig. 7A, B.** RNA synthesis in developing cotyledons of *Vicia faba* L. var. *minor* (PÜCHEL et al. 1979). **A** fresh weight increase (●) length of cotyledons (○) storage protein accumulation: globulins (▲), legumin (△), and vicilin (×). **B** Concentration of RNA per cotyledon (▲), incorporation of radioactive precursors of RNA into cotyledons (○), and specific labelling of RNA (●). In labelling experiments isolated cotyledons were aseptically incubated in nutrient solution containing [$^3$H]-labelled uridine at 25 °C for 3 h in the dark

1965, JOHARI et al. 1977), but opaque-2 mutant maize loses all RNA from the endosperm during late maturation stages (MEHTA et al. 1972). In the endosperm of mature castor beans only approximately 38% of the maximum RNA content of stage 2 were measured (STURANI and COCUCCI 1965).

Since continued storage protein synthesis has frequently been observed after the amount of RNA per storage cell remains unchanged or has even started to decrease, special efforts have been made to elucidate whether late protein biosynthesis is based on long-lived RNA or depends on continuous RNA synthesis which represents a part of RNA turnover.

WALBOT (1973) reported that in developing cotyledons of *Phaseolus vulgaris* L. the incorporation of $^{32}$P into RNA ceases when the constant level of RNA in the cotyledon has been reached. A similar incorporation pattern has been observed with embryonic axes of these beans (WALBOT 1971). During embryogenesis of *Pisum sativum* L. seeds the incorporation of [$^3$H]-cytidine into RNA still takes place after the RNA quantity per cotyledon reaches its constant level, but RNA labelling decreases and stops before the main period of storage protein formation is finished (POULSON and BEEVERS 1973). Both findings led to the suggestion that protein biosynthesis at late stages of cotyledon development is based on long-lived RNA. PÜCHEL et al. (1979) working with *Vicia faba* L. var. *minor* demonstrated that the incorporation of [$^3$H]-uridine into cotyledons of seeds of several harvests after constant RNA levels had been reached led to continued and constant rates of labelling of the organ and to a constant specific activity of cotyledonary RNA, thus indicating an equilibrium of RNA synthesis and degradation until seeds dessicate and protein formation stops (Fig. 7). In *Sorghum vulgare* $^{32}$P incorporation into RNA has been measured in parallel with the increasing amount of RNA per total grain. Unfortunately, the endosperm was not investigated separately, but RNA labelling paralleled the main period of storage protein accumulation (JOHARI et al. 1977). The maize embryo still shows an increase of the RNA content after maximum RNA amount per endosperm was reached and this increase continued up to the beginning of stage 3 (GALITZ and HOWELL 1965, INGLE et al. 1965, MEHTA et al. 1972).

Depending on the method used, rRNA has been reported to account at maximum for 68% in *Zea mays* L. (METHA et al. 1972), 85% in *Pisum sativum* L. (DAVIES and BREWSTER 1975), 84% in *Phaseolus vulgaris* L. (WALBOT 1973), and 85% in *Vicia faba* L. var. *minor* (PÜCHEL, unpublished). WALBOT (1973) claimed that approximately 20% of the total rRNA from *Phaseolus* cotyledons was of organellar origin, whereas POULSON and BEEVERS (1973) have calculated that the low molecular RNA fractions, comprising 5.8S, 5S, and 4S RNA, represent approximately 19% of the total cotyledonary RNA in developing pea cotyledons. The ratio of the different RNA fractions remains nearly constant through all stages of cotyledon development in *Pisum sativum* L. (POULSON and BEEVERS 1973, DAVIES and BREWSTER 1975). The same result was reported for *Phaseolus vulgaris,* although in the second half of embryogenesis the percentage of cytoplasmic rRNA decreases a little, whereas that of the organellar rRNA increased (WALBOT 1973). Poly(A)$^+$-RNA (mRNA) accounts for approximately 1%–2% of the total cellular RNA in storage tissue; if poly(A)$^+$-RNA with "long" poly(A) tails only is taken into account this value is approximately 1% in legume cotyledons as well as in barley endosperm (PÜCHEL et al. 1979 for *Vicia faba* L. var. *minor;* EVANS et al. 1979 for *Pisum sativum* L., calculation for *Phaseolus vulgaris* L. from RNA determinations of WALBOT 1973 and poly(A)$^+$-RNA measurements of HALL et al. 1980, BRANDT and INGVERSEN 1978 for *Hordeum vulgare* L.).

Although poly(A)$^+$-RNA from developing *Vicia faba* cotyledons only represents 1–2% of the total RNA and no significant change in this proportion has been observed during stage 2 of seed development, it was 13 times more strongly labelled with [$^3$H]-uridine during 3-h incorporation experiments using detached cotyledons in aseptic nutrient solution than was poly(A)$^-$-RNA. Preferential labelling of poly(A)$^+$-RNA indicates therefore a rapid turnover of this RNA fraction. The precursor incorporation into poly(A)$^+$-RNA can be 80%

inhibited by α-amanitin at concentrations which only slightly affect the labelling of poly(A)$^{-}$-RNA (PÜCHEL et al. 1979).

Continuous RNA degradation represents part of the RNA turnover which has been demonstrated in cotyledons of *Vicia faba* L. var. *minor* during developmental stage 2 (PÜCHEL et al. 1979). The decrease in RNA amounts per organ which has been observed with several storage tissues from different plants could be the result of RNA degradation. RNase activities have been measured in developing maize kernels and *Sorghum* grains (MEHTA et al. 1973, JOHARI et al. 1977). The endosperm of opaque-2 maize develops much higher RNase activity during the second half of the grain-filling period than does normal maize; opaque-2 maize endosperm loses nearly all RNA at this time, whereas normal endosperm still contains large amounts of RNA at maturity. In *Sorghum* grains RNase activity doubles during late maturation stages on a fresh weight basis, as well as in relation to mg protein. Surprisingly, no concomitant decrease in RNA content per seed was registered. Total RNase activity is much lower in *Sorghum* grains than in maize kernels, at the same developmental stages.

In agreement with the general nucleotide composition of plant RNA, the sum of G+C was higher than of A+U in rRNA of *Sorghum* grains (JOHARI et al. 1977), in total RNA from whole seeds, testa, and cotyledons plus embryo axis, of *Vicia faba* L. var. *major* (WHEELER and BOULTER 1967), and green beans of *Coffea arabica* (DE et al. 1972). In cotyledons of *Vicia faba* L., the purine to pyrimidine ratio was approximately 1 throughout seed development. In *Sorghum,* values above unity have been found in rRNA of grains at early developmental stages, whereas rRNA from grains of late maturation stages approximated 1.

### 3.3.3 Poly(A)$^{+}$-RNA from Developing Seeds

The discovery that most of the eukaryotic mRNA's bear a poly(A)-tail at their 3′-terminus (HADJIVASSILIOU and BRAWERMAN 1966, BRAWERMAN 1974, BRAWERMAN et al. 1972) led to the development of affinity chromatographic isolation methods based on the pairing of the poly(A)-sequences with poly(U) or oligo-dT bound to Sepharose and cellulose, respectively. Using these methods poly(A)$^{+}$-RNA has been isolated recently from several kinds of developing seeds started with total cellular RNA or with polysomal RNA.

Poly(A)$^{+}$-RNA which can direct in vitro biosynthesis of polypeptides in cell-free translation systems has already been isolated from cotyledons of seeds of *Glycine max* L. Merr. (CHANDRA and ABDUL-BAKI 1977, BEACHY et al. 1978, 1980, MORI et al. 1978), *Pisum sativum* L. (EVANS et al. 1979, CROY et al. 1980a, 1980b), *Phaseolus vulgaris* L. (HALL et al. 1978, 1980), *Vicia faba* L. var. *minor* (PÜCHEL et al. 1979), from *Zea mays* kernels (BURR et al. 1978, LARKINS and HURKMAN 1978, VIOTTI et al. 1978b, WIENAND and FEIX 1978, MELCHER 1979), endosperm of grains of *Hordeum vulgare* L. (BRANDT and INGVERSEN 1978, MIFLIN et al. 1980), and groats of grains of *Avena sativa* L. (LUTHE and PETERSON 1977).

In cotyledons of *Pisum sativum* L., *Vicia faba* L. and *Glycine max* L. Merr. the prominent fraction of poly(A)$^{+}$-RNA exhibits sedimentation constants between 18 and 20S (EVANS et al. 1979, PÜCHEL et al. 1979, BEACHY et al. 1980, CROY et al. 1980a, PÜCHEL et al. 1981, SCHMIDT et al. 1981). In *Phaseolus* cotyledons the main poly(A)$^+$-RNA is approximately 16S (HALL et al. 1978, 1980), in barley 11S, and in *Zea mays* 13S (BRANDT and INGVERSEN 1978, LARKINS and HURKMAN 1978). The length of the poly(A)-tail from cotyledons of *Vicia faba* L. var. *minor* is approximately 160 nucleotides (PÜCHEL et al. 1979), and $m^7$GTP inhibition of in vitro-translation of poly(A)$^+$-RNA from *Zea mays* L., *Phaseolus vulgaris* L., *Vicia faba* L. var. *minor, Glycine max* L. Merr. and *Hordeum vulgare* L. points towards the presence of cap structures at the 5′-terminus of these mRNA (BEACHY et al. 1978, BRANDT and INGVERSEN 1978, BURR et al. 1978, MASON and LARKINS 1979, NICHOLS 1979, TAKAGI and MORI 1979, MÜNTZ et al. 1981). Indications have been obtained by chemical modification that soyabean (*Glycine max*) poly(A)$^+$-RNA possesses the following 5′-terminal capped structure: $m^7G(5')ppp\text{-}(5')N$ and $m^7G(5')N^m$ (TAKAGI and MORI 1979).

Long-lived mRNA has been found in seeds (see reviews by PAYNE 1976, PEUMANS and CARLIER 1977, PEUMANS et al. 1979, 1981, DURE 1977, and also Chap. 15, this Vol.).

### 3.3.4 cDNA Complementary to Poly(A)$^+$-RNA from Seeds and the Molecular Cloning of DNA

Since cDNA preparation and cloning became a powerful tool for the analysis of the regulation of gene expression and gene structure, isolated and purified poly(A)$^+$-RNA from developing seeds actively synthesizing storage proteins has been used to obtain single and double-stranded cDNA (BURR 1979a, BURR and BURR 1980, VIOTTI et al. 1979, WIENAND et al. 1979, LARKINS et al. 1980, PEDERSON et al. 1981, all with *Zea mays* L., BRANDT 1979, INGVERSEN et al. 1981, with *Hordeum vulgare* L.; EVANS et al. 1980, with *Pisum sativum* L.; HALL et al. 1980, with *Phaseolus vulgaris* L.; MÜNTZ et al. 1981, PÜCHEL et al. 1981, with *Vicia faba* L. var. *minor;* FUKAZAWA et al. 1981, with *Glycine max* L. Merr.). Cloned cDNA fractions have been selected and proven to be complementary to zein mRNA (PEDERSON and LARKINS 1981), hordein mRNA (BRANDT 1979), and phaseolin (syn. G 1 protein, glycoprotein II) mRNA from kidney beans (*Phaseolus vulgaris*) (HALL et al. 1980). A gene coding for phaseolin, the major storage protein of *Phaseolus vulgaris* L., has been selected from a genomic DNA library of this plant. Comparison of the nucleotide sequence of part of this gene with that of the corresponding cloned cDNA revealed the presence of three intervening sequences (SUN et al. 1981).

### 3.3.5 tRNA and the Biosynthesis of Storage Proteins During Seed Development

The amino acid composition of cereal prolamin is characterized by extremely low percentages of lysine and tryptophan and very high proportions of glutamic acid (and amides), proline, and leucine. Several authors have investigated the total tRNA content, the tRNA composition and presence of isoaccepting tRNA's as well as the amino acyl-tRNA synthetase complement of developing cereal grains in the expectation that compositional or quantitative changes in the components of the amino acid-activating system would reflect a control of storage protein synthesis at the translational level. NORRIS et al. (1973, 1975) found the total synthetase activity in wheat endosperm increased up to week 5 after pollination followed by a rapid decline. A continuous decrease in total synthetase activity was found during caryopsis formation in testa plus pericarp, whereas in the developing embryo a steady increase was observed. A decreasing ratio of lysyl-tRNA synthetase to total synthetase activity has been found during endosperm development, but this was not specific for the lysine-activating enzyme, since several other amino acyl-tRNA synthetases showed the same pattern of activity change. In contrast, an increase in the proportion of prolyl, glutamyl, and glutamine activation was found, but other synthetases also showed similar activity changes. No indication has yet been found for translational controls on prolamin formation in wheat grain endosperm.

VIOTTI et al. (1978a) investigated the endosperm tRNA population in comparison with that of the embryo in maize kernels 30 days after anthesis, when active zein formation was taking place. The endosperm tRNA had a higher activity for glutamine, leucine, and alanine than the embryo tRNA. Similar differences between the two kernel tissues were not found for valine, tyrosine, and serine tRNA, which had been included as controls into these experiments. Higher proline and lower lysine-accepting activities which might have been expected in the endosperm compared with the embryo, were not found. The increase in activity for glutamine, leucine, and alanine was accompanied by changes in the distribution of isoaccepting tRNA species in the endosperm. Therefore, since glutamine, leucine, and alanine together represent 55% of the total amino acid content of zein and activation activities for these three amino acids show quantitative and qualitative changes correlated with prolamin synthesis, the authors interpreted their results to indicate a functional adaptation of the tRNA population to the changing pattern of protein synthesis during kernel development.

MERRICK and DURE (1973a, b) and DURE (1973) did not find any changes in the population of 15 different cytoplasmic tRNA during seed formation and germination of seeds of *Gossypium hirsutum* L.; in addition no differences could be found between the tRNA populations of seeds and roots as was to be expected if in seeds an adaptation of tRNA to the needs of storage protein had occurred. The question remains open whether or not storage protein synthesis is under translation control at the level of amino acid activation.

## 4 Ribosomes and Polysomes

Membrane-bound polysomes increase dramatically in amount at the onset of storage protein synthesis as determined both in the electron microscope and biochemically using isolated preparations which were fractionated on isokinetic sucrose gradients. By the end of stage 3 all polysome peaks disappear and

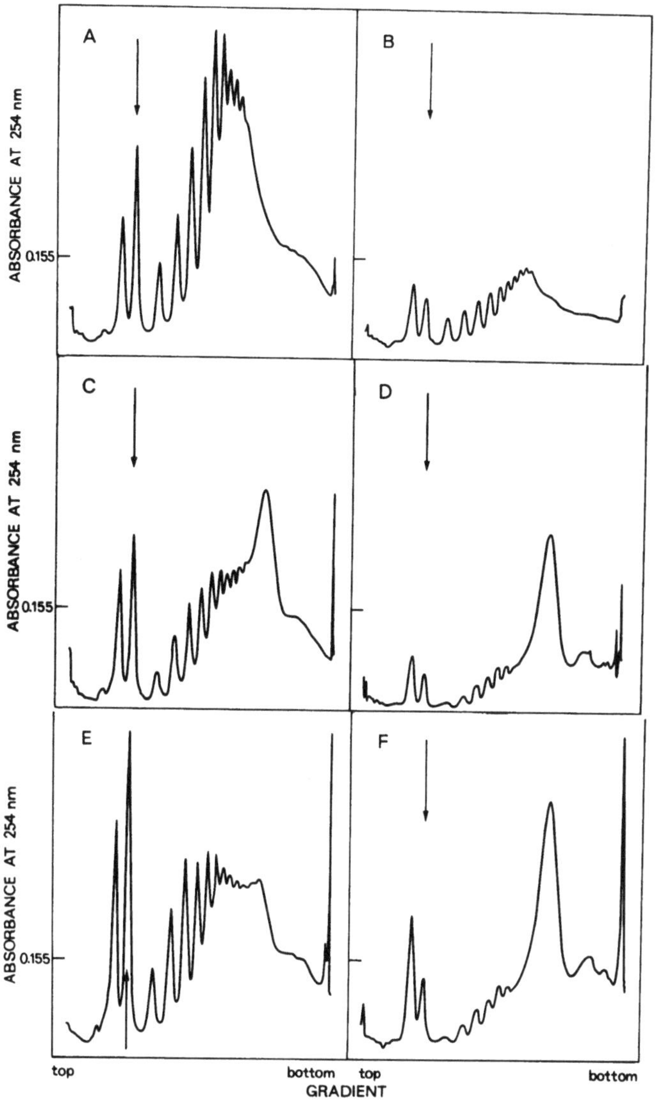


**Fig. 8 A–F.** Changing polysome profiles in developing field bean cotyledons (SCHMIDT et al. 1981). Polysomes have been isolated from cotyledons at day 33 (**A, B**), 38 (**C, D**), and 52 (**E, F**) after flowering and fractionated into free (**A, C** and **E**) and mb polysomes (**B, D** and **F**). The *arrow* indicates the monoribosome peak; the main mb polysome peak corresponds to 16–17 ribosomes per polysome; the ratios of mb/free polysomes are 0.46, 0.65 and 0.62 at the three different harvests

monosomes remain in the cotyledons (PAYNE et al. 1971 a, POULSON and BEEVERS 1973, SCHMIDT et al. 1981), whereas the total disappearance of ribosomes was reported in the endosperm of *Ricinus communis* (STURANI and COCUCCI 1965). The ratio of free and membrane-bound polysomes decreases as the amount of membrane-bound polysomes rises in developing cotyledons of *Vicia faba* L.

(PAYNE and BOULTER 1969, PÜCHEL et al. 1979, SCHMIDT et al. 1981). Precursor incorporation studies with [$^3$H]-uridine have shown a preferential labelling of membrane-bound polysomes at this stage, indicating de novo synthesis of rRNA and subunits of the membrane-bound polysomes (PAYNE and BOULTER 1969). The following ratios of free to membrane-bound polysomes were measured in different seeds at the stage of active synthesis of storage proteins: *Hordeum vulgare* L. 0.6 (BRANDT and INGVERSEN 1976), *Vicia faba* L. var. *major* 0.5–0.6 (PAYNE and BOULTER 1969), *Vicia faba* var. *minor* 0.87 (SCHMIDT et al. 1981), and *Phaseolus vulgaris* L. 0.66 (BOLLINI and CHRISPEELS 1979). Determinations of the length of membrane-bound polysomes were based on electron microscopy or gradient centrifugation and led to values of 16–17 ribosomes per polysome (Fig. 8) in *Vicia faba* L. var. *minor* (SCHMIDT et al. 1981), 14–15 in *Vicia faba* L. var *major* (BRIARTY et al. 1969), 7 in the endosperm of developing grains of *Hordeum vulgare* L. (BRANDT and INGVERSEN 1976), and 25 in the endosperm of kernels of *Zea mays* L. (LARKINS et al. 1976). These data indicate that polysomes contain mRNA sufficiently long to be a template for the corresponding storage protein polypeptides.

# 5 Protein Synthesis During Seed Development and Its Regulation

Numerous descriptive papers have been published on the time course and pattern of the accumulation of different seed proteins during embryogenesis, nearly all of which are on storage proteins. Recently the use of polysome or poly(A)$^+$-RNA-directed cell-free translation systems has given some information on the possible regulation mechanism.

## 5.1 Pattern of in Vivo Protein Formation

### 5.1.1 Albumins and Globulins from Cereals and Legumes

#### *5.1.1.1 Albumins*

Albumin synthesis predominates during early cotyledon development before storage protein synthesis starts. In most legume and cereal seeds the absolute amount of albumin per organ remains more or less constant during seed development, but its concentration decreases as the amount of storage compounds accumulated in the seeds increase. Albumins represent a complex mixture of several biologically active proteins, mainly enzymes, and these undergo protein turnover (at least in *Zea mays* L., SODEK and WILSON 1970) but so far it has not been demonstrated for storage proteins (MADISON et al. 1981). During stage 2 the percentage of albumins relative to the total amount of protein decreases (e.g., RAACKE 1957b, *Pisum sativum* L.; SODEK and WILSON 1970, *Zea mays* L.; PALMIANO et al. 1968, *Oryza sativa* L., IVANKO 1971, BRANDT 1975, *Hordeum vulgare* L.; DONOVAN et al. 1977, *Triticum aestivum* L.).

*5.1.1.2 The Salt-Soluble Fraction of Cereal Grain Proteins*

Cereal globulins are generally considered to represent a part of the metabolically active proteins, mainly enzyme proteins, although in oat grains large amounts of storage globulins are accumulated from day 4 to day 16 after anthesis. At day 12 after anthesis the oat (*Avena sativa*) grain globulins represent approximately 50% of the storage proteins consisting of two electrophoretically separated major components (PETERSON and SMITH 1976, LUTHE and PETERSON 1977). Frequently, the albumin and globulin fractions of cereal grains are extracted together as the so-called "salt-soluble" fraction, which represents mainly enzyme and other biologically active proteins, whereas prolamins and glutelins are taken together as storage proteins in the restricted sense. During grain development of cereals other than oats the ratio of salt-soluble proteins to storage proteins decreases continuously once storage protein synthesis has been initiated (DONOVAN et al. 1977, *Triticum aestivum* L.; EWART 1966, *Hordeum vulgare* L.; PALMIANO et al. 1968, *Oryza sativa* L.), although salt-soluble proteins remain constant on a per-grain basis (EWART 1966, *Hordeum vulgare* L.). Where the globulin fraction has been determined separately to the albumins, a slight increase in the percentage per total protein was registered during the initial phase of storage protein accumulation (Fig. 9) followed by a decrease (IVANKO 1971, *Hordeum*

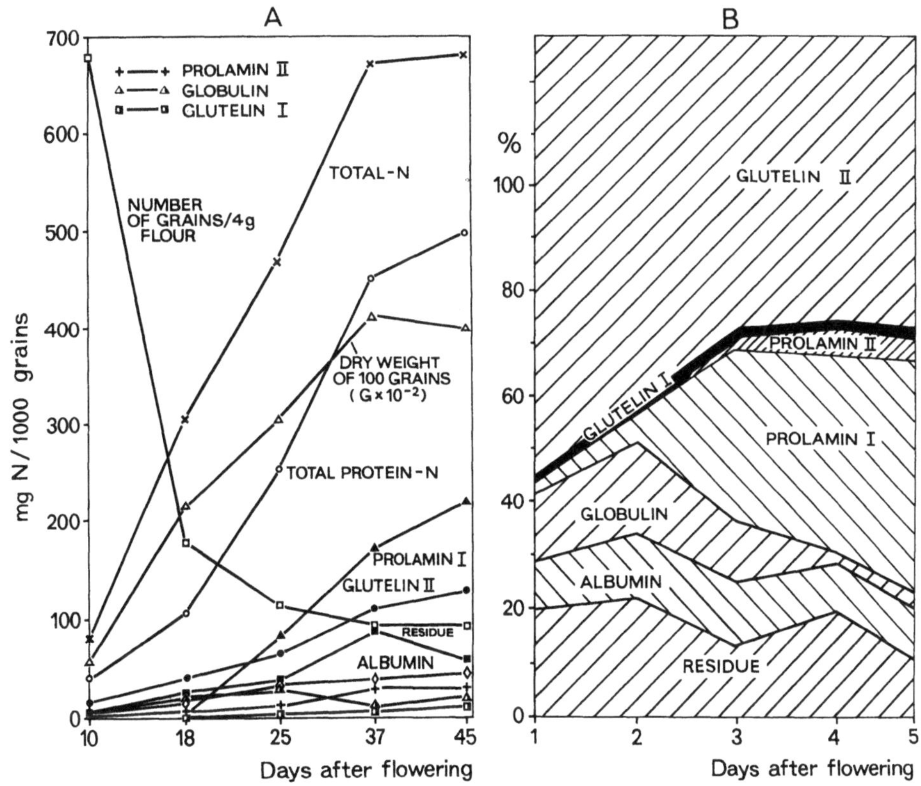


**Fig. 9 A, B.** Growth and accumulation of different protein fractions in developing barley grains (*Hordeum vulgare* L.) **A** on a per-grain basis; **B** expressed as percent of the total grain nitrogen (IVANKO 1971)

*vulgare* L.; IVANKO 1978, *Zea mays* L.), other authors, however, find a small but continuous increase with the same cereals (MEHTA et al. 1972, *Zea mays* L.; BRANDT 1975, *Hordeum vulgare* L.). In rice the salt-soluble fraction reached maximum in the grains at day 12 after anthesis and afterwards decreased (CAGAMPANG et al. 1976, TRAVIESO 1977).

### *5.1.1.3 Legume Globulins*

Legume cotyledons are characterized by the presence of two major classes of storage globulins, the 7S and the 11–13S globulins, typical representatives of which are the vicilin and legumin of *Pisum sativum* L. and *Vicia faba* L. (DERBYSHIRE et al. 1976, see Chap. 8, this Vol.). Biosynthesis of both types of globulins appears to start after cell divisions have stopped in storage parenchyma of the cotyledons (see Sect. 2.1 this Chap.). In some cases 11–13S globulin synthesis has apparently been shown to start later than the formation of the 7S globulins (*Pisum sativum* L.: MILLERD et al. 1975, MILLERD et al. 1978; *Vicia faba* L.: WRIGHT and BOULTER 1972; *Glycine max* L. Merr.: OCHIAI-YANAGI et al. 1978; *Vigna unguiculata* L. Walp.: CARASCO et al. 1978: *Phaseolus vulgaris* L.: SUN et al. 1978). In developing soyabeans the appearance of the 7S and 11–13S globulins is preceded by the accumulation of 2.2S protein belonging to the globulin fraction (HILL and BREIDENBACH 1974, OCHIAI-YANAGI et al. 1978). Working with field beans (MANTEUFFEL et al. 1976, WEBER et al. 1981) and with soyabeans (HILL and BREIDENBACH 1974), the presence of 11–13S proteins was already detected at the time when biosynthesis and accumulation of the 7S globulin commences. In developing soyabeans, analysis of the timing of globulin synthesis showed at the onset of the synthesis already both classes of storage globulins when ultracentrifugation was used (HILL and BREIDENBACH 1974, OCHIAI-YANAGI et al. 1978), whereas immunological investigations led to the conclusion that 11–13S globulin formation started later than 7S globulin synthesis (OCHIAI-YANAGI et al. 1978). This difference was explained by suggesting that changes in the immunological reactivity of the two major globulins took place during development. Quantitative immunological determination of *Vicia faba* L. legumin indicated a late accumulation during cotyledon development, but with small amounts present at very early stages of development (MILLERD et al. 1971). These results agree with the finding that in vivo incorporation of labelled amino acids into immunoprecipitated legumin of field beans always occurred parallel with the beginning of precursor incorporation into vicilin, the 7S globulin (WEBER et al. 1981). Thus, the type and sensitivity of the different methods used strongly influence the results obtained. Though no final decision can be taken as to whether a sequential or simultaneous start to 7S and 11–13S globulin formation takes place, there seems to be no doubt that the 7S proteins are at first synthesized with a higher rate than the 11–13S globulin (Fig. 10), but that later during stage 2 an increased rate of legumin synthesis leads in many cases to the final prevalence of 11–13S globulin (DANIELSSON 1952, RAACKE 1957b, WRIGHT and BOULTER 1972, MILLERD et al. 1975, MANTEUFFEL et al. 1976, WEBER et al. 1981). Since a wide range of final ratios of 7S to 11–13S globulins is found in mature legume seeds, different ratios

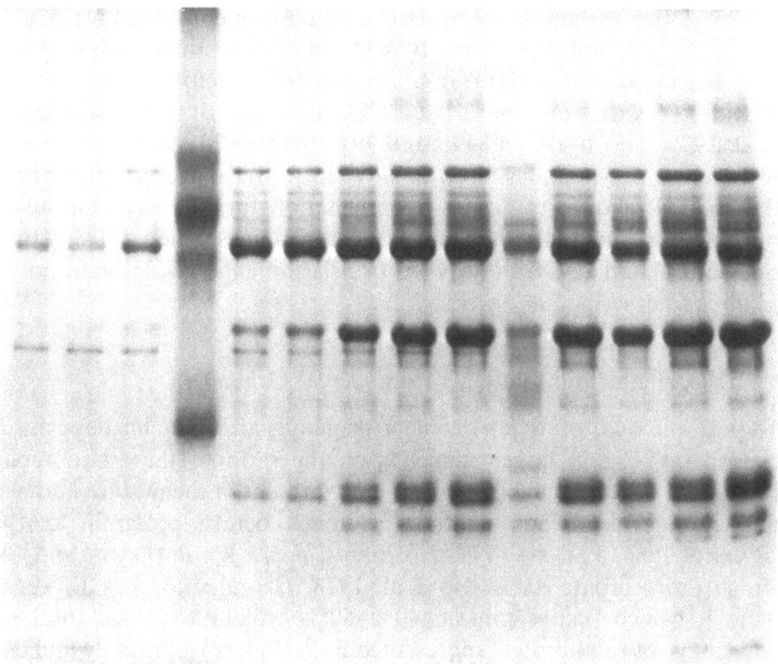


**Fig. 10.** Gel electrophoresis of globulin extracts from developing field bean cotyledons (*Vicia faba* L. var. *minor*) under denaturing conditions. Time scale is given below the electropherogram to characterize the stages of seed development; the *fourth* and the *tenth lanes* represent molecular weight standards and globulins from mature seeds

of the corresponding rates of formation and accumulation, respectively, are found, e.g., *Vicia faba* L. with a final ratio of vicilin to legumin of 1:4, which during the late developmental period 2 nearly exclusively synthesizes legumin (WRIGHT and BOULTER 1972, WEBER et al. 1981), and *Phaseolus vulgaris* L. with the 7S globulin (synonymously called phaseolin or G1 protein by the group of Hall (see SUN et al. 1974, SUN and HALL 1975) and glycoprotein II by PUSZTAI and WATT 1970) as the predominant storage protein (DERBYSHIRE and BOULTER 1976), where the biosynthesis of this protein seems to prevail throughout the whole developmental stage 2 (SUN et al. 1978). Genetically fixed differences in the accumulation pattern of different components of the globulin fraction have been observed in three pea cultivars that differed greatly in the vicilin to legumin ratio in mature seeds (MILLERD et al. 1978, THOMPSON and DOLL 1979). Although the 11–13S storage proteins exhibit micro-heterogeneity (e.g., NIELSEN et al. 1981 with *Glycine max* L. Merr.) no changes in the ratio of the acidic to the basic subunit have so far been reported during seed development (e.g., HILL and BREIDENBACH 1974, *Glycine max* L. Merr.; WRIGHT and BOULTER 1972, *Vicia faba* L.; MILLERD et al. 1978, *Pisum sativum* L.). In contrast, the 7S globulin fraction appears to be inhomogeneous in addition to the heterogeneity of individual proteins. It is not surprising that quantitative differences have been observed in the formation and accumulation of its constituent polypeptides (WRIGHT and BOULTER 1972, HILL and BREIDENBACH 1974, MILLERD et al. 1978, THOMPSON and DOLL 1979). Pattern differences in the accumulation of different polypeptides have already led to the suggestion that they are under the control of different genes. Confirmation for this assumption comes from breeding experiments with legumes and cereal lines which differ in their polypeptide pattern and also from in vitro translation experiments with poly(A)$^+$-RNA as already described in Section 3.2.

### 5.1.2 Prolamins and Glutelins from Cereal Grains

Though the quantitative pattern of prolamin and glutelin deposition in cereal endosperm (Fig. 9) largely depends on the ratio of these two typical storage proteins to other protein fractions finally present in the mature kernels, generally the endosperm already contains glutelins before prolamin synthesis starts (EWARTS 1966, IVANKO 1971, *Hordeum vulgare* L.; PETERSON and SMITH 1976, *Avena sativa* groat; PALMIANO et al. 1978, *Oryza sativa* L.). In kernels of *Zea mays* L. which finally contain 40%–50% prolamins in the total protein, the percentage of total protein represented by zein increases rapidly and continuously once its synthesis has been started. Although the amount of glutelin increases alongside zein, its percentage of the total grain protein decreases with time, since the accumulation rate of the prolamins is higher than that of the glutelins (IVANKO 1978). Normal maize kernels start the biosynthesis of zein between day 12 and 16 after anthesis, and at day 28 approximately 50% of the final quantity of zein has already been accumulated (JONES et al. 1977b). Barley grains exhibit patterns of storage protein accumulation similar to maize (IVANKO 1971). Since in rice grains glutelins represent approximately 80% of the total protein and prolamins only at maximum approximately 5%, glutelin synthesis and accumulation predominates throughout the whole period of storage protein formation (PALMIANO et al. 1968, JULIANO 1972, CAGAMPANG et al. 1976, TRAVIESO 1977). In oat groats the percentage of glutelins in the total proteins continuously decreases from day 4 after anthesis, whereas the percentage of prolamins increases; the glutelin content per gram dry weight remains more or less constant, but that of prolamins increases during the grain-filling period (PETERSON and SMITH 1976).

Extensive work concerning compositional changes during wheat grain development has been published (GRAHAM and MORTON 1963, GRAHAM et al. 1963, JENNINGS and MORTON 1963a), but fractionation procedures different to those of the Osborne scheme have been applied and therefore the results are difficult to compare with those reported for other cereals. The Na-pyrophosphate-soluble protein fraction seems to correspond to the total salt-soluble fraction composed of the albumins and globulins, but results obtained with further fractionation using successive acetic acid and sodium hydroxide extractions cannot be compared with classical prolamin and glutelin proteins. The Na-pyrophosphate-soluble proteins increase up to day 25 to 30 days after anthesis and then remain constant on a per-grain basis, whereas acetic acid- and sodium hydroxide-soluble proteins, which is the storage protein fraction, continuously increase per grain up to day 45 after anthesis. Like all other cereals, different cultivars of wheat exhibit some differences in their pattern of protein accumulation, which is also modified by environmental influences.

The changes in the pattern of synthesis of different classes of storage proteins are reflected in the pattern of incorporation of labelled amino acid. Labelling experiments with [$^{14}$C]-lysine led to the preferential incorporation of this amino acid into albumins, globulins, and glutelins of developing barley endosperm. Only at late maturation stages does the percentage of label incorporated into glutelins decrease somewhat, whereas the increased prolamin synthesis is indicated by an increasing percentage of the label appearing in the hordein fraction. The low percentage of lysine incorporation into prolamins of the barley mutant Risø 1508 in comparsion with the barley line Bomi reflects the deficiency of this mutant in prolamin synthesis (BRANDT 1975). Similar experiments have been performed using normal and opaque-2 maize kernels (SODEK and WILSON 1970), and similar results have been obtained. In developing wheat grains the incorporation ratio of [$^{3}$H]-lysine to [$^{14}$C]-leucine decreased from day 8–28 after anthesis, reflecting the increasing synthesis of proteins poor in lysine. In addition GUPTA et al. (1976) assumed a decrease in the formation of lysine-rich proteins, which should furthermore undergo a preferential degradation by a special protease appearing at the respective developmental stage (GOSH and DAS 1980). FLINT et al. (1975) have incubated developing wheat grains from seven different harvests between day 5 and 26 after anthesis with [$^{14}$C]-labelled amino acids, and after isolation of the endosperm the protein extracts, which have been prepared with 8 M urea, 2% SDS and 5% 2-mercaptoethanol under denaturing conditions without previous fractionation, were electrophoretically analyzed by scanning of the stained bands and scintillation counting of solubilized gel slices. Two main labelling periods have been detected, the first up to day 5 after anthesis, when profound changes in labelling profiles took place, and the second starting with day 5 after anthesis, characterized by more stable labelling profiles. The techniques used in these experiments do not permit a comparison of the polypeptides with those of the Osborne classes of seed proteins, but the first 5 days should correspond to the period of preferential albumin plus globulin synthesis, whereas the later period should correspond to the period of storage protein formation.

### 5.2 In Vitro Biosynthesis of Seed Proteins – a Tool to Investigate Regulatory Processes of Storage Protein Formation

Polysomes, total polysomal RNA, RNA fractions from methylated albumin absorbed on kieselguhr column (MAK) chromatography and recently poly(A)$^{+}$-RNA isolated from storage tissues at the most active stages of protein formation have been used to direct in vitro biosynthesis of seed proteins in cell-free reticulocyte and wheat germ systems. The large capacity of the ribosomal machinery from embryonic tissue of dry seeds to perform active translation processes has been used to develop several in vitro translation systems using cell-free extracts of wheat germ (MARCUS et al. 1974), of rye (*Secale cereale*) germ (CARLIER and PEUMANS 1976, PEUMANS et al. 1980a), of dry pea primary axis (PEUMANS et al. 1980b), and of dry and germinated mung bean (*Phaseolus aureus*) primary axis (CARLIER et al. 1978). All these systems have been shown to translate exogenous mRNA more or less efficiently from various sources including viral RNA. The cell-free wheat germ system is now one of the standard in vitro translation systems used in protein biosynthesis research generally. WELLS and BEEVERS (1973, 1974, 1975) and the group of Boulter (YARWOOD et al. 1971a, b, PAYNE et al. 1971b, LIDDELL and BOULTER 1974) have investigated in detail the processes of initiation, elongation and amino acid activation in cell-free translation systems derived from cotyledons of pea, broad bean, and mung bean, respectively, using viral RNA and synthetic messengers, e.g., poly(U).

The early in vitro translation experiments suffered from two main constraints, (1) the lack of purified mRNA fractions which only became available from plants after the introduction of poly(A)-affinity chromatography, and (2) from the lack of reliable specific methods for identification of in vitro translation products (*Glycine max* L. Merr.: MORI and MATSUSHITA 1970; *Arachis hypogaea* L.: JACHYMCZYK and CHERRY 1968; *Zea mays* L.: RABSON et al. 1961, MEHTA et al. 1973; *Triticum aestivum* L.: SCHULTZ et al. 1972; *Vicia faba* L.: PAYNE et al. 1971b). The specific solubility of cereal prolamins in 50%–70% ethanol, n- or iso-propanol, and the characteristic ratio of leucine to lysine in these storage proteins offered the first tool for more specific product identification after in vitro translation of cereal polysomes, the isolation of which was very much improved by the methods developed by LARKINS and DAVIES (1975).

### 5.2.1 Cereal Prolamins

#### *5.2.1.1 Zein of Maize*

In 1975 LARKINS and DALBY demonstrated preferential in vitro biosynthesis of zein-like polypeptides in the cell-free wheat germ system directed by mb polysomes in comparison to free polysomes isolated from the endosperm of *Zea mays* L. 23 days after pollination (see also LARKINS et al. 1976). Using similar techniques JONES et al. (1976) incubated free and mb polysome preparations, respectively, from *Zea mays* L. W 22 line and from the mutant opaque-2 W $22_{02}$ with the cell-free wheat germ system. Previous analysis of the polysome profiles had shown an identical pattern of free polysomes from both types of maize, but showed that the mutant lacked mb polysomes with more than 8 ribosomes. The quantity of mb polysomes of opaque-2 kernels was approximately 27% lower than in line W 22, and these polysomes directed a 30% decreased synthesis of zein-1 polypeptides in the cell-free translation system. The reduction in zein-1 formation which can be calculated from the lower in vitro synthesis activity and from the decreased amount of heavy mb polysomes amounts to 37%–40% compared with the normal maize kernels and corresponds well with the in vivo situation (JONES et al. 1977a, b). Consequently, reduced prolamin synthesis represents the basis of increased lysine concentration in the mutant maize endosperm. MEHTA et al. (1972) observed a total disappearance of RNA accompanied by a large rise in RNase activity of later stages of opaque-2 maize kernel development in comparison to the normal line of *Zea mays* L. Thus premature RNA degradation may also contribute to the low prolamin quantity in mutant maize endosperm (MEHTA et al. 1973). This interpretation has been supported by the experiments of JONES et al. (1976). Corresponding investigations with floury-2 maize mutants also revealed decreased amounts of mb polysomes in comparison to normal maize endosperm, but differences in the polysome profiles similar to those of opaque-2 were not observed. The presence of increasing numbers of floury-2 alleles in the endosperm decreased the recovery of mb polysomes in a stepwise fashion and this progressive reduction in mb polysomes correlated linearly with the decrease of in vitro biosynthesis of zein and in vivo zein accumulation. Zein-1 and zein-2 polypeptides were proportionally diminished, disproportional changes were found with the minor zein bands (JONES 1978).

Based on the observation of KHOO and WOLF (1970), that in the endosperm of *Zea mays* L. polysomes are attached to the exterior protein body membranes, BURR and BURR (1976) used polysomes detached from isolated maize protein bodies as a source of nearly pure zein mRNA (BURR et al. 1978, BURR 1979a, b). After further purification by sucrose gradient centrifugation zein mRNA was characterized as consisting of 1,230 nucleotides including a poly(A)-tail and the 5′-terminal cap structure. In vitro translation products were 2,000 and 1,100 larger in molecular weight on SDS polyacrylamide gels than the 22,500 und 19,000 molecular weight subunits, respectively, of zein preparations from mature endosperm (Fig. 11). The additional part of the polypeptides is probably a N-terminal extension in accordance with the signal hypothesis of Blobel (BLOBEL and DOBBERSTEIN 1975a, b, BLOBEL 1977, BLOBEL et al. 1979). Zein-1 and zein-2 were translated from separate monocistronic mRNA (BURR et al. 1978, WIENAND and FEIX 1978), which can be fractionated by gel electrophoresis. Only 50% of the nucleotides appear to code for the polypeptide chains including the signal peptide. The growing zein polypeptides are vectorially segregated from the mb polysomes into microsomal vesicles in vitro, and the processing of the precursor zein polypeptides takes place inside these vesicles, including selective cleavage of the signal peptide (LARKINS and HURKMAN 1978) and glycosylation (BURR 1979b). Electronmicrographs of cells from developing maize endosperm showed continuities between RER and protein body membranes. After cell fractionation RER and protein body membranes showed identical marker enzyme characteristics and both types of membranes directed zein synthesis in cell-free translation experiments (LARKINS and HURKMAN 1978). After injection into *Xenopus* oocytes zein mRNA directed the formation of processed mature zein polypeptides which were deposited in protein body-like membrane bounded granules (HURKMAN et al. 1979, 1981, LARKINS et al. 1979). Protein bodies of maize endosperm arise directly from the RER by a process leading to RER vesicles with polysomes that are only specialized to zein formation attached to the membrane (LARKINS et al. 1980, PEDERSEN and LARKINS 1981).

In vitro biosynthesis of a methionine-rich polypeptide belonging to the zein fraction which was translated from mRNA isolated from protein body membranes has been reported (MELCHER 1979). Storage globulin polypeptides have also been synthesized in cell-free translation systems from *Avena sativa* L. (LUTHE and PETERSON 1977).

### 5.2.1.2 Barley Hordein

Similar attempts to those described above have been made to investigate the hordein formation in developing grains of *Hordeum vulgare* L. The in vitro biosynthesis of hordein peptide groups by mb polysomes has been demonstrated by BRANDT and INGVERSEN (1976). Poly(A)$^+$-RNA which had been isolated from total cellular RNA and from mb polysomes directed the formation of hordein polypeptides along with other products in cell-free translation experiments. These hordein polypeptides had molecular weights 2000–2400 larger than hordein preparations from mature endosperm. The hordein polypeptides that were formed by cell-free translation with microsome preparations must have been segregated into the microsomal vesicles, since these polypeptides were resistant to tryptic cleavage, but were made susceptable to proteolytic degradation after detergent treatment (BRANDT and INGVERSEN 1978, CAMERON-MILLS and INGVERSEN 1978, CAMERON-MILLS et al. 1978, HOLDER and INGVERSEN 1978, INGVERSEN et al. 1981). Products of polysome fractions and poly(A)$^+$-RNA from wild-type Bomi and mutant Risø 1508 barley endosperm have been compared in cell-free translation experiments. Cells of the mutant endosperm contained double the amount of RNA than did the cells from wild-type endosperm, but the percentage of poly(A)$^+$-RNA in the total cellular RNA was only 50% of that in the total RNA of the wild-type endosperm. In cell-free translation experiments the specific activity of mutant poly(A)$^+$-RNA amounted to only 50% in comparison to wild-type poly(A)$^+$-RNA. These differences became even more pronounced when poly(A)$^+$-RNA isolated from mb polysomes of the mutant and wild-type barley endosperms have been compared (BRANDT and INGVERSEN 1978). Only one type of the normal hordein precursor polypeptides was formed by cell-free translation of the mutant mRNA. Polypeptide patterns of hordein formed in vitro by the translation of different normal and mutant poly(A)$^+$-RNA and polysome preparations, respectively, reflect the in vivo situation.

### 5.2.1.3 Wheat Proteins

In vitro translation experiments corresponding to those of *Zea mays* L. and *Hordeum vulgare* L. have not been reported with wheat. Earlier results, which suggested that the protein bodies of wheat contained a complete protein synthesis system leading to storage protein formation (MORTON and RAISON 1963, 1964, MORTON et al. 1964) have been criticized since experiments with maize protein bodies (WILSON 1966) have demonstrated that bacterial contaminations may have caused the protein biosynthesis activity in the investigations of Morton et al.

### 5.2.2 Intracellular Site of Storage Protein Formation

Except in maize, the protein bodies of other cereals and of dicotyledonous plants appear to be free of polysomes on their membrane, although in wheat and barley, endosperm RER membranes may be partially attached to the surface of protein bodies (MIFLIN et al. 1980). Membrane-bound polysomes have been demonstrated to be the intracellular site of storage protein biosynthesis in legumes (MÜNTZ 1978a, PÜCHEL et al. 1979, *Vicia faba* L. var. *minor:* BOLLINI and CHRISPEELS 1979, *Phaseolus vulgaris* L.) as well as in cereals.

### 5.2.3 Legume Globulins

In vitro translation experiments have been performed with polysomes and poly(A)$^+$-RNA of *Glycine max* L. Merr. *Phaseolus vulgaris* L., *Pisum sativum* L. and *Vicia faba* L. Immunological techniques are normally needed to isolate and identify the in vitro translation products.

MORI et al. (1978) first identified glycinin polypeptides belonging to the 11S storage protein of *Glycine max* L. Merr. after incubation of poly(A)$^+$-RNA from dry cotyledons with the cell-free wheat germ system. Using polysomes from developing soyabean cotyledons in vitro translation products were obtained which could be reconstituted to give an oligomeric protein exhibiting the same electrophoretic mobility and identical immunological specificity as glycinin preparations from mature seeds (MORI et al. 1979). After a period with contradictory results concerning the size of the primary glycinin-specific cell-free translation product obtained with poly(A)$^+$-RNA from developing soyabean cotyledons (BEACHY et al. 1977, 1978, 1980, FUKAZAWA et al. 1981) recently NIELSEN et al. (1981) isolated a glycinin-specific polypeptide with molecular weights approximately 60,000 using immunoaffinity chromatography. Since on reduction 40,000 acidic and 20,000 basic subunits were not found, as is the case with in vivo glycinin, the authors concluded that the primary in vitro translation product was a common precursor of both subunits. Similar results were first demonstrated for *Pisum sativum* L. and *Vicia faba* L. by CROY et al. 1980a (see Fig. 11). Since the molecular weight of the glycinin precursor is somewhat larger than that corresponding to the sum of the acidic plus basic subunits a two-step processing of the primary translation products was suggested, i.e., the splitting of a presumed signal polypeptide and the separation of the large from the small subunit by specific proteolytic cleavage (CROY et al. 1980a, NIELSEN et al. 1981).

The second type of legume storage globulins, the 7S proteins, seems to be formed as pre-polypeptides in accordance with the signal hypothesis of BLOBEL (1977), since in vitro formation of immunologically isolated 7S protein-specific polypeptides with relative molecular weights larger than known for corresponding polypeptides of mature seeds has been reported with peas (CROY et al. 1980b, HIGGINS and SPENCER 1980) and with soyabeans (BEACHY et al. 1980). Since this type of globulin in most cases is a glycoprotein, co- or post-translational glycosylation is involved in polypeptide processing prior to the final storage protein deposition. Authentic preparations of the kidney bean 7S globulin,

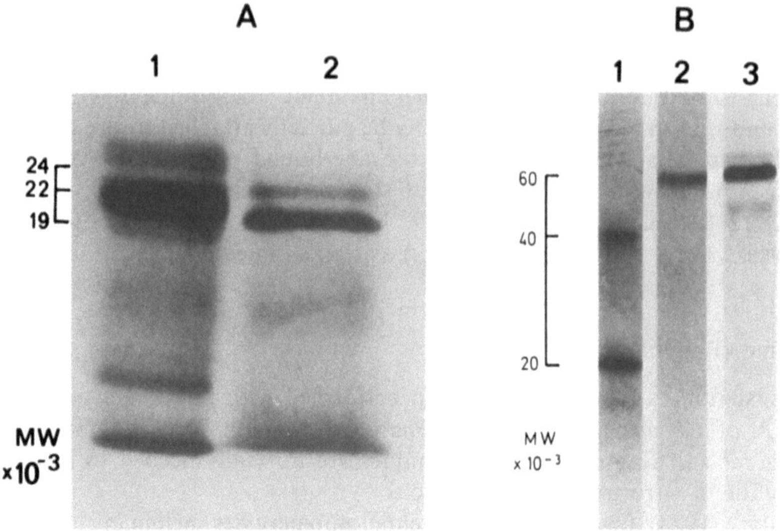


**Fig. 11A, B.** Storage protein precursors in *Zea mays* L. and *Pisum sativum* L. cotyledons. **A** autoradiographs of electrophoretically fractionated zein polypeptides after in vitro-translation using detached mb polysomes (*1*) and microsomes (*2*) indicating the precursor molecules of zein (*1*) with mol · wt 24,000 and 21,000 and the mature zein polypeptides (*2*) with mol · wt 22,000 and 19,000 (LARKINS and HURKMAN 1978); **B** autoradiographs of electrophoretically fractionated legumin subunits from mature seeds after denaturation under reducing (*1*) and non-reducing conditions (*2*) in comparison to legumin produced by in vitro translation of poly(A)-containing RNA from developing pea cotyledons, under denaturation and reducing conditions (*3*). Even under reducing conditions the legumin precursor (*3*) has a mol · wt of approx. 60,000 corresponding to the mol · wt of the non-reduced standard legumin preparation (*2*), which after reduction (*1*) dissociates into the acid 40,000 and the basic 20,000 sub-units. (CROY et al. 1980a)

G1 protein of Hall (SUN and HALL 1975) synonymous with glycoprotein II of Pusztai (PUSZTAI and WATT 1970), give three bands in polyacrylamide gel electrophoresis under denaturing conditions (SUN et al. 1974, 1975, HALL et al. 1980), named $\alpha$ molecular weight 53,000, $\beta$ molecular weight 47,000, and $\gamma$ molecular weight 43,000, respectively. The same pattern of polypeptides, all of which are glycosylated, has been obtained by mb polysome-directed in vitro formation of G1 polypeptides, but cell-free translation of 16S poly(A)$^+$-RNA, that preferentially leads to the formation of G1 polypeptides, only resulted in two bands with molecular weights 47,000 and 43,000 on SDS gels after immunological isolation of G1-specific proteins from the incubation mixture (HALL et al. 1978). Tunicamycin inhibition of in vivo glycosylation in cotyledon slices changed the pattern of authentic G1 polypeptides in *Phaseolus* cotyledons from three to only two subunits, the latter with molecular weight similar to the in vitro translation products obtained with poly(A)$^+$-RNA. The lack of $\alpha$-polypeptide formation by cell-free translation of poly(A)$^+$-RNA from developing *Phaseolus* cotyledons could not be explained by a failure of immunoreaction or a lack of the poly(A)-tail in the corresponding mRNA, since the specific antibodies precipitated all three polypeptides in control experiments with polysome-directed G1 protein formation and no G1 proteins could be detected in translation products of poly(A)$^-$-RNA. Therefore HALL et al. (1980) assume the lack of glycosylation in experiments with cell-free translation of poly(A)$^+$-RNA to be the reason for the differing polypeptide pattern: Non-glycosylated $\alpha$-polypeptides should have the same molecular weight as glycosylated $\beta$-polypeptides, whereas the molecular weight of non-glycosylated $\beta$-chain coincides with that of the glycosylated $\gamma$-polypeptide. Since the $\gamma$-chain is only poorly glycosylated, the position of the non-glycosylated in vitro translation product from poly(A)$^+$-RNA should not be different from that of the glycosylated form.

### 5.2.4 Changes in Storage Protein mRNA's During Seed Development

Polysomes and poly(A)$^+$-RNA, isolated from different developmental stages of the endosperm of *Zea mays* L. and cotyledons of *Phaseolus vulgaris* L. and *Vicia faba* L., were used to investigate the changing pattern of mRNA coding for storage proteins during seed ontogenesis. Provided equal amounts of polysomes or poly(A)$^+$-RNA from the different harvests are incubated with a standardized cell-free translation system from wheat germs or reticulozytes, and providing the procedures give more or less 100% yield, then the radioactivity of storage proteins which are specifically immunoprecipitated from the incubation mixture should reflect the changes in the ratio of their corresponding mRNA's to other mRNA's in the poly(A)$^+$-RNA fraction.

Starting from day 12 after pollination, mb polysome profiles of normal maize kernels show an increase of large-sized polysome fractions which accompanies the rising capacity of normal maize endosperm to synthesize zein whereas mb polsomes from opaque-2 mutant endosperm did not show this increase. In vitro translation experiments with mb polysome preparations from both types of maize demonstrated phase-dependent changes in the amount as well as in the biosynthetic capacity of the different polysome preparations that paralleled the in vivo situation. These results were taken as an indication that mRNA availability could qualitatively, as well as quantitatively, regulate the pattern of zein formation during endosperm development, though without experiments using isolated poly(A)$^+$-RNA to direct cell-free translation, the level of regulation, e.g., during mRNA formation or during mRNA translation (JONES et al. 1977a, b), cannot be defined.

Analogous experiments have been performed with poly(A)$^+$-RNA isolated from five different harvests of *Vicia faba* cotyledons at day 31 to day 54 after anthesis. Since only one cotyledon was used for the preparation of mRNA, the other could be used to prepare extracts for rocket immunoelectrophoresis. Legumin and vicilin polypeptides have been specifically immunoprecipitated from the in vitro translation mixtures and the incorporated radioactivity was quantitatively determined by liquid scintillation counting. There was only a small increase in vicilin labelling but a large increase in legumin over the same period as in vivo legumin accumulation indicating a corresponding increase in the percentage of legumin-specific mRNA in the total poly(A)$^+$-RNA (BASSÜNER et al. 1981, MÜNTZ et al. 1981). These results, taken together with the previous findings that poly(A)$^+$-RNA undergoes a rapid turnover at this time and that α-amanitin selectively inhibits precursor incorporation into poly(A)$^+$-RNA as well as into globulins (PÜCHEL et al. 1979), indicate that at least legumin formation is probably quantitatively regulated at the transcription level at this stage of cotyledon development (MÜNTZ et al. 1981).

Similar experiments with developing cotyledons of *Phaseolus vulgaris* L. (SUN et al. 1978) gave only tentative results, indicating parallel rises of in vivo and in vitro globulin formation since no specific protein isolation procedure of cell-free translation products was used and in vitro translation was directed by polysomes only.

### 5.3 Glycosylation of Storage Proteins During Seed Development

Since the storage proteins of legumes have often been reported to be glycosylated (PUSZTAI 1964, MILLERD 1975, DERBYSHIRE et al. 1976) glycosylation must be an important post- or co-translational event during storage protein synthesis and accumulation. Few publications are available concerning cereal storage proteins, only BURR (1979b) reported zein polypeptides to be glycosylated. This topic is dealt with in detail by BEEVERS (Chap. 4, this Vol.) and will not be considered here.

## 6 Protein Body Formation

Convincing evidence has been presented by several authors working with different cereals and grain legumes that mb polysomes represent the major or even sole intracellular site of reserve protein polypeptide formation in storage tissues (see Sect. 5.2.1.1 of this Chap.). Therefore these polypeptides must pass from the RER to the protein bodies. Two hypotheses have been developed to explain protein body formation: (a) Protein bodies are of vacuolar origin, and (b) protein bodies arise from the ER. Except with some cereals, no convincing evidence is at present available whether protein bodies, especially those of dicotyledons, develop according hypothesis (a) or (b).

Protein bodies of the endosperm of *Zea mays* L. have polysomes attached to the exterior surface of the protein body membrane (KHOO and WOLF 1970, BURR and BURR 1976). These mb polysomes have been shown to direct zein formation in vitro, suggesting a direct segregation of the nascent zein polypeptide chains into the protein bodies. Membrane continuity has been demonstrated between RER and protein bodies of the endosperm of *Zea mays* L., and mb polysomes of the RER synthesize zein in cell-free translation systems as do those from the protein body membrane. In addition the primary products of zein are pre-proteins according to Blobel's signal hypothesis and maturation of these pre-proteins occurs inside the RER vesicles (LARKINS and HURKMAN 1978, see Sect. 5.2.1.1 of this Chap.). The formation of maize endosperm protein bodies by the filling of RER vesicles seems to be convincingly demonstrated.

Close connections between RER membranes and protein bodies seem also to be part of storage protein deposition in *Oryza sativa* L. (HARRIS and JULIANO 1977), *Hordeum vulgare* L. (VON WETTSTEIN 1975, MIFLIN et al. 1980), and *Triticum sativum* L. endosperm (MIFLIN et al. 1980). In mature barley and wheat grains the protein bodies do not appear to be bounded by a single membrane as in the case of maize endosperm, but rather lumps of storage protein are deposited from the RER outside the ER vesicles (MIFLIN et al. 1980). Hence, direct deposition of the storage proteins from the site of biosynthesis, the mb polysomes, into the site of deposition, the protein bodies, mediated by contact of the RER with the protein bodies seems to be a general feature in cereal endosperm (MIFLIN et al. 1980).

In contrast mb polysomes attached to the exterior protein body membrane have never been found in grain legume cotyledons or those of other dicotyledonous plants (for reviews see DIECKERT and DIECKERT 1972, 1976, 1979, BRIARTY 1978, PERNOLLET 1978, WEBER and NEUMANN 1980, BOULTER 1981). Immunohistochemical evidence has been presented that storage proteins occur in the protein bodies of developing broad beans (GRAHAM and GUNNING 1970) as well as in mature broad beans and peas (GRAHAM and GUNNING 1970, CRAIG et al. 1979b respectively). Sometimes the appearance of protein inside membrane-bound cytoplasmic granules (e.g., DIECKERT and DIECKERT 1972) or ER vesicles (e.g., NEUMANN and WEBER 1978) has been related to the transfer of storage protein polypeptides from the site of synthesis to the site of deposition, but the identity of these proteins with storage proteins has never been demonstrated by immunological methods to be similar to those within protein bodies. In the case of *Sinapis alba* L. (BERGFELD et al. 1980, SCHOPFER et al. 1980, BERGFELD and SCHOPFER 1981), where electron micrographs most convincingly agree with the hypothesis of the vacuolar origin of the globulin-containing protein bodies, no evidence has been presented that the growing protein lumps to be seen in the vacuoles are identical with storage globulins of the white mustard.

For the first time protein lumps that are attached to the inner surface of the tonoplast in vacuoles of developing *Pisum sativum* cotyledons have been identified to contain vicilin and legumin by immuno-histochemical methods using monospecific antibodies (CRAIG et al. 1980a, b).

Pulse-chase experiments with broad bean cotyledons have demonstrated that the radioactivity of [$^{14}$C]-leucine which had been accumulated at the RER during the labelling period was mainly transferred into the protein bodies during a chase period (BAILEY et al. 1970). Since connections have been observed between ER and the Golgi apparatus when semi-thin sections of broad bean cotyledons were investigated by high voltage electron microscopy, HARRIS (1979) and BOULTER (1979, 1981) suggested the possible participation of dictyosomes in the processing, presumably glycosylation, and transfer of storage polypeptides into protein bodies. BRIARTY et al. (1969) from the same group favoured the hypothesis of the vacuolar origin of *Vicia faba* protein bodies. DIECKERT and DIECKERT (1972, 1976, 1979) demonstrated Golgi-derived vesicles which fused with the vacuoles in developing cotton cotyledons, and the content of these electron-dense vesicles as well as the "lumps" in vacuoles were shown to be protein by pronase digestion in ultra-thin sections. The segregation of Golgi vesicles has been reported by many authors working with different storage tissues (see Sect. 2.3.1, this Chap.), including *Vicia faba* L. var. *major* and *var. minor* (BRIARTY et al. 1969, NEUMANN and WEBER 1978, WEBER et al. 1981). Since connections between Golgi apparatus and protein bodies have never been observed, the transfer of glycosylated storage polypeptides into the vacuoles which later could become protein bodies has been assumed to take place via dictyosome-derived vesicles (DIECKERT and DIECKERT 1972). Evidence is accumulating, however, that glycosylation of storage proteins occurs at least in part in the RER of legume cotyledons (see Chap. 4, this Vol.). Furthermore, legumin of *Vicia faba* L. var. *minor* does not seem to be glycosylated (WEBER et al. 1981), and electron microscopic analysis of developing cotyledons of *Vicia faba* L. var. *minor* led NEUMANN and WEBER (1978, WEBER and NEUMANN 1980, WEBER et al. 1981) to the conclusion that at least in part field bean protein bodies must be derived directly from the ER. Since these contradictory results also reflect the situation with other dicotyledons, and recent reviews are available (BRIARTY 1978, PERNOLLET 1978, WEBER and NEUMANN 1980, BOULTER 1981), the discussion of these problems will not be extended here.

# 7 The Influence of Environmental Factors on Protein Synthesis and Accumulation During Seed Development

Morphology and biochemistry of seed development follow a species-specific pattern within genetically fixed limits. The qualitative composition of seed storage proteins is also genetically determined, but the time-course of seed development, including the pattern of storage protein deposition, the total amount of storage proteins finally present in the respective seed and the quantitative composition of the storage protein complex may be strongly influenced by environmental factors, e.g., nitrogen and sulphur supply, temperature and water regime, or the length of the vegetation period. Once the period of storage protein synthesis and deposition has been initiated during seed development, these factors act preferentially on the formation and accumulation of storage material, since these are the major metabolic activities during this period.

In cereals where prolamins poor in the limiting essential amino acids lysine and tryptophan represent major storage protein fractions, late nitrogen supply is mainly channelled into the formation of these nutritionally unbalanced proteins, thus lowering the nutritional quality of the total grain protein, though the total amount of protein formed usually rises under the influence of the nitrogen fertilization (Fig. 12). Thus nitrogen fertilization has frequently been reported to induce negatively correlated changes of grain protein content and nutritional quality in cereals (e.g., WITEHOUSE 1973). This negative correlation is observed with different genotypes when large collections of wheat and barley are screened for protein and lysine content (HAGBERG and KARLSSON 1969, JOHNSON et al. 1973, LEHMANN et al. 1978a, b, MÜNTZ et al. 1979). The negative correlation between grain protein content and lysine concentration may become

Fig. 12. Changes in the composition of proteins in relation to increasing nitrogen content of barley grains. The percentage of the three protein fractions is expressed as a percentage of total protein nitrogen. (PRÉAUX and LONTIE 1975). The salt-soluble fraction corresponds to albumins plus globulins

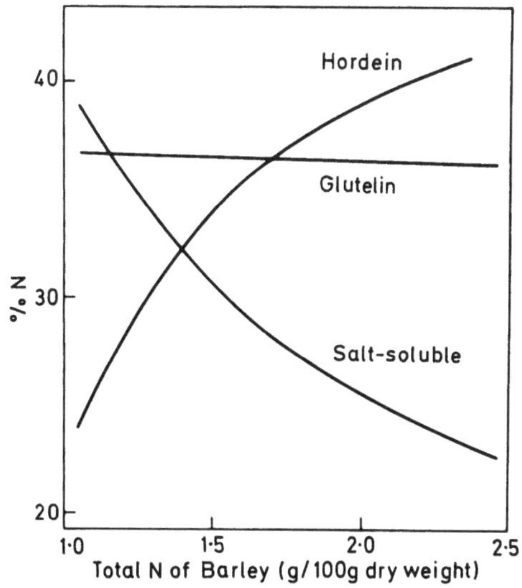


weaker if the grain protein content rises above 15% in wheat (JOHNSON et al. 1979).

Nitrogen fertilization generally increases cereal grain yields and also increases the grain protein content, leading to an increased protein yield (e.g., WHITEHOUSE 1973, RÖBBELEN 1976). The percentage increase in non-protein compounds induced by the nitrogen supply is generally higher than the rise in protein content. Consequently, in modern cereal varieties, grain yield and protein yield increases that have been induced by nitrogen fertilization are positively correlated, though a negative correlation governs the relation between grain yield and protein content per mass unit of the grain. These correlations are also usually found when different genotypes are investigated. Successful attempts have been made to overcome these negative correlations by breeding, and wheat strains have been obtained in which high levels of grain protein content are combined with excellent productivity, and lysine per unit of grain protein is maintained (JOHNSON et al. 1979).

Energy balance calculations which have been based on the work of PENNING DE VRIES et al. (1974) indicated that much more carbohydrate can be formed per unit of glucose than protein or lipids. Seeds with low protein content like cereals exhibit a much higher biomass productivity, defined as gram seed mass produced per glucose equivalent, than do high protein seeds like legumes (SINCLAIR and DE WIT 1975). Furthermore, MITRA et al. (1977) have demonstrated that histidine, arginine, lysine, and tryptophan formation require nearly double the amount of glucose equivalents than does glutamic acid. Consequently, the synthesis of prolamins, which are poor in lysine but rich in glutamic acid, needs less assimilates than does the formation of glutelins, albumins, and globulins in wheat, maize, rice, and oats. Therefore it was suggested that limitations of energy supply may lead to the negative correlations found between grain yield and grain protein content as well as between grain protein content and the amount of lysine in the grain protein (SINCLAIR and DE WIT 1975, BHATIA and RABSON 1976). MIFLIN (1979) has critically discussed these calculations, since they did not directly apply to the special situation of autotrophic $CO_2$-fixing and nitrate-assimilating organisms like higher plants but started with glucose as the primary energy and carbon source. Under field conditions the availability of nitrogen represents the first-order factor limiting grain protein yield, and energy availability should only act as limiting factor under conditions where photosynthesis or the supply of carbohydrates for heterotrophic nitrogen metabolism are minimal. In general, during vegetative stages of plant growth, nitrogen supply stimulates leaf development, thus generating the basis for elevated assimilate formation concomitant with the formation of a source for subsequent nitrogen transfer from leaves to grains when storage protein accumulation takes place (BOULTER and GATEHOUSE 1978, MÜNTZ 1980).

The existence of a negative correlation between grain protein content and the percentage of limiting essential amino acids per unit protein (e.g., methionine and cysteine in legumes) has also been established in *Vicia faba* L. (HANELT et al. 1978). Different legumes react differently to late nitrogen application. *Vicia faba* L. does not show increased seed yield and seed protein production when additional external nitrogen is applied during seed development since symbiotic nitrogen fixation supplies sufficient quantities of nitrogen for grain filling (SCHILLING 1977). In contrast, peas and white lupin (*Lupinus albus*) react to late dressed nitrogen with increased seed and seed protein yield (SCHILLING and SCHALLDACH 1966, SCHILLING et al. 1967). The dependence of storage protein deposition in the seeds on the supply of carbon skeletons is reflected by competition success of the developing new sink on root nodule nitrogen fixation when generative organs start to grow rapidly in soybeans (HARDY and HAVELKA 1976, PATE 1976).

Since sulphur-containing amino acids represent the limiting essential amino acids in legume storage proteins some work has been done on the influence of sulphur deficiency and supply on the storage protein composition in peas (RANDALL et al. 1979) and lupins (BLAGROVE et al. 1976, GILLESPIE et al. 1976, 1978). Lupins, grown with normal sulphur supply during vegetative development, were transferred to nutrient solutions containing low, normal and high amounts of sulphate at the time of first flowering. Plants exhibiting sulphur deficiency characteristics at low sulphur supply nevertheless developed viable seeds which did not show differences in the protein nitrogen per unit dry matter in comparison with the normal and high level sulphur variants, the protein nitrogen quantity per unit dry matter even increased in the seeds of one of the lupin varieties used in these investigations. Non-protein nitrogen was somewhat increased in sulphur-deficient seeds. Electrophoretic investigation of the storage proteins revealed that in the low sulphur variant those components of the globulins which contain sulfur, e.g., $\alpha$-conglutin representing the 11S component and $\gamma$-conglutin, have nearly disappeared whereas the sulphur-free 7S globulin, called $\beta$-conglutin, increased. Subunit analysis of storage proteins formed under conditions of sulphur deficiency indicated the appearance of polypeptides not found in seeds with normal or high sulphur supply, but it remained unclear whether these polypeptides of the 7S globulin fraction represent only an increased amount of normally undetected minor components or were newly formed. In *Lupinus angustifolius* L. the percentage of different components in the storage globulin fraction undergoes dramatic changes under sulphur-deficiency conditions without affecting the amount of protein deposited, thus indicating that the constraints governing the proportion and types of storage proteins can only be relatively weak. Increased sulphur fertilization does not seriously influence the seed globulins, except for a small increase in the amount of total protein sulphur and non-protein organic sulphur per unit dry matter.

Experiments with peas (RANDALL et al. 1979) led also to decreased proportions of the sulphur-containing legumin and increased quantities of sulphur-free vicilin under low sulphur supply. Some components of the vicilin fraction were preferentially stimulated, e.g., the main 7S globulin fraction. In some cases a reduction of total TCA-insoluble nitrogen per unit dry matter has been observed by the same authors. The effect of sulphur application on the total amount of sulphur amino acids in seeds has been investigated with *Vicia faba* L. (EPPENDORFER 1971), *Phaseolus aureus* L. (ARORA and LUTHRA 1971), *Glycine max* L. Merr. (SHARMA and BRADFORD 1973), and *Vigna unguiculata* (EVANS et al. 1977).

The effect of other mineral deficiences on the globulins of developing peas has also been studied (RANDALL et al. 1979). Since in mature cotyledons of *Vicia faba* L. the 11S globulin, legumin, represents about three quarters of the total globulin, and in addition this protein is preferentially formed and deposited during the second half of stage 2 factors such as elevated temperature and drought, which prematurely terminate seed growth induce elevated vicilin to legumin ratios and may even lead to higher quantities of vicilin than legumin at maturity (MANTEUFFEL et al. 1976). The influence of different temperatures on DNA-template and RNA-polymerase activity in developing peas has already been mentioned previously in this chapter.

## References


Aldana AB, Fites RC, Pattee HE (1972) Changes in nucleic acids, protein and ribonuclease activity during maturation of peanut seeds. Plant Cell Physiol 13:515–525

Arora SK, Luthra YP (1971) Relationship between sulphur content of leaf with methionine, cysteine and cystine content in the seeds of *Phaseolus aureus* L. as affected by S, P and N application. Plant Soil 34:91–96

Bailey CJ, Cobb A, Boulter D (1970) A cotyledon slice system for the electron autoradiographic study of the synthesis and intracellular transport of the seed storage protein of *Vicia faba*. Planta 95:103–118

Bain JM, Mercer FV (1966) Subcellular organization of the developing cotyledons of *Pisum sativum* L. Aust J Biol Sci 19:49–67

Bassüner R (1981) *In vitro* Translationsversuche zur Analyse der Biosynthese von Reserveproteinen in reifenden Samen der Ackerbohne (*Vicia faba* L.). Ph D Thesis, Gatersleben

Beachy RN, Thompson JF, Madison JT (1977) Isolation of polyribosomes from immature soybeans and their translation *in vitro*. Plant Physiol 59:106

Beachy RN, Thompson JF, Madison JT (1978) Isolation of polyribosomes and messenger RNA active in *in vitro* synthesis of soybean seed protein. Plant Physiol 61:139–144

Beachy RN, Barton KA, Madison JT, Thompson JF, Jarvis N (1980) The mRNAs that code for soybean seed proteins. In: Leaver CJ (ed) Genome organisation and expression in plants. Plenum Press, New York London pp 273–282
Bergfeld R, Schopfer P (1981) Fine structural aspects of storage protein formation in developing cotyledons of *Sinapis alba* L. In: Müntz K, Manteuffel R, Scholz G (eds) Proc 2nd Symp Seed Protein. Regulation of protein biosynthesis and degradation during embryogenesis and germination of plant seeds. Akademie Verlag, Berlin p 235
Bergfeld R, Kühnl T, Schopfer P (1980) Formation of protein storage bodies during embryogenesis in cotyledons of *Sinapis alba* L. Planta 148:146–156
Bhatia CR, Rabson R (1976) Bioenergetic considerations in cereal breeding for protein improvement. Science 194:1418–1421
Bils RF, Howell RW (1963) Biochemical and cytological changes in developing soybean cotyledons. Crop Sci 3:304–308
Blagrove RJ, Gillespie JM, Randall PJ (1976) Effect of sulphur supply on the seed globulin composition of *Lupinus angustifolius*. Aust J Plant Physiol 3:173–184
Blobel G (1977) Synthesis and segregation of secretory proteins: the signal hypothesis. In: Brinkley BR, Porter KR (eds) Int Cell Biol 1st Congress. Rockefeller Univ Press, New York, pp 318–325
Blobel G, Dobberstein B (1975a) Transfer of proteins across membranes. I. Presence of proteolytically processed and unprocessed nascent immunoglobulin light chains on membrane-bound ribosomes of Murine myeloma. J Cell Biol 67:835–851
Blobel G, Dobberstein B (1975b) Transfer of protein across membranes. II. Reconstitution of functional rough microsomes from heterologous components. J Cell Biol 67:852–862
Blobel G, Walter P, Chung Nan Chang, Goldman AH, Erickson AS, Lingappa VE (1979) Translocation of protein across membranes: The signal hypothesis and beyond. Symp Soc Exp Biol 33:9–36
Bollini R, Chrispeels MJ (1979) The rough endoplasmic reticulum is the site of reserve protein synthesis in developing *Phaseolus vulgaris* cotyledons. Planta 146:487–501
Boulter D (1979) Structure and biosynthesis of legume storage proteins. In: Seed protein improvement in cereals and grain legumes. Proc Int Symp FAO/LAEA. IAEA, Vienna, pp 125–133
Boulter D (1981) The ultrastructure and composition of protein bodies. In: Müntz K, Manteuffel R, Scholz G (eds) Proc 2nd Seed Protein Symp: Regulation of protein biosynthesis and degradation during embryogenesis and germination of plant seeds. Akademie Verlag, Berlin, pp 95–102
Boulter D, Davis OJ (1968) Nitrogen metabolism in developing seeds of *Vicia faba*. New Phytol 67:935–946
Boulter D, Gatehouse JA (1978) The efficiency of plant protein synthesis in nature. Proc FEBS-Meet 1977, Copenhagen 44:181–192
Brady T (1973) Feulgen cytophotometric determination of the DNA content of the embryo proper and suspensor cells of *Phaseolus coccineus*. Cell Differ 2:65–75
Brandt AB (1975) *In vivo* incorporation of $^{14}C$-lysine into the endosperm proteins of wild type and high-lysine barley. FEBS Lett 52:288–291
Brandt A (1979) Cloning of double stranded DNA coding for hordein polypeptides. Carlsberg Res Commun 44:255–267
Brandt A, Ingversen J (1976) *In vitro* synthesis of barley endosperm proteins on wild type and mutant templates. Carlsberg Res Commun 41:311–320
Brandt A, Ingversen J (1978) Isolation and translation of hordein messenger RNA from wild type and mutant endosperms in barley. Carlsberg Res Commun 43:451–469
Brawerman G (1974) Eukaryotic messenger RNA. Annu Rev Biochem 43:621–642
Brawerman G, Mendecki J, Lee SY (1972) A procedure for the isolation of mammalian messenger ribonucleic acid. Biochemistry 11:637–641
Briarty LG (1973) Stereology in seed development studies: Some preliminary work. Caryologica 25:289–301
Briarty LG (1978) The mechanism of protein body deposition in legumes and cereals. Proc Easter School Agric Sci Univ Nottingham 24:81–106

Briarty LG, Coult DA, Boulter D (1969) Protein bodies of developing seeds of *Vicia faba*. J Exp Bot 20:358–372
Briarty LG, Hughes CE, Evers AD (1979) The developing endosperm of wheat – a stereological analysis. Ann Bot 44:641–658
Broekaert D, Van Parijs R (1978) The relationship between the endomitotic cell cycle and the enhanced capacity for protein synthesis in *Leguminosae* embryogeny. Z Pflanzenphysiol 86:165–175
Brown MM (1917) The development of the embryo sac and the embryo in *Phaseolus vulgaris*. Bull Torrey Bot Club 44:535–544
Burr B (1979a) Identification of zein structural genes in the maize genome. In: Seed protein improvement in cereals and grain legumes. Proc Int Symp FAO/IAEA. IAEA, Vienna, pp 175–177
Burr FA (1979b) Zein synthesis and processing on zein protein body membranes. In: Seed protein improvement in cereals and grain legumes. Proc Int Symp FAO/IAEA. IAEA, Vienna, pp 159–162
Burr B, Burr FA (1976) Zein synthesis in maize endosperm by polyribosomes attached to protein bodies. Proc Natl Acad Sci USA 73:515–519
Burr FA, Burr B (1980) The cloning of zein sequences and an approach to zein genetics. In: Leaver CJ (ed) Genome organisation and expression in plants. Plenum Press, New York London, pp 227–232
Burr B, Burr FA, Rubenstein I, Simon MN (1978) Purification and translation of zein messenger RNA from maize endosperm protein bodies. Proc Natl Acad Sci USA 75:696–700
Buttrose MS (1963a) Ultrastructure of developing wheat endosperm. Aust J Biol Sci 16:305–307
Buttrose MS (1963b) Ultrastructure of the developing aleurone cells of wheat grain. Aust J Biol Sci 16:768–774
Cagampang G, Perdon A, Juliano BO (1976) Changes in salt-soluble proteins of rice during grain development. Phytochemistry 15:1425–1429
Cameron-Mills V, Ingversen J (1978) *In vitro* synthesis and transport of barley endosperm proteins: Reconstitution of functional rough microsomes from polyribosomes and stripped microsomes. Carlsberg Res Commun 43:471–489
Cameron-Mills V, Ingversen J, Brandt A (1978) Transfer of *in vitro* synthesized barley endosperm proteins into the lumen of the endoplasmatic reticulum. Carlsberg Res Commun 43:91–102
Carasco JF, Croy R, Derbyshire E, Boulter D (1978) The isolation and characterization of the major polypeptides of the seed globulin of cowpea (*Vigna unguiculata* L.) and their sequential synthesis in developing seeds. J Exp Bot 29:309–324
Carlier AR, Peumans WJ (1976) The rye embryo system as an alternative to the wheat system for protein biosynthesis *in vitro*. Biochim Biophys Acta 447:436–444
Carlier AR, Peumans WJ, Manickam A (1978) Cell-free translation in extracts from dry and germinated mung bean primary axes. Plant Sci Lett 11:207–216
Carr DJ, Skene KG (1961) Diauxic growth curves of seeds, with special reference to french beans (*Phaseolus vulgaris* L.). Aust J Biol Sci 14:1–12
Castroviejo M, Tharaud D, Tarrago-Litvak L, Litvak S (1979) Multiple deoxyribonucleic acid polymerase from quiescent wheat embryos. Biochem J 181:183–191
Chandra GR, Abdul-Baki A (1977) Separation of poly(A)-RNA's synthesized by soybean embryos. Plant Cell Physiol 18:271–275
Cooper DC (1938a) Embryology of *Pisum sativum*. Bot Gaz 100:123–132
Cooper GO (1938b) Cytological investigations of *Pisum sativum*. Bot Gaz 99:584–591
Craig S, Goodchild DJ, Hardham AR (1979a) Structural aspects of protein accumulation in developing pea cotyledons. I Qualitative and quantitative changes in parenchyma cell vacuoles. Aust J Plant Physiol 6:81–98
Craig S, Goodchild DJ, Millerd A (1979b) Immunofluorescence localization of pea storage proteins in glycol methacrylate embedded tissue. J Histochem Cytochem 27:1312–1316
Craig S, Goodchild DJ, Miller C (1980a) Structural aspects of protein accumulation in

developing pea cotyledons. II. Three-dimensional reconstructions of vacuoles and protein bodies from serial sections. Aust J Plant Physiol 7:329–337
Craig S, Millerd A, Goodchild DJ (1980b) Structural aspects of protein accumulation in developing pea cotyledons. III. Immunocytochemical localization of legumin and vicilin using antibodies shown to be specific by the enzyme-linked immunosorbent assay (ELISA). Aust J Plant Physiol 7:339–351
Croy RRD, Gatehouse JA, Evans IM, Boulter D (1980a) Characterisation of the storage protein subunits synthesized *in vitro* by polyribosomes and RNA from developing pea (*Pisum sativum* L.) I. Legumin. Planta 148:49–56
Croy RRD, Gatehouse JA, Evans IM, Boulter D (1980b) Characterisation of the storage protein subunits synthesized *In vitro* by polyribosomes and RNA from developing pea (*Pisum sativum* L.). II. Vicilin. Planta 148:57–63
Cullis CA (1976) Chromatin bound DNA dependent RNA-polymerase in developing pea cotyledons. Planta 131:293–208
Cullis CA (1978) Chromatin-bound DNA-dependent RNA polymerase in developing pea cotyledons. II. Polymerase activity and template availability under different growth conditions. Planta 144:57–62
Cullis CA, Davies DR (1975) Ribosomal DNA amounts in *Pisum sativum*. Genetics 81:485–492
Danielsson CE (1952) A contribution to the study of the synthesis of the reserve proteins in ripening pea seeds. Acta Scand Chem 6:149–159
Davey JE, Van Staden J (1978) Ultrastructural aspects of reserve protein deposition during cotyledonary cell development in *Lupinus albus*. Z Pflanzenphysiol 89:259–271
Davies DR (1973) Differential activation of maternal and paternal loci in seed development. Nature (London) New Biol 245:30–32
Davies DR (1976) DNA and RNA contents in relation to cell and seed weight in *Pisum sativum*. Plant Sci Lett 7:17–25
Davies DR (1977) DNA contents and cell number in relation to seed size in the genus *Vicia*. Heredity 39:153–163
Davies DR, Brewster V (1975) Studies of seed development in *Pisum sativum*. II. Ribosomal RNA contents in reciprocal crosses. Planta 124:303–309
De GN, Gosh JJ, Bhattacharyya (1972) Ribonucleic acid from coffee beans. Phytochemistry 11:3349–3353
Derbyshire E, Boulter D (1976) Isolation of legumin-like protein from *Phaseolus aureus* and *Phaseolus vulgaris*. Phytochemistry 15:411–414
Derbyshire E, Wright DJ, Boulter D (1976) Legumin and vicilin, storage proteins of legume seeds. Phytochemistry 15:3–24
Dieckert JW, Dieckert MC (1972) The deposition of vacuolar proteins in oilseeds. In: Inglett GE (ed) Symposium: Seed proteins. Avi Publ Comp, Westport, pp 52–85
Dieckert JW, Dieckert MC (1976) The chemistry and cell biology of vacuolar proteins of seeds. J Food Sci 41:475–482
Dieckert JW, Dieckert MC (1979) The comparative anatomy of the principal reserve proteins of seeds. In: Müntz K (ed) Proc symp seed proteins of dicotyledonous plants. Akademie Verlag, Berlin, pp 73–86
Doll H, Køie B, Eggum BO (1974) Induced high lysine mutants in barley. Radiat Bot 14:73–80
Donovan GR (1979) Relationship between grain nitrogen, non-protein nitrogen and nucleic acids during wheat grain development. Aust J Plant Physiol 6:449–458
Donovan GR, Hill RD, Lee JW (1977) Compositional changes in the developing grain of high and low protein wheats I. Chemical composition. Cereal Chem 54:638–645
Duda CT (1976) Plant RNA polymerases. Annu Rev Plant Physiol 27:119–132
Dure LS (1973) Regulation of protein synthesis in cotton seed embryogenesis and germination. Biochem Soc Symp 38:217–234
Dure LS (1975) Seed formation. Annu Rev Plant Physiol 26:259–278
Dure LS (1977) Stored messenger ribonucleic acid and seed germination. In: Khan AA (ed) Physiol biochem seed dormancy germination. North-Holland, Amsterdam, pp 335–345

Engleman EM (1966) Ontogeny of aleurone grains in cotton embryo. Am J Bot 53:231–237
Eppendorfer WH (1971) Effects of S, N and P on amino acid composition of field beans (*Vicia faba*) and responses of the biological value of the seed protein to S-amino acid content. J Sci Food Agric 22:501–505
Evans IM, Boulter D, Fox RL, Kang BT (1977) The effect of sulfur fertilisers on the content of sulpho-amino acids in seeds of cowpea (*Vigna unguiculata*). J Sci Food Agric 28:161–166
Evans MI, Croy RRD, Hutchinson P, Boulter D, Payne PJ, Gordon ME (1979) Cell free synthesis of some storage protein subunits by polyribosomes and RNA isolated from developing seeds of pea (*Pisum sativum* L.). Planta 144:455–462
Evans IM, Croy RDD, Brown PA, Boulter D (1980) Synthesis of complementary DNAs to partially purified mRNAs coding for storage proteins of Pisum sativum L. Biochim Biophys Acta 610:81–95
Evers AD (1970) Development of the endosperm of wheat. Ann Bot 34:547–555
Ewart JAD (1966) Fingerprinting of glutenin and gliadin. J Sci Food Agric 17:30–33
Finlayson AJ, Christ CM (1971) Changes in the nitrogenous components of maturing rapeseed (*Brassica napus*). Can J Bot 49:1733–1735
Flinn AM, Pate IS (1968) Biochemical and phytological changes during maturation of fruit of the field pea (*Pisum arvense* L.). Ann Bot 32:479–495
Flint D, Ayers GS, Ries SK (1975) Synthesis of endosperm proteins in wheat seed during maturation. Plant Physiol 56:381–384
Fukazawa C, Udaka K, Kainuma K (1981) Characterization of glycinin mRNA and reiteration frequency of glycinin DNA. In: Müntz K, Manteuffel R, Scholz G (eds) Proc. 2nd Seed Protein Symp. Regulation of protein biosynthesis and degradation during embryogenesis and germination of plant seeds. Akademie Verlag, Berlin, pp 217–218
Galitz D, Howell RW (1965) Measurement of ribonucleic acids and total free nucleotides of developing soybean seeds. Physiol Plant 18:1018–1021
Ghosh M, Das HK (1980) Role of proteolytic enzymes in the development of amino acid imbalance in wheat endosperm proteins. Phytochemistry 19:2535–2540
Gillespie JM, Blagrove RJ, Randall PJ (1976) Regulation of the globulins of lupine seeds. In: Eval seed protein alterations mutant breed. Proc 3rd Res Co-ord Meet Seed Protein Improv Programme. IAEA, Vienna, pp 151–156
Gillespie JM, Blagrove RJ, Randall PJ (1978) Effect of sulfur supply on the seed globulin composition of various species of lupin. Aust J Plant Physiol 5:641–650
Gottschalk W, Müller H (1970) Monogenic alteration of seed protein content and protein pattern in X-ray-induced *Pisum* mutants. In: Improving plant protein by nuclear techniques. Proc Symp Int At Energy Agency, Vienna, pp 201–215
Graham JSD, Morton RK (1963) Studies of proteins of developing wheat endosperm: Separation by starch-gel electrophoresis and incorporation of $^{35}$S-sulfate. Aust J Biol Sci 16:357–365
Graham JSD, Morton RK, Simmonds DH (1963) Studies of proteins of developing wheat endosperm. Fractionation by ion-exchange chromatography. Aust J Biol Sci 16:350–356
Graham TA, Gunning BES (1970) Localization of legumin and vicilin in bean cotyledon cells using fluorescent antibodies. Nature (London) 228:81–82
Gupta RK, Tiwari OP, Gupta AK, Das HK (1976) Synthesis and degradation of protein during wheat endosperm development. Phytochemistry 15:1101–1104
Hadjivassiliou A, Brawerman G (1966) Polyadenylic acid in the cytoplasm of rat liver. J Mol Biol 20:1–7
Hagberg A, Karlsson KE (1969) Breeding for high protein content and quality in barley. In: New Approaches Breed Improv Plant Protein, Proc Panel Meet. IAEA, Vienna, pp 17–21
Hahn H, Servos D (1980) The isolation of multiple forms of DNA-dependent RNA polymerases from nuclei of quiescent wheat embryos by affinity chromatography. Z Pflanzenphysiol 97:43–57
Hall TC, McLeester RC, Bliss FA (1977) Equal expression of the maternal and paternal alleles for the polypeptide subunits of the major storage protein of the bean *Phaseolus vulgaris* L. Plant Physiol 59:1122–1124

Hall TC, Buchbinder BU, Payne JW, Sun SM (1978) Messenger RNA for G1 protein of french bean seeds: Cell-free translation and product characterization. Proc Natl Acad Sci USA 75:3196–3200

Hall TC, Sun SM, Buchbinder BU, Payne JW, Bliss FA, Kemp JD (1980) Bean seed globulin mRNA: Translation, characterization, and its use as a probe towards genetic engineering of crop plants. In: Leaver CJ (ed) Genome organisation and expression in plants. Plenum Press, New York, London, pp 259–272

Handa AK, Harmodson M, Tsai CY, Larkins BA (1979) Analysis of isoelectric heterogeneity among zein proteins. Plant Physiol 63:94

Hanelt P, Rudolph A, Hammer K, Jank HW, Müntz K, Scholz F (1978) Eiweißuntersuchungen am Getreide- und Leguminosen-Sortiment Gatersleben. Teil 3: Gehalt an Rohprotein und schwefelhaltigen Aminosäuren der Ackerbohnen (*Vicia faba* L.) Kulturpflanze 24:183–212

Hardy RWF, Havelka UD (1976) Photosynthate as a major factor limiting nitrogen fixation by field-grown legumes with emphasis on soybeans. In: Nutman PS (ed) Symbiontic nitrogen fixation in plants. Cambridge Univ Press, Cambridge London New York Melbourne, pp 421–439

Harris N (1978) Nuclear pore distribution and relation to adjacent cytoplasmic organelles in cotyledon cells of developing *Vicia faba*. Planta 141:121–128

Harris N (1979) Endoplasmic reticulum in developing seeds of *Vicia faba*. A high voltage electron microscope study. Planta 146:63–70

Harris N, Juliano BO (1977) Ultrastructure of endosperm protein bodies in developing rice grains differing in protein content. Ann Bot 41:1–6

Higgins TJV, Spencer D (1980) Biosynthesis of pea proteins: Evidence for precursor forms from *in vivo* and *in vitro* studies. In: Leaver CJ (ed) Genome organisation and expression in plants. Plenum Press, New York London, pp 245–258

Hill JE, Breidenbach RW (1974) Proteins of soybean seeds. II. Accumulation of the major components during seed development and maturation. Plant Physiol 53:747–751

Holder A, Ingversen J (1978) Peptide mapping of the major components of *in vitro* synthesized barley hordein: Evidence of structural homology. Carlsberg Res Commun 43:177–184

Hurkman WJ, Pedersen K, Smith LD, Larkins BA (1979) Synthesis and processing of maize storage proteins in *Xenopus* oocytes injected with zein mRNA. Plant Physiol 63:94

Hurkman WJ, Smith LD, Richter J, Larkins BA (1981) Subcellular compartmentalization of maize storage proteins in *Xenopus* oocytes injected with zein messenger RNAs. J Cell Biol 89:292–299

Ingle J, Beitz D, Hagemann RH (1965) Changes in the composition during development and maturation of maize seeds. Plant Physiol 40:835–839

Ingversen J, Brandt A, Cameron-Mills V (1981) The structure, biosynthesis and intracellular transport of barley endosperm reserve proteins. In: Müntz K, Manteuffel R, Scholz G (eds) Proc 2nd Seed Protein Symp. Regulation of protein biosynthesis and degradation during embryogenesis and germination of plant seeds. Akademie Verlag, Berlin, pp 41–56

Ivanko S (1971) Changeability of protein fractions and their amino acid composition during maturation of barley grain. Biol Plant 13:155–164

Ivanko S (1978) Die Bildung des Proteinkomplexes in Maiskörnern während der Reifung (Slovakian). Biol Práce 4:XXIV, pp 1–107

Jachymczyk WJ, Cherry JH (1968) Studies on messenger RNA from peanut plants: *in vitro* polyribosome formation and protein synthesis. Biochim Biophys Acta 157:368–377

Jendrisak J (1980) Purification, structure and functions of the nuclear RNA polymerases of higher plants. In: Leaver CJ (ed) Genome organization and expression in plants. Plenum Press, New York London, pp 77–92

Jennings AC, Morton RK (1963a) Changes in carbohydrate, protein, and non-protein nitrogenous compounds of developing wheat grain. Aust J Biol Sci 16:318–331

Jennings AC, Morton RK (1963b) Changes in nucleic acids and other phosphorus-containing compounds of developing wheat grain. Aust J Biol Sci 16:322–341

Jennings AC, Morton RK (1963c) Amino acid and protein synthesis in developing wheat endosperm. Aust J Biol Sci 16:384–394

Johari RP, Mehta SL, Naik MS (1977) Protein synthesis and changes in nucleic acids during grain development of *Sorghum*. Phytochemistry 16:19–24

Johnson VA, Mattern PJ, Schmidt JW, Stroike JE (1973) Genetic advances in wheat protein quantity and composition. In: Sears ER, Sears LMS (eds) Proc 4th Int Wheat Genet Symp, pp 547–556

Johnson VA, Mattern PJ, Kuhr SL (1979) Genetic improvement of wheat protein. In: Seed protein improvement in cereals and grain legumes. Proc Int Symp FAO/IAEA. IAEA, Vienna, pp 165–178

Jones RA (1978) Effects of floury-2-locus on zein accumulation and RNA metabolism during maize endosperm development. Biochem Genet 16:27–28

Jones RA, Larkins BA, Tsai CY (1976) Reduced synthesis of zein *in vitro* by a high lysine mutant of maize. Biochem Biophys Res Commun 69:404–410

Jones RA, Larkins BA, Tsai CY (1977a) Storage protein synthesis in maize. II. Reduced synthesis of a major zein component by the opaque-2 mutant of maize. Plant Physiol 59:525–529

Jones RA, Larkins BA, Tsai CY (1977b) Storage protein synthesis in maize. III. Developmental changes in membrane-bound polyribosome composition and *in vitro* protein synthesis of normal and opaque-2 maize. Plant Physiol 59:733–737

Juliano BO (1972) Studies on protein quality and quantity of rice. In: Inglett GE (ed) Symposium: Seed proteins. Avi Publ Comp, Westport, Connecticut, pp 114–125

Kasarda DD, Bernardin JE, Qualset CO (1976) Relationship of gliadin protein components to chromosomes in hexaploid wheats (*Triticum aestivum* L.). Proc Natl Acad Sci USA 73:3646–3650

Khoo U, Wolf MJ (1970) Origin and development of protein granules in maize endosperm. Am J Bot 57:1042–1050

Køie B, Ingversen J, Andersen AJ, Doll H, Eggum BO (1976) Composition and nutritional quality of barley protein. In: Evaluation of seed protein alterations by mutation breeding. Proc Meet IAEA. IAEA, Vienna pp 55–61

Larkins BA, Dalby A (1975) *In vitro* synthesis of zein-like protein by maize polyribosomes. Biochem Biophys Res Commun 66:1048–1054

Larkins BA, Davies E (1975) Polyribosomes from peas. V An attempt to characterize the total free and membrane bound polysomal population. Plant Physiol 55:749–756

Larkins BA, Hurkman WJ (1978) Synthesis and deposition of zein in protein bodies of maize endosperm. Plant Physiol 62:256–263

Larkins BA, Bracker CE, Tsai CY (1976) Storage protein synthesis in maize. Isolation of zein-synthesizing polysomes. Plant Physiol 57:740–745

Larkins BA, Pedersen K, Handa AK, Hurkman WJ, Smith LD (1979) Synthesis and processing of maize storage proteins in *Xenopus laevis* oocytes. Proc Natl Acad Sci USA 76:6448–6452

Larkins BA, Pedersen K, Hurkman WJ, Handa AK, Mason AC, Tsai CY, Hermodson MA (1980) Maize storage proteins: Characterization and biosynthesis. In: Leaver CJ (ed) Genome organization and expression in plants. Plenum Press, New York London, pp 203–218

Lehmann CO, Rudolph A, Hammer K, Meister A, Müntz K, Scholz F (1978a) Eiweißuntersuchungen am Getreide- und Leguminosen-Sortiment Gatersleben. Teil 1: Gehalt an Rohprotein und Lysin von Weizen sowie von Weizen-Art- und Gattungsbastarden. Kulturpflanze 24:133–161

Lehmann CO, Rudolph A, Hammer K, Meister A, Müntz K, Scholz F (1978b) Eiweißuntersuchungen am Getreide- und Leguminosen-Sortiment Gatersleben. Teil 2: Gehalt an Rohprotein und Lysin von Gersten. Kulturpflanze 24:163–181

Liddell JW, Boulter D (1974) Amino acid incorporation on 80S plant ribosomes: The complete system. Phytochemistry 13:2065–2069

Lin CY, Guilfoyle TJ, Chen YM, Key JL (1975) Isolation of nucleoli and localization of ribonucleic acid polymerase I from soybean hypocotyl. Plant Physiol 56:850–852

Loewenberg JR (1955) The development of bean seeds (*Phaseolus vulgaris*). Plant Physiol 30:244–250
Luthe DS, Peterson DH (1977) Cell-free synthesis of globulin by developing oat (*Avena sativa* L.) seeds. Plant Physiol 59:836–841
Madison JT, Thompson JF, Muenster AE (1981) Turnover of storage proteins in seeds of soyabean and pea. Ann Bot 47:65–74
Manteuffel R, Müntz K, Püchel M, Scholz G Phase-dependent changes of DNA, RNA and protein accumulation during ontogenesis of broad bean seeds. Biochem Physiol Pflanz 169:595–605
Marcus A, Efron D, Weeks DP (1974) The wheat embryo cell-free system. In: Moldave K, Grossman L (eds) Methods in enzymology, Vol 30. Academic Press, London New York, pp 749–754
Mares D, Jeffery C, Norstog K (1976) The differentiation of the modified aleurone of wheat endosperm: Nuclear membrane-mediated endoplasmic reticulum development. Plant Sci Lett 7:305–311
Marinos NG (1970) Embryogenesis of the pea (*Pisum sativum*). I. The cytological environment of the developing embryo. Protoplasma 70:261–279
Marks GE, Davies DR (1969) The cytology of cotyledon cells and the induction of giant polytene chromosomes in *Pisum sativum*. Protoplasma 101:73–80
Mason AC, Larkins BA (1979) Preferential translation of zein mRNAs in vitro. Plant Physiol 63:94
Matile P (1968) Aleuron vacuoles as lysosomes. Z Pflanzenphysiol 58:365–368
Matile P (1975) The lytic compartment of plant cells. In: Cell biology monographs, Vol I. Springer, Wien New York, pp 46–50
Mazuś B, Buchowicz J (1972) RNS-Polymeraseaktivität in ruhenden und keimenden Weizensamen. Phytochemistry 11:2443–2446
McKee HS (1958) Protein metabolism in ripening and dormant seeds and fruits. In: Ruhland W (ed) Handbuch der Pflanzenphysiologie, Vol 8. Springer, Berlin Heidelberg New York, pp 581–609
McKee HS, Robertson RN, Lee JB (1955a) Physiology of pea fruits. I. The developing fruit. Aust J Biol Sci 8:137–163
McKee HS, Nestel L, Robertson RN (1955b) Physiology of pea fruits. II Soluble nitrogenous constituents in the developing fruit. Aust J Biol Sci 8:467–475
Mecham DK, Kasarda DD, Qualset CO (1978) Genetic aspects of wheat gliadin proteins. Biochem Genet 16:831–853
Mehta SL, Srivastava KN, Mali PC, Naik MS (1972) Changes in the nucleic acid and protein fractions in opaque-2 maize kernels during development. Phytochemistry 11:937–942
Mehta SL, Lodha ML, Mali PC, Singh J, Naik MS (1973) Characterization of polysomes and incorporation *in vitro* of leucine and lysine in normal and opaque-2 *Zea mays* endosperm during development. Phytochemistry 12:2815–2820
Melcher U (1979) *In vitro* synthesis of a precursor to the methionine-rich polypeptide of the zein fraction of corn. Plant Physiol 63:354–358
Melcher U (1980) Heterogeneity of *Zea mays* protein body messenger RNA. Plant Sci Lett 18:133–141
Melcher U, Fraij B (1979) Zein heterogeneity: Isoelectric focusing and nucleic acid hybridization. Plant Physiol 63:94
Merrick WC, Dure LS (1973a) Developmental biochemistry of cottonseed embryogenesis and germination. Preferential charging of cotton chloroplastic transfer ribonucleic acid by *Escherichia coli* enzymes. Biochemistry 12:629–635
Merrick WC, Dure LS (1973b) The developmental biochemistry of cotton. Seed embryo genesis and germination. IV. Levels of cytoplasmic and chloroplastic transfer ribonucleic acid species. J Biol Chem 247:7988–7999
Miflin BJ (1979) Energy considerations in nitrogen metabolism. In: Miflin BJ, Zoschke M (eds) Carbohydrate and protein synthesis. Commission of European Communities, Luxembourg, pp 13–31

Miflin BJ, Shewry PR (1979) The biology and biochemistry of cereal seed prolamine. Seed protein improvement in cereal and grain legumes. Proc Int Symp FAO/IAEA. IAEA, Vienna, pp 137–158
Miflin BJ, Matthews JA, Burgess SR, Faulks AJ, Shewry PR (1980) The synthesis of barley storage proteins. In: Leaver CJ (ed) Genome organisation and expression in plants. Plenum Press, New York London, pp 233–244
Millerd A (1973) DNA and RNA synthesis during growth by cell expansion in *Vicia faba* cotyledons. In: Pollack JK, Lee JW (eds) The biochemistry of gene expression in higher organisms. Proc Symp IUB, Sydney, Australien. Reidel Publ Comp, Dordrecht-Boston, pp 357–366
Millerd A (1975) Biochemistry of legume seed proteins. Annu Rev Plant Physiol 26:53–72
Millerd A, Spencer D (1974) Changes in RNA-synthesizing activity and template activity in nuclei from cotyledons of developing pea seeds. Aust J Plant Physiol 1:331–341
Millerd A, Whitfeld PR (1973) Deoxyribonucleic acid and ribonucleic acid synthesis during the cell expansion phase of cotyledon development in *Vicia faba* L. Plant Physiol 51:1005–1010
Millerd A, Simon M, Stern H (1971) Legumin synthesis in developing cotyledons of *Vicia faba* L. Plant Physiol 48:419–425
Millerd A, Spencer D, Dudman WF, Stiller M (1975) Growth of immature pea cotyledons in culture. Aust J Plant Physiol 2:51–60
Millerd A, Thomson JA, Schroeder HE (1978) Cotyledonary storage proteins in *Pisum sativum*. III. Patterns of accumulation during development. Aust J Plant Physiol 5:1–16
Mitra RK, Bhatia CR, Rabson R (1977) Bioenergetic cost of altering the amino acid composition of cereal grains. Proc 62nd Ann Meet Am Assoc Cereal Chem, San Francisco
Mori T, Matsushita A (1970) Isolation of ribonucleic acids having template activities from particulate components of soybean seeds. Agric Biol Chem 34:1004–1008
Mori T, Wakabayashi Y, Takagi S (1978) Occurrence of mRNA for storage protein in dry soybean seeds. J Biochem 84:1103–1111
Mori T, Takagi S, Utsumi A (1979) Synthesis of glycinin in a wheat germ cell-free system. Biochem Biophys Res Commun 87:43–49
Morton RK, Raison JK (1963) A complete intracellular unit for incorporation of amino acid into storage protein utilizing adenosine triphosphate generated from phytate. Nature (London) 200:429–433
Morton RK, Raison JK (1964) The separation incorporation of amino acids into storage and soluble protein catalysed by two independent systems isolated from developing wheat endosperm. Biochem J 91:528–539
Morton RK, Raison JK, Smeaton JR (1964) Enzymes and ribonucleic acid associated with the incorporation of amino acids into proteins of wheat endosperm. Biochem J 91:539–546
Munck L, Karlsson KE, Hagberg A (1969) High nutritional value of cereal protein. Sver Utsaedesfoeren Tidskr 79:196–205
Müntz K (1978a) Cell specialisation processes during embryogenesis and storage of proteins in plant seeds. In: Schütte HR, Gross D (eds) Regulation of developmental processes in plants. Proc Int Conf, Halle. VEB Fischer, Jena, pp 70–97
Müntz K (1978b) Biosynthese von Speicherproteinen und Proteinspeicherung während der Entwicklung pflanzlicher Samen. In: Nover L, Luckner M, Parthier B (eds) Zelldifferenzierung – molekulare Grundlagen, Regulation, Probleme. VEB Fischer, Jena, pp 369–394
Müntz K (1980) Stoffwechselphysiologische Grundlagen für die Bildung des Sameneiweißertrages bei Kulturpflanzen. In: Aus d Arb Plenum u Kl AdW d DDR 5:30–57
Müntz K, Silhengst P (1979) Protein pattern in reciprocal crosses of Faba beans. In: Müntz K (ed) Proc symp seed proteins of dicotyledonous plants. Akademie Verlag, Berlin, p 229
Müntz K, Hammer K, Lehmann C, Meister A, Rudolph A, Scholz F (1979) Variability of protein and lysine content in barley and wheat specimens from the world collection

of cultivated plants at Gatersleben. In: Seed protein improvement in cereals and grain legumes. Proc Int Symp FAO/IAEA. IAEA, Vienna, pp 183–200

Müntz K, Bassüner R, Bäumlein H, Manteuffel R, Püchel M, Schmidt P, Wobus U (1981) The biosynthesis of storage proteins in developing field bean seeds. In: Müntz K, Manteuffel R, Scholz G (eds) Proc 2nd Seed Protein Symp: Regulation of protein biosynthesis and degradation during embryogenesis and germination of plant seeds. Akademie Verlag, Berlin, pp 57–72

Nagl W (1974) The *Phaseolus* suspensor and its polytene chromosomes. Z Pflanzenphysiol 73:1–44

Neumann D, Weber E (1978) Formation of protein bodies in ripening seeds of *Vicia faba*. Biochem Physiol Pflanz 173:167–180

Nichols JL (1979) 'CAP' structures in maize poly(A)-containing RNA. Biochim Biophys Acta 563:490–495

Nielsen NC, Moreira M, Staswick P, Hermodson MA, Tumer N, Than VH (1981) The structure of glycinin from soybeans. In: Müntz K, Manteuffel R, Scholz G (eds) Proc 2nd Seed Protein Symp: Regulation of protein biosynthesis and degradation during embryogenesis and germination of plant seeds. Akademie Verlag, Berlin, pp 73–94

Norris RD, Lea PJ, Fowden L (1973) Aminoacyl-tRNA synthetase in *Triticum aestivum* L. during seed development and germination. J Exp Bot 24:615–625

Norris RD, Lea PJ, Fowden L (1975) tRNA species in the developing grain of *Triticum aestivum*. Phytochemistry 14:1683–1686

Norton G, Harris JF (1975) Compositional changes in developing rape seed. Planta 123:163–174

Norton G, Harris JF, Tomlinson A (1978) Development and deposition of protein in oilseeds. Proc Easter School Agric Sci Univ Nottingham 24:59–79

Ochiai-Yanagi S, Fukazawa C, Harada K (1978) Formation of storage protein components during soybean seed development. Agric Biol Chem 42:697–716

Öpik H (1965) The form of nuclei in the storage cells of the cotyledons of germinating seeds of *Phaseolus vulgaris*. Exp Cell Res 38:517–522

Öpik H (1968) Development of cotyledon cell structure in ripening *Phaseolus vulgaris* seeds. J Exp Bot 19:64–76

Oram RN, Doll H, Køie B (1975) Genetics of two storage protein variants in barley. Hereditas 80:53–58

Osborne TB (1924) The vegetable proteins. In: Plimmer RHA, Hopkins FG (eds) Monographs in Biochemistry. Longmans, Green and Co, London

Palmiano EP, Almazan AM, Juliano BO (1968) Physicochemical properties of protein of developing and mature rice grain. Cereal Chem 45:1–12

Pate JS (1976) Physiology of the reaction of nodulated legumes to environment. In: Nutman PS (ed) Symbiontic nitrogen fixation in plants. Cambridge Univ Press, Cambridge, New York Melbourne, pp 335–360

Payne PI (1976) The long-lived messenger ribonucleic acid of flowering-plant seeds. Biol Rev 51:329–363

Payne PI, Boulter D (1969) Free and membrane bound ribosomes of the cotyledons of *Vicia faba* (L.). I. Seed development. Planta 84:263–271

Payne ES, Brownrigg A, Yarwood A, Boulter D (1971a) Changing protein synthetic machinery during development of seeds of *Vicia faba*. Phytochemistry 10:2299–2303

Payne ES, Boulter D, Brownrigg A, Lonsdale D, Yarwood A, Yarwood JN (1971b) A polyuridylic acid-directed cell-free system from 60 day-old developing seeds of *Vicia faba*. Phytochemistry 10:2293–2298

Pederson K, Larkins BA (1981) Factors regulating zein biosynthesis during maize endosperm development. In: Müntz K, Manteuffel R, Scholz G (eds) Proc 2nd Seed Protein Symp: Regulation of protein biosynthesis and degradation during embryogenesis and germination of plant seeds. Akademie Verlag, Berlin, pp 31–40

Penning de Vries FWT, Brunstieg AHM, Van Laar HH (1974) Products, requirements and efficiency of biosynthesis: a quantitative approach. J Theor Biol 45:339–377

Pernollet JC (1978) Protein bodies of seeds: Ultrastructure, biochemistry, biosynthesis and degradation. Phytochemistry 17:1473–1480

Peterson DM, Smith D (1976) Changes in nitrogen and carbohydrate fractions in developing oats groats. Crop Sci 16:67–71
Peumans WJ, Carlier AR (1977) Messenger ribonucleoprotein particles in dry wheat and rye embryos. Planta 136:195–201
Peumans WJ, Caers LI, Carlier AR (1979) Some aspects of the synthesis of long-lived messenger ribonucleoproteins in developing rye embryos. Planta 144:485–490
Peumans WJ, Carlier AR, Caers LI (1980a) Botanical aspects of cell-free protein synthesizing systems from cereal embryos. Planta 147:307–311
Peumans WJ, Carlier AR, Schreurs J (1980b) Cell free translation of exogenous mRNA in extracts of dry pea primary axes. Planta 147:302–306
Peumans WJ, Carlier AR, Manickam A, Delaey BH (1981) Protein biosynthesis at the beginning of embryo development in germinating seeds. In: Müntz K, Manteuffel R, Scholz G (eds) Proc 2nd Seed Protein Symp: Regulation of protein biosynthesis and degradation during embryogenesis and germination of plant seeds. Akademie Verlag, Berlin, pp 163–180
Poulson R, Beevers L (1973) RNA metabolism during the development of cotyledons of *Pisum sativum* L. Biochem Biophys Acta 308:381–389
Préaux G, Lontie R (1975) The proteins of barley. In: Harborne JB, Van Sumere C (eds) Chemistry and biochemistry of plant proteins. Academic Press, London New York, pp 89–111
Prokofiev AA, Sveshnikova IN, Sobolev AM (1967) Changes in the structure and composition of aleuron grains in maturing seeds of castor-bean plants (Russian). Fiziol Rast 16:889–897
Püchel M, Müntz K, Parthier B, Aurich O, Bassüner R, Manteuffel R, Schmidt P (1979) RNA metabolism and membrane bound polysomes in relation to globulin biosynthesis in cotyledons of developing field beans. Eur J Biochem 96:321–329
Püchel M, Bäumlein H, Silhengst P, Wobus U (1981) Cloning of double stranded cDNA derived from polysomal poly(A)-containing RNA of field bean cotyledons (*Vicia faba* L.) In: Müntz K, Manteuffel R, Scholz G (eds) Proc 2nd Seed Protein Symp: Regulation of protein biosynthesis and degradation during embryogenesis and germination of plants seeds. Akademie Verlag, Berlin, pp 227–228
Pusztai A (1964) Hexosamines in the seeds of higher plants (spermatophytes). Nature (London) 201:1328–1329
Puzsztai A, Watt WB (1970) Glycoprotein II. Isolation and characterization of a major antigenic and nonhemagglutinating glycoprotein from *Phaseolus vulgaris*. Biochim Biophys Acta 207:413–431
Raacke ID (1957a) Protein synthesis in ripening pea seeds. I. Analysis of whole seeds. Biochem J 66:101–110
Raacke ID (1957b) Protein synthesis in ripening pea seeds. II. Development of embryos and seed coats. Biochem J 66:110–113
Rabson R, Mans RJ, Novelli GD (1961) Changes in cell-free amino acid incorporating activity during maturation of maize kernels. Arch Biochem Biophys 93:555–562
Randall PJ, Thomson JA, Schroeder HE (1979) Cotyledonary storage proteins in *Pisum sativum*. IV. Effect of sulphur, phosphorus, potassium and magnesium deficiencies. Aust J Plant Physiol 6:11–24
Reeve RM (1948) Late embryology and histogenesis in *Pisum*. Am J Bot 35:591–602
Rest JA, Vaughan JG (1972) The development of protein and oil bodies in the seed of *Sinapis alba* L. Planta 105:245–262
Righetti PG, Gianazza A, Viotti A, Soave C (1977) Heterogeneity of storage proteins in maize. Planta 136:115–123
Röbbelen G (1976) Besseres Pflanzeneiweiß durch Anbaumaßnahmen und Züchtung. Ernährungsumschau 23:49–57
Romero J, Sun SMM, McLeester RC, Bliss FA, Hall TC (1975) Heritable variation in a polypeptide subunit of the major storage protein of the bean, *Phaseolus vulgaris* L. Plant Physiol 56:776–779
Scharpé A, Van Parijs R (1973) The formation of polyploid cells in ripening cotyledons of *Pisum sativum* L. in relation to ribosome and protein synthesis. J Exp Bot 24:216–222

Scharpé A, Van Parijs R (1974) The metabolism of nucleic acids and proteins in cotyledons of *Pisum sativum* L. during seed formation. Meded Rijksfac Landbouwwet Gent 38:322–342

Schilling G (1977) Einige Probleme bei der Anwendung hoher Stickstoffdüngergaben in der industriemäßigen Pflanzenproduktion und Wege zu ihrer Lösung. Arch Acker Pflanzenbau Bodenkd 21:175–189

Schilling G, Schalldach I (1966) Untersuchungen über Transport, Einbau und Verwertung spät gedüngten Mineralstickstoffs bei *Pisum sativum* (L.) Thaer-Arch 10:895–907

Schilling G, Schalldach I, Polz S (1967) Neue Ergebnisse über die N-Düngung zu Leguminosen, erzielt auf der Grundlage von Versuchen mit doppelt N-markiertem $NH_4NO_3$. Wiss Z Friedrich-Schiller-Univ Jena, Math-Naturwiss Reihe 16:385–389

Schmidt P, Bassüner R, Jank HW, Manteuffel R, Püchel M (1981) Polysomes and poly(A)-containing RNA of membrane-bound polysomes from developing broad bean seeds (*Vicia faba* L., var. *minor*). In: Müntz K, Manteuffel R, Scholz G (eds) Proc 2nd Seed Protein Symp: Regulation of protein biosynthesis and degradation during embryogenesis and germination of plant seeds. Akademie Verlag, Berlin, pp 229–230

Schopfer P, Bergfeld R, Kühnl T (1980) Development of cytoplasmic organelles (aleurone bodies, oleosomes) from the endoplasmic reticulum in plants. In: Nover L, Lynen F, Mothes K (eds) Cell compartmentation and metabolic channeling. VEB Fischer, Jena, pp 345–357

Schultz GA, Chen C, Katchalski E (1972) Localization of a messenger RNA in a ribosomal fraction from ungerminated wheat embryos. J Mol Biol 66:379–390

Setterfield G, Stern H, Johnston FB (1959) Fine structure in cells of pea and wheat embryos. Can J Bot 37:65–74

Sharma GC, Bradford RR (1973) Effect of sulphur on yield and amino acids of soybeans. Commun Soil Sci Plant Anal 4:77–82

Shepherd KW (1968) Chromosomal control of endosperm proteins in wheat and rye. Proc 3rd Int Wheat Genet Symp. Aust Acad Sci, Canberra, pp 86–96

Shewry PR, Pratt HM, Finch RA, Miflin BJ (1978) Genetic analysis of hordein polypeptides from single seeds of barley. Heredity 40:463–466

Sinclair TR, De Wit CT (1975) Photosynthate and nitrogen requirements for seed production by various crops. Science 189:565–567

Smith DL (1973) Nucleic acid, protein and starch synthesis in developing cotyledons of *Pisum arvense* L. Ann Bot 37:795–804

Soave C, Suman N, Viotti A, Salamini F (1978) Linkage relationships between regulatory and structural gene loci involved in zein biosynthesis in maize. Theor Appl Genet 52:263–267

Soave C, Viotti A, Di Fonzo N, Salamini F (1979) Maize prolamin: Synthesis and genetic regulation. In: Seed protein improvement in cereals and grain legumes. Proc Int Symp FAO/IAEA. IAEA, Vienna, pp 165–174

Sobolev AM, Sveshnikova IN, Ivanova AJ (1968) The appearance of aleuronic grains in the endosperm of ripening ricinus seed (Russian). Dokl Akad Nauk SSSR 181:1503–1505

Sodek L, Wilson CM (1970) Incorporation of leucine-14-C and lysine-14-C into protein in the developing endosperm of normal and opaque-2 corn. Arch Biochem Biophys 140:29–38

Sturani E, Cocucci S (1965) Changes of the RNA system in the endosperm of ripening castor bean seeds. Life Sci 4:1937–1944

Sun SM, Hall TC (1975) Solubility characteristics of globulins from *Phaseolus* seeds in regard to their isolation and characterization. J Agric Food Chem 23:184–189

Sun SM, McLeester RC, Bliss FA, Hall TC (1974) Reversible and irreversible dissociation of globulins from *Phaseolus vulgaris*. J Biol Chem 249:2118–2121

Sun SMM, Buchbinder BU, Hall TC (1975) Cell-free synthesis of the major storage protein of the bean, *Phaseolus vulgaris* L. Plant Physiol 56:780–785

Sun SM, Mutschler M, Bliss FA, Hall TC (1978) Protein synthesis and accumulation in bean cotyledons during growth. Plant Physiol 61:918–923 (1978)

Sun SM, Slightom JL, Hall TC (1981) Intervening sequences in a plant gene-comparison of the partial sequence of cDNA and genomic DNA of French bean phaseolin. Nature (London) 289:37–41

Swift H (1950) The constancy of DNA in plant nuclei. Proc Natl Acad Sci USA 36:643–645

Takagi S, Mori T (1979) The 5′-terminal structure of poly(A)-containing RNA of soybean seeds. J Biochem 86:231–238

Thomson JA, Doll H (1979) Genetics and evolution of seed storage proteins. In: Seed protein improvement in cereals and grain legumes. Proc Int Symp FAO/IAEA. IAEA, Vienna, pp 109–124

Thomson JA, Schroeder HE (1978) Cotyledonary storage proteins in *Pisum sativum*. II. Hereditary variation in components of the legumin and vicilin fractions. Austral J Plant Physiol 5:281–294

Travieso A (1977) Untersuchungen über Sorten-, Umwelt- und Entwicklungsanhängigkeit des Gehaltes, der Zusammensetzung und des Ertrages von Proteinen beim Reis (*Oryza sativa* L.) unter den Klimabedingungen in Kuba. Ph D Thesis, Gatersleben

Tsai CY (1979) Tissue-specific zein synthesis in maize kernel. Biochem Genet 17:1109–1119

Vecher AS, Matoshko IV (1965) Accumulation of a nucleic and other forms of phosphorus, total and protein nitrogen in maturing seeds of *Lupinus* (Russian). Biokhimiya 30:939–946

Viotti A, Balducci C, Weil JH (1978a) Adaptation of the tRNA population of maize endosperm for zein synthesis. Biochim Biophys Acta 517:125–132

Viotti A, Sala E, Albergi P, Soave C (1978b) Heterogeneity of zein synthesized in vitro. Plant Sci Lett 13:365–375

Viotti A, Sala E, Marotta R, Albergi P, Balducci C, Soave C (1979) Genes and mRNAs coding for zein polypeptides in *Zea mays*. Eur J Biochem 102:211–222

Walbot V (1971) RNA metabolism during embryo development and germination of *Phaseolus vulgaris*. Dev Biol 26:369–379

Walbot V (1973) RNA metabolism in developing cotyledons of *Phaseolus vulgaris*. New Phytol 72:479–483

Walbot V, Brady T, Clutter M, Sussex I (1972) Macromolecular synthesis during plant embryogeny: Rates of RNA synthesis in *Phaseolus coccineus* embryos and suspensors. Dev Biol 29:104–111

Wardlaw CW (1955) Embryogenesis in plants. Wiley, New York

Weber E, Neumann D (1980) Protein bodies as storage organelles in plant seeds. Biochem Physiol Pflanz 175:279–306

Weber E, Neumann D, Manteuffel R (1981) Storage protein accumulation and protein body formation in *Vicia* seeds. In: Müntz K, Manteuffel R, Scholz G (eds) Proc 2nd Seed Protein Symp: Regulation of protein biosynthesis and degradation during embryogenesis and germination of plant seeds. Akademie Verlag, Berlin, pp 103–120

Wells GN, Beevers L (1973) Protein synthesis in cotyledons of *Pisum sativum* L. II. The requirements for initiation with plant messenger RNA. Plant Sci Lett 1:281–286

Wells GN, Beevers L (1974) Protein synthesis in the cotyledons of *Pisum sativum* L. Protein factors involved in the binding of phenylalanyl-transfer ribonucleic acid to ribosomes. Biochem J 139:61–69

Wells GN, Beevers L (1975) Protein synthesis in the cotyledons of *Pisum sativum* L. Messenger RNA independent formation of a methionyl-tRNA initiation complex. Arch Biochem Biophys 170:384–391

Wettstein D von (1975) Grundforskning og Målforskning i Genetik. Naturens Verden 12

Wheeler CT, Boulter D (1967) Nucleic acids of developing seeds of *Vicia faba* L. J Exp Bot 18:229–240

Whitehouse RNH (1973) The potential of cereal grain crops for protein production. In: Jones JGW (ed) The biological efficiency of protein production. Univ Press, Cambridge, pp 83–99

Wienand U, Feix G (1978) Electrophoretic fractionation and translation *in vitro* of poly(A)-containing RNA from maize endosperm. Europ J Biochem 92:605–611

Wienand U, Brüschke C, Feix G (1979) Cloning of double stranded DNAs derived from polysomal mRNA of maize endosperm and characterisation of zein clones. Nucleic Acid Res 6:2707–2715

Wilden van der W, Herman EM, Chrispeels MJ (1980) Protein bodies of mung bean cotyledons as autophagic organelles. Proc Natl Acad Sci USA 77:428–432

Wilson CM (1966) Bacteria, antibodies and amino acid incorporation into maize endosperm protein bodies. Plant Physiol 41:325–327

Wollgiehn R (1960) Untersuchungen über den Zusammenhang zwischen Nukleinsäure- und Eiweißstoffwechsel in reifenden Samen. Flora 148:479–483

Wright DJ, Boulter D (1972) The characterization of vicilin during seed development in *Vicia faba* L. Planta 105:60–65

Wrigley CW, Shepherd KW (1973) Electrofocusing of grain proteins from wheat genotypes. Ann NY Acad Sci 209:154–162

Yarwood A, Boulter D, Yarwood JN (1971 a) Methionyl-tRNAs and initiation of protein synthesis in *Vicia faba* L. Biochem Biophys Res Commun 44:353–361

Yarwood A, Payne ES, Yarwood JN, Boulter D (1971 b) Aminoacyl-tRNA binding and peptide chain elongation on 80 S plant ribosomes. Phytochemistry 10:2305–2311

Yatsu LY (1965) The ultrastructure of cotyledonary tissue from *Gossypium hirsutum* L. seeds. J Cell Biol 25:193–199

# 15 Protein and Nucleic Acid Synthesis During Seed Germination and Early Seedling Growth

J.D. BEWLEY

## 1 Introduction

Our understanding of the mechanism of protein synthesis in plants has resulted largely from studies on in vitro systems prepared from dry, quiescent seeds – for details see Chapter 3, this Volume.

When water is added to a dry seed, a variety of metabolic processes quickly resume, including RNA and protein synthesis. In this chapter I will describe the events that are intimately associated with the resumption of protein synthesis, and in particular I will consider the dependence of protein synthesis upon the synthesis of the RNA components that make up the protein-synthesizing complex. First, however, it is necessary to define some of the terms that will be used to describe the physiological states of the rehydrated seed. Uptake

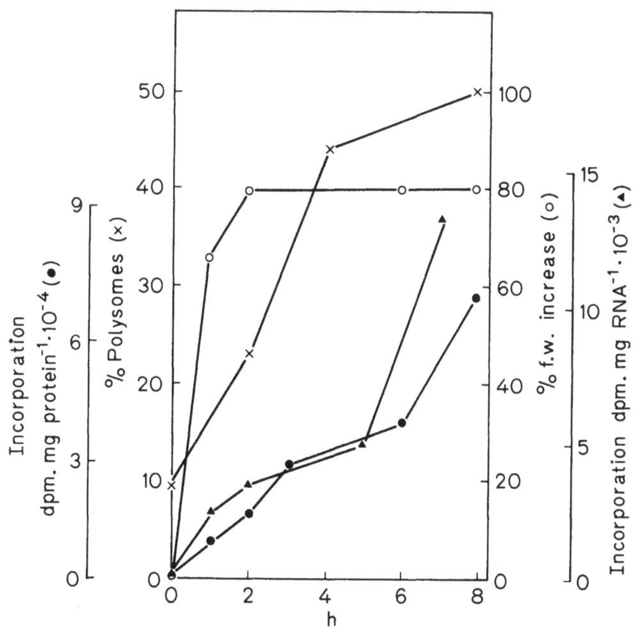


**Fig. 1.** Imbibition, polysome content and protein synthesis by embryos of oat (*Avena sativa* cv. Harmon). ○—○ Fresh weight (% increase of 500 mg dry weight samples of embryos). ●—● Leucine incorporation into protein by 25 embryos. ×—× Percent polysome content of 300 mg dry weight embryos. ▲—▲ Uridine incorporation into RNA by 25 embryos. (AKALEHIYWOT and BEWLEY: previously unpublished data)

of water by mature dry seeds is triphasic (BEWLEY and BLACK 1978). The initial phase is imbibition, when the fresh weight of the seed increases appreciably (Fig. 1). Imbibition can last from several minutes to many hours and is determined by such factors as seed size, and seed coat and tissue permeability. When imbibition is completed there is a period when no increase in fresh weight occurs (Fig. 1). This lag phase is followed by a further increase in fresh weight as the radicle elongates and growth of the seedling commences. Germination is that period from the start of imbibition to the elongation of the radicle to penetrate the structures surrounding the embryo, e.g., testa, pericarp, etc. Events occurring after radicle protrusion are considered as seedling growth.

There can be little doubt that protein synthesis is a pre-requisite for germination to be completed. It is reasonable to assume that germination-related protein synthesis occurs within the embryonic axis itself, since the termination of germination is marked by further growth of the radicle. Hence it is important to distinguish between studies on protein synthesis in the axis and those on protein synthesis in storage organs. It is likely that protein synthesis in the latter is unrelated to germination per se, but is related instead to the mobilization of stored reserves which is a post-germination phenomenon.

## 2 Protein Synthesis in Imbibing Embryos and Axes

Protein synthesis does not occur within the mature dry seed, and polysomes are usually absent. Sometimes, seeds harvested prior to completion of maturation and then rapidly air-dried by commercial procedures may contain a few polysomes, but these are absent from naturally dried seeds. Isolated axes of *Phaseolus lunatus* (KLEIN et al. 1971), and *Phaseolus vulgaris* (GILLARD and WALTON 1973), and isolated embryos of rye *Secale cereale* (SEN et al. 1975), rice *Oryza sativa* (BHAT and PADAYATTY 1974), wheat *Triticum durum* (MARCUS et al. 1966) and oats *Avena sativa* (Fig. 1) commence protein synthesis within 30–60 min of the start of imbibition. In some other species protein synthesis has not been detected until after several hours. However, the techniques used for the detection of in vivo protein synthesis depend upon the incorporation of exogenously supplied radioactive amino acids into nascent polypeptide chains, and hence failure to detect early protein synthesis is most likely a consequence of insufficient uptake of radioactive precursor into the sites of synthesis within the seed.

It is now well accepted that ribosomes are present within mature dry seeds, this having been demonstrated in a number of species, e.g., embryos of castor bean *Ricinus communis* (BEWLEY and LARSEN 1979), wheat (MARCUS et al. 1973), rye (CARLIER and PEUMANS 1976), *Pinus lambertiana* (BARNETT et al. 1974), axes of peanut *Arachis hypogaea* (MARCUS and FEELEY 1964), pea *Pisum arvense* (BRAY and CHOW 1976), and *Phaseolus vulgaris* (BEWLEY unpublished) and whole lettuce *Lactuca sativa* seeds (FOUNTAIN and BEWLEY 1973). Upon imbibition these ribosomes combine with mRNA, and in the presence of various other cytoplasmic components (e.g., initiation and elongation factors, tRNA, amino acids, aminoacyl tRNA synthetases, ATP and GTP) commence protein synthesis. A question that has aroused considerable interest over the past decade is: are the components required for initial protein synthesis conserved within the dry seed, and are they sufficient for all the protein synthesis required for germination to be completed? The answers are by no means unequivocal. However, as will be seen in the following sections, there is reasonable evidence that components of the protein-synthesizing complex are conserved within mature dry seeds and that they appear to be present in sufficient quantities at least for protein synthesis to commence.

## 2.1 Messenger RNA: Conserved, Synthesized, or Both?

A topic of considerable controversy that has occupied the literature in recent years is whether protein synthesis can commence in the imbibing seed without the necessity for prior RNA synthesis. Several researchers claim to have demonstrated that incorporation of radioactive amino acids into protein precedes the incorporation of radioactive precursors into RNA. Consequently, their contention is that protein synthesis is directed by preformed mRNA which has conserved in the dry seed. Nowadays, however, it is generally recognized that RNA synthesis and protein synthesis are concomitant events commencing at the earliest times of imbibition (e.g., Fig. 1), and failure to show this can be attributed to inadequate techniques (PAYNE 1976).

Isolated wheat embryos synthesize rRNA and mRNA very soon after they start to imbibe water (SPIEGEL et al. 1975). Moreover, analysis of the extracted ribosomal/polysomal pellet from wheat embryos imbibed for 0.5–1.5 h in the presence of radioactive uridine shows newly synthesized mRNA associated with polysomes at this time. Rye embryos also synthesize poly(A)-containing RNA, which is probably mRNA, during the first hour from the start of imbibition (PAYNE 1977). Synthesis of RNA, although not necessarily mRNA, has been demonstrated in other seeds as well [e.g., radish (*Raphanus sativus*) embryo axes (DELSENY et al. 1977a)]. Despite these observations, however, new mRNA synthesis might not be essential for the resumption of protein synthesis upon imbibition. During the first hour of imbibition of wheat embryos, polysomes are still formed when over 90% of the initial RNA synthesis, including mRNA, is prevented by chemical inhibitors (Fig. 2).

The addition of a sequence of poly(A) is a necessary step in the formation of most mRNA's prior to their involvement in cellular protein synthesis. Cordycepin (3′-deoxyadenosine) inhibits RNA synthesis (Fig. 2) and suppresses the polyadenylation reaction. Since imbibition of wheat embryos in the presence of effective concentrations of cordycepin does not prevent rapid polysome formation (Fig. 2), this suggests that mRNA is conserved in the dry embryo and does not require further adenylation before being utilized for early protein synthesis. While protein synthesis can commence in wheat embryos without a requirement for new RNA synthesis it is entirely possible that protein synthesis becomes dependent upon RNA synthesis before germination is completed (CAERS et al. 1979, CHEUNG et al. 1979).

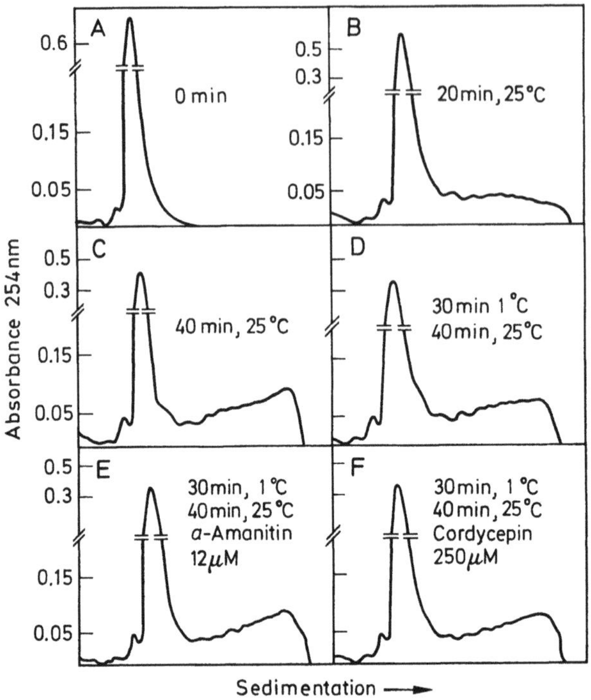


**Fig. 2A–F.** Polysome formation during early germination of wheat embryos and the effect of RNA synthesis inhibitors. Embryos were imbibed in water at 25 °C for **A** 0 min; **B** 20 min; **C** 40 min; **D** 1 °C 30 min, then for 40 min at 25 °C, both in water or **E** 1 °C for 30 min, then for 40 min at 25 °C, both in 12 μM α-amanitin; **F** 1 °C for 30 min, then for 40 min at 25 °C, both in 250 μM cordycepin. Inhibition by 250 μM cordycepin of uridine incorporation into RNA: 83%, and of adenosine incorporation: 82%. Inhibition by 12 μM α-amanitin of uridine incorporation into RNA: 96%, and of adenosine incorporation: 93%. (After SPIEGEL and MARCUS 1975)

Only about 25–40% of the conserved mRNA in mature dry wheat embryos moves into polysomes when protein synthesis is initiated (BROOKER et al. 1978). However, on the basis of hybridization studies, there does not appear to be any difference between the mRNA's that become associated with polysomes and those that do not. Moreover, mRNA in polysomes does not appear to change appreciably between 45 min and 5 h from the start of imbibition, even though at 5 h the mRNA in polysomes might be that which is newly synthesized and at 45 min that which was conserved in the dry seed (BROOKER et al. 1978). However, it is debatable whether the hybridization technique used in these studies is sufficiently discriminating to be certain that there were no qualitative changes in mRNA. An analysis of the proteins synthesized in vitro on the mRNA's present on polysomes at 45 min and 5 h from the start of imbibition has still to be done. By 6 h it is likely that most of the utilized mRNA's are newly synthesized, and it has been confirmed independently that the proteins synthesized by the new mRNA's are not substantially different from those coded for by conserved mRNA's (see later).

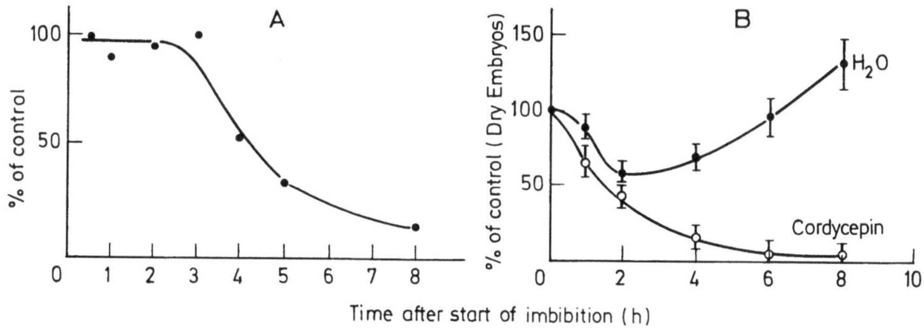


**Fig. 3. A** Decline in the amount of in vivo protein synthesis coded for by stored mRNA of radish (*Raphanus sativus*). Embryonic axes were germinated on water (control) or on cordycepin (200 μg/ml) for various times, and the effects of this inhibitor on RNA synthesis or subsequent protein synthesis over a 30-min period was determined. **B** Poly(A)-containing RNA content of radish embryos during germination in the presence (○) or absence (●) of cordycepin (200 μg/ml). (After DELSENY et al. 1977b)

Protein synthesis in radish embryonic axes is insensitive to cordycepin for the first 2–3 h after the start of imbibition (Fig. 3A), indicating that new RNA synthesis, presumably including mRNA synthesis, is not required during this time. Poly(A)-containing RNA is lost from embryos germinating on cordycepin during the first few hours, although the total poly(A) content of embryos increases after an initial decline (Fig. 3B). Thus it appears that there is a loss of conserved mRNA during the first few hours of germination, and an increase in new messages, with protein synthesis becoming increasingly dependent upon the latter with time of germination.

Protein synthesis in the embryos of cotton seeds is unaffected by concentrations of cordycepin that reduce poly(A)-containing RNA synthesis by 75% (HAMMETT and KATTERMAN 1975). Hence neither new mRNA synthesis nor processing of conserved mRNA's appears to be required for early protein synthesis.

In germinating embryos of *Agrostemma githago,* protein synthesis during the first 3 h from the start of imbibition is reduced considerably by incubation on cordycepin or on α-amanitin (HECKER and BERNHARDT 1976a, HECKER et al. 1976). During these early times there is considerable synthesis of poly(A) sequences (HECKER et al. 1976) in newly formed mRNA (HECKER 1977). This had led to the conclusion that early protein synthesis is dependent upon transcriptional events. However, the results of experiments upon which this conclusion is based are open to an alternative interpretation. Since the effects of inhibitors were measured over a 3-h period, it is possible that very early protein synthesis was dependent upon conserved mRNA's, but that sometime during the first 3 h the embryo required newly synthesized mRNA. The latter process would have been sensitive to the inhibitors, giving an overall lowered protein synthesis over a 3-h period, and hence masking any earlier, inhibitor-insensitive protein synthesis. Going back to the wheat embryo experiments: there it has been found also that inhibitors of mRNA synthesis and processing can lower the intensity of early protein synthesis if applied during the first 3 h (CHEUNG et al. 1979), but at the same time it is known that conserved mRNA's are used for protein synthesis at the earliest times.

It has been claimed that protein synthesis in pea embryonic axes commences after 2 h from the start of imbibition but that RNA synthesis does not start until after 24 h (SIELIWANOWICZ and CHMIELEWSKA 1973, 1974), although the results presented do not appear to bear this claim out. Moreover, the sensitivity of the techniques used to measure RNA and protein synthesis can be questioned. Indeed, more recently others have shown that RNA synthesis occurs in imbibed pea axes at considerably earlier times than 24 h (TANIFUJI et al. 1969, ROBINSON and BRYANT 1975, TAKAHASHI et al. 1977), even as early as during the first hour, when protein synthesis occurs also (BRAY and DASGUPTA 1976). It has been claimed further that there is no poly(A)-containing RNA in mature dry pea embryos but that it appears during the first 2 h of imbibition (SIELIWANOWICZ et al. 1977). The contention is that mRNA exists in the dry seed in a non-adenylated state and must be activated before it can be utilized; when it becomes adenylated (during the first 2 h), protein synthesis can commence. On the basis of what is known about other seeds, and about the work of others on pea axes themselves, this course of events is unusual, and independent confirmation is appropriate.

The contention that protein synthesis commences in rice grains within 30 min of imbibition, but RNA synthesis is negligible for the first 9 h (BHAT and PADAYATTY 1974) also requires confirmation before being accepted.

While the utilization of conserved messages during germination has been studied in a number of species, the presence of mRNA in dried seeds has also been recorded in several others. Not all claims for conserved mRNA's can be regarded as having adequate supportive evidence, and a comprehensive review of the parameters required as evidence of conserved mRNA's has been outlined (PAYNE 1976). Where the evidence is adequate, it is apparent that dry seeds contain quite a number of mRNA's. This is shown in Fig. 4, where the in vitro translation products are displayed of an mRNA fraction extracted from dry rye embryos and from dry whole seeds of pea and rape. The heterogeneity of stored poly(A)-containing RNA has been demonstrated also for wheat embryos (PEUMANS and CARLIER 1977, CUMING and LANE 1978, CAERS et al. 1979), rye embryos (PEUMANS and CARLIER 1977), cotton (*Gossypium hirsutum*) embryos (HAMMETT and KATTERMAN 1975), radish seeds (possibly) (FERRER et al. 1979), and it has been inferred for rape seed (*Brassica napus*) cotyledons (PAYNE et al. 1976).

Since mature dry seeds contain mRNA's, it is pertinent to consider how and where they are stored within the cell. It is probably reasonable to assume that many of the mRNA's required for the synthesis of proteins during seed development (and in particular those involved in storage product synthesis) are destroyed prior to the completion of seed maturation. This can be inferred from the results in Fig. 4, for the translation products of mRNA from the mature pea seed (slot 8) do not include storage protein polypeptides. Others not required for germination might be conserved but destroyed quickly upon imbibition (DURE 1979). How, then, are some messages conserved and utilized

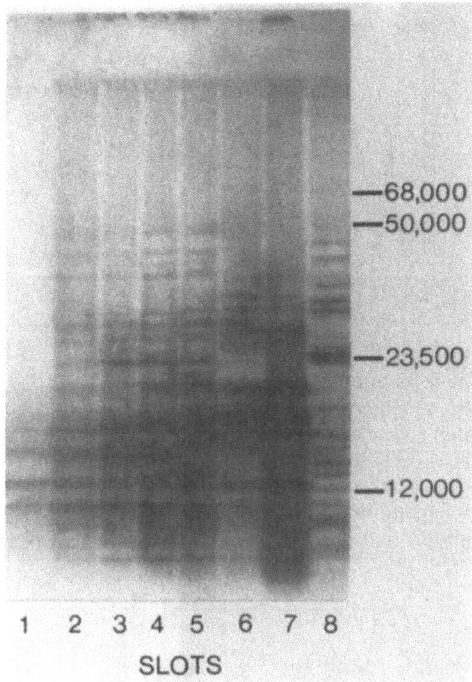


**Fig. 4.** Fractionation by polyacrylamide gel electrophoresis of the products of poly(A)-containing RNA from dry seeds. The poly(A)-rich fractions [II and III, obtained by oligo (dT) cellulose column chromatography] were translated in an in vitro protein-synthesizing system containing [$^{35}$S] methionine, and polypeptide products were separated by sodium dodecylsulphate polyacrylamide gel electrophoresis (PAGE) and incorporated radioactivity detected by autoradiography. *Numbers on the right* are the positions to which marker proteins of known molecular weight migrated. *1* Endogenous activity of the in vitro system minus mRNA; *2* High molecular weight RNA from rye embryo; *3, 4* Fractions II and III respectively, of poly(A)-containing RNA from an oligo (dT) cellulose column; *5* Fraction I (95% of total RNA) from rye embryos reintroduced onto an oligo (dT) cellulose column to obtain Fraction III; *6* Fraction II from rape seeds; *7* Fraction III from rape seeds; *8* Fraction II from pea. There are 23 polypeptide bands synthesized by polyadenylated messages from rye embryos, 15 from rape seeds, and 20 from pea seeds. (GORDON and PAYNE 1976)

during germination while others are being degraded? One possible reason is that mRNA's to be conserved in the dry seed are protected by being stored in the nucleus during maturation, either in their activated (polyadenylated) state, or ready to be activated. There is some evidence that in cotton embryos (HAMMETT and KATTERMAN 1975) and in rape seed (PAYNE et al. 1976, 1977) the subcellular location of the polyadenylated messenger fraction is within the nucleus. On the other hand, the accumulation of mRNA's in cytoplasmic informosomes during the development of wheat embryos has been advocated (AJTKHOZHIN et al. 1976). In dry rye embryos most of the stored mRNA is present in the form of cytoplasmic mRNP's, ranging in size from 25S to 104S (Fig. 5A and B). Removal of the protein moiety from the mRNP's reveals three peaks

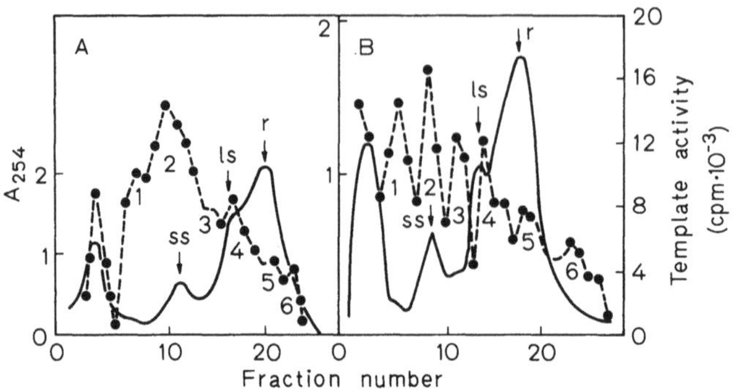


**Fig. 5A, B.** Heterogeneity of endogenous mRNP's in mature dry rye embryos as shown by separation on sucrose gradients. Absorbance at 254 nm by the fractions is shown by the *solid line* and their template activity by the *dotted lines*. **A** Separation was by centrifugation using a 10–30% linear sucrose gradient. **B** Separation was by centrifugation using a 14–38% hyperbolic gradient. *ss* position of small ribosomal subunit; *ls* position of large ribosomal subunit; *r* position of ribosomes. Peaks *1, 2, 3, 4, 5* and *6* correspond to sedimentation values of 25S, 35S, 54S, 64S, 85S and 104S, respectively. (After PEUMANS and CARLIER 1977)

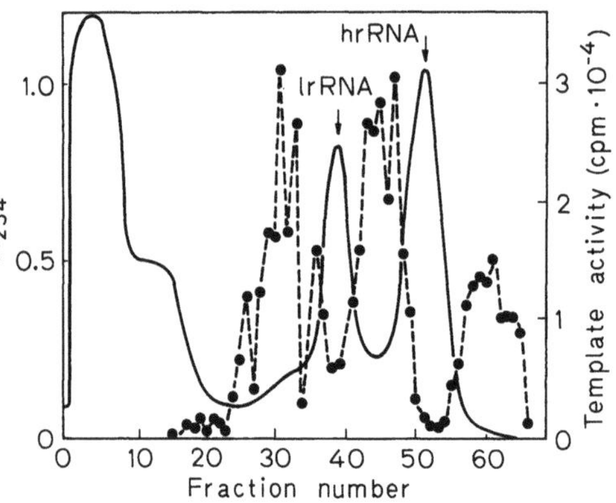


**Fig. 6.** Localization of the mRNA in the total RNA from dry rye embryos. Protein-free fractions were separated on a 14%–30% hyperbolic gradient. Absorbance at 254 nm is shown by the *solid line*, and template activity by the *dotted lines*. The position of ribosomal RNA's is marked by: *lr RNA* light ribosomal; *hr RNA* heavy ribosomal RNA. (After PEUMANS and CARLIER 1977)

of messenger activity on subsequent analysis using sucrose gradients (Fig. 6). However, this distribution of activity does not reflect the real mRNA distribution pattern, since fractions containing rRNA have reduced template activity because of interference with the in vitro assay by this contaminant (PEUMANS and CARLIER 1977). What can be concluded is that the size of the mRNA's varies between 8S and 35S. In wheat embryos the sizes of the mRNP particles and of the poly(A)-containing RNA's are purported to be considerably smaller (FILIMONOV et al. 1977, ISKHAKOV et al. 1978).

It is interesting to note in relation to the distribution of conserved mRNA's or mRNP particles within the cell that variation in technique can cause variation in apparent distribution. For example, the normally soluble mRNP's of dry rye embryos can be converted to an insoluble form at high $Mg^{2+}$ or $Ca^{2+}$ concentrations, or at low pH (PEUMANS 1978).

As in radish embryos (Fig. 3), the conserved mRNA's in wheat embryos are lost during the first few hours of germination and new ones are synthesized (Caers et al. 1979). But despite this changeover in source of mRNA the same proteins are coded for by the messenger fraction from dry embryos as by that from 1-, 2-, 4-, 6- and 8-h germinating embryos (Caers et al. 1979). Similarly, in rye embryos the same proteins are synthesized in dry embryos and up to 8 h of germination (Peumans 1978). Analysis by hybridization techniques of the prevalent classes of mRNA in wheat embryos has shown no differences between the messages associated with polysomes after 40 min of imbibition and with those associated after 5 h (Brooker et al. 1978) – although see earlier comments. Big differences were evident between the translated messages in 5-h- and 2-day-incubated embryos, but this is not surprising since the latter would have grown into a sizeable seedling. Contrasting results have been reported for imbibing and germinated rape seed (Gordon and Payne 1977). At 8, 12, 16 and 24 h the in vitro translation products of extracted poly(A)-containing RNA are similar, but unlike those coded for by messages extracted from the dry seed. In rape seed, however, the bulk of mRNA is probably extracted from the cotyledons and not from the germinating axis, and hence the relationship of such changes to germination per se is obscure.

In summary, there can be little doubt that mRNA's are conserved in mature dry seeds. They are in a protected form, perhaps in the nucleus or in the cytoplasm associated with proteins as mRNP particles. Some of the conserved mRNA it utilized at the earliest times of imbibition and germination to direct protein synthesis. There is an excess of mRNA in relation to the early requirements of the cells, but the messages that are not utilized might not be qualitatively different from those that are. Within a few hours there is synthesis of new mRNA's, and continued protein synthesis probably becomes increasingly dependent upon them. De novo synthesized mRNA's appear to be similar to those conserved within the dry seed, although subtle but important changes in protein synthesis could occur as a consequence of changes to minor mRNA species that are undetectable using currently available techniques. At the present time it is not possible to identify any proteins that are essential for completion of the germination process, nor to determine if such proteins (assuming that they exist) require the synthesis of mRNA's at any time during germination. The identification of such proteins and mRNA's may remain beyond the resolution of available techniques for some time – changes of a few proteins within a few cells of the radicle may be all that is required for the initiation of radicle elongation.

## 3 RNA Synthesis in Imbibing Embryos and Axes

It is evident from the foregoing section that RNA synthesis occurs very soon after imbibition commences. The involvement of newly synthesized mRNA's in protein synthesis has been discussed, so in this section the synthesis of other

RNA components involved in protein synthesis (tRNA and rRNA) will be considered.

### 3.1 Ribosomal RNA

Synthesis of rRNA in imbibing wheat embryos has been detected as early as that of mRNA, and within the first 1–2 h of imbibition (DOSHCHANOV et al. 1975, SPIEGEL et al. 1975). The rate of rRNA synthesis increases during and after imbibition (SPIEGEL et al. 1975), with an overall increase that is at least twofold during the first 5.5 h, with about half of that increase occurring up to 3.5 h (HUANG et al. 1980). This new rRNA is associated with ribosomes as early as 1–1.5 h (DOSHCHANOV et al. 1975, HUANG et al. 1980), contrary to a previous suggestion that assembly into mature ribosomes takes as long as 12 h (CHEN et al. 1971). Processing of the 26S rRNA of the larger ribosomal subunit appears to be slower than that of the 18S rRNA of the smaller subunit during the early time period.

It has been suggested that synthesis and processing of RNA occurs more slowly in intact wheat grains than in isolated embryos (GRZELCZAK and BUCHOWICZ 1977), for it was found that radioactive pre-rRNA and mature rRNA species appeared faster in the isolated embryo than in the embryo of the intact wheat grain. While this observation might have some significance, it must be remembered that imbibition and germination of isolated embryos is faster, and hence it might be expected that metabolic events occur sooner.

Maturation of rRNA in rye embryos occurs quickly after imbibition, the pattern of synthesis being (SEN et al. 1975):

rDNA ---→ 31S RNA (mol. wt. $2.3 \times 10^6$) → 25S RNA (mol. wt. $1.3 \times 10^6$)

31S RNA (mol. wt. $2.3 \times 10^6$) → 22S RNA (mol. wt. $1 \times 10^6$) ——→ 18S RNA (mol. wt. $0.7 \times 10^6$)

Twenty minutes after the start of imbibition synthesis of a heterogeneous nuclear RNA (possibly a precursor mRNA), 4S RNA (possibly tRNA) and 5S RNA (the low mol. wt. RNA of ribosomes) is underway. After 40 min of imbibition the 31S rRNA precursor (pre-rRNA) is evident and by 60 min it has started to mature into the 25S rRNA of the larger ribosomal subunit and the 18S rRNA of the smaller subunit, the latter via the immature 22S rRNA form. Between the 3rd and the 6th h from the start of imbibition the processing of the rRNA is virtually complete and large amounts of 25S and 18S rRNA are present within the embryos. It has still to be established when the newly synthesized rRNA becomes associated with ribosomes participating in cellular protein synthesis.

Synthesis of rRNA in germinating soyabean (*Glycine max*) axes is evident within the first 2 h from the start of imbibition and it increases up to the

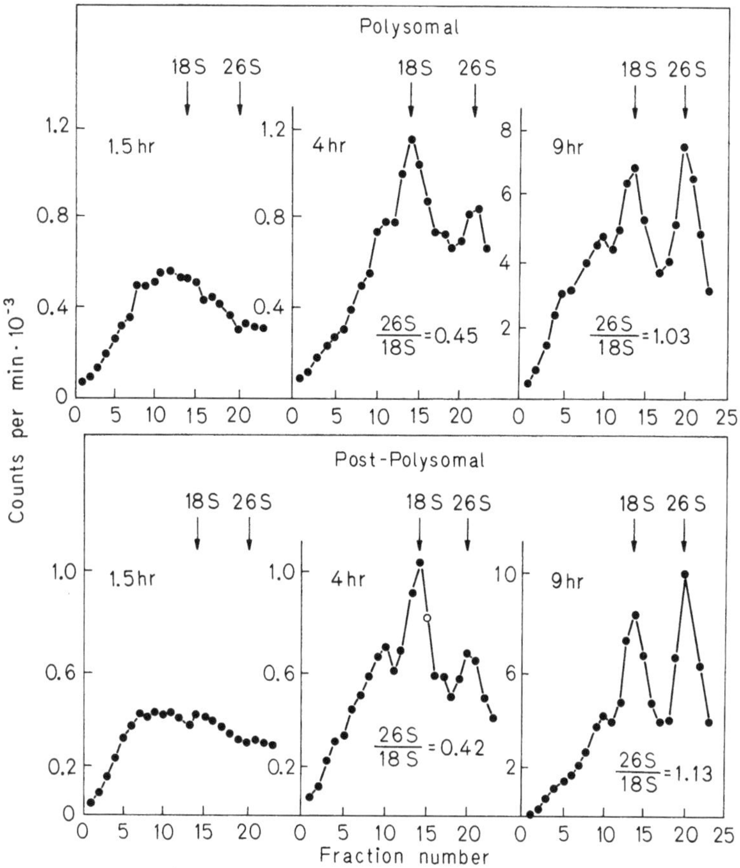


**Fig. 7.** The synthesis and maturation of rRNA in the polysomal and post-polysomal fractions of germinating soyabean axes. Soyabean axes were incubated for 80 min (1 h to 2 h 20 min; 3 h 20 min to 4 h 40 min; 8 h 20 min to 9 h 40 min) in [$^3$H]-uridine. RNA then was extracted from polysomal and post-polysomal (which includes supernatant, ribosomes and subunits) fractions and the non-adenylated (i.e., non-template active) RNA was subjected to sucrose gradient analysis. The 26S/18S ratios were determined from the sum of radioactivity under each peak. (After HUANG et al. 1980)

10th h (HUANG et al. 1980). This newly synthesized rRNA is associated both with the polysomal and ribosomal fractions at these times, suggesting that it is involved in protein synthesis early in imbibition. Processing of pre-rRNA into mature 18S and 26S forms occurs most rapidly after 9 h, maturation of the 26S rRNA being considerably slower at 4 h, and processing occurs only at a low rate at 1.5 h (Fig. 7). Methylation of pre-rRNA is an essential step in its processing to the mature forms, and the control of methylation of newly synthesized rRNA is a possible mechanism whereby processing and utilization is controlled. However, the methylation of soyabean axes rRNA is as prevalent at 2–5 h from the start of imbibition as it is at later times (7.5–10.5 h). Thus the capacity for methylation keeps pace with the capacity for rRNA transcription, and seems not to be the rate-limiting process (HUANG et al. 1980).

**Table 1.** rRNA genes during development and germination of wheat embryo

| Source of DNA | % DNA hybridized | |
|---|---|---|
| | $1.3 \times 10^6$ rRNA | $0.70 \times 10^6$ rRNA |
| Ovule, 5 days after pollination | 0.101 | 0.066 |
| Embryo, 10 days after pollination | 0.086 | 0.067 |
| Embryo, 15 days after pollination | 0.096 | 0.069 |
| Embryo, from dry grain | 0.087 | 0.068 |
| Embryo, 3 days after imbibition | 0.089 | 0.068 |

DNA was prepared from complete ovules 5 days after pollination, from the embryos 10 and 15 days after pollination, from the embryo of the dry grain, and 72 h after imbibition. RNA was prepared from pea roots, which had grown in the presence of 100 μCi/ml [$^{32}$P] orthophosphate for 24 h. The specific activity of the RNA used was 58,000 cpm/μg, and the values are calculated from means of triplicate samples. $1.3 \times 10^6$ rRNA is 25S RNA from the large ribosomal subunit and $0.7 \times 10^6$ rRNA is 18S RNA from the small subunit. (After INGLE and SINCLAIR 1972)

Studies on rRNA synthesis in pea embryos have shown that a large RNA precursor (approx. 31S) accumulates in the nucleolus prior to migration into the cytoplasm (TANIFUJI et al. 1970). Within 4–7 h from imbibition starting, newly synthesized 25S and 17S rRNA has been detected, which possibly matures from a 33S–35S precursor (TAKAIWA and TANIFUJI 1979). Synthesis of pre-rRNA has been observed after 2 h in isolated embryos of radish, and processing into mature rRNA species occurs within 5–6 h, which is also when they are associated with the ribosomes in polysomes (DELSENY et al. 1977a). Synthesis of rRNA in other species of seeds has been recorded, although some studies have only provided information on events occurring after long incubation times, sometimes long after germination is completed, e.g., in *Vicia faba* (FUKUEI et al. 1977), *Phaseolus vulgaris* (WALBOT 1972), *Vigna unguiculata* (CHAKRAVORTY 1969), and in cotton (WATERS and DURE 1966). More recently, though, rRNA synthesis has been detected in whole cotton seeds within 8 h of soaking (CLAY et al. 1975).

Amplification of rRNA genes occurs during certain developmental processes in animals. Since seeds undergo dramatic metabolic changes during development, maturation and subsequent germination, the possibility must be considered that gene amplification occurs in them also. However, on the basis of fragmentary evidence, this seems unlikely. In developing, germinating and growing wheat embryos the percentage of DNA capable of hybridizing with the large ribosomal subunit RNA and with the small subunit RNA remains the same (Table 1) showing that, contrary to a previous report (CHEN and OSBORNE 1970a), there is no amplification or deletion of rRNA genes. No gross amplification of genes occurs in axes of *Vicia faba* either (FUKUEI et al. 1975).

To briefly summarize this section: Mature dry embryos contain ribosomes that are demonstrably active under in vitro protein-synthesizing conditions, and yet rRNA synthesis is an early event during germination. Ribosomes containing de novo synthesized rRNA may become associated with the protein-synthesizing complex early during germination, but it remains to be determined if new rRNA's in new ribosomes are essential for early protein synthesis or, for that matter, for protein synthesis at any stage during germination.

### 3.2 Transfer RNA Synthesis

Transfer RNA's and their aminoacylating enzymes are present within dry seeds, and in vitro protein-synthesizing systems have been developed from wheat embryos (MARCUS et al. 1968) and from rye embryos (CARLIER and PEUMANS 1976) without any requirement for addition of these components. Synthesis of tRNA, however, begins within 20 min in imbibed rye embryos (SEN et al. 1975) and within 1 h in dissected *Phaseolus vulgaris* axes (WALBOT 1972), but its involvement in protein synthesis remains to be determined.

In isolated axes of *P. vulgaris* the total amount of low molecular weight RNA (tRNA and 5S RNA in molar proportions of 15:1) remains nearly constant during the first 15 h from imbibition. An increase occurs between the 18th and 25th to coincide with the initiation of cell division in the axis and with expansion of the hypocotyl. A very different pattern of events has been claimed to occur in imbibing wheat embryos (VOLD and SYPHERD 1968). Here, tRNA levels decline substantially during the first 15 h after imbibition starts and only regain the levels present in the dry embryo by the 20th h, increasing thereafter. It is difficult to reconcile the decline in tRNA (i.e., that there is an excess) with the fact that synthesis of proteins and other components of the protein-synthesizing complex (mRNA and rRNA) is increasing in the embryos at the same time. If, as in rye embryos, there is early tRNA synthesis, then turnover of tRNA's needs to be determined. The presence of aminoacyl tRNA synthetases in dry wheat embryos has been reported, and some of these appear to be contained within high molecular weight complexes (QUINTARD et al. 1978).

No more than 10% of the tRNA molecules in dry lupin *Lupinus luteus* seeds are aminoacylated, and about 40% of the axis tRNA is degraded at the $-C-C-A_{OH}$ end (KEDZIERSKI and PAWELKIEWICZ 1977, and quoted in DZIEGIELEWSKI et al. 1979). Following imbibition, tRNA aminoacylation increases and defective tRNA molecules disappear, presumably because their defects are made good. The enzyme responsible for repair of tRNA, nucleotidyltransferase, is present in the dry axis and its activity increases after imbibition (DZIEGIELEWSKI et al. 1979). No significant changes occur in the size of the tRNA population during germination, and probably not in the levels of aminoacyl-tRNA synthetases either. The scheme of events presented here for lupin must be regarded as being tentative, for many observations were made on dry seeds and one-day-incubated seeds only and not during germination itself. Moreover, some experiments involved the use of whole seeds, others axes, and yet others only cotyledons. Nucleotidyltransferase activity is present in dry wheat embryos, but its activity remains low until after 12 h. It increases in activity therafter, and presumably this increase is associated with growth of the seedling (BALLIVIAN et al. 1977).

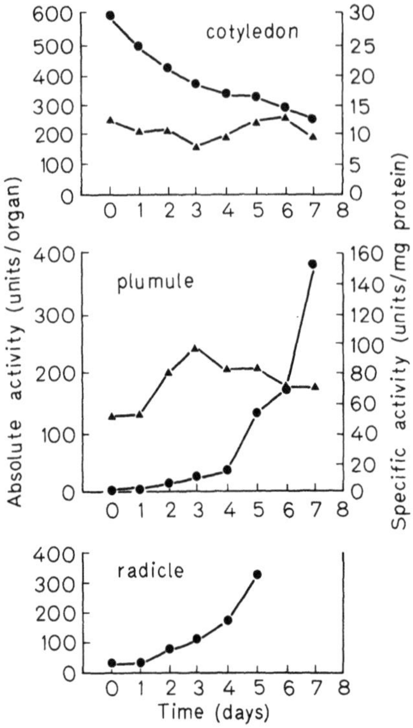


**Fig. 8.** Activity of the aminoacyl tRNA synthetases in various organs of *Phaseolus vulgaris* during germination and growth. ●–● Activity per organ. ▲–▲ Specific activity. (After ANDERSON and FOWDEN 1969)

In an exhaustive study of the activities of the aminoacyl tRNA synthetases for 20 tRNA species from the radicle, plumule and cotyledons of *P. vulgaris* it was found that specific enzyme activity remains unchanged prior to growth of the axis, but activity in the plumule and radicle increases thereafter for several days (Fig. 8). It is likely that enzymes present in the dry seed are sufficient for aminoacylation at all stages of germination, and that increased enzyme activity is necessary only when seedling growth commences. The eventual decline in the cotyledons is related to its senescence following expansion of the first leaf.

### 3.3 The Sequence of RNA Synthesis During Germination

The general pattern of RNA and protein synthesis that is emerging is that embryos can synthesize all types of RNA during germination, and in some seeds this synthesis commences very early upon imbibition.

There are several suggestions in the literature that transcription of different RNA species occurs in a sequential, or cascade, manner. It has been claimed, for example, that mRNA synthesis in wheat embryos commences within the first hour of imbibition, protein synthesis some 2 h later, rRNA synthesis after 12 h and tRNA synthesis after 18 h (DOBRZAŃSKA et al. 1973). But in the light of substantial evidence that mRNA, rRNA and protein are all synthesized

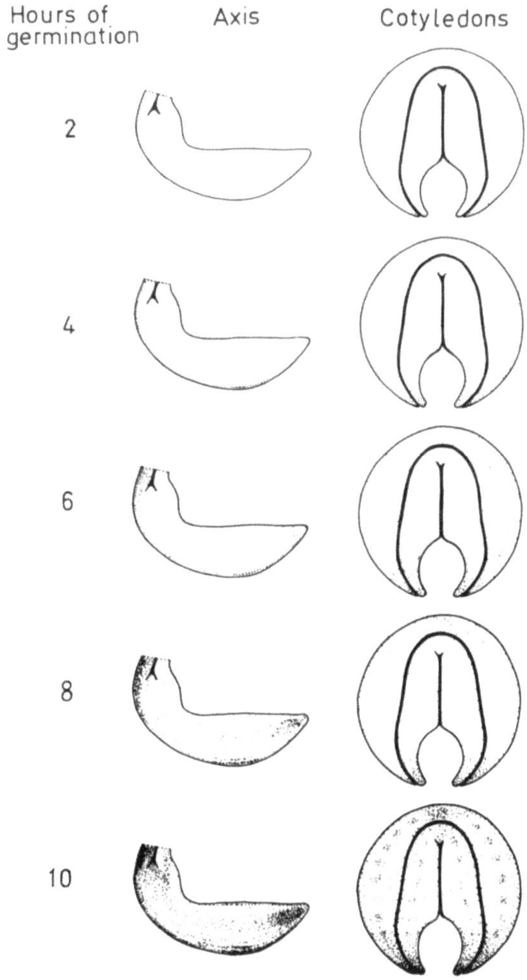


**Fig. 9.** Distribution of newly synthesized RNA in germinating rape seed (*Brassica napus*) embryos. The radioactively-labelled areas, indicated by the *stippling*, are typical of the central region of the cotyledon and the median plane of the embryo axis at each time of germination. Cotyledons were sectioned transversely at right angles to the midribs, and embryos were sectioned longitudinally. (PAYNE et al. 1978)

very early after imbibition starts (e.g., SPIEGEL et al. 1975, HUANG et al. 1980), this claim is not acceptable. Early, and more or less concomitant synthesis of 4–5S RNA, rRNA's and poly(A)-containing RNA occurs within rye embryos, all RNA species being synthesized within the first hour of imbibition (SEN et al. 1975, PAYNE 1977). Both the synthesis of polyadenylated RNA and of rRNA species occurs within the first 2 h of imbibition of radish embryonic axes (DELSENY et al. 1977a).

The synthesis of pre-rRNA, 25S rRNA, 18S rRNA and poly(A)-rich RNA has been detected in the axes of rape seed by the end of the imbibition period (4–6 h from planting) (PAYNE et al. 1978). This observation came from experiments where incorporation of radioactive precursors was followed into RNA. Autoradiographic studies on the same seed have shown even earlier RNA synthesis (by 2 h from the start of imbibition), although this was confined to the peripheral regions of the axis (Fig. 9). Thus RNA synthesis appears initially to be localized within a few cells in one region of the axis, and failure to detect this by precursor incorporation techniques could be because any newly formed radioactive RNA is effectively diluted by unlabelled RNA in the rest of the axis. The order in which cells commence RNA synthesis will be determined by the course of water uptake (and its detection by the course of radioactive precursor uptake). Hence in large seeds, or in intact seeds (as compared to isolated embryos or axes) detectable RNA synthesis might be evident only when a large number of cells are hydrated. Since the embryonic axes are comprised of a variety of cell types, even within the fully imbibed tissues, it can be expected that some cells will synthesize more RNA, and/or more of one type of RNA, than others.

### 3.4 Enzymes and Precursors of RNA Synthesis

Synthesis of RNA during germination requires the involvement of RNA polymerase. There are probably three RNA polymerases required: (1) RNA polymerase I, a nuclear polymerase that transcribes genes coding for 18S and 25S rRNA; (2) RNA polymerase II, localized within the nucleoplasm and responsible for transcription of Hn RNA and presumably mRNA, (3) RNA polymerase III, found within the nucleoplasm (and possibly the cytoplasm), it transcribes the genes coding for tRNA and 5S RNA. The presence and function of these polymerases in seeds remains to a large extent to be demonstrated, however.

Dry wheat germ (JENDRISAK and BECKER 1973), rye embryos (FABISZ-KIJOWSKA et al. 1975) and soyabean axes (GUILFOYLE and KEY 1977) contain DNA-dependent RNA polymerases, and it is presumed that these enzymes are available in sufficient quantities to catalyze RNA synthesis as soon as is necessary upon imbibition. Two RNA polymerase activities have been defined in dry wheat germ. RNA polymerase II activity is present in the embryos of dry grains (JENDRISAK and BURGESS 1975, JENDRISAK 1977), and it remains unchanged during germination (MAZUŚ and BRODNIEWICZ-PROBA 1976). RNA polymerase I activity is low initially and increases slowly after 6 h by de novo synthesis. There is a contradictory claim that RNA polymerase activity declines in isolated wheat embryos after 6 h (MAZUŚ 1973): rightly or wrongly this contradiction has been attributed to differences between experiments using isolated embryos and those using embryos attached to the grain.

Dry soyabean axes contain RNA polymerase II (GUILFOYLE and KEY 1977). The predominant form of this polymerase in dry axes is called IIA (over 95% is in this form). It is a soluble enzyme. With increasing time after imbibition a smaller molecular weight form called IIB which is tightly bound to the chromatin template, increases. It is not clear, however, whether this change occurs after germination is completed, or before. The IIB form has been found in rapidly proliferating soyabean seedling tissues (GUILFOYLE and JENDRISAK 1978). It has been suggested that the larger IIA form is cleaved to the smaller IIB form by proteolysis, and that this cleavage is necessary before the polymerase can become active in transcription. This remains to be confirmed, but even

**Table 2.** Cellular levels of ribonucleoside triphosphates in early germination of wheat embryos. (After CHEUNG and SUHADOLNIK 1978)

| Germination time | ATP | GTP | UTP | CTP |
|---|---|---|---|---|
| | pmol per mg of embryo | | | |
| 0 | 8 | 46 | <1 | <1 |
| 40 min | 1288 | 320 | 132 | 84 |
| 3 h | 2000 | 304 | 420 | 200 |
| 5.5 h | 2072 | 425 | 540 | 320 |

so it must be assumed that there is sufficient active polymerase during early times of germination to support mRNA synthesis.

While it is evident that protein and RNA synthesis commence rapidly during germination, the mechanisms whereby these processes are regulated remain to be elucidated. Initial protein synthesis depends upon the reconstitution of the conserved components of the protein-synthesizing complex, but later synthesis involves newly formed RNA's. The initiation and continuity of RNA synthesis is dependent upon a ready supply of ribonucleoside triphosphate precursors, and it has been suggested (CHEUNG and SUHADOLNIK 1978) that the initial rise in RNA synthesis is dependent upon a rise in all nucleoside triphosphates. Certainly, ATP, UTP, GTP and CTP are present in very low quantities in a dry wheat embryo (Table 2), and all four increase rapidly during the first 40 min of imbibition. The rise in the pyrimidine nucleoside triphosphates (UTP and CTP) is slower, however, and these increase substantially also between 40 min and 5.5 h. This has led to the further suggestion (CHEUNG and SUHADOLNIK 1978) that RNA synthesis at these later times is controlled by pyrimidine nucleoside levels alone. Other studies on wheat embryos (HUANG et al. 1980) have shown an approximate, but by no means concrete correlation between increasing UTP levels and increasing RNA synthesis, with ATP levels increasing earlier than either of these. In soyabean axes the rise in UTP precedes most of the increase in rate of RNA synthesis (HUANG et al. 1980), and hence it might not be the regulating agent – another ribonucleoside triphosphate might be important in this respect, though.

The possible nature of qualitative control mechanisms for RNA synthesis is unknown. Changes in transcriptional activity at the chromatin level have been followed in wheat embryos, but these occur many hours after germination is completed (YOSHIDA et al. 1979).

## 4 Protein and RNA Synthesis in Storage Organs

Whereas protein and RNA synthesis in embryonic axes precedes and accompanies cell elongation, division and differentiation, the development of these synthetic activities within storage organs occurs in the absence of such cellular changes.

Nevertheless, protein and RNA metabolism is a complex process, for synthesis and/or activation of hydrolytic enzymes to mobilize stored reserves is accompanied by their catabolism, and also that of RNA. It is important to note that changes in metabolism within storage organs are to a very large extent unrelated to germination; they are almost exclusively related to the mobilization of stored reserves, which is a post-germination phenomenon associated with seedling growth.

The mobilization of stored reserves is an event that requires the participation of many enzymes, a substantial number of which must be synthesized de novo. It is beyond the intended scope of this chapter to detail the processes associated with the synthesis of the many enzymes involved with stored reserve mobilization and utilization in different seeds. Instead, a general outline of RNA synthesis and turnover will be outlined, and a few examples of synthesis of hydrolytic enzymes will be given – particularly in relation to the dependence of this process on new or conserved mRNA's.

### 4.1 General RNA Metabolism

Total RNA levels in imbibing storage organs may rise before declining, e.g., peanut cotyledons (CHERRY et al. 1965), cucumber (*Cucumis sativus*) cotyledons (BECKER et al. 1978), castor bean endosperms (ROBERTS and LORD 1979), or may not, e.g., *Pisum sativum* (BEEVERS and GUERNSEY 1966), *Vicia faba* (PAYNE and BOULTER 1969); the decline is associated with organ senescence. Not all regions of a storage organ necessarily exhibit the same pattern of RNA metabolism during and after imbibition. Nuclear RNA appears to increase in the outer storage tissues of *Pisum arvense* cotyledons during the first 5 days after imbibition and then declines to a low level before any changes occur within the nuclei of the inner storage tissues (SMITH 1971). RNA synthesis has been detected in the cotyledons of rape seed as early as 2 h from the start of imbibition, but only in localized regions (Fig. 9). More regions commence RNA synthesis with time. It appears that in these regions the synthesis of pre-rRNA, rRNA, tRNA and poly(A)-containing RNA begins almost simultaneously (PAYNE et al. 1978).

Mature dry storage organs of many seeds contain ribosomes that can catalyze protein synthesis in an in vitro system, and it is now widely assumed that other components of the protein-synthesizing complex are present in the dry storage organs too. It has been claimed, however, that ribosomes are lost during late maturation of castor bean endosperms (STURANI 1968), and that there is a massive resynthesis of rRNA and of other RNA species following the first day after imbibition (MARRÉ et al. 1965). While the large increase in total RNA content after the first day has been confirmed (ROBERTS and LORD 1979), the claim that there are no ribosomes in dry castor bean endosperms has been refuted (BEWLEY and LARSEN 1979). In another endosperm, that of onion (*Allium cepa*), RNA precursors become associated with the nucleolus of some cells within 15 min of imbibition starting, and with that of most cells by the second hour (PAYNE and BAL 1972). This observation is suggestive of rRNA synthesis occurring at these early times.

Dry cotyledons appear to contain all of the tRNA species and appropriate aminoacyl tRNA synthetases necessary for protein synthesis to commence upon hydration. Several synthetases have been isolated from dry cotyledons of *Aesculus* species (ANDERSON and FOWDEN 1970), and there is substantial evidence that dry cotton cotyledons contain a full complement of tRNA species (including isoacceptors) capable of being aminoacylated (MERRICK and DURE 1972). Six leucyl and three tyrosyl tRNA's have been extracted from 2-day-imbibed soyabean seeds. Between the 2nd and 15th day the amounts of two of the leucyl tRNA's increase about threefold, while the level of others fluctuates only slightly. Of the tyrosyl tRNA's, two decline and one rises over the same time period (BICK et al. 1970). The significance of these changes is open to speculation, but it is possible that protein synthesis in the cotyledons and other seedling parts is regulated at the translational level by the availability of certain aminoacyl tRNA's. In differentiated seedlings of pea and soyabean there are differences in the synthetases and tRNA species (ANDERSON and CHERRY 1969, KANABUS and CHERRY 1971, WRIGHT et al. 1974), but there is no evidence available yet to suggest that a cell's capacity for protein synthesis can be controlled in this manner.

### 4.2 Synthesis of Enzymes Involved in Reserve Mobilization in Relation to Their mRNA's

The involvement of mRNA's in the synthesis of proteins associated with reserve mobilization is a subject that has attracted much interest and has stirred up considerable controversy. There can be no doubt that a vital post-germination event is a relatively massive synthesis of enzymes responsible for the hydrolysis of stored reserves and for their conversion into metabolites of use to the growing seedling.

#### 4.2.1 α-Amylase Synthesis in the Barley Aleurone Layer

The aleurone layers from dry, de-embryonated barley grains contain an appreciable number of poly(A)-containing RNA's that can be extracted and translated in vitro. After 3 days incubation in water the aleurone layers still contain the same complement of translatable template RNA (JONES and JACOBSEN 1978). The mRNA's present are presumably for constitutive proteins that are required for the normal metabolism of the aleurone cells. Within 2–4 h of the addition of gibberellic acid ($GA_3$) to the aleurone layers of hydrated de-embryonated grains there are qualitative changes in protein metabolism, including an increase in α-amylase synthesis (Fig. 10). Some workers have claimed that the number of ribosomes increases in barley aleurone cells in response to $GA_3$ (JONES 1969, EVINS 1971), although others have shown no $GA_3$-stimulated synthesis of rRNA, or of tRNA (JACOBSEN and ZWAR 1974a, VAN ONCKELEN et al. 1974). Since rRNA synthesis can be stopped in $GA_3$-treated aleurone layers without any

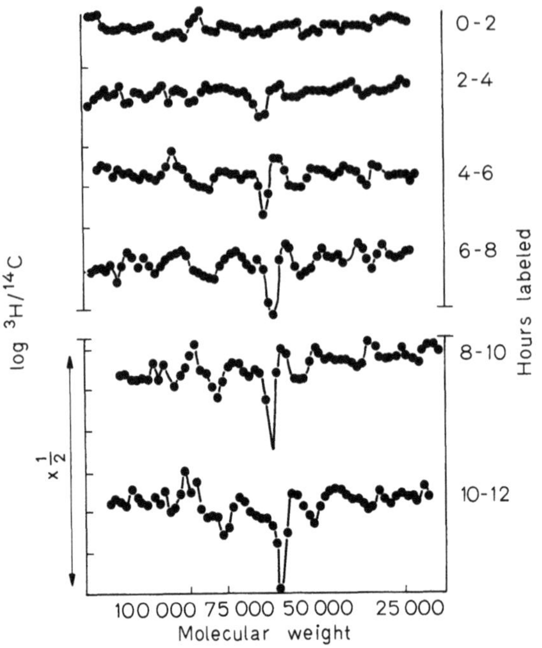


**Fig. 10.** Salt-soluble proteins extracted from control ($^3$H-labelled) and $GA_3$-treated ($^{14}$C-labelled) barley aleurone layers incubated for 2–12 h. Proteins were extracted from aleurone layers, subjected to each treatment, mixed, and fractionated by PAGE. Relative synthetic activities are expressed as a $^3$H/$^{14}$C ratio. The minimum at approx. mol. wt. 50,000 is α-amylase. (After VARNER et al. 1976)

significant effect on either α-amylase synthesis or cellular RNA levels (JACOBSEN and ZWAR 1974a), it is unlikely that the production of ribosomes is necessary for enzyme synthesis. An increase in the number of polysomes in $GA_3$-stimulated aleurone cells has been reported, commencing 3–4 h after addition of the hormone (EVINS 1971). This may indicate that aleurone cells are becoming increasingly active at synthesizing proteins, although there is some question that the polysome extraction techniques used could have led to spurious increases. On balance, it is more likely that $GA_3$ acts by effecting a qualitative change in protein synthesis (Fig. 10) rather than a quantitative one.

One action of $GA_3$ in the isolated barley aleurone is to enhance the synthesis of poly(A)-containing RNA (HO and VARNER 1974, JACOBSEN and ZWAR 1974b, HIGGINS et al. 1976, MUTHURKRISHNAN et al. 1979). When this RNA is placed in an in vitro protein-synthesizing system several proteins are produced, including α-amylase (Fig. 11) (MUTHURKRISHNAN et al. 1979). The amount of mRNA for α-amylase increases with time after $GA_3$ addition, and after about 15 h it accounts for 15%–25% of the total translatable mRNA (JONES and JACOBSEN 1978). The way in which $GA_3$ induces the increase in α-amylase mRNA is still open to debate. It may occur by de novo synthesis, and $GA_3$ may be an activator of transcription (perhaps by activation of some GA–receptor complex). Alternatively, $GA_3$, or a product of $GA_3$ action, might protect mRNA from degradation, so that mRNA for α-amylase could be destroyed in aleurone layers incubated in water but not in those incubated in $GA_3$. A third possibility is that $GA_3$ could somehow enhance the translational capacity of preformed mRNA for α-amylase by effecting its processing, activation, or release from an inactive form. Lastly, the increase in α-amylase activity might simply occur

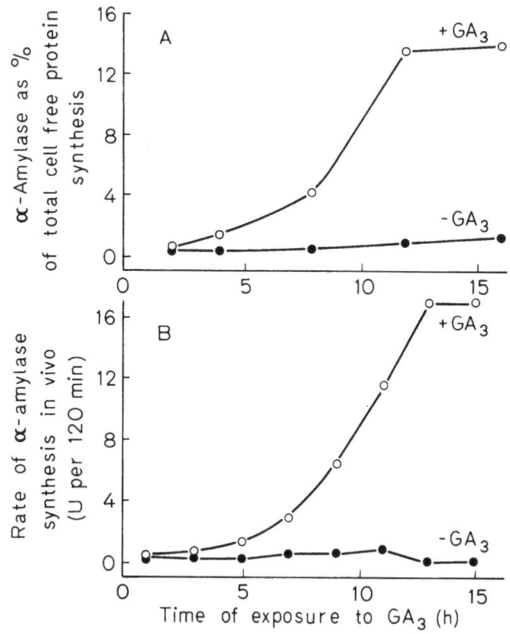


**Fig. 11A, B.** The increase with time in **A** the level of translatable mRNA for α-amylase and **B** the increase in rate of synthesis in vivo in response to $GA_3$. *U* units of α-amylase activity. For **A** the poly(A)-containing RNA was extracted from aleurone layers treated with $GA_3$ for different lengths of time and used to support α-amylase synthesis in vitro. (After HIGGINS et al. 1976)

as a result of a chain of processes set in motion by $GA_3$, and induction of synthesis of this enzyme might be only indirectly related to the primary action of the hormone (i.e., $GA_3$ might not directly affect transcription or translation of α-amylase mRNA per se). In relation to this last possibility, it has been suggested that prior to the appearance of mRNA for α-amylase there is a requirement for the synthesis of other proteins, and that the primary action of $GA_3$ is related to the induction of synthesis of these proteins (MUTHURKRISHNAN et al. 1979). But the evidence for this sequence of events is indirect, and further confirmation is required.

### 4.2.2 Castor Bean Endosperm and Cucumber Cotyledons

Both of these storage organs contain considerable amounts of lipid, the catabolism of which requires the participation of numerous enzymes. Fatty acids released from lipids are converted to sugars, and part of this process occurs within a newly synthesized organelle, the glyoxysome, in which enzymes of the glyoxylate cycle are sequestered. There are few studies on the synthesis of individual enzymes associated with lipid catabolism, and whether they are synthesized from conserved or from newly synthesized mRNA's has not been determined.

The mature dry castor bean endosperm contains a little poly(A)-RNA, the level of which declines initially before there is a marked rise on the second day (Fig. 12B). The mRNA fraction present in the dry endosperm might include residual messages left over from seed development, which are destroyed early upon imbibition. The large increase in RNA and mRNA (Fig. 12A and B) up to 4 days coincides with an increase in the enzymes of lipid catabolism

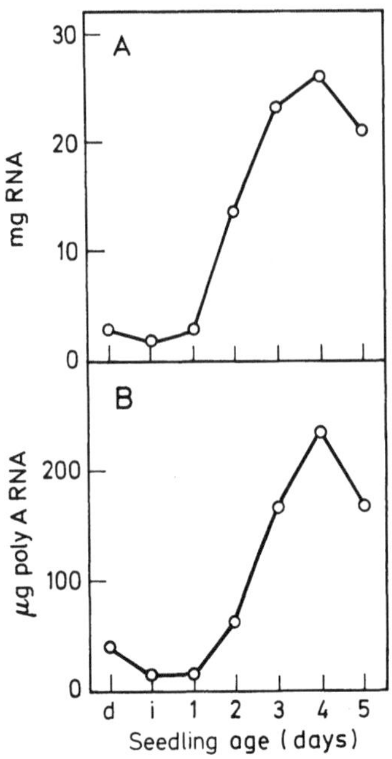


**Fig. 12A, B.** Changes to **A** total RNA and **B** poly(A)-containing RNA in the endosperms of castor bean. *d* and *i* signify dry and imbibed seeds, respectively. (After ROBERTS and LORD 1979)

and in glyoxysome formation (GERHARDT and BEEVERS 1970, BEEVERS 1979). It is reasonable to expect that at least some of the mRNA's are for enzymes involved in mobilization and utilization of lipid reserves.

The amount of poly(A)-containing RNA in the cotyledons of mature dry cucumber seeds is also low, but it increases within 24 h of the start of imbibition (WEIR et al. 1979). Very low levels of translatable messages for the glyoxysomal enzymes isocitrate lyase and malate synthetase are present initially, but in cotyledons of seedlings grown under a 12-h photoperiod they increase dramatically within 3 days to a maximum which coincides with the peak of lipid metabolism. An increase in glyoxysomal enzymes is also accompanied by an increase in other cytoplasmic RNA species (particularly rRNA), indicating a generally enhanced capacity for protein synthesis by the cotyledons (BECKER et al. 1978).

### 4.2.3 Cotton Seed Cotyledons: The Carboxypeptidase Message

DURE (1979) has proposed that there could be several mRNA sub-sets involved in protein synthesis in cotton cotyledons during and after germination:

a) Residual mRNA's – these are mRNA's that are produced during seed development and are not destroyed during late maturation and desiccation. They are not important for germination and may be destroyed early during imbibition. An example of this type of mRNA might be the storage protein mRNA present in dry soyabean seeds (MORI et al. 1978).

b) Stored or conserved mRNA's – these are translated into proteins that are an integral part of germination, or of post-germination events such as reserve mobilization.

c) Newly synthesized mRNA's – these are for proteins unique to germination or post-germination, without which these events could not take place.

d) Newly synthesized, constitutive mRNA's – for enzymes of intermediary metabolism, i.e., for essential cellular processes not necessarily unique to germination or growth.

Of these subsets, the one that has aroused the most interest is (b), the conserved mRNA's. The concept of stored mRNA's for the hydrolytic enzymes involved in reserve hydrolysis is one that received quite widespread support and acceptance in the late 1960's and early 1970's, but in recent years it has been subjected to considerable criticism.

A proteinase (carboxypeptidase C), which is probably involved in the hydrolysis of storage protein, arises in the cotyledons of cotton seeds by de novo synthesis after germination is completed. It has been claimed (IHLE and DURE 1969, DURE 1975) that the mRNA for this enzyme is conserved in the cotyledons of the mature dry seed and is activated and translated following germination. Preliminary experiments on the mature dry cotton seed showed that protein synthesis (DURE and WATERS 1965), including synthesis of the proteinase (IHLE and DURE 1969) is not inhibited by actinomycin D applied during the first 3 days after sowing. It was inferred from this observation that the synthesis of the enzyme does not depend upon prior RNA synthesis (including mRNA synthesis). Evidence was then sought to show that production of mRNA for the proteinase occurred during seed development and was stored in the dry seed.

Immature embryos of cotton germinate precociously when excised from the ovule during their development. Before the final desiccation phase the mature embryos weigh up to 125 mg (IHLE and DURE 1969, 1972, DURE 1975). Developing embryos, weighing as little as 95 mg, germinate when they are removed from the ovule and placed on agar gel; like the more mature embryos they produce proteinase also. Dissected, developing embryos placed on actinomycin D also germinate, but those at a stage of development where they weigh less than 100 mg fail to produce the enzyme. This has been taken to show that mRNA for proteinase is produced within the embryo sometime between the 95-mg and 100-mg stage of development, approximately 15–20 days before the maturation process is completed. It is important to note that an assumption behind these early experiments was that actinomycin D inhibited DNA-dependent mRNA synthesis in the developing and germinating embryos. More recently, it has been shown that in mature germinating/germinated cotton embryos actinomycin D inhibits $^{32}PO_4$ incorporation into adenylated mRNA by 62%, into non-adenylated RNA by 70%, and into poly(A) by only 30% (HARRIS and DURE 1978). The effects of actinomycin D upon RNA synthesis in developing embryos have not been studied, and it is necessary to assume that this drug behaves likewise there.

It has been suggested that upon imbibition of the mature dry seed the putative stored proteinase message has to be processed before it can be translated. This might involve post-transcriptional addition of adenylic acid residues to the 3′OH end of the mRNA. Polyadenylation is inhibited by cordycepin, and this compound also inhibits the production of proteinase (WALBOT et al. 1974). It is possible that over 50% of the total mass of mRNA polyadenylated during the first day after imbibition pre-exists in the mature dry seed (HARRIS and DURE 1978).

While the cotton carboxypeptidase mRNA story is an attractive one, not all the conclusions drawn are justified on the basis of the types of experiments that have been carried out (PAYNE 1976). Even the more recent studies on the effects of inhibitors on mRNA synthesis and processing are subject to various interpretations (HARRIS and DURE 1978) and a variety of new approaches are required (DURE 1979). It is possible to concede that stored non-adenylated messages exist within mature dry cotton cotyledons, and that some of them might be processed for use during or after germination. On the other hand, the evidence that any one of these mRNA's is for the proteinase, carboxypeptidase C, is extremely tenuous and no direct link has been forged between conserved mRNA or mRNA activation, and de novo synthesis of this enzyme.

Another enzyme that arises in germinated cotton seed cotyledons by de novo synthesis is isocitrate lyase. One group of workers claim that it is synthesized on conserved messages (IHLE and DURE 1972), but others have made more substantial claims that it is not (SMITH et al. 1974).

## 5 Protein and RNA Synthesis in Relation to Dormancy Breaking

In the previous sections changes in protein and RNA synthesis during germination have been surveyed in non-dormant seeds, i.e., in those species of seeds that will germinate when introduced to sufficient water at a moderate temperature. But many seeds initially require an extra environmental stimulus, e.g., light or cool temperatures, when in the imbibed state, before germination can be promoted. Alternatively, the requirement for an external stimulus can often be replaced by imbibing the seeds in a solution of phytohormone (usually a gibberellin). Several studies have been initiated to elucidate the mechanisms whereby an environmental stimulus or an applied hormone can release seeds from their dormant condition. Unfortunately, very little is known about the process of germination per se, and hence workers have concentrated their efforts on the effects of dormancy-breaking agents on metabolic processes that are accepted to be pre-requisites for successful radicle emergence – including nucleic acid and protein synthesis. For it has been suggested that dormancy could result from an impediment to one or both of these processes.

It is important to note that dormancy should not be equated with metabolic quiescence, for there is ample evidence that dormant seeds are capable of a variety of metabolic reactions. For example, dormant lettuce seeds are capable of appreciable RNA (FRANKLAND et al. 1971, FOUNTAIN 1974) and protein (FOUNTAIN and BEWLEY 1973, 1976) synthesis, as are dormant seeds of *Agrostem-*

*ma githago* (HECKER and BERNHARDT 1976a), *Melandrium noctiflorum* (HECKER and BERNHARDT 1976b), *Paulownia tomentosa* (GRUBISIĆ et al. 1978), *Xanthium pennsylvanicum* (SATOH and ESASHI 1979) and *Fraxinus excelsior* (VAN DE WALLE and FORGEUR 1977). Thus any promotive action of hormones or environmental stimuli is likely to entail some modification or enchancement of this "basal" metabolism.

### 5.1 Dormancy Breaking by Hormones and Light

On the premise that germination might require the synthesis of specific enzymes/ proteins, attempts have been made to determine if applied hormones, and in particular gibberellins, can modulate transcription or translation to bring about quantative and qualitative shifts in protein synthesis.

A considerable body of work has been carried out to determine the effects of RNA and protein synthesis inhibitors on $GA_3$-induced germination. The results obtained have, however, been very variable, the effects of the inhibitors ranging from completely inhibitory to completely ineffective. Conclusions drawn from inhibitor studies should be treated with caution, however, for: (a) proof is often lacking that inhibitors are taken up by the seed and that they do indeed inhibit protein or RNA synthesis; (b) very high concentrations of some inhibitors are needed for them to be effective, perhaps because of their limited ability to penetrate the seeds. In the tissues that they do penetrate there is a possibility of side-effects due to their high concentration; (c) the specificity of RNA and protein synthesis inhibitors is not as narrow as was once assumed, e.g., actinomycin D is not a specific mRNA synthesis inhibitor; (d) to date, no evidence has been provided that when an inhibitor is effective it is preferentially eliminating hormone-induced RNA or protein synthesis. It may be extremely difficult to obtain such evidence, but without it the possibility exists that the inhibitors affected "basal" metabolism as well as, or instead of, specifically hormone-induced processes. Reduction of basal metabolism, which presumably is essential for the metabolic well-being of the seed, could result indirectly in reduced germination.

Several biochemical studies have been made on the effects of $GA_3$ on protein and RNA synthesis during germination (reviewed by BEWLEY 1979). In embryonic axes of hazel (*Corylus avellana*) there is an increase in total RNA 2–3 days after the start of imbibition in the presence of $GA_3$, compared to water controls (JARVIS et al. 1968a). But this increase was found to be very small prior to the observed increase in axis fresh weight, which occurred 3–4 days after the start of imbibition and which was the first physical indication of a $GA_3$ effect (JARVIS et al. 1968a). Axis elongation commenced after 5 days. Synthesis of RNA increased within 16–24 h from the start of imbibition of $GA_3$, and to a somewhat greater extent than in the axes of water-imbibed seeds, although the largest increase was after the increase in fresh weight (JARVIS et al. 1968b). Little or no incorporation of radioactive precursors into RNA was detected before 15 h – this is rather disturbing in the light of more recent observations that RNA synthesis in seeds generally commences soon after imbibition (Sect. 3).

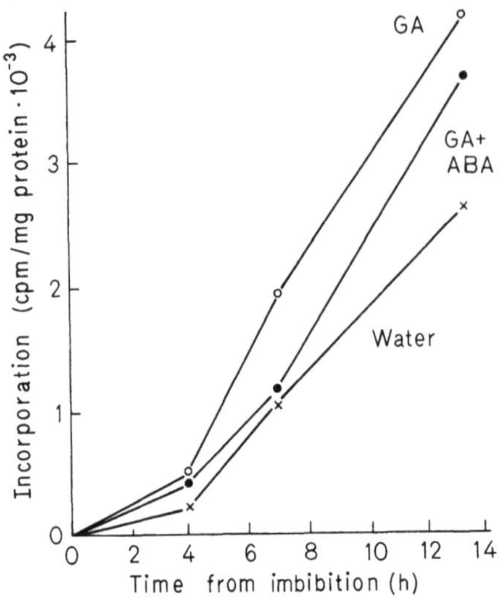


**Fig. 13.** The effects of $GA_3$ and $GA_3$+ABA on the incorporation of [$^3$H]-leucine into protein by intact lettuce seeds. (Based on FOUNTAIN and BEWLEY 1976 and FOUNTAIN 1974)

Also in the experiments with hazel cotyledons, the possibility was not eliminated that $GA_3$ facilitated the uptake of radioactive precursor, thus simply making more available for incorporation than in water-imbibed cotyledons.

$GA_3$ enhances uridine incorporation into RNA, and leucine incorporation into proteins of wild oat (*Avena fatua*) embryos prior to radicle expansion (CHEN and PARK 1973). From this observation it is implied that the hormone stimulates transcription, and hence subsequent translation. This remains to be proven.

In wheat embryos no effects of $GA_3$ upon RNA synthesis were detected during the first 12 h after imbibition starts (in fact, no RNA synthesis was detected on $GA_3$- or water-imbibed seeds) but an enhancement of protein synthesis was detected by the 12th h (CHEN and OSBORNE 1970b). However, it is not possible to tell from those results whether the $GA_3$-induced increase in protein synthesis occurred prior to radicle elongation (i.e., during germination) or later, as a consequence of growth of the radicle. It is disturbing also that no RNA synthesis was detected during the first 12 h, for it is known that wheat embryos conduct considerable RNA synthesis during this time (see Sect. 2.1 and 3). Maize embryos imbibed in $GA_3$ synthesize 20%–40% more RNA over the first 4 h than do water controls (WIELGAT et al. 1974). Interestingly, though, it is the scutellum (a non-germinating tissue) that responds most positively to $GA_3$. Neither wheat nor maize embryos are dormant and they do not require $GA_3$ to stimulate germination, and hence the hormone might stimulate protein and RNA synthesis that is not important for germination.

$GA_3$ promotes polysome formation and protein synthesis in lettuce seeds (FOUNTAIN and BEWLEY 1976) (Fig. 13), but not RNA synthesis (FOUNTAIN 1974) above the levels in water controls, and prior to completion of germination. Similarly, charlock (*Sinapis arvensis*) shows increases in amino acid and protein synthesis in the presence of $GA_3$ (EDWARDS 1976), and autoradiography has shown that much of the synthesis is confined to the apical meristems. In neither seed is it known if the increased protein synthesis is essentially linked to germination. Prevention of $GA_3$-induced lettuce seed germination by abscisic acid (ABA) is accompanied by a partial reversal of $GA_3$-stimulated protein synthesis (Fig. 13). It may be argued that ABA is inhibiting germination by reversing $GA_3$-induced protein synthesis that is essential for dormancy breaking. On the other hand, the possibility that ABA is simply reducing basal protein synthesis and indirectly affecting seed germinability cannot be ignored. Indeed, studies on isolated lettuce embryos have shown that ABA suppresses polysome formation and protein synthesis below the levels found in water-imbibed control embryos (FOUNTAIN and BEWLEY 1976).

Inhibition of germination of a number of seed species by ABA is accompanied by a general reduction in nucleic acid synthesis, e.g., in wheat (CHEN and OSBORNE 1970b), lettuce (FOUNTAIN 1974), pear (*Pyrus communis*) (KHAN and HEIT 1968) and ash (*Fraxinus excelsior*) (VILLIERS 1968) embryos, and in germinating bean axes (WALBOT et al. 1975). Cytokinins stimulate the germination of very few seed species, although they reverse the inhibitory effects of ABA. In bean axes (SUSSEX et al. 1975) and pear embryos (KHAN and HEIT 1968) this reversal is accompanied by an increase in RNA synthesis. In lettuce embryos the cytokinin, benzyladenine, overcomes the inhibition of germination by ABA and also induces increased polysome formation and protein synthesis (FOUNTAIN 1974, FOUNTAIN and BEWLEY 1976).

In brief summary, gibberellin appears to promote RNA and/or protein synthesis in some seeds, and ABA is inhibitory. This latter inhibition can be overcome by cytokinin. As yet, however, there is no compelling evidence that any of these hormones act directly at the level of transcription and translation or regulate the synthesis of proteins essential for germination – if, indeed, such proteins exist.

The same conclusions can be drawn from experiments on the effects of light, mediated via the phytochrome system, upon protein and RNA synthesis during the breaking of dormancy. For example, embryos of dormant lettuce seeds irradiated with red light show no stimulation of $^{32}PO_4$ incorporation into RNA before germination is completed (FRANKLAND et al. 1971), although polysome levels increase (KHAN et al. 1974). Nor does red light enhance, or far-red light reduce [$^3$H]-uracil incorporation into RNA of *Melandrium noctiflorum* seeds (HECKER and BERNHARDT 1976b). These observations are indications that dormancy breaking by light, as by hormones, does not involve the synthesis of new RNA molecules. However, the level of detection was such that subtle changes in RNA synthesis might not have been detected, particularly if there were changes to only a few specific mRNA's for proteins essential for germination. Moreover, the response to light or to hormones might occur only in the radicle, and maybe within only a few cells of this organ, thus making them beyond the resolution of available techniques. This argument can be applied quite widely when it is not possible to detect any light- or hormone-induced response – testing its validity is the problem!

# References

Ajtkhozhin MA, Doschanov KhJ, Akhanov AJ (1976) Informosomes as a stored form of mRNA in wheat embryos. FEBS Lett 66:124–126
Anderson JW, Fowden L (1969) A study of the aminoacyl-sRNA synthetases of *Phaseolus vulgaris* in relation to germination. Plant Physiol 44:60–68
Anderson JW, Fowden L (1970) Properties and substrate specificities of the phenylalany-transfer-ribonucleic acid synthetases of *Aesculus* species. Biochem J 119:677–690
Anderson MB, Cherry JH (1979) Differences in leucyl-transfer RNA's and synthetase in soybean seedlings. Proc Natl Acad Sci USA 62:202–209
Ballivian L, Julius D, Litvak S (1977) The purification and properties of wheat embryo tRNA nucleotidyl transferase. CNRS Colloq acides nucleiques et synthèse des protéines chez les végétaux, Strasbourg pp 265–271
Barnett LB, Adams RE, Ramsey JA (1974) The effect of stratification on in vitro protein synthesis in seeds of *Pinus lambertiana*. Life Sci 14:653–658
Becker WM, Leaver CJ, Weir EM, Riezman H (1978) Regulation of glyoxysomal enzymes during germination of cucumber. I. Developmental changes in cotyledonary protein, RNA, and enzyme activities during germination. Plant Physiol 62:542–549
Beevers H (1979) Microbodies in higher plants. Annu Rev Plant Physiol 30:159–193
Beevers L, Guernsey FS (1966) Changes in some nitrogenous components during germination of pea seeds. Plant Physiol 41:1455–1458
Bewley JD (1979) Dormancy breaking by hormones and other chemicals – action at the molecular level. In: Rubenstein I, Phillips RL, Green CE, Gegenbach BE (eds) The plant seed: Development, preservation and germination. Academic Press, London New York, pp 219–239
Bewley JD, Black M (1978) Physiology and biochemistry of seeds in relation to their germination, vol I. Development, germination and growth. Springer, Berlin Heidelberg New York, pp 306
Bewley JD, Larsen KM (1979) Endosperms and embryos of mature dry castor bean seeds contain active ribosomes. Phytochemistry 18:1617–1619
Bhat SP, Padayatty JD (1974) Presence of conserved messenger RNA in rice embryos. Indian J Biochem Biophys 11:47–50
Bick HD, Liebke H, Cherry JH, Strehler BL (1970) Changes in leucyl- and tyrosyl-tRNA of soybean cotyledons during plant growth. Biochim Biophys Acta 204:175–182
Bray CM, Chow T-Y (1976) Lesions in post-ribosomal supernatant fractions associated with loss of viability in pea (*Pisum arvense*) seed. Biochim Biophys Acta 442:1–13
Bray CM, Dasgupta J (1976) Ribonucleic acid synthesis and loss of viability in pea seed. Planta 132:103–108
Brooker JD, Tomaszewski M, Marcus A (1978) Preformed messenger RNAs and early wheat embryo germination. Plant Physiol 61:145–149
Caers LI, Peumans WJ, Carlier AC (1979) Preformed and newly synthesized messenger RNA in germinating wheat embryos. Planta 144:491–496
Carlier AR, Peumans WJ (1976) The rye embryo as an alternative to the wheat system for protein synthesis in vitro. Biochim Biophys Acta 447:436–444
Chakravorty AK (1969) Ribosomal RNA synthesis in the germinating black pea (*Vigna unguiculata*). II. The synthesis and maturation of ribosomes in the later stages of germination. Biochim Biophys Acta 179:83–96
Chen D, Osborne DJ (1970a) Ribosomal genes and DNA replication in germinating wheat embryos. Nature (London) 225:336–340
Chen D, Osborne DJ (1970b) Hormones in the translational control of early germination in wheat embryos. Nature (London) 226:1157–1160
Chen D, Schultz G, Katchalski E (1971) Early ribosomal RNA transcription and appearance of cytoplasmic ribosomes during germination of the wheat embryo. Nature (London) New Biol 231:69–72
Chen SCC, Park WM (1973) Early actions of gibberellic acid on the embryo and on the endosperm of *Avena fatua* seeds. Plant Physiol 52:174–176

Cherry JH, Chiroboczek H, Carpenter WJG, Richmond A (1965) Nucleic acid metabolism in peanut cotyledons. Plant Physiol 40:582–587
Cheung CP, Suhadolnik RJ (1978) Regulation of RNA synthesis in early germination of isolated wheat (*Triticum aestivum* L.) embryo. Nature (London) 271:357–358
Cheung CP, Wu J, Suhadolnik RJ (1979) Dependence of protein synthesis on RNA synthesis during the early hours of germination of wheat embryos. Nature (London) 277:66–67
Clay WF, Katterman FRH, Hammett JR (1975) Nucleic acid metabolism during germination of pima cotton (*Gossypium barbadense*). Plant Physiol 55:231–236
Cuming AC, Lane BG (1978) Wheat embryo ribonucleates. XI. Conserved mRNA in dry wheat embryos and its relation to protein synthesis during early imbibition. Can J Biochem 56:365–369
Delseny M, Aspart L, Got A, Cooke R, Guitton Y (1977a) Early synthesis of polyadenylic acid, polyadenylated and ribosomal nucleic acids in germinating radish embryo. Physiol Veg 15:413–428
Delseny M, Aspart L, Guitton Y (1977b) Disappearance of stored polyadenylic acid and mRNA during early germination of radish (*Raphanus sativus* L.) embryo axes. Planta 135:125–128
Dobrzańska M, Tomaszewski M, Grzelczak Z, Rejman E, Buchowicz J (1973) Cascade activation of genome transcription in wheat. Nature (London) 244:507–508
Doshchanov KI, Ajtkhozhin MA, Darkanbaev TB (1975) Studies of cytoplasmic RNAs at the early stages of germination of wheat embryos. Sov Plant Physiol 22:305–310
Dure LS III (1975) Seed formation. Annu Rev Plant Physiol 26:259–278
Dure LS III (1979) Role of stored messenger RNA in late embryo development and germination. In: Rubenstein I, Phillips RL, Green CE, Gegenbach BE (eds) The plant seed: Development, preservation and germination. Academic Press, London New York, 113–127
Dure LS III, Waters LC (1965) Long-lived messenger RNA: evidence from cotton seed germination. Science 147:410–412
Dziegielewski T, Kedzierski W, Pawelkiewicz J (1979) Levels of aminoacyl-tRNA synthetases, tRNA nucleotidyltransferase and ATP in germinating lupin seeds. Biochim Biophys Acta 564:37–42
Edwards M (1976) Dormancy in seeds of charlock (*Sinapis arvensis* L.): Early effects of gibberellic acid on the synthesis of amino acids and proteins. Plant Physiol 58:626–630
Evins WH (1971) Enhancement of polyribosome formation and induction of tryptophan-rich proteins by gibberellic acid. Biochemistry 10:4295–4303
Fabisz-Kijowska A, Dullin P, Walerych W (1975) Isolation and purification of RNA polymerases from rye embryos. Biochim Biophys Acta 390:105–116
Ferrer A, Delseny M, Guitton Y (1979) Isolation and characterization of sub-ribosomal ribonucleo-protein particles from radish seeds and seedlings. Plant Sci Lett 14:31–42
Filimonov NG, Ajtkhozhin MA, Tarantul VZ, Gasaryan KG (1977) Free poly (A) tracts complexed with protein in the cytoplasm of dried wheat embryos. FEBS Lett 79:348–352
Fountain DW (1974) On the physiology and biochemistry of dormancy and germination in light-sensitive lettuce seeds. PhD Thesis, Univ Calgary
Fountain DW, Bewley JD (1973) Polyribosome formation and protein synthesis in imbibed but dormant lettuce seeds. Plant Physiol 52:604–607
Fountain DW, Bewley JD (1976) Lettuce seed germination: Modulation of pregermination protein synthesis by gibberellic acid, abscisic acid, and cytokinin. Plant Physiol 58:530–536
Frankland B, Jarvis BC, Cherry JH (1971) RNA synthesis and the germination of light-sensitive lettuce seeds. Planta 97:39–49
Fukuei K, Sakamaki T, Takahashi N, Tanifuji S (1975) Stability of rRNA genes during germination of *Vicia faba* seeds. Plant Cell Physiol 16:387–394
Fukuei K, Sakamaki T, Takahashi N, Takaiwa F, Tanifuji S (1977) RNA synthesis required for DNA replication in *Vicia* seed embryos. Plant Cell Physiol 18:173–180
Gerhardt BP, Beevers H (1970) Developmental studies on glyoxysomes in *Ricinus* endosperm. J Cell Biol 44:94–102

Gillard DF, Walton DC (1973) Germination in *Phaseolus vulgaris*. IV. Patterns of protein synthesis in excised axes. Plant Physiol 51:1147–1149
Gordon ME, Payne PI (1976) In vitro translation of the long-lived messenger ribonucleic acid of dry seeds. Planta 130:269–273
Gordon ME, Payne PI (1977) In vitro translation of poly A-rich RNA from dry and imbibing seeds of rape. In: Legocki AB (ed) Translation of natural and synthetic polynucleotides. Agric Univ Press, Poznań, pp 228–231
Grubisić D, Petrović J, Konjević R (1978) Biosynthesis of nucleic acids and proteins in dormant and light-stimulated *Paulownia tomentosa* seeds. Biochem Physiol Pflanz 173:333–339
Grzelczak Z, Buchowicz J (1977) A comparison of the activation of ribosomal RNA synthesis during germination of isolated and non-isolated embryos of *Triticum aestivum* L. Planta 134:263–265
Guilfoyle TJ, Jendrisak JJ (1978) Plant DNA-dependent RNA polymerases: subunit structures and enzymatic properties of the class II enzymes from quiescent and proliferating tissues. Biochemistry 17:1860–1866
Guilfoyle TJ, Key JL (1977) The subunit structures of soluble and chromatin-bound RNA polymerase II from soybean. Biochem Biophys Res Commun 74:308–313
Hammett JR, Katterman FR (1975) Storage and metabolism of poly (adenylic acid)-mRNA in germinating cotton seeds. Biochemistry 14:4375–4379
Harris B, Dure LS III (1978) Developmental regulation in cotton seed germination: Polyadenylation of stored messenger RNA. Biochemistry 17:3250–3256
Hecker M (1977) Polyadenylation of long-lived mRNA at the early imbibition phases of *Agrostemma githago* seeds? Biol Rundsch 15:58–61
Hecker M, Bernhardt D (1976a) Proteinbiosynthesen in dormanten und nachgereiften Embryonen und Samen von *Agrostemma githago*. Phytochemistry 15:1105–1109
Hecker M, Bernhardt D (1976b) Untersuchungen über protein und RNA-synthese in dormanten und nachgereiften samen von *Melandrium noctiflorum*. Biochem Physiol Pflanz 169:417–426
Hecker M, Weidmann M, Köhler K-H, Serfling E (1976) Studies on early RNA and protein synthesis in imbibing embryos of *Agrostemma githago*. Biochem Physiol Pflanz 169:427–436
Higgins TJV, Zwar JA, Jacobsen JV (1976) Gibberellic acid enhances the level of translatable mRNA for α-amylase in barley aleurone layers. Nature (London) 260:166–169
Ho DT-H, Varner JE (1974) Hormonal control of messenger ribonucleic acid metabolism is barley aleurone layers. Proc Natl Acad Sci USA 71:4783–4786
Huang B, Rodaway S, Wood A, Marcus A (1980) RNA synthesis in germinating embryos of soybean and wheat. Plant Physiol 65:1155–1159
Ihle JN, Dure LS III (1969) Synthesis of a protease in germinating cotton cotyledons catalysed by mRNA synthesised during embryogenesis. Biochem Biophys Res Commun 36:705–710
Ihle JN, Dure LS III (1972) The developmental biochemistry of cottonseed embryogenesis and germination. III. Regulation of the biosynthesis of enzymes utilized in germination. J Biol Chem 247:5048–5055
Ingle J, Sinclair J (1972) Ribosomal RNA genes and plant development. Nature (London) 235:30–32
Iskhakov BK, Filimonov NG, Ajtkhozhin MA (1978) Proteins bound to poly (A) sequences of polyribosomes from germinating wheat embryos. Biochim Biophys Acta 521:470–475
Jacobsen JV, Zwar JA (1974a) Gibberellic acid and RNA synthesis in barley aleurone layers: metabolism of rRNA and tRNA and of RNA containing polyadenylic acid sequences. Aust J Plant Physiol 1:343–356
Jacobsen JV, Zwar JA (1974b) Gibberellic acid causes increased synthesis of RNA which contains poly (A) in barley aleurone tissue. Proc Natl Acad Sci USA 71:3290–3293
Jarvis BC, Frankland B, Cherry JH (1968a) Increase in nucleic acid synthesis in relation to the breaking of dormancy of hazel seed by gibberellic acid. Planta 83:257–266
Jarvis BC, Frankland B, Cherry JH (1968b) Increased DNA template and RNA polymerase associated with the breaking of seed dormancy. Plant Physiol 43:1734–1736

Jendrisak JJ (1977) RNA polymerase II from wheat germ: purification and subunit structure. CNRS Colloq acides nucleiques et synthèse des proteins chez les végétaux, Strasbourg, pp 179–185
Jendrisak JJ, Becker WM (1973) Isolation, purification and characterization of RNA polymerases from wheat germ. Biochim Biophys Acta 319:48–54
Jendrisak JJ, Burgess RR (1975) A new method for the large-scale purification of wheat germ DNA-dependent RNA polymerase II. Biochemistry 14:4639–4654
Jones RL (1969) Gibberellic acid and the fine structure of barley aleurone cells. I. Changes during the lag phase of α-amylase synthesis. Planta 87:119–133
Jones RL, Jacobsen JV (1978) Membrane and RNA metabolism in the response of aleurone cells to GA. In: Controlling factors in plant development. Bot Mag Spec Ed 1:83–99
Kanabus J, Cherry JH (1971) Isolation of an organ-specific leucyl tRNA synthetase from soybean seedling. Proc Natl Acad Sci USA 68:873–876
Kedzierski W, Pawelkiewicz J (1977) Effect of seed germination on levels of tRNA aminoacylation. Phytochemistry 16:503–504
Khan AA, Heit CE (1968) Selective effect of hormones on nucleic acid metabolism during germination of pear embryos. Biochem J 113:707–712
Khan AA, Tao K-L, Stone MA (1974) Polyribosome formation during light- and gibberellic acid-induced germination of lettuce seeds. In: Plant growth substances 1973. Hirokawa Publ Co, Tokyo, pp 608–615
Klein S, Barenholz H, Budnik A (1971) The initiation of growth in isolated lima bean axes. Physiological and fine structural effects of actinomycin D, cycloheximide and chloramphenicol. Plant Cell Physiol 12:41–60
Marcus A, Feeley J (1964) Activation of protein synthesis in the imbibition phase of seed germination. Proc Natl Acad Sci USA 51:1075–1079
Marcus A, Feeley J, Volcani T (1966) Protein synthesis in imbibed seeds. III. Kinetics of amino acid incorporation, ribosome activation and polysome formation. Plant Physiol 41:1167–1172
Marcus A, Luginbill B, Feeley J (1968) Polysome formation with tobacco mosaic virus RNA. Proc Natl Acad Sci USA 59:1243–1250
Marcus A, Weeks DP, Seal SN (1973) Protein chain initiation in wheat embryo. Biochem Soc Symp 38:97–109
Marré E, Cocucci S, Sturani E (1965) On the development of the ribosomal system in the endosperm of germinating castor bean seeds. Plant Physiol 40:1162–1170
Mazuś B (1973) RNA polymerase activity in isolated *Triticum aestivum* embryos during germination. Phytochemistry 12:2809–2813
Mazuś B, Brodniewicz-Proba T (1976) RNA polymerases I and II in germinating wheat embryo. Acta Biochim Pol 23:261–267
Merrick WC, Dure LS III (1972) The developmental biochemistry of cotton seed embryogenesis and germination. IV. Levels of cytoplasmic and chloroplastic transfer ribonucleic acid species. J Biol Chem 247:7988–7999
Mori T, Takagi S, Utsuni A (1979) Synthesis of glycininin wheat germ cotyledons. Biochem Biophys Res Commun 87:43–49
Mori T, Waskabayashi Y, Takagi S (1978) Occurrence of mRNA for storage protein in dry soybean seeds. J Biochem (Tokyo) 84:1103–1111
Muthurkrishnan S, Chandra GR, Maxwell ES (1979) Hormone-induced increase in levels of functional mRNA and α-amylase mRNA in barley aleurones. Proc Natl Acad Sci USA 76:6181–6185
Onckelen Van HA, Verbeek R, Khan AA (1974) Relationship of ribonucleic acid metabolism in embryo and aleurone to α-amylase synthesis in barley. Plant Physiol 53:562–568
Payne JF, Bal AK (1972) RNA polymerase activity in germinating onion seeds. Phytochemistry 11:3105–3110
Payne PI (1976) The long-lived messenger ribonucleic acid of flowering plant seeds. Biol Rev 51:329–363
Payne PI (1977) Synthesis of poly A-rich RNA in embryos of rye during imbibition and early germination. Phytochemistry 16:431–434

Payne PI, Boulter D (1969) Free and membrane-bound ribosomes of the cotyledons of *Vicia faba* (L.). II. Seed germination. Planta 87:63–68

Payne PI, Gordon ME, Dobrzańska M, Parker ML, Barlow PW (1975) The long-lived mRNA of dry seeds. CNRS Colloq acides nucleiques et synthèse des proteines chez les végétaux, Strasbourg, pp 487–499

Payne PI, Gordon ME, Barlow PW, Parker ML (1977) The subcellular location of the long-lived messenger RNA of rape seed. In: Legocki AB (ed) Translation of natural and synthetic polynucleotides. Agric Univ Press, Poznań, pp 224–227

Payne PI, Dobrzańska M, Barlow PW, Gordon ME (1978) The synthesis of RNA in imbibing seed of rape (*Brassica napus*) prior to the onset of germination: A biochemical and cytological study. J Exp Bot 29:73–88

Peumans WJ (1978) Study of cell-free translation in extracts from cereal embryos and some aspects of the long-lived mRNPs in rye embryos. PhD Thesis, Katholieke Univ Leuven, pp 191

Peumans WJ, Carlier AR (1977) Messenger ribonucleoprotein particles in dry wheat and rye embryos. In vitro translation and size distribution. Planta 136:195–201

Quintard B, Mouricout M, Carias JR, Julien R (1978) Occurrence of aminoacyl-tRNA synthetase complexes in quiescent wheat germ. Biochem Biophys Res Commun 85:999–1006

Roberts LM, Lord JM (1979) Developmental changes in the activity of messenger RNA isolated from germinating castor bean endosperm. Plant Physiol 64:630–634

Robinson NE, Bryant JA (1975) Onset of nucleic acid synthesis during germination of *Pisum sativum*. Planta 127:63–68

Satoh S, Esashi Y (1979) Protein synthesis in dormant and non-dormant cocklebur seed segments. Physiol Plant 47:229–234

Sen S, Payne PI, Osborne DJ (1978) Early ribonucleic acid synthesis during the germination of rye (*Secale cereale*) embryos and the relationship to early protein synthesis. Biochem J 148:381–387

Sieliwanowicz B, Chmielewska I (1973) Studies on the initiation of protein synthesis in the course of germination of pea seeds. Bull Acad Sci Pol 21:399–404

Sieliwanowicz B, Chmielewska I (1974) Synthesis of pea embryo axes proteins directed by preexisting mRNA. Bull Acad Sci Pol 22:159–162

Sieliwanowicz B, Kalinowska M, Chmielewska I (1977) Appearance of poly (A)-rich RNA in germinating pea seeds. Acta Biochim Pol 24:59–64

Smith DL (1971) Nuclear changes in the cotyledons of *Pisum arvense* L. during germination. Ann Bot 35:511–521

Smith RA, Schubert AM, Benedict CR (1974) The development of isocitric lyase activity in germinating cotton seed. Plant Physiol 54:197–200

Spiegel S, Marcus A (1975) Polyribosome formation in early wheat embryo germination independent of either transcription or polyadenylation. Nature (London) 256:228–230

Spiegel S, Obendorf RL, Marcus A (1975) Transcription of ribosomal and messenger RNAs in early wheat embryo germination. Plant Physiol 56:502–507

Sturani E (1968) Protein synthesis activity of ribosomes from developing castor bean endosperm. Life Sci 7:527–537

Sussex I, Clutter M, Walbot V (1975) Benzyladenine reversal of abscisic acid inhibition of growth and RNA synthesis in germinating pea axes. Plant Physiol 56:575–578

Takahashi N, Takaiwa F, Fukuei K, Sakamaki T, Tanifuji S (1977) Appearance of newly formed mRNA and rRNA as ribonucleoprotein-particles in the cytoplasmic subribosomal fraction of pea embryos. Plant Cell Physiol 18:235–246

Takaiwa F, Tanifuji S (1979) RNA synthesis in embryo axes of germinating pea seeds. Plant Cell Physiol 20:875–884

Tanifuji S, Asamizu T, Sakaguchi K (1969) DNA-like RNA synthesized in pea embryos at very early stage of germination. Bot Mag 82:56–58

Tanifuji S, Higo M, Shimada T, Higo S (1970) High molecular weight RNA synthesized in nucleoli of higher plants. Biochim Biophys Acta 217:418–425

Varner JE, Flint D, Mitra R (1976) Characterization of protein metabolism in cereal grains. In: Genetic improvement of seed proteins. Natl Acad Sci Natl Res Counc, pp 301–328

Villiers TA (1968) An autoradiographic study of the effect of the plant hormone abscisic acid on nucleic acid and protein metabolism. Planta 82:342–354

Vold BS, Sypherd PS (1968) Changes in soluble RNA and ribonuclease activity during germination of wheat. Plant Physiol 43:1221–1226

Walbot V (1972) Rate of RNA synthesis and tRNA end-labeling during early development of *Phaseolus*. Planta 108:161–171

Walbot V, Capdevila A, Dure LS III (1974) Action of 3′d adenosine (cordycepin) and 3′d cytidine on the translation of the stored mRNA of cotton cotyledons. Biochem Biophys Res Commun 60:103–110

Walbot V, Clutter M, Sussex I (1975) Effects of abscisic acid on growth, RNA metabolism, and respiration in germinating bean axes. Plant Physiol 56:570–574

Walle Van de C, Forgeur G (1977) RNA synthesis during imbibition of a dormant seed. CNRS colloq acides nucleiques et synthèse des protèines chez les végétaux, Strasbourg, pp 515–519

Waters LC, Dure LS III (1966) Ribonucleic acid synthesis in germinating cotton seeds. J Mol Biol 19:1–27

Weir EM, Leaver CJ, Riezman H, Becker WM, Grienenberger JM (1979) Gene expression during germination of the cucumber (*Cucumis sativus*). Abstract: Genome organisation and expression in plants. NATO Adv Stud Inst, p 91

Wielgat B, Wasilewska LD, Kleczkowski K (1974) RNA synthesis in germinating maize seeds under hormonal control. In: Plant growth substances 1973. Hirokawa Publ Co, Tokyo, pp 593–598

Wright RD, Kanabus J, Cherry JH (1974) Multiple leucyl tRNA synthetases in pea seedlings. Plant Sci Lett 2:347–355

Yoshida K, Sugita M, Sasaki K (1979) Involvement of nonhistone chromosomal proteins in transcriptional activity of chromatin during wheat germination. Plant Physiol 63:1016–1021

# 16 Leaf Senescence

J.L. STODDART and H. THOMAS

## 1 Introduction

For some time it has been conventional to regard leaf senescence as a species of catastrophe in which the activities of the mature, carbon-exporting leaf lose both their integrity and coordination in an essentially unregulated manner. This view assumes parallels with gerontological changes occurring in humans and other animals. However, it is becoming increasingly apparent that leaf senescence is least understood when considered in these terms and that degenerative sequences in plants and animals, whilst having underlying similarities, also have fundamental differences.

We now understand that foliar senescence is part of a pre-programmed progression of development capable of modulation by environment or growth status and conditioned, in certain respects, by the previous life history of the plant. Whilst, for the purposes of this chapter, senescence can be defined as the sequence of events concerned with cellular disassembly in the leaf and the mobilization of released materials, it is important to recognize that the process has significance at higher orders of organization. These are exemplified by effects on growth correlation at the whole plant level or aspects of flexibility and survival in populations.

General reviews of the extensive literature covering leaf senescence have been provided by LEOPOLD (1961), WOOLHOUSE (1974) and THOMAS and STODDART (1980) and we, therefore, will be more specifically concerned with the role played by protein turnover in the initiation and regulation of leaf disassembly and the way in which other modifying factors may act upon the requisite transcription and translation processes. In addition, because of the central importance of nitrogen mobilization occurring during senescence, consideration will be given to this process and its consequences at the cell, whole plant and crop levels.

## 2 Senescence as a Developmental Event

During its life span the leaf passes through three recognizable developmental phases. Initially it is a net carbon-importing structure and remains so until the full development of photosynthetic capacity and the subsidence of the peak demand for carbohydrate for the assembly of mature cells and their walls. There then follows a period, of variable duration, when the leaf becomes an asset to the carbon economy of the plant and this continues until the advent of internal or external conditions which initiate senescence. From this point the leaf progresses into a period of massive mobilization and export of carbon, nitrogen and minerals. In considering senescence we are dealing therefore, not only with the gradual decline of photosynthetic capacity in the mature leaf, but also with the mobilization processes involved in cellular disassembly. Factors operative during leaf expansion or early maturity may have strong conditioning effects upon these events and, in addition, we must take account of the relationships between the activities of various sub-cellular components during development, such as changes in the co-operative activities of the plastid, nuclear and, possibly, mitochondrial genomes (see Sect. 2.4).

### 2.1 Changes in Photosynthetic Capacity

As the leaf expands and chloroplasts are assembled there is a rapid increase in the rate of carbon fixation (CATSKÝ et al. 1976). Thereafter, photosynthesis declines steadily with time (HOPKINSON 1964, WOOLHOUSE and BATT 1976, CATSKÝ et al. 1976). The reason for the slow decline in capacity is unclear but WOOLHOUSE and BATT (1976) have suggested that the underlying cause may be the cessation of chloroplast protein and nucleic acid synthesis which, when coupled with continued protein turnover, leads to declining photosynthetic enzyme activities. The primary carboxylating enzyme ribulose-1,5-bisphosphate carboxylase (RuBPC) is the only example, at present, where we can compare the reduction in catalytic activity with the loss of enzyme protein during ageing of attached leaves. Detailed data, based upon immunological techniques, is available for *Triticum* (HALL et al. 1978) and in this case the fall in carboxylation capacity *precedes* any drop in the level of enzyme protein. However, RuBPC may be unusual in so far as it is synthesized in a massive burst at the time of expansion and, thereafter, the turnover rate is very low indeed (HUFFAKER and PETERSON 1974). WOOLHOUSE and BATT (1976) measured the activities of several other photosynthetic enzymes during leaf development in *Perilla* and found that ribose-5-phosphate isomerase, phosphoribulokinase and NADP triosephosphate dehydrogenase began to decline in activity at the completion of expansion (30 days), whereas phosphoglycerokinase and fructose-1,6-bisphosphatase were retained at a high level until leaf yellowing became apparent (40–45 days). In this species a clear distinction exists, therefore, between enzymes synthesized on 70S chloroplast ribosomes, which decline at an early stage of leaf development, and those formed on 80S cytoplasmic ribosomes, which persist up to the point of senescence. However, these studies measured only catalytic activity and there is no information on the amount of enzyme protein present at each point. When enzymes are known to be regulated by feed-back mechanisms or allosteric effectors (e.g., RICARD et al. 1977) measurements of activity alone do not give reliable information on turnover. Data available for changes in rRNA and polysome content in plastids and cytoplasm (CALLOW et al. 1972, TREHARNE et al. 1970) do, nevertheless, suggest that the total activity measurements made on these stroma enzymes are a reasonable reflection of their rate of synthesis and that the suggested correlation between turnover and declining photosynthetic activity could be sound.

## 2.2 The Disassembly Processes

### 2.2.1 Changes in Stroma and Thylakoid Functions

The onset of senescence is accompanied by progressive modifications of chloroplast integrity and performance.

RuBPC content declines rapidly (PETERSON and HUFFAKER 1975, THOMAS 1976b, HALL et al. 1978) and chlorophyll degradation proceeds on a similar time course, but is not obligately synchronized with protein mobilization. This latter point is well illustrated by the studies of THOMAS and STODDART (1975) on a *Festuca* genotype (NY) with a nuclear gene mutation concerned with chlorophyll breakdown. In this material senescence, measured in terms of total protein, RuBPC, total RNA and plastid structural degradation, proceeds at the same rate as in control genotypes, but the chlorophyll content of the leaf, when corrected for moisture loss, remains close to the original value. Enzymes activated during senescence, such as peroxidase and certain esterase isoenzymes (THOMAS and BINGHAM 1977), also increase in a normal manner in the mutant. The non-yellowing lesion is, thus, apparently restricted to the turnover of chlorophyll and is not expressed as a general disruption of senescence. Indeed, the organizational complexity of the process has been emphasized by further studies which suggest that chlorophyll persists in the NY mutant because it remains inaccessible to the degrading enzymes (THOMAS 1977). Ultrastructural studies on the plastids from this material indicate that the initial unstacking of membranes occurs in the mutant but subsequent steps, involving degradation of the unit membrane structure which sandwiches the chlorophyll molecules, are arrested. The plastids become filled with whorled membrane figures and the buildup of lipid globules normally characteristic of senescence is reduced (THOMAS 1977). In this context it should be mentioned that whilst individuals in a population of normal plastids may senesce at different rates (Fig. 1, d and f), the end-point of the sequence, as in Fig. 1h, is always the same. Illustrations of sequential ultrastructural changes in chloroplasts from normal and NY leaves are given in Fig. 1.

More than 30 polypeptide species, with molecular weights (mol. wt.) in the 10,000 to 70,000 range, have been identified by SDS-polyacrylamide gel electrophoresis of *Festuca* plastid membrane preparations. Five major components with mol. wt. of 15,000, 19,000, 23,000, 32,000 and 54,000 accounted for more than half the membrane protein (THOMAS 1977). A marked reduction in total protein occurred between days 3 and 6 after leaf excision, with individual components disappearing at differential rates. For example, the 15,000, 19,000 and 23,000 mol. wt. components declined most rapidly. However, in the NY mutant, these were conserved.

Thylakoid membrane proteins can be classified as either structural (intrinsic) or weakly bound to the stroma-facing membrane surface (extrinsic). The major extrinsic protein is the photophosphorylation coupling factor and plastid mem-

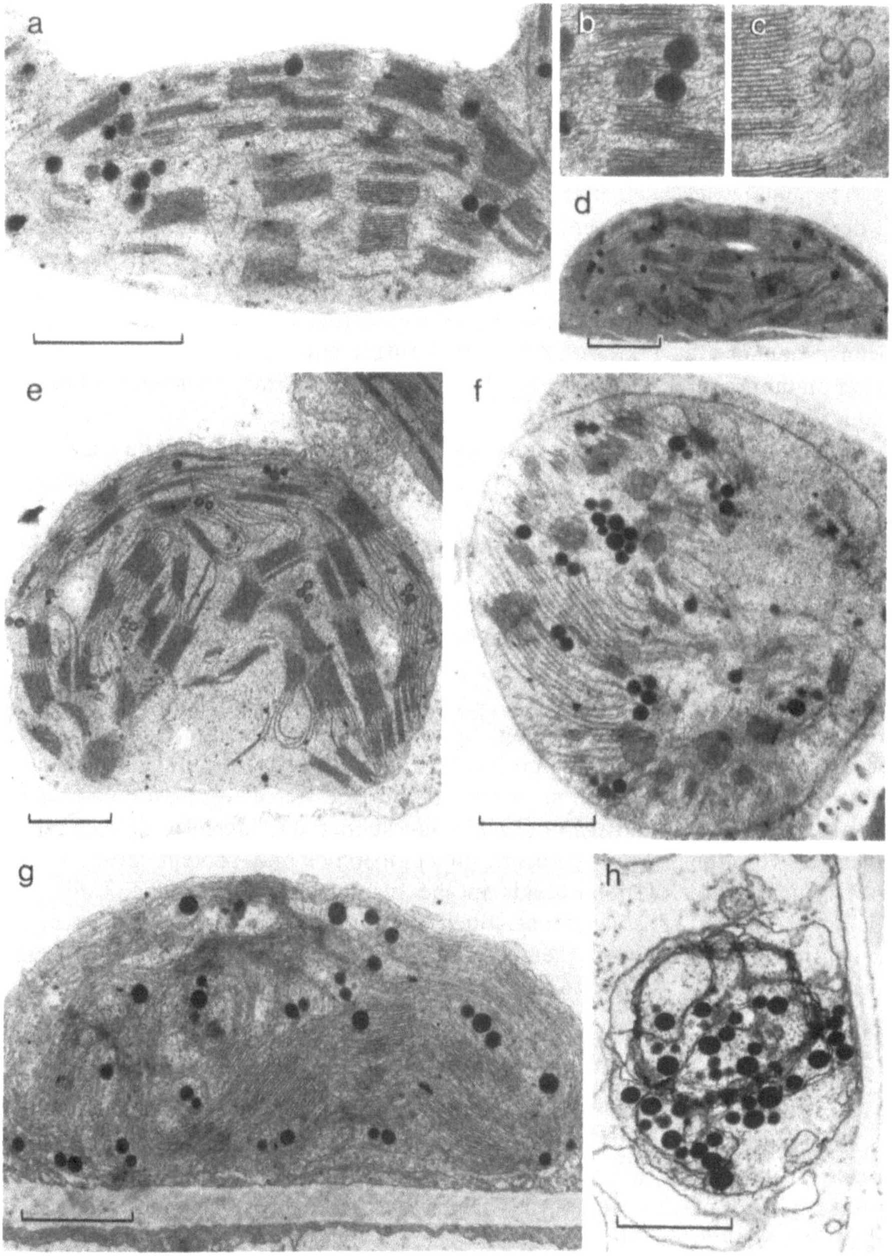


**Fig. 1a–h.** Electron micrographs of mesophyll chloroplasts from *Festuca pratensis* at different times after the initiation of senescence (THOMAS 1977). **a** Normal plastid, day 0. **b** Plastoglobuli in normal plastid, day 0. **c** Plastoglobuli in mutant, day 0. **d** Normal plastid, day 3. **e** Mutant plastid, day 3. **f** Rapidly senescing, normal plastid, day 3. **g** Mutant plastid, day 6. **h** Normal plastid, day 6. Horizontal bar = 1 μm. Magnification of **b** and **c** approx. × 60,000

brane polypeptides with mol. wt. of 32,000 and 54,000 have been identified as subunits of this protein (HENRIQUES and PARK 1976). The apoprotein of the chlorophyll–protein light-harvesting complex, with a mol. wt. of 25,000, is the predominant intrinsic thylakoid protein. It has been suggested that the differential decay kinetics of prominent extrinsic and intrinsic membrane constituents may reflect the operation of separate degrading enzyme systems.

The membrane-associated electron transport activities of photosystems I and II decline steadily during senescence at a rate which matches the loss of stroma and extrinsic thylakoid proteins (THOMAS 1977) and there is also evidence to indicate that the individual components of the chloroplast electron transport chain are inactivated during senescence in a sequential, rather than a simultaneous, manner (BISWAL and MOHANTY 1976). Changes also occur in the non-polar membrane fractions as shown by the buildup, during senescence, of osmiophilic globular bodies (Fig. 1) which result from the removal of lipid components from the thylakoids (BUTLER and SIMON 1970). The changing physical state of lipid in chloroplasts of senescing leaves of *Phaseolus* has been studied by X-ray diffraction (MCKERSIE and THOMPSON 1978). In young leaves the phase-transition temperature for thylakoid membrane lipids is below $-30$ °C but, at late maturity, there is an abrupt change to a value of around $+30$ °C and the transition temperature continues to increase thereafter with leaf age. The onset of plastid degradation is correlated with the time at which the chloroplast lipid transition temperature enters the physiological range. Shifts in phase-transition temperature are not thought to be merely reflections of the saturation level of membrane lipids, although alterations in composition are known to occur during yellowing (FONG and HEATH 1977). The largest change in *Phaseolus* plastid membrane constitution occurs in the free sterol to phospholipid ratio and MCKERSIE and THOMPSON (1978) suggest that this increase in free sterols, together with the senescent decline in chlorophyll and protein levels, results in a redistribution of polar lipids in the plane of the membrane, leading to the formation of gel-state zones. Such inhomogeneities may cause localized loss of membrane integrity, thereby contributing to the breakdown of compartmentalization which is one of the characteristics of senescence.

### 2.2.2 Changes in the Chloroplast Envelope

A separation of the inner and outer layers of the chloroplast envelope is one of the earliest observable consequences of senescence and a selective change in the properties of this membrane system may be a crucial initiating event. Disassembly of plastid contents has been shown to be due to enzymes synthesized in the cytoplasm (see Sect. 2.4.3) and there is reason to believe that these enzymes, or the RNA's from which they are translated, are present from an early stage in leaf development (THOMAS and STODDART 1980). Thus, the envelope may constitute a barrier preventing ingress of degradative agents and its effectiveness may be determined by structural changes occurring during progressive development. Such changes are also likely to be important in the context of metabolite exchange between plastid and cytoplasm.

The inner envelope membrane is thought to be the main osmotic barrier to the transport of metabolites. A range of distinguishable protein translocators are inserted into this membrane (CHUA and SCHMIDT 1979) and enzymes such as acyl CoA synthetase, acylase, phosphatidic acid phosphatase and galactosyl transferase are also bound to the inner envelope (JOYARD and DOUCE 1977). In contrast, the outer membrane seems to be a relatively inert structure. Denaturing polyacrylamide gel electrophoresis resolves at least 30 polypeptides from *Spinacia* envelopes, all with molecular weights in the 30,000 to 100,000 range. However, isolated chloroplasts synthesize only three of these in vitro and the remainder are thought to be of cytoplasmic origin (MORGENTHALER and MENDIOLA-MORGENTHALER 1976). Thus, if plastid envelope proteins turn over at a significant rate in the mature leaf, and this remains to be established, then the protein-synthesizing activities of the cytoplasm are likely to be of major importance in maintaining the selective and structural properties of this barrier.

Changes in the transport rates of small molecules (sucrose, malate, α-ketoglutarate, glutamate, glycine) into plastids during greening have been recorded for *Spinacia* (GIMMLER et al. 1974) and *Avena* (HAMPP and SCHMIDT 1976). Protein synthesis in isolated *Phaseolus* etioplasts has also been shown to be accompanied by pool-independent increase in leucine uptake (DRUMM and MARGULIES 1970). The establishment of such selective properties has clear implications for the regulation of the plastid/cytoplasm balance via the exchange of key metabolites and an associated property may be the maintenance of a barrier to the ingress of hydrolytic agents active in the cytoplasmic turnover of proteins and nucleic acids.

Disassembly of the plastid by cytoplasmic agents implies either a loss of envelope integrity, thus allowing indiscriminate incursion by cytoplasmic components, or a directed synthesis of specific hydrolytic enzymes with the capacity to cross the envelope inner membrane using the normal transport mechanisms. The possibility that additional N or C-terminal amino acid "signal" sequences may direct to the plastid proteins synthesized on free polysomes, or facilitate passage of the proteins into this organelle, has been examined for the small subunit of RuBPC. When synthesized by cell-free systems this polypeptide has a molecular weight 4,000 larger than the final incorporated subunit (HIGHFIELD and ELLIS 1978) but the extra sequences are lost during, or immediately after, transport into the chloroplast (CHUA and SCHMIDT 1978). Certain thylakoid membrane proteins, such as ferredoxin, are known to exhibit a similar behaviour. Thus, proteinase ingress may depend upon the synthesis of species with appropriate attachments to allow successful transport across the intact inner envelope. The general area of transport of proteins into organelles has been reviewed by CHUA and SCHMIDT (1979).

Information on envelope properties during senescence is sparse and is restricted to observations on the loss of activity of enzymes such as galactosyl transferase (DALGARN et al. 1979) and, possibly, alanine aminotransferase (THOMAS 1975). This is clearly a critical area for future studies on the regulation of plastid senescence.

### 2.2.3 Extraplastidic Changes

Advancing senescence is paralleled by marked physical and compositional changes in microsomal membranes. Detailed studies have been carried out on the rough and smooth components of microsomal vesicles derived from the endoplasmic reticulum (ER) of *Phaseolus* cotyledons. Senescence in this tissue proceeds with advancing seedling development and the activities of the membrane-associated enzymes, NADPH- and NADH-cytochrome c oxidoreductase, glucose-6-phosphatase and 5′nucleotidase have been determined at various points during the ageing process (MCKERSIE and THOMPSON 1975). Activities of all the enzymes declined, indicating changes in membrane organization. Phospholipid levels in both the smooth and rough vesicle fractions also fell in a co-ordinated manner although disparities in associated enzyme profiles suggested they were not being dismantled in an identical fashion. As in plastid studies (Sect. 2.2.1) the *Phaseolus* microsomal membranes exhibited a dramatic shift in phase transition temperature with increasing age, changing from 22° to 38 °C in the period between 2 and 3 weeks after emergence. The free sterol to phospholipid ratio increased threefold during the same period, reflecting either a selective loss of polar lipids or some conversion of sterol esters and glycosides to the free form during senescence. Polyribosome profiles are modified with advancing leaf age and the proportion of higher polymers in the total ribosome population falls progressively (CALLOW et al. 1972). This process correlates with the rise in ribonuclease activity in the senescent leaf (SACHER and DAVIES 1974).

Ultrastructural studies have shown that mitochondria remain intact, except for some swelling or distortion of the cristae, up to a very late stage. This is supported by a number of studies on respiration and mitochondrial enzyme activity during ageing in sections and intact leaves which indicate a slow decline without any abrupt features (HARDWICK et al. 1968, TETLEY and THIMANN 1974, LLOYD 1980). Intact nuclei are also recoverable from leaf cells throughout senescence, but nucleoli become less frequent with ageing and nuclear contents tend to aggregate (BUTLER and SIMON 1970).

Cells become progressively more vacuolate with age and the surviving organelles are confined within a diminishing rim of cytoplasm. The outer protoplast membrane, or plasmalemma, retains a normal appearance at all stages, surviving up to the point where the cytoplasm has almost disappeared. Ultimately, however, it does assume a somewhat corrugated appearance in electron micrographs (BUTLER and SIMON 1970). Changes in the permeability of the tonoplast membrane, which bounds the vacuole, could allow transfer of vacuolar materials to the cytoplasm where they might act to lower pH to a value favouring the operation of acid hydrolases. At the same time, loss of tonoplast integrity could permit migration of the enzymes themselves from vacuole to cytoplasm or chloroplast (PEOPLES et al. 1980). A range of hydrolases, including proteinase, β-glucosidase, phosphatase and nuclease, has been localized in vacuoles prepared from leaf protoplasts (BOLLER and KENDE 1979, HECK et al. 1981). The enzyme profile of the central vacuole of plant cells resembles that described for animal lysosomes (e.g., STRAUS 1967) but the validity of the implied functional comparison has been questioned (LEIGH 1979). The timing of vacuolar rupture during senescence, occurring as it does when plastid degradation is already well advanced, suggests that release of vacuolar contents into the cytoplasm is a relatively late event. However, we should be careful to appreciate that electron micrographs give little direct information on changes in the permeability or selectivity of the tonoplast membrane and substantial transfers could occur before visible rupture.

These data convey a general impression of differential lability at the organelle level with the chloroplast and endomembrane systems exhibiting the highest susceptibility to degradative influences. The resistance shown by mitochondria and the bounding membranes of the cell (and by nuclei to a lesser extent) hint at differences in their molecular architecture which allow them to continue functioning, as essential components of the mobilization apparatus, in the presence of increasing cytoplasmic degeneration.

### 2.2.4 Changes in Respiration and Energy Supply

Measurements of $O_2$ uptake by *Avena* leaf sections (TETLEY and THIMANN 1974) have shown that, after an initial 24 h lag, respiration rate rises rapidly to achieve values 2.5 times higher than those measured at excision. The rise, and chlorophyll loss, could be completely prevented by kinetin application. These authors suggest that approximately one-fifth of the respiratory rise could be attributed to the consumption of the free amino acids liberated during senescence and that a further, smaller, amount resulted from sugar utilization. Uncoupling of respiration and phosphorylation was considered to be the major component of the increase. A somewhat different pattern was obtained by measuring $CO_2$ evolution in attached *Perilla* leaves (HARDWICK et al. 1968), where the respiratory burst occurred only during the most advanced stages of senescence.

Closer agreement with the leaf section findings was obtained using attached fourth leaves of *Lolium temulentum*, labelled with $^{14}C$ during expansion by root-feeding with sucrose (LLOYD 1980). Senescence was induced by either maintaining in darkness or by removing the $CO_2$ supply in the light. Under both conditions $^{14}CO_2$ evolution rose by a factor of 5, when compared to controls, over a period of 4 to 6 days and, thereafter, diminished steadily with the progress of senescence (Fig. 2). This respiratory surge was associated with a rapid transfer of $^{14}C$ between the insoluble and soluble fractions as well as with the rate of [$^{14}C$]-translocation. Interpretation of these data does, however, rest upon the assumption that no major changes occur in the specific activity of the respiratory substrate pool(s).

The respiratory quotient (RQ) in *Avena* leaves fell from 0.84 to 0.7 during yellowing but could be restored to unity by the addition of glucose at 0.3 M; an unphysiologically high concentration (TETLEY and THIMANN 1974). A low value in senescent leaves may reflect an increasing use, as substrates, of lipids mobilized during plastid degeneration or may suggest incomplete oxidation of amides. Keto-acids generated as a result of the extensive glutamate-forming transaminations occurring during senescence (Sect. 2.3.3, Fig. 3) are also likely to be major substrates at this time. A final rise in RQ in *Polygonum, Rumex* and *Triticum* to values as high as 1.3 has been reported by JAMES (1953), who

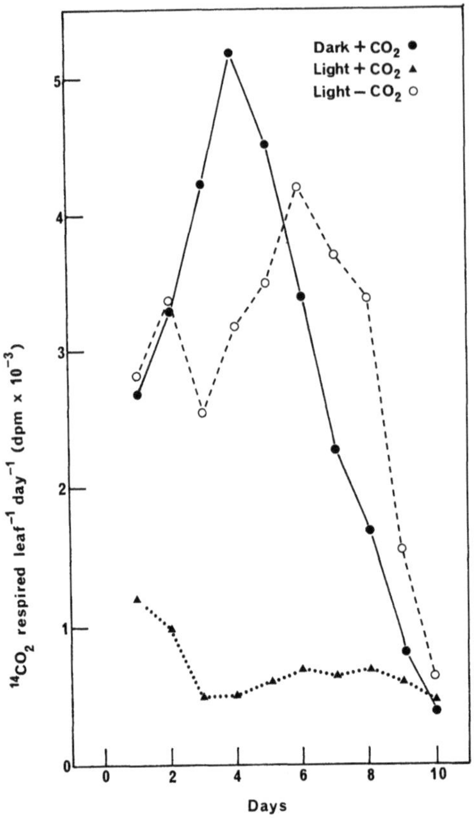


**Fig. 2.** Effect of light and $CO_2$ supply on the evolution of respiratory $^{14}CO_2$ from attached fourth leaves of *Lolium temulentum*. (Redrawn from LLOYD 1980)

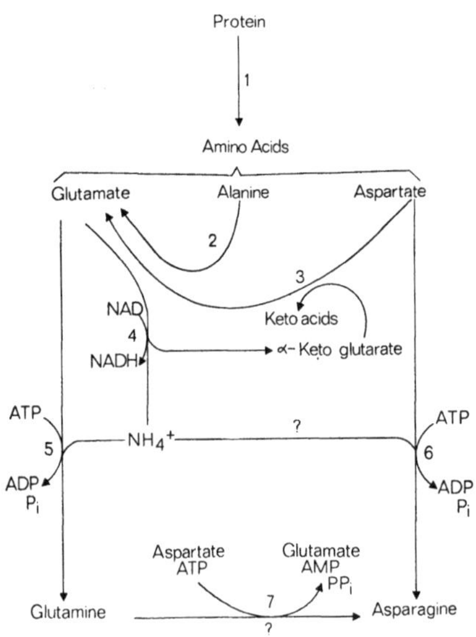


**Fig. 3.** Proposed pathway for the conversion of leaf protein to amides during senescence (THOMAS 1978). *1* endo- and exo-peptidases; *2* alanine aminotransferase; *3* aspartate aminotransferase; *4* glutamate dehydrogenase; *5* glutamine synthetase; *6* asparagine synthetase; *7* glutamine-dependent asparagine synthetase

reasoned that this late change could be due to a higher consumption of organic acids in the first two taxa or a further level of oxidation of amides and amino acids in *Triticum*. What might be the significance of enhanced respiratory rates during senescence? It is evident that nutrient mobilization processes will have an energy demand which must be satisfied via the respiratory pathway, but there is also the possibility that this process might provide a means for the removal of an untranslocatable excess of substrates, which would otherwise make the senescing leaf an attractive site for invasion by pathogens.

A measure of the energy available for metabolism in a system is provided by the "energy charge" (EC) value (ATKINSON 1968) which can be computed from the following relationship:

$$EC = ([ATP] + \tfrac{1}{2}[ADP])/([ATP] + [ADP] + [AMP])$$

When the proportions of adenosine phosphates are such that a value in excess of 0.5 is obtained then ATP-requiring systems can operate, whereas EC values below this threshold are indicative of zero-growth or senescence. Actively dividing bacteria (CHAPMAN et al. 1971) or germinating seeds (CHING and CHING 1972, SIMMONDS and DUMBROFF 1974) typically have an EC of around 0.8, whilst dormant axes of *Acer* give values as low as 0.15 (SIMMONDS and DUMBROFF 1974). The regulatory aspects of the adenylate balance have been demonstrated in *Lactuca* seeds where the depression of adenylate kinase activity by anaerobiosis has been shown to limit both germination and seedling growth (BOMSEL and PRADET 1968).

Energy balances of this type may be capable of regulating metabolism in a subtle and effective manner but, to date, there are no detailed studies of EC changes in relation to leaf ageing. The crossing of the 0.5 threshold could well be related to the sensing or triggering mechanisms associated with senescence.

## 2.3 The Senescent Leaf as a Source of Mineral Elements

### 2.3.1 Mobilization of Nitrogen

Senescent leaves are net exporters of mobilized minerals (WILLIAMS 1955, THROWER 1967, BRADY 1973, DERMAN et al. 1978, THOMAS and STODDART 1980). Some elements, such as iron and calcium, are not mobile and leaves continue to accumulate these elements until they die. Symptoms of their deficiency become apparent first in young tissues and growing points (WALLACE 1961, SPRAGUE 1964, EPSTEIN 1972, THOMAS and STODDART 1980). The major mineral elements, N, P, K and, possibly, S and Mg, are readily withdrawn from ageing leaves and redistributed to younger tissues (WILLIAMS 1955, HOPKINSON 1966, FRIEDRICH and SCHRADER 1979). The fact that leaf senescence may be prevented by removing or inactivating organs or tissues which receive the minerals mobilized from senescing leaves suggests that the senescence signal originating in young tissue and transmitted to older leaves (LINDOO and NOODÉN 1977) is related to, and possibly identical with, the mobilization signal.

For most of the mineral elements little is known of the mechanism of mobilization during leaf senescence, or of how this process is regulated. However, recent interest in the role of senescence in the internal N economy of crop plants has led to a better understanding of the factors determining redistribution of this element from old leaves. Nitrogen reutilization is discussed here as an example of a mobilization process initiated at the onset of leaf senescence.

Some plant species accumulate nitrate. It has long been known that this stored nitrate can be used for protein synthesis in developing seeds (BERTHELOT and ANDRÉ 1884). Maturing seeds are able to synthesize amino acids de novo if supplied with nitrate and an organic carbon source such as sucrose (LEWIS et al. 1970). Nevertheless, most of the N reaching developing seeds and other sinks is reduced; and, although it varies with growth conditions and from species to species, it is genrally true that much or most of this reduced N has been remobilized from older vegative tissues (WILLIAMS 1955, NEALES et al. 1963, DALLING et al. 1976, SINCLAIR and DE WIT 1975, STOREY and BEEVERS 1977, HUFFAKER and RAINS 1978).

The largest source of mobilized reduced N in the leaf is RuBPC (HUFFAKER and MILLER 1978). During senescence both the amount of RuBPC protein and the carboxylating activity of the enzyme decrease (PETERSON and HUFFAKER 1975, WOOLHOUSE and BATT 1976, HALL et al. 1978, THOMAS 1976b, 1977, WITTENBACH 1979, FRIEDRICH and HUFFAKER 1980). RuBPC is hydrolyzed by one or more proteolytic enzymes in the leaf and the amino acids released by proteolysis are loaded into the translocation system, either directly or following some degree of metabolic interconversion (THOMAS 1978).

### 2.3.2 Proteolysis

Proteolytic activity in relation to leaf senescence has been studied in many species including: *Nicotiana* (ANDERSON and ROWAN 1965); *Triticum* (DALLING et al. 1976, PEOPLES and DALLING 1978, WITTENBACH 1978, 1979); *Hordeum* (PETERSON and HUFFAKER 1975, HUFFAKER and MILLER 1978, FRIEDRICH and HUFFAKER 1980); *Glycine* (RAGSTER and CHRISPEELS 1979); *Pisum* (STOREY and BEEVERS 1977); *Lolium* (THOMAS 1978); *Zea* (FELLER et al. 1977, REED et al. 1980); and *Avena* (MARTIN and THIMANN 1972). The enzymes of *Avena, Hordeum* and *Triticum* leaves have been extensively purified and their enzymic properties described (DRIVDAHL and THIMANN 1977, 1978; FRITH et al. 1978a, b, c, PEOPLES et al. 1979, HUFFAKER and MILLER 1978). In addition to enzymes that hydrolyze protein substrates, leaves contain exopeptidase-like activities which hydrolyze artificial esters such as the p-nitroanilide and β-naphthylamide derivatives of amino acids, but it is not certain whether these enzymes function in protein degradation in vivo (THOMAS 1978).

The endopeptidases of ageing leaves have the following general properties. They have an acid pH optimum, generally around 5.0 and require a sulphydryl reagent, or a chelator such as EDTA, or both, for full activity. The substrate specificity is low and in most studies they have been assayed using an animal protein substrate such as gelatin, haemoglobin, casein or azocasein. Soluble leaf protein is readily hydrolyzed by leaf proteinase in vitro (VON ABRAMS 1974,

**Fig. 4.** Distribution of radioactivity in SDS-gel electrophoretograms of $^{14}C$-RuPBC incubated, at pH 5.0, with a crude extract of 12-day-old barley primary leaves. Incubation times (h): *A* 0; *B* 1.5; *C* 3.0; *D* 6.0. Molecular weight scale determined by reference to known values for large and small subunits (THOMAS and HUFFAKER 1981)

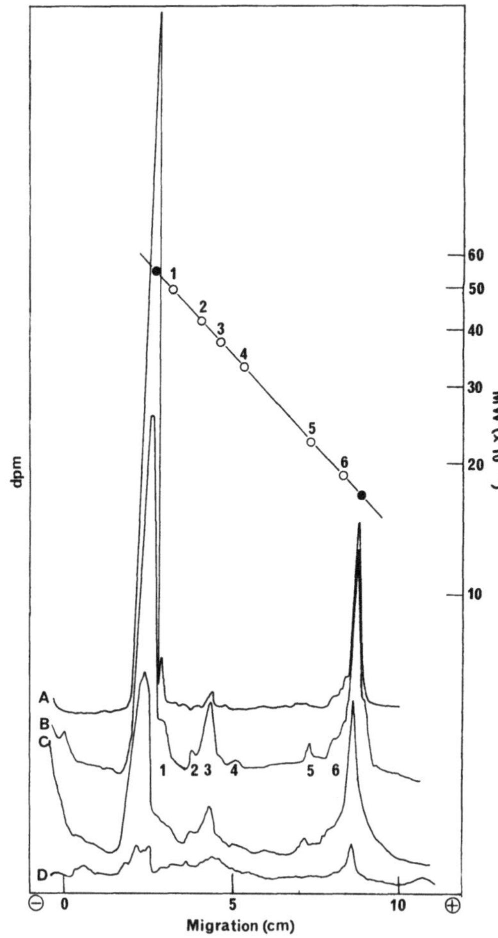


STOREY and BEEVERS 1977). The enzyme also has a high affinity for RuBPC, presumably its natural substrate (WITTENBACH 1978). Using purified [$^{14}C$]-labelled RuBPC as a substrate for the assay of endopeptidase activity in crude extracts of *Hordeum* primary leaves, high rates of hydrolysis, up to 4% per hour with extracts of senescent tissue, have been obtained. The products of endopeptidic cleavage of RuBPC were visualized by SDS gel electrophoresis and fluorography (Fig. 4). The large subunit (mol. wt. 55,000) was cleaved into a series of fragments, including major polypeptides with molecular weights of 50,000, 42,200, 37,300, 33,100, 22,400 and 18,800. No evidence of a comparable breakdown pattern for the small subunit of RuBPC was obtained in these experiments (THOMAS and HUFFAKER 1981).

As well as acid endopeptidases, there are reports of leaf proteolytic enzymes with higher pH optima (DRIVDAHL and THIMANN 1977, FELLER et al. 1977, THOMAS 1978). It is possible that these enzymes are the products of self-hydrolysis or ageing of the acid endopeptidase (THOMAS 1978); novel proteinases appear to arise in this way during enzyme purification or storage of enzyme extracts

(DRIVDAHL and THIMANN 1978, HUFFAKER and MILLER 1978). On the other hand the multiplicity of proteinases found in leaf tissue may reflect a corresponding variety of function and substrate specificty.

Activation of acid endopeptidase apparently related to degradation of leaf protein has been observed both in detached leaves (PETERSON and HUFFAKER 1975, MARTIN and THIMANN 1972, THOMAS 1978) and in leaves attached to the plant (FELLER et al. 1977, DALLING et al. 1976, STOREY and BEEVERS 1977, FRIEDRICH and HUFFAKER 1980). The fact that inhibitors of protein synthesis, such as cycloheximide and D-MDMP, prevent leaf senescence, loss of protein and proteinase activation is further evidence that protein mobilization is initiated by an increase in endopeptidase activity (perhaps the result of de novo enzyme synthesis) at the start of senescence (PETERSON and HUFFAKER 1975, MARTIN and THIMANN 1972, THOMAS 1976b, HUFFAKER and MILLER 1978).

On the other hand, there is increasing evidence that proteinase activation is not a prerequisite for protein mobilization in senescent leaves. Protein degradation in the absence of increased proteinase activity occurs in leaves of *Nicotiana* (ANDERSON and ROWAN 1965) and *Tropaeolum* (BEEVERS 1968). Even species such as *Triticum* and *Hordeum*, in which activation has been shown to occur, may be made to senesce under conditions which promote a decrease in leaf protein but no accompanying change, or even a reduction, in endopeptidase activity (WITTENBACH 1978, 1979, HUFFAKER and MILLER 1978, MILLER and HUFFAKER 1979). It is a typical feature of acid hydrolases that levels of these enzymes present in the pre-senescent leaf are more than adequate to degrade their in vivo substrate at the rate observed during senescence without the necessity for further activation (BEEVERS 1976, POLLOCK and LLOYD 1978).

If protein mobilization is not initiated by an increase in proteinase activity, it may be the consequence of an increase in the accessibility of the substrate. It has been suggested that proteinases cannot penetrate the plastid envelope and hydrolyze RuBPC and other chloroplast proteins until the envelope has been modified (perhaps by a membrane-specific enzyme synthesized de novo during senescence) or has aged and become leaky (HUFFAKER and MILLER 1978). Alternatively stroma proteins may cross the membrane to the cytoplasm or vacuole when this happens. It is also possible that the chloroplast fuses with or is engulfed by the vacuole, a mechanism that would be consistent with the latter's lysosome-like enzyme complement (see Sect. 2.2.3).

It may be, however, that endopeptidases localized outside the chloroplast do not function at all in degradation of the organelle. The chloroplast itself may contain its own proteinase(s) which remain inactive until senescence is initiated. The work of CHOE and THIMANN (1977), showing functional and structural stability of chloroplasts isolated from leaves early in senescence and cultured in vitro, argues against this view. On the other hand, PEOPLES and DALLING (1979) have recently reported that appreciable acid proteinase activity equilibrates with chloroplast marker enzymes in sucrose gradient fractionations of *Triticum* leaf organelles; and THOMAS and HUFFAKER (1981) found that RuBPC from entire *Hordeum* leaves and from isolated chloroplasts exhibited a measurable level of associated endopeptidase even when purified to apparent homogenity. HUFFAKER and MILLER (1978) suggest that conflicting reports concerning the intracellular location of RuBPC hydrolase may arise out of the use of isolation procedures which favour the extraction of healthy presenescent chloroplasts and the discarding of more fragile senescing organelles in which proteolysis in underway (see Fig. 1d and f). Furthermore, it would be difficult to assay, by conventional procedures, a chloroplast-associated proteinase in the presence of high concentrations of RuBPC with which it appears to associate with high affinity. The question remains as to how hydrolysis of chloroplast proteins by such an intrachloroplastic proteinase is prevented until the appropriate time in leaf development. Regulation by pH is possible (PEOPLES and DALLING 1978). An acid proteinase would be inoperative in photosynthesizing chloroplasts, where the pH of the stroma is maintained at about 8, but would become increasingly active as the rate of photosynthesis and stroma pH fell in ageing leaves.

Regulation by proteinase inhibitors is a possibility. These are present in leaves and in at least one species, *Lycopersicon,* have been localized in the vacuole (WALKER-SIMMONS and RYAN 1977); but generally they are far less effective against plant proteinases than against those of animal origin and appear to function in wound reactions and other defence mechanisms.

### 2.3.3 Metabolism and Loading

The primary structure of RuBPC is highly conserved across the range of plant groups. Glutamate, aspartate, glycine, leucine and alanine are the five major constituent amino acids, occurring in the approximate molar ratios 1:0.85:0.75:0.74:0.64 respectively (data from KUNG 1976). It is probable that some or all of the glutamate and aspartate normally exists in the amide form in the intact enzyme. If the amino acid products of proteolysis occurring during senescence are loaded into the translocation system without further metabolism, then the amino acid composition of phloem sap leaving the leaf or arriving at the sink should resemble that of RuBPC, the major mobilized leaf protein. A study of C and N translocation in *Lupinus* (PATE et al. 1979) includes figures for the amino acid fraction of the phloem exudate from petioles of primary leaves sampled at senescence, when both total leaf N and the ratio of sucrose to amino acid exported were falling. The approximate molar ratios of glutamate-glutamine, aspartate-asparagine, glycine, leucine and alanine were respectively, 1:2.8:0 (not reported):0.3:0.3. These values are very different from those predicted by the amino acid composition of the major leaf protein. The amino acid complement of phloem sap obtained from legume fruits also contrasts markedly with that of leaf protein (PATE et al. 1974).

The major organic forms of N in phloem are amides, notably asparagine and glutamine (ZIMMERMANN 1960, PATE et al. 1979, CARR and PATE 1967). N can be moved in amide form from source to sink through intervening tissues which might have a high demand for N but to which amide N is unavailable because these tissues lack the necessary enzymes of amide metabolism – glutamate synthetase (MIFLIN and LEA 1977, STOREY and BEEVERS 1977, 1978); asparaginase (ATKINS et al. 1975) or asparagine transaminase (STREETER 1977). In this respect amides are "protected" metabolites, analogous to sucrose, which is the major translocated form of C and which is also unavailable to tissues lacking the appropriate invertase or sucrose synthetase activity (ARNOLD 1968, POLLOCK 1976, POLLOCK and LLOYD 1978).

The capacity of plants to make amides is prodigious. CHIBNALL (1939) describes how high levels of ammonia fertilizer applied to *Lolium* plants resulted in the exudation of glutamine in such high concentrations that it crystallized out on the leaves. Detached leaves convert almost all their protein N to amides during senescence (CHIBNALL 1939, YEMM 1950, MCKEE 1962, THOMAS 1978).

The available information adds up to a good case for believing that, before they reach the sinks, the amino acids released by proteolysis during leaf senescence are subjected to further metabolism, particularly interconversion to form amides, either before being loaded into the transport system, or in the phloem itself, or both (ZIEGLER 1974, THOMAS 1978, STOREY and BEEVERS 1978). THOMAS (1978) has proposed a scheme for the formation of amides during senescence, based on a study of protein breakdown in detached *Lolium* leaves (Fig. 3). Each of the enzymes measured (transaminases, glutamate dehydrogenase, glutamine synthetase) was present at levels in excess of the rate of conversion of protein N to amide N. STOREY and BEEVERS (1978) also found that the catalytic activity of glutamine synthetase in ageing *Pisum* leaves is more than adequate to account for the observed rate of glutamine formation.

The mechanism of asparagine synthesis in senescent leaves is unknown. Asparagine synthetase is probably not responsible for accumulation of the amide in detached *Lolium* leaves (THOMAS 1978). Synthesis from serine and cyanide is a possibility (CASTRIC et al. 1972) but endogenous levels of cyanide in senescent leaves appear to be insufficient to support asparagine synthesis at the observed rate (SADLER and SCOTT 1974). Alternatively, asparagine may be synthesized from aspartate using glutamine as the amide donor (LEA and FOWDEN 1975).

Many gaps remains in our knowledge of N redistribution during senescence. What is the rate-determining step? The absolute activities of proteinases and the enzymes of amide synthesis do not seem to be limiting. Compartmentation or substrate-level regulation of enzyme activity are likely areas where the rate of the process is controlled. Is the export of mobilized N from senescent leaves limited by selectivity in the translocation system, or by the operation of a specific active loading mechanism like that for sucrose? Information on this point is meagre. JOY and ANTCLIFF (1966) supplied different radioactive amino acids to *Beta* leaves of various ages and measured the extent of translocation to young tissues. They found a marked increase in amino acid translocation at the onset of senescence and a reduction in the ability of the transport system to discriminate between protein and non-protein amino acids. More recently, experiments on phloem loading in the petioles of *Ricinus* leaves have shown the existence of an active uptake mechanism for amino acids, with evidence of proton co-transport in competition with sugars (BAKER et al. 1980). In studying the route taken by mobilized amino acids as they move from the senescent leaf to the young sink, THIMANN et al. (1974) detected a pronounced basipetal transport of mobilized amino acids in leaves of intact oat seedlings senescing in the dark. They suggest that the amino acids so transported are accumulated in the roots. Transfer to xylem and retranslocation upward to young sinks may occur in the root, although interchange of metabolites between xylem and phloem is known to occur in the shoot too (SHARKEY and PATE 1975, PATE et al. 1975). Amino acids are also redistributed by direct translocation from older leaves to young sinks through the phloem (WALLACE and PATE 1967, OGHOGHORIE and PATE 1972). For most plant species the relative contribution of each route to the flow of N between old leaves and sinks is unknown. Also largely uncomputed are the energy costs of recovering N from senescent leaves. Only when coordinated studies, both at the molecular and at the structural level, of proteolysis, amino acid metabolism, phloem loading, transport and unloading at the sink have been carried out can firm answers to these questions be expected.

## 2.4 Characteristics of the Senescence Switch

### 2.4.1 Senescence is Genetically Programmed

We have seen that leaf development is a genetically programmed sequence of biochemical and physiological changes, some of which are summarized in Fig. 5, and that the initiation of senescence is marked by a transition in the status of the leaf from being a source of photosynthate to becoming a source of mobilized N and other mineral elements. The sequence itself seems to be qualitatively invariable for a given species; but the rate at which the leaf pro-

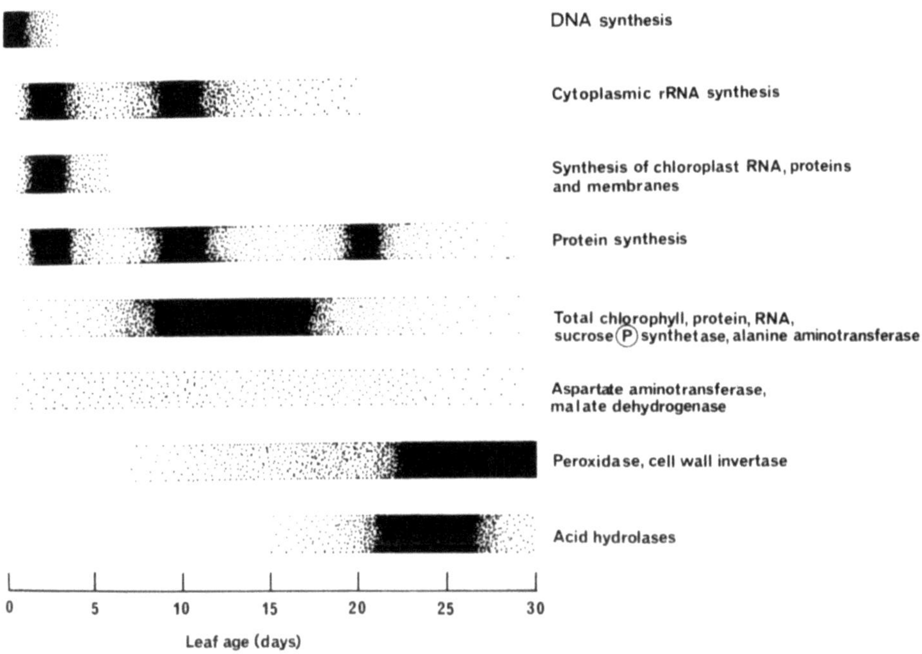


**Fig. 5.** Metabolic events during development of a grass leaf from emergence to senescence. *Density of shading* indicates the level of the constituent or intensity of activity

gresses through it and the relative duration of each phase are determined by environmental and correlative factors. Senescence comes at its due time in leaf development. Generally speaking, factors that initiate senescence do so by compressing or terminating previous phases in the developmental sequence, and those that delay senescence act by prolonging the pre-senescent condition. For example, senescence in *Lolium temulentum* is sensitive to daylength: short days delay senescence and long days promote it by respectively extending and curtailing the previous phase of leaf development (HEDLEY and STODDART 1972). Temperature influences senescence in a similar way, as illustrated by inhibition of leaf expansion in high altitude ecotypes of *Festuca pratensis* in response to low temperatures during winter, which leads to a rapid senescence of newly formed leaves, a high rate of leaf turnover and, consequently, an apparent dormancy condition (POLLOCK et al. 1980).

Direct evidence that leaf senescence is a programmed developmental process is provided by studies showing that genetic variation exists for expression of the syndrome as a whole (ABU-SHAKRA et al. 1978, BOYD and WALKER 1972, KAHANAK et al. 1978) and also for individual components of the syndrome, notably chlorophyll degradation (THOMAS and STODDART 1975, 1980, THOMAS 1976a, 1977). Furthermore, the work of YOSHIDA (1961) with nucleate and enucleate protoplasts of *Elodea* clearly shows that chloroplast breakdown is under direct nuclear control.

It may be concluded, therefore, that leaf development is regulated by a gene complex which includes a programme specifically regulating senescence. Expression of this programme is blocked until the appropriate point in the developmental sequence. The initiating event in senescence – the switch – is the unblocking of the senescence programme (THOMAS and STODDART 1980). The question of where the blockage is located raises, in turn, the question of the relative contributions of chloroplast and nuclear genomes to the regulation of senescence.

### 2.4.2 Activity of the Chloroplast Genome

Chloroplasts are assembled in the early part of leaf ontogeny during growth and expansion (Fig. 5). Plastid rRNA and proteins coded by chloroplast DNA (for example, the large subunit of RuBPC) are also synthesized during the growth phase of leaf development. Once full expansion has been achieved, chloroplast protein and nucleic acid synthesis are shut down (WOOLHOUSE and BATT 1976, THOMAS and STODDART 1980, NESS and WOOLHOUSE 1980) and the rate of photosynthesis begins to decline steadily (see Sect. 2.1). It appears that repression of the chloroplast genome contributes to the initiation of leaf senescence through its function in terminating the phase of chloroplast assembly during leaf growth.

### 2.4.3 Requirement for Cytoplasmic Protein Synthesis

Measurements of protein synthesis during the life of the leaf show three bursts of incorporation of labelled amino acids; the first associated with leaf expansion and representing the synthesis of proteins participating in chloroplast assembly, the second occurring in the mature, photosynthate-exporting leaf, and the third coinciding with senescence (HEDLEY and STODDART 1972, STODDART 1972, Fig. 5). Treatments which delay senescence, such as daylength or cytokinins, also delay the onset of the third phase of protein synthesis. So, too, do inhibitors of protein synthesis including cycloheximide and D-MDMP (STODDART 1972, THOMAS 1975, 1976a, b, THOMAS and STODDART 1980). Inhibitors of chloroplast protein synthesis are not effective in preventing senescence (THOMAS 1976a, THOMAS and STODDART 1980). Thus synthesis of proteins by cytoplasmic ribosomes appears to be a positive requirement for leaf senescence. This conclusion is supported by studies of the stability of chloroplasts isolated from leaves at incipient senescence (CHOE and THIMANN 1977) and by the observation that cytoplasmic polyribosomes (but not chloroplastic) may be recovered from leaves up to an advanced stage of yellowing (CALLOW et al. 1972, EILAM et al. 1971, JUPP 1980).

### 2.4.4 Transcription

Proteins are synthesized during senescence. These proteins function in the degradative processes that occur in ageing leaves. Presumably, therefore, they are the translation products of transcripts of the senescence programme in the nuclear genome; but there are few convincing reports of a transcription requirement for leaf senescence. The rate of RNA synthesis in senescent leaves is usually low, and such RNA as is made is largely cytoplasmic rRNA. Actinomycin-D and other inhibitors of transcription do not prevent yellowing and loss of protein (THOMAS and STODDART 1980). Fluorouracil, which inhibits post-transcriptional processing of rRNA precursor, is also ineffective in preventing senescence (PARANJOTHY and WAREING 1971). Senescence thus seems not to require the continuous production of mature rRNA.

On the other hand, cordycepin, an inhibitor of post-transcriptional polyadenylation of mRNA, has been reported to inhibit leaf senescence (TAKEGAMI and YOSHIDA 1975, THOMAS and STODDART 1980). Incorporation of radioactively labelled adenine into poly-(A)RNA is detectable in leaf tissue during senescence (JUPP 1980). Latent mRNA transcripts of the senescence programme may become active as a result of polyadenylation in the nucleus in older leaves.

Existing information does not provide unequivocal support for an exclusively post-transcriptional mode of genetic regulation controlling the initiation and progress of leaf senescence; but it is a testable working hypothesis that, except for some rRNA synthesis, the nuclear genome is repressed fairly early in leaf development – perhaps at the same time as chloroplast transcription and translation close down. Further development of the leaf, including senescence, runs on RNA transcripts amassed before the nucleus becomes inactive and on translation products of these transcripts, some of which may be in a latent or masked form (THOMAS 1976a, THOMAS and STODDART 1977, 1980).

### 2.4.5 Reversibility of Senescence

Treatments which delay senescence are also able to reverse the process in many species if applied to the leaves once senescence is underway. The most extreme examples of senescence reversal are seen amongst the dicotyledonous plants, notably *Nicotiana* (MOTHES 1960, WOLLGIEHN 1961, THOMAS and STODDART 1980), *Perilla* (CALLOW and WOOLHOUSE 1973) and *Phaseolus* (NESS and WOOLHOUSE 1980). By removing some or all of the upper shoot, lower leaves that have become senescent may be stimulated to re-green. The process is enhanced by feeding fertilizer to the leaves or, in some cases, treating with cytokinins (AVERY 1934, DYER and OSBORNE 1971, MOTHES 1960, MOTHES and BAUDISCH 1958, WOLLGIEHN 1961, BÖTTGER and WOLLGIEHN 1958, WOOLHOUSE 1967). It has been shown that diverting the transpiration stream from upper leaves by enclosing the young shoot in a plastic bag is enough to bring about re-greening of older leaves (NESS and WOOLHOUSE 1980). The ability of senescent leaves to re-green in response to a reduction in the demands of younger organs is much more limited among monocotyledonous plants, although under certain conditions cereal leaves may be made to retreat from senescence to a small extent (WALKLEY 1940).

Senescence reversibility has also been studied using plants that have been induced to senesce by placing them in the dark. By re-exposing such plants to light after various lengths of time it has been possible to identify an initial phase when senescence is readily reversible, leading to a second stage when transfer back to light is no longer able to prevent senescence from reaching completion. Cytokinins are effective in extending the reversible phase (WITTENBACH 1977, VONSHAK and RICHMOND 1975).

The general characteristics of senescence reversibility may be summarized (THOMAS and STODDART 1980). In continuous darkness the ability to re-green is lost. It is not directly dependent on retention of chloroplast structure and function. It requires that all the information for reconstructing the chloroplast and other cellular systems is stable and available until an advanced stage in senescence. It follows that leaf senescence is probably not a consequence of irreversible degradation of, or damage to, the genome, in contrast to other forms of ageing such as loss of seed viability, gerontological changes in animals and so on. Finally, studies on reversibility suggest that the onset of senescence is a sequential process during which a developmental threshold is crossed and the syndrome becomes irreversibly established.

## 3 Senescence as a Growth Correlation

### 3.1 Correlative Regulation of Senescence

#### 3.1.1 Competition Between Organs

The senescence of a leaf is influenced by other organs on the same plant. It is useful here to think of an individual plant as a population of parts (HARPER 1977). Just as the members of a plant community interact by crowding, shading, competing for nutrients and exuding allelopathic compounds, so too do the structural units of a single plant modify the course of each other's development through physical, nutritional and hormonal influences. The role of plant growth regulators in leaf senescence is discussed in Section 3.2. Here we examine the correlative regulation of senescence in terms of competition between organs for space, light and nutrients.

### 3.1.2 Space

THOMAS and STODDART (1980) have described how expanding axillary shoots in dicotyledonous plants, and increasing numbers of young leaves produced within the oldest leaf sheaths in monocotyledonous plants, might trigger senescence in older leaves simply by imposing a physical constraint. One can conceive of a sequence of events in which the stress experienced by a petiole in the region of the axil where a secondary shoot is increasing in girth leads to the initiation of an abscission layer, possibly mediated by ethylene, and the consequent senescence of the lamina. With the exception of a few species, notably *Molinia* (JEFFERIES 1916) true leaf abscission does not occur in the Gramineae; but recent studies (LEE 1979, J.A. Pearson unpublished) have identified a distinct zone in the ligule region of *Festuca* leaves across which transport of dyes and radioactively labelled compounds introduced into the xylem becomes restricted with increasing age. It is possible that sequential senescence in grasses is at least in part the result of vascular blockage in the ligule-sheath region of the older leaf in reaction to the increasing bulk of the young expanding tissue it encloses.

### 3.1.3 Light

Young leaves generally occupy the upper region of the canopy. By intercepting light they may impose a pattern of progressive senescence on the leaves below them. The effects of light on senescence are extremely complex. Alterations in the duration, intensity, quality and interaction of light with other environmental variables can bring about changes in the senescence pattern. In general, darkness induces senescence and sustains the process once it is underway, whereas light delays and slows senescence. But extended daylengths can cause enhanced senescence compared with short days, both in intact plants, where it is often related to the induction of flowering (HEDLEY and STODDART 1972, SCHWABE 1970) and in isolated leaf tissue (CARVER et al. 1979). Moreover, light can interact with factors such as the presence of chelators and other chemicals (HOLDEN 1972) or low temperatures (VAN HASSELT and STRIKWERDA 1976) to bring about degradation of chlorophyll by photooxidation.

The senescence-retarding effect of light may be exerted through its role in maintaining photosynthesis (GOLDTHWAITE and LAETSCH 1967, THIMANN et al. 1977). Recent evidence in support of this proposal comes from work on *Lolium temulentum* (LLOYD 1980). During senescence of the 4th leaf in the dark there was a fall in total chlorophyll and protein and a rise in free amino acids (Table 1) and rate of respiration (Fig. 2). These changes did not occur in leaves of an equivalent age maintained in the light and a normal atmosphere; but in leaves exposed to light in $CO_2$-free air, yellowing, proteolysis and respiration followed

**Table 1.** Chlorophyll, soluble protein and amino acid contents of mature attached 4th leaves of *L. temulentum* senesced under light (with and without $CO_2$) and in the dark. (LLOYD 1980)

| Days after induced senescence | Chlorophyll (mg leaf$^{-1}$) Treatment | | | Protein (mg leaf$^{-1}$) Treatment | | | Amino acid (mEq. leucine) Treatment | | |
|---|---|---|---|---|---|---|---|---|---|
| | A | B | C | A | B | C | A | B | C |
| 0 | 0.21 | – | – | 1.5 | – | – | 15.2 | – | – |
| 3 | 0.21 | 0.13 | 0.10 | 1.0 | 1.3 | 1.0 | 9.6 | 21.6 | 43.2 |
| 5 | 0.16 | 0.08 | 0.02 | 1.4 | 0.8 | 0.4 | 14.4 | 46.2 | 46.2 |
| 7 | 0.16 | 0.02 | 0.01 | 1.1 | 0.2 | 0.2 | 9.6 | 44.6 | 52.2 |

A = under lights + $CO_2$; B = under lights − $CO_2$; C = in darkness

the pattern of darkened leaves. On the other hand, the action spectrum for light inhibition of senescence and the inability of DCMU to overcome the light effect suggest that maintenance of photosynthesis is not the whole story (HABER et al. 1969). There are reports of red light inhibition of senescence reversible by far-red implying the participation of the phytochrome system (DE GREEF et al. 1971, SUGIURA 1963, MISHRA and PRADHAN 1973, BISWAL and SHARMA 1976). It is known that the canopy acts as a filter so that as light penetrates to lower leaves it decreases not only in intensity but also in red: far-red ratio (HOLMES and SMITH 1977). Upper leaves may thus influence the senescence pattern of lower leaves by altering both the quantity and the quality of incoming light.

### 3.1.4 Nutrients

During senescence nutrients are removed from the old leaf and transferred to younger tissues. MOLISCH (1938) observed that leaf senescence is often associated with reproduction and proposed that the transfer of nutrients to flowers and fruits killed the leaves by starving them. If plants are maintained under conditions that do not induce flowering, or if pollination is prevented, or if flowers and developing fruits and seeds are removed, it is possible to delay senescence almost indefinitely in many species (HEDLEY and STODDART 1972, LEOPOLD 1961, LEOPOLD et al. 1959, MOLISCH 1938, SCHWABE 1970, LINDOO and NOODÉN 1977, MURNEEK 1926, WAREING and SETH 1967, WOOLHOUSE 1967, WOOLHOUSE 1978). Molisch's concept of Erschöpfungstod (death by exhaustion) also seems to apply in cases of progressive or sequential senescence in the vegetative condition where removal of young leaves halts or even reverses senescence of the old leaves (Sect. 2.4.5).

On the other hand, there are species (e.g., *Perilla*) in which the leaves of fruiting plants live longer and are often bigger than comparable leaves on vegetative individuals (WOOLHOUSE 1974). Furthermore, removing developing fruits has the effect of accelerating foliar senescence in *Capsicum* (HALL and BRADY 1977), *Zea* (ALLISON and WEINMANN 1970) and *Hordeum* (MANDAHAR and GARG 1975).

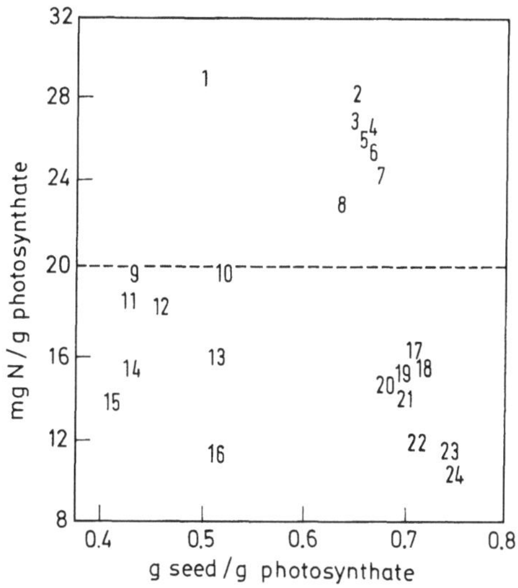


**Fig. 6.** Nitrogen requirement and seed biomass yield per g available photosynthate for 24 crop species. ----- requirement when supply of available N and photosynthate is 5 g and 250 kg/ha/day respectively. *1* Soyabean; *2* Lentil; *3* Pea; *4* Mung bean; *5* Cowpea; *6* Pigeon pea; *7* Lima bean; *8* Chickpea; *9* Hemp; *10* Cotton; *11* Peanut; *12* Flax; *13* Sunflower; *14* Rape; *15* Sesame; *16* Safflower; *17* Wheat; *18* Rye; *19* Oat; *20* Popcorn; *21* Sorghum; *22* Corn; *23* Barley; *24* Rice. (Redrawn from SINCLAIR and DE WIT 1975)

THOMAS and STODDART (1980) suggest that the degree to which reproduction is linked to leaf senescence may vary in different species with the relative requirements of the developing sinks for fixed C on the one hand and mobilized N on the other. Protein reserves account for about 10%, whereas starch and fat make up 80%, of the dry weight of *Zea* and *Hordeum* grains (SINCLAIR and DE WIT 1975). Although some of the carbohydrate of the grain may have been mobilized from sites of storage in the vegetative parts (YOSHIDA 1972) most of it appears to come from current photosynthesis (GIFFORD 1974). Thus grain filling is dependent more on the photosynthetic phase of leaf development than on the nutrient-mobilizing senescence phase. Removing developing grains depresses the rate of photosynthesis in the source leaves (KING et al. 1967, NEALES and INCOLL 1968) and this may be the signal to initiate senescence (THOMAS and STODDART 1980).

By contrast, seeds of legume species such as *Glycine* have a much higher proportion of protein (25%–40% of dry weight) relative to fat and carbohydrate (SINCLAIR and DE WIT 1975). In these species the demand for current photosynthate is supplemented by a high demand for N; so high that the presence of developing seeds may invoke the transition from mature C-exporting to senescent N-exporting function in source leaves. Removing fruits will delay the onset of mobilization and, pari passu, the initiation of senescence. Analysis of the relationship between crop yields and requirements for assimilated C and N (SINCLAIR and DE WIT 1975, 1976) tells a similar story: to meet the N demands of the developing seeds, *Glycine* and the other pulses must "self-destruct" – that is, they must mobilize the N in vegetative tissues by senescing. When N requirements and biomass yield are plotted on a unit photosynthate basis for the major seed crops (Fig. 6) the pulses and *Glycine* form a quite distinct group from the cereals and the oilseeds. It would be interesting to establish, by a systematic study, whether these groupings reflect differing relationships between fruiting and senescence as proposed above.

Clearly, C and N requirements of developing structures are matched closely with the maturity-senescence sequence in source leaves, but there is no general answer as to how supply and demand are coordinated. In some circumstances regulation of source activity by the sink through a feedback mechanism seems to apply, but just as frequently the reverse appears to be true (NEALES and INCOLL 1968, EVANS 1975, GIFFORD 1974). One is bound to question also whether the nutrient demands of developing flowers are adequate to bring about leaf senescence via a simple source-sink feedback system. Flowering in dioecious species such as *Cannabis* and *Spinacia* is correlated with leaf senescence not only in female individuals, which carry the burden of demand from the developing fruits, but also in males (LEOPOLD et al. 1959, WOOLHOUSE 1974, MOTHES 1960). Such observations raise the possibility that growth regulators play a coordinating role in the induction of leaf senescence in response to changes elsewhere on the plant.

## 3.2 The Role of Growth Regulators

It is probable that some form of correlative influence, originating outside the leaf, is implicated in the initiation of senescence and it is generally considered that the "signal" involved is hormonal in nature. Clearly, there are a number of other possibilities, exemplified by subtle changes in the nutrient composition of the translocation stream or the transmission of a specific mobile nucleic acid, but there is little direct evidence to support such suggestions. The various structural classes of known plant growth regulators remain, therefore, as the most probable candidates and, in assessing their effects upon foliar senescence, emphasis will be placed upon possible relationships to the processes involved in protein turnover.

### 3.2.1 Cytokinins

Chlorophyll retention is the most obvious visible effect of cytokinin treatment (RICHMOND and LANG 1957, BEEVERS 1968) but there are also effects upon a range of other cellular components and activities. Polysome aggregates are stabilized (BERRIDGE and RALPH 1969), RNA and soluble protein levels are maintained (RICHMOND and LANG 1957), changes in respiration rate and mitochondrial coupling associated with senescence are suppressed (TETLEY and THIMANN 1974) and the decline of a range of membrane-associated activities in the chloroplast is postponed (THOMAS 1975, WOOLHOUSE and BATT 1976).

This type of cytokinin response is, however, largely confined to excised tissues and only small effects have been described for attached leaves (MULLER and LEOPOLD 1966). The inference is that endogenous sources normally fulfil any growth regulator requirement and that exogenous supply to excised material replaces imports from other parts of the plant. Where might such a source be located? CHIBNALL (1939) observed that leaf senescence was apparently sensitive to influences emanating from the root system in cereals and this finding has been supported by studies showing that root initiation on cuttings or excised leaves leads to a postponement of senescence (WOLLGIEHN 1961). Root exudates and extracts are known to contain significant amounts of cytokinin-like compounds (KULAEVA 1962, KENDE 1965) and the level has been shown to decrease with plant age (BEEVER and WOOLHOUSE 1973) and in response to drought or water-logging (ITAI et al. 1968, BURROWS and CARR 1969), nutrient deficiency (WAREING 1976) and photoperiod (HENSON and WAREING 1974). Application of bleeding sap from roots has been shown to delay senescence of excised leaf tissue from the same species (KULAEVA 1962).

The finding that root tips contain high concentrations of free cytokinin-like compounds (WEISS and VAADIA 1965) added the final piece to the hypothesis that leaf senescence is controlled by the supply of cytokinins from the root system. This is a scheme which accounts for the rapid yellowing of detached leaves and the ineffectiveness of applied cytokinins in delaying senescence of attached tissue. It is still not clear, however, whether the influence of roots on leaf senescence is explicable solely in terms of cytokinins or whether a more complex mixture of factors must be invoked.

Further support for the implication of endogenous cytokinins in regulating the onset of senescence has been derived from studies on the so-called "green islands" which are observed in yellowing leaves infected with various pathogens (WOOD 1967, MOTHES 1970). Chlorophyll, plastid integrity and full photosynthetic function are preserved in a distinct zone surrounding each infection locus (SHAW 1963) and it has been postulated that this may be due to the radial diffusion of a cytokinin-like substance produced by the pathogen. Spot applications of cytokinin solutions to yellowing leaves result in green zones mimicking those produced by fungal infection but this, of course, is not rigorous proof of cytokinin implication in the expression of infection symptoms.

A related concept is that of hormone-directed transport' exemplified by the observation that localized applications of cytokinins to leaves result in a preferential accumulation of translocated metabolites in those areas (e.g., MOTHES 1960). Whilst the nominal inference is that the cytokinin directs this accumulation, it is more strictly correct to say that the important factor is a higher rate of metabolism in these zones as a result of delaying the onset of senescence (i.e., metabolism-directed transport). In a system where the principal cytokinin supply is thought to come from the root system, the highest concentrations, and the largest proportion of other translocated metabolites, would be anticipated in the most actively growing tissues and, thus, the cause and effect relationship between leaf cytokinin levels and propensity to senesce is difficult to establish. Are cytokinin levels high because the tissues are growing (and importing) or are they growing because of the presence of cytokinin? Undoubtedly, the most momentous consequences of excision are the loss of both a sink for photosynthate and an external supply of minerals, including N, to sustain the synthetic capacity of the leaf. The metabolic imbalance between plastid and cytoplasm resulting from this correlative trauma may well be of greater significance in initiating chloroplast disassembly than the reduction in cytokinin supply.

If, however, we accept that a prima facie case exists for the involvement of cytokinins, it is relevant to ask how they might operate in the leaf, at a sub-cellular level, to produce a retardation of senescence. Studies on the rate of protein synthesis in leaves of *Lolium temulentum* have shown that, during the period from emergence to full senescence, three rate maxima of amino-acid incorporation can be detected (Sect. 2.4.3) and the third rate increase is causally related to yellowing. Cytokinins supplied after the commencement of the final phase of protein synthesis have little effect upon chlorophyll retention but, if given immediately before or after excision, are able to maintain the tissue in a green state for an extended period. These findings suggest that the senescence trigger is, in some way, sensitive to the decline of "peak 2"-type protein synthesis and that disassembly processes are not activated for as long as the rate is kept above an undefined minimum. Consequently, the role of cytokinin is seen in terms of the maintenance of protein synthetic rate above a threshold related to the senescence trigger rather than as a specific inhibitory influence on one or more of the degradative processes. This interpretation implies that the primary sub-cellular action of the growth regulator must be concerned with the operation of the protein synthetic apparatus. Current theories on the biological mode of action of cytokinins are indeed almost exclusively oriented towards one or more facets of protein synthesis. For example, $N^6$-isopentenyl adenosine, or its hydroxylated derivative zeatin, have been isolated as constituents of plant transfer RNA(tRNA) (ARMSTRONG et al. 1969, BURROWS et al. 1971). Incorporated cytokinins are located at the 3′ end of the anticodon sequence in a position which is believed to exert influence on the configuration of that part of the tRNA molecule and additional detail on this point may be found elsewhere (e.g., STODDART and VENIS 1980). However, there is considerable doubt as to whether such incorporation reflects the primary growth-regulatory action of the cytokinin molecule. For example, when tissue cultures are supplied with exogenous synthetic cytokinin (benzyladenine, BA) in large excess, only a small proportion of the tRNA cytokinins are subsequently recovered as BA, suggesting that the large growth effect produced in response to BA is not mediated through the incorporation process (BURROWS et al. 1971). The total amount of cytokinin recoverable from tRNA never represents more than a small fraction of the total cellular content of this growth regulator (STODDART and VENIS 1980).

A more promising line of investigation is concerned with the discovery that proteins with structure-specific binding sites for cytokinins can be extracted from ribosomes (BERRIDGE et al. 1970a, b, FOX and ERION 1975, 1977). High and low affinity binding sites can be solubilized with 0.5 M KCl and the active component has been shown to have a mean molecular weight of 93,000. The kinetics of binding for this protein indicate a dissociation constant for the high affinity site which is similar to the saturating concentration for biological effectiveness of exogenous cytokinin. However, an inconsistency exists in so far as the excellent correlation between binding at this site and biological effectiveness over a range of cytokinin structures is marred by a low affinity for zeatin, the most widely distributed natural cytokinin. This problem could be an expression of the need for additional factors to facilitate the interaction.

Thus cytokinins may be generally implicated in the operation of leaf protein synthesis; a process which includes the formation of those elements necessary for the continued health of cellular membranes and the maintenance of the inactivated/compartmentalized state of the senescence information. There is no obligation to postulate a unique mode of action for cytokinins in the regulation of senescence, but the manner and rate of their supply to the leaf tissue may be a strong correlative influence in determining the life span of that organ.

It is implicit in the foregoing discussion that the leaf protein population changes qualitatively with age and that the products formed during senescence, in a cellular environment where hydrolytic agents are continuously present and compartmental separation is decreasing, must have properties and/or synthetic routes which differ in some respects from those encountered during preceding developmental phases. The generalization that cytokinins act to delay senescence by maintaining protein synthesis and that decay of this function is synonymous with disassembly (OSBORNE 1967) can only be true in a limited sense as it demonstrably does not apply to the protein formation accompanying senescence. There is a clear inference of changes in the properties or operation of the protein synthetic apparatus at this stage, amongst the consequences of which are the removal of the requirement for a continuous supply of cytokinins.

### 3.2.2 Gibberellins, Growth Retardants and Auxins

Senescence in some species, notably *Rumex* (WHYTE and LUCKWILL 1966), *Taraxacum* (FLETCHER and OSBORNE 1966) and *Tropaeolum* (BEEVERS 1968) is retarded by gibberellin (GA) as well as by cytokinins. In contrast to cytokinins chlorophyll loss in *Rumex* leaf discs was retarded even when GA was added during the logarithmic phase of yellowing. Protein degradation was also arrested and pulse-chase experiments indicated that the effect of GA gradually diminished after removal of the exogenous source (GOLDTHWAITE and LAETSCH 1967).

Senescence can also, apparently paradoxically, be delayed by quaternary ammonium growth retardants such as CCC or AMO 1618 (HALEVY et al. 1966) which act by blocking GA synthesis (LANG 1970). Clearly, this is at variance with the observed effects of GA applications and, in *Tropaeolum*, both CCC and GA act as senescence retardants (BEEVERS 1968) whereas the related inhibitor B995 delays senescence at low concentrations and accelerates the process at higher dose rates. If we add to this the observation that, at high concentrations growth retardants have more general effects upon metabolism (HEATHERBELL et al. 1966), it seems unlikely that these compounds function exclusively via their influence on the cyclization step in the GA biosynthetic pathway. Their mode of action is probably indirect and may involve a more general interference with protein synthesis; a conclusion which is consistent with their ability to affect senescence when added in mid-course.

Gibberellins probably exert their effect at a different level of cellular activity but we need more precise knowledge on their biological mode of action before we can devise schemes to explain their influence on senescence. Recent work by SILK and JONES (1975), STUART and JONES (1977) and STODDART (1979a, b) raise the possibility that GA acts by influencing cell wall extensibility. In the presence of the growth regulator wall plasticity is increased in lettuce hypocotyl tissue and the ageing processes which normally limit the duration of the expansion phase are postponed. It may be that the decline of wall turnover processes in leaves and the consequent feedback to general metabolism through a reduced demand for polysaccharides operates as a facet of the senescence trigger. However, at present, such a mechanism must be entirely speculative.

Promotion of senescence by the synthetic auxins 2,4-dichlorophenoxyacetic acid (2,4 D) and 2,4,5-trichlorphenoxyacetic acid (2,4,5 T) has been reported for some species (OSBORNE and HALLAWAY 1961, OSBORNE 1967) whilst, in *Euonymus* 2,4 D has been found to retard yellowing (OSBORNE and HALLAWAY 1961). Both indoleacetic acid and 1-naphthylacetic acid reduce the rate of senescence in *Rhoeo* leaf sections (SACHER 1967, DELEO and SACHER 1970).

There are indications that auxin-binding sites with high specificity and affinity exist on some classes of sub-cellular membranes (for details see STODDART and VENIS 1980) and an interaction of auxins with the protein synthetic apparatus is, therefore, a possibility.

### 3.2.3 Abscisic Acid

Applied abscisic acid (ABA) can induce premature yellowing in leaf tissue from a diversity of species ranging from deciduous trees (EAGLES and WAREING 1964) to herbaceous plants such as *Rhoeo* (DELEO and SACHER 1970) and *Tropaeolum* (BEEVERS 1968). The effect is, however, generally greatest in excised tissue and intact leaves sprayed in situ have been shown to be less sensitive (EL-ANTABLY et al. 1967, SLOGER and CALDWELL 1970). Where present, the effect of ABA can be often counteracted, in a competitive manner, by suitable concentrations of cytokinins (WOOLHOUSE and BATT 1976, ASPINALL et al. 1967). Variable evidence is available on possible interactions between ABA and GA, ranging from zero effects in *Nasturtium* (BEEVERS 1968) or *Rumex* (MANOS and GOLDTHWAITE 1975) to substantial reversal of ABA effects in *Rhoeo* (DELEO and SACHER 1970).

Increased levels of endogenous ABA have been measured in senescing leaves of *Coleus* (BÖTTGER 1970) *Glycine* (LINDOO and NOODÉN 1978), *Nasturtium* (CHIN and BEEVERS 1970) and *Nicotiana* (EVEN-CHEN and ITAI 1975). A contradictory pattern was found by OSBORNE et al. (1972) in *Phaseolus* leaves where the levels of free ABA were less than those determined in non-senescing tissue. However, bound ABA levels did correlate with senescence. In interpreting such data we must keep in mind the large body of evidence which shows dramatic rises in endogenous ABA levels in leaves subjected to drought stress or injury (WRIGHT and HIRON 1969), as senescence may be caused, or accompanied, by exposure to such conditions. Correlative evidence which supports an indirect role for ABA has been quoted by LINDOO and NOODÉN (1978) for *Glycine*. These data suggest that ABA does not, itself, have a senescence-initiating role but must act in concert with other factors.

There is almost no hard information on the sub-cellular mode of action of ABA. Some workers have suggested a possible role in the regulation of transcription (e.g., PEARSON and WAREING 1969) and binding to a soluble subcellular component has also been detected (HOCKING et al. 1978). The structural and biosynthetic affinities between ABA and GA, as well as their antagonistic action in a number of systems (e.g., EVINS and VARNER 1971) leaves open the possibility that they may operate at related cellular sites.

### 3.2.4 Ethylene

Ethylene has been implicated in both fruit-ripening processes and flower senescence in various species (BURG and BURG 1965, NICHOLS 1968). The correlation between ethylene production and corolla senescence has been extensively investigated (e.g., KENDE and HANSON 1976).

Recent studies with *Beta, Nicotiana* and *Phaseolus* leaves (AHARONI et al. 1979, AHARONI and LIEBERMAN 1979) have shown that ethylene production increases during the phase of rapid chlorophyll loss and then declines in a manner similar to that described for $CO_2$ evolution (Sect. 2.2.4). Exogenous ethylene supplied during the first 24 h of senescence-accelerated chlorophyll loss did not affect the timing of the evolution curve. The maximum rate of evolution was, however, enhanced. Whilst ethylene is undoubtedly effective in regulating senescence, the timing of the surge of endogenous production strongly suggests that this process does not form a part of the primary regulating signal. There may be, however, a role for ethylene in transducing a senescence response to adverse conditions affecting remote plant parts such as, for example, waterlogging of the root system (KAWASE 1974). Here, ethylene generation would occur at the site of stress but would initiate a response only after transfer to the leaves.

The recent findings of HALL et al. (1980) which point to an ethylene action site on the endomembrane system suggest that this regulator may also exert its effects by modulating some aspect of protein synthesis.

### 3.2.5 Other Compounds

Studies on the senescence-influencing properties of specific molecules have tended, for obvious reasons, to be predominantly concerned with compounds falling within the broad structural groupings of the established growth-regulator classes. There are, however, occasional reports of sparing or promoting activities associated with amino acids and other metabolites.

SHIBAOKA and THIMANN (1970) noted that L-serine specifically enhanced senescence in *Avena* leaf sections, especially in the presence of kinetin and, on the basis of inhibitor and time-course studies (MARTIN and THIMANN 1972) it was postulated that L-serine acted by being incorporated into proteinases synthesized during senescence, the required specificity resulting from occupation of a critical site near the active centre of the enzyme. Serine was effective only in the presence of cytokinin, suggesting that proteinase formation could be limited by a shortage of critical amino acids, with the growth regulator acting to divert supplies to other categories of protein synthesis. Such an explanation raises interesting questions about the compartmentation of protein synthesis within the cell and implies that senescence-associated protein formation is physically separated from normal synthetic activities.

A converse effect was noted with the basic amino acids L-ornithine, L-lysine and L-diaminobutyric acid (VON ABRAMS 1974) which strongly retarded both chlorophyll and protein loss in *Avena* leaf sections. Within the reservations applicable to cell-free systems it was shown that ornithine did not act as a direct inhibitor of proteinase action and it seems more probable that it functions by repressing synthesis of the enzyme or indirectly modifying its effectiveness.

Within this general category it is also appropriate to include undefined entities which have been attributed a possible controlling role in leaf senescence. The major item in this list must be the "senescence factor" (SF) described by OSBORNE et al. (1972). Diffusates from yellowing petioles of a range of herbaceous and arboreal species were found to accelerate abscission of *Phaseolus* leaves and similar preparations obtained from young leaves were ineffective. In view of the tissue of origin and the assay system used it might, however, be more accurate to describe the principle as "abscission factor". The active substance, separated by column and thin layer chromatography, was found to differ from both ABA and IAA and, when applied to non-senescent leaf tissue, it stimulated a rise in endogenous ethylene production. The authors suggested that leaf senescence was accompanied by a loss of membrane-dependent compartmentalization, one of the consequences of which was a release of SF. This, in turn, resulted in a rise in ethylene levels which triggered leaf senescence. In the absence of any information on the identitiy of SF it is impossible to comment on a possible controlling function in foliar senescence and the suspicion must remain that a large number of metabolites extractable from mature leaves could have a non-specific yellowing effect when reapplied at non-physiological levels.

There have been suggestions (e.g., NOODÉN and LINDOO 1978) that senescence, especially in annual plants, may results from the synthesis and transmission of a "death signal" or "killer factor" originating in the reproductive structures. No single factor has so far been isolated which could be responsible and, in most cases, it is difficult to separate the postulated effects of such a compound from the consequences of massive diversion of nutrients to the apex or seed tissues. The theory also has to account for monocarpic plants, such as *Xanthium*, where removal of the reproductive structures does not prevent co-ordinated foliar senescence (KRIZEK et al. 1966). If such a signal exists it is more likely to be a summation of a complex series of organ interactions involving transport, to and fro, of a wide range of metabolic components. Until a specific compound with the required properties is isolated we are probably justified in regarding the concept of a "death signal" as a convenient umbrella to cover the whole range of metabolic intercommunication between seeds/flowers and vegetative tissues.

# 4 Senescence as an Adaptation

## 4.1 Senescence and the Plant Life Cycle

An organism meets its commitment to survival in two ways. It persists; and it reproduces. Some plants, trees for example, place the accent on growth and survival of the individual over a long period, devoting only a relatively small proportion of their resources, in terms of time, material and energy, to reproduction: they are, to use the terminology of COLE (1954), iteroparous, reproducing repeatedly throughout their lives. At the other extreme are the annuals, in which the vegetative individual exists only to manufacture progeny, and to die in the process. COLE (1954) calls this kind of life-cycle semelparous. Between the extreme perennials and the ephemerals is a spectrum of plant life-cycle with a corresponding gamut of variation in longevity and fecundity. Full discussion of the allocation of resources between reproduction and vegetative growth may be found in HARPER (1977).

Semelparity subordinates survival of the individual to the cause of survival of the population; iteroparity is the reverse. Discussion of the survival value of programmed senescence and its contribution to the adaptive tactics and strategies employed by plants may be conveniently divided along the line separating semelparity from iteroparity.

## 4.2 Survival of the Individual

### 4.2.1 Tactical Senescence

Senescence may be induced by a random or unpredictable environmental stimulus such as shading by other plants, shortages of essential mineral nutrients, unseasonal water stress or extremes of temperature (THOMAS and STODDART 1980). Included here are mechanical damage (including detachment or excision, widely employed to initiate senescence in studies of the process – SIMON 1967, POLLOCK and LLOYD 1978, THIMANN et al. 1974, SPENCER and TITUS 1973, LEWINGTON et al. 1967) and infection by pathogens (URITANI 1971, FINNEY 1979, THOMAS and STODDART 1980). Under these circumstances senescence may be regarded as a tactic; a rapid response to adverse conditions which has the effect of limiting the damage to disposable tissue and of pruning the plant down to a healthy or resistant configuration from which it can grow anew when things improve. The latent, post-transcriptional nature of senescence regulation fits it well for such a tactical function.

In many species, the ability to re-green is an important element in recovery from unforeseen, transient stresses. Young tissue is often the first to suffer when the plant is subject to extremes of temperature (LARCHER 1973), inadequate water supply (GATES 1968) or attack by predators (HARPER 1977). Rejuvenescent older leaves may make a significant contribution to survival and re-growth in these circumstances.

### 4.2.2 Strategic Senescence

As well as senescing in response to current environmental conditions, plants deploy programmed senescence in a strategic role for avoiding or withstanding periods of seasonal adversity. The growth cycles of long-lived plants from regions subject to wide seasonal fluctuations in climate are closely synchronized with the regular ebb and flow of the environment. The senescence pattern is a characteristic manifestation of this synchrony, seen most dramatically in the deciduous senescence of trees and the top senescence of herbaceous perennials (Leopold 1961, Simon 1967). Even the chloroplasts of evergreen leaves go through a regular cycle of partial disassembly and renewal related to winter survival (Öquist et al. 1978).

A common feature of responses of this kind is that they are initiated by a different environmental variable from that which becomes limiting during the resting season. Thus deciduous senescence in temperate trees is a response to daylength, but is part of an adaptive mechanism that avoids low winter temperatures (Spencer and Titus 1972, Perry 1971, Pollock et al. 1980). In this context it is important to note that a particular influence may initiate senescence but may have quite distinct effects on the process as a whole and on its component activities once senescence is underway. For example, chilling induces leaf senescence in winter-inactive grass ecotypes (Pollock et al. 1980); but the rate at which chlorophyll and protein are lost from senescing leaves decreases markedly with decreasing temperature (Thomas et al. 1980). Similarly, detaching a leaf will initiate senescence; but the act of excision, and the isolated condition of the tissue, also result in distorted and artefactual responses (such as wound reactions and accumulation of amino acids and other metabolites normally exported from attached leaves during senescence) which are superimposed on the senescence process itself.

## 4.3 Senescence and Survival of the Population

### 4.3.1 Monocarpic Senescence

Plants with a semelparous life-cycle are termed monocarpic. The monocarpic condition appears to have evolved independently in many different taxonomic groups (Noodén and Lindoo 1978). Although characteristic of annuals and biennials, monocarpic senescence also occurs in certain perennials, such as *Agave,* which lives vegetatively for about 30 years, flowers once, and dies. It is significant that the major food crops are monocarpic. Agricultural practice exploits the massive transfer of nutrients from the vegetative parts of the fruit which occurs during monocarpic reproduction and which concentrates a high proportion of crop biomass in the harvestable fraction.

In terms of life-cycle, there is a fundamental difference between semelparity and iteroparity. From the point of view of seasonal adaptation, however, the monocarpic character of annuals and biennials is an extreme kind of stress-avoiding stratagem essentially identical to that employed by the iteroparous perennials. Preparation for surviving winter cold or summer drought and so on consists of a programmed senescence of susceptible parts leaving only resistant vegetative organs or seeds (Leopold 1961, Larcher 1973).

### 4.3.2 Turnover

Leopold (1978) has argued that two important factors determine the rate of evolutionary change in a population, namely, its genetic variability and its turnover rate, and that monocarpy catalyzes evolutionary adaptation by ensuring a high rate of turnover. Monocarpic senescence is part of a syndrome of population behaviour that combines maintenance of equilibrium between the population and the existing environment with the potential for rapid readjustment to new conditions and colonization of new habitats. Indeed, so flexible is the semelparous life-cycle that it has prompted the question as to why iteroparity should have developed at all in organism life-cycles (COLE 1954). Discussion of this point is beyond the scope of this article and the reader is referred to CHARNOV and SCHAFFER (1973) for further information.

The turnover of organs, individuals and populations is a part of a continuum of dynamic, cyclical change extending from metabolic substrates all the way up the hierarchy of biological organization to entire floras (LEOPOLD 1975). Turnover has long been considered to be the key to the regulation and biological significance of senescence. However, at most levels of organization, turnover is extremely difficult to measure and interrelationships between, for example, protein turnover on the one hand and leaf turnover on the other, or turnover of the constructional units of the individual plant and turnover of the population, are not easy to understand in most cases (THOMAS and STODDART 1980).

## 4.4 Senescence and Crop Yield

### 4.4.1 Objectives in Crop Improvement

How, when and to what extent the canopy senesces has a major influence on yield. Crop improvement may be achieved in several ways: by increasing the total productivity (biomass per unit area) and the efficiency of the crop as an energy converter; by raising the harvestable fraction of the crop biomass; by improving the quality of the product; and, particularly in the case of perennial crops, by altering the pattern of seasonal growth. Programmed leaf senescence is, or should be, an important consideration in each of these breeding objectives, although in many instances it is not certain whether its contribution is a positive or negative one. In the following sections we discuss examples of particular crops in which the part played by leaf senescence has been the subject of direct study.

### 4.4.2 Soyabeans

The self-destructive characteristic limits yield in *Glycine*. Studies of the N budget of fruiting plants (THIBODEAU and JAWORSKI 1975) and mathematical simulation of *Glycine* growth (SINCLAIR and DE WIT 1976) agree that the limiting factor is N availability within the plant during pod fill. In the period of vegetative growth preceding flowering, nitrate reduction predominates over N fixation as the source of N for further growth. At the onset of fruiting, nitrate reduction falls markedly and N fixation increases to a maximum before declining as the beans mature. THIBODEAU and JAWORSKI (1975) suggest that N fixation and nitrate reduction are in competition for photosynthate during vegetative growth and that N reduction and seed maturation compete during pod fill.

The availability of reduced C during fruiting is further restricted by declining photosynthesis associated with leaf senescence and nutrient mobilization from the vegetative parts.

The imbalance between the demand for reduced N and photosynthate and the intake and supply of C and N could be redressed if there were some way of getting N into the plant during pod fill. A promising method is foliar feeding. GARCIA and HANWAY (1976) have shown that up to four applications of a mixture containing NPKS during the period of pod fill gave marked improvements in yield. The beans of treated plants were slightly smaller than those of untreated, but there were more of them. In this respect it is significant that SHIBLES et al. (1975) have identified increasing seed set as a major objective for *Glycine* improvement.

Extending the period of canopy photosynthesis by exploiting genetic variation in the correlation between vegetative growth and reproduction has great attractions as an economic alternative to ferilizer treatments. ABU-SHAKRA et al. (1978) measured N fixation and RuBPC content in a population of *Glycine* plants selected for leaf greenness at pod maturity. Of six $F_3$ plants derived from an $F_2$ plant exhibiting the delayed senescence character, three were found to retain chlorophyll, total leaf protein, RuPBC protein and activity and nodule acetylene-reducing capacity at high levels compared with the remaining three. In progeny of the elite green $F_3$ individual, leaf chlorophyll, protein and RuBPC levels at pod maturity were 60% to 210% higher than in progeny of a normal $F_3$ plant and, unlike those of the normal population, root nodules retained the ability to reduce acetylene. ABU-SHAKRA et al. conclude by proposing that incorporation of the heritable delayed senescence character into a favourable genetic background may give increases in yield by maintaining photosynthesis and N assimilation during the reproductive phase of *Glycine* growth.

### 4.4.3 Wheat

"In spite of the very great differences in conditions during grain filling, about half of the variation in (wheat) yield due to climate, agronomic practice, and variety is related to variation in leaf area duration". These words of EVANS et al. (1975) serve to focus attention on the onset of senescence, which determines the duration of canopy photosynthesis and is thus a major limiting factor in the productivity of this species. Leaf area duration is usually measured as green area duration (D) and high correlations have been obtained between D and yield for the total canopy (SPIERTZ et al. 1971) and for the flag leaf (STAMP and HERZOG 1976).

*Triticum,* like *Glycine,* is monocarpic, but the relatively low demand of the developing seeds for N means more room for manoeuvre in the manipulation of the reproduction-senescence correlation by applying fertilizer during fill. Fertilizer supplied to *Triticum* plants during the reproductive phase enhances their N content and increases D (LANGER and LIEW 1973, THOMAS and THORNE 1975); but many *Triticum* varieties, particularly the modern non-lodging types, do not show yield increases commensurate with extended D (FIDDIAN 1970). This is because there is a fall in the photosynthetic efficiency of the canopy, possibly as a result of an increase in mutual leaf shading (THORNE and BLACKLOCK 1971). Thus the use of fertilizer to uncouple reproduction and vegetative senescence may require a restructuring of the canopy to be fully effective. Although the flag leaf is the major source of photosynthate for the developing grain, shading and senescence limit the contribution of leaves below the flag leaf to total crop photosynthesis. Significant gains might be expected if senescence could be delayed in lower levels of the canopy. AUSTIN et al. (1976) observed such an improvement in the duration of lower leaves in a genotype with erect flag and penultimate leaves. This improvement may have resulted from a reduction in the amount of shading in this genotype as compared with varieties possessing normal, lax leaves; but these workers did not rule out an indirect pleiotropic effect of leaf posture on senescence rates.

A problem associated with long D is that a considerable proportion of N taken in is committed to the non-harvestable fraction of the crop. In extreme cases this must severely reduce the efficiency of the crop as an energy converter. Under the best circumstances there should be a balance struck between the duration of the photosynthetic and N-mobilizing phases of canopy development to obtain optimal yield and product quality. In this connection, the *Triticum* ideotype might include a rapid and efficient system for N redistribution during the later part of grain maturation, when N availability has its major influence on grain protein content (LANGER and LIEW 1973). Good correlations have been observed between leaf proteolytic activity, rate of N loss from leaves and grain protein content (RAO and CROY 1972, DALLING et al. 1976); but we need to know a lot more about the regulation of N mobilization and its genetic background (Sect. 2.3) before these observations can be put to practical use. Nevertheless, simple measurement of senescence offers a potentially useful screening procedure for characteristics associated with yield and quality in *Triticum*. A study of genetic variation in the rate of yellowing of excised flag leaf tissue was reported by BOYD and WALKER (1972). Of the 19 varieities measured, 6 retained 40% or more of flag leaf chlorophyll after incubating for 4 days in the dark. A partial diallel cross involving three slow-yellowing and one fast-yellowing genotype indicated heritable control of these characters. In general, fast yellowing occurred in varieties known to have a high grain protein content and slow yellowing in low protein varieities.

### 4.4.4 Grasses

In perennials the role of leaf senescence in determining yield is less clear-cut. There is little evidence to suggest that during the period of active growth, the rate of leaf turnover is a major determinant of yield but, where a dense canopy is produced, the rapid senescence of leaves in unfavourable positions for light interception could be an important mechanism for maximizing photosynthetic efficiency. Tiller formation can also act as a senescence trigger and the subtending leaf often begins to yellow when active growth is established in the subsidiary axis. This may reflect nutrient diversion to the new growing point or may be a programmed correlative event providing an enhanced N supply for the developing tiller.

The perennial habit, in colder regions, is associated with some form of strategy for winter survival and this is accompanied by the conservation of a proportion of the fixed carbon in the form of reserves. In grasses two types of behaviour predominate. Either growth ceases below a threshold temperature and the canopy is conserved, or as in the case of Scandinavian ecotypes of *Dactylis* (POLLOCK et al. 1980), there is a gradual, temperature- and daylength-triggered senescence of the entire canopy. In the latter case senescence is preceded by a phase in which massive amounts of current photosynthate are committed to root and leaf base reserves (POLLOCK 1979) and canopy death is accompanied by mobilization of N from the leaves (MORTON 1977). Where the canopy is conserved, growth in the following season relies heavily on new tiller formation and here senescence of existing leaves can be viewed as an important means of providing substrates for the early growth of the new axes.

Traditionally, a considerable proportion of the yield from a grass crop is conserved to provide winter feed. This involves cutting followed by air-drying or anaerobic storage. Excision triggers the leaf senescence syndrome so that mobilization of cell material and carbon loss via respiration occurs during the conservation process. Two consequential problems arise. Firstly, there is a loss in total dry matter and, secondly, proteolysis during ensiling contributes to unsuitably high levels of soluble nitrogenous compounds which interfere with N utilization in the rumen (CARPINTERO et al. 1979). It is in these areas that delayed senescence might be expected to produce the greatest economic benefit. A genetic lesion which interfered with protein mobilization in a manner similar to the chlorophyll-preserving mutant already isolated in *Festuca* (THOMAS and STODDART 1975) could result in a significant improvement in the quality of conserved herbage.

*Acknowledgements.* We wish to thank Professor J.P. Cooper, F.R.S., Director of the Welsh Plant Breeding Station, for his continued interest in leaf senescence studies. Thanks are also due to Mair Evans, B.Sc for help in organizing the bibliography and to various colleagues for providing information on unpublished studies in this subject area.

## References

Abrams von GJ (1974) An effect of ornithine on degradation of chlorophyll and protein in excised leaf tissue. Z Pflanzenphysiol 72:410–421

Abu-Shakra SS, Phillips DA, Huffaker RC (1978) Nitrogen fixation and delayed leaf senescence in soybeans. Science 199:973–975

Aharoni N, Lieberman M (1979) Ethylene as a regulator of senescence in tobacco leaf discs. Plant Physiol 64:801–804

Aharoni N, Lieberman M, Sisler HD (1979) Patterns of ethylene production in senescing leaves. Plant Physiol 64:796–800

Allison JCS, Weinmann H (1970) Effect of absence of developing grain on carbohydrate content and senescence of maize leaves. Plant Physiol 46:435–436

Anderson JW, Rowan KS (1965) Activity of peptidase in tobacco-leaf tissue in relation to senescence. Biochem J 97:741–746

Armstrong DJ, Burrows WJ, Skoog F, Roy KL, Soll D (1969) Cytokinins: distribution in transfer RNA species of *Escherichia coli*. Proc Natl Acad Sci USA 63:834–841

Arnold WN (1968) The selection of sucrose as the translocate of higher plants. J Theor Biol 21:13–20

Aspinall D, Paleg L, Addicott F (1967) Abscisin II and some hormone-regulated plant responses. Aust J Biol Sci 20:869–882

Atkins CA, Pate JS, Sharkey PS (1975) Asparagine metabolism-key to the nitrogen nutrition of developing legume seeds. Plant Physiol 56:807–812

Atkinson DE(1968) The energy charge of the adenylate pool as a regulatory parameter. Interaction with feedback modifiers. Biochemistry 7:4030–4034

Austin RB, Ford MA, Edrich JA, Hooper BE (1976) Some effects of leaf posture on photosynthesis and yield in wheat. Ann Appl Biol 83:425–446

Avery GS (1934) Structural responses to the practice of topping tobacco plants: a study of cell size, cell number, leaf size and veinage of leaves at different levels on the stalk. Bot Gaz 96:314–329

Baker DA, Malek F, Dehvar FD (1980) Phloem loading of amino acids from the petioles of *Ricinus* leaves. Ber Dtsch Bot Ges 93:203–209

Beever JF, Woolhouse HW (1973) Increased export from root system of *Perilla frutescens* associated with flower and fruit development. Nature (London) New Biol 246:31–32

Beevers L (1968) Growth regulator control of senescence in leaf discs of Nasturtium (*Tropaeolum majus*). In: Wightman F, Setterfield G (eds) Biochemistry and physiology of plant growth substances. Runge Press, Ottawa, pp 1417–1434

Beevers L (1976) Nitrogen metabolism in plants. Elsevier, Amsterdam New York, pp 276–289

Berridge MV, Ralph RK (1969) Some effects of kinetin on floated Chinese Cabbage leaf discs. Biochim Biophys Acta 182:266–269

Berridge MV, Ralph RK, Letham DS (1970a) The binding of cytokinin to plant ribosomes. Biochem J 119:75–84

Berridge MV, Ralph RK, Letham DS (1970b) On the significance of cytokinin binding to plant ribosomes. In: Carr DJ (ed) Plant growth substances 1970. Springer, Berlin Heidelberg New York, pp 248–255

Berthelot M, André G (1884) Les azotates dans les plantes, aux diverses périodes de la végétation. CR Acad Sci 99:550–568

Biswal UC, Mohanty P (1976) Ageing induced changes in photosynthetic electron transport of detached barley leaves. Plant Cell Physiol 17:323–331

Biswal UC, Sharma R (1976) Phytochrome regulation of senescence in detached barley leaves. Z Pflanzenphysiol 80:71–73

Boller T, Kende H (1979) Hydrolytic enzymes in the central vacuole of plant cells. Plant Physiol 63:1123–1132

Bomsel J, Pradet A (1968) Study of adenosine 5′-mono- di- and triphosphates in plant tissues. IV. Regulation of the level of nucleotides in vitro by adenylate kinase: theoretical and experimental study. Biochim Biophys Acta 162:230–242

Böttger I, Wollgiehn R (1958) Untersuchungen über den Zusammenhang zwischen Nucleinsäure- und Eiweißstoffwechsel in grünen Blättern höherer Pflanzen. Flora (Jena) 146:302–320

Böttger M (1970) Die hormonale Regulation des Blattfalls bei *Coleus rehneltianus* Berger. II. Die natürliche Rolle von Abscisinsäure im Blattfallprozeß. Planta 93:205–213

Boyd WJR, Walker MG (1972) Variation in chlorophyll a content and stability in wheat flag leaves. Ann Bot 36:87–92

Brady CJ (1973) Changes accompanying growth and senescence and effect of physiological stress. In: Butler GW, Bailey RW (eds) Chemistry and biochemistry of herbage, vol II. Academic Press, London New York, pp 317–351

Burg SP, Burg EA (1965) Ethylene action and the ripening of fruits. Science 148:1190–1196

Burrows WJ, Carr DJ (1969) Effects of flooding the root system of sunflower plants on the cytokinin content in the xylem sap. Physiol Plant 22:1105–1112

Burrows WJ, Skoog F, Leonard NJ (1971) Isolation and identification of cytokinins located in the transfer ribonucleic acid of tobacco callus grown in the presence of 6-benzylaminopurine. Biochemistry 10:2189–2194

Butler RD, Simon EW (1970) Ultrastructural aspects of senescence in plants. Adv Gerontol Res 3:73–129
Callow JA, Callow ME, Woolhouse HW (1972) In vitro protein synthesis, ribosomal RNA synthesis and polyribosomes in senescing leaves of *Perilla*. Cell Differ 1:79–90
Callow ME, Woolhouse HW (1973) Changes in nucleic-acid metabolism in regreening leaves of *Perilla*. J Exp Bot 24:285–294
Carpintero CM, Henderson AR, McDonald P (1979) The effect of some pre-treatments on proteolysis during the ensiling of herbage. Grass Forage Sci 34:311–315
Carr DJ, Pate JS (1967) Ageing in the whole plant. Symp Soc Exp Biol 21:559–599
Carver TLW, Jones IT, Thomas H (1979) Influence of photoperiod on mildew development and its relationship with senescence. Rep Welsh Plant Breed Stn 1978:188–189
Castric PA, Farnden KJF, Conn EE (1972) Cyanide metabolism in higher plants. V. The formation of asparagine from β-cyanoalanine. Arch Biochem Biophys 152:62–69
Catský J, Tichá I, Solárová J (1976) Ontogenetic changes in the internal limitations to bean leaf photosynthesis. I. Carbon dioxide exchange and conductances of carbon dioxide transfer. Photosynthetica 10:394–402
Chapman AG, Fall L, Atkinson DE (1971) Adenylate energy charge in *Escherichia coli* during growth and starvation. J Bacteriol 108:1072–1086
Charnov EL, Schaffer WM (1973) Life history consequences of natural selection: Cole's result revisited. Am Nat 107:791–793
Chibnall AC (1939) Protein metabolism in the plant. Yale Univ, New Haven, pp 1–306
Chin T, Beevers L (1970) Changes in endogenous growth regulators in Nasturtium leaves during senescence. Planta 92:178–188
Ching TM, Ching KK (1972) Content of adenosine phosphate and adenylate energy charge in germinating ponderosa pine seeds. Plant Physiol 50:536–540
Choe HT, Thimann KV (1977) The retention of photosynthetic activity by senescing chloroplasts of oat leaves. Planta 135:101–107
Chua NH, Schmidt GW (1978) Post-translational transport into intact chloroplasts of a precursor to the small subunit of ribulose-1,5-bisphosphate carboxylase. Proc Natl Acad Sci USA 75:6110–6114
Chua NH, Schmidt GW (1979) Transport of proteins into mitochondria and chloroplasts. J Cell Biol 81:461–483
Cole LC (1954) The population consequences of life-history phenomena. Q Rev Biol 29:103–137
Dalgarn D, Miller P, Bricker T, Speer N, Jaworski JG, Newman DW (1979) Galactosyl transferase activity of chloroplast envelopes from senescent soybean cotyledons. Plant Sci Lett 14:1–6
Dalling MJ, Boland G, Wilson JH (1976) Relation between acid proteinase activity and redistribution of nitrogen during grain development in wheat. Aust J Plant Physiol 3:721–730
de Greef JA, Butler WL, Roth TF, Fredericq H (1971) Control of senescence in *Marchantia* by phytochrome. Plant Physiol 48:407–412
DeLeo P, Sacher JA (1970) Control of ribonuclease and acid phosphatase by auxin and abscisic acid during senescence of *Rhoeo* leaf sections. Plant Physiol 46:806–811
Derman BD, Rupp DC, Noodén LD (1978) Mineral distribution in relation to fruit development and monocarpic senescence in Anoka soybeans. Am J Bot 65:205–213
Drivdahl RH, Thimann KV (1977) The proteases of senescing oat leaves. I. Purification and general properties. Plant Physiol 59:1059–1063
Drivdahl RH, Thimann KV (1978) The proteases of senescing oat leaves. II. Reaction to substrates and inhibitors. Plant Physiol 61:501–505
Drumm HE, Margulies MM (1970) In vitro protein synthesis by plastids of *Phaseolus vulgaris*. IV. Amino-acid incorporation by etioplasts and effects of illumination of leaves on incorporation by plastids. Plant Physiol 45:442–453
Dyer TA, Osborne DJ (1971) Leaf nucleic acids. II. Metabolism during senescence and the effect of kinetin. J Exp Bot 22:552–560
Eagles CF, Wareing PF (1964) The role of growth substances in the regulation of bud dormancy. Physiol Plant 17:697–709

Eilman Y, Butler RD, Simon EW (1971) Ribosomes and polysomes in cucumber leaves during growth and senescence. Plant Physiol 47:317–323
El-Antably HMM, Wareing PF, Hillman J (1967) Some physiological responses to DL-abscisin (dormin). Planta 73:74–90
Epstein E (1972) Mineral nutrition of plants: Principles and perspectives. Wiley, New York, pp 1–412
Evans LT (1975) The physiological basis of crop yield. In: Evans LT (ed) Crop physiology. Univ Press, Cambridge, pp 327–355
Evans LT, Wardlaw IF, Fischer RA (1975) Wheat. In: Evans LT (ed) Crop physiology. Univ Press, Cambridge, pp 101–149
Even-Chen Z, Itai C (1975) The role of abscisic acid in senescence of detached tobacco leaves. Physiol Plant 34:97–100
Evins WH, Varner JE (1971) Hormone controlled synthesis of endoplasmic reticulum in barley aleurone cells. Proc Natl Acad Sci USA 68:1631–1633
Feller UK, Soong T-ST, Hageman RH (1977) Proteolytic activities and leaf senescence during grain development of field-grown corn (*Zea mays* L.). Plant Physiol 59:290–294
Fiddian WEH (1970) Cereal variety and nitrogen fertilizer relationships. J Natl Inst Agric Bot (GB) 12:57–64
Finney ME (1979) The influence of infection by *Erysiphe graminis* D.C. on the senescence of the first leaf of barley. Physiol Plant Pathol 14:31–36
Fletcher RA, Osborne DJ (1966) Gibberellin as a regulator of protein and ribonucleic acid synthesis. Can J Bot 44:739–745
Fong F, Heath RL (1977) Age dependent changes in phospholipids and galactolipids in primary bean leaves (*Phaseolus vulgaris*). Phytochemistry 16:215–217
Fox JE, Erion JL (1975) A cytokinin binding protein from higher plant ribosomes. Biochem Biophys Res Commun 64:694–700
Fox JE, Erion JL (1977) Cytokinin binding proteins in higher plants. In: Pilet PE (ed) Plant growth regulation. Springer, Berlin Heidelberg New York, pp 139–146
Friedrich JW, Huffaker RC (1980) Photosynthesis, leaf resistances, and ribulose-1,5-bisphosphate carboxylase degradation in senescing barley leaves. Plant Physiol 65:1103–1107
Friedrich JW, Schrader LE (1979) N and S accumulation in maize is influenced by N deprivation during grain filling. II. Remobilization of $^{15}N$ and $^{35}S$ and the relationship between N and S accumulation. Agron J 71:466–472
Frith GJT, Gordon KHJ, Dalling MJ (1978a) Proteolytic enzymes in green wheat leaves. I. Isolation on DEAE-cellulose of several proteinases with acid pH optima. Plant Cell Physiol 19:491–500
Frith GJT, Peoples MB, Dalling MJ (1978b) Proteolytic enzymes in green wheat leaves. III. Inactivation of acid proteinase II by diazoacetyl-DL-norleucine methyl ester and 1,2-epoxy-3-(p-nitrophenoxy)-propane. Plant Cell Physiol 19:819–824
Frith GJT, Swinden LB, Dalling MJ (1978c) Proteolytic enzymes in green wheat leaves. II. Purification by affinity chromatography and some properties of proteinases with acid pH optima. Plant Cell Physiol 19:1029–1041
Garcia R, Hanway JJ (1976) Foliar fertilization of soybeans during the seed-filling period. Agron J 68:653–657
Gates CT (1968) Water deficits and growth of herbaceous plants. In: Kozlowski TT (ed) Water deficits and plant growth, vol II. Academic Press, London New York, pp 135–190
Gifford RM (1974) Photosynthetic limitations to cereal yield. Bull R Soc NZ 12:887–893
Gimmler M, Schafer G, Kraminer H, Heber U (1974) Amino acid permeability of the chloroplast envelope as measured by light scattering, volumetry and amino-acid uptake. Planta 120:47–61
Goldthwaite JJ, Laetsch WM (1967) Regulation of senescence in bean leaf discs by light and chemical growth regulators. Plant Physiol 42:1757–1762
Haber AH, Thompson PJ, Walne PL, Triplett LL (1969) Nonphotosynthetic retardation of chloroplast senescence by light. Plant Physiol 44:1619–1628
Halevy AH, Dilley DR, Wittwer SH (1966) Senescence inhibition and respiration induced by growth retardants and $N^6$-benzyladenine. Plant Physiol 41:1085–1089
Hall AJ, Brady CJ (1977) Assimilate source – sink relationships in *Capsicum annuum*

L. II. Effects of fruiting and defloration on the photosynthetic capacity and senescence of the leaves. Aust J Plant Physiol 4:771–783
Hall MA, Acaster MA, Bengochea T, Dodds JH, Evans DE, Jones JF, Jerie PH, Mutumba GC, Niepel B, Shaari AR (1980) Ethylene and Seeds. In: Skoog F (ed) Plant growth substances 1979. Springer, Berlin Heidelberg New York, pp 199–207
Hall NP, Keys AJ, Merrett MJ (1978) Ribulose-1,5-diphosphate carboxylase protein during flag leaf senescence. J Exp Bot 29:31–37
Hampp R, Schmidt HW (1976) Changes in envelope permeability during chloroplast development. Planta 129:69–73
Hardwick K, Wood M, Woolhouse HW (1968) Photosynthesis and respiration in relation to leaf age in *Perilla fructescens* (L) Britt. New Phytol 67:79–86
Harper JL (1977) Population biology of plants. Academic Press, London New York, pp 1–892
Hasselt van TPR, Strikwerda JT (1976) Pigment degradation in discs of the thermophilic *Cucumis sativus* as affected by light, temperature, sugar application and inhibitors. Physiol Plant 37:253–257
Heatherbell DA, Howard BH, Wicken AJ (1966) The effects of growth retardants on the respiration and coupled phosphorylation of preparations from etiolated pea seedlings. Phytochemistry 5:635–642
Heck U, Martinoia E, Matile P (1981) Subcellular localization of acid proteinase in barley mesophyll protoplasts. Planta 151:198–200
Hedley CL, Stoddart JL (1972) Patterns of protein synthesis in *Lolium temulentum* L. I. Changes occurring during leaf development. J Exp Bot 23:490–501
Henriques F, Park RB (1976) Identification of chloroplast membrane peptides with subunits of coupling factor and ribulose-1,5-diphosphate carboxylase. Arch Biochem Biophys 176:472–478
Henson I, Wareing PF (1974) Cytokinins in *Xanthium strumarium*: a rapid response to short-day treatments. Physiol Plant 32:185–187
Highfield PE, Ellis RJ (1978) Synthesis and transport of the small subunit of chloroplast ribulose bisphosphate carboxylase. Nature (London) 271:420–424
Hocking TJ, Clapham J, Cattell KJ (1978) Abscisic acid binding to subcellular fractions from leaves of *Vicia faba*. Planta 138:303–304
Holden M (1972) Effects of EDTA and other compounds on chlorophyll breakdown in detached leaves. Phytochemistry 11:2393–2402
Holmes MG, Smith H (1977) The function of phytochrome in the natural environment. II. The influence of vegetation canopies on the spectral energy distribution of natural daylight. Photochem Photobiol 25:539–545
Hopkinson JM (1964) Studies on the expansion of the leaf surface. IV. The carbon and phosphorus economy of a leaf. J Exp Bot 15:125–137
Hopkinson JM (1966) Studies on the expansion of the leaf surface. VI. Senescence and the usefulness of old leaves. J Exp Bot 17:762–770
Huffaker RC, Miller BL (1978) Reutilization of ribulose bisphosphate carboxylase. In: Siegelman HW (ed.) Photosynthetic carbon assimilation. Plenum Press, New York, pp 139–152
Huffaker RC, Peterson LW (1974) Protein turnover in plants and possible means of its regulation. Annu Rev Plant Physiol 25:363–392
Huffaker RC, Rains DW (1978) Factors influencing nitrate acquisition by plants; assimilation and fate of reduced nitrogen. In: Nielsen DR, MacDonald JG (eds) Nitrogen in the environment, vol II. Academic Press, London New York, pp 1–43
Itai C, Richmond A, Vaadia Y (1968) Effect of water stress on root exudation of cytokinin-like compounds. Isr J Bot 17:187–193
James WO (1953) Plant respiration. Oxford Univ Press, Oxford
Jefferies TA (1916) The vegetative anatomy of *Molinia caerulea*, the purple heath grass. New Phytol 15:49–71
Joy KW, Antcliff AJ (1966) Translocation of amino acids in sugar beet. Nature (London) 211:210–211

Joyard J, Douce R (1977) Site of synthesis of phosphatidic acid and diacylglycerol in spinach chloroplasts. Biochim Biophys Acta 486:273–285
Jupp DJ (1980) Nucleic acids and proteins during development of leaves of *Festuca pratensis*. Ph D Thesis, Univ Durham
Kahanak GM, Okatan Y, Rupp DC, Noodén LD (1978) Hormonal and genetic alteration of monocarpic senescence in soybeans. Plant Physiol 61:Suppl 26
Kawase M (1974) Role of ethylene in induction of flooding damage in sunflower. Physiol Plant 31:29–38
Kende H (1965) Kinetin-like factors in the root exudate of sunflowers. Proc Natl Acad Sci USA 53:1302–1307
Kende H, Hanson AD (1976) Relationship between ethylene evolution and senescence in morning glory flower tissue. Plant Physiol 57:523–527
King RW, Wardlaw IF, Evans LT (1967) Effect of assimilate utilization on photosynthetic rate in wheat. Planta 77:261–276
Krizek DT, McIlrath WJ, Vergara BS (1966) Photoperiodic induction of senescence in *Xanthium* plants. Science 151:95–96
Kulaeva O (1962) The effect of roots on leaf metabolism in relation to the action of kinetin on leaves. Sov Plant Physiol 9:182–189
Kung S-D (1976) Tobacco fraction I protein: a unique genetic marker. Science 191:429–434
Lang A (1970) Gibberellins: Structure and metabolism. Annu Rev Plant Physiol 21:537–570
Langer RHM, Liew FKY (1973) Effects of varying nitrogen supply at different stages of the reproductive phase on spikelet and grain production and on grain nitrogen in wheat. J Agric Res 24:647–656
Larcher W (1973) Temperature resistance and survival. In: Precht H, Christopherson J, Hensel H, Larcher W (eds) Temperature and life. Springer, Berlin Heidelberg New York, pp 203–231
Lea PJ, Fowden L (1975) The purification and properties of glutamine-dependent asparagine synthetase isolated from *Lupinus albus*. Proc R Soc London Ser B 192:13–26
Lee MJ (1979) Senescence studies of *Festuca pratensis*. M Sc Thesis, Univ Durham
Leigh RA (1979) Do plant vacuoles degrade cytoplasmic components? Trends Biochem Sci 4:N37–N38
Leopold AC (1961) Senescence in plant development. Science 134:1727–1732
Leopold AC (1975) Aging, senescence and turnover in plants. Bioscience 25:659–662
Leopold AC (1978) The biological significance of death in plants. In: Behnke JA, Finch CE, Moment GB (eds) The biology of aging. Plenum Press, New York, pp 101–114
Leopold AC, Niedergang-Kamien E, Janick J (1959) Experimental modification of plant senescence. Plant Physiol 34:570–573
Lewington RJ, Talbot M, Simon EW (1967) The yellowing of attached and detached cucumber cotyledons. J Exp Bot 18:526–534
Lewis OAM, Nieman E, Munz A (1970) Origin of amino acids in *Datura stramonium* seeds. Ann Bot 34:843–848
Lindoo SJ, Noodén LD (1977) Studies on the behaviour of the senescence signal in Anoka soybeans. Plant Physiol 59:1136–1140
Lindoo SJ, Noodén LD (1978) Correlation of cytokinins and abscisic acid with monocarpic senescence in soybeans. Plant Cell Physiol 19:997–1006
Lloyd EJ (1980) The effects of leaf age and senescence on the distribution of carbon in *Lolium temulentum*. J Exp Bot 31:1067–1079
Mandahar CL, Garg ID (1975) Effect of ear removal on sugars and chlorophylls of barley leaves. Photosynthetica 9:407–409
Manos PJ, Goldthwaite J (1975) A kinetic analysis of the effects of gibberellic acid, zeatin and abscisic acid on leaf tissue senescence in *Rumex*. Plant Physiol 55:192–198
Martin C, Thimann KV (1972) The role of protein synthesis in the senescence of leaves. I. The formation of protease. Plant Physiol 49:64–71
McKee HS (1962) Nitrogen metabolism in plants. Univ Press, Oxford, pp 1–728
McKersie BD, Thompson JE (1975) Cytoplasmic membrane senescence in bean cotyledons. Phytochemistry 14:1485–1491

McKersie BD, Thompson JE (1978) Phase behaviour of chloroplast and microsomal membranes during leaf senescence. Plant Physiol 61:639–643
Miflin BJ, Lea PJ (1977) Amino acid metabolism. Annu Rev Plant Physiol 28:299–329
Miller BL, Huffaker RC (1979) Changes in endopeptidases, protein and chlorophyll in senescing barley leaves. Agron Abstr Am Soc Agron 71st Annu Meet Fort Collins, Colorado, p 91
Mishra D, Pradhan PK (1973) Regulation of senescence in detached rice leaves by light, benzimidazole and kinetin. Exp Gerontol 8:153–155
Molisch H (1938) The longevity of plants. Science Press, Lancaster Pa, pp 1–226
Morgenthaler JJ, Mendiola-Morgenthaler L (1976) Synthesis of soluble, thylakoid and envelope membrane proteins by spinach chloroplasts purified from gradients. Arch Biochem Biophys 172:51–58
Morton AJ (1977) Mineral nutrient pathways in a molinetum in autumn and winter. J Ecol 65:993–999
Mothes K (1960) Über das Altern der Blätter und die Möglichkeit ihrer Wiederverjüngung. Naturwissenschaften 47:337–351
Mothes K (1970) Über grüne Inseln. Leopoldina 15:171–172
Mothes K, Baudisch W (1958) Untersuchungen über die Reversibilität der Ausbleichung grüner Blätter. Flora (Jena) 146:521–531
Muller K, Leopold AC (1966) The mechanism of kinetin-induced transport in corn leaves. Planta 68:186–205
Murneek AE (1926) Effects of correlation between vegetative and reproductive functions in the tomato (*Lycopersicon esculentum* Mill). Plant Physiol 1:3–56
Neales TF, Incoll LD (1968) The control of leaf photosynthesis rate by the level of assimilate concentration in the leaf: a review of the hypothesis. Bot Rev 34:107–125
Neales TF, Anderson MJ, Wardlaw IF (1963) The role of leaves in the accumulation of nitrogen by wheat during ear development. Aust J Agric Res 14:725–736
Ness PJ, Woolhouse HW (1980) RNA synthesis in *Phaseolus* chloroplasts. II. Ribonucleic acid synthesis in chloroplasts from developing and senescing leaves. J Exp Bot 31:235–245
Nichols R (1968) The response of carnation (*Dianthus caryophyllus*) to ethylene. J Hortic Sci 43:335–349
Noodén LD, Leopold AC (1978) Phytohormones and the regulation of senescence and abscission. In: Letham DS, Goodwin PB, Higgins TJV (eds) Phytohormones and related compounds: A comprehensive treatise, vol II. Elsevier, Amsterdam New York, pp 329–369
Noodén LD, Lindoo SJ (1978) Monocarpic senescence. What's New Plant Physiol 9:25–28
Oghoghorie CGO, Pate JS (1972) Exploration of the nitrogen transport system of a nodulated legume using $^{15}N$. Planta 104:35–49
Öquist G, Bjorn M, Martensson O, Christersson L, Malmberg G (1978) Effects of season and low temperature on polypeptides from thylakoids isolated from chloroplasts of *Pinus sylvestris*. Physiol Plant 44:300–306
Osborne DJ (1967) Hormonal regulation of leaf senescence. Symp Soc Exp Biol 21:305–322
Osborne DJ, Hallaway M (1961) Auxin control of protein levels in detached autumn leaves. Nature (London) 188:240–241
Osborne DJ, Jackson M, Milborrow BV (1972) Physiological properties of abscission accelerator from senescent leaves. Nature (London) New Biol 240:98–101
Paranjothy K, Wareing PF (1971) The effect of abscisic acid, kinetin and 5-fluorouracil on ribonucleic acid and protein synthesis in senescing radish leaf discs. Planta 99:112–119
Pate JS, Sharkey PJ, Lewis OAM (1974) Phloem bleeding from legume fruits – a technique for study of fruit nutrition. Planta 120:229–243
Pate JS, Sharkey PJ, Lewis OAM (1975) Xylem to phloem transfer of solutes in fruiting shoots of legumes, studied by a phloem bleeding technique. Planta 122:11–26
Pate JS, Atkins CA, Hamel K, McNeil DL, Layzell DB (1979) Transport of organic solutes in phloem and xylem of a nodulated legume. Plant Physiol 63:1082–1088
Pearson JA, Wareing PF (1969) Effect of abscisic acid on activity of chromatin. Nature (London) 221:672–673

Peoples MB, Dalling MJ (1978) Degradation of ribulose-1,5-bisphosphate carboxylase by proteolytic enzymes from crude extracts of wheat leaves. Planta 138:153–160

Peoples MB, Dalling MJ (1979) Intracellular localization of acid peptide hydrolases in wheat leaves. Plant Physiol 63:Suppl 159

Peoples MB, Frith GJT, Dalling MJ (1979) Proteolytic enzymes in green wheat leaves. IV. Degradation of ribulose-1,5-bisphosphate carboxylase by acid proteases isolated on DEAE cellulose. Plant Cell Physiol 20:253–258

Peoples MB, Beilharz VC, Waters SP, Simpson RJ, Dalling MJ (1980) Nitrogen redistribution during grain growth in wheat (*Triticum aestivum* L.) II. Chloroplast senescence and the degradation of ribulose-1,5-bisphosphate carboxylase. Planta 149:241–251

Perry TO (1971) Dormancy of trees in winter. Science 171:29–36

Peterson LW, Huffaker RC (1975) Loss of ribulose-1,5-diphosphate carboxylase and increase in proteolytic activity during senescence of detached primary barley leaves. Plant Physiol 55:1009–1015

Pollock CJ (1976) Changes in the activity of sucrose-synthesizing enzymes in developing leaves of *Lolium temulentum*. Plant Sci Lett 7:27–31

Pollock CJ (1979) Pathway of fructosan synthesis in leaf bases of *Dactylis glomerata*. Phytochemistry 18:777–779

Pollock CJ, Lloyd EJ (1977) The distribution of acid invertase in developing leaves of *Lolium temulentum* L. Planta 133:197–200

Pollock CJ, Lloyd EJ (1978) Acid invertase activity during senescence of excised leaf tissue of *Lolium temulentum*. Z Pflanzenphysiol 90:79–84

Pollock CJ, Riley GJP, Stoddart JL, Thomas H (1980) The biochemical basis of plant response to temperature limitations. Rep Welsh Plant Breed Stn 1979:227–246

Ragster L, Chrispeels MJ (1979) Azocoll-digesting proteinases in soybean leaves. Characteristics and changes during leaf maturation and senescence. Plant Physiol 64:857–862

Rao SC, Croy LI (1972) Protease and nitrate reductase seasonal patterns and their relation to grain protein production of 'high' versus 'low' protein wheat varieties. J Agric Food Chem 20:1138–1141

Reed AJ, Below FE, Hageman RH (1980) Grain protein accumulation and the relationship between leaf nitrate reductase and protease activities during grain development in maize (*Zea mays* L.). I. Variation between genotypes. Plant Physiol 66:164–170

Ricard J, Nari J, Buc J, Meunier JC (1977) Conformational changes and modulation of enzyme catalysis. In: Smith H (ed) Regulation of enzyme synthesis and activity in higher plants, ch 9. Academic Press, London New York, pp 156–174

Richmond AE, Lang A (1957) Effect of kinetin on protein content and survival of detached *Xanthium* leaves. Science 125:650–651

Sacher JA (1967) Studies of permeability, RNA and protein turnover during aging of fruit and leaf tissues. Symp Soc Exp Biol 21:269–303

Sacher JA, Davies DD (1974) Demonstration of de novo synthesis of RNase in *Rhoeo* leaf sections by deuterium oxide labelling. Plant Cell Physiol 15:157–162

Sadler R, Scott KJ (1974) Nitrogen assimilation and metabolism in barley leaves infected with the powdery mildew fungus. Physiol Plant Pathol 4:235–247

Schwabe WW (1970) The control of leaf senescence in *Kleinia articulata* by photoperiod. Ann Bot 34:43–55

Sharkey PJ, Pate JS (1975) Selectivity in xylem to phloem transfer of amino acids in fruiting shoots of white lupin (*Lupinus albus* L.). Planta 127:251–262

Shaw M (1963) The physiology and host-parasite relations of the rusts. Annu Rev Phytopathol 1:259–294

Shibaoka H, Thimann KV (1970) Antagonisms between kinetin and amino-acids. Plant Physiol 46:212–220

Shibles R, Anderson IC, Gibson AH(1975) Soybean. In: Evans LT (ed) Crop physiology. Univ Press, Cambridge, pp 151–189

Silk WK, Jones RL (1975) Gibberellin response in lettuce hypocotyl sections. Plant Physiol 56:267–272

Simon EW (1967) Types of leaf senescence. Symp Soc Exp Biol 21:215–230

Simmonds JA, Dumbroff EB (1974) High energy charge as a requirement for axis elongation

in response to gibberellic acid and kinetin during stratification of *Acer saccharinum* seeds. Plant Physiol 53:91–95
Sinclair TR, de Wit CT (1975) Photosynthate and nitrogen requirements for seed production by various crops. Science 189:565–567
Sinclair TR, de Wit CT (1976) Analysis of the carbon and nitrogen limitations to soybean yield. Agron J 68:319–324
Sloger C, Caldwell BE (1970) Response of cultivars of soybean to synthetic abscisic acid. Plant Physiol 46:634–635
Spencer PW, Titus JS (1972) Biochemical and enzymatic changes in apple leaf tissue during autumnal senescence. Plant Physiol 49:746–750
Spencer PW, Titus JS (1973) Apple leaf senescence: leaf disc compared to attached leaf. Plant Physiol 51:89–92
Spiertz HJH, Ten Hag BA, Kupers LPJ (1971) Relation between green area duration and grain yield in some varieties of spring wheat. Neth J Agric Sci 19:211–222
Sprague HB (ed) (1964) Hunger signs in crops 3rd edn. McKay, New York, pp 1–461
Stamp P, Herzog H (1976) Untersuchungen zur Fahnenblattalterung und zum Kornwachstum einiger deutscher Sommerweizensorten (*Triticum aestivum* L.). Z Pflanzenzuecht 77:330–338
Stoddart JL (1972) Protein synthesis during leaf and seed development. Rep Welsh Plant Breed Stn 1971:107–116
Stoddart JL (1979a) Interaction of [$^3$H]-gibberellin $A_1$ with a sub-cellular fraction from lettuce (*Lactuca sativa*) hypocotyls. I. Kinetics of labelling. Planta 146:353–361
Stoddart JL (1979b) Interaction of [$^3$H]-gibberellin $A_1$ with a sub-cellular fraction from lettuce (*Lactuca sativa*) hypocotyls. II. Stability and properties of the association. Planta 146:363–368
Stoddart JL, Venis MA (1980) Molecular and subcellular aspects of plant growth regulator action. In: MacMillan J (ed) Encyclopaedia of plant physiology. New series. Plant hormones, vol IX. Springer, Berlin Heidelberg New York, pp 445–510
Storey R, Beevers L (1977) Proteolytic activity in relationship to senescence and cotyledonary development in *Pisum sativum* L. Planta 137:37–44
Storey R, Beevers L (1978) Enzymology of glutamine metabolism related to senescence and seed development in the pea (*Pisum sativum* L.). Plant Physiol 61:494–500
Straus W (1967) Lysosomes, phagosomes and related particles. In: Roodyn DR (ed) Enzyme cytology. Academic Press, London New York, pp 239–306
Streeter JG (1977) Asparaginase and asparagine transaminase in soybean leaves and root nodules. Plant Physiol 60:235–239
Stuart DA, Jones RL (1977) The roles of extensibility and turgor in gibberellin- and dark-stimulated growth. Plant Physiol 59:61–68
Sugiura M (1963) Effect of red and far-red light on protein and phosphate metabolism in tobacco leaf discs. Bot Mag 76:174–180
Takegami T, Yoshida K (1975) Remarkable retardation of the senescence of tobacco leaf discs by cordycepin, an inhibitor of RNA polyadenylation. Plant Cell Physiol 16:1163–1166
Tetley RM, Thimann KV (1974) The metabolism of oat leaves during senescence. I. Respiration, carbohydrate metabolism and the action of cytokinins. Plant Physiol 54:294–303
Thibodeau PS, Jaworski EG (1975) Patterns of nitrogen utilization in the soybean. Planta 127:133–147
Thimann KV, Tetley RM, Thann TV (1974) The metabolism of oat leaves during senescence. III. Senescence in leaves attached to the plant. Plant Physiol 54:859–862
Thimann KV, Tetley RM, Krivak BM (1977) Metabolism of oat leaves during senescence. V. Senescence in light. Plant Physiol 59:448–454
Thomas H (1975) Regulation of alanine aminotransferase in leaves of *Lolium temulentum* during senescence. Z Pflanzenphysiol 74:208–218
Thomas H (1976a) Leaf growth and senescence in grasses. Rep Welsh Pl Breed Stn 1975:133–138
Thomas H (1976b) Delayed senescence in leaves treated with the protein synthesis inhibitor MDMP. Plant Sci Lett 6:369–377

Thomas H (1977) Ultrastructure, polypeptide composition and photochemical activity of chloroplasts during foliar senescence of a non-yellowing genotype of *Festuca pratensis* Huds. Planta 137:53–60

Thomas H (1978) Enzymes of nitrogen mobilization in detached leaves of *Lolium temulentum* during senescence. Planta 142:161–169

Thomas H, Bingham MJ (1977) Isoenzymes of naphthylacetate esterase in senescing leaves of *Festuca pratensis*. Phytochemistry 16:1887–1889

Thomas H, Huffaker RC (1981) Hydrolysis of radioactively-labelled ribulose-1,5-bisphosphate carboxylase by an endopeptidase from the primary leaf of barley seedlings. Plant Sci Lett 20:251–262

Thomas H, Stoddart JL (1975) Separation of chlorophyll degradation from other senescence processes in leaves of a mutant genotype of meadow fescue (*Festuca pratensis*). Plant Physiol 56:438–441

Thomas H, Stoddart JL (1977) Biochemistry of leaf senescence in grasses. Ann Appl Biol 85:461–463

Thomas H, Stoddart JL (1980) Leaf senescence. Annu Rev Plant Physiol 31:83–111

Thomas H, Stoddart JL, Potter JF (1980) Temperature responses of membrane-associated activities from spring and winter oats. Plant Cell Environ 3:271–277

Thomas SM, Thorne GN (1975) Effect of nitrogen fertilizer on photosynthesis and ribulose-1,5-diphosphate carboxylase activity in spring wheat in the field. J Exp Bot 26:43–51

Thorne GN, Blacklock JC (1971) Effects of plant density and nitrogen fertilizer on growth and yield of short varieties of wheat derived from Norin 10. Ann Appl Biol 68:93–111

Thrower SL (1967) The pattern of translocation during leaf ageing. Symp Soc Exp Biol 21:483–506

Treharne KJ, Stoddart JL, Pughe J, Paranjothy K, Wareing PF (1970) Effects of gibberellin and cytokinins on the activity of photosynthetic enzymes and plastid ribosomal RNA synthesis in *Phaseolus vulgaris* L. Nature (London) 228:129–131

Uritani I (1971) Protein changes in diseased plants. Annu Rev Phytopathol 9:211–234

Vonshak A, Richmond AE (1975) Initial states in the onset of senescence in tobacco leaves. Plant Physiol 55:786–790

Walker-Simmons M, Ryan CA (1977) Immunological identification of proteinase inhibitors I and II in isolated tomato leaf vacuoles. Plant Physiol 60:61–63

Walkley J (1940) Protein synthesis in mature and senescent leaves of barley. New Phytol 39:362–369

Wallace T (1961) Mineral deficiencies in plants. HMSO London

Wallace W, Pate JS (1967) Nitrate assimilation in higher plants with special reference to the cocklebur, *Xanthium pennsylvanicum* Wallr. Ann Bot 31:213–228

Wareing PF (1976) Endogenous cytokinins as growth regulators. In: Sunderland N (ed) Perspectives in experimental biology. Pergamon Press, Oxford, pp 103–109

Wareing PF, Seth AK (1967) Ageing and senescence in the whole plant. Symp Soc Exp Biol 21:543–558

Weiss C, Vaadia Y (1965) Kinetin-like activity in root apices of sunflower plants. Life Sci 4:1323–1326

Whyte P, Luckwill LC (1966) Sensitive bioassays for gibberellins based upon retardation of leaf senescence in *Rumex obtusifolius*. Nature (London) 210:1360

Williams RF (1955) Redistribution of mineral elements during development. Annu Rev Plant Physiol 6:25–42

Wittenbach VA (1977) Induced senescence of intact wheat seedlings and its reversibility. Plant Physiol 59:1039–1042

Wittenbach VA (1978) Breakdown of ribulose bisphosphate carboxylase and change in proteolytic activity during dark-induced senescence of wheat seedlings. Plant Physiol 62:604–608

Wittenbach VA (1979) Ribulose bisphosphate carboxylase and proteolytic activity in wheat leaves from anthesis through senescence. Plant Physiol 64:884–887

Wollgiehn R (1961) Untersuchungen über den Zusammenhang zwischen Nukleinsäure- und Eiweißstoffwechsel in grünen Blättern. Flora (Jena) 150:117–127

Wood RKS (1967) Physiological plant pathology. Blackwell,Oxford Edinburgh, pp 393–397

Woolhouse HW (1967) The nature of senescence in plants. Symp Soc Exp Biol 21:179–214
Woolhouse HW (1974) Longevity and senescence in plants. Sci Prog (London) 61:123–147
Woolhouse HW (1978) Cellular and metabolic aspects of senescence in higher plants. In: Behnke JA, Finch CE, Moment GB (eds) The biology of aging. Plenum Press, New York, pp 83–99
Woolhouse HW, Batt T (1976) The nature and regulation of senescence in plastids. In: Sunderland N (ed) Perspectives in experimental biology. Pergamon Press, Oxford, pp 163–175
Wright STC, Hiron RWP (1969) (+)-Abscisic acid, the growth inhibitor induced in detached wheat leaves by a period of wilting. Nature (London) 224:5220–5221
Yemm EW (1950) Respiration of barley plants. IV. Protein catabolism and the formation of amides in starving leaves. Proc R Soc London Ser B 136:632–649
Yoshida S (1972) Physiological aspects of crop yield. Annu Rev Plant Physiol 23:437–464
Yoshida Y (1961) Nuclear control of chloroplast activity in *Elodea* leaf cells. Protoplasma 54:476–492
Ziegler H (1974) Biochemical aspects of phloem transport. Symp Soc Exp Biol 28:43–62
Zimmermann MH (1960) Transport in the phloem. Annu Rev Plant Physiol 11:167–190

# 17 Macromolecular Aspects of Cell Wall Differentiation

D.H. NORTHCOTE

## 1 Introduction

The development of a structure containing macromolecules such as a plant cell wall depends on the coordination and ordered sequence of a series of synthetic and transport processes within the cell throughout its development. These processes arise by changes in the enzymic activities of the cell whereby the synthetic or degradative pathways of those polymers making up the structure or taking part in the organised transport are modified. It is possible therefore to limit the consideration of the enzymic systems that are involved in the control of differentiation by knowing the chemical changes that take place and the mechanism of the transport of the materials involved. Some of these enzymic steps are the effectors whereby the control is monitored and they will respond to signals such as a hormonal supply or change in environment of the cell or tissue. This review will consider these two aspects separately, first the enzyme systems involved in the control of the differentiation and then the signals which bring about the changes in enzymic activitity.

### 1.1 Control of Enzyme Activity

The flux through a metabolic sequence of biochemical reactions is controlled by the regulation of particular enzymic activities. This is achieved by two distinct methods. One is by operation of a biochemical control on a rate-limiting enzymic step when, for instance, products of the enzymic sequence can act as modulators of this particular enzymic activitity. In this way a constant monitoring of the metabolic flow is possible and control is swift and responsive to the immediate metabolic requirements of the cell (NEWSHOLME and START 1973). The other method of control that is exerted during a developmental sequence operates so that amounts of enzymes of a metabolic route are modified and the route can be switched on or repressed or its speed regulated by increasing or decreasing the amount of enzyme in the steady-state system. These processes are mechanisms which work at the level of mRNA transcription or its translation, mRNA processing or post-translational modification.

### 1.2 The Cell Wall

The wall of a plant cell is initially formed at cytokinesis in telophase when the cell plate is formed. At this early stage the plane of division is established and the cell plate is subsequently modified, elaborated, thickened and built upon to give the mature cell wall. The sequence of organization and transport within the cell and the sequence of the synthetic steps involved during the deposition of the cell wall represents a complex differentiation process from cell division to final mature cell; the results of which are all recorded in the chemical nature and form of the wall.

The essential chemical changes which occur during wall development are the deposition of pectin during the early stage of primary growth as the cell surface increases in area, the cessation of pectin deposition during secondary thickening and the increase in hemicellulose and cellulose deposition at this secondary stage together with the initiation of lignin synthesis (THORNBER and NORTHCOTE 1961). These changes in the synthetic capabilities of the cell are accomplished by the activation and inactivation of certain key enzymes, the induction and repression of other enzymes and the control of transport and organization processes.

## 2 The Enzyme Systems

### 2.1 Pectin Polysaccharides

The pectins are made up of acidic polysaccharides composed of polygalacturonic acid chains to which are attached varying amounts of arabinogalactan moieties, and separate neutral polysaccharides consisting of arabinogalactans (NORTHCOTE 1972). The relative amounts of these different polysaccharides and their composition change during the cell plate formation and primary growth and these changes are reflected in the physical nature and extensibility of the wall during the period of its increase in surface area (NORTHCOTE 1977).

The scheme whereby carbohydrate flows through the metabolic routes to form the nucleoside diphosphate sugar precursors for pectin and hemicellulose synthesis is shown in Fig. 1. The main flow is from UDP glucose to UDP glucuronic acid and UDP xylose by means of a dehydrogenase and a decarboxylase to form the donors of hemicellulose synthesis or via epimerases from each of these three uridine diphosphate sugars to form UDP galactose, UDP galacturonic acid and UDP arabinose (HASSID et al. 1959, NORTHCOTE 1962). These last three UDP sugars form the donors for pectin synthesis. Obvious points of control are therefore at the level of these epimerase reactions which would effectively control the flow of carbohydrate either into pectin or hemicellulose. It would be expected that the epimerase actitivities would be shut down during secondary growth so that no pectin would be formed. However, when these enzymes are extracted from cells undergoing either primary or secondary growth of their walls, there is little difference between the activities of the extracts (Table 1) (DALESSANDRO and NORTHCOTE 1977a, b), so that unless there is biochemical modulation of the epimerases there is no control at these sites and certainly there is no repression of the synthesis of the epimerases during secondary growth. The cell at all stages of growth therefore is capable of main-

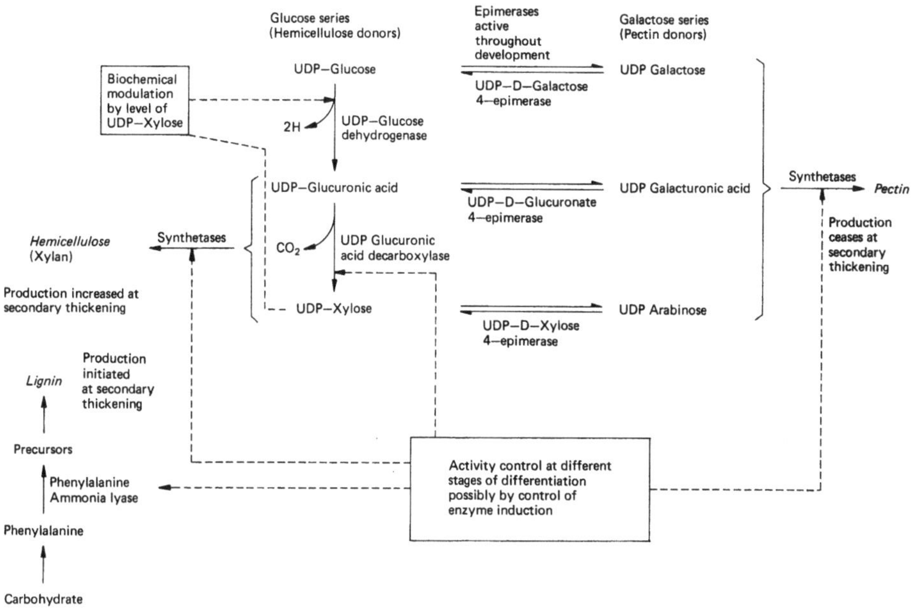


**Fig. 1.** Diagram to show the possible control of enzyme activities of a plant cell during cell wall development

**Table 1.** Specific activities and units of enzyme activities per cell of the enzymes of UDP-sugar interconversions during differentiation in sycamore stems

| | Cambial cells | | Differentiating xylem cells | | Differentiated xylem cells | |
|---|---|---|---|---|---|---|
| | Specific activity (n mol min$^{-1}$ per mg protein) | Units of enzyme activity per cell (n mol min$^{-1}$ per cell) | Specific activity (n mol min$^{-1}$ per mg protein) | Units of enzyme activity per cell (n mol min$^{-1}$ per cell) | Specific activity (n mol min$^{-1}$ per mg protein) | Units of enzyme activity per cell (n mol min$^{-1}$ per cell) |
| Enzyme | | | | | | |
| UDP-D-glucose dehydrogenase | 6.4 | $7.9 \times 10^{-7}$ | 10.0 | $15.8 \times 10^{-7}$ | 18.0 | $35.8 \times 10^{-7}$ |
| UDP-D-glucuronic acid decarboxylase | 132.8 | $16.5 \times 10^{-6}$ | 313.5 | $49.7 \times 10^{-6}$ | 303,7 | $60.4 \times 10^{-6}$ |
| UDP-D-galactose 4-epimerase | 16.3 | $2.0 \times 10^{-6}$ | 19.9 | $3.1 \times 10^{-6}$ | 12.1 | $2.4 \times 10^{-6}$ |
| UDP-D-xylose 4-epimerase | 1.0 | $12.7 \times 10^{-8}$ | 1.3 | $20.1 \times 10^{-8}$ | 0.9 | $17.0 \times 10^{-8}$ |

taining a pool of all the uridine diphosphate sugars even if some of these are not used to any extent for wall formation at secondary thickening.

The structure of the pectin is important all through primary growth and it varies in its composition so that there are changes in the relative amounts of arabinose, galactose and galacturonic acid in the polymers at any particular time. The different types of pectin alter the texture of the wall and the charge at the surface of the plasmamembrane. During primary growth while the cell is growing rapidly in surface area and while the pectin composition is varying, the activities of the epimerases do fluctuate and these slight changes may well exert some control upon the flow of arabinose, galactose and galacturonic acid into the pectin substances (DALESSANDRO and NORTHCOTE 1977c).

However, the main control of pectin synthesis, and especially the signal for the shut-down of its formation at secondary thickening, must be at sites other than the epimerases and it is probably exerted at some aspect of the synthetase system. The synthetases form the polysaccharides from the pool of nucleoside diphosphate sugar donors by tranglycosylation reactions which result in polymerization of the glycosyl monomers (NORTHCOTE 1969).

## 2.2 Hemicellulose

In angiosperms the main hemicellulose is xylan. During differentiation of a xylem cell of a dicotyledon, such as sycamore (*Acer pseudoplatanus*), the activity of UDP glucuronic acid decarboxylase, that produces UDP xylose (Fig. 1), increases significantly (Table 1) (DALESSANDRO and NORTHCOTE 1977a). Thus the flow of glucose towards the nucleoside diphosphate donor for xylan synthesis increases. The UDP glucose dehydrogenase (Fig. 1) has always a lower activitiy than the decarboxylase and is probably a rate-limiting step for the flow to UDP-xylose. This enzyme is inhibited by an increased concentration of UDP-xylose so that the dehydrogenase can monitor the flow to the xylan precursor by a biochemical feed-back mechanism (NEUFELD and HALL 1965, DALESSANDRO and NORTHCOTE 1977a). UDP glucuronic acid can also arise from glucose via myo-inositol (LOEWUS et al. 1973) and this route will by-pass the modulation at the dehydrogenase so that an effective control of UDP-xylose production by the decarboxylase can occur. The main control during differentiation can, however, be clearly seen if the activity of the synthetase system is measured, since this increases in cells which are differentiating into xylem compared with the cambium initials from which they arise (Table 2) (NORTHCOTE 1979b, DALESSANDRO and NORTHCOTE 1981).

Callose is a $(1 \rightarrow 3)$-$\beta$ glucan which the plant cell forms at particular times during development (ASPINALL and KESSLER 1957). It can be rapidly synthesized and rapidly removed. It is used to plug wounds in the cell wall (BRETT 1978) and it occurs at specific sites of pores in sieve tubes at the sieve plate and at special pores between sieve tubes and companion cells (KESSLER 1958, ESAU et al. 1962, WOODING and NORTHCOTE 1965). Glucans containing a mixture of $(1 \rightarrow 3)$-$\beta$ and $(1 \rightarrow 4)$-$\beta$ linkages also occur in some storage and hemicellulose or matrix polysaccharides of the cell wall (BUCHALA and WILKIE 1974). During

**Table 2.** Xylan synthetase activity in cambial cells, differentiating and differentiated xylem cells in sycamore trees

| Source of enzyme fraction | Xylan synthetase activity (nmol/min/mg of protein) |
|---|---|
| Cambial cells | 0.7 |
| Differentiating xylem cells | 1.6 |
| Differentiated xylem cells | 4.2 |

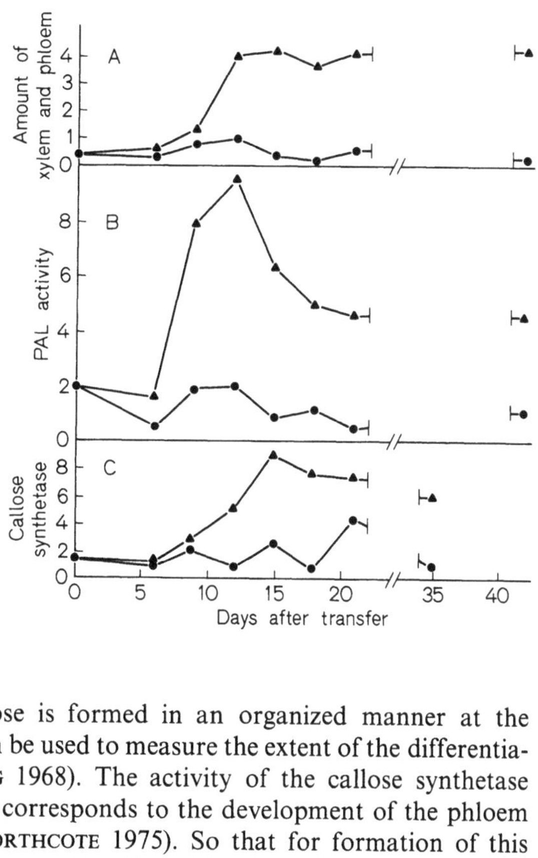


**Fig. 2A–C.** The time course of differentiation in a bean (*Phaseolus vulgaris*) callus after three transfers to maintenance medium (sucrose 2%; 2:4 dichlorophenoxyacetic acid 2 mg/l: kinetin 0.0 mg/l) and then either to maintenance medium (-●-) or to induction medium (-▲-) (sucrose 3%; NAA 1 mg/l, kinetin 0.2 mg/l). **A** xylem and phloem concentration in nodules **B** phenylalanine ammonia lyase activity **C** callose synthetase activity. The amounts in each case (**A, B, C**) are expressed per g fresh weight of tissue in arbitrary units

phloem differentiation the callose is formed in an organized manner at the sieve plate and its appearance can be used to measure the extent of the differentiation (Northcote and Wooding 1968). The activity of the callose synthetase increases over a time course and corresponds to the development of the phloem tissue (Fig. 2) (Haddon and Northcote 1975). So that for formation of this polysaccharide as for the xylan, control is effected by the activity of the synthetase and it probably represents an increase in the amount of the enzyme and an induction of its synthesis at the commencement of the differentiation.

## 2.3 Cellulose

Very little information is known about the synthetase system for cellulose in higher plants. It is probably complex and lipid and protein carriers of the glucan chain may be involved (Hopp et al. 1978). However, since a satisfactory assay system for the synthetase complex is not available, no information on its induction or variations in its activity during development is known. The organizational changes at cell membranes associated with cellulose microfibril formation are to some extent better understood and these will be discussed later together with the transport and organization of the hemicellulose and pectin matrix material of the wall.

### 2.4 Lignin

The onset of lignification of the secondary thickening wall can be seen by electron microscopic examination of differentiating xylem cells within a stem (HEPLER and NEWCOMB 1963, WOODING and NORTHCOTE 1964). It occurs at a very early stage of secondary thickening, one or two cells away from the dividing cambial initial (WOODING and NORTHCOTE 1964). The lignification starts at the middle lamella (outer region of the wall) and progresses inwards towards the developing secondary thickening and longitudinally between them down the middle lamella (KERR and BAILEY 1934, BAILEY 1936).

The final stage of the polymerization processes is initiated by an oxidation of phenolic alcohols producing free radicals which polymerize (FREUDENBERG 1968). This oxidation is probably brought about by a peroxidase present in the wall (HARKIN and OBST 1973). The peroxidase activity of several tissues has been examined and the level is high even in tissues that are not undergoing lignification (YUNG and NORTHCOTE 1975, HADDON and NORTHCOTE 1976b) so that this step is not rate-limiting or a site of control. One control site of the lignification is probably at the reactions which produce the precursors (phenolic alcohols) or their derivatives within the cytoplasm of the cell. These precursors are p-coumaryl alcohol, coniferyl alcohol and sinapyl alcohol, all of which can arise from cinnamic acid, which is produced from phenylalanine by phenylalanine ammonia lyase (NEISH 1961, KOUKOL and CONN 1961, FREUDENBERG 1965). Thus phenylalanine ammonia lyase is a key enzyme for the production and possible regulation of the supply of lignin precursors. It is a highly regulated enzyme in plant tissues and its production is controlled by light (BETZ et al. 1978, HAHLBROCK and GRISEBACH 1979, BETZ and HAHLBROCK 1979) and by plant growth hormones (HADDON and NORTHCOTE 1975, 1976a, b, KUBOI and YAMADA 1978c, BEVAN and NORTHCOTE 1979a, b). In tissue which is differentiating to xylem, induction of phenylalanine ammonia lyase activity is correlated with the increase in xylem formation measured by direct visual assessment of the extent of the differentiation (Fig. 2) (RUBERY and NORTHCOTE 1968, HADDON and NORTHCOTE 1975).

## 3 Site of Synthesis, Transport and Organization of Material During Differentiation

The formation of the wall gives information about aspects of differentiation other than that of control of protein synthesis. For instance, mechanisms for intra- and inter-cellular transport of materials which are an essential part of the development of a eukaryotic cell can be studied.

The organization of transport within the cell establishes the plane of division and hence the formation of patterns of cells which make up the shape of the tissue in a plant. It is also concerned with the movement of material synthesized within the cell to the wall outside the plasmamembrane. Polysaccharides are deposited so that either an evenly thickened cell wall is formed or the characteristic spiral or reticulate thickening of the xylem vessels is obtained.

Transport of material between cells brings about coordinated growth within a tissue for the formation of organs such as stomata. These are produced by discrete patterns of cells established by control of the plane of cell division together with a related growth in size and shape between the individual cells (PICKETT-HEAPS and NORTHCOTE 1966b, STEBBINS and SHAH 1968). Similarly the vascular cambium is maintained while it produces different types of cells which differentiate into xylem or phloem on either side of it. These interrelated aspects of growth need an exchange of information between the cells in the form of organized transport of materials such as plant hormones, sucrose and ions.

## 3.1 Polysaccharide Synthesis and Movement

### 3.1.1 Sites of Synthesis of Pectin and Hemicelluloses

The polysaccharides of the matrix of the wall are elaborated within the lumen of the membrane system that consists of the endoplasmic reticulum, the Golgi apparatus, vesicles and plasmamembrane (DE DUVE 1969, NORTHCOTE 1970). It has been shown for some glycoproteins that glyco-lipid intermediates are concerned with the glycosylation mechanism (PARODI and LELOIR 1979). The lipids are polyisoprenyl phosphates. The sugar chain is built up from nucleoside diphosphate sugar donors or polyisoprenyl monophosphate sugar donors on to a polyisoprenyl diphosphate oligosaccharide chain (HEMMING 1978). This sugar chain is then transferred to the protein. No isoprenoid intermediates have been found to be used directly either as donors or carriers for polysaccharide formation (NORTHCOTE 1979a). However, some polysaccharides have been shown to be carried on proteins before they are released and with these the intermediate glycoprotein may be formed via the isoprenoid compounds (GREEN and NORTHCOTE 1978, 1979a, b). Thus two transglycosylases, a nucleoside diphosphate sugar: isoprenyl phosphate transglycosylase and an isoprenyl phosphate sugar: protein transglycosylase may be involved in the synthetase system and either of these could be rate-limiting and the control step of the synthetic reactions (NORTHCOTE 1979a, b).

If proteins are involved these will be synthesized and inserted into the membranes at the endoplasmic reticulum so that glycosylation may be started at this site (MOLNAR 1976). However, the bulk of the glycosylation of the polysaccharide occurs at the membrane of the Golgi apparatus which receives any initiators by membrane flow and vesicle fusion from the endoplasmic reticulum

(NORTHCOTE 1974a, b). The completed polysaccharide is packaged into the vesicles formed by the dispersal of the Golgi cisternae and passed across the plasmamembrane by membrane fusion and reverse pinocytosis (MOLLENHAUER and WHALEY 1963, NORTHCOTE and PICKETT-HEAPS 1966, NORTHCOTE 1970, WHALEY 1975).

### 3.1.2 Changes in the Products Synthesized by the Membrane System During Differentiation

The distance between cells which are actively dividing and the regions of different differentiating tissues is very small. In a vascular cambium the cells destined to become phloem and those which are on the pathway to xylem with its different synthetic capabilities arise from the same actively dividing zone of cells and are separated one from another by only a ring of two or four cambial cells. In a root tip, distinct cells in the cap arise from the same meristem which forms the vascular tissue and cortex of the growing root (ESAU 1965).

The differentiated cells of the outer root cap can be clearly distinguished because they produce a slime (WRIGHT and NORTHCOTE 1974, 1975, BARLOW 1975). In *Zea mays* this slime consists of polysaccharide that contains fucose and this sugar is only present in the polysaccharides formed by these cap cells so that it is a characteristic product (HARRIS and NORTHCOTE 1970, KIRBY and ROBERTS 1971, WRIGHT and NORTHCOTE 1976). The slime, like the polysaccharides of the hemicellulose and pectin of the matrix of the wall, is synthesized, elaborated and packed in the Golgi cisternae and vesicles and is then transported across the plasmamembrane (BOWLES and NORTHCOTE 1972, 1974, 1976). The synthetases forming the slime polysaccharides are therefore unique to the membrane system of the root cap cells and they are not found in the regions of the root a few millimetres away from the tip where no cap cells are being differentiated.

The transferases which probably make up the synthetase system have been detected in the membranes of the root-tip cells and they are not present in the membranes from cells further up the root. Two transglycosylases have been found, one occurs in the membranes of the endoplasmic reticulum and transfers fucose to a polyprenyl phosphate compound which may also carry other sugars (GREEN and NORTHCOTE 1979a), and the other transfers fucose to a polysaccharide or glycoprotein and occurs mainly in the Golgi apparatus but also in the endoplasmic reticulum (GREEN and NORTHCOTE 1978, 1979b, JAMES and JONES 1979). Whether this second enzyme uses the lipid intermediate as a substrate at either location is not known. Nevertheless these synthetases are obviously induced and inserted into the membranes of the Golgi and endoplasmic reticulum only at the region of the root which is differentiated to carry out the particular function of slime production. Thus the metabolism of the organelle in the various regions of the root is determined and can be altered by the insertion of different glycosyl transferases which are the control points of the synthetic reactions.

### 3.1.3 Movement of Vesicles During Secondary Thickening

Since the wall can be thickened in a spiral or reticulate fashion the vesicles, containing the hemicellulose, which are packed into the wall at definite localized positions, must be moved to particular areas of the wall in an organized manner. This is in part achieved by two mechanisms, an organized directed movement of the vesicles in association with microtubules to the area of the plasmamembrane at which fusion occurs (WOODING and NORTHCOTE 1964, PICKETT-HEAPS and NORTHCOTE 1966c) and an elaboration of the membranes both chemically and by ultrastructural changes so that fusion is possible at the required site (NORTHCOTE 1979a). The control is therefore exerted by the aggregation of microtubules in definite groups and directions within the cytoplasm and by modifications to the distribution of proteins within the fluid lipid bilayer of the membrane. How these controls are achieved is not known, but the level of ions and in particular $Ca^{2+}$ ions play some plart.

### 3.1.4 Formation of Cell Plate and Orientation of Cell Division

The distribution of microtubules, their aggregation and disaggregation have a major role in establishing the plane of division in a plant cell. The plane of the cell plate is partly determined very early in mitosis by the orientation of the preprophase nucleus in a definite direction and within a definite region of the cytoplasm (PICKETT-HEAPS and NORTHCOTE 1966a, b, BURGESS and NORTHCOTE 1967). This movement of the nucleus is associated with the preprophase band of micro-tubules that encircles the nucleus. The micro-tubules are generally located at right angles to the direction of the mitotic spindle at a position near the plasmalemma where the future cell plate will join the mother cell wall. The microtubules of the preprophase band are dispersed as the nuclear membrane is disaggregated at prophase.

The organelles of the cytoplasm are distributed fairly equally to the two mitotic poles at prophase, again under the influence of microtubules that are directed longitudinally in the cytoplasm outside the nuclear plasm, in the direction of the mitotic spindle (ROBERTS and NORTHCOTE 1971). At the time of cell plate formation which is elaborated by vesicle fusion there is a further involvement of microtubules to direct the vesicles towards the growing edge of the plate as it extends to the mother cell wall (ESAU and GILL 1965, PICKET-HEAPS and NORTHCOTE 1966a, b, ROBERTS and NORTHCOTE 1970).

### 3.1.5 Microfibrillar Deposition

Although the mechanics of hemicellulose and pectin deposition in the wall are fairly well established, they by no means account for the laying down of microfibrils of cellulose. But again it is apparent from electron microscope observation that the orientation of the microfibrils within the wall is in some way related to the orientation of the microtubules within the cytoplasm but no direct connection has been observed (WOODING and NORTHCOTE 1964, PICKETT-HEAPS and NORTHCOTE 1966c). There is, however, an ever-increasing amount of evidence to suggest that the microfibrils are spun out into the wall from the plasmalemma surface (BOWLES and NORTHCOTE 1972, NORTHCOTE 1974a, b) in association with a specific substructure of granules (probably proteinaceous) (BROWN and MONTEZINOS 1976, MONTEZINOS and BROWN 1976, WILLISON and BROWN 1978). The granules may be distributed in a definite form at the origin of the microfibril on membrane from which it grows (GIDDINGS et al. 1980, MUELLER and BROWN 1980). However, care has to be taken with the interpretation of these studies which are based on freeze-fracture experiments since the effects of plasmolysis on the structure of the plasmamembrane can give rise to ultrastructural changes in particle distribution (WILKINSON and NORTHCOTE 1980).

### 3.2 Lignification

The localization of the deposition of lignin in the wall is difficult to explain since the precursors arise in the cytoplasm and the polymerization takes place in the wall, by the formation of short-lived free radicals which not only establish covalent linkages between themselves but also between the phenolic radicals and the polysaccharides of the wall (MEREWETHER et al. 1972, HARRIS and HARTLEY 1976, WHITMORE 1978). The lignification is initiated at discrete zones at the outside of the wall and progresses inwards. It is even extended in the walls of cells which have little or no contents and for which the precursors must have been produced by nearby cells that are at earlier stages in the differentiation process. Thus the precursors are freely diffusible within the unlignified wall and are protected from the dehydrogenation step which produces the free radicals until they arrive at the specific site of lignification. Coniferin, the glucoside of coniferyl alcohol and other glycosides of phenol alcohols may be used as lignin precursors (FREUDENBERG 1959). Glycosylation increases the solubility of the phenolic alcohols and protects them from dehydrogenation by peroxidases. The sugar must be removed to give the free alcohol, by $\beta$-glycosidases which are present in the xylem tissue, before oxidation and polymerization can occur. Thus regulation of the formation of these glycosidases and their location within the wall provides a mechanism for the control of the rate and site of intercellular lignification. How these hydrolytic enzymes are transported and anchored at the sites of lignification is not clear.

## 4 Possible Control Signals for the Changes in Cell Wall Synthesis

The mechanism for polysaccharide deposition in the cell wall depends on two different biological processes. One is the biochemical synthesis of the polymers and the other is transport of vesicles and membrane fusion. The control for the macromolecular assembly of the cell wall can operate at both. The cell wall continually changes throughout growth and can change in response to alterations in the external environment. The growth of the cell wall is limited and comes to an end. There must therefore be information that is passed back from the wall to the cytoplasm, so that these changes can take place and be monitored by the cytoplasmic machinery which produces the changes.

### 4.1 Signals at the Cell Surface

By incubating normal and freshly plasmolyzed tobacco mesophyll cells (as leaf discs) with radioactive glucose, it is possible to compare the pattern of polysaccharide synthesis in the two types of cells. Plasmoylsis generally decreases the incorporation of sugars into all the polysaccharide fractions of the wall, but affects the incorporation of arabinose into pectin material in a different way from its effect on other sugars. In freshly plasmolyzed tissue, the incorporation of arabinose into pectin is stimulated, whereas the incorporation into other sugars and uronic acids is decreased (Table 3) (Boffey and Northcote 1975).

The process of plasmolysis involves several interdependent events. One consequence is the decrease in total surface area of the plasmalemma, and this must cause the general decrease in incorporation of radioactivity from glucose into the sugars of the polysaccharide that are formed after plasmolysis. However, this cannot explain the increase in arabinose incorporation. Loss of contact between much of the cell wall and the plasmalemma must bring about significant changes in the environment of the membrane. For example, the ionic atmosphere at the outer surface of the membrane changes and this could result in disruption and alteration of those syntheses occurring at the cell surface. The removal of the acidic pectin molecules of the cell wall from the membrane surface could also alter the charge distribution across the membrane and the distribution

**Table 3.** Radioactive incorporation into the sugars of pectins extracted from leaf discs stripped of lower epidermis, incubated with D-[U-$^{14}$C] glucose for 5 h after 1 h in non-radioactive media

| Sugar | Radioactivity (%) | | | | Ratio of incorporation (plasmolyzed/ unplasmolyzed) |
|---|---|---|---|---|---|
| | Unplasmolyzed | | Plasmolysed (0.7 M sorbitol) | | |
| | Sample 1 | Sample 2 | Sample 1 | Sample 2 | |
| Galacturonic acid | 28.3 | 34.2 | 32.8 | 31.8 | 0.3 |
| Galactose | 43.2 | 33.9 | 30.9 | 34.2 | 0.2 |
| Glucose | 18.1 | 14.4 | 9.2 | 13.4 | 0.2 |
| Arabinose | 2.5 | 3.7 | 13.3 | 12.8 | 1.2 |
| Rhamnose | 5.4 | 5.6 | 5.0 | 2.6 | 0.2 |
| $10^{-3}$ radioactivity in pectin (c.p.m.) | 34 | 28 | 13 | 3 | 0.3 |

and orientation of the membrane constituents, so that vesicle fusion from the Golgi bodies is changed and control of membrane flow from the endomembrane system influenced. Plasmolysis causes the disappearance of dictyosomes from onion root cells, and this could result in a control of those polysaccharides synthesized and transported by the Golgi bodies (PRAT 1972).

Another result of plasmolysis is the reduction of cell–cell interaction that occurs when an organized tissue is used, especially on the breakage of the plasmodesmata which link adjacent cells. The effects of plasmolysis might therefore be due to interference with transport from cell to cell of substances that control the differentiation of the tissue and which will therefore alter cell wall synthesis.

There are, however, several experiments that indicate that the response to plasmolysis and the accompanying alteration of cell wall synthesis is due primarily to the withdrawal of the membrane from the immediate vincinity of the wall and the consequent change in the environment of the cell membrane (BOFFEY and NORTHCOTE 1975). The effects of plasmolysis on the polysaccharide synthesis pattern of separated cells compared with those of intact leaf discs are very similar. The effects of plasmolysis on the incorporation of radioactivity are partially reversed when the plasmolyzed tissue is placed in a solution that causes deplasmolysis. During plasmolysis of the cells in a leaf disc, the alteration in the pattern of pectin synthesis is brought about immediately, as soon as the cells are placed in solutions just above the iso-osmotic point, as the plasmalemma begins to be withdrawn from the wall and the plasmodesmata are not broken at this stage. This immediate effect is much greater than the subsequent response to decreasing the plasmalemma surface as the osmotic pressure of the plasmolyzing solution is progressively increased above the isotonic point (NORTHCOTE 1977).

Thus the loss of contact between much of the cell wall and the plasmalemma occurs simultaneously with a change in the synthesis occurring within the cytoplasm. It is possible that this change is brought about because the ionic atmosphere at the outer surface of the membrane has changed. A possible control point in the secretion of polysaccharides by the endomembrane system is therefore the rate of fusion of the vesicles with the plasmamembrane. Among many other factors the fusion is dependent upon the charge at the membrane surface and the presence of $Ca^{2+}$ or $Mg^{2+}$ ions (LOISTER and LAYTER 1973, MAEDA and OHNISHI 1974, GRATZL and DAHL 1976, 1978, DAHL et al. 1978) and by specific proteins as part of a characteristic distribution of granules within the membrane (SATIR et al. 1973, SATIR 1974, BURWEN and SATIR 1977, DA SILVA and NOGUEIRA 1977). Some of these factors are probably related and are parts of the same control mechanism.

Membrane fractions enriched in Golgi apparatus or plasmamembrane (BAYDOUN and NORTHCOTE 1980a) prepared from maize root tips are made radioactive if the roots are pre-incubated with radioactive glucose or choline. When in vitro radioactive and non-radioactive membrane preparations are mixed, the amount of membrane fusion under specified conditions is related to the transfer of radioactivity between them (BAYDOUN and NORTHCOTE 1980b); a quantitative assay of membrane fusion can therefore be made. $Ca^{2+}$ ions are necessary for fusion. Furthermore, a specific, integral protein of the membranes is removed by treatment with trypsin and this protein, which is soluble in deoxycholate, is a $Ca^{2+}$-activated ATPase. The presence of this protein is necessary for the $Ca^{2+}$-dependent fusion of the membranes (BAYDOUN and NORTHCOTE 1980b, 1981).

Suspension-cultured sycamore cells secrete soluble polysaccharides and the steady-state rate of secretion is substantially and very rapidly increased by addition of various electrolytes to the medium. The increased secretion is induced, from within the cell, by the presence of cations including $Ca^{2+}$ and $Mg^{2+}$ ions, primarily at the outer surface of the plasmamembrane. The increase in rate of secretion is brought about by a stimulation of the normal mechanism of polysaccharide secretion, that is, it is the fusion of membrane-bounded vesicles from the Golgi apparatus with the plasmamembrane that is activated, so that this step is a rate-limiting process in polysaccharide secretion and a potential control point (MORRIS and NORTHCOTE 1977).

Control of exocytosis occurs at the outer surface of the plasmamembrane and the immediate effect is on the last stage of the mechanism of polysaccharide secretion. A further control of the biochemical synthetic steps within the membrane system at the level of the endoplasmic reticulum and the Golgi apparatus must be coordinated with these surface regulatory mechanisms.

## 4.2 Plant Growth Hormones as Signals

The plant growth hormones, indole acetic acid (or artificial compounds like 2,4 dichlorophenoxyacetic acid or naphthylacetic acid), cytokinin (or kinetin), gibberellic acid and abscisic acid, together with sucrose, applied to plant tissue cultures, either as suspension cultures or on solid media, influence the differentiation of the callus cells to xylem and phloem (SKOOG and MILLER 1957, WETMORE and RIER 1963, JEFFS and NORTHCOTE 1966, 1967, WRIGHT and NORTHCOTE 1972, HADDON and NORTHCOTE 1976a). These hormones and possibly sucrose might therefore act in part to influence the induction of particular enzymes such as xylan synthetase, phenylalanine ammonia lyase and callose synthetase which are necessary for the formation of xylem and phloem cells. The induction of differentiation of a callus transferred to an induction medium is measured by the increase in activities of these enzymes (HADDON and NORTHCOTE 1975). Usually a ratio of NAA to kinetin of about 5:1 (1 mg/l NAA, 0.2 mg/l kinetin) will bring about differentiation in the callus in about 6–12 days and it will induce the enzyme activities just prior to the onset of visual differentiation of the cells. Gibberellic acid in the induction medium (one containing the required ratio of NAA to kinetin) will influence the type of differentiation and may slightly delay its onset and the induction of the enzyme activities, while the presence of abscisic acid will inhibit both the enzyme induction and the differentiation (HADDON and NORTHCOTE 1976a).

The induction of PAL and callose synthetase activitity is shown in Fig. 2 and the increase of these two activities can be directly related, in a series of experiments, to the induction of the differentiation of xylem and phloem

(HADDON and NORTHCOTE 1975). It is difficult to prove that the induced activities are due to a de novo increase or onset of synthesis of enzyme rather than an activation of enzyme previously synthesized but inactive. However, when the enzymes are assayed they are isolated and purified to an extent that small molecular weight activators and repressors will be removed from the preparations, so that the measurement of activities under the optimum kinetic conditions probably represents the amount of enzyme present. If this is correct then the induction medium has brought about an increase, or an initiation in some cases, of the transcription and translation of the mRNA from the plant cell genome and the control exerted by the application of the exogenous plant growth factors is operated at the stage of transcription, RNA processing, translation or post-translational modification of the enzyme.

Inhibitors of transcription (actinomycin D) and translation [D-2-(4-methyl-2,6-dinitroanilo)-N-methylpropionamide, MDMP] applied to bean suspension culture cells grown in induction media inhibit the rising phase of PAL activity. This indicates that both transcription and translation is required for the increase in PAL activity. Superinduction by actinomycin D but not by MDMP during the falling phase of PAL activity suggests that availability of PAL mRNA for translation may be controlled to produce the PAL activity response to plant hormones (JONES and NORTHCOTE 1981). A theoretical model based on de novo synthesis of PAL mRNA has been shown to apply for other stimuli which bring about an increase in PAL activity (HAHLBROCK and GRISEBACH 1979).

With bean tissue, the increase of phenylalanine ammonia lyase activity that is associated with cell differentiation depends upon the presence of NAA and kinetin. The NAA needs to be present at least 48 h prior to the increase in PAL activity while the kinetin can be applied immediately prior to the expected rise in PAL activity (BEVAN and NORTHCOTE 1979a). Therefore it is probable that the hormones are acting at two different sites to control the production of the enzyme.

During continued subculture on a maintenance medium containing a ratio of auxin:kinetin which does not induce differentiation, callus cells lose their potential to differentiate when they are subsequently transferred to an induction medium (HADDON and NORTHCOTE 1975, BEVAN and NORTHCOTE 1979b). This loss in potential for differentiation is measured by a corresponding loss in the capacity of the tissue for the induction of PAL and is progressive during the continued subculture on the maintenance medium. The loss in ability to differentiate is not due to cell selection on the maintenance medium, nor does it seem to be a consequence of the loss of the genetic machinery necessary to produce some of the necessary enzymes. The culture tissue no longer responds to the ratio of exogenous hormones because of a change in its metabolism (BEVAN and NORTHCOTE 1979b), such as the rate of formation and level of endogenous hormones; brought about by a change in their synthesis or rate of degradation monitored by a biochemical feed-back control by the pool-size of the growth factors within the cell. This pool-size is in part dependent on the external supply.

The work with the tissue cultures indicates that the ratio of the various growth factors at the cell, together with nutrients such as sucrose, determine the path of differentiation which the cell takes. The plant cell in situ receives these growth factors by two routes, one by synthesis within the cell and secondly by transport from other cells. In the intact plant, differentiation and the establishment of tissues and organs depends therefore upon the synthesis, degradation and transport of the various growth factors such as auxin, cytokinin, gibberellic acid and nutrients such as sucrose.

The sites of synthesis of the growth factors are not known with certainty. But during the controlled breakdown of the contents of a cell such as a xylem tracheid, as it differentiates to maturity, much of its contents including the protein is autolyzed liberating free amino acids such as tryptophan. Tryptophan can be metabolized to give indole acetic acid by a great variety of animal and plant tissues. It is possible therefore that one of the main sites of auxin production within the plant is the controlled differentiation of xylem and phloem tissue on either side of the ring of cambium (SHELDRAKE and NORTHCOTE 1968a, b, c, SHELDRAKE 1973). This developing vascular tissue could supply growth factors, such as indole acetic acid, because of the autolysis of the tissue which is part of the differentiating process, to the meristemic cells at the phloem or xylem side of the cambium. The amount of auxin transported from the xylem would be different from that from the phloem. The distribution of sucrose, since it is carried in the phloem tissue, would also be different on both sides of the cambium. It is possible therefore that the developing cambium is maintained so that it differentiates phloem on one side and xylem on the other by the presence of existing differentiating cells already on either side of it which supply different amounts of growth factors to the cambial cells. A biostatic mechanism is therefore possible between the dividing and differentiating cells which produce the shape and form of the organ (stem and root etc.) and the shape and pattern of cells which make up the organ and which control the supply and distribution of growth factors to the meristemic cells which are forming the pattern of cells.

*Editors' Comment*

Although this paper could have been expected to have appeared as a contribution to Vol. 13B of this Encyclopedia which deals with extracellular carbohydrates (W. Tanner and F.A. Loewus, eds.), the control of enzymes involved in cell wall formation can be considered also in the present volume as a kind of model system. For further and more detailed information the reader is referred to the relevant chapters of the volume mentioned.

## References

Aspinal GO, Kessler G (1957) The structure of callose from the grape vine. Chem Ind NY p 1296

Bailey AJ (1936) Lignin in Douglas fir. Composition of the middle lamella. Ind Eng Chem (Anal) 8:52–55

Barlow P (1975) In: Lorrey JG, Clarkson DC (eds) The development and function of roots. The Root Cap. Academic Press, New York, pp 22–54

Baydoun EAH, Northcote DH (1980a) Isolation and characterization of membranes from the cells of maize root tips. J Cell Sci 45:147–167

Baydoun EAH, Northcote DH (1980b) Measurement and characteristics of fusion of isolated membrane fractions from maize root tips. J Cell Sci 45:169–186

Baydoun EAH, Northcote DH (1981) The extraction from maize (*Zea mays*) root cells of membrane-bound protein with $Ca^{2+}$-dependent ATPase activity and its possible role in membrane fusion *in vitro*. Biochem J 193:781–791
Betz B, Hahlbrock K (1979) Identity of differently-induced phenylalanine ammonia lyases from cell suspension cultures of *Petroselinum hortense*. FEBS Lett 107:233–236
Beth B, Schafer E, Hahlbrock K (1978) Light-induced phenyalanine ammonia-lyase in cell-suspension cultures of *Petroselinum hortense*. Arch Biochem Biophys 190:126–135
Bevan M, Northcote DH (1979a) The interaction of auxin and cytokinin in the induction of phenylalanine ammonia-lyase in suspension cultures of *Phaseolus vulgaris*. Planta 147:77–81
Bevan M, Northcote DH (1979b) The loss of morphogenetic potential and induction of phenylalanine ammonia-lyase in suspension cultures of *Phaseolus vulgaris*. J Cell Sci 39:339–353
Boffey SA, Northcote DH (1975) Pectin synthesis during the wall regeneration of plasmolysed tobacco leaf cells. Biochem J 150:433–440
Bowles DJ, Northcote DH (1972) The sites of synthesis and transport of extracellular polysaccharides in the root tissues of maize. Biochem J 130:1133–1145
Bowles DJ, Northcote DH (1974) The amounts and rates of export of polysaccharides found within the membrane system of maize root cells. Biochem J 142:139–144
Bowles DJ, Northcote DH (1976) The size and distribution of polysaccharides during their synthesis within the membrane system of maize root cells. Planta 128:101–106
Brett CT (1978) Synthesis of $\beta$-(1-3)-glucan from extracellular uridine diphosphate glucose as a wound response in suspension-cultured soybean cells. Plant Physiol 62:377–382
Brown R, Montezinos D (1976) Cellulose microfibrils: Visualisation of biosynthetic and orienting complexes in association with the plasma membrane. Proc Natl Acad Sci USA 73:143–147
Buchala AJ, Wilkie KCB (1974) Total hemicelluloses from *Hordeum vulgare* plants at different stages of maturity. Phytochemistry 13:1347–1351
Burgess J, Northcote DH (1967) A function of the preprophase band of microtubules in *Phleum pratense*. Planta 75:319–326
Burwen SJ, Satir BH (1977) A freeze-fracture study of early membrane events during mast cell secretion. J Cell Biol 73:660–671
Dahl G, Schudt C, Gratzl M (1978) Fusion of isolated myoblast plasma membranes: an approach to the mechanism. Biochem Biophys Acta 514:105–116
Dalessandro G, Northcote DH (1977a) Changes in enzymic activities of nucleoside diphosphate sugar interconversions during differentiation of cambium to xylem in sycamore and poplar. Biochem J 162:267–279
Dalessandro G, Northcote DH (1977b) Changes in enzymic activities of nucleoside diphosphate sugar interconversions during differentiation of cambium to xylem in pine and fir. Biochem J 162:281–288
Dalessandro G, Northcote DH (1977c) Possible control sites of polysaccharide synthesis during cell growth and wall expansion of pea seedlings (*Pisum sativum* L.). Planta 134:39–44
Dalessandro G, Northcote DH (1981) Increase of xylan synthetase activity during xylem differentiation of the vascular cambium of sycamore and poplar trees. Planta 151:61–67
Da Silva PP, Nogueira L (1977) Membrane fusion during secretion. A hypothesis based on electron microscope observation of *Phytophthora palmivora* zoospores during encystment. J Cell Biol 73:161–181
De Duve C (1969) The lysosome in retrospect. In: Dingle JT, Fell HB (eds) Lysosomes in biology and pathology. North Holland, Amsterdam, pp 3–41
Esau K (1965) Plant Anatomy, 2nd edn. Wiley and Sons, New York
Esau K, Gill RH (1965) Observations on cytokinesis. Planta 67:168–181
Esau K, Cheadle VI, Risley EB (1962) Development of sieve-plate pores. Bot Gaz 123:233–243
Freudenberg K (1959) Biosynthesis and constitution of lignin. Nature (London) 183:1152–1155
Freudenberg K (1965) Lignin: its constitution and formation from p-hydroxycinnamyl alcohols. Science 148:595–600

Freudenberg K (1968) The constitution and biosynthesis of lignin. In: Kleinzeller A (ed) Constitution and biosynthesis of lignin. Mol Biol Biochem and Biophys Vol 2. Springer, Berlin Heidelberg New York, pp 45–122

Giddings TH, Brower DL, Staehelin LA (1980) Visualization of particle complexes in the plasma membrane of *Micrasterias denticulata* associated with the formation of cellulose fibrils in primary and secondary cell walls. J Cell Biol 84:327–339

Gratzl M, Dahl G (1976) $Ca^{2+}$-induced fusion of Golgi-derived secretory vesicles isolated from rat liver. FEBS Lett 62:142–145

Gratzl M, Dahl G (1978) Fusion of secretory vesicles isolated from rat liver. J Membr Biol 40:343–364

Green JR, Northcote DH (1978) The structure and function of glycoproteins synthesised during slime-polysaccharide production by membranes of the root-cap cells of maize (*Zea mays*). Biochem J 170:599–608

Green JR, Northcote DH (1979a) Polyprenyl phosphate sugar synthesised during slime-polysaccharide production by membranes of the root-cap cells of maize (*Zea mays*) Biochem J 178:661–671

Green JR, Northcote DH (1979b) Location of fucosyl transferases in the membrane system of maize root cells. J Cell Sci 40:235–244

Haddon LE, Northcote DH (1975) Quantitative measurement of the course of bean callus differentiation. J Cell Sci 17:11–26

Haddon L, Northcote DH (1976a) The influence of gibberellic acid and abscissic acid on cell and tissue differentiation of bean callus. J Cell Sci 20:47–55

Haddon L, Northcote DH (1976b) Correlation of the induction of various enzymes concerned with phenylpropanoid and lignin synthesis during differentiation of bean callus (*Phaseolus vulgaris* L). Planta 128:255–262

Hahlbrock K, Grisebach H (1979) Enzymic controls in the biosynthesis of lignin and flavonoids. Annu Rev Plant Physiol 30:105–130

Harris PJ, Hartley RD (1976) Detection of bound ferulic acid in cell walls of the *Gramineae* by ultraviolet fluorescence microscopy. Nature (London) 259:508–510

Harris PJ, Northcote DH (1970) Patterns of polysaccharide biosynthesis in differentiating cells of maize root-tips. Biochem J 120:479–491

Harkin JM, Obst JR (1973) Lignification in trees: indication of exclusive peroxidase participation. Science 180:296–297

Hassid WZ, Neufeld EF, Feingold DS (1959) Sugar nucleotides in the interconversion of carbohydrates in higher plants. Proc Natl Acad Sci USA 45:905–915

Hemming FW (1978) Polyprenyl phosphates as coenzymes in protein and oligosaccharide glycosylation. Philos Trans R Soc London Ser B 284:559–568

Hepler PK, Newcomb EH (1963) The fine structure of young tracheary xylem elements arising by redifferentiation of parenchyma in wounded *Coleus stem*. J Exp Bot 14:496–502

Hopp HE, Romero PA, Daleo GR, Pont Lezica R (1978) Synthesis of cellulose precursors. Eur J Biochem 84:561–571

James DW, Jones RL (1979) Intracellular localization of GDP fucose. Polysaccharide fucoxyl transferase in corn roots (*Zea mays* L.). Plant Physiol 64:914–918

Jeffs RA, Northcote DH (1966) Experimental induction of vascular tissue in an undifferentiated plant callus. Biochem J 101:146–152

Jeffs RA, Northcote DH (1967) The influence of indol-3yl acetic acid and sugar on the pattern of induced differentiation in plant tissue culture. J Cell Sci 2:77–88

Jones DH, Northcote DH (1981) Induction by hormones of phenylalanine ammonia-lyase in bean-cell suspension cultures. Inhibition and superinduction by actionomycin D. Eur J Biochem 116:117–125

Kerr T, Bailey IW (1934) Structure, optical properties and chemical composition of the so-called middle lamella. J Arnold Arbor 15:327–349

Kessler G (1958) Zur Charakterisierung der Silberröhrenkallose. Ber Schweiz Bot Ges 68:5–43

Kirby KG, Roberts RM (1971) The localized incorporation of $^{3}$H-L-fucose into cell-wall polysaccharides of the cap and epidermis of corn roots. Planta 99:211–221

Koukol J, Conn EE (1961) The metabolism of aromatic compounds in higher plants. IV purification and properties of the phenylalanine deaminase of *Hordeum vulgare*. J Biol Chem 236:2692–2698
Kuboi T, Yamada Y (1978) Regulation of the enzyme activities related to lignin synthesis in cell aggregates of tobacco cell culture. Biochem Biophys Acta 542:181–190
Loewus F, Chen M-S, Loewus MF (1973) Biogenesis of plant cell wall polysaccharides. Academic Press, New York, pp 1–27
Loister Z, Layter A (1973) Mechanism of cell fusion. II formation of chicken erythrocyte polykaryons. J Biol Chem 248:422–432
Maeda T, Ohnishi S-I (1974) Membrane fusion. Transfer of phospholipid molecules between phospholipid bilayer membranes. Biochem Biophys Res Commun 60:1509–1516
Merewether JWT, Samsuzzaman LAM, Cooke RG (1972) Studies on a lignin-carbohydrate complex III. Nature of the complex. Holzforschung 26:193
Mollenhauer HH, Whaley WG (1963) An observation on the functioning of the Golgi apparatus. J Cell Biol 17:222
Molnar J (1976) Role of endoplasmic reticulum and Golgi apparatus in the biosynthesis of plasma glycoproteins. In: Martonosi A (ed) Enzymes of biological membranes Vol 2. Wiley and Sons, London New York, pp 385–419
Montezinos D, Brown RM (1976) Surface architecture of the plant cell: biogenesis of the cell wall with special emphasis on the role of the plasma membrane in cellulose biosynthesis. J Supramol Struct 5:277–290
Morris MR, Northcote DH (1977) Influence of cations at the plasma membrane in controlling polysaccharide secretion from sycamore suspension cells. Biochem J 166:603–618
Mueller SC, Brown RM (1980) Evidence for an intramembrane component associated with a cellulose microfibril-synthesising complex in higher plants. J Cell Biol 84:315–326
Neish AC (1961) Formation of m- and p-coumaric acids by enzymatic deamination of the corresponding isomers of tyrosine. Phytochemistry 1:1–24
Neufeld EF, Hall CW (1965) Inhibition of UDP-D-glucose dehydrogenase by UDP-D-xylose: a possible regulatory mechanism. Biochem Biophys Res Commun 19:456–461
Newsholme EA, Start C (1973) Regulation in metabolism. Wiley and Sons, London New York
Northcote DH (1962) The nature of plant cell surfaces. Biochem Soc Symp 22:105–125
Northcote DH (1969) The synthesis and metabolic control of polysaccharides and lignin during the differentiation of plant cells. Essays in Biochem 5:90–137
Northcote DH (1970) The Golgi apparatus. Endeavour 30:26–33
Northcote DH (1972) Chemistry of the plant cell wall. Annu Rev Plant Physiol 23:113–132
Northcote DH (1974a) Membrane systems of plant cells. Philos Trans R Soc London Ser B 268:119–128
Northcote DH (1974b) Sites of synthesis of the polysaccharides of the cell wall. In: Pridham JP (ed) Plant carbohydrate biochemistry. Academic Press, London New York, pp 165–181
Northcote DH (1977) The synthesis and assembly of plant cell walls: possible control mechanisms. In: Poste G, Nicholson GL (eds) The synthesis, assembly and turnover of cell surface components. Elsevier/North-Holland Biomedical Press 4:717–739
Northcote DH (1979a) The involvement of the Golgi apparatus in the biosynthesis and secretion of glycoproteins and polysaccharides. In: Manson LA (ed) Biomembranes Vol 10. Plenum Press, New York, pp 51–76
Northcote DH (1979b) In: George EC (ed) Biochemical mechanisms involved in plant morphogenesis. British Plant Growth Regulator Group monograph 3:11–20
Northcote DH, Pickett-Heaps JP (1966) A function of the Golgi apparatus in polysaccharide synthesis and transport in the root-cap cells of wheat. Biochem J 98:159–167
Northcote DH, Wooding FBP (1968) The structure and function of phloem tissue. Sci Prog 56:35–58
Parodi AJ, Leloir LF (1979) The role of lipid intermediates in the glycosylation of proteins in the eucaryotic cell. Biochem Biophys Acta 559:1–37
Pickett-Heaps JD, Northcote DH (1966a) Organisation of microtubules and endoplasmic reticulum during mitosis and cytokinesis in wheat meristems. J Cell Sci 1:109–120

Pickett-Heaps JD, Northcote DH (1966b) Cell division in the formation of the stomatal complex of the young leaves of wheat. J Cell Sci 1:121–128
Pickett-Heaps JD, Northcote DH (1966c) Relationships of cellular organelles to the formation and development of the plant cell wall. J Exp Bot 17:20–26
Prat R (1972) Plant protoplasts. Effect of the isolation procedure on cell structure. J Microscopie (Paris) 14:85–114
Roberts K, Northcote DH (1970) The structure of sycamore callus cells during division in a partially synchronised suspension culture. J Cell Sci 6:299–321
Roberts K, Northcote DH (1971) Ultrastructure of the nuclear envelope; structural aspects of the interphase nucleus of sycamore suspension culture cells. Microsc Acta 71:102–120
Rubery PH, Northcote DH (1968) Site of phenylalanine ammonia-lyase activity and synthesis of lignin during xylem differentiation. Nature (London) 219:1230–1234
Satir B (1974) Ultrastructural aspects of membrane fusion. J Supramol Struct 2:529–537
Satir B, Schooley C, Satir P (1973) Membrane fusion in a model system. Mucocyst secretion in Tetrahymena. J Cell Biol 56:153–176
Sheldrake AR (1973) The production of hormones in higher plants. Biol Rev 48:509–529
Sheldrake AR, Northcote DH (1968a) The production of auxin by autolysing tissues. Planta 80:227–236
Sheldrake AR, Northcote DH (1968b) The production of auxin by tobacco internode tissues. New Phytol 67:1–13
Sheldrake AR, Northcote DH (1968c) Some constituents of xylem sap and their possible relationship to xylem differentiation. J Exp Bot 19:681–689
Skoog F, Miller CO (1957) Chemical regulation of growth and organ formation in plant tissue cultured *in vitro*. Symp Soc Exp Biol 11:118–131
Stebbins GL, Shah SS (1968) Development studies of cell differentiation in the epidermis of monocotyledons. II Cytological features of stomatal development in the Gramineae. Dev Biol 2:477–500
Thornber JP, Northcote DH (1961) Changes in the chemical composition of a cambial cell during its differentiation into xylem and phloem tissue in trees. 1. Main components. Biochem J 81:449–455
Wetmore RH, Rier JP (1963) Experimental induction of vascular tissues in callus of angiosperms. Am J Bot 50:418–430
Whaley WG (1975) The Golgi apparatus. Cell Biology Mono Vol 2. Springer, Wien New York
Whitmore FW (1978) Lignin-arbohydrate complex formed in isolated cell walls of callus. Phytochemistry 17:421–425
Wilkinson MJ, Northcote DH (1980) Plasma membrane ultrastructure during plant protoplast plasmolysis, isolation and wall regeneration: a freeze fracture study. J Cell Sci 42:401–415
Willison JHM, Brown RM (1978) Cell wall structure and deposition in *Glaucocystis*. J Cell Biol 78:103–119
Wooding FBP, Northcote DH (1964) The development of the secondary wall of the xylem in *Acer pseudoplatanus*. J Cell Biol 23:327–337
Wooding FBP, Northcote DH (1965) The fine structure and development of the companion cell of the phloem of *Acer pseudoplatanus*. J Cell Biol 24:117–128
Wright K, Northcote DH (1972) Induced root differentiation in sycamore callus. J Cell Sci 11:319–337
Wright K, Northcote DH (1974) The relationship of root-cap slime to pectins. Biochem J 139:525–534
Wright K, Northcote DH (1975) An acidic oligosaccharide from maize slime. Phytochemistry 14:1793–1798
Wright K, Northcote DH (1976) Identification of $\beta 1 \rightarrow 4$ glucan chains as part of a fraction of slime synthesised within the dictyosomes of maize root caps. Protoplasma 88:225–239
Yung K-H, Northcote DH (1975) Some enzymes present in the walls of mesophyll cells of tobacco leaves. Biochem J 151:141–144

# Author Index

Page numbers in *italics* refer to the references

# Plant Name Index

# Subject Index

# Encyclopedia of Plant Physiology

**New Series**
**Editors: A. Pirson, M. H. Zimmermann**

Volume 1

## Transport in Plants I

**Phloem Transport**
**Editors: M. H. Zimmermann, J. A. Milburn**
With contributions by numerous experts
1975. 93 figures. XIX, 535 pages
ISBN 3-540-07314-0

Volume 2

## Transport in Plants II

**Editors: U. Lüttge, M. G. Pitman**
Part A: Cells
With contributions by numerous experts
1976. 97 figures, 64 tables. XVI, 419 pages
ISBN 3-540-07452-X

Part B: Tissues and Organs
With contributions by numerous experts
1976. 129 figures, 45 tables. XII, 475 pages
ISBN 3-540-07453-8

Volume 3

## Transport in Plants III

**Intracellular Interactions and Transport Processes**
**Editors: C. R. Stocking, U. Heber**
With contributions by numerous experts
1976. 123 figures. XXII, 517 pages
ISBN 3-540-07818-5

Volume 4

## Physiological Plant Pathology

**Editors: R. Heitefuss, P. H. Williams**
With contributions by numerous experts
1976. 92 figures. XX, 890 pages
ISBN 3-540-07557-7

Volume 5

## Photosynthesis I

**Photosynthetic Electron Transport and Photophosphorylation**
**Editors: A. Trebst, M. Avron**
With contributions by numerous experts
1977. 128 figures. XXIV, 730 pages
ISBN 3-540-07962-9

Volume 6

## Photosynthesis II

**Photosynthetic Carbon Metabolism and Related Processes**
**Editors: M. Gibbs, E. Latzko**
With contributions by numerous experts
1979. 75 figures, 27 tables. XX, 578 pages
ISBN 3-540-09288-9

Volume 7

## Physiology of Movements

**Editors: W. Haupt, M. E. Feinleib**
With contributions by numerous experts
1979. 185 figures, 19 tables. XVII, 731 pages
ISBN 3-540-08776-1

Volume 8

## Secondary Plant Products

**Editors: E. A. Bell, B. V. Charlwood**
With contributions by numerous experts
1980. 176 figures, 44 tables and numerous schemes and formulas. XVI, 674 pages
ISBN 3-540-09461-X

Volume 9

## Hormonal Regulation of Development I

**Molecular Aspects of Plant Hormones**
**Editor: J. MacMillan**
With contributions by numerous experts
1980. 126 figures. XVII, 681 pages
ISBN 3-540-10161-6

Springer-Verlag
Berlin
Heidelberg
New York

Volume 10

## Hormonal Regulation of Development II

**From the Cell to the Whole Plant**
**Editor: T. K. Scott**
With contributions by numerous experts
ISBN 3-540-10196-9
In preparation

Volume 11

## Hormonal Regulation of Development III

**Role of Environmental Factors**
**Editors: R. P. Pharis, D. M. Reid**
With contributions by numerous experts
ISBN 3-540-10197-7
In preparation

## Physiological Plant Ecology

**Editors of Volume 12A – D: O. L. Lange, P. S. Nobel, C. B. Osmond, H. Ziegler**

Volume 12 A

## Responses to the Physical Environment

1981. 109 figures, 40 tables.
Approx. 670 pages
ISBN 3-540-10763-0

Volume 12 B

## Water Relations and Carbon Assimilation

ISBN 3-540-10906-4
In preparation

Volume 12 C

## Responses to the Chemical and Biological Environment

ISBN 3-540-10907-2
In preparation

Volume 12 D

## Ecosystem Processes: Mineral Cycling, Productivity and Man's Influence

ISBN 3-540-10908-0
In preparation

Volume 13 A

## Plant Carbohydrates I

**Intracellular Carbohydrates**
**Editors: F. A. Loewus, W. Tanner**
1982. Approx. 103 figures. approx. 98 tables.
Approx. 900 pages
ISBN 3-540-11060-7

Volume 13 B

## Plant Carbohydrates II

**Extracellular Carbohydrates**
**Editors: W. Tanner, F. A. Loewus**
1981. Approx. 124 figures, approx. 49 tables.
Approx. 790 pages
ISBN 3-540-11007-0

Volume 14 B

## Nucleic Acids and Proteins in Plants II

**Structure, Biochemistry and Physiology of Nucleic Acids**
ISBN 3-540-11140-9
In preparation

Springer-Verlag
Berlin
Heidelberg
New York